Organic Chemistry

Organic Chemistry

Sixth Edition

Herbert Meislich, PhD
Professor Emeritus of Chemistry, City College of CUNY

Howard Nechamkin, EdD
Professor Emeritus of Chemistry, Trenton State College

Jacob Sharefkin, PhD
Professor Emeritus of Chemistry, Brooklyn College of CUNY

George J. Hademenos, PhD
Former Visiting Assistant Professor, Department of Physics, University of Dallas

Schaum's Outline Series

New York Chicago San Francisco Athens
London Madrid Mexico City Milan
New Delhi Singapore Sydney Toronto

HERBERT MEISLICH holds a BA degree from Brooklyn College and an MA and PhD from Columbia University. He is a professor emeritus from the City College of CUNY, where he taught Organic and General Chemistry for 40 years at both the undergraduate and doctoral levels. He received the Outstanding Teacher award in 1985 and has coauthored eight textbooks, three laboratory manuals in General and Organic Chemistry, and 15 papers on his research interests.

HOWARD NECHAMKIN is Professor Emeritus of Chemistry at Trenton State College; for 11 years of his tenure he served as Department Chairman. His Bachelor's degree is from Brooklyn College, his Master's from the Polytechnic Institute of Brooklyn, and his Doctorate in Science Education from New York University. He is the author or coauthor of 53 papers and 6 books in the areas of inorganic, analytical, and environmental chemistry.

JACOB SHAREFKIN is Professor Emeritus of Chemistry at Brooklyn College. After receiving a BS from City College of New York, he was awarded an MA from Columbia University and a PhD from New York University. His publications and research interest in Qualitative Organic Analysis and organic boron and iodine compounds have been supported by grants from the American Chemical Society, for whom he has also designed national examinations in Organic Chemistry.

GEORGE J. HADEMENOS was a Visiting Assistant Professor of Physics at the University of Dallas. He received his BS with a combined major of physics and chemistry from Angelo State University, and his MS and PhD in physics from the University of Texas at Dallas, and he has completed postdoctoral fellowships in nuclear medicine at the University of Massachusetts Medical Center and in radiological sciences/biomedical physics at UCLA Medical Center. His research interests have involved biophysical and biochemical mechanisms of disease processes, particularly cerebrovascular diseases and stroke. He has published his work in journals such as *American Scientist, Physics Today, Neurosurgery,* and *Stroke.* In addition, he is the author or coauthor of three books: *Physics of Cerebrovascular Diseases: Biophysical Mechanisms of Development, Diagnosis,* and *Therapy,* published by Springer-Verlag; *McGraw Hill's MCAT,* and *Schaum's Outline of Biology,* both published by McGraw Hill.

1 2 3 4 5 6 7 8 9 10 LON 28 27 26 25 24 23

ISBN 978-1-265-51332-0
MHID 1-265-51332-5

e-ISBN 978-1-265-51379-5
e-MHID 1-265-51379-1

To Amy Nechamkin, Belle D. Sharefkin,
John B. Sharefkin,
Kelly Hademenos, and Alexandra Hademenos

Preface

The beginning student in Organic Chemistry is often overwhelmed by facts, concepts, and new language. Each year, textbooks of Organic Chemistry grow in quantity of subject matter and in level of sophistication. This Schaum's Outline was undertaken to give a clear view of first-year Organic Chemistry through the careful detailed solution of illustrative problems. Such problems make up over 80% of the book, the remainder being a concise presentation of the material. Our goal is for students to learn by thinking and solving problems rather than by merely being told.

This book can be used in support of a standard text, as a supplement to a good set of lecture notes, as a review for taking professional examinations, and as a vehicle for self-instruction.

The second edition has been reorganized by combining chapters to emphasize the similarities of functional groups and reaction types as well as the differences. Thus, polynuclear hydrocarbons are combined with benzene and aromaticity. Nucleophilic aromatic displacement is merged with aromatic substitution. Sulfonic acids are in the same chapter with carboxylic acids and their derivatives, and carbanion condensations are in a separate new chapter. Sulfur compounds are discussed with their oxygen analogs. This edition has also been brought up-to-date by including solvent effects, CMR spectroscopy, an elaboration of polymer chemistry, and newer concepts of stereochemistry, among other material.

HERBERT MEISLICH
HOWARD NECHAMKIN
JACOB SHAREFKIN
GEORGE J. HADEMENOS

Contents

Structure and Properties of Organic Compounds

1.1 Carbon Compounds

Organic chemistry is the study of carbon (C) compounds, all of which have **covalent bonds**. Carbon atoms can bond to each other to form **open-chain** compounds, Fig. 1.1(*a*), or **cyclic** (**ring**) **compounds**, Fig. 1.1(*c*). Both types can also have branches of C atoms, Fig. 1.1(*b*) and (*d*). **Saturated** compounds have C atoms bonded to each other by **single bonds,** C—C; **unsaturated** compounds have C's joined by **multiple bonds**. Examples with **double bonds** and **triple bonds** are shown in Fig. 1.1(*e*). Cyclic compounds having at least one atom in the ring other than C (a **heteroatom**) are called **heterocyclics**, Fig. 1.1(*f*). The heteroatoms are usually oxygen (O), nitrogen (N), or sulfur (S).

Problem 1.1 Why are there so many compounds that contain carbon?

Bonds between C's are covalent and strong, so that C's can form long chains and rings, both of which may have branches. C's can bond to almost every element in the periodic table. Also, the number of isomers increases as the organic molecules become more complex.

Problem 1.2 Compare and contrast the properties of ionic and covalent compounds.

Ionic compounds are generally inorganic, have high melting and boiling points due to the strong electrostatic forces attracting the oppositely charged ions, are soluble in water and insoluble in organic solvents, are hard to burn, and involve reactions that are rapid and simple. Also, bonds between like elements are rare, with isomerism being unusual.

Covalent compounds, on the other hand, are commonly organic; have relatively low melting and boiling points because of weak intermolecular forces; are soluble in organic solvents and insoluble in water; burn readily and are thus susceptible to oxidation because they are less stable to heat, usually decomposing at temperatures above 700°C; and involve reactions that are slow and complex, often needing higher temperatures and/or catalysts, yielding *mixtures* of products. Also, bonds between carbon atoms are typical, with isomerism being common.

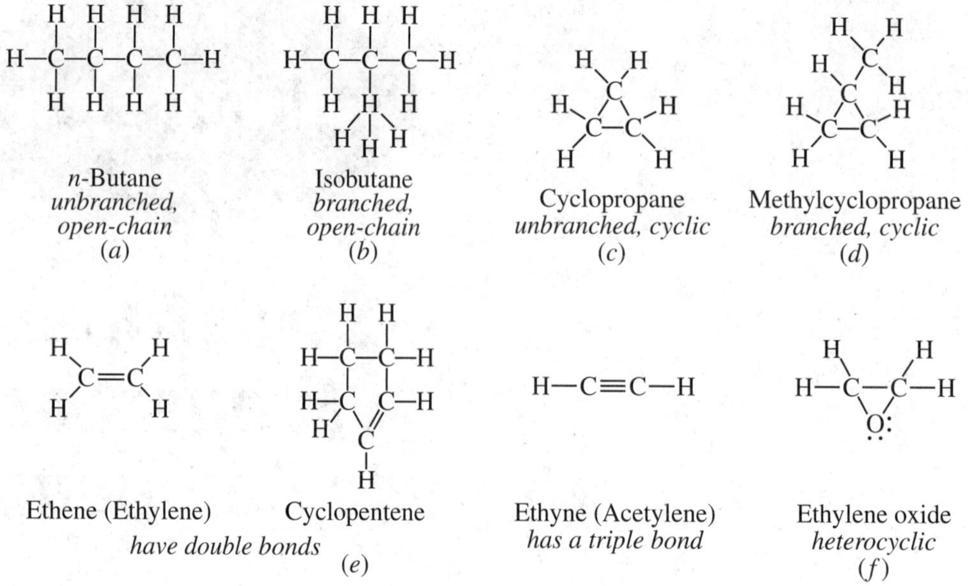

n-Butane
unbranched, open-chain
(*a*)

Isobutane
branched, open-chain
(*b*)

Cyclopropane
unbranched, cyclic
(*c*)

Methylcyclopropane
branched, cyclic
(*d*)

Ethene (Ethylene)   Cyclopentene
have double bonds
(*e*)

Ethyne (Acetylene)
has a triple bond

Ethylene oxide
heterocyclic
(*f*)

Figure 1.1

1.2 Lewis Structural Formulas

Molecular formulas merely include the kinds of atoms and the number of each in a molecule (as C_4H_{10} for butane). **Structural formulas** show the arrangement of atoms in a molecule (see Fig. 1.1). When unshared electrons are included, the latter are called **Lewis (electron-dot) structures** [see Fig. 1-1(*f*)]. **Covalences** of the common elements—the numbers of covalent bonds they usually form—are given in Table 1.1; these help us to write Lewis structures. Multicovalent elements such as C, O, and N may have multiple bonds, as shown in Table 1.2. In **condensed** structural formulas, all H's and branched groups are written immediately after the C atom to which they are attached. Thus, the condensed formula for isobutane [Fig. 1-1(*b*)] is $CH_3CH(CH_3)_2$.

Problem 1.3 (*a*) Are the covalences and **group numbers** (numbers of **valence electrons**) of the elements in Table 1.1 related? (*b*) Do all the elements in Table 1.1 attain an octet of valence electrons in their bonded states? (*c*) Why aren't Group 1 elements included in Table 1.1?

(*a*) Yes. For the elements in Groups 4 through 7, Covalence = 8 − (Group number).

(*b*) No. The elements in Groups 4 through 7 do attain the octet, but the elements in Groups 2 and 3 have less than an octet. (The elements in the third and higher periods, such as Si, S, and P, may achieve more than an octet of valence electrons.)

(*c*) They form ionic rather than covalent bonds. (The heavier elements in Groups 2 and 3 also form mainly ionic bonds. In general, as one proceeds down a group in the periodic table, ionic bonding is preferred.)

Most carbon-containing molecules are three-dimensional. In methane, the bonds of C make equal angles of 109.5° with each other, and each of the four H's is at a vertex of a regular tetrahedron whose center is occupied by the C atom. The spatial relationship is indicated as in Fig. 1.2(*a*) (Newman projection) or in Fig. 1.2(*b*) ("wedge" projection). Except for ethene, which is planar, and ethyne, which is linear, the structures in Fig. 1.1 are all three-dimensional.

Organic compounds show a widespread occurrence of **isomers**, which are compounds having the same molecular formula but different structural formulas, and therefore possessing different properties. This phenomenon of **isomerism** is exemplified by isobutane and *n*-butane [Fig. 1.1(*a*) and (*b*)]. The number of isomers increases as the number of atoms in the organic molecule increases.

TABLE 1.1 Covalences of H and Second-Period Elements in Groups 2 through 7

Group	1	2	3	4	5	6	7
Lewis Symbol	H·	·Be·	·Ḃ·	·Ċ·	·N̈·	·Ö·	·F̈:
Covalence	1	2	3	4	3	2	1
Compounds with H	H—H Hydrogen	H—Be—H Beryllium hydride	H—B(—H)—H Boron hydride*	CH_4 Methane	H—N̈(—H)—H Ammonia	H—Ö—H Water	H—F̈: Hydrogen fluoride

* Exists as B_2H_6.

TABLE 1.2 Normal Covalent Bonding

BONDING FOR C				BONDING FOR N			BONDING FOR O	
—C— (four single bonds)	—C=	=C=	—C≡	—N̈— (three single bonds)	—N̈=	N̈≡	—Ö—	Ö=
as in	*as in*	*as in*	*as in*	*as in*	*as in*	*as in*	*as in*	*as in*
CH_4	$H_2C{=}CH_2$	:Ö=C=Ö:	H—C≡C—H	H—N̈(—H)—H	H—Ö—N̈=Ö:	:N≡C—H	H—Ö—H	Ö=C(H)H
Methane	Ethene (Ethylene)	Carbon dioxide	Ethyne (Acetylene)	Ammonia	Nitrous acid	Hydrogen cyanide	Water	Formaldehyde

Problem 1.4 Write structural and condensed formulas for (*a*) three isomers with molecular formula C_5H_{12} and (*b*) two isomers with molecular formula C_3H_6.

(*a*) Carbon forms four covalent bonds; hydrogen forms one. The carbons can bond to each other in a chain:

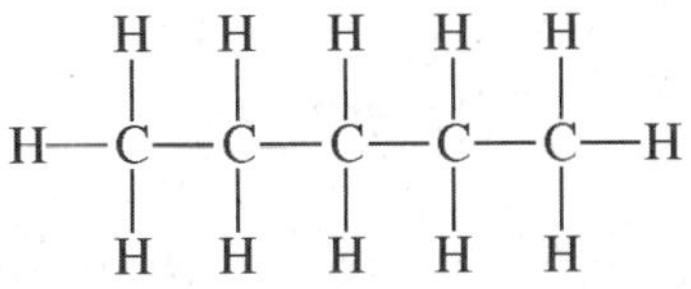

H—C—C—C—C—C—H, with H above and below each C (structural formula)

$CH_3(CH_2)_3CH_3$ (condensed formula)

n-Pentane

or there can be "branches" (shown circled in Fig. 1.3) on the linear backbone (shown in a rectangle).

(*b*) We can have a double bond or a ring.

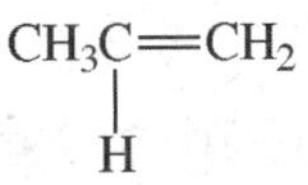

$CH_3C(H){=}CH_2$

Propene (Propylene)

$(CH_2)_3$ ring, each C bearing two H

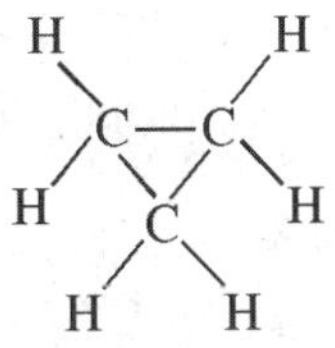

Cyclopropane

H^f's project toward viewer
H^b's project away from viewer

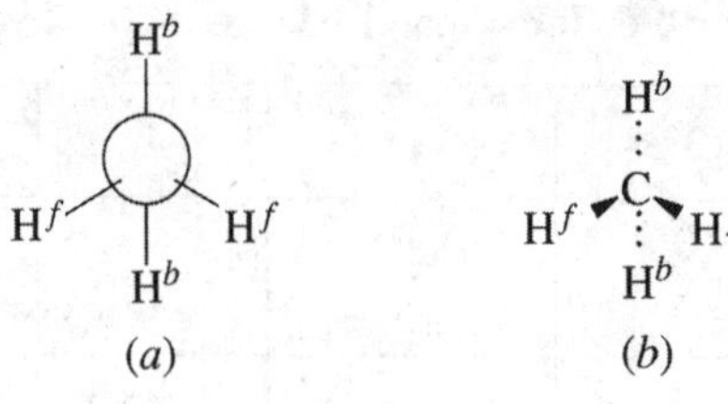

... projects into plane of paper away from reader
projects out of plane of paper toward reader

Figure 1.2

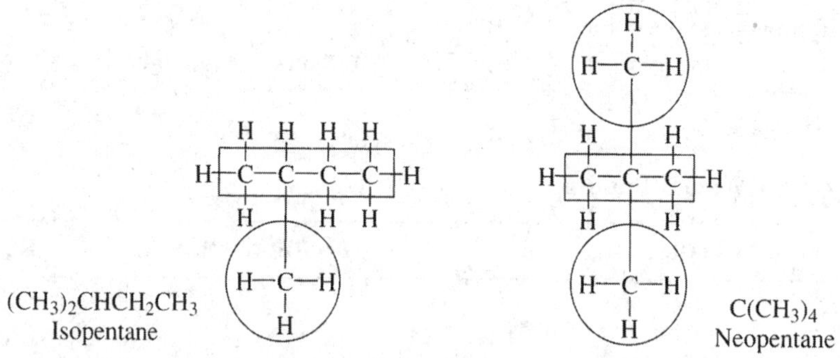

$(CH_3)_2CHCH_2CH_3$
Isopentane

$C(CH_3)_4$
Neopentane

Figure 1.3

Problem 1.5 Write Lewis structures for (*a*) hydrazine, N_2H_4; (*b*) phosgene, $COCl_2$; and (*c*) nitrous acid, HNO_2.

In general, first bond the multicovalent atoms to each other and then, to achieve their normal covalences, bond them to the univalent atoms (H, Cl, Br, I, and F). If the number of univalent atoms is insufficient for this purpose, use multiple bonds or form rings. In their bonded state, the second-period elements (C, N, O, and F) should have eight (an octet) electrons but not more. Furthermore, the number of electrons shown in the Lewis structure should equal the sum of all the valence electrons of the individual atoms in the molecule. Each bond represents a shared pair of electrons.

(*a*) N needs three covalent bonds, and H needs one. Each N is bonded to the other N and to two H's.

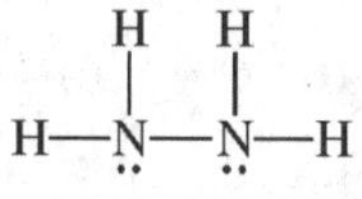

H—N̈—N̈—H (each N also bonded to an H)

(*b*) C is bonded to O and to each Cl. To satisfy the tetravalence of C and the divalence of O, a double bond is placed between C and O.

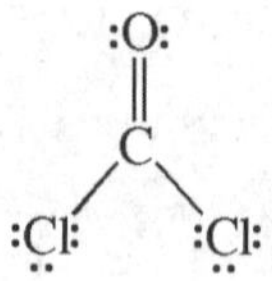

:Ö: double-bonded to C; C bonded to :C̈l: and :C̈l:

(*c*) The atom with the higher covalence, in this case the N, is usually the more central atom. Therefore, each O is bonded to the N. The H is bonded to one of the O atoms, and a double bond is placed between the N and the other O. (Convince yourself that bonding the H to the N would not lead to a viable structure.)

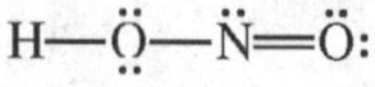

H—Ö—N̈=Ö:

Problem 1.6 Why are none of the following Lewis structures for $COCl_2$ correct?

(*a*) :C̈l—C̈=Ö—C̤̈l: (*b*) :C̤̈l—C≡O—C̤̈l: (*c*) :C̈l=C=Ö—C̤̈l: (*d*) :C̈l=C=Ö̤—C̤̈l:

The total number of valence electrons that must appear in the Lewis structure is 24, from [2 × 7](2Cl's) + 4(C) + 6(O). Structures (*b*) and (*c*) can be rejected because they each show only 22 electrons. Furthermore, in (*b*), O has 4 rather than 2 bonds, and in (*c*), one Cl has 2 bonds. In (*a*), C and O do not have their normal covalences. In (*d*), O has 10 electrons, though it cannot have more than an octet.

Problem 1.7 Use the Lewis-Langmuir octet rule to write Lewis electron-dot structures for: (*a*) HCN, (*b*) CO_2, (*c*) CCl_4, and (*d*) C_2H_6O.

(*a*) Attach the H to the C, because C has a higher covalence than N. The normal covalences of N and C are met with a triple bond. Thus, H—C≡N: is the correct Lewis structure.

(*b*) The C is bonded to each O by double bonds to achieve the normal covalences.

:O̤=C=O̤:

(*c*) Each of the four Cl's is singly bonded to the tetravalent C to give:

```
       :C̈l:
        |
  :C̤̈l—C—C̤̈l:
        |
       :C̤l:
```

(*d*) The three multicovalent atoms can be bonded as C—C—O or as C—O—C. If the six H's are placed so that C and O acquire their usual covalences of 4 and 2, respectively, we get two correct Lewis structures (isomers):

```
    H  H               H     H
    |  |               |     |
 H—C—C—Ö̤—H          H—C—Ö̤—C—H
    |  |               |     |
    H  H               H     H

   Ethanol          Dimethyl ether
```

Problem 1.8 Determine the positive or negative charge, if any, on:

```
         H               H                 H  H
         |               |                 |  |
 (a) H—C—Ö̤:     (b) H—C=Ö:     (c) H—C—C·
         |                                 |  |
         H                                 H  H

         H                 :C̈l:
         |                  |
 (d) H—N—Ö̤—H      (e)    :C—C̤̈l:
         |                  |
         H                 :C̤l:
```

The charge on a species is numerically equal to the total number of valence electrons of the unbonded atoms minus the total number of electrons shown (as bonds or dots) in the Lewis structure.

(*a*) The sum of the valence electrons (6 for O, 4 for C, and 3 for three H's) is 13. The electron-dot formula shows 14 electrons. The net charge is 13 − 14 = −1, and the species is the methoxide anion, $CH_3\ddot{\underset{..}{O}}:^-$.

(*b*) There is no charge on the formaldehyde molecule, because the 12 electrons in the structure equals the number of valence electrons—that is, 6 for O, 4 for C, and 2 for two H's.
(*c*) This species is neutral, because there are 13 electrons shown in the formula and 13 valence electrons: 8 from two C's and 5 from five H's.
(*d*) There are 15 valence electrons: 6 from O, 5 from N, and 4 from four H's. The Lewis dot structure shows 14 electrons. It has a charge of $15 - 14 = +1$ and is the hydroxylammonium cation, $[H_3NOH]^+$.
(*e*) There are 25 valence electrons, 21 from three Cl's and 4 from C. The Lewis dot formula shows 26 electrons. It has a charge of $25 - 26 = -1$ and is the trichloromethide anion, $:CCl_3^-$.

1.3 Types of Bonds

Covalent bonds, the mainstays of organic compounds, are formed by the sharing of pairs of electrons. Sharing can occur in two ways:

$$(1)\quad A\cdot + \cdot B \rightarrow A:B$$

$$(2)\quad \underset{acceptor}{A} + \underset{donor}{:B} \rightarrow A:B \text{ coordinate covalent}$$

In method (1), each atom brings an electron for the sharing. In method (2), the donor atom (B:) brings both electrons to the "marriage" with the acceptor atom (A); in this case, the covalent bond is termed a **coordinate covalent** bond.

Problem 1.9 Each of the following molecules and ions can be thought to arise by coordinate covalent bonding. Write an equation for the formation of each one and indicate the donor and acceptor molecule or ion: (*a*) NH_4^+; (*b*) BF_4^-; (*c*) $(CH_3)_2OMgCl_2$; and (*d*) $Fe(CO)_5$.

	acceptor	*donor*		
(*a*)	H^+	$+ :NH_3$	$\longrightarrow$	NH_4^+ (All N—H bonds are alike.)
(*b*)	F_3B	$+ :\ddot{\underset{..}{F}}:^-$	$\longrightarrow$	BF_4^- (All B—F bonds are alike.)
(*c*)	$:\ddot{\underset{..}{Cl}}—Mg—\ddot{\underset{..}{Cl}}:$	$+ :\ddot{O}—CH_3$ (with CH_3 on O)	$\longrightarrow$	$(CH_3)_2\ddot{O}—MgCl_2$
(*d*)	Fe	$+ 5:C\equiv O:$	$\longrightarrow$	$Fe(C\equiv O:)_5$

Notice that in each of the products there is at least one element that does not have its usual covalence—this is typical of coordinate covalent bonding.

Recall that an ionic bond results from a *transfer* of electrons ($M\cdot + A\cdot \rightarrow M^+ + :A^-$). Although C usually forms covalent bonds, it sometimes forms an ionic bond (see Section 3.2). Other organic ions, such as CH_3COO^- (acetate ion), have charges on heteroatoms.

Problem 1.10 Show how the ionic compound Li^+F^- forms from atoms of Li and F.

These elements react to achieve a stable noble-gas electron configuration (NGEC). Li(3) has one electron more than He and loses it. F(9) has one electron less than Ne and therefore accepts the electron from Li.

$$Li\cdot + \cdot\ddot{\underset{..}{F}}: \longrightarrow Li^+ :\ddot{\underset{..}{F}}:^- \quad \text{(or simply LiF)}$$

1.4 Functional Groups

Hydrocarbons contain only C and hydrogen (H). H's in hydrocarbons can be replaced by other atoms or groups of atoms. These replacements, called **functional groups**, are the reactive sites in molecules. The C-to-C double and triple bonds are considered to be functional groups. Some common functional groups are given in Table 1.3.

Compounds with the same functional group form a **homologous series** having similar characteristic chemical properties and often exhibiting a regular gradation in physical properties with increasing molecular weight.

Problem 1.11 Methane, CH_4; ethane, C_2H_6; and propane, C_3H_8, are the first three members of the alkane homologous series. By what structural unit does each member differ from its predecessor?

These members differ by a C and two H's; the unit is —CH_2— (a methylene group).

Problem 1.12 (*a*) Write possible Lewis structural formulas for (1) CH_4O; (2) CH_2O; (3) CH_2O_2; (4) CH_5N; and (5) CH_3SH. (*b*) Indicate and name the functional group in each case.

The atom with the higher valence is usually the one to which most of the other atoms are bonded.

(*a*) (1) $H_3C—\ddot{O}—H$ (2) $H_2C{=}\ddot{O}{:}$ (3) $H—C({=}\ddot{O}{:})—\ddot{O}—H$ (4) $H_3C—\ddot{N}H_2$ (5) $H_3C—\ddot{S}H$

(*b*) (1) $—\ddot{O}H$ alcohol (2) $HC{=}\ddot{O}{:}$ aldehyde (3) $—C({=}\ddot{O}{:})—\ddot{O}—H$ carboxylic acid (4) $—\ddot{N}H_2$ amine (5) $—\ddot{S}H$ thiol

1.5 Formal Charge

The formal charge on a covalently bonded atom equals the number of valence electrons of the unbonded atom (the group number) minus the number of electrons assigned to the atom in its bonded state. The assigned number is one-half the number of shared electrons plus the total number of unshared electrons. The sum of all formal charges in a molecule equals the charge on the species. In this outline, formal charges and actual ionic charges (e.g., Na^+) are both indicated by the signs + and −.

Problem 1.13 Determine the formal charge on each atom in the following species: (*a*) H_3NBF_3; (*b*) $CH_3NH_3^+$; and (*c*) SO_4^{2-}.

(*a*) $H_3N^+—B^-(\ddot{F}{:})_3$

	GROUP NUMBER	−	[UNSHARED ELECTRONS	+	1/2 SHARED ELECTRONS]	=	FORMAL CHARGE
H atoms	1	−	0	+	1	=	0
F atoms	7	−	6	+	1	=	0
N atom	5	−	0	+	4	=	+1
B atom	3	−	0	+	4	=	−1

The sum of all formal charges equals the charge on the species. In this case, the +1 on N and the −1 on B cancel

(*b*) $[H_3C—N^+H_3]^+$

and the species is an unchanged molecule:

	GROUP NUMBER	−	[UNSHARED ELECTRONS	+	1/2	SHARED ELECTRONS]	=	FORMAL CHARGE
C atoms	4	−	0	+		4	=	0
N atoms	5	−	0	+		4	=	+1
H atoms	1	−	0	+		1	=	0
						Net charge on species	=	+1

(*c*)

$$\left[\begin{array}{c} :\ddot{O}:^{-} \\ | \\ :\ddot{O}^{-}—S^{2+}—\ddot{O}:^{-} \\ | \\ :\ddot{O}:^{-} \end{array}\right]^{2-}$$

	GROUP NUMBER	−	[UNSHARED ELECTRONS	+	1/2	SHARED ELECTRONS]	=	FORMAL CHARGE
S atoms	6	−	0	+		4	=	+2
each O atom	6	−	6	+		1	=	−1
						Net charge is + 2 + 4(−1)	=	−2

These examples reveal that formal charges appear on an atom that does not have its usual covalence and does not have more than an octet of valence electrons. Formal charges always occur in a molecule or ion that can be conceived to be formed as a result of coordinate covalent bonding.

Problem 1.14 Show how (*a*) H_3NBF_3 and (*b*) $CH_3NH_3^+$ can be formed from coordinate covalent bonding. Indicate the donor and acceptor, and show the formal charges.

donor *acceptor*

(*a*) $H_3N: + BF_3 \longrightarrow H_3\overset{+}{N}—\overset{-}{B}F_3$

(*b*) $CH_3\ddot{N}H_2 + H^+ \longrightarrow [CH_3\overset{+}{N}H_3]$

SUPPLEMENTARY PROBLEMS

Problem 1.15 Why are the compounds of carbon covalent rather than ionic?

With four valence electrons, it would take too much energy for C to give up or accept four electrons. Therefore, carbon shares electrons and forms covalent bonds.

Problem 1.16 Classify the following as (i) branched chain, (ii) unbranched chain, (iii) cyclic, (iv) multiple bonded, or (v) heterocyclic:

(*a*) cyclopentene ring: CH_2 bonded to H_2C and CH_2; H_2C—HC, CH_2—CH; $HC=CH$

(*b*) $CH_3—C(CH_3)_2—CH_3$

(*c*) CH_3 on $H_2C—CH_2—CH_2$, with CH_3 on the last CH_2

(*d*) ring of H_2C, H_2C and NH

(*e*) $CH_2=HC$, HC bonded to $H_2C—CH_3$

(*a*) (iii) and (iv); (*b*) (i); (*c*) (ii); (*d*) (v); (*e*) (iv) and (ii).

TABLE 1.3 Some Common Functional Groups

FUNCTIONAL GROUP	GENERAL FORMULA	GENERAL NAME	EXAMPLE		
			FORMULA	IUPAC NAME[1]	COMMON NAME
None	C_nH_{2n+2}	Alkane	CH_3CH_3	Eth*ane*	Ethane
$>C=C<$	C_nH_{2n}	Alkene	$H_2C=CH_2$	Eth*ene*	Ethylene
$-C\equiv C-$	C_nH_{2n-2}	Alkyne	$HC\equiv CH$	Eth*yne*	Acetylene
$-Cl$	$R-Cl$	Chloride	CH_3CH_2Cl	*Chloro*ethane	Ethyl chloride
$-Br$	$R-Br$	Bromide	CH_3Br	*Bromo*methane	Methyl bromide
$-OH$	$R-OH$	Alcohol	CH_3CH_2OH	Ethan*ol*	Ethyl alcohol
$-O-$	$R-O-R$	Ether	$CH_3CH_2OCH_2CH_3$	*Ethoxy*ethane	Diethyl ether
$-NH_2$	RNH_2	Amine[2]	$CH_3CH_2CH_2NH_2$	1-*Amino*propane[3]	Propylamine
$-NR_3^+X^-$	$R_4N^+X^-$	Quaternary ammonium salt	$CH_3(CH_2)_9N(CH_3)_3^+Cl^-$	Decyltrimethyl-*ammonium* chloride	Decyltrimethyl-ammonium chloride
$-\underset{\mid}{\overset{}{C}}(H)=O$	$R-C(H)=O$	Aldehyde	$CH_3CH_2CH(H)=O$	Propan*al*	Propionaldehyde
$-\underset{\mid}{C}=O$	$R-C(R)=O$	Ketone	$CH_3CH_2C(CH_3)=O$	2-Butan*one*	Methyl ethyl ketone
$-C(=O)-OH$	$R-C(=O)-OH$	Carboxylic acid	$CH_3-C(=O)-OH$	Ethan*oic acid*	Acetic acid

(continued)

TABLE 1.3 *(continued)*

FUNCTIONAL GROUP	GENERAL FORMULA	GENERAL NAME	EXAMPLE		
			FORMULA	IUPAC NAME[1]	COMMON NAME
$-C(=O)-OR'$	$R-C(=O)-OR'$	Ester	$CH_3-C(=O)-OC_2H_5$	Ethyl ethan*oate*	Ethyl acetate
$-C(=O)-NH_2$	$R-C(=O)-NH_2$	Amide	$CH_3-C(=O)-NH_2$	Ethan*amide*	Acetamide
$-C(=O)-Cl$	$R-C(=O)-Cl$	Acid chloride	$CH_3-C(=O)-Cl$	Ethan*oyl chloride*	Acetyl chloride
$-C(=O)-O-C(=O)-$	$R-C(=O)-O-C(=O)-R$	Acid anhydride	$CH_3-C(=O)-O-C(=O)-CH_3$	Ethan*oic anhydride*	Acetic anhydride
$-C\equiv N$	$R-C\equiv N$	Nitrile	$CH_3C\equiv N$	Ethane*nitrile*	Acetonitrile
$-NO_2$	$R-NO_2$	Nitro	CH_3-NO_2	*Nitro*methane	Nitromethane
$-SH$	$R-SH$	Thiol	CH_3-SH	Methane*thiol*	Methyl mercaptan
$-S-$	$R-S-R$	Thioether (sulfide)	CH_3-S-CH_3	Dimethyl *thioether*	Dimethyl sulfide
$-S-S-$	$R-S-S-R$	Disulfide	$CH_3-S-S-CH_3$	Dimethyl *disulfide*	Dimethyl disulfide
$-S(=O)_2-OH$	$R-S(=O)_2-OH$	Sulfonic acid	$CH_3-S(=O)_2-OH$	Methane*sulfonic acid*	Methanesulfonic acid
$-S(=O)-$	$R-S(=O)-R$	Sulfoxide	$CH_3-S(=O)-CH_3$	Dimethyl *sulfoxide*	Dimethyl sulfoxide
$-S(=O)_2-$	$R-S(=O)_2-R$	Sulfone	$CH_3-S(=O)_2-CH_3$	Dimethyl *sulfone*	Dimethyl sulfone

[1] The italicized portion indicates the group.
[2] A primary (1°) amine; there are also secondary (2°) R_2NH, and tertiary (3°) R_3N amines.
[3] Another name is propanamine.

Problem 1.17 Refer to a periodic chart and predict the covalences of the following in their hydrogen compounds: (*a*) O; (*b*) S; (*c*) Cl; (*d*) C; (*e*) Si; (*f*) P; (*g*) Ge; (*h*) Br; (*i*) N; (*j*) Se.

The number of covalent bonds typically formed by an element is 8 minus the group number. Thus: (*a*) 2; (*b*) 2; (*c*) 1; (*d*) 4; (*e*) 4; (*f*) 3; (*g*) 4; (*h*) 1; (*i*) 3; (*j*) 2.

Problem 1.18 Which of the following are isomers of 2-hexene, $CH_3CH{=}CHCH_2CH_2CH_3$?

(*a*) $CH_3CH_2CH{=}CHCH_2CH_3$ (*b*) $CH_2{=}CHCH_2CH_2CH_2CH_3$
(*c*) $CH_3CH_2CH_2CH{=}CHCH_3$ (*d*) cyclopentane ring: C(H)(CH₃), H_2C, CH_2, $H_2C{-}CH_2$ (*e*) cyclobutane ring: $H_2C{-}C(H){-}CH_3$ / $H_2C{-}C(H){-}CH_3$

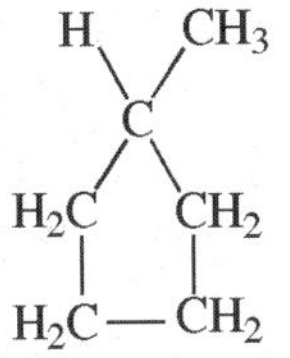

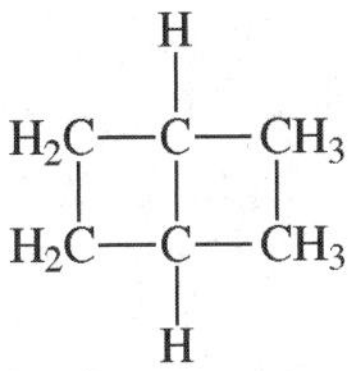

All but (*c*), which is 2-hexene itself.

Problem 1.19 Find the formal charge on each element of

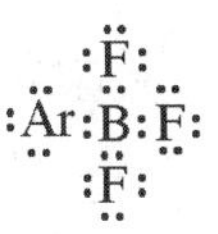

$$:\ddot{Ar}:B:\ddot{F}: \text{ with } :\ddot{F}: \text{ above and below B}$$

and the net charge on the species (BF_3Ar).

ATOM	GROUP NUMBER	−	[# UNSHARED ELECTRONS	+	1/2 # SHARED ELECTRONS]	=	FORMAL CHARGE OF ATOM
each F	7	−	6	+	1		0
B	3	−	0	+	4		−1
Ar	8	−	6	+	1		+1
							0 = net charge

Problem 1.20 Write Lewis structures for the nine isomers having the molecular formula C_3H_6O, in which C, H, and O have their usual covalences; name the functional group(s) present in each isomer.

One cannot predict the number of isomers by mere inspection of the molecular formula. A logical method runs as follows. First write the different bonding skeletons for the multivalent atoms; in this case, the three C's and the O. There are three such skeletons:

(*i*) C—C—C—O (*ii*) C—O—C—C (*iii*) C—C—C with O attached to the middle C

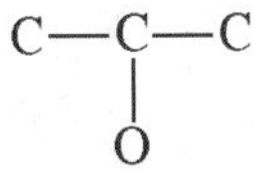

To attain the covalences of 4 for C and 2 for O, eight H's are needed. Since the molecular formula has only six H's, a double bond or ring must be introduced onto the skeleton. In (i), the double bond can be situated three ways, between either pair of C's or between the C and O. If the H's are then added, we get three isomers: (1), (2), and (3). In (ii), a double bond can be placed only between adjacent C's to give (4). In (iii), a double bond can be placed between a pair of C's or C and O, giving (5) and (6), respectively.

(1) $H_2C{=}CHCH_2OH$ (2) $CH_3CH{=}CHOH$ — *alkene alcohols (enols)*
(3) $CH_3CH_2CH_2CH{=}O$ — *an aldehyde*
(4) $CH_3OCH{=}CH_2$ — *an alkene ether*

(5) $H_2C{=}C(\ddot{O}H)CH_3$ — *an enol*

(6) $CH_3C(={\ddot{O}}:)CH_3$ — *a ketone*

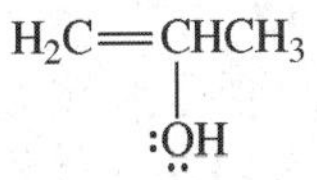

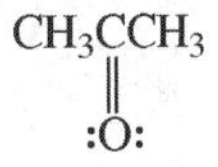

In addition, three ring compounds are possible:

(7) 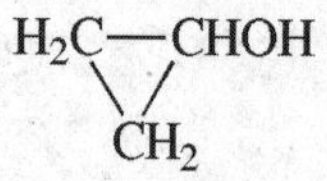 $H_2C—CHOH$ (ring closed through CH_2)

a cyclic alcohol

(8) 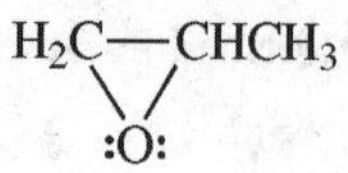 $H_2C—CHCH_3$ (ring closed through $:\ddot{O}:$)

(9) 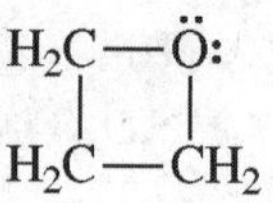 $H_2C—\ddot{O}:$ / $H_2C—CH_2$ (four-membered ring)

heterocyclic ethers

Bonding and Molecular Structure

2.1 Atomic Orbitals

An **atomic orbital** (AO) is a region of space about the nucleus in which there is a high probability of finding an electron. An electron has a given energy, as designated by (*a*) the principal energy level (quantum number) n, related to the size of the orbital; (*b*) the sublevel *s*, *p*, *d*, *f*, or *g*, related to the shape of the orbital; (*c*) except for the *s*, each sublevel having some number of equal-energy (**degenerate**) orbitals differing in their spatial orientation; and, (*d*) the electron spin, designated ↑ or ↓. Table 2.1 shows the distribution and designation of orbitals.

TABLE 2.1

Principal energy level, n	1	2	3	4
Maximum no. of electrons, $2n^2$	2	8	18	32
Sublevels [n in number]	$1s$	$2s$, $2p$	$3s$, $3p$, $3d$	$4s$, $4p$, $4d$, $4f$
Maximum electrons per sublevel	2	2, 6	2, 6, 10	2, 6, 10, 14
Designations of filled orbitals	$1s^2$	$2s^2$, $2p^6$	$3s^2$, $3p^6$, $3d^{10}$	$4s^2$, $4p^6$, $4d^{10}$, $4f^{14}$
Orbitals per sublevel	1	1, 3	1, 3, 5	1, 3, 5, 7

The *s* orbital is a sphere around the nucleus, as shown in cross section in Fig. 2.1(*a*). A *p* orbital is two spherical lobes touching on opposite sides of the nucleus. The three *p* orbitals are labeled p_x, p_y, and p_z because they are oriented along the *x*-, *y*-, and *z*-axes, respectively [Fig. 2.1(*b*)]. In a *p* orbital, there is no chance of finding an electron at the nucleus—the nucleus is called a **node point**. Regions of an orbital separated by a node are assigned + and − signs. These signs are *not associated with electrical or ionic charges*. The *s* orbital has no node and is usually assigned a +.

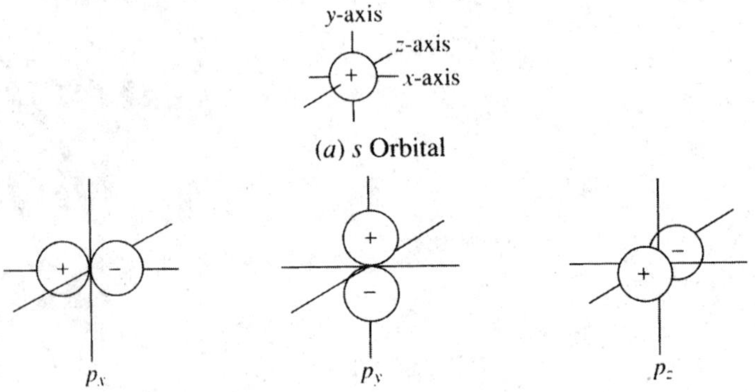

y-axis
z-axis
x-axis

(*a*) *s* Orbital

p_x p_y p_z

(*b*) *p* Orbitals

Figure 2.1

Three principles are used to distribute electrons in orbitals:

1. **"Aufbau" or building-up principle.** Orbitals are filled in order of increasing energy: 1*s*, 2*s*, 2*p*, 3*s*, 3*p*, 4*s*, 3*d*, 4*p*, 5*s*, 4*d*, 5*p*, 6*s*, 4*f*, 5*d*, 6*p*, and so on.
2. **Pauli exclusion principle.** No more than two electrons can occupy an orbital and then only if they have opposite spins.
3. **Hund's rule.** One electron is placed in each equal-energy orbital so that the electrons have parallel spins, before pairing occurs. (Substances with unpaired electrons are **paramagnetic**—they are attracted to a magnetic field.)

Problem 2.1 Show the distribution of electrons in the atomic orbitals of (*a*) carbon and (*b*) oxygen.

A dash represents an orbital; a horizontal space between dashes indicates an energy difference. Energy increases from left to right.

(*a*) Atomic number of C is 6.

$$\underline{\uparrow\downarrow}\;\underline{\uparrow\downarrow}\;\underline{\;\uparrow\;}\;\underline{\;\uparrow\;}\;\underline{\;\;\;}$$
$$1s\;\;2s\;\;2p_x\;\;2p_y\;\;2p_z$$

The two 2*p* electrons are unpaired in each of two *p* orbitals (Hund's rule).

(*b*) Atomic number of O is 8.

$$\underline{\uparrow\downarrow}\;\underline{\uparrow\downarrow}\;\underline{\uparrow\downarrow}\;\underline{\;\uparrow\;}\;\underline{\;\uparrow\;}$$
$$1s\;\;2s\;\;2p_x\;\;2p_y\;\;2p_z$$

2.2 Covalent Bond Formation—Molecular Orbital (MO) Method

A covalent bond forms by overlap (fusion) of two AO's—one from each atom. This overlap produces a new orbital, called a **molecular orbital (MO),** which embraces both atoms. The interaction of two AO's can produce two kinds of MO's. If orbitals with like signs overlap, a **bonding MO** results which has a high electron density between the atoms and therefore has a lower energy (greater stability) than the individual AO's. If AO's of unlike signs overlap, an **antiboding MO*** results which has a node (site of zero electron density) between the atoms and therefore has a higher energy than the individual AO's. An asterisk indicates antibonding.

Head-to-head overlap of AO's gives a **sigma** (σ) **MO**—the bonds are called $\boldsymbol{\sigma}$ **bonds**, [Fig. 2.2(*a*)]. The corresponding antibonding MO* is designated σ^* [Fig. 2.2(*b*)]. The imaginary line joining the nuclei of the bonding atoms is the **bond axis**, whose length is the **bond length**.

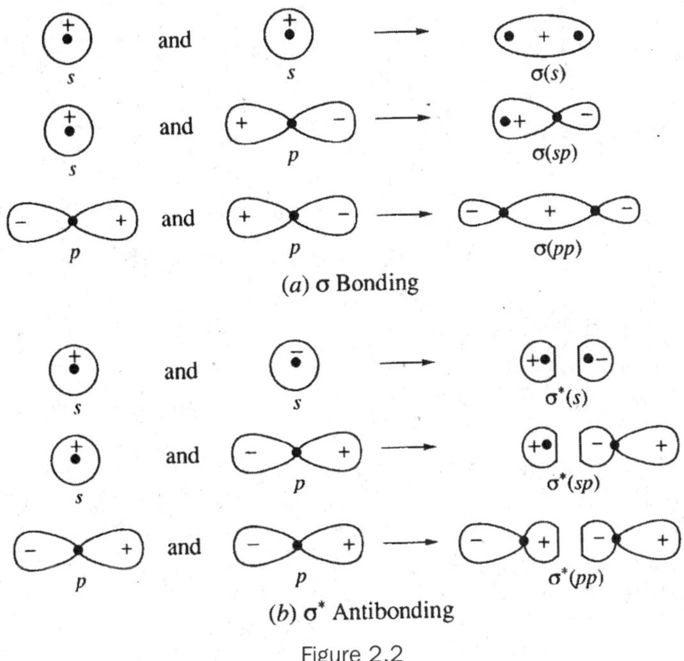

(*a*) σ Bonding

(*b*) σ* Antibonding

Figure 2.2

Two parallel *p* orbitals overlap side by side to form a **pi** (π) bond [Fig. 2.3(*a*)] or a π^* bond [Fig. 2.3(*b*)]. The bond axis lies in a nodal plane (plane of zero electronic density) perpendicular to the cross-sectional plane of the π bond.

Single bonds are σ bonds. A double bond is one σ and one π bond. A triple bond is one σ and two π bonds (a π_z and a π_y, if the triple bond is taken along the *x*-axis).

Although MO's encompass the entire molecule, it is best to visualize most of them as being localized between pairs of bonding atoms. This description of bonding is called **linear combination of atomic orbitals (LCAO)**.

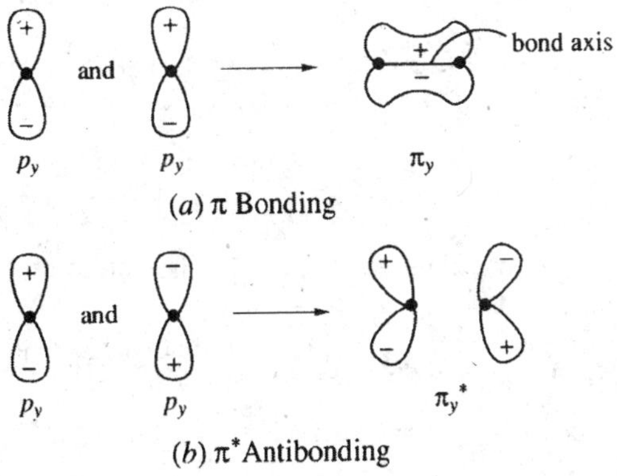

p_y and p_y → π_y (bond axis)

(*a*) π Bonding

p_y and p_y → π_y^*

(*b*) π* Antibonding

Figure 2.3

Problem 2.2 What type of MO results from side-to-side overlap of an *s* and a *p* orbital?

The overlap is depicted in Fig. 2.4. The bonding strength generated from the overlap between the $+s$ AO and the $+$ portion of the *p* orbital is canceled by the antibonding effect generated from overlap between the $+s$ and the $-$ portion of the *p*. The MO is **nonbonding** (*n*); it is no better than two isolated AO's.

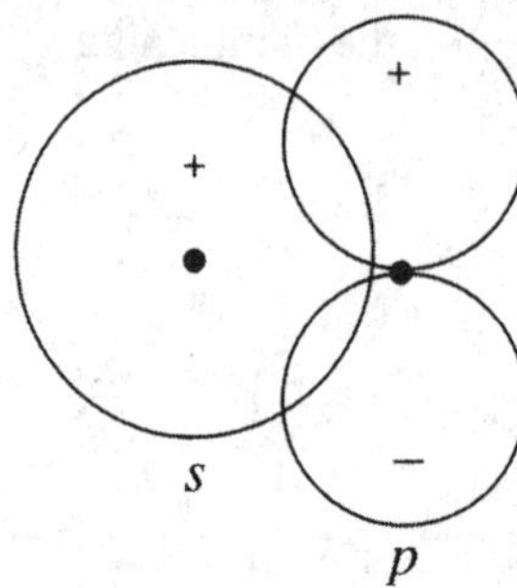

Figure 2.4

Problem 2.3 List the differences between a σ bond and a π bond.

σ **Bond**	π **Bond**
1. Formed by head-to-head overlap of AO's.	1. Formed by lateral overlap of *p* orbitals (or *p* and *d* orbitals).
2. Has cylindrical charge symmetry about bond axis.	2. Has maximum charge density in the cross-sectional plane of the orbitals.
3. Has free rotation.	3. Does not have free rotation.
4. Has lower energy.	4. Has higher energy.
5. Only one bond can exist between two atoms.	5. One or two bonds can exist between two atoms.

Problem 2.4 Show the electron distribution in MO's of (*a*) H_2, (*b*) H_2^+, (*c*) H_2^-, (d) He_2. Predict which are unstable.

Fill the lower-energy MO first with no more than two electrons.

(*a*) H_2 has a total of two electrons, therefore:

$$\underset{\sigma}{\underline{\uparrow\downarrow}}\ \underset{\sigma^*}{\underline{\quad}}$$

Stable (excess of two bonding electrons).

(*b*) H_2^+, formed from H^+ and H·, has one electron:

$$\underset{\sigma}{\underline{\uparrow}}\ \underset{\sigma^*}{\underline{\quad}}$$

Stable (excess of one bonding electron). Has less bonding strength than H_2.

(*c*) H_2^-, formed theoretically from $H:^-$ and H·, has three electrons:

$$\underset{\sigma}{\underline{\uparrow\downarrow}}\ \underset{\sigma^*}{\underline{\uparrow}}$$

Stable (has net bond strength of one bonding electron). The antibonding electron cancels the bonding strength of one of the bonding electrons.

(*d*) He_2 has four electrons, two from each He atom. The electron distribution is

$$\underset{\sigma}{\underline{\uparrow\downarrow}}\ \underset{\sigma^*}{\underline{\uparrow\downarrow}}$$

Not stable (antibonding and bonding electrons cancel, and there is no net bonding). Two He atoms are more stable than a He_2 molecule.

Problem 2.5 Since the σ MO formed from 2*s* AO's has a higher energy than the σ^* MO formed from 1*s* AO's, predict whether (*a*) Li_2, (*b*) Be_2 can exist.

The MO levels are as follows: $\sigma_{1s}\sigma^*_{1s}\sigma_{2s}\sigma^*_{2s}$, with energy increasing from left to right.

(*a*) Li_2 has six electrons, which fill the MO levels to give

$$\frac{\uparrow\downarrow}{\sigma_{1s}}\ \frac{\uparrow\downarrow}{\sigma^*_{1s}}\ \frac{\uparrow\downarrow}{\sigma_{2s}}\ \frac{\quad}{\sigma^*_{2s}}$$

designated $(\sigma_{1s})^2(\sigma^*_{1s})^2(\sigma_{2s})^2$. Li_2 has an excess of two electrons in bonding MO's and therefore can exist; it is by no means the most stable form of lithium.

(*b*) Be_2 would have eight electrons:

$$\frac{\uparrow\downarrow}{\sigma_{1s}}\ \frac{\uparrow\downarrow}{\sigma^*_{1s}}\ \frac{\uparrow\downarrow}{\sigma_{2s}}\ \frac{\uparrow\downarrow}{\sigma^*_{2s}}$$

There are no net bonding electrons, and Be_2 does not exist.

Stabilities of molecules can be qualitatively related to the **bond order,** defined as

$$\text{Bond order} \equiv \frac{(\text{Number of valence electrons in MO's}) - (\text{Number of valence electrons in MO*'s})}{2}$$

The bond order is usually equal to the number of σ and π bonds between two atoms—in other words, 1 for a single bond, 2 for a double bond, 3 for a triple bond.

Problem 2.6 The MO's formed when the two sets of the three 2*p* orbitals overlap are

$$\pi_{2p_y}\pi_{2p_z}\sigma_{2p_x}\pi^*_{2p_y}\pi^*_{2p_z}\sigma^*_{2p_x}$$

(the π and π^* pairs are degenerate). (*a*) Show how MO theory predicts the paramagnetism of O_2. (*b*) What is the bond order in O_2?

The valence sequence of MO's formed from overlap of the $n = 2$ AO's of diatomic molecules is

$$\sigma_{2s}\sigma^*_{2s}\pi_{2p_y}\pi_{2p_z}\sigma_{2p_x}\pi^*_{2p_y}\pi^*_{2p_z}\sigma^*_{2p_x}$$

O_2 has 12 elecrons to be placed in these MO's, giving

$$(\sigma_{2s})^2(\sigma^*_{2s})^2(\pi_{2p_y})^2(\pi_{2p_z})^2(\sigma_{2p_x})^2(\pi^*_{2p_y})^1(\pi^*_{2p_z})^1$$

(*a*) The electrons in the two, equal-energy, π^* MO*'s are unpaired; therefore, O_2 is paramagnetic.

(*b*) Electrons in the first two molecular orbitals cancel each other's effect. There are 6 electrons in the next 3 bonding orbitals and 2 electrons in the next 2 antibonding orbitals. There is a net bonding effect due to 4 electrons. The bond order is 1/2 of 4, or 2; the two O's are joined by a net double bond.

2.3 Hybridization of Atomic Orbitals

A carbon atom must provide four equal-energy orbitals in order to form four equivalent σ bonds, as in methane, CH_4. It is assumed that the four equivalent orbitals are formed by blending the 2*s* and the three 2*p* AO's give four new **hybrid orbitals**, called sp^3 HO's (Fig. 2.5). The shape of an sp^3 HO is shown in Fig. 2.6. The larger lobe, the "head," having most of the electron density, overlaps with an orbital of its bonding mate to form the bond. The smaller lobe, the "tail," is often omitted when HO's are depicted (see Fig. 2.11). However, at times the tail plays an important role in an organic reaction.

The AO's of carbon can hybridize in ways other than sp^3, as shown in Fig. 2.7. Repulsion between pairs of electrons causes these HO's to have the maximum bond angles and geometries summarized in Table 2.2. The sp^2 and sp HO's induce geometries about the C's as shown in Fig. 2.8. Only σ bonds, not π bonds, determine molecular shapes.

Energy

$2p$ ↑ ↑ — $2sp^3$ ↑ ↑ ↑ ↑

$2s$ ↑↓

ground-state valence shell of carbon atom

sp^3-hybridized state of carbon atom (before bonding)

Figure 2.5

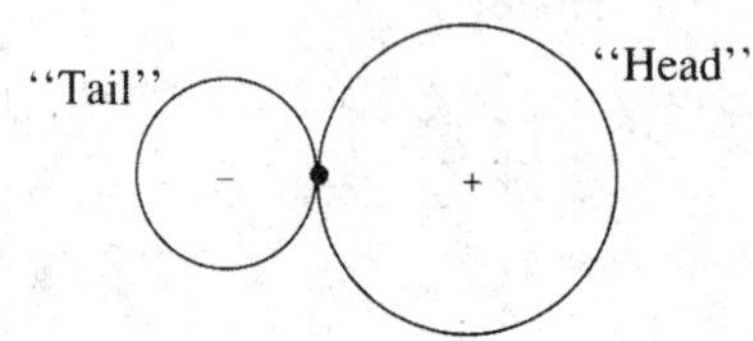

Figure 2.6

$2p$ ↑ ↑ — | ↑ ↑ ↑ ↑ sp^3 | ↑ ↑ ↑ sp^2 ↑ | ↑ ↑ sp ↑ ↑

$2s$ ↑↓

Ground state

Hybridized states

Figure 2.7

TABLE 2.2

TYPE	BOND ANGLE	GEOMETRY	NUMBER OF REMAINING p's	TYPE OF BOND FORMED
sp^3	109.5°	Tetrahedral*	0	σ
sp^2	120°	Trigonal planar	1	σ
sp	180°	Linear	2	σ

* See Fig. 1.2.

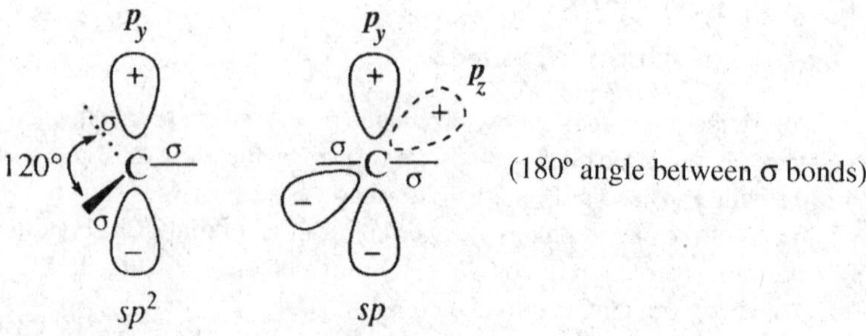

Figure 2.8

Problem 2.7 The H_2O molecule has a bond angle of 105°. (*a*) What type of AO's does O use to form the two equivalent σ bonds with H? (*b*) Why is this bond angle less than 109.5°?

(*a*)

$$_8O = \frac{\uparrow\downarrow}{1s}\ \frac{\uparrow\downarrow}{2s}\ \frac{\uparrow\downarrow}{2p_x}\ \frac{\uparrow}{2p_y}\ \frac{\uparrow}{2p_z}\quad \text{(ground state)}$$

O has two degenerate orbitals, the p_y and p_z, with which to form two equivalent bonds to H. However, if O used these AO's, the bond angle would be 90°, which is the angle between the *y*- and *z*-axes. Since the angle is actually 105°, which is close to 109.5°, O is presumed to use sp^3 HO's.

$$_8O = \frac{\uparrow\downarrow}{1s}\ \underbrace{\uparrow\downarrow\ \ \uparrow\downarrow\ \ \uparrow\ \ \uparrow}_{2sp^3}\quad (sp^3\ \text{HO's})$$

(*b*) Unshared pairs of electrons exert a greater repulsive force than do shared pairs, which causes a contraction of bond angles. The more unshared pairs there are, the greater is the contraction.

Problem 2.8 Each H—N—H bond angle in $:NH_3$ is 107°. What type of AO's does N use?

$$_7N = \frac{\uparrow\downarrow}{1s}\ \frac{\uparrow\downarrow}{2s}\ \frac{\uparrow}{2p_x}\ \frac{\uparrow}{2p_y}\ \frac{\uparrow}{2p_z}\quad \text{(ground state)}$$

If the ground-state N atom were to use its three equal-energy *p* AO's to form three equivalent N—H bonds, each H—N—H bond angle would be 90°. Since the actual bond angle is 107° rather than 90°, N, like O, uses sp^3 HO's:

$$_7N = \frac{\uparrow\downarrow}{1s}\ \underbrace{\uparrow\downarrow\ \ \uparrow\ \ \uparrow\ \ \uparrow}_{2sp^3}\quad (sp^3\ \text{HO's})$$

Apparently, for atoms in the second period forming more than one covalent bond (Be, B, C, N, and O), *a hybrid orbital must be provided for each* σ *bond and each unshared pair of electrons*. Atoms in higher periods also often use HO's.

Problem 2.9 Predict the shape of (*a*) the boron trifluoride molecule (BF_3) and (*b*) the boron tetrafluoride anion (BF_4^-). All bonds are equivalent.

(*a*) The HO's used by the central atom, in this case B, determine the shape of the molecule:

$$_5B = \frac{\uparrow\downarrow}{1s}\ \frac{\uparrow\downarrow}{2s}\ \frac{\uparrow}{2p_x}\ \frac{\ }{2p_y}\ \frac{\ }{2p_z}\quad \text{(ground state)}$$

There are three sigma bonds in BF_3 and no unshared pairs; therefore, three HO's are needed. Hence, B uses sp^2 HO's, and the shape is trigonal planar. Each F—B—F bond angle is 120°.

$$_5B = \frac{\uparrow\downarrow}{1s}\ \underbrace{\uparrow\ \ \uparrow\ \ \uparrow}_{2sp^2}\ \frac{\ }{2p_z}\quad (sp^2\ \text{hybrid state})$$

The empty p_z orbital is at right angles to the plane of the molecule.

(*b*) B in BF_4^- has four σ bonds and needs four HO's. B is now in an sp^3 hybrid state:

$$_5B = \frac{\uparrow\downarrow}{1s}\quad \underbrace{\uparrow\ \ \uparrow\ \ \uparrow\ \ -}_{2sp^3}\quad (sp^3\ \text{hybrid state})$$

used for bonding

The empty sp^3 hybrid orbital overlaps with a filled orbital of F^-, which holds two electrons:

$$:\ddot{\underset{..}{F}}:^- + BF_3 \longrightarrow BF_4^- \quad \text{(coordinate covalent bonding)}$$

The shape is tetrahedral; the bond angles are 109.5°.

Problem 2.10 Arrange the *s*, *p*, and the three *sp*-type HO's in order of decreasing energy.

The more *s* character in the orbital, the lower the energy. Therefore, the order of decreasing energy is

$$p > sp^3 > sp^2 > sp > s$$

Problem 2.11 What effect does hybridization have on the stability of bonds?

Hybrid orbitals can (*a*) overlap better and (*b*) provide greater bond angles, thereby minimizing the repulsion between pairs of electrons and making for great stability.

By use of the generalization that *each unshared and σ-bonded pair of electrons needs a hybrid orbital, but π bonds do not*, the number of hybrid orbitals **(HON)** needed by C or any other central atom can be obtained as

HON = (Number of σ bonds) + (Number of *unshared pairs* of electrons)

The hybridized state of the atom can then be predicted from Table 2.3. If more than four HO's are needed, *d* orbitals are hybridized with the *s* and the three *p*'s. If five HO's are needed, as in PCl_5, one *d* orbital is included to give **trigonal-bipyramidal** sp^3d HO's [Fig. 2.9(*a*)]. For six HO's, as in SF_6, two *d* orbitals are included to give **octahedral** sp^3d^2 HO's [Fig. 2.9(*b*)].

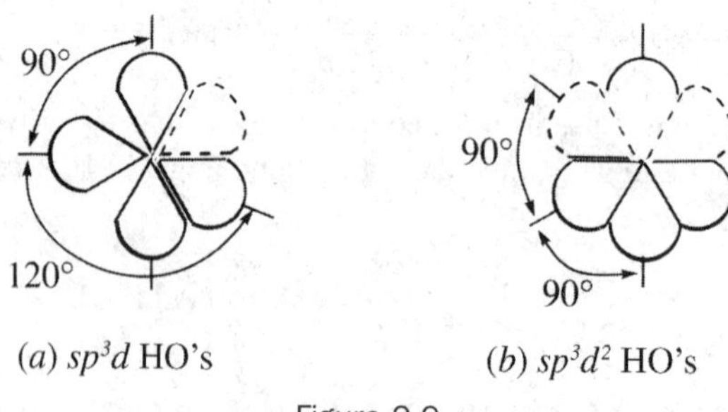

Figure 2.9

TABLE 2.3

HON	PREDICTED HYBRID STATE
2	sp
3	sp^2
4	sp^3
5	sp^3d
6	sp^3d^2

The preceding method can also be used for the multicovalent elements in the second period and, with few exceptions, in the higher periods of the periodic table.

Problem 2.12 Use the HON method to determine the hybridized state of the underlined elements:

(*a*) $\underline{C}HCl_3$ (*b*) $H_2C{=}\underline{C}H_2$ (*c*) $O{=}\underline{C}{=}O$ (*d*) $H\underline{C}{\equiv}\underline{N}{:}$ (*e*) $H_3\underline{\ddot{O}}^+$

	NUMBER OF σ BONDS +	NUMBER OF UNSHARED ELECTRON PAIRS	= HON	HYBRID STATE
(*a*)	4	0	4	sp^3
(*b*)	3	0	3	sp^2
(*c*)	2	0	2	sp
(*d*) C	2	0	2	sp
(*d*) N	1	1	2	sp
(*e*)	3	1	4	sp^3

2.4 Electronegativity and Polarity

The relative tendency of a bonded atom in a molecule to attract electrons is expressed by the term **electronegativity**. The higher the electronegativity, the more effectively does the atom attract and hold electrons. A bond formed by atoms of dissimilar electronegativities is called **polar**. A **nonpolar** covalent bond exists between atoms having a very small or zero difference in electronegativity. A few relative electronegativities are

$$F(4.0) > O(3.5) > Cl, N(3.0) > Br(2.8) > S, C, I(2.5) > H(2.1)$$

The more electronegative element of a covalent bond is relatively negative in charge, while the less electronegative element is relatively positive. The symbols $\delta+$ and $\delta-$ represent partial charges (**bond polarity**). These partial charges should not be confused with ionic charges. Polar bonds are indicated by $+\!\!\longrightarrow$; the head points toward the more electronegative atom.

The vector sum of all individual bond moments gives the net **dipole moment** of the molecule.

Problem 2.13 What do the molecular dipole moments $\mu = 0$ for CO_2 and $\mu = 1.84$ D for H_2O tell you about the shapes of these molecules?

In CO_2:

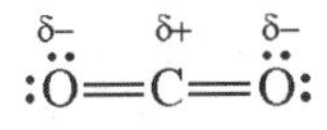

$$\overset{\delta-}{:\ddot{O}}=\overset{\delta+}{C}=\overset{\delta-}{\ddot{O}:}$$

O is more electronegative than C, and each C—O bond is polar as shown. A zero dipole moment indicates a symmetrical distribution of $\delta-$ charges about the $\delta+$ carbon. The geometry must be linear; in this way, individual bond moments cancel:

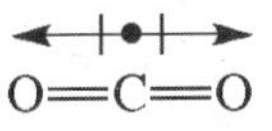

$$\overset{\longleftarrow\!+\bullet+\!\longrightarrow}{O=C=O}$$

H_2O also has polar bonds. However, since there is a net dipole moment, the individual bond moments do not cancel, and the molecule must have a *bent* shape:

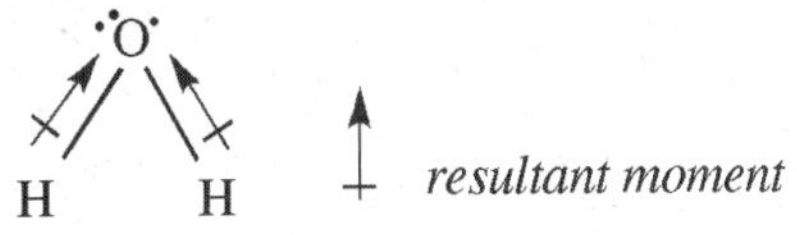

H—Ö—H (bent), resultant moment ↑ *resultant moment*

2.5 Oxidation Number

The oxidation number (ON) is a value assigned to an atom based on relative electronegativities. It equals the group number minus the number of assigned electrons, when the bonding electrons are assigned to the more electronegative atom. The sum of all (ON)'s equals the charge on the species.

Problem 2.14 Determine the oxidation number of each C, $(ON)_C$, in: (*a*) CH_4, (*b*) CH_3OH, (*c*) CH_3NH_2, (*d*) $H_2C{=}CH_2$. Use the data $(ON)_N = -3$; $(ON)_H = 1$; $(ON)_O = -2$.

All examples are molecules; therefore, the sum of all (ON) values is 0.

(*a*) $(ON)_C + 4(ON)_H = 0;\quad (ON)_C + (4 \times 1) = 0;\quad (ON)_C = -4$
(*b*) $(ON)_C + (ON)_O + 4(ON)_H = 0;\quad (ON)_C + (-2) + 4 = 0;\quad (ON)_C = -2$
(*c*) $(ON)_C + (ON)_N + 5(ON)_H = 0;\quad (ON)_C + (-3) + 5 = 0;\quad (ON)_C = -2$
(*d*) Since both C atoms are equivalent,

$$2(ON)_C + 4(ON)_H = 0;\qquad 2(ON)_C + 4 = 0;\qquad (ON)_C = -2$$

2.6 Intermolecular Forces

(a) **Diplole-dipole** interaction results from the attraction of the $\delta+$ end of one polar molecule for the $\delta-$ end of another polar molecule.

(b) **Hydrogen-bond.** X—H and :Y may be bridged X—H---:Y if X and Y are small, highly electronegative atoms such as F, O, and N. H-bonds also occur intramolecularly.

(c) **London (van der Waals) forces.** Electrons of a nonpolar molecule may momentarily cause an imbalance of charge distribution in neighboring molecules, thereby inducing a temporary dipole moment. Although constantly changing, these induced dipoles result in a weak net attractive force.

The greater the molecular weight of the molecule, the greater the number of electrons and the greater these forces.

The order of attraction is

H-bond >> dipole-dipole > London forces

Problem 2.15 Account for the following progressions in boiling point. (*a*) CH_4, −161.5°C; Cl_2, −34°C; CH_3Cl, −24°C. (*b*) CH_3CH_2OH, 78°C; CH_3CH_2F, 46°C; $CH_3CH_2CH_3$, −42°C.

The greater the intermolecular force, the higher the boiling point. Polarity and molecular weight must be considered.

(*a*) Only CH_3Cl is polar, and it has the highest boiling point. CH_4 has a lower molecular weight (16 g/mole) than has Cl_2 (71 g/mole) and therefore has the lowest boiling point.

(*b*) Only CH_3CH_2OH has H-bonding, which is a stronger force of intermolecular attraction than the dipole-dipole attraction of CH_3CH_2F. $CH_3CH_2CH_3$ has only London forces, the weakest attraction of all.

Problem 2.16 The boiling points of *n*-pentane and its isomer neopentane are 36.2°C and 9.5°C, respectively. Account for this difference. (See Problem 1.4 for the structural formulas.)

These isomers are both nonpolar. Therefore, another factor, the shape of the molecule, influences the boiling point. The shape of *n*-pentane is rodlike, whereas that of neopentane is spherelike. Rods can touch along their entire length; spheres touch only at a point. The more contact between molecules, the greater the London forces. Thus, the boiling point of *n*-pentane is higher.

2.7 Solvents

The oppositely charged ions of salts are strongly attracted by electrostatic forces, thereby accounting for the high melting and boiling points of salts. These forces of attraction must be overcome in order for salts to dissolve in a solvent. **Nonpolar solvents** have a zero or very small dipole moment. **Protic solvents** are highly polar molecules that have an H that can form an H-bond. **Aprotic solvents** are highly polar molecules that do not have an H that can form an H-bond.

Problem 2.17 Classify the following solvents: (*a*) $(CH_3)_2S{=}O$, dimethyl sulfoxide; (*b*) CCl_4, carbon tetrachloride; (*c*) C_6H_6, benzene; (*d*) HC(=O)N$(CH_3)_2$ Dimethylformamide; (*e*) CH_3OH, methanol; (*f*) liquid NH_3.

Nonpolar: (*b*) Because of the symmetrical tetrahedral molecular shape, the individual C—Cl bond moments cancel. (*c*) With few exceptions, hydrocarbons are nonpolar. *Protic*: (*e*) and (*f*). *Aprotic*: (*a*) and (*d*). The S=O and C=O groups are strongly polar, and the H's attached to C do not typically form H-bonds.

Problem 2.18 Mineral oil, a mixture of high-molecular-weight hydrocarbons, dissolves in *n*-hexane but not in water or ethyl alcohol, CH_3CH_2OH. Explain.

Attractive forces between nonpolar molecules such as mineral oil and *n*-hexane are very weak. Therefore, such molecules can mutually mix and solution is easy. The attractive forces between polar H_2O or C_2H_5OH molecules are strong H-bonds. Most nonpolar molecules cannot overcome these H-bonds and therefore do not dissolve in such **polar protic** solvents.

Problem 2.19 Explain why CH_3CH_2OH is much more soluble in water than is $CH_3(CH_2)_3CH_2OH$.

The OH portion of an alcohol molecule tends to interact with water—it is **hydrophilic**. The hydrocarbon portion does not interact. Rather, it is repelled—it is **hydrophobic**. The larger the hydrophobic portion, the less soluble in water is the molecule.

Problem 2.20 Explain why NaCl dissolves in water.

Water, a protic solvent, helps separate the strongly attracting ions of the solid salt by **solvation**. Several water molecules surround each positive ion (Na^+) by an **ion-dipole** attraction. The O atoms, which are the negative ends of the molecular dipole, are attracted to the cation. H_2O typically forms an H-bond with the negative ion (in this case Cl^-).

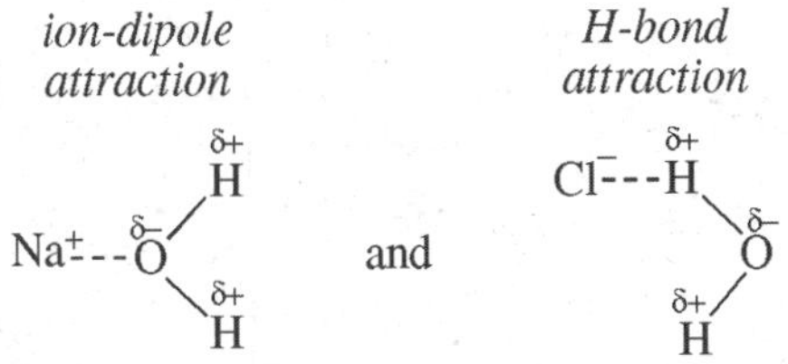

ion-dipole attraction $Na^+ \text{- - } \overset{\delta-}{O}(\overset{\delta+}{H})_2$ and *H-bond attraction* $Cl^- \text{- - } \overset{\delta+}{H}\text{—}\overset{\delta-}{O}\text{—}\overset{\delta+}{H}$

Problem 2.21 Compare the ways in which NaCl dissolves in water and in dimethyl sulfoxide.

The way in which NaCl, a typical salt, dissolves in water, a typical protic solvent, was discussed in Problem 2.20. Dimethyl sulfoxide also solvates positive ions by an ion-dipole attraction; the O of the S=O group is attracted to the cation. However, since this is an aprotic solvent, there is no way for an H-bond to be formed and the *negative ions are not solvated when salts dissolve in aprotic solvents*. The S, the positive pole, is surrounded by the methyl groups and cannot get close enough to solvate the anion.

The bare negative ions discussed in Problem 2.21 have a greatly enhanced reactivity. The small amounts of salts that dissolve in nonpolar or weakly polar solvents exist mainly as **ion-pairs** or ion-clusters, where the oppositely charged ions are close to each other and move about as units. **Tight** ion-pairs have no solvent molecules between the ions; **loose** ion-pairs are separated by a small number of solvent molecules.

2.8 Resonance and Delocalized π Electrons

Resonance theory describes species for which a single Lewis electron structure cannot be written. As an example, consider dinitrogen oxide, N_2O:

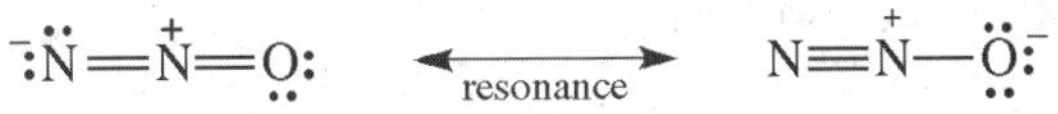

$$:\overset{..}{N}^{-}=\overset{+}{N}=\underset{..}{O}: \quad \xleftrightarrow{\text{resonance}} \quad N\equiv\overset{+}{N}—\overset{..}{\underset{..}{O}}:^{-}$$

Calculated Bond Length	0.120 0.115	0.110 0.147
Observed Bond Length	0.112 0.119	0.112 0.119

A comparison of the calculated and observed bond lengths show that neither structure is correct. Nevertheless, these **contributing** (**resonance**) structures tell us that the actual **resonance hybrid** has some double-bond character between N and O, and some triple-bond character between N and N. This state of affairs is described by the non-Lewis structure:

$$:\overset{\delta-}{N}\equiv\overset{+}{N}=\overset{\delta-}{O}:$$

in which broken lines stand for the partial bonds in which there are delocalized p electrons in an extended π bond created from overlap of p orbitals on each atom. See also the orbital diagram in Fig. 2.10. The symbol $\leftrightarrow$ denotes resonance, *not equilibrium*.

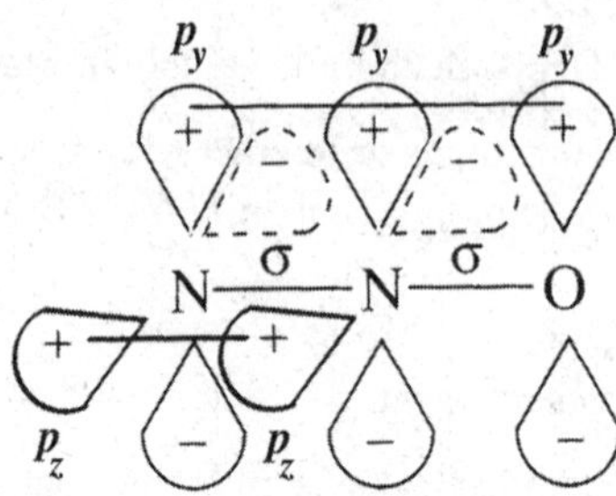

Figure 2.10

The energy of the hybrid, E_h, is always less than the calculated energy of any hypothetical contributing structure, E_c. The difference between these energies is the **resonance (delocalization) energy**, E_r:

$$E_r = E_c - E_h$$

The more nearly equal in energy the contributing structures, the greater the resonance energy and the less the hybrid looks like any of the contributing structures. When contributing structures have dissimilar energies, the hybrid looks most like the lowest-energy structure.

Contributing structures (a) differ only in positions of electrons (atomic nuclei must have the same positions) and (b) must have the same number of paired electrons. Relative energies of contributing structures are assessed by the following rules:

1. Structures with the greatest number of covalent bonds are most stable. However, for second-period elements (C, O, N), the octet rule must be observed.
2. With a few exceptions, structures with the least amount of formal charges are most stable.
3. If all structures have formal charge, the most stable (lowest energy) one has − on the more electronegative atom and + on the more electropositive atom.
4. Structures with like formal charges on adjacent atoms have very high energies.
5. Resonance structures with electron-deficient, positively charged atoms have very high energy and are usually ignored.

Problem 2.22 Write contributing structures, showing formal charges when necessary, for (a) ozone, O_3; (b) CO_2; (c) hydrazoic acid, HN_3; (d) isocyanic acid, HNCO. Indicate the most and least stable structures and give reasons for your choices. Give the structure of the hybrid.

(a) $:\ddot{O}=\overset{+}{\ddot{O}}-\ddot{\underset{..}{O}}:$ $\longleftrightarrow$ $:\ddot{\underset{..}{O}}-\overset{+}{\ddot{O}}=\ddot{O}:$ (equal-energy structures). The hybrid is $:\overset{\delta-}{\ddot{O}}\!=\!\overset{+}{\ddot{O}}\!=\!\overset{\delta-}{\ddot{O}}:$.

(b) $\ddot{O}=C=\ddot{O}:$ $\longleftrightarrow$ $:\overset{-}{\ddot{\underset{..}{O}}}-C\equiv\overset{+}{O}:$ $\longleftrightarrow$ $:\overset{+}{O}\equiv C-\overset{-}{\ddot{\underset{..}{O}}}:$

(1) (2) (3)

(1) is most stable; it has no formal charge. (2) and (3) have equal energy and are least stable because they have formal charges. In addition, in both (2) and (3), one O, an electronegative element, bears a + formal charge. Since (1) is so much more stable than (2) and (3), the hybrid is $:\underset{..}{O}=C=\underset{..}{O}:$, which is just (1).

(*c*) $\mathrm{H{-}\ddot{N}{=}\overset{+}{N}{=}\overset{-}{\ddot{N}}{:}}$ ⟷ $\mathrm{H{-}\overset{-}{\underset{\cdot\cdot}{\ddot{N}}}{-}\overset{+}{N}{\equiv}N{:}}$ ⟷ $\mathrm{H{-}\overset{+}{N}{\equiv}\overset{+}{N}{=}\overset{2-}{\underset{\cdot\cdot}{\ddot{N}}}{:}}$ ⟷ $\mathrm{H{-}\overset{2+}{N}{=}\ddot{N}{-}\overset{2-}{\underset{\cdot\cdot}{\ddot{N}}}{:}}$

(1) (2) (3) (4)

(1) and (2) have about the same energy and are the most stable, since they have the least amount of formal charge. (3) has a very high energy, since it has + charge on adjacent atoms and, in terms of absolute value, a total formal charge of 4. (4) has a very high energy because the N bonded to H has only six electrons. The hybrid, composed of (1) and (2), is as follows:

$$\mathrm{H{-}\overset{\delta-}{\ddot{N}}{=\!=}\overset{+}{N}{\equiv\!\equiv}\overset{\delta-}{N}{:}}$$

(*d*) $\mathrm{H{-}\ddot{N}{=}C{=}\ddot{O}{:}}$ ⟷ $\mathrm{H{-}\overset{-}{\underset{\cdot\cdot}{\ddot{N}}}{-}C{\equiv}\overset{+}{O}{:}}$ ⟷ $\mathrm{H{-}\overset{+}{N}{\equiv}C{-}\overset{-}{\underset{\cdot\cdot}{\ddot{O}}}{:}}$

(1) (2) (3)

(1) has no formal charge and is most stable. (2) is least stable since the − charge is on N rather than on the more electronegative O as in (3). The hybrid is $\mathrm{H{-}\underset{\cdot\cdot}{N}{=}C{=}\underset{\cdot\cdot}{O}{:}}$, which is the same as (1), the most stable contributing structure.

Problem 2.23 (*a*) Write contributing structures and the delocalized structure for (i) NO_2^- and (ii) NO_3^-. (*b*) Use *p* AO's to draw a structure showing the delocalization of the *p* electrons in an extended π bond for (i) and (ii). (*c*) Compare the stability of the hybrids of each.

(*a*) (i) $\mathrm{{:}\overset{-}{\underset{\cdot\cdot}{\ddot{O}}}{-}\ddot{N}{=}\ddot{O}{:}}$ ⟷ $\mathrm{{:}\ddot{O}{=}N{-}\overset{-}{\underset{\cdot\cdot}{\ddot{O}}}{:}}$ or $\left[\mathrm{{:}\overset{-\frac{1}{2}}{\ddot{O}}{=\!=}N{=\!=}\overset{-\frac{1}{2}}{\underset{\cdot\cdot}{O}}{:}}\right]$

The − is delocalized over both O's so that each can be assumed to have a $-\frac{1}{2}$ charge. Each N—O bond has the same bond length.

(ii) $\mathrm{{:}\overset{-}{\underset{\cdot\cdot}{\ddot{O}}}{-}\overset{+}{N}({=}\ddot{O}{:})({-}\ddot{O}{:}^-)}$ ⟷ $\mathrm{{:}\ddot{O}{=}\overset{+}{N}({-}\ddot{O}{:}^-)({-}\ddot{O}{:}^-)}$ ⟷ $\mathrm{{:}\overset{-}{\underset{\cdot\cdot}{\ddot{O}}}{-}\overset{+}{N}({-}\ddot{O}{:}^-)({=}\ddot{O}{:})}$ or $\left[\mathrm{{:}\overset{-\frac{2}{3}}{\ddot{O}}{=\!=}\overset{+}{N}({=\!=}\overset{-\frac{2}{3}}{\ddot{O}}{:})({=\!=}\overset{-\frac{2}{3}}{\ddot{O}}{:})}\right]^-$

The – charges are delocalized over three O's so that each has a $-\frac{2}{3}$ charge.

(*b*) See Fig. 2.11.

(*c*) We can use resonance theory to compare the stability of these two ions because they differ in only one feature—the number of O's on each N, which is related to the oxidation numbers of the N's. We could not, for example, compare NO_3^- and HSO_3^-, since they differ in more than one way; N and S are in different groups and periods of the periodic table. NO_3^- is more stable than NO_2^- since the charge on NO_3^- is delocalized (dispersed) over a greater number of O's and since NO_3^- has a more extended π bond system.

Problem 2.24 Indicate which one of the following pairs of resonance structures is the less stable and is an unlikely contributing structure. Give reasons in each case.

(*a*) $:\ddot{O}^{-}—\overset{2+}{N}(—\ddot{O}:^{-})_2$ $\longleftrightarrow$ $:\ddot{O}^{-}—\overset{+}{N}(—\ddot{O}:^{-})(=\ddot{O}:)$ (*b*) $H_2\overset{+}{C}—\ddot{O}:^{-}$ $\longleftrightarrow$ $H_2\overset{-}{C}—\ddot{O}:^{+}$

I II III IV

(*c*) $H_2C{=}CH—\overset{-}{C}H_2$ $\longleftrightarrow$ $H_2\overset{+}{C}—\overset{-}{C}H—\overset{-}{C}H_2$ (*d*) $H—\overset{-}{C}::\overset{+}{N}:$ $\longleftrightarrow$ $H—C:::N:$

V VI VII VIII

(*e*) $H_3C—\ddot{Cl}:$ $\longleftrightarrow$ $H_3\overset{-}{C}{=}\overset{+}{Cl}:$

IX X

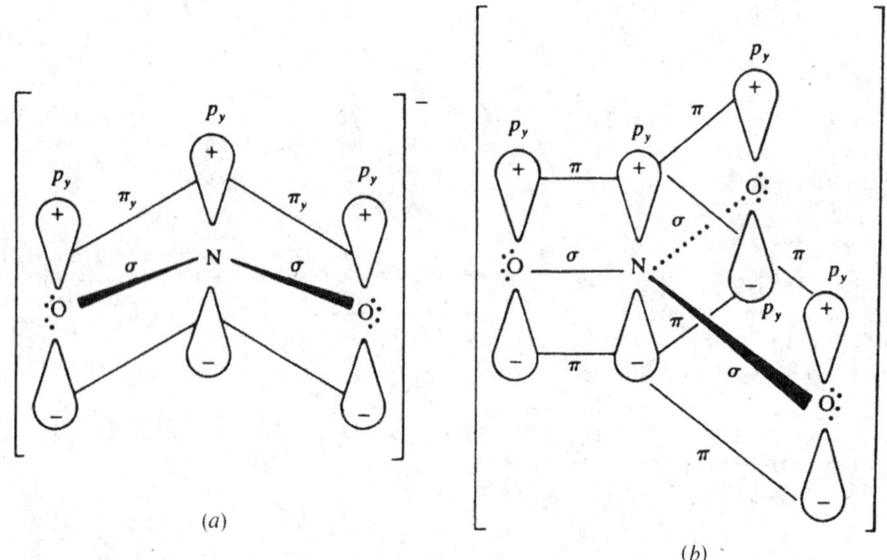

Figure 2.11

(*a*) I has fewer covalent bonds, more formal charge, and an electron-deficient N.
(*b*) IV has + on the more electronegative O.
(*c*) VI has similar − charges on adjacent C's, fewer covalent bonds, more formal charge, and an electron-deficient C.
(*d*) VII has fewer covalent bonds and a + on the more electronegative N, which is also electron-deficient.
(*e*) C in X has 10 electrons; this is not possible with the elements of the second period.

SUPPLEMENTARY PROBLEMS

Problem 2.25 Distinguish between an AO, an HO, an MO and a localized MO.

An AO is a region of space in an *atom* in which an electron may exist. An HO is mathematically fabricated from some number of AO's to explain equivalency of bonds. An MO is a region of space about the *entire molecule* capable of accommodating electrons. A localized MO is a region of space between a pair of bonded atoms in which the bonding electrons are assumed to be present.

Problem 2.26 Show the orbital population of electrons for unbonded N in (*a*) ground state, and for (*b*) sp^3, (*c*) sp^2, and (*d*) sp hybrid states.

(*a*)	$\underset{1s}{\underline{\uparrow\downarrow}}\ \underset{2s}{\underline{\uparrow\downarrow}}\ \underset{2p}{\underline{\uparrow}\ \underline{\uparrow}\ \underline{\uparrow}}$	(*c*)	$\underset{1s}{\underline{\uparrow\downarrow}}\ \underset{2sp^2}{\underline{\uparrow\downarrow}\ \underline{\uparrow}\ \underline{\uparrow}}\ \underset{2p}{\underline{\uparrow}}$
(*b*)	$\underset{1s}{\underline{\uparrow\downarrow}}\ \underset{2sp^3}{\underline{\uparrow\downarrow}\ \underline{\uparrow}\ \underline{\uparrow}\ \underline{\uparrow}}$	(*d*)	$\underset{1s}{\underline{\uparrow\downarrow}}\ \underset{2sp^2}{\underline{\uparrow\downarrow}\ \underline{\uparrow}}\ \underset{2p}{\underline{\uparrow}\ \underline{\uparrow}}$

Note that since the energy difference between hybrid and p orbitals is so small, Hund's rule prevails over the Aufbau principle.

Problem 2.27 (*a*) NO_2^+ is linear, (*b*) NO_2^- is bent. Explain in terms of the hybrid orbitals used by N.

(*a*) NO_2^+, $:\ddot{O}{=}\overset{+}{N}{=}\ddot{O}:$. N has two σ bonds, no unshared pairs of electrons and therefore needs two hybrid orbitals. N uses *sp* hybrid orbitals and the σ bonds are linear. The geometry is controlled by the arrangement of the sigma bonds.

(*b*) NO_2^-, $:\ddot{O}{=}\ddot{N}{:}\ddot{\underset{..}{O}}:^-$. N has two σ bonds and one unshared pair of electrons and, therefore, needs three hybrid orbitals. N uses sp^2 hybrid HO's, and, the bond angle is about 120°.

Problem 2.28 Draw an orbital representation of the cyanide ion, $:C{\equiv}N:^-$.

See Fig. 2.12. The C and N each have one σ bond and one unshared pair of electrons, and therefore each needs two *sp* hybrid HO's. On each atom, one *sp* hybrid orbital forms a σ bond, while the other has the unshared pair. Each atom has a p_y AO and a p_z AO. The two p_y orbitals overlap to form a π_y bond in the *xy*-plane; the two p_z orbitals overlap to form a π_z bond in the *xz*-plane. Thus, two π bonds at right angles to each other and a σ bond exist between the C and N atoms.

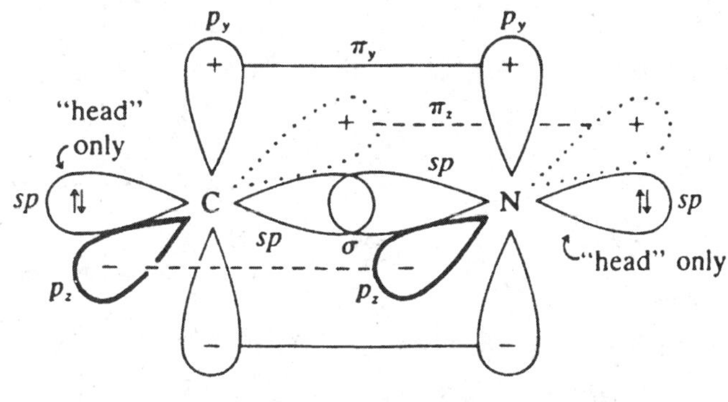

Figure 2.12

Problem 2.29 (*a*) Which of the following molecules possess polar bonds: F_2, HF, BrCl, CH_4, $CHCl_3$, CH_3OH? (*b*) Which are polar molecules?

(*a*) HF, BrCl, CH_4, $CHCl_3$, CH_3OH.
(*b*) HF, BrCl, $CHCl_3$, CH_3OH. The symmetrical individual bond moments in CH_4 cancel.

Problem 2.30 Considering the difference in electronegativity between O and S, would H_2O or H_2S exhibit greater (*a*) dipole-dipole attraction, (*b*) H-bonding?

(*a*) H_2O (*b*) H_2O.

Problem 2.31 Nitrogen trifluoride (NF_3) and ammonia (NH_3) have an electron pair at the fourth corner of a tetrahedron and have similar electronegativity *differences* between the elements (1.0 for N and F and 0.9 for N and H). Explain the larger dipole moment of ammonia (1.46 D) as compared with that of NF_3 (0.24 D).

The dipoles in the three N—F bonds are toward F, see Fig. 2.13(*a*), and oppose and tend to cancel the effect of the unshared electron pair on N. In NH_3, the moments for the three N—H bonds are toward N, see Fig. 2.13(*b*), and add to the effect of the electron pair.

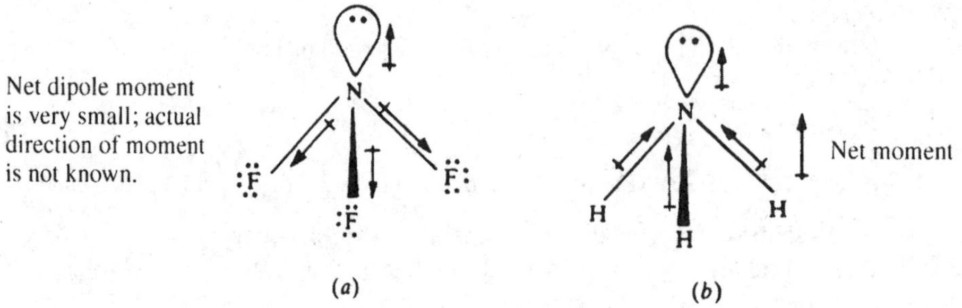

Net dipole moment is very small; actual direction of moment is not known.

(*a*) (*b*) Net moment

Figure 2.13

Problem 2.32 NH_4^+ salts are much more soluble in water than are the corresponding Na^+ salts. Explain.

Na^+ is solvated merely by an ion-dipole interaction. NH_4^+ is solvated by H-bonding

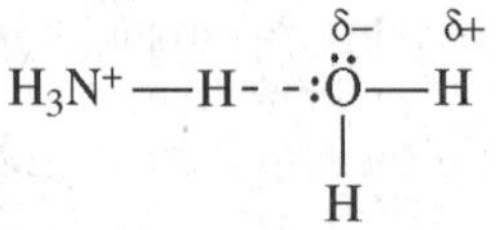

$$\overset{}{H_3N^+}\text{—H- -:}\overset{\delta-}{\ddot{O}}\text{—}\overset{\delta+}{H}$$
$$\qquad\qquad\quad | $$
$$\qquad\qquad\quad H$$

which is a stronger attractive force.

Problem 2.33 The F^- of dissolved NaF is more reactive in dimethyl sulfoxide:

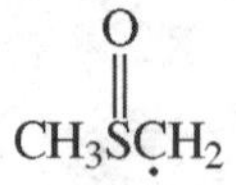

$$CH_3\overset{\overset{O}{\|}}{\underset{\cdot}{S}}CH_3$$

and in acetonitrile, $CH_3C{\equiv}N$, than in CH_3OH. Explain.

H-bonding prevails in CH_3OH (a protic solvent), $CH_3OH\text{---}F^-$, thereby decreasing the reactivity of F^-. CH_3SOCH_3 and CH_3CN are aprotic solvents; their C—H H's do not H-bond.

Problem 2.34 Find the oxidation of the C in (*a*) CH_3Cl, (*b*) CH_2Cl_2, (*c*) H_2CO, (*d*) HCOOH, and (*e*) CO_2, if $(ON)_{Cl} = -1$.

From Section 2.5:

(*a*) $(ON)_C + (3 \times 1) + (-1) = 0$; $(ON)_C = -2$

(*b*) $(ON)_C + (2 \times 1) + [2(-1)] = 0$; $(ON)_C = 0$

(*c*) $(ON)_C + (2 \times 1) + [1(-2)] = 0$; $(ON)_C = 0$

(*d*) $(ON)_C + 2 + (-4) = 0$; $(ON)_C = 2$

(*e*) $(ON)_C + (-4) = 0$; $(ON)_C = 4$

Problem 2.35 Give a True or False answer to each question and justify your answer. (*a*) Since in polyatomic anions XY_m^{n-} (such as SO_4^{2-} and BF_4^-), the central atom X is usually less electronegative than the peripheral atom Y, it tends to acquire a positive oxidation number. (*b*) Oxidation numbers tend to be smaller values than formal charges. (*c*) A bond between dissimilar atoms always leads to nonzero oxidation numbers. (*d*) Fluorine never has a positive oxidation number.

(*a*) True. The bonding electrons will be allotted to the more electronegative peripheral atoms, leaving the central atoms with a positive oxidation number.
(*b*) False. In determining formal charges, an electron of each shared pair is assigned to each bonded atom. In determining oxidation numbers, pairs of electrons are involved, and more electrons are moved to or away from an atom. Hence, larger oxidation numbers result.
(*c*) False. The oxidation numbers will be zero if the dissimilar atoms have the same electronegativity, as in PH_3.
(*d*) True. F is the most electronegative element; in F_2, it has a zero oxidation number.

Problem 2.36 Which of the following transformations of organic compounds are oxidations, which are reductions, and which are neither?

(*a*) $H_2C{=}CH_2 \longrightarrow CH_3CH_2OH$ (*c*) $CH_3CHO \longrightarrow CH_3COOH$ (*e*) $HC{\equiv}CH \longrightarrow H_2C{=}CH_2$
(*b*) $CH_3CH_2OH \longrightarrow CH_3CH{=}O$ (*d*) $H_2C{=}CH_2 \longrightarrow CH_3CH_2Cl$

To answer the question, determine the average oxidation numbers (ON) of the C atoms in reactant and in product. An increase (more positive or less negative) in ON signals an oxidation; a decrease (more negative or less positive) signals a reduction; no change means neither.

(*a*) and (*d*) are *neither*, because $(ON)_C$ is invariant at -2. (*b*) and (*c*) are *oxidations*, the respective changes being from -2 to -1 and from -1 to 0. (*e*) is a reduction, the change being from -1 to -2.

Problem 2.37 Irradiation with ultraviolet (uv) light permits rotation about a π bond. Explain in terms of bonding and antibonding MO's.

Two *p* AO's overlap to form two pi MO's, π (bonding) and π^* (antibonding). The two electrons in the original *p* AO's fill only the π MO (ground state). A photon of UV causes excitation of one electron from π to π^* (excited state).

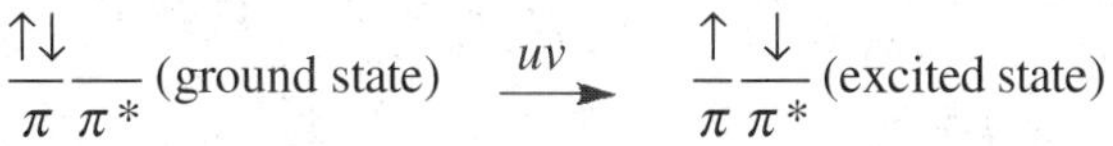

$$\underset{\pi}{\underline{\uparrow\downarrow}}\ \underset{\pi^*}{\underline{\quad}}\ \text{(ground state)} \xrightarrow{uv} \underset{\pi}{\underline{\uparrow}}\ \underset{\pi^*}{\underline{\downarrow}}\ \text{(excited state)}$$

(Initially, the excited electron does not change its spin.) The bonding effects of the two electrons cancel. There is now only a sigma bond between the bonded atoms, and rotation about the bond can occur.

Problem 2.38 Write the contributing resonance structures and the delocalized hybrid for (*a*) BCl_3, (*b*) H_2CN_2 (diazomethane).

(*a*) Boron has six electrons in its outer shell in BCl_3 and can accommodate eight electrons by having a B—Cl bond assume some double-bond character.

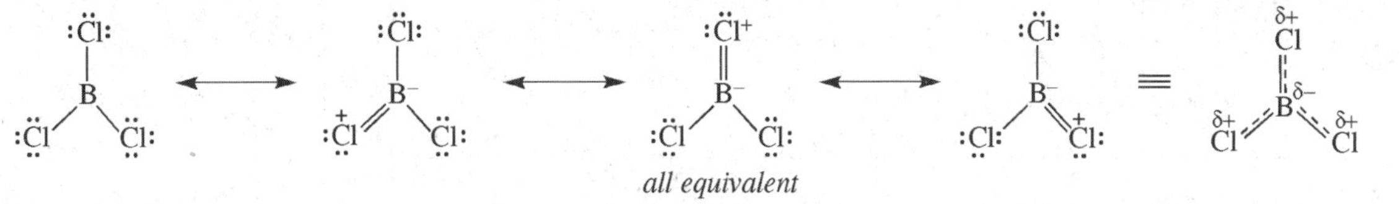

:C̈l: / B / :C̈l C̈l: ⟷ :C̈l: / B⁻ / :C̈l⁺ (=B) C̈l: ⟷ :Cl⁺ (=B) / B⁻ / :C̈l C̈l: ⟷ :C̈l: / B⁻ / :C̈l: Cl⁺ (=B) ≡ δ+ Cl / δ− B / δ+ Cl δ+ Cl

all equivalent

(*b*) $H_2C{=}\overset{+}{N}{=}\overset{-}{\ddot{N}}{:} \longleftrightarrow H_2\overset{-}{\ddot{C}}{-}\overset{+}{N}{\equiv}N{:} \equiv \overset{\delta-}{C}{-}{-}\overset{+}{N}{\equiv}{\equiv}\overset{\delta-}{N}{:}$

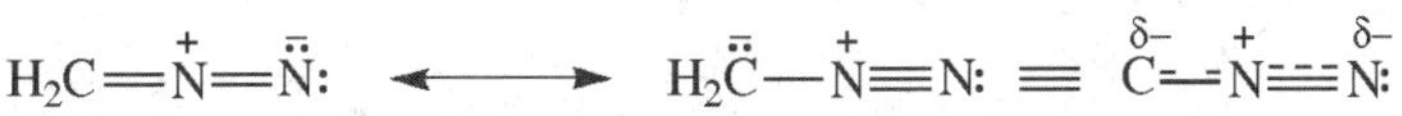

Problem 2.39 Arrange the contributing structures for (*a*) vinyl chloride, $H_2C{=}CHCl$, and (*b*) formic acid, HCOOH, in order of increasing importance (increasing stability) by assigning numbers starting with 1 for most important and stable.

(*a*) $H_2C{=}CH{-}\ddot{\underset{..}{Cl}}{:} \longleftrightarrow H_2\overset{-}{\ddot{C}}{-}CH{=}\overset{+}{\underset{..}{Cl}}{:} \longleftrightarrow H_2\overset{+}{C}{-}CH{=}\overset{..-}{\underset{..}{Cl}}{:}$

I II III

I is most stable because it has no formal charge. III is least stable since it has an electron-deficient C. In III, Cl uses an empty 3*d* orbital to accommodate a fifth pair of electrons. Fluorine could not do this. The order of stability is

$$I(1) > II(2) > III(3)$$

(*b*) $H{-}\underset{\|}{\overset{:O:}{C}}{-}\ddot{\underset{..}{O}}{-}H \longleftrightarrow H{-}\overset{:\ddot{O}:^-}{\underset{|}{C}}{=}\overset{+}{\underset{..}{O}}{-}H \longleftrightarrow H{-}\overset{:\ddot{O}:^-}{\underset{+}{C}}{-}\ddot{\underset{..}{O}}{-}H \longleftrightarrow H{-}\overset{:\overset{+}{O}:}{\underset{..}{C}}{-}\ddot{\underset{..}{O}}{-}H$

V VI VII VIII

V and VI have the greater number of covalent bonds and are more stable than either VII or VIII. V has no formal charge and is more stable than VI. VIII is less stable than VII, since VIII's electron deficiency is on O, which is a more electronegative atom than the electron-deficient C of VII. The order of stability is

$$V(1) > VI(2) > VII(3) > VIII(4)$$

Problem 2.40 What is the difference between isomers and contributing resonance structures?

Isomers and *real compounds* differ in the arrangement of their atoms. Contributing structures have the same arrangement of atoms; they differ only in the distribution of their electrons. Their *imaginary structures* are written to give some indication of the electronic structure of certain species for which a typical Lewis structure cannot be written.

Problem 2.41 Use the HON method to determine the hybridized state of the underlined elements:

(*a*) $H\underline{C}{\equiv}CH$ (*b*) $H_2\underline{C}{=}O$ (*c*) $HC{\equiv}\underline{C}^-$ (*d*) $\underline{Al}Cl_6^{3-}$ (*e*) $\underline{P}F_5$ (*f*) $CH_3\underline{\ddot{O}}CH_3$

	NUMBER OF σ BONDS	+ NUMBER OF UNSHARED ELECTRON PAIRS	= HON	HYBRID STATE
(*a*)	2	0	2	*sp*
(*b*)	3	0	3	sp^2
(*c*)	1	1	2	*sp*
(*d*)	6	0	6	sp^3d^2
(*e*)	5	0	5	sp^3d
(*f*)	2	2	4	sp^3

Chemical Reactivity and Organic Reactions

3.1 Reaction Mechanism

The way in which a reaction occurs is called a **mechanism**. A reaction may occur in one step or, more often, by a sequence of several steps. For example, A + B → X + Y may proceed in two steps:

$$(1)\ \mathrm{A} \longrightarrow \mathrm{I} + \mathrm{X} \qquad \text{followed by} \qquad (2)\ \mathrm{B} + \mathrm{I} \longrightarrow \mathrm{Y}$$

Substances such as I, formed in intermediate steps and consumed in later steps, are called **intermediates**. Sometimes the same reactants can give different products via different mechanisms.

3.2 Carbon-Containing Intermediates

Carbon-containing intermediates often arise from two types of bond cleavage:

Heterolytic (polar) cleavage. Both electrons go with one group, for example:

$$\mathrm{A{:}B} \longrightarrow \mathrm{A}^+ + {:}\mathrm{B}^- \quad \text{or} \quad \mathrm{A{:}}^- + \mathrm{B}^+$$

Homolytic (radical) cleavage. Each separating group takes one electron, for example:

$$\mathrm{A{:}B} \longrightarrow \mathrm{A}\cdot + \cdot\mathrm{B}$$

1. **Carbocations** are positively charged ions containing a carbon atom having only six electrons in three bonds:

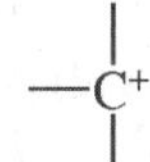

2. **Carbanions** are negatively charged ions containing a carbon atom with three bonds and an unshared pair of electrons:

$$-\overset{|}{\underset{|}{C}}:^-$$

3. **Radicals** (or **free radicals**) are species with at least one unparied electron. This is a broad category in which carbon radicals,

$$-\overset{|}{\underset{|}{C}}\cdot$$

are just one example.

4. **Carbenes** are neutral species having a carbon atom with two bonds and two electrons. There are two kinds: **singlet**

$$-\overset{\uparrow\downarrow}{C}-$$

in which the two electrons have opposite spins and are paired in one orbital, and **triplet**

$$-\overset{\uparrow}{\underset{\uparrow}{C}}-$$

in which the two electrons have the same spin and are in different orbitals.

Problem 3.1 Determine the hybrid orbital number (HON) of the five C-containing intermediates tabulated below, and give the hybrid state of the C atom. Unpaired electrons do *not* require an HO and should not be counted in your determination.

	NUMBER OF σ BONDS	+	NUMBER OF UNSHARED ELECTRON PAIRS	=	HON	HYBRID STATE
(*a*) carbocation	3		0		3	sp^2
(*b*) carbanion	3		1		4	sp^3
(*c*) radical	3		0		3	sp^2
(*d*) singlet carbene	2		1		3	sp^2
(*e*) triplet carbene	2		0		2	sp

Recall that the two unshared electrons of the triplet carbene are not paired and, hence, are not counted; they are in different orbitals.

Problem 3.2 Give three-dimensional representations for the orbitals used by the C's of the five carbon intermediates of Problem 3.1. Place all unshared electrons in the appropriate orbitals.

(*a*) A carbocation has three trigonal planar sp^2 HO's to form three σ bonds, and an empty *p* AO perpendicular to the plane of the σ bonds. See Fig. 3.1(*a*).

(*b*) A carbanion has four tetrahedral sp^3 HO's; three form three σ bonds and one has the unshared pair of electrons. See Fig. 3.1(*b*).

(*c*) A radical has the same orbitals as the carbocation. The difference lies in the presence of the odd electron in the *p* orbital of the radical. See Fig. 3.1(*c*).

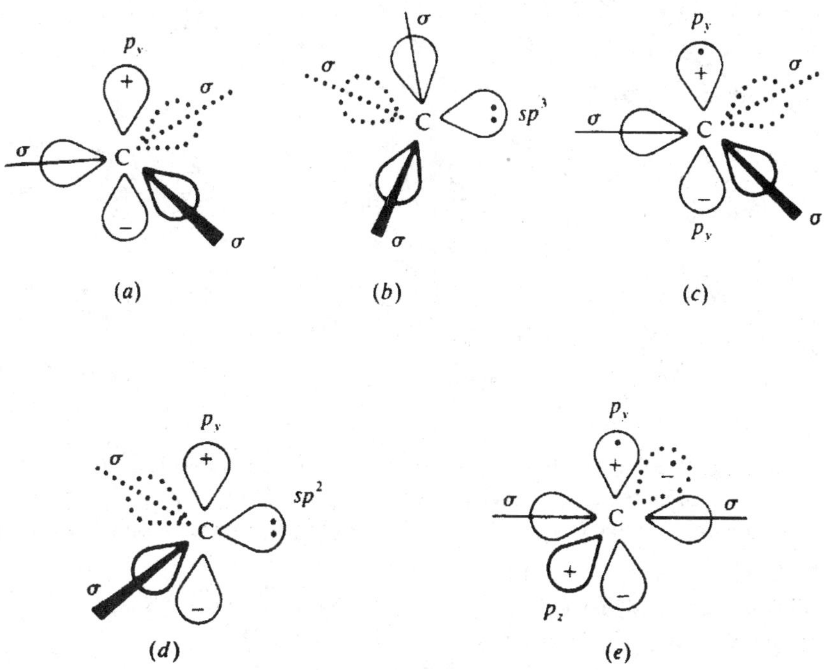

Figure 3.1

(*d*) A singlet carbene has three sp^2 HO's; two form two σ bonds and the third holds the unshared pair of electrons. It also has an empty *p* AO perpendicular to the plane of the sp^2 HO's. See Fig. 3.1(*d*).

(*e*) A triplet carbene uses two *sp* HO's to form two linear σ bonds. Each of the two unhybridized *p* AO's has one electron. See Fig. 3.1(*e*).

Problem 3.3 Write formulas for the species resulting from the (*a*) homolytic cleavage and (*b*) heterolytic cleavage of the C—C bond in ethane, C_2H_6, and classify these species.

(*a*) $\underset{\text{Ethane}}{H_3C:CH_3} \rightarrow \underset{\text{Methyl radicals}}{H_3C\cdot + \cdot CH_3}$

(*b*) $\underset{\text{Ethane}}{H_3C:CH_3} \rightarrow \underset{\text{Carbocation}}{H_3C^+} + \underset{\text{Carbanion}}{^-:CH_3}$

3.3 Types of Organic Reactions

1. **Displacement (substitution).** An atom or group of atoms in a molecule or ion is replaced by another atom or group.

2. **Addition.** Two molecules combine to yield a single molecule. Addition frequently occurs at a double or triple bond and sometimes at small-size rings.

3. **Elimination.** This is the reverse of addition. Two atoms or groups are removed from a molecule. If the atoms or groups are taken from adjacent atoms (**β-elimination**), a multiple bond is formed; if they are taken from other than adjacent atoms, a ring results. Removal of two atoms or groups from the same atom (**α-elimination**) produces a carbene.

4. **Rearrangement.** Bonds in the molecule are scrambled, converting it to its isomer.

5. **Oxidation-reduction (redox).** These reactions involve transfer of electrons or change in oxidation number. A decrease in the number of H atoms bonded to C and an increase in the number of bonds to other atoms such as C, O, N, Cl, Br, F, and S signals oxidation.

Problem 3.4 The following represents the steps in the mechanism for chlorination of methane:

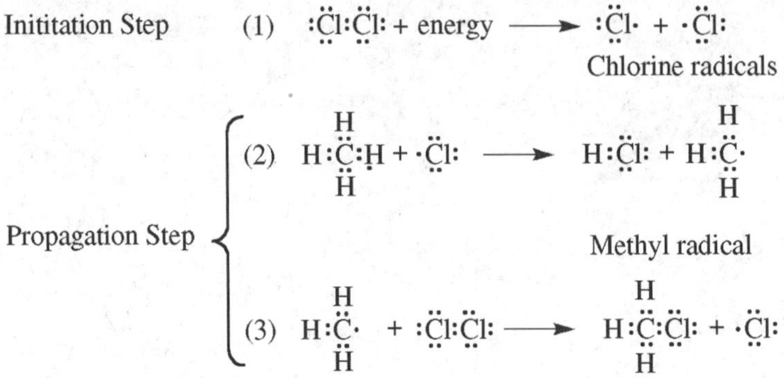

Initiation Step (1) $:\ddot{\underset{..}{Cl}}:\ddot{\underset{..}{Cl}}: + \text{energy} \longrightarrow :\ddot{\underset{..}{Cl}}\cdot + \cdot\ddot{\underset{..}{Cl}}:$ Chlorine radicals

Propagation Step

(2) $H:\underset{H}{\overset{H}{\ddot{C}}}:H + \cdot\ddot{\underset{..}{Cl}}: \longrightarrow H:\ddot{\underset{..}{Cl}}: + H:\underset{H}{\overset{H}{\ddot{C}}}\cdot$ Methyl radical

(3) $H:\underset{H}{\overset{H}{\ddot{C}}}\cdot + :\ddot{\underset{..}{Cl}}:\ddot{\underset{..}{Cl}}: \longrightarrow H:\underset{H}{\overset{H}{\ddot{C}}}:\ddot{\underset{..}{Cl}}: + \cdot\ddot{\underset{..}{Cl}}:$

The propagation steps constitute the overall reaction. (*a*) Write the equation for the overall reaction. (*b*) What are the intermediates in the overall reaction? (*c*) Which reactions are homolytic? (*d*) Which is a displacement reaction? (*e*) In which reaction is addition taking place? (*f*) The collision of which species would lead to side products?

(*a*) Add Steps 2 and 3: $CH_4 + Cl_2 \longrightarrow CH_3Cl + HCl$.
(*b*) The intermediates formed and then consumed are the $H_3C\cdot$ and radicals.
(*c*) Each step is homolytic. In Steps 1 and 3, Cl_2 cleaves; in Step 2, CH_4 cleaves.
(*d*) Step (3) involves the displacement of one $\cdot\ddot{\underset{..}{Cl}}:$ of Cl_2 by a $\cdot CH_3$ group. In Step 2, $\cdot\ddot{\underset{..}{Cl}}:$ displaces a $\cdot CH_3$ group from an H.
(*e*) None.
(*f*) $H_3C\cdot + \cdot CH_3 \longrightarrow H_3CCH_3$ (ethane)

Problem 3.5 Identify each of the following as (1) carbocations, (2) carbanions, (3) radicals, or (4) carbenes:

(*a*) $(CH_3)_2C:$ (*d*) $(CH_3)_3C:^-$ (*g*) $C_6H_5\dot{C}HCH_3$
(*b*) $(CH_3)_3C\cdot$ (*e*) $(CH_3)CH_2\dot{C}H_2$ (*h*) $CH_3\ddot{C}H$
(*c*) $(CH_3)_3C^+$ (*f*) $CH_3CH{=}CH$

(1) (*c*), (*f*). (2) (*d*). (3) (b), (e), (g). (4) (*a*), (*h*).

Problem 3.6 Classify the following as substitution, addition, elimination, rearrangement, or redox reactions. (A reaction may have more than one designation.)

(*a*) $CH_2{=}CH_2 + Br_2 \longrightarrow CH_2BrCH_2Br$
(*b*) $C_2H_5OH + HCl \longrightarrow C_2H_5Cl + H_2O$
(*c*) $CH_3CHClCHClCH_3 + Zn \longrightarrow CH_3CH{=}CHCH_3 + ZnCl_2$
(*d*) $NH_4^+(CNO)^- \longrightarrow H_2N\overset{\overset{\large O}{\|}}{C}NH_2$ (C=O)
(*e*) $CH_3CH_2CH_2CH_3 \longrightarrow (CH_3)_3CH$
(*f*) $H_2C{-}CH_2$ (ring with CH_2) $+ Br_2 \longrightarrow BrCH_2CH_2CH_2Br$
(*g*) $3CH_3CHO + 2MnO_4^- + OH^- \xrightarrow{\Delta} 3CH_3COO^- + 2MnO_2 + 2H_2O$ (Δ means heat.)
(*h*) $HCCl_3 + OH^- \rightarrow :CCl_2 + H_2O + Cl^-$
(*i*) $BrCH_2CH_2CH_2Br + Zn \longrightarrow H_2C{-}CH_2$ (ring with CH_2) $+ ZnBr_2$

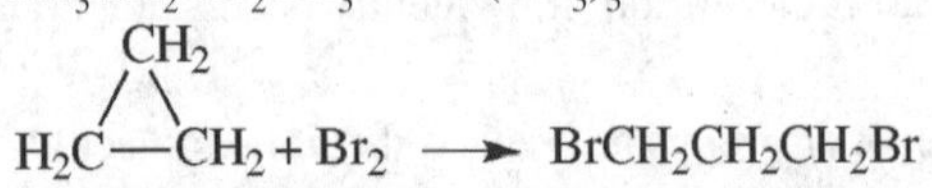

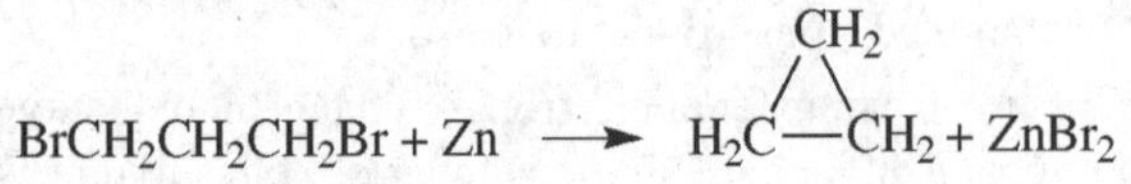

(*a*) Addition and redox. In this reaction, the two Br's add to the two double-bonded C atoms (1,2-addition). The oxidation number (ON) for C has changed from $4 - 2(2) - 2 = -2$ to $4 - 2(2) - 1 = -1$; (ON) for Br has changed from $7 - 7 = 0$ to $7 - 8 = -1$.
(*b*) Substitution of a Cl for an OH.
(*c*) Elimination and redox. Zn removes two Cl atoms from adjoining C atoms to form a double bond and $ZnCl_2$ (a β-elimination). The organic compound is reduced and Zn is oxidized.
(*d*) Rearrangement (isomerization).
(*e*) Rearrangement (isomerization).
(*f*) Addition and redox. The Br's add to two C atoms of the ring. These C's are oxidized, and the Br's are reduced.
(*g*) Redox. CH_3CHO is oxidized and MnO_4^- is reduced.
(*h*) Elimination. An H^+ and Cl^- are removed from the same carbon (α-elimination).
(*i*) Elimination and redox. The two Br's are removed from nonadjacent C's, giving a ring [see (*c*)].

3.4 Electrophilic and Nucleophilic Reagents

Reactions generally occur at the reactive sites of molecules and ions. These sites fall mainly into two categories. One category has a high electron density because the site (*a*) has an unshared pair of electrons or (*b*) is the $\delta-$ end of a polar bond or (*c*) has C=C π electrons. Such electron-rich sites are **nucleophilic** and the species possessing such sites are called **nucleophiles** or **electron-donors**. The second category (*a*) is capable of acquiring more electrons or (*b*) is the $\delta+$ end of a polar bond. These electron-deficient sites are **electrophilic**, and the species possessing such sites are called **electrophiles** or **electron-acceptors**. Many reactions occur by coordinate covalent bond formation between a nucleophilic and an electrophilic site.

$$\text{Nu:} + \text{E} \longrightarrow \text{Nu:E}$$

Problem 3.7 Classify the following species as being (1) nucleophiles or (2) electrophiles, and give the reason for your classification: (*a*) $H\ddot{\underset{\cdot\cdot}{O}}:^-$, (*b*) $:C{\equiv}N:^-$, (*c*) $:\underset{\cdot\cdot}{Br}^+$, (*d*) BF_3, (*e*) $H_2\ddot{O}:$, (*f*) $AlCl_3$, (*g*) $:NH_3$, (*h*) $H_3C:^-$ (a carbanion), (*i*) SiF_4, (*j*) Ag^+, (*k*) H_3C^+ (a carbocation), (*l*) $H_2C:$ (a carbene), (*m*) $:\ddot{\underset{\cdot\cdot}{I}}:^-$.

(1) (*a*), (*b*), (*e*), (*g*), (*h*), and (*m*). They all have unshared pairs of electrons. All anions are potential nucleophiles.
(2) (*d*) and (*f*) are molecules whose central atoms (B and Al) have only six electrons rather than the more desirable octet; they are electron-deficient. (*c*), (*j*), and (*k*) have positive charges and therefore are electron-deficient. Most cations are potential electrophiles. The Si in (*i*) can acquire more than eight electrons by utilizing its *d* orbitals; for example:

$$SiF_4 \text{ (an electrophile)} + 2:\ddot{\underset{\cdot\cdot}{F}}:^- \longrightarrow SiF_6^{2-}$$

Although the C in (*l*) has an unshared pair of electrons, (*l*) is electrophilic because the C has only six electrons.

Problem 3.8 Why is the reaction $CH_3Br + OH^- \longrightarrow CH_3OH + Br^-$ a nucelophilic displacement?

The $:\underset{\cdot\cdot}{\dot{O}}H^-$ has unshared electrons and is a nucleophile. Because of the polar nature of the C—Br bond:

$$\overset{\delta+}{\text{C}}\text{—}\overset{\delta-}{\text{Br}}$$

C acts as an electrophilic site. The displacement of Br^- by OH^- is initiated by the nucleophile $H\ddot{\underset{\cdot\cdot}{O}}:^-$.

Problem 3.9 Indicate whether reactant (1) or (2) is the nucleophile or electrophile in the following reactions:

(*a*) $H_2C{=}CH_2$ (1) + Br_2 (2) $\longrightarrow$ $BrCH_2$—CH_2Br

(*b*) $CH_3NH_3^+$ (1) + CH_3COO^- (2) ⟶ CH_3NH_2 + CH_3COOH

(*c*) $CH_3C(=O)—\ddot{Cl}:$ (1) + $AlCl_3$ (2) ⟶ $CH_3C^+(=O)$ + $AlCl_4^-$

(*d*) $CH_3CH{=}O$ (1) + $:SO_3H^-$ (2) ⟶ $CH_3—CH(O^-)SO_3H$

	(a)	(b)	(c)	(d)
Nucleophile	(1)	(2)	(1)	(2)
Electrophile	(2)	(1)	(2)	(1)

3.5 Thermodynamics

The thermodynamics and the rate of a reaction determine whether the reaction proceeds. The thermodynamics of a system is described in terms of several important functions:

(1) ΔE, the change in **energy**, equals q_V, the heat transferred to or from a system at constant volume: $\Delta E = q_V$.

(2) ΔH, the change in **enthalpy**, equals q_p, the heat transferred to or from a system at constant pressure: $\Delta H = q_p$. Since most organic reactions are performed at atmospheric pressure in open vessels, ΔH is used more often than is ΔE. For reactions involving only liquids or solids, $\Delta E = \Delta H$. ΔH of a chemical reaction is the difference in the enthalpies of the products, H_p, and the reactants, H_R:

$$\Delta H = H_P - H_R$$

If the bonds in the products are more stable than the bonds in the reactants, energy is released, and ΔH is negative. The reaction is **exothermic**.

(3) ΔS is the change in **entropy**. Entropy is a measure of randomness. The more the randomness, the greater is S; the greater the order, the smaller is S. For a reaction:

$$\Delta S = S_P - S_R$$

(4) $\Delta G = G_P - G_R$ is the change in **free energy**. At constant temperature:

$$\Delta G = \Delta H - T\Delta S \ (T = \text{Absolute temperature})$$

For a reaction to be spontaneous, ΔG must be negative.

Problem 3.10 State whether the following reactions have a positive or negative ΔS, and give a reason for your choice.

(*a*) $H_2 + H_2C{=}CH_2$ ⟶ H_3CCH_3

(*b*) cyclo-CH_2–$H_2C—CH_2$ (cyclopropane) $\xrightarrow{\Delta}$ $H_3C—CH{=}CH_2$

(*c*) $CH_3COO^-(aq) + H_3O^+(aq)$ ⟶ $CH_3COOH + H_2O$

(*a*) Negative. Two molecules are changing into one molecule and there is more order (less randomness) in the product ($S_P < S_R$).
(*b*) Positive. The rigid ring opens to give compounds having free rotation about the C—C single bond. There is now more randomness ($S_P > S_R$).
(*c*) Positive. The ions are solvated by more H_2O molecules than is CH_3COOH. When ions form molecules, many of these H_2O molecules are set free and therefore have more randomness ($S_P > S_R$).

Problem 3.11 Predict the most stable state of H_2O (steam, liquid, or ice) in terms of (*a*) enthalpy, (*b*) entropy, and (*c*) free energy.

(*a*) Gas → Liquid → Solid are exothermic processes and, therefore, ice has the least enthalpy. For this reason, ice should be most stable.
(*b*) Solid → Liquid → Gas shows increasing randomness and therefore increasing entropy. For this reason, steam should be most stable.
(*c*) Here, the trends to lowest enthalpy and highest entropy are in opposition; neither can be used independently to predict the favored state. Only G, which gives the balance between H and S, can be used. The state with lowest G or the reaction with the most negative ΔG is favored. For H_2O, this is the liquid state, a fact which cannot be predicted until a calculation is made using the equation $G = H - TS$.

3.6 Bond-Dissociation Energies

The **bond-dissociation energy**, ΔH, is the energy needed for the endothermic homolysis of a covalent bond A:B → A· + ·B; ΔH is positive. Bond *formation*, the reverse of this reaction, is exothermic and the ΔH values are negative. The more positive the ΔH value, the stronger is the bond. The ΔH of reaction is the sum of all the (positive) ΔH values for bond cleavages *plus* the sum of all the (negative) ΔH values for bond formations.

Problem 3.12 Calculate ΔH for the reaction $CH_4 + Cl_2 \longrightarrow CH_3Cl + HCl$. The bond-dissociation energies, in kJ/mol, are 427 for C—H, 243 for Cl—Cl, 339 for C—Cl, and 431 for H—Cl.

The values are shown under the bonds involved:

$$\underbrace{H_3C{-}H + Cl{-}Cl}_{\substack{427 + 243 \\ \text{cleavage (endothemic)}}} \longrightarrow \underbrace{H_3C{-}Cl + H{-}Cl}_{\substack{(-339) + (-431) \\ \text{formations (exothermic)}}}$$

$$427 + 243 \quad + \quad (-339) + (-431) = -100$$

The reaction is exothermic, with $\Delta H = -100$ kJ/mol.

Problem 3.13 Compare the strengths of bonds between similar atoms having: (*a*) single bonds between atoms with and without unshared electron pairs; (*b*) triple, double, and single bonds.

(*a*) Bonds are weaker between atoms with unshared electron pairs because of interelectron repulsion.
(*b*) Overlapping of p orbitals strengthens bonds, and bond energies are greatest for triple and smallest for single bonds.

3.7 Chemical Equilibrium

Every chemical reaction can proceed in either direction, $dA + eB \rightleftharpoons fX + gY$, even if it goes in one direction to a microscopic extent. A **state of equilibrium** is reached when *the concentrations of A, B, X, and Y no longer change* even though the reverse and forward reactions are taking place.

Every reversible reaction has an equilibrium expression in which K_e, the **equilibrium constant**, is defined in terms of molar concentrations (mol/L), as indicated by the square brackets:

$$dA + eB \rightleftharpoons fX + gY$$

$$K_e = \frac{[X]^f[Y]^g}{[A]^d[B]^e} \quad \begin{array}{l}\text{Products favored; } K_e \text{ is large}\\ \text{Reactants favored; } K_e \text{ is small}\end{array}$$

K_e varies only with temperature.

The ΔG of a reaction is related to K_e by

$$\Delta G = -2.303\, RT \log K_e$$

where R is the molar gas constant ($R = 8.314$ J mol^{-1} K^{-1}) and T is the absolute temperature (in K).

Problem 3.14 Given the reversible reaction

$$C_2H_5OH + CH_3COOH \rightleftharpoons CH_3COOC_2H_5 + H_2O$$

what changes could you make to increase the yield of $CH_3COOC_2H_5$?

The equilibrium must be shifted to the right, the side of the equilibrium where $CH_3COOC_2H_5$ exists. This is achieved by any combination of the following: adding C_2H_5OH, adding CH_3COOH, removing H_2O, removing $CH_3COOC_2H_5$.

Problem 3.15 Summarize the relationships between the signs of ΔH, $T\Delta S$, and ΔG, and the magnitude of K_e, and state whether a reaction proceeds to the right or to the left for the reaction equation as written.

See Table 3.1.

Problem 3.16 At 25°C, the following reactions have the indicated K_e values:

(*a*) $CH_3CH_2OH + CH_3COOH \rightleftharpoons CH_3COOCH_2CH_3 + H_2O \quad K_e = 4.0$

Ethanol, Acetic acid, Ethyl acetate, Water

(*b*) $H_2C(COOH)$–CH_2–H_2C–OH (ring-open hydroxy acid) $\rightleftharpoons$ H_2C–CH_2–$C{=}O$ / H_2C–O (five-membered ring) $+ H_2O \quad K_e = 1000$

a lactone

Since the changes in bonding in these reactions are similar, both reactions have about the same ΔH. Use thermodynamic functions to explain the large difference in the magnitude of K_e.

TABLE 3.1

ΔH −	$T\Delta S$ =	ΔG	Reaction Direction	K_e
−	+	−	Forward → right	> 1
+	−	+	Reverse → left	< 1
−	−	Usually − if $\Delta H < -60$ kJ/mol	Depends on conditions	?
+	+	Usually + if $\Delta H > +60$ kJ/mol	Depends on conditions	?

A larger K_e means a more negative ΔG. Since ΔH is about the same for both reactions, a more negative ΔG means that ΔS for this reaction is more positive. A more positive ΔS (greater randomness) is expected in reaction (*b*) because one molecule is converted into two molecules, whereas in reaction (*a*) two molecules are changed into two other molecules. When two reacting sites, such as OH and COOH, are in the same molecule, the reaction is **intramolecular**. When reaction sites are in different molecules, as in (*a*), the reaction is **intermolecular**. Intramolecular reactions often have a more positive ΔS than similar intermolecular reactions.

3.8 Rates of Reactions

The rate of a reaction is how fast reactants disappear or products appear. For the general reaction $d\text{A} + e\text{B} \rightarrow f\text{C} + g\text{D}$, the rate is given by a rate equation

$$\text{Rate} = k[\text{A}]^x [\text{B}]^y$$

where k is the **rate constant** at the given temperature, T, and [A] and [B] are molar concentrations. Exponents x and y may be integers, fractions, or zero; their sum defines the **order** of the reaction. The values of x and y are found experimentally, and they *may differ from* the stoichiometric coefficients d and e.

Experimental conditions, other than concentrations, that affect rates of reactions are as follows:

Temperature. A rough rule is that the value of k doubles for every rise in temperature of 10°C.

Particle size. Increasing the surface area of solids by pulverization increases the reaction rate.

Catalysts and inhibitors. A catalyst is a substance that increases the rate of a reaction but is recovered unchanged at the end of the reaction. Inhibitors decrease the rate.

At a given set of conditions, the factors that determine the rate of a given reaction are as follows:

1. **Number of collisions per unit time.** The greater the chances for molecular collision, the faster the reaction. Probability of collision is related to the number of molecules of each reactant and is proportional to the molar concentrations.
2. **Enthalpy of activation (activation energy, E_{act})** ($\Delta H^\ddagger$). Reaction may take place only when colliding molecules have some enthalpy content, $\Delta H^\ddagger$, in excess of the average. *The smaller the value of $\Delta H^\ddagger$, the more successful will be the collisions and the faster the reaction.* ($\Delta H^\ddagger = E_{act}$ at constant pressure.)
3. **Entropy of activation** ($\Delta S^\ddagger$), also called the **probability factor**. Not all collisions between molecules possessing the requisite $\Delta H^\ddagger$ result in reaction. Often collisions between molecules must also occur *in a certain orientation*, reflected by the value of $\Delta S^\ddagger$. *The more organized or less random the required orientation of the colliding molecules, the lower the entropy of activation and the slower the reaction.*

Problem 3.17 Predict the effect on the rate of a reaction if a change in the solvent causes: (*a*) an increase in $\Delta H^\ddagger$ and a decrease in $\Delta S^\ddagger$, (*b*) a decrease in $\Delta H^\ddagger$ and an increase in $\Delta S^\ddagger$, (*c*) an increase in $\Delta H^\ddagger$ and in $\Delta S^\ddagger$, (*d*) a decrease in $\Delta H^\ddagger$ and in $\Delta S^\ddagger$.

(*a*) Decrease in rate. (*b*) Increase in rate. (*c*) The increase in $\Delta H^\ddagger$ tends to decrease the rate but the increase in $\Delta S^\ddagger$ tends to increase the rate. The combined effect is unpredictable. (*d*) The trends here are opposite to those in part (*c*); the effect is also unpredictable. In many cases, the change in $\Delta H^\ddagger$ is more important than the change in $\Delta S^\ddagger$ in affecting the rate of reaction.

3.9 Transition-State Theory and Enthalpy Diagrams

When reactants have collided with sufficient enthalpy of activation ($\Delta H^\ddagger$) and with the proper orientation, they pass through a hypothetical **transition state** in which some bonds are breaking while others may be forming.

The relationship of the transition state (TS) to the reactants (R) and products (P) is shown by the enthalpy (energy) diagram, Fig. 3.2, for a one-step *exothermic* reaction A + B → C + D. At equilibrium, formation of

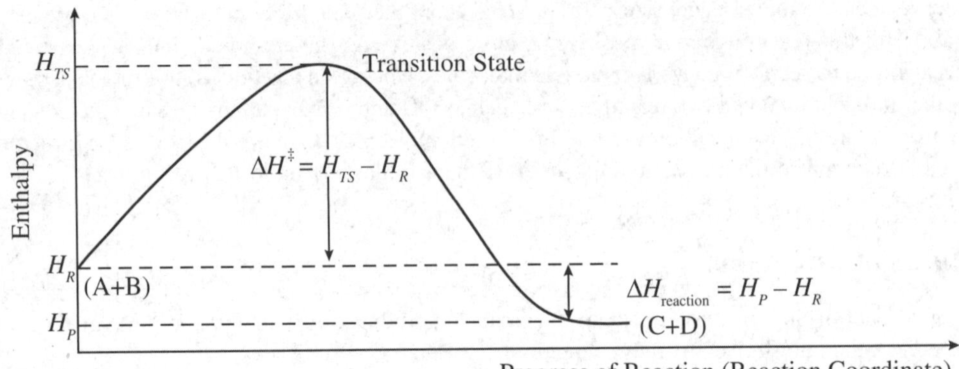

Figure 3.2

molecules with lower enthalpy is favored—that is, C + D. However, this applies only if ΔH of reaction predominates over $T\,\Delta S$ of reaction in determining the equilibrium state. The reaction rate is actually related to the free energy of activation, $\Delta G^{\ddagger}$, where $\Delta G^{\ddagger} = \Delta H^{\ddagger} - T\Delta S^{\ddagger}$.

In multistep reactions, each step has its own transition state. The step with the highest-enthalpy transition state is the slowest step and determines the overall reaction rate.

The number of species colliding in a step is called the **molecularity** (of that step). If only one species breaks down, the reaction is **unimolecular**. If two species collide and react, the reaction is **bimolecular**. Rarely do three species collide (**termolecular**) at the same instant.

The rate equation gives molecules and ions and the number of each (from the exponents) involved in the slow step and in any preceding fast step(s). Intermediates do not appear in rate equations, although products occasionally do appear.

Problem 3.18 Draw an enthalpy diagram for a one-step endothermic reaction. Indicate ΔH of reaction and $\Delta H^{\dagger}$.

See Fig. 3.3.

Problem 3.19 Draw a reaction-enthalpy diagram for an exothermic two-step reaction in which (*a*) the first step is slow and (*b*) the second step is slow. Why do these reactions occur?

See Fig. 3.4, in which R = reactants, I = intermediates, P = products, TS_1 = transition state of first step, TS_2 = transition state of second step. Because the reactions are exothermic, $H_P < H_R$.

Problem 3.20 In Problem 3.19(*b*), the first step is not only fast but is also reversible. Explain.

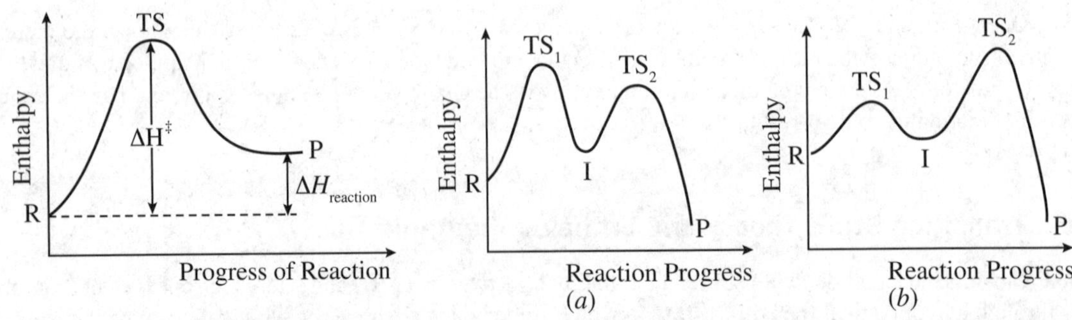

Figure 3.3

Figure 3.4

The $\Delta H^{\ddagger}$ for I to revert to reactants R is less than that for I to form the products P. Therefore, most I's re-form R, so that the first step is fast and reversible. A few I's have enough enthalpy to go through the higher-enthalpy TS_2 and form the products. The $\Delta H^{\ddagger}$ for P to revert to I is prohibitively high; hence, the products accumulate, and the second step is at best insignificantly reversible.

Problem 3.21 Catalysts generally speed up reactions by lowering $\Delta H^{\ddagger}$. Explain how this occurs in terms of ground-state and transition-state enthalpies (H_R and H_{TS}).

$\Delta H^{\ddagger}$ can be decreased by (*a*) lowering H_{TS}, (*b*) raising H_R or (*c*) both of these.

Problem 3.22 The reaction A + B → C + D has (*a*) rate = k[A][B], or (*b*) rate = k[A]. Offer possible mechanisms consistent with these rate expressions.

(*a*) Molecules A and B must collide in a bimolecular rate-controlling step. Since the balanced chemical equation calls for the reaction of one A molecule with one B molecule, the reaction must have a single (concerted) step.
(*b*) The rate-determining step is unimolecular and involves only an A molecule. There can be no prior fast steps. Molecule B reacts in the second step, which is fast. Following is a possible two-step mechanism:

$$\text{Step 1:}\quad \text{A} \longrightarrow \text{C} + \text{I} \quad (\text{I} = \text{intermediate})$$
$$\text{Step 2:}\quad \text{B} + \text{I} \longrightarrow \text{D}$$

Adding the two steps gives the balanced chemical equation: A + B → C + D.

Problem 3.23 For the reaction 2A + 2B → C + D, rate = $k[A]^2[B]$. Give a mechanism using only unimolecular or bimolecular steps.

One B molecule and two A molecules are needed to give the species for the slow step. The three molecules do not collide simultaneously, since we are disregarding the very rare termolecular steps. There must then be some number of prior fast steps to furnish at least one intermediate needed for the slow step. The second B molecule that appears in the reaction equation must be consumed in a fast step following the slow step.

Mechanism 1

$$\text{A} + \text{B} \xrightarrow{\text{fast}} \text{AB} \quad (\text{intermediate})$$
$$\text{AB} + \text{A} \xrightarrow{\text{slow}} \text{A}_2\text{B} \quad (\text{intermediate})$$
$$\text{A}_2\text{B} + \text{B} \xrightarrow{\text{fast}} \text{C} + \text{D}$$

Mechanism 2

$$\text{A} + \text{A} \xrightarrow{\text{fast}} \text{A}_2 \quad (\text{intermediate})$$
$$\text{A}_2 + \text{B} \xrightarrow{\text{slow}} \text{A}_2\text{B} \quad (\text{intermediate})$$
$$\text{A}_2\text{B} + \text{B} \xrightarrow{\text{fast}} \text{C} + \text{D}$$

Problem 3.24 For the reaction A + 2B ⟶ C + D, rate = $k[A][B]^2$. Offer a mechanism in which the rate-determining step is unimolecular.

The slow step needs an intermediate formed from one A molecule and two B molecules. Since the rate expression involves the same kinds and numbers of molecules as does the chemical equation, there are no fast steps following the slow step.

Mechanism 1

$$\text{A} + \text{B} \xrightarrow{\text{fast}} \text{AB}$$
$$\text{AB} + \text{B} \xrightarrow{\text{fast}} \text{AB}_2$$
$$\text{AB}_2 \xrightarrow{\text{slow}} \text{C} + \text{D}$$

Mechanism 2

$$\text{B} + \text{B} \xrightarrow{\text{fast}} \text{B}_2$$
$$\text{B}_2 + \text{A} \xrightarrow{\text{fast}} \text{A}_2\text{B}$$
$$\text{A}_2\text{B} \xrightarrow{\text{slow}} \text{C} + \text{D}$$

Notice that often the rate expression is insufficient to allow the suggestion of an unequivocal mechanism. More experimental information is often needed.

3.10 Brönsted Acids and Bases

In the Brönsted definition, an *acid donates a proton* and a *base accepts a proton*. The strengths of acids and bases are measured by the extent to which they lose or gain protons, respectively. In these reactions, acids are converted to their **conjugate bases** and bases to their **conjugate acids**. Acid-base reactions go in the direction of *forming the weaker acid and the weaker base*.

Problem 3.25 Show the conjugate acids and bases in the reaction of H_2O with gaseous (*a*) HCl, (*b*) $:NH_3$.

(*a*) H_2O, behaving as a Brönsted base, accepts a proton from HCl, the Brönsted acid. They are converted to the conjugate acid H_3O^+ and the conjugate base Cl^-, respectively.

$$\underset{\substack{\text{acid}_1\\ \text{(stronger)}}}{HCl} + \underset{\substack{\text{base}_2\\ \text{(stronger)}}}{H_2O} \rightleftharpoons \underset{\substack{\text{acid}_2\\ \text{(weaker)}}}{H_3O^+} + \underset{\substack{\text{base}_1\\ \text{(weaker)}}}{Cl^-}$$

The conjugate acid-base pairs have the same subscript and are bracketed together. This reaction goes almost to completion because HCl is a good proton donor and hence a strong acid.

(*b*) H_2O is **amphoteric** and can also act as an acid by donating a proton to $:NH_3$. H_2O is converted to its conjugate base, OH^-, and $:NH_3$ to its conjugate acid, NH_4^+.

$$\underset{\text{acid}_1}{H{:}OH} + \underset{\text{base}_2}{NH_3} \rightleftharpoons \underset{\text{acid}_2}{NH_4^+} + \underset{\text{base}_1}{:OH^-}$$

$:NH_3$ is a poor proton acceptor (a weak base); the arrows are written to show that the equilibrium lies mainly to the left.

To be called an acid, the species must be more acidic than water and be able to donate a proton to water. Some compounds, such as alcohols, are not acidic toward water but have an H which is acidic enough to react with very strong bases or with Na.

Problem 3.26 Write an equation for the reaction of ethanol with (*a*) NH_2^-, a very strong base; (*b*) Na.

(*a*) $CH_3CH_2\ddot{\underset{..}{O}}H + :\ddot{N}H_2^- \longrightarrow CH_3CH_2\ddot{\underset{..}{O}}:^- + :NH_3$

(*b*) $2CH_3CH_2\ddot{\underset{..}{O}}H + 2Na \longrightarrow 2CH_3CH_2\ddot{\underset{..}{O}}:^- + 2Na^+ + H_2$

Relative quantitative strengths of acids and bases are given either by their ionization constants, K_a and K_b, or by their pK_a and pK_b values as defined by:

$$pK_a = -\log K_a \quad pK_b = -\log K_b$$

The stronger an acid or base, the larger its ionization constant and the smaller its p*K* value. The strengths of bases can be evaluated from those of their conjugate acids; the strengths of acids can be evaluated from those of their conjugate bases. The *strongest acids* have the *weakest conjugate bases* and the *strongest bases* have the *weakest conjugate acids*. This follows from the relationships

$$K_w = (K_a)(K_b) = 10^{-14} \quad pK_a + pK_b = pK_w = 14$$

in which K_w, the ion-product of water = $[H_3O^+][OH^-]$.

3.11 Basicity (Acidity) and Structure

The basicity of a species depends on the reactivity of the atom with the unshared pair of electrons, this atom being the **basic site** for accepting the H^+. The more spread out (dispersed, delocalized) is the electron density on the basic site, the less basic is the species.

The acidity of a species can be determined from the basicity of its conjugate base.

Delocalization Rules

(1) For bases of binary acids (H_nX) of elements in the same Group, the larger the basic site X, the more spread out is the charge. Compared bases must have the same charge.

(2) For like-charged bases of binary acids in the same period, the more unshared pairs of electrons the basic site has, the more delocalized is the charge.

(3) The more *s* character in the orbital with the unshared pair of electrons, the more delocalized the electronic charge.

(4) Extended *p–p* π bonding between the basic site and an adjacent π system (resonance) delocalizes the electronic charge.

(5) Extended *p–d* π bonding with adjacent atoms that are able to acquire more than an octet of electrons delocalizes the electronic charge.

(6) Delocalization can occur via the **inductive effect,** whereby an electronegative atom transmits its electron-withdrawing effect through a chain of σ bonds. Electropositive groups are electron-donating and localize more electron density on the basic site.

With reference to (4) and (5), some common π systems that participate in extended π bonding are

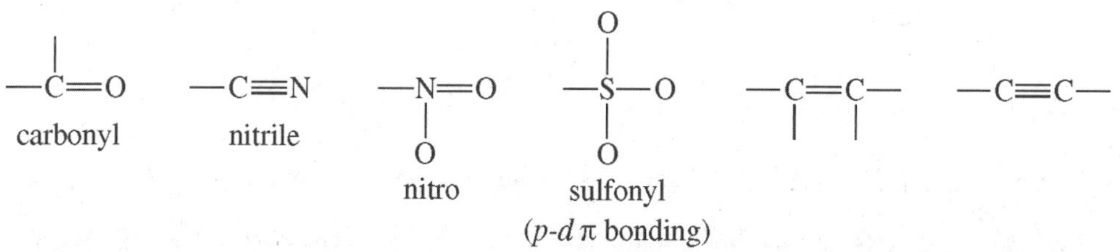

$-\overset{\mid}{C}=O$	$-C\equiv N$	$-\underset{\mid}{N}=O$ (with O below N)	$-S(=O)(=O)-O$ (O above and below S)	$-\underset{\mid}{C}=\underset{\mid}{C}-$	$-C\equiv C-$
carbonyl	nitrile	nitro	sulfonyl (*p-d* π bonding)		

Problem 3.27 Compare the basicities of the following pairs: (*a*) RS^- and RO^-; (*b*) $:NH_3$ (N uses sp^3 HO's) and $:PH_3$ (P uses *p* AO's for its three bonds); (*c*) NH_2^- and OH^-.

(*a*) S and O are in the same periodic group, but S is larger and its charge is more delocalized. Therefore, RS^- is a weaker base than RO^- and RSH is a stronger acid than ROH.

(*b*) Since the bonding pairs of electrons of $:PH_3$ are in *p* AO's, the unshared pair of electrons is in an *s* AO. In $:NH_3$, the unshared pair is in an sp^3 HO. Consequently, the orbital of P with the unshared electrons has more *s* character and PH_3 is the weaker base. In water, PH_3 has no basic property.

(*c*) $:OH^-$ has more unshared pairs of electrons than $:NH_2^-$, its charge is more delocalized, and it is the weaker base.

Problem 3.28 Compare and account for the acidity of the underlined H in:

(*a*) $R-O-\underline{H}$ and $R-\underset{\parallel}{C}-O-\underline{H}$ (C=O)

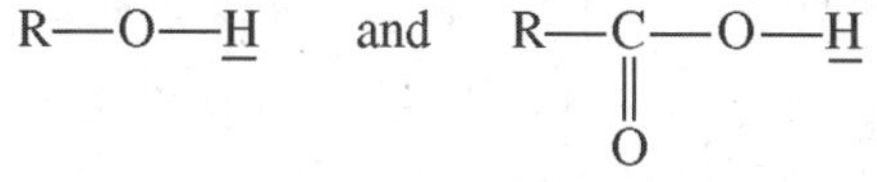

(*b*) $\underline{H}-CH_2-C(CH_3)=CH_2$ (i) $\underline{H}-CH_2-CH(CH_3)-CH_3$ (ii) $\underline{H}-CH_2-C(=O)=CH_3$ (iii)

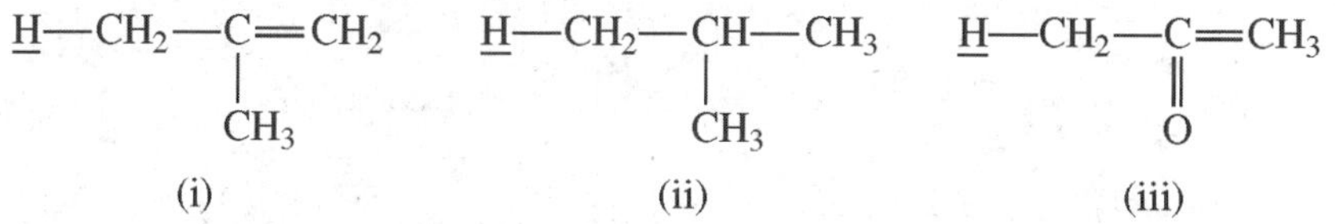

Compare the stability of the conjugate bases in each case.

(a) In $R{-}C({-}O^-)({=}O) \longleftrightarrow R{-}C({=}O)({-}O^-)$ or $\left\{R{-}C(\cdots O)(\cdots O)\right\}^-$

the C and O's participate in extended π bonding so that the − is distributed to each O. In RO^-, the − is localized on the O. Hence, $RCOO^-$ is a weaker base than RO^-, and the RCOOH is a stronger acid than ROH.

(b) The stability of the carbanions and relative acidity of these compounds is

$$(III) > (I) > (II)$$

Both (I) and (III) have a double bond not present in (II) that permits delocalization by extended π bonding. Delocalization is more effective in (III) because charge is delocalized to the electronegative O.

(I) $B{:} + H{-}CH_2{-}C(CH_3){=}CH_2 \rightleftharpoons B{:}H^+ + \left[{:}\bar{C}H_2{-}C(CH_3){=}CH_2 \longleftrightarrow H_2C{=}C(CH_3){-}\bar{C}H_2{:}\right]$

(II) $B{:} + H{-}CH_2{-}CH(CH_3){-}CH_3 \rightleftharpoons B{:}H^+ + {:}\bar{C}H_2{-}CH(CH_3)CH_3$ (− localized on one C)

(III) $B{:} + H{-}CH_2{-}C({=}\ddot{O}{:}){-}CH_3 \rightleftharpoons B{:}H^+ + \left[{:}\bar{C}H_2{-}C({=}\ddot{O}{:}){-}CH_3 \longleftrightarrow H_2C{=}C({-}{:}\ddot{O}{:}^-){-}CH_3\right]$

Problem 3.29 Account for the decreasing order of basicity in the following amines: $CH_3NH_2 > NH_3 > NF_3$.

The F's are very electronegative and, by their inductive effects, they delocalize electron density from the N atom. The N in NF_3 has less electron density than the N in NH_3; NF_3 is a weaker base than NH_3. The CH_3 group, on the other hand, is electron-donating and localizes more electron density on the N of CH_3NH_2, making CH_3NH_2 a stronger base than NH_3.

Problem 3.30 Account for the decreasing order of acidity: $HC{\equiv}CH > H_2C{=}CH_2 > CH_3CH_3$.

We apply the HON rule to the basic site C of the conjugate bases. For $HC{\equiv}C{:}^-$, the HON is 2 and the C uses *sp* HO's. For $H_2C{=}\ddot{C}H^-$, the HON is 3 and the C uses sp^2 HO's. For $CH_3\ddot{C}H_2^-$, the HON is 4 and the C uses sp^3 HO's. As the *s* character decreases, the basicities increase and the acidities decrease.

3.12 Lewis Acids and Bases

A **Lewis acid** (**electrophile**) shares an electron pair furnished by a **Lewis base** (**nucleophile**) to form a covalent (coordinate) bond. The Lewis concept is especially useful in explaining the acidity of an aprotic acid (no available proton), such as BF_3.

$$\underset{\text{Lewis base}}{H_3N{:}} + \underset{\text{Lewis acid}}{BF_3} \longrightarrow H_3\overset{+}{N}{:}\overset{-}{B}F_3$$

The three types of nucleophiles are listed in Section 3.4.

Problem 3.31 Given the following Lewis acid-base reactions:

Lewis base	+ Lewis acid	→	Product
$4H_3N:$	$+ Cu^{2+}$	→	$Cu(NH_3)_4^{2+}$
$2:\ddot{F}:^-$	$+ SiF_4$	→	SiF_6^{2-}
$H:\ddot{O}:^-$	$+ :\ddot{O}=C=\ddot{O}:$	→	$^-:\ddot{O}-C(OH)=\ddot{O}:$
$H_2C=CH_2$	$+ H^+$ (from a Brönsted acid)	→	$H_2\overset{+}{C}-CH_2-H$
$H_2C=\ddot{O}:$	$+ BF_3$	→	$H_2C-\overset{+}{\ddot{O}}:\overline{B}F_3$

(*a*) Group the bases as follows: (1) anions, (2) molecules with an unshared pair of electrons, (3) negative end of a π bond dipole, and (4) available π electrons. (*b*) Group the acids as follows: (1) cations, (2) species with electron-deficient atoms, (3) species with an atom capable of expanding an octet, and (4) positive end of a π bond dipole.

(*a*) (1) OH^-, F^- (2) $:NH_3$, $H_2C=\ddot{O}$ (3) $H_2\overset{\delta+}{C}=\overset{\delta-}{\ddot{O}}:$ (4) $H_2C=CH_2$

(*b*) (1) Cu^{2+}, H^+ (2) BF_3 (3) SiF_4 (4) $:\overset{\delta-}{\ddot{O}}=\overset{\delta+}{C}=\overset{\delta-}{\ddot{O}}:$

SUPPLEMENTARY PROBLEMS

Problem 3.32 Which of the following reactions can take place with carbocations? Give examples when reactions do occur: (*a*) acts as an acid; (*b*) reacts as an electrophile; (*c*) reacts as a nucleophile; (*d*) undergoes rearrangement.

Carbocations may undergo (*a*), (*b*), and (*d*).

(*a*) $CH_3-C^+(CH_3)_2 \xrightarrow{-H^+} CH_2=C(CH_3)_2$ (*b*) $CH_3-C^+(CH_3)_2 + :\ddot{O}H_2 \longrightarrow CH_3-C(CH_3)_2-\overset{+}{\ddot{O}}H_2$

(*d*) $CH_3-C(CH_3)(\ddot{H})-\overset{+}{C}HCH_3 \xrightarrow{\sim:H} CH_3-\overset{+}{C}(CH_3)-CH_2CH_3$

Problem 3.33 Give three-dimensional representations for the orbitals used by the C's in (*a*) $H_2C=CH^+$ and (*b*) $H_2C=\ddot{C}H^-$.

In both (*a*) and (*b*), the C of the CH_2 group uses sp^2 HO's to form σ bonds with two H's and the other C. In (*a*), the charged C uses two *sp* HO's to form two σ bonds, one with H and one with the other C. One *p* AO forms a π bond with the other C, and the second unhybridized *p* AO has no electrons [see Fig. 3.5(*a*)]. In (*b*), the charged C uses three sp^2 HO's: two form σ bonds with the H and the other C, and the third has the unshared pair of electrons [see Fig. 3.5(*b*)].

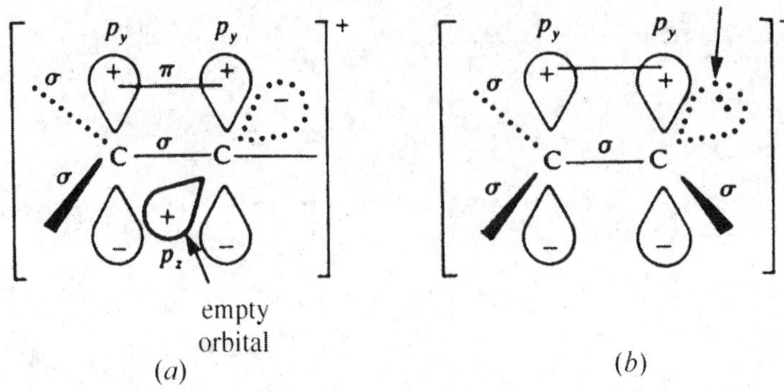

Figure 3.5

Problem 3.34 Write the formula for the carbon intermediate indicated by ?, and label as to type.

(*a*) $H_3C—\ddot{N}=\ddot{N}—CH_3 \longrightarrow ? + :N\equiv:N$
(*b*) $(CH_3)_2Hg \longrightarrow ? + \cdot Hg\cdot$
(*c*) $H_2\ddot{\bar{C}}—\overset{+}{N}\equiv N: \longrightarrow ? + N_2$
(*d*) $(CH_3)_3C\ddot{O}H + H^+ \longrightarrow ? + H_2\ddot{O}:$
(*e*) $H—C\equiv C—H + Na\cdot \longrightarrow ? + Na^+ + \frac{1}{2}H_2$
(*f*) $H_2C=CH_2 + D—Br \longrightarrow ? + Br^-$
(*g*) $H_2CI_2 + Zn: \longrightarrow ? + Zn_2^+ + 2I^-$
(*h*) $(CH_3)_3C—Cl + AlCl_3 \longrightarrow ? + AlCl_4^-$

(*a*) and (*b*) $H_3C\cdot$, a radical. (*c*) and (*g*) $H_2C:$, a carbene. (*d*) and (*h*) $(CH_3)_3C^+$, a carbocation. (*e*) $H—C\equiv C:^-$, a carbanion. (*f*) $H_2\overset{+}{C}—CH_2—D$, a carbocation.

Problem 3.35 Classify the following reactions by type.

(*a*) $H_2C(OH)—CH_2—Br + OH^- \longrightarrow H_2C—CH_2(O) + H_2O + Br$ (epoxide: $H_2C—CH_2$ bridged by O)
(*b*) $(CH_3)_2CHOH \xrightarrow{\text{Cu. heat}} (CH_3)_2C=O + H_2$
(*c*) $H_2C—CH_2(CH_2)$ (cyclopropane) $\xrightarrow{\text{heat}} H_2C=CHCH_3$
(*d*) $H_3C—CH_2Br + :H^- \longrightarrow H_3C—CH_3 + :Br^-$
(*e*) $H_2C=CH_2 + H_2 \xrightarrow{\text{Pt}} H_3C—CH_3$
(*f*) $C_6H_6 + HNO_3 \xrightarrow{H_2SO_4} C_6H_5NO_2 + H_2O$
(*g*) $HCOOH \xrightarrow{\text{heat}} H_2O + CO$
(*h*) $CH_2=C=O + H_2O \longrightarrow CH_3COOH$
(*i*) $H_2C=O + 2Ag(NH_3)_2^+ + 3OH^- \longrightarrow HCOO^- + 2Ag + 4NH_3 + 2H_2O$
(*j*) $H_2C=O + HCN \longrightarrow H_2C(OH)CN$

(*a*) Elimination and an intramolecular displacement; a C—O bond is formed in place of a C—Br bond. (*b*) Elimination and redox; the alcohol is oxidized to a ketone. (*c*) Rearrangement. (*d*) Displacement and redox; H_3CCH_2Br is reduced. (*e*) Addition and redox; $H_2C=CH_2$ is reduced. (*f*) Substitution. (*g*) Elimination. (*h*) Addition. (*i*) Redox. (*j*) Addition.

Problem 3.36 Which of the following species behave as (1) a nucleophile, (2) an electrophile, (3) both, or (4) neither?

(*a*) $:\ddot{\underset{..}{Cl}}:^-$	(*d*) $AlBr_3$	(*g*) Br^+	(*j*) NO_2^+	(*m*) $CH_2{=}CHCH_2^+$
(*b*) $H_2\ddot{O}:$	(*e*) $CH_3\ddot{\underset{..}{O}}H$	(*h*) Fe^{3+}	(*k*) $H_2C{=}\ddot{\underset{..}{O}}$	(*n*) CH_4
(*c*) H^+	(*f*) $BeCl_2$	(*i*) $SnCl_4$	(*l*) $CH_3C{\equiv}\ddot{N}$	(*o*) $H_2C{=}CHCH_3$

(1): (*a*), (*b*), (*e*), (*o*). (2): (*c*), (*d*), (*f*), (*g*), (*h*), (*i*), (*j*), (*m*). (3): (*k*) and (*l*) (because carbon is electrophilic; oxygen and nitrogen are nucleophilic). (4): (*n*).

Problem 3.37 Formulate the following as a two-step reaction and label nucleophiles and electrophiles.

$$H_2C{=}CH_2 + Br_2 \longrightarrow \underset{Br}{H_2C}{-}\underset{Br}{CH_2}$$

Step 1

$$\underset{\text{nucleophile}_1}{H_2C{=}CH_2} + \underset{\text{electrophile}_2}{Br{-}Br} \longrightarrow \underset{\text{electrophile}_1}{\underset{Br}{H_2C}{-}\overset{+}{C}H_2} + \underset{\text{nucleophile}_2}{Br^-}$$

Step 2

$$\underset{\text{electrophile}_1}{\underset{Br}{H_2C}{-}\overset{+}{C}H_2} + \underset{\text{nucleophile}_2}{Br^-} \longrightarrow \underset{Br}{H_2C}{-}\underset{Br}{CH_2}$$

Problem 3.38 The addition of 3 mol of H_2 to 1 mol of benzene, C_6H_6,

$$3H_2 + C_6H_6 \underset{300^\circ C}{\overset{Pd(rt)}{\rightleftharpoons}} C_6H_{12}$$

occurs at room temperature (rt); the reverse elimination reaction proceeds at 300°C. For the addition reaction, ΔH and ΔS are both negative. Explain in terms of thermodynamic functions: (*a*) why ΔS is negative and (*b*) why the addition doesn't proceed at room temperature without a catalyst.

(*a*) A negative ΔH tends to make ΔG negative, but a negative ΔS tends to make ΔG positive. At room temperature, ΔH exceeds $T\Delta S$, and therefore, ΔG is negative. At the higher temperature (300°C), $T\Delta S$ exceeds ΔH, and ΔG is then positive. ΔS is negative because four molecules become one molecule, thereby reducing the randomness of the system.

(*b*) The addition has a very high $\Delta H^\ddagger$, and the rate without catalyst is extremely slow.

Problem 3.39 The reaction $CH_4 + I_2 \rightarrow CH_3I + HI$ does not occur as written because the equilibrium lies to the left. Explain in terms of the bond dissociation energies, which are 427, 151, 222, and 297 kJ/mol for C—H, I—I, C—I, and H—I, respectively.

The endothermic, bond-breaking energies are +427(C—H) and +151(I—I), for a total of +578 kJ/mol. The exothermic, bond-forming energies are −222(C—I) and −297(H—I), for a total of −519 kJ/mol. The net ΔH is

$$+578 - 519 = +59 \text{ kJ/mol}$$

and the reaction is endothermic. The reactants and products have similar structures, so the ΔS term is insignificant. Reaction does not occur because ΔH and ΔG are positive.

Problem 3.40 Which of the isomers ethyl alcohol, H_3CCH_2OH, or dimethyl ether, H_3COCH_3, has the lower enthalpy? The bond dissociation energies are 356, 414, 360, and 464 kJ/mol for C—C, C—H, C—O and O—H, respectively.

C_2H_5OH has 1 C—C bond (356 kJ/mol), 5 C—H bonds (5 × 414 kJ/mol), one C—O bond (360 kJ/mol) and one O—H bond (464 kJ/mol), giving a total energy of 3250 kJ/mol. CH_3OCH_3 has 6 C—H bonds (6 × 414 kJ/mol) and 2 C—O bonds (2 × 360 kJ/mol), and the total bond energy is 3,204 kJ/mol. C_2H_5OH has the higher total bond energy and therefore the lower enthalpy of formation. More energy is needed to decompose C_2H_5OH into its elements.

Problem 3.41 Consider the following sequence of steps:

$$(1)\ A \longrightarrow B \qquad (2)\ B + C \xrightarrow{\text{slow}} D + E \qquad (3)\ E + A \longrightarrow 2F$$

(*a*) Which species may be described as (i) reactant, (ii) product, and (iii) intermediate? (*b*) Write the net chemical equation. (*c*) Indicate the molecularity of each step. (*d*) If the second step is rate-determining, write the rate expression. (*e*) Draw a plausible reaction-enthalpy diagram.

(*a*) (i) A, C; (ii) D, F; (iii) B, E.
(*b*) $2A + C \rightarrow D + 2F$ (add steps 1, 2, and 3).
(*c*) (1) unimolecular, (2) bimolecular, (3) bimolecular.
(*d*) Rate = k[C][A], since A is needed to make the intermediate, B.
(*e*) See Fig. 3.6.

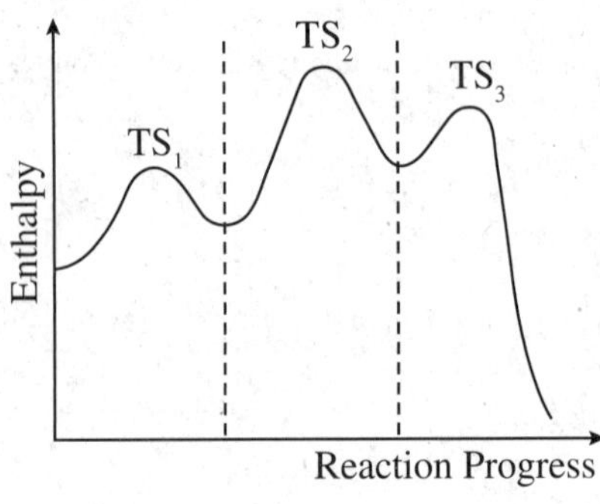

Figure 3.6

Problem 3.42 A minor step in Problem 3.41 is $2E \rightarrow G$. What is G?

A side product.

Problem 3.43 The rate expression for the reaction

$$(CH_3)_3C\text{—}Br + CH_3COO^- + Ag^+ \longrightarrow CH_3COOC(CH_3)_3 + AgBr$$

is

$$\text{Rate} = k[(CH_3)_3C\text{—}Br][Ag^+]$$

Suggest a plausible two-step mechanism showing the reacting electrophiles and nucleophiles.

The rate-determining step involves only $(CH_3)_3C$—Br and Ag^+. The acetate ion CH_3COO^- must participate in an ensuing fast step.

Step 1 $(CH_3)_3\overset{\delta+}{C}\text{—}\overset{\delta-}{\ddot{Br}}: + Ag^+ \xrightarrow{\text{slow}} (CH_3)_3C^+ + AgBr$

(Br: nucleophilic site; Ag^+: electrophile)

Step 2 $CH_3COO^- + \overset{+}{C}(CH_3)_3 \xrightarrow{\text{fast}} CH_3COOC(CH_3)_3$

nucleophile electrophile

Problem 3.44 Give the conjugate acid of (*a*) CH_3NH_2, (*b*) CH_3O^-, (*c*) CH_3OH, (*d*) $:\dot{H}^-$, (*e*) $:CH_3^-$, (*f*) $H_2C{=}CH_2$.

(*a*) $CH_3NH_3^+$, (*b*) CH_3OH, (*c*) $CH_3OH_2^+$, (*d*) H_2, (*e*) CH_4, (*f*) $H_3CCH_2^+$.

Problem 3.45 What are the conjugate bases, if any, for the substances in Problem 3.44?

(*a*) $CH_3\ddot{N}H^-$, (*b*) $:CH_2O^{2-}$, (*c*) CH_3O^-, (*d*) none, (*e*) $H_2\ddot{C}{:}^{2-}$, (*f*) $H_2C{=}\ddot{C}H^-$. The bases in (*b*) and (*e*) are extremely difficult to form; from a *practical* point of view, CH_3O^- and $H_3C{:}^-$ have no conjugate bases.

Problem 3.46. Are any of the following substances amphoteric? (*a*) H_2O, (*b*) NH_3, (*c*) NH_4^+, (*d*) Cl^-, (*e*) HCO_3^-, (*f*) HF.

(*a*) Yes, gives H_3O^+ and OH^-. (*b*) Yes, gives NH_4^+ and H_2N^-. (*c*) No, cannot accept H^+. (*d*) No, cannot donate H^+. (*e*) Yes, gives H_2CO_3 ($CO_2 + H_2O$) and CO_3^{2-}. (*f*) Yes, gives H_2F^+ and F^-.

Problem 3.47 Account for the fact that acetic acid, CH_3COOH, is a stronger acid in water than in methanol, CH_3OH.

The equilibrium

$$CH_3COOH + H_2O \rightleftharpoons CH_3COO^- + H_3O^+$$

lies more to the right than does

$$CH_3COOH + CH_3OH \rightleftharpoons CH_3COO^- + CH_3OH_2^+$$

This difference could result if CH_3OH were a weaker base than H_2O. However, this might not be the case. The significant difference arises from solvation of the ions. Water solvates ions better than does methanol; thus, the equilibrium is shifted more toward the right to form ions that are solvated by water.

Problem 3.48 Refer to Fig. 3.7, the enthalpy diagram for the reaction A $\rightarrow$ B. (*a*) What do states 1, 2, and 3 represent? (*b*) Is the reaction exothermic or endothermic? (*c*) Which is the rate-determining step, A $\rightarrow$ 2 or 2 $\rightarrow$ B? (*d*) Can substance 2 ever be isolated from the mixture? (*e*) What represents the activation enthalpy of the overall reaction A $\rightarrow$ B? (*f*) Is step A $\rightarrow$ 2 reversible?

(*a*) 1 and 3 are transition states; 2 is an intermediate.
(*b*) Since the overall product, B, is at lower energy than reactant, A, the reaction is exothermic.
(*c*) The rate-determining step is the one with the higher enthalpy of activation, 2 $\rightarrow$ *B*.
(*d*) Yes. The activation enthalpy needed for 2 to get through transition state 3 may be so high that 2 is stable enough to be isolated.
(*e*) The $\Delta H^\ddagger$ is represented by the difference in enthalpy between A, the reactant, and the higher transition state, 3.
(*f*) The $\Delta H^\ddagger$ for 2 $\rightarrow$ *A* is less than the $\Delta H^\ddagger$ for 2 $\rightarrow$ *B*; therefore, 2 returns to A more easily than it goes on to B. The step *A* $\rightarrow$ 2 is fast and reversible.

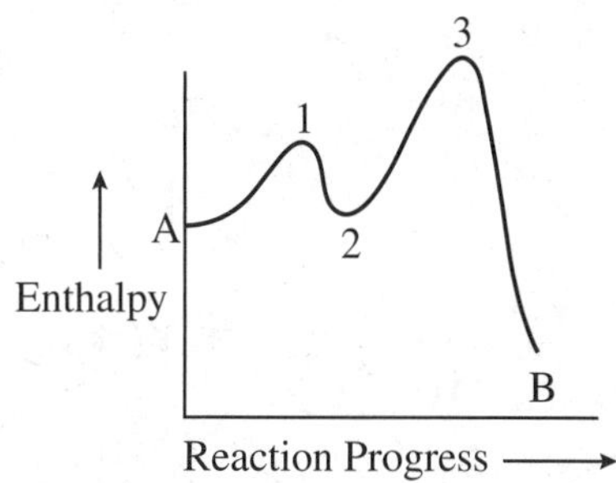

Figure 3.7

CHAPTER 4

Alkanes

4.1 Definition

Alkanes are open-chain (acyclic) hydrocarbons constituting the homologous series with the general formula C_nH_{2n+2}, where n is an integer. They have only single bonds and therefore are said to be **saturated**.

Problem 4.1 (*a*) Use the superscripts *1*, *2*, *3*, and so on to indicate the different kinds of equivalent H atoms in propane, $CH_3CH_2CH_3$. (*b*) Replace one of each kind of H by a CH_3 group. (*c*) How many isomers of butane, C_4H_{10}, exist?

(*a*) $CH_3^1CH_2^2CH_3^1 \equiv \begin{array}{c} \quad H^1 \quad H^2 \quad H^1 \\ H^1\,C{-}C{-}CH^1 \\ \quad H^1 \quad H^2 \quad H^1 \end{array}$

(*b*) $CH_3^1CH_2^2CH_3^1 \xrightarrow[+\boxed{CH_3}]{-H^1} CH_3CH_2CH_2\boxed{CH_3}$ *n*-Butane

$CH_3^1CH_2^2CH_3^1 \xrightarrow[+\boxed{CH_3}]{-H^2} \begin{array}{c} \boxed{CH_3} \\ | \\ CH_3CHCH_3 \end{array}$ Isobutane

(*c*) Two: *n*-butane and isobutane.

Problem 4.2 (*a*) Use the superscripts *1*, *2*, *3*, *4*, and so on, to indicate the different kinds of equivalent H's in (1) *n*-butane and (2) isobutane. (*b*) Replace one of each kind of H in the two butanes by a CH_3. (*c*) Give the number of isomers of pentane, C_5H_{12}.

(1) $CH_3^1CH_2^2CH_2^2CH_3^1$ or $\begin{array}{c} \quad H^1 \quad H^2 \quad H^2 \quad H^1 \\ H^1C{-}C{-}C{-}CH^1 \\ \quad H^1 \quad H^2 \quad H^2 \quad H^1 \end{array}$ (2) $\begin{array}{c} \overset{3}{CH_3}\overset{4}{CH}\overset{3}{CH_3} \\ |\,3 \\ CH_3 \end{array}$ or $\overset{4}{CH}(\overset{3}{CH_3})_3$ or $\begin{array}{c} \overset{3}{CH_3} \\ {}_3H \quad | \quad H_3 \\ HC{-}C{-}CH \\ H^3 \quad H^4 \quad H^3 \end{array}$

(*b*) $CH_3^1CH_2^2CH_2^2CH_3^1 \xrightarrow[+\boxed{CH_3}]{-H^1} CH_3CH_2CH_2CH_2\boxed{CH_3}$

n-Pentane

$CH_3^1CH_2^2CH_2^2CH_3^1 \xrightarrow[+\boxed{CH_3}]{-H^2} CH_3\overset{\boxed{CH_3}}{\overset{|}{C}}HCH_2CH_3$

Isopentane

$CH_3^3\overset{CH_3^3}{\overset{|}{C}}H^4CH_3^3 \xrightarrow[+\boxed{CH_3}]{-H^3} CH_3\overset{\boxed{CH_3}}{\overset{|}{C}}H_2CH_2\boxed{CH_3}$

Isopentane

$CH_3^3\overset{CH_3^3}{\overset{|}{C}}CH_3^3\ (H^4) \xrightarrow[+\boxed{CH_3}]{-H^4} CH_3\overset{CH_3}{\overset{|}{\underset{CH_3}{\underset{|}{C}}}}CH_3$

Neopentane

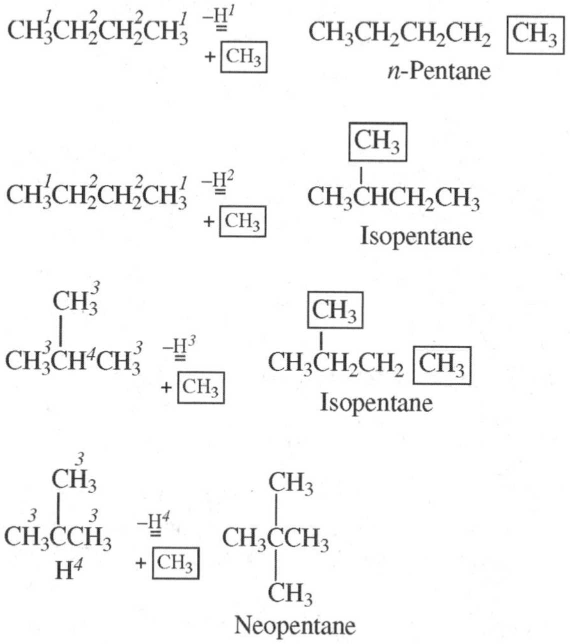

(*c*) Three: *n*-pentane, isopentane, and neopentane.

Sigma-bonded C's can rotate about the C—C bond, and hence, a chain of singly bonded C's can be arranged in any zigzag shape (**conformation**). Two such arrangements, for four consecutive C's, are shown in Fig. 4.1. *Since these conformations cannot be isolated, they are not isomers.*

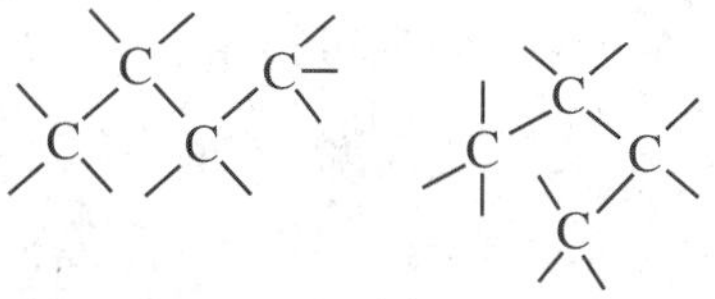

Figure 4.1

The two extreme conformations of ethane—called **eclipsed** [Fig. 4.2(*a*)] and **staggered** [Fig. 4.2(*b*)]—are shown in the "wedge" and Newman projections. With the Newman projection, we sight along the C—C bond, so that the back C is hidden by the front C. The circle aids in distinguishing the bonds on the front C (touching at the center of the circle) from those on the back C (drawn to the circumference of the circle). In the eclipsed conformation, the bonds on the back C are, for visibility, slightly offset from a truly eclipsed view. The angle between a given C—H bond on the front C and the closest C—H bond on the back C is called the **dihedral (torsional) angle** (θ). The θ values for the closest pairs of C—H bonds in the eclipsed and staggered conformations are 0° and 60°, respectively. All intermediate conformations are called **skew**; their θ values lie between 0° and 60° (see Fig. 4.2).

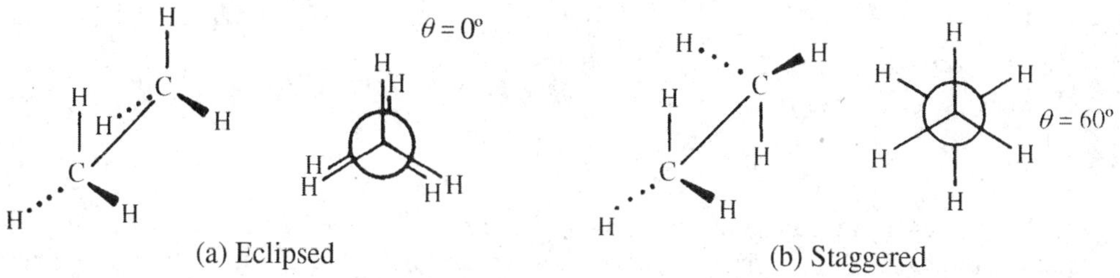

$\theta = 0°$

(a) Eclipsed

$\theta = 60°$

(b) Staggered

Figure 4.2

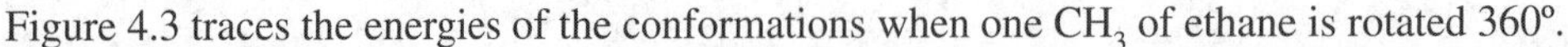

Figure 4.3 traces the energies of the conformations when one CH_3 of ethane is rotated 360°.

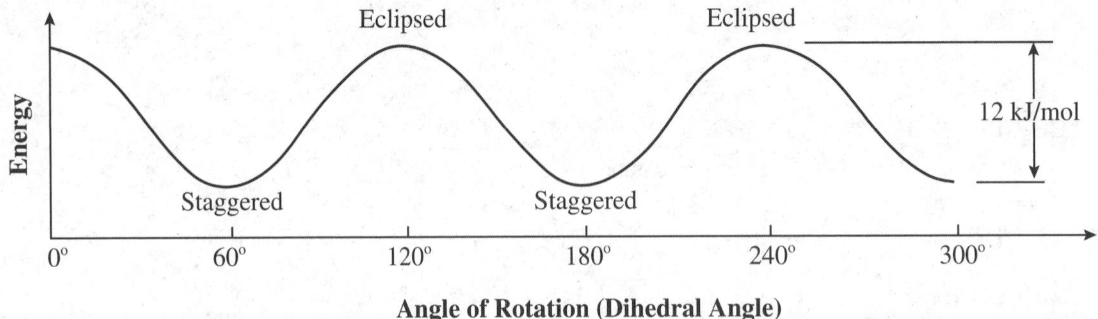

Figure 4.3

Problem 4.3 (*a*) Are the staggered and eclipsed conformations the only ones possible for ethane? (*b*) Indicate the preferential conformation of ethane molecules at room temperature. (*c*) What conformational changes occur as the temperature rises? (*d*) Is the rotation about the σ C—C bond, as in ethane, really "free"?

(*a*) No. There is an infinite number with energies between those of the staggered and eclipsed conformations. For simplicity, we are concerned only with conformations at minimum and maximum energies.
(*b*) The staggered form has the minimal energy and hence is the preferred conformation.
(*c*) The concentration of eclipsed conformations increases.
(*d*) There is an energy barrier of 12 kJ/mol (enthalpy of activation) for one staggered conformation to pass through the eclipsed conformation to give another staggered conformation. Therefore, rotation about the sigma C—C bond in ethane is somewhat restricted rather than "free."

Problem 4.4 How many distinct compounds do the following structural formulas represent?

```
(a) CH3—CH—CH2—CH—CH2—CH3          (b) CH2—CH—CH2—CH—CH3
        |         |                     |   |        |
        CH3       CH3                   CH3 CH3      CH3

        CH3                                     CH3
        |                                       |
(c) CH3—CH—CH2—CH—CH2—CH3          (d) CH3—CH—CH—CH2—CH2
                 |                          |            |
                 CH3                        CH3          CH3

        CH3                                CH3
        |                                  |
(e) CH3—CH—CH2                     (f) CH3—CH    H
            |                              |     |
  CH3—CH2—C—CH3                           CH2—C—CH3
            |                                    |
            H                                    CH2
                                                 |
                                                 CH3
```

Two. (*a*), (*b*), (*c*), (*e*), and (*f*) are conformations of the same compound. This becomes obvious when the longest chain of carbons, in this case six, is written in a linear fashion. (*d*) represents a different compound.

Problem 4.5 (*a*) Which of the following compounds can exist in different conformations? (1) hydrogen peroxide, HOOH; (2) ammonia, NH_3; (3) hydroxylamine, H_2NOH; (4) methyl alcohol, H_3COH. (*b*) Draw two structural formulas for each compound in (*a*) possessing conformations.

(*a*) A compound must have a sequence of at least three consecutive single bonds, with no π bonds, in order to exist in different conformations. (1), (3), and (4) have such a sequence. In (2):

H—N(—H)—H

the three single bonds are not consecutive.

(*b*) (1) H—O—O—H (eclipsed); H—O—O—H (staggered) (3) H₂N—O—H (eclipsed); H₂N—O—H (staggered)

(4) H_3C—O—H (eclipsed); H_3C—O—H (staggered)

The first-drawn structure in each case is the eclipsed conformation; the second one is staggered.

Problem 4.6 Explain the fact that the calculated entropy for ethane is much greater than the experimentally determined value.

The calculated value incorrectly assumes unrestricted free rotation so that all conformations are equally probable. Since most molecules of ethane have the staggered conformation, the structural randomness is less than calculated, and the actual observed entropy is less. This discrepancy led to the concept of conformations with different energies.

Figure 4.4 shows extreme conformations of *n*-butane. The two eclipsed conformations, I and II, are least stable. The totally eclipsed structure I, having eclipsed CH_3's, has a higher energy than II, in which CH_3 eclipses H. Since the other three conformations are staggered, they are at energy minima and are the stable conformations (**conformers**) of butane. The *anti* conformer, having the CH_3's farthest apart, has the lowest energy, is the most stable, and constitutes the most numerous form of butane molecules. In the two, higher-energy, staggered, *gauche* conformers the CH_3's are closer than they are in the more stable *anti* form.

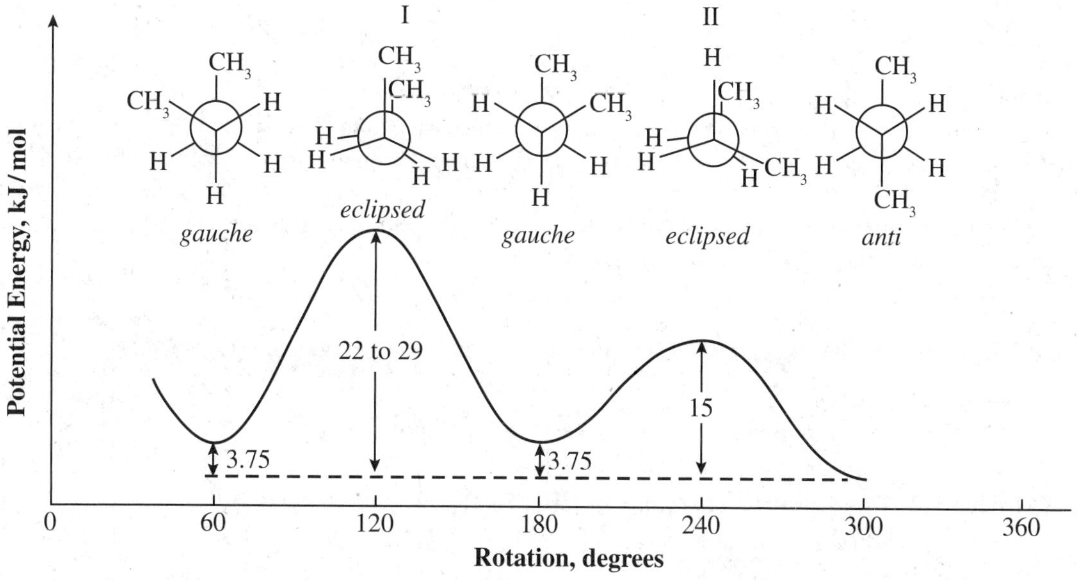

Figure 4.4

Problem 4.7 Give two factors that account for the resistance to rotation through the high-energy eclipsed conformation.

Torsional strain arises from repulsion between the bonding pairs of electrons, which is greater in the eclipsed form because the electrons are closer. **Steric strain** arises from the proximity and bulkiness of the bonded atoms or group of atoms. This strain is greater in the eclipsed form because the groups are closer. The larger the atom or group, the greater the steric strain.

Problem 4.8 How does the relative population of an eclipsed and a staggered conformation depend on the energy difference between them?

The greater the energy difference, the more the population of the staggered conformer exceeds that of the eclipsed.

Problem 4.9 Draw a graph of potential energy plotted against angle of rotation for conformations of (*a*) 2,3-dimethylbutane, (*b*) 2-methylbutane. Point out the factors responsible for energy differences.

Start with the conformer having a pair of CH_3's *anti*. Write the conformations resulting from successive rotations about the central bond of 60°.

(*a*) As shown in Fig. 4.5(*a*), structure IV has each pair of CH_3's eclipsed and has the highest energy. Structures II and VI have the next highest energy; they have only one pair of eclipsed CH_3's. The stable conformers at energy minima are I, III, and V. Structure I has both pairs of CH_3 groups *anti* and has the lowest energy. Structures III and V have one pair of CH_3's *anti* and one pair *gauche*.

(*b*) As shown in Fig. 4.5(*b*), the conformations in decreasing order of energy are:

1. IX and XI; have eclipsing CH_3's.
2. XIII; CH_3 and H eclipsing.
3. X; CH_3's are all *gauche*.
4. VIII and XII; have a pair of *anti* CH_3's.

4.2 Nomenclature of Alkanes

The letter *n* (for *normal*), as in *n*-butane, denotes an unbranched chain of C atoms. The prefix *iso-*(*i-*) indicates a CH_3 branch on the second C from the end. For example, isopentane is

$$\begin{array}{l}CH_3\underset{\displaystyle CH_3}{\underset{|}{C}}HCH_2CH_3\end{array}$$

Alkyl groups, such as methyl (CH_3) and ethyl (CH_3CH_2), are derived by removing one H from alkanes.

The prefixes *sec-* and *tert-* before the name of the group indicate that the H was removed from a **secondary** or **tertiary** C, respectively. A secondary C has bonds to two other C's, a tertiary to three other C's, and a **primary** either to three H's or to two H's and *one* C.

The H's attached to these types of carbon atoms are also called *primary*, *secondary*, and *tertiary* (1°, 2°, and 3°), respectively. A **quaternary** C is bonded to four other C's.

The letter R is often used to represent any alkyl group.

Problem 4.10 Name the alkyl groups originating from (*a*) propane, (*b*) *n*-butane, (*c*) isobutane.

(*a*) $CH_3CH_2CH_2—$ is *n*-propyl (*n*-Pr); $CH_3\overset{|}{C}HCH_3$ is isopropyl (*i*-Pr).

(*b*) $CH_3CH_2CH_2CH_2—$ is *n*-butyl (*n*-Bu); $CH_3\overset{|}{C}HCH_2CH_3$ is *sec*-butyl (*s*-Bu).

(*c*) $CH_3—\underset{\displaystyle CH_3}{\underset{|}{C}}HCH_2—$ is isobutyl (*i*-Bu); $CH_3\underset{\displaystyle CH_3}{\underset{|}{\overset{|}{C}}}CH_3$ is *tert*-butyl (*t*-Bu).

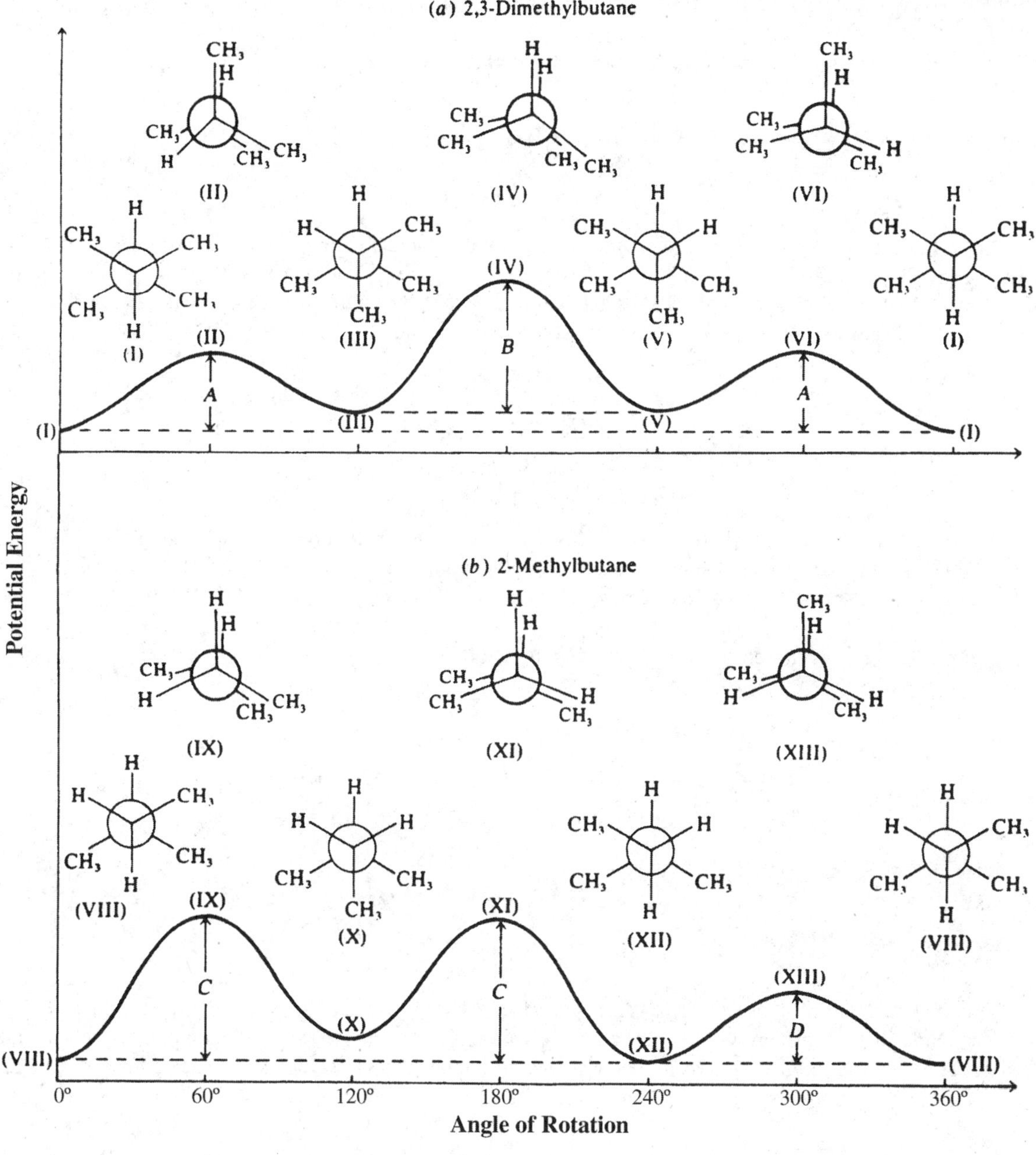

Figure 4.5

Problem 4.11 Use numbers *1*, *2*, *3*, and *4* to designate the 1°, 2°, 3°, and 4° C's, respectively, in $CH_3CH_2C(CH_3)_2CH_2CH(CH_3)_2$. Use letters *a*, *b*, *c*, and so on to indicate the different kinds of 1° and 2° C's.

```
              1b
              CH3          H
 1a   2a      |4    2b     |3    1c
 CH3CH2—C—CH2—C—CH3
              |            |
              CH3          CH3
              1b           1c
```

Problem 4.12 Name by the IUPAC system the isomers of pentane derived in Problem 4.2.

(*a*) $CH_3CH_2CH_2CH_2CH_3$ Pentane (IUPAC does not use *n*)

(*b*) The longest consecutive chain in

$$\begin{array}{c} \quad CH_3 \\ \quad | \\ \overset{1}{C}H_3\overset{2}{C}H\overset{3}{C}H_2\overset{4}{C}H_3 \end{array}$$

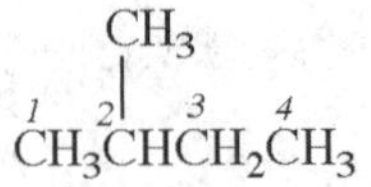

has 4 C's, and therefore, the IUPAC name is a substituted butane. Number the C's as shown so that the branch CH_3 is on the C with the lower number; in this case C^2. The name is 2-methylbutane and not 3-methylbutane. Note that numbers are separated from letters by a hyphen and words are run together.

(*c*) The longest consecutive chain in

$$\begin{array}{ccc} & CH_3 & \\ & | & \\ CH_3- & C & -CH_3 \\ & | & \\ & CH_3 & \end{array}$$

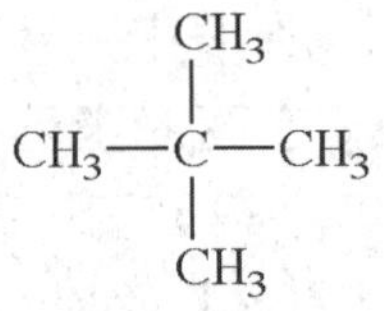

has three C's; the parent is propane. The IUPAC name is 2,2-dimethylpropane. Note the use of the prefix *di-* to show two CH_3 branches, and the repetition of the number 2 to show that both CH_3's are on C^2. *Commas separate numbers and hyphens separate numbers and words.*

Problem 4.13 Name the compound in Fig. 4.6(*a*) by the IUPAC system.

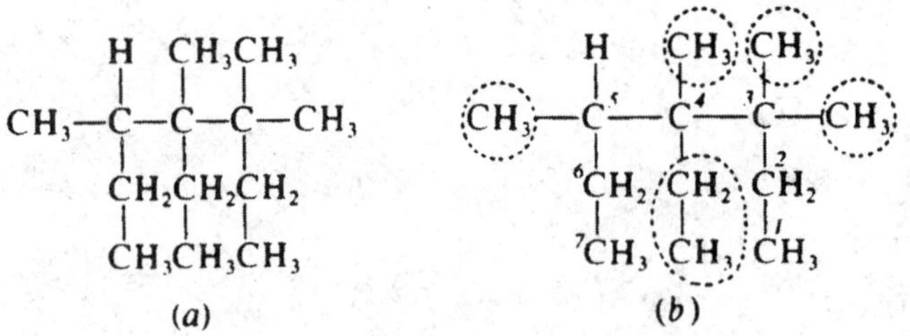

$$\begin{array}{ccccccc} & & H & & CH_3 & & CH_3 \\ & & | & & | & & | \\ CH_3 & - & C & - & C & - & C & -CH_3 \\ & & | & & | & & | \\ & & CH_2 & & CH_2 & & CH_2 \\ & & | & & | & & | \\ & & CH_3 & & CH_3 & & CH_3 \end{array}$$

(*a*)

$$\begin{array}{ccccccc} & & H & & (CH_3) & & (CH_3) \\ & & | & & | & & | \\ (CH_3) & - & \overset{5}{C} & - & \overset{4}{C} & - & \overset{3}{C} & -(CH_3) \\ & & | & & | & & | \\ & & {}^6CH_2 & & (CH_2 & & {}^2CH_2 \\ & & | & & | & & | \\ & & {}^7CH_3 & & CH_3) & & {}^1CH_3 \end{array}$$

(*b*)

Figure 4.6

The longest chain of consecutive C's has 7 C's [see Fig. 4.6(*b*)], and so the compound is named as a heptane. Note that, as written, this longest chain is bent and not linear. Circle the branch alkyl groups, and consecutively number the C's in the chain so that the lower-numbered C's hold the most branch groups. The name is 3,3,4,5-tetramethyl-4-ethylheptane.

4.3 Preparation of Alkanes

Reactions with No Change in Carbon Skeleton

1. **Reduction of Alkyl Halides** (RX, X = Cl, Br, or I) **(Substitution of Halogen by Hydrogen)**

(*a*) $RX + Zn + H^+ \longrightarrow RH + Zn^{2+} + X^-$

(*b*) $4RX + LiAlH_4 \longrightarrow 4RH + LiX + AlX_3$

or $RX + H{:}^- \longrightarrow RH + X^-$ ($H{:}^-$ comes from $LiAlH_4$)

(*c*) $RX + (n\text{-}C_4H_9)_3\,SnH \longrightarrow RH + (n\text{-}C_4H_9)_3\,SnX$

(*d*) Via organometallic compounds (Grignard reagent). Alkyl halides react with either Mg or Li in dry ether to give **organometallics** having a basic carbanionic site.

$$RX + 2Li \xrightarrow{\text{dry ether}} \underset{\text{Alkyllithium}}{\bar{R}{:}Li^+ + LiX} \quad \text{then} \quad \bar{R}{:}\overset{+}{Li} + H_2O \longrightarrow RH + \overset{+}{Li}\overline{OH}$$

$$RX + Mg \xrightarrow{\text{dry ether}} \underset{\text{Grignard reagent}}{\bar{R}{:}(MgX)^+} \quad \text{then} \quad \bar{R}(\overset{+}{Mg}X) + H_2O \longrightarrow RH + (MgX)^+(OH)^-$$

The net effect is replacement of X by H.

2. Hydrogenation of $\gt C{=}C\lt$ (alkenes) or $-C{\equiv}C-$ (alkynes)

$$\underset{\text{Isobutylene}}{CH_3-\overset{\overset{CH_3}{|}}{C}{=}CH_2} + H_2 \xrightarrow[\text{or Ni}]{\text{Pt}} \underset{\text{Isobutane}}{CH_3-\overset{\overset{CH_3}{|}}{CH}-CH_3}$$

$$\underset{\text{Propyne}}{CH_3-C{\equiv}C-H} + 2H_2 \xrightarrow[\text{or Ni}]{\text{Pt}} \underset{\text{Propane}}{CH_3-CH_2-CH_3}$$

Products with More Carbons Than the Reactants

Two alkyl groups can be **coupled** by indirectly reacting two molecules of RX, or RX with R′X, to give R—R or R—R′, respectively. The preferred method is the **Corey-House synthesis**, which uses the organometallic **lithium dialkylcuprates**, R_2CuLi, as intermediates.

$$2R-Li + CuI \xrightarrow{\text{ether}} R_2CuLi + LiI \quad \text{(Most R groups are possible.)}$$

$$R_2CuLi + R'-X \xrightarrow{\text{ether}} R-R' + RCu + LiX \quad \text{(All groups except 3°.)}$$

$$\underset{\text{Lithium diisopropylcuprate}}{[(CH_3)_2\underset{\underset{H}{|}}{C}]_2CuLi} + Br-CH_2CH_2CH_3 \longrightarrow \underset{\text{2-Methylpentane}}{(CH_3)_2\underset{\underset{H}{|}}{C}-CH_2CH_2CH_3} + (CH_3)_2CHCu + LiBr$$

Problem 4.14 Write equations to show the products obtained from the reactions:

(*a*) 2-Bromo-2-methylpropane + magnesium in dry ether
(*b*) Product of (*a*) + H_2O
(*c*) Product of (*a*) + D_2O

(*a*) $$CH_3-\overset{\overset{CH_3}{|}}{\underset{\underset{CH_3}{|}}{C}}-Br + Mg \longrightarrow CH_3-\overset{\overset{CH_3}{|}}{\underset{\underset{CH_3}{|}}{C}}-MgBr \quad \textit{tert}\text{-Butylmagnesium bromide}$$

(*b*) $$\underset{\text{base}_1}{CH_3-\overset{\overset{CH_3}{|}}{\underset{\underset{CH_3}{|}}{C}}{:}^-(MgBr)^+} + \underset{\text{acid}_2}{HOH} \longrightarrow \underset{\text{acid}_1}{CH_3-\overset{\overset{CH_3}{|}}{\underset{\underset{CH_3}{|}}{C}}-H} + \underset{\text{base}_2}{(MgBr^+)(OH^-)}$$

(*c*) The *t*-butyl carbanion accepts a deuterium cation to form 2-methyl-2-deuteropropane, $(CH_3)_3CD$.

Problem 4.15 Use 1-bromo-2-methylbutane and any other one- or two-carbon compounds, if needed, to synthesize the following with good yields:

(*a*) 2-methylbutane (*b*) 3,6-dimethyloctane (*c*) 3-methylhexane

(*a*)
$$BrCH_2CH(CH_3)CH_2CH_3 \xrightarrow{Mg} BrMgCH_2CH(CH_3)CH_2CH_3 \xrightarrow{H_2O} CH_3CH(CH_3)CH_2CH_3$$

(*b*)
$$BrCH_2CH(CH_3)CH_2CH_3 \xrightarrow[\text{ether}]{Li} LiCH_2CH(CH_3)CH_2CH_3 \xrightarrow{CuI} CuLi(CH_2CH(CH_3)CH_2CH_3)_2$$
$$\xrightarrow{BrCH_2CH(CH_3)CH_2CH_3} CH_3CH_2CH(CH_3)CH_2\text{—}CH_2CH(CH_3)CH_2CH_3$$

(*c*)
$$CuLi(CH_2CH(CH_3)CH_2CH_3)_2 \xrightarrow{CH_3CH_2Br} CH_3CH_2CH_2CH(CH_3)CH_2CH_3$$

Problem 4.16 Give the different combinations of RX and R'X that can be used to synthesize 3-methylpentane. Which synthesis is "best"?

From the structural formula

$$CH_3\overset{1}{—}CH_2\overset{2}{—}\underset{\underset{CH_3}{|3}}{CH}\overset{2}{—}CH_2\overset{1}{—}CH_3$$

we see that there are three kinds of C—C bonds, labeled *1*, *2*, *3*. The combinations are as follows: for bond *1*, CH_3X and $XCH_2CH(CH_3)CH_2CH_3$; for bond *2*, CH_3CH_2X and $XCH(CH_3)CH_2CH_3$; for bond *3*, CH_3X and $XCH(CH_2CH_3)_2$. The chosen method utilizes the simplest, and least expensive, alkyl halides. On this basis, bond *2* is the one to form on coupling.

4.4 Chemical Properties of Alkanes

Alkanes are unreactive except under vigorous conditions.

1. **Pyrolytic Cracking** [heat (Δ) in absence of O_2; used in making gasoline]

$$\text{Alkane} \xrightarrow{\Delta} \text{mixture of smaller hydrocarbons}$$

2. **Combustion**

$$CH_4 + 2O_2 \xrightarrow{600°C} CO_2 + 2H_2O \qquad \Delta H \text{ of combustion} = -809.2\ \text{kJ/mol}$$

Problem 4.17 (*a*) Why are alkanes inert? (*b*) Why do the C—C rather than the C—H bonds break when alkanes are pyrolyzed? (*c*) Although combustion of alkanes is a strongly exothermic process, it does not occur at moderate temperatures. Explain.

(*a*) A reactive site in a molecule usually has one or more unshared pairs of electrons, a polar bond, an electron-deficient atom, or an atom with an expandable octet. Alkanes have none of these.

(*b*) The C—C bond has a lower bond energy ($\Delta H = +347$ kJ/mol) than the C—H bond ($\Delta H = +414$ kJ/mol).
(*c*) The reaction is very slow at room temperature because of a very high $\Delta H^{\ddagger}$.

3. Halogenation

$$RH + X_2 \xrightarrow[\text{or } \Delta]{\text{uv}} RX + HX$$

(Reactivity of X_2: $F_2 > Cl_2 > Br_2$. I_2 does not react; F_2 destroys the molecule)
Chlorination (and bromination) of alkanes such as methane, CH_4, has a radical-chain mechanism, as follows:

Initiation Step

$$Cl{:}Cl \xrightarrow[\text{or } \Delta]{\text{uv}} 2Cl\cdot \qquad \Delta H = +243 \text{ kJ/mol}$$

The required enthalpy comes from ultraviolet (uv) light as heat.

Propagation Steps

(i) $H_3C{:}H + Cl\cdot \longrightarrow H_3C\cdot + H{:}Cl \qquad \Delta H = -4$ kJ/mol (rate-determining)

(ii) $H_3C\cdot + Cl{:}Cl \longrightarrow H_3C{:}Cl + Cl\cdot \qquad \Delta H = -96$ kJ/mol

The sum of the two propagation steps is the overall reaction:

$$CH_4 + Cl_2 \longrightarrow CH_3Cl + HCl \qquad \Delta H = -100 \text{ kJ/mol}$$

In propagation steps, the same free-radical intermediates, here $Cl\cdot$ and $H_3C\cdot$, are being formed and consumed. Chains terminate on those rare occasions when two free-radical intermediates form a covalent bond:

$$Cl\cdot + Cl\cdot \longrightarrow Cl_2, \quad H_3C\cdot + Cl\cdot \longrightarrow H_3C{:}Cl, \quad H_3C\cdot + \cdot CH_3 \longrightarrow H_3C{:}CH_3$$

Inhibitors stop chain propagation by reacting with free-radical intermediates, for example:

$$H_3C\cdot + \cdot\ddot{\underset{\cdot\cdot}{O}}—\ddot{\underset{\cdot\cdot}{O}}\cdot \longrightarrow H_3C\ddot{\underset{\cdot\cdot}{O}}—\ddot{\underset{\cdot\cdot}{O}}\cdot$$

The inhibitor—here O_2—must be consumed before chlorination can occur.

In more complex alkanes, the abstraction of each different kind of H atom gives a different isomeric product. Three factors determine the relative yields of the isomeric product:

(1) **Probability factor.** This factor is based on the number of each kind of H atom in the molecule. For example, in $CH_3CH_2CH_2CH_3$, there are *six* equivalent 1° H's and *four* equivalent 2° H's. The odds on abstracting a 1° H are thus 6 to 4, or 3 to 2.

(2) **Reactivity of H.** The order of reactivity of H is 3° > 2° > 1°.

(3) **Reactivity of X·**. The more reactive Cl· is less selective and more influenced by the probability factor. The less reactive Br· is more selective and less influenced by the probability factor. As summarized by the **reactivity-selectivity principle**: If the attacking species is more reactive, it will be less selective, and the yields will be closer to those expected from the probability factor.

Problem 4.18 (*a*) List the monobromo derivatives of (i) $CH_3CH_2CH_2CH_3$ and (ii) $(CH_3)_2CHCH_3$. (*b*) Predict the predominant isomer in each case. The order of reactivity of H for bromination is

$$3° (1600) > 2° (82) > 1° (1)$$

(*a*) There are two kinds of H's, and there are two possible isomers for each compound: (i) $CH_3CH_2CH_2CH_2Br$ and $CH_3CHBrCH_2CH_3$; (ii) $(CH_3)_2CHCH_2Br$ and $(CH_3)_2CBrCH_3$.

(*b*) In bromination, in general, the difference in reactivity completely overshadows the probability effect in determining product yields. (i) $CH_3CHBrCH_2CH_3$ is formed by replacing a 2° H; (ii) $(CH_3)_2CBrCH_3$ is formed by replacing the 3° H and predominates.

Problem 4.19 Using the bond dissociation energies for X_2,

X_2	F_2	Cl_2	Br_2	I_2
ΔH, kJ/mol	+155	+243	+193	+151

show that the initiation step for halogenation of alkanes,

$$X_2 \xrightarrow[\text{or } \Delta]{\text{uv}} 2X\cdot$$

is not rate-determining.

The enthalpy $\Delta H^\ddagger$ (Section 3.8) is seldom related to ΔH of the reaction. In this reaction, however, $\Delta H^\ddagger$ and ΔH are identical. In simple homolytic dissociations of this type, the free radicals formed have the same enthalpy as does the transition state. On this basis alone, iodine, having the smallest ΔH and $\Delta H^\ddagger$, should react fastest. Similarly, chlorine, with the largest ΔH and $\Delta H^\ddagger$, should react slowest. But the actual order of reaction rates is

$$F_2 > Cl_2 > Br_2 > I_2$$

Therefore, the initiation step is not rate-determining; the rate is determined by the first propagation step, H-abstraction.

Problem 4.20 Draw the reactants, transition state, and products for

$$Br\cdot + CH_4 \longrightarrow HBr + \cdot CH_3$$

In the transition state, Br is losing radical character while C is becoming a radical; both atoms have partial radical character as indicated by $\delta\cdot$. The C atom undergoes a change in hybridization as indicated:

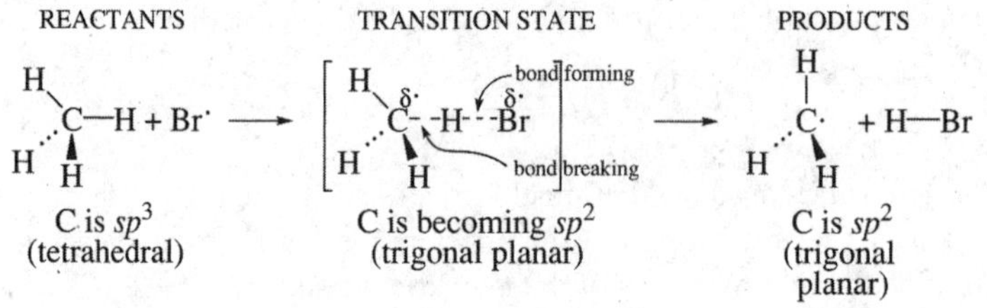

REACTANTS

$H_3C{-}H + Br\cdot \longrightarrow$

C is sp^3 (tetrahedral)

TRANSITION STATE

$[H_3C^{\delta\cdot}\cdots H\cdots Br^{\delta\cdot}] \longrightarrow$ (bond forming; bond breaking)

C is becoming sp^2 (trigonal planar)

PRODUCTS

$H_3C\cdot + H{-}Br$

C is sp^2 (trigonal planar)

Problem 4.21 Bromination of methane, like chlorination, is exothermic, but it proceeds at a slower rate under the same conditions. Explain in terms of the factors that affect the rate, assuming that the rate-controlling step is

$$X\cdot + CH_4 \longrightarrow HX + \cdot CH_3$$

Given the same concentration of CH_4 and Cl· or Br·, the frequency of collisions should be the same. Because of the similarity of the two reactions, $\Delta S^\ddagger$ for each is about the same. The difference must be due to the $\Delta H^\ddagger$, which is less (17 kJ/mol) for Cl· than for Br· (75 kJ/mol).

Problem 4.22 2-Methylbutane has 1°, 2°, and 3° H's as indicated:

$$\overset{1^\circ}{(CH_3)_2}\overset{3^\circ}{CH}\overset{2^\circ}{CH_2}\overset{1^\circ}{CH_3}$$

(*a*) Use enthalpy-reaction progress diagrams for the abstraction of each kind of hydrogen by ·X. (*b*) Summarize the relationships of relative (i) stabilities of transition states, (ii) $\Delta H^\ddagger$ values, (iii) stabilities of alkyl radicals, and (iv) rates of H-abstraction.

(*a*) See Fig. 4.7.
(*b*) (i) 3° > 2° > 1° since the enthalpy of the $TS_{1°}$ is the greatest and the enthalpy of the $TS_{3°}$ is the smallest. (ii) $\Delta H^{\ddagger}_{3°} < \Delta H^{\ddagger}_{2°} < \Delta H^{\ddagger}_{1°}$. (iii) 3° > 2° > 1°. (iv) 3° > 2° > 1°.

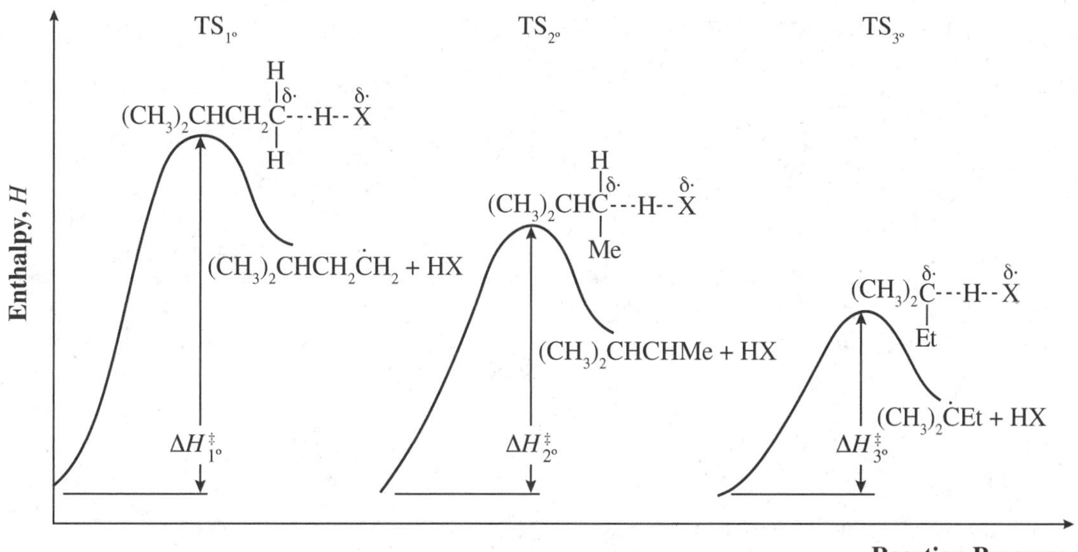

Figure 4.7

Problem 4.23 List and compare the differences in the properties of the transition states during chlorination and bromination that account for the different reactivities for 1°, 2°, and 3° H's.

The differences may be summarized as follows:

	Chlorination	Bromination
1. Time of formation of transition state	Earlier in reaction	Later in reaction
2. Amount of breaking of C—H bond	Less, H_3C---H-----Cl	More, H_3C----H---Br
3. Free-radical character ($\delta\cdot$) of carbon	Less	More
4. Transition state more closely resembles	Reactants	Products

These show that the greater selectivity in bromination is attributable to the greater free-radical character of carbon. With greater radical character, the differences in stability between 1°, 2°, and 3° radicals become more important, and reactivity of H (3° > 2° > 1°) also become more significant.

4. Isomerization

$$CH_3CH_2CH_2CH_3 \underset{}{\overset{AlCl_3,\ HCl}{\rightleftharpoons}} CH_3-\underset{}{\overset{\overset{CH_3}{|}}{C}H}-CH_3$$

4.5 Summary of Alkane Chemistry

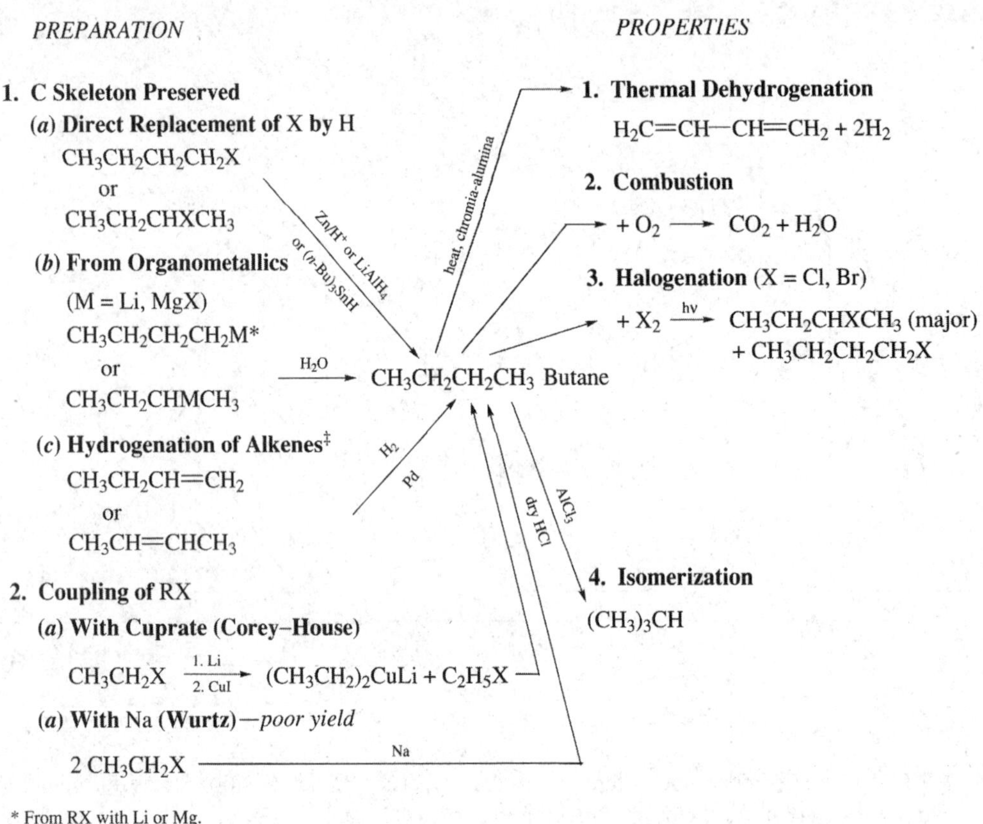

* From RX with Li or Mg.
‡ Also from corresponding alkynes.

SUPPLEMENTARY PROBLEMS

Problem 4.24 Assign numbers, ranging from (1) for LOWEST to (3) for HIGHEST, to the boiling points of the following hexane isomers: 2,2-dimethylbutane, 3-methylpentane, and *n*-hexane. Do not consult any tables for data.

Hexane has the longest chain, the greatest intermolecular attraction, and, therefore, the highest boiling point, (3). 2,2-Dimethylbutane is (1), since, with the most spherical shape, it has the smallest intermolecular contact and attraction. This leaves 3-methylpentane as (2).

Problem 4.25 Write structural formulas for the five isomeric hexanes, and name them by the IUPAC system.

The isomer with the longest chain is hexane, $CH_3CH_2CH_2CH_2CH_2CH_3$. If we use a five-carbon chain, a CH_3 may be placed on either C^2 or C^4 to produce 2-methylpentane, or on C^3 to give another isomer, 3-methylpentane.

$$\overset{1}{C}H_3\overset{2}{C}H(CH_3)\overset{3}{C}H_2\overset{4}{C}H_2\overset{5}{C}H_3$$

2-Methylpentane

$$\overset{1}{C}H_3\overset{2}{C}H_2\overset{3}{C}H(CH_3)\overset{4}{C}H_2\overset{5}{C}H_3$$

3-Methylpentane

With a four-carbon chain, either a CH_3CH_2 or two CH_3's must be added as branches to give a total of 6 C's. Placing CH_3CH_2 anywhere on the chain is ruled out because it lengthens the chain. Two CH_3's are added, but only to central C's to avoid extending the chain. If both CH_3's are introduced on the same C, the isomer is 2,2-dimethylbutane. Placing one CH_3 on each of the central C's gives the remaining isomer, 2,3-dimethylbutane.

$$\begin{array}{ccc} & CH_3 & \\ & | & \\ CH_3— & C & —CH_2CH_3 \\ & | & \\ & CH_3 & \end{array}$$

2,2-Dimethylbutane

$$\begin{array}{cccc} & CH_3 & CH_3 & \\ & | & | & \\ CH_3— & C— & C & —CH_3 \\ & | & | & \\ & H & H & \end{array}$$

2,3-Dimethylbutane

Problem 4.26 Write the structural formulas for (*a*) 3,4-dichloro-2,5-dimethylhexane; (*b*) 5-(1,2-dimethylpropyl)-6-methyldodecane. (Complex branch groups are usually enclosed in parentheses.)

(*a*)

$$\begin{array}{cccccc} & CH_3 & Cl & Cl & CH_3 & \\ & | & | & | & | & \\ CH_3— & CH— & CH— & CH— & CH— & CH_3 \end{array}$$

(*b*) The group in parentheses is bonded to the fifth C. It is a propyl group with CH_3's on its first and second C's (denoted *1′* and *2′*) counting from the attached C.

$$\begin{array}{ccc} & H & CH_3 \\ & | & | \\ CH_3CH_2CH_2CH_2— & \overset{5}{C}— & CHCH_2CH_2CH_2CH_2CH_2CH_3 \\ & |_{1'} & \\ & HC—CH_3 & \\ & |_{2'} & \\ & HC—CH_3 & \\ & |_{3'} & \\ & CH_3 & \end{array}$$

Problem 4.27 Write the structural formulas and give the IUPAC names for all monochloro derivatives of (*a*) isopentane, $(CH_3)_2CHCH_2CH_3$; (*b*) 2,2,4-trimethylpentane, $(CH_3)_3CCH_2CH(CH_3)_2$.

(*a*) Since there are four kinds of equivalent H's:

$$(\overset{1}{CH_3})_2\overset{2}{C}H\overset{3}{C}H_2\overset{4}{C}H_3$$

there are four isomers:

$$\begin{array}{c} H \\ | \\ ClCH_2CCH_2CH_3 \\ | \\ H \end{array}$$

1-Chloro-2-methylbutane

$$\begin{array}{ccc} & CH_3 & \\ & | & \\ CH_3— & C & —CH_2CH_2 \\ & | & \\ & Cl & \end{array}$$

2-Chloro-2-methylbutane

$$\begin{array}{ccc} & CH_3 & \\ & | & \\ CH_3— & C— & CHCH_3 \\ & | & | \\ & H & Cl \end{array}$$

2-Chloro-3-methylbutane

$$\begin{array}{ccc} & CH_3 & \\ & | & \\ CH_3— & C & —CH_2CH_2Cl \\ & | & \\ & H & \end{array}$$

1-Chloro-3-methylbutane

(*b*) There are four isomers because there are four kinds of H's $(\overset{1}{CH_3})_3C\overset{2}{C}H_2\overset{3}{C}H(\overset{4}{CH_3})_2$.

$$\begin{array}{ccccc} & CH_3 & & CH_3 & \\ & | & & | & \\ ClCH_2— & C— & CH_2— & C & —CH_3 \\ & | & & | & \\ & CH_3 & & H & \end{array}$$

1-Chloro-2,2,4-trimethylpentane

$$\begin{array}{ccccc} & CH_3 & H & CH_3 & \\ & | & | & | & \\ CH_3— & C— & C— & C & —CH_3 \\ & | & | & | & \\ & CH_3 & Cl & H & \end{array}$$

3-Chloro-2,2,4-trimethylpentane

$$\begin{array}{ccccc} & CH_3 & & CH_3 & \\ & | & & | & \\ CH_3— & C— & CH_2— & C & —CH_3 \\ & | & & | & \\ & CH_3 & & Cl & \end{array}$$

2-Chloro-2,2,4-trimethylpentane

$$\begin{array}{ccccc} & CH_3 & & CH_3 & \\ & | & & | & \\ CH_3— & C— & CH_2— & C & —CH_2Cl \\ & | & & | & \\ & CH_3 & & H & \end{array}$$

1-Chloro-2,4,4-trimethylpentane

Problem 4.28 Give topological structural formulas for (*a*) propane, (*b*) butane, (*c*) isobutane, (*d*) 2,2-dimethylpropane, (*e*) 2,3-dimethylbutane, (*f*) 3-ethylpentane, (*g*) 1-chloro-3-methylbutane, (*h*) 2,3-dichloro-2-methylpentane, (*i*) 2-chloro-2,4-trimethylpentane.

In this method, one writes only the C—C bonds and all functional groups bonded to C. The approximate bond angles are used

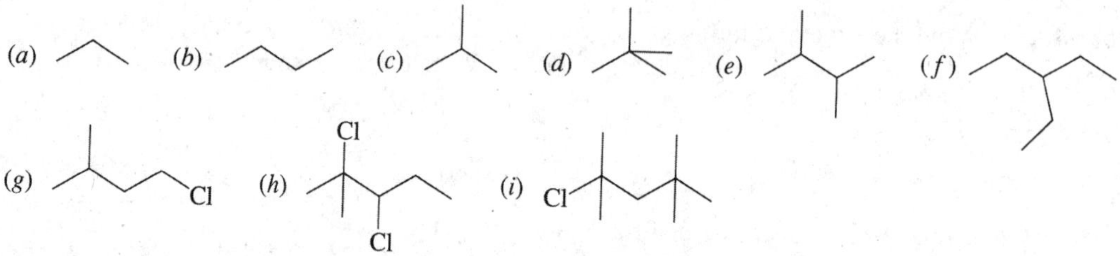

(*a*) (*b*) (*c*) (*d*) (*e*) (*f*)

(*g*) Cl (*h*) Cl Cl (*i*) Cl

Problem 4.29 Synthesize (*a*) 2-methylpentane from $CH_3CH{=}CH{-}CH(CH_3)_2$, (*b*) isobutane from isobutyl chloride, (*c*) 2-methyl-2-deuterobutane from 2-chloro-2-methylbutane. Show all steps.

(*a*) The alkane and the starting compound have the same carbon skeleton.

$$CH_3{-}CH{=}CH{-}\underset{}{\overset{CH_3}{\overset{|}{C}}}H{-}CH_3 \xrightarrow{H_2/Pt} CH_3{-}CH_2{-}CH_2{-}\overset{CH_3}{\overset{|}{C}}H{-}CH_3$$

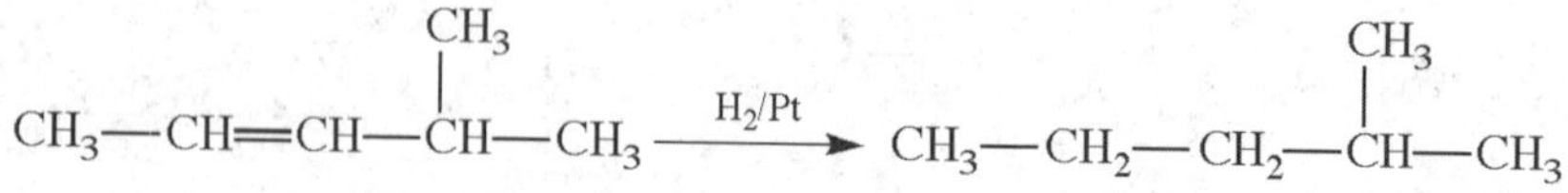

(*b*) The alkyl chloride and alkane have the same carbon skeleton.

$$CH_3\overset{CH_3}{\overset{|}{C}}HCH_2{-}Cl \xrightarrow[\text{or } LiAlH_4]{Zn,HCl} CH_3\overset{CH_3}{\overset{|}{C}}HCH_2{-}H$$

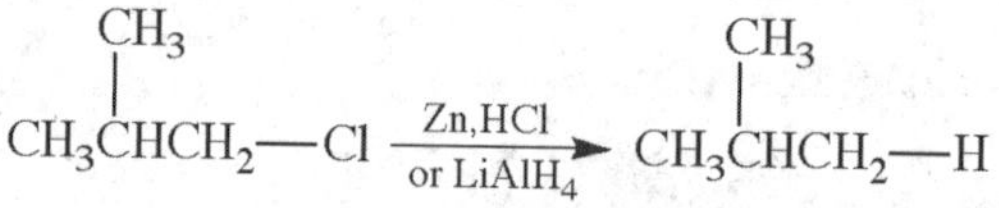

(*c*) Deuterium can be bonded to C by the reaction of D_2O with a Grignard reagent.

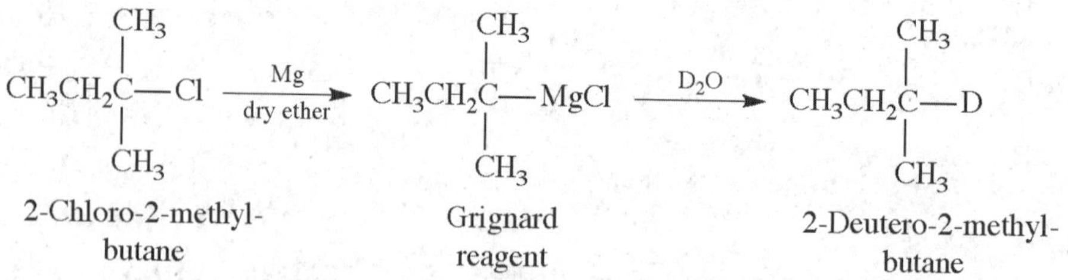

$$CH_3CH_2\underset{CH_3}{\underset{|}{\overset{CH_3}{\overset{|}{C}}}}{-}Cl \xrightarrow[\text{dry ether}]{Mg} CH_3CH_2\underset{CH_3}{\underset{|}{\overset{CH_3}{\overset{|}{C}}}}{-}MgCl \xrightarrow{D_2O} CH_3CH_2\underset{CH_3}{\underset{|}{\overset{CH_3}{\overset{|}{C}}}}{-}D$$

2-Chloro-2-methylbutane — Grignard reagent — 2-Deutero-2-methylbutane

Problem 4.30 RCl is treated with Li in ether solution to form RLi. RLi reacts with H_2O to form isopentane. Using the Corey-House method, RCl is coupled to form 2,7-dimethyloctane. What is the structure of RCl?

To determine the structure of a compound from its reactions, the structures of the products are first considered and their formation is then deduced from the reactions. The coupling product must be a symmetrical molecule whose carbon-to-carbon bond was formed between C^4 and C^5 of 2,7-dimethyloctane. The only RCl which will give this product is isopentyl chloride:

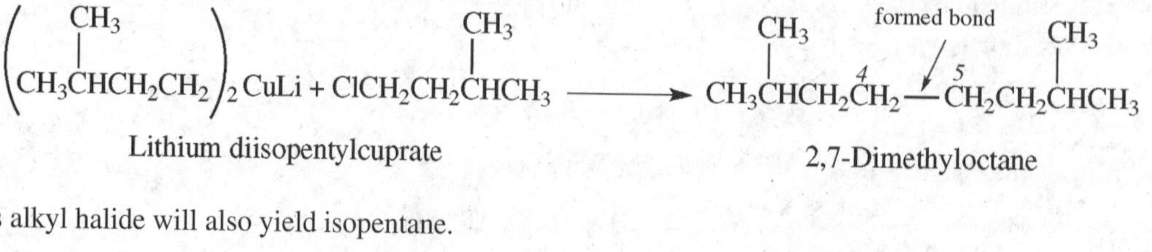

$$\left(CH_3\overset{CH_3}{\overset{|}{C}}HCH_2CH_2\right)_2CuLi + ClCH_2CH_2\overset{CH_3}{\overset{|}{C}}HCH_3 \longrightarrow CH_3\overset{CH_3}{\overset{|}{C}}HCH_2\overset{4}{C}H_2{-}\overset{5}{C}H_2CH_2\overset{CH_3}{\overset{|}{C}}HCH_3$$

Lithium diisopentylcuprate — 2,7-Dimethyloctane (formed bond between C^4 and C^5)

This alkyl halide will also yield isopentane.

$$\underset{\text{}}{CH_3\overset{\overset{CH_3}{|}}{C}HCH_2CH_2Li} \xrightarrow{H_2O} \underset{\text{(Isopentane)}}{CH_3\overset{\overset{CH_3}{|}}{C}HCH_2CH_3}$$

Problem 4.31 Give steps for the following syntheses: (*a*) propane to $(CH_3)_2CHCH(CH_3)_2$, (*b*) propane to 2-methylpentane, (*c*) $^{14}CH_3Cl$ to $^{14}CH_3{}^{14}CH_2{}^{14}CH_2{}^{14}CH_3$.

(*a*) The symmetrical molecule is prepared by coupling an isopropyl halide. Bromination of propane is preferred over chlorination because the ratio of isopropyl to *n*-propyl halide is 96%/4% in bromination and only 56%/44% in chlorination.

$$CH_3CH_2CH_3 \xrightarrow{Br_2\ (127°C)} (CH_3)_2CHBr \xrightarrow[\text{2. CuI}\quad \text{3. } CH_3CHBrCH_3]{\text{1. Li}} CH_3\overset{\overset{CH_3}{|}}{C}H{-}\overset{\overset{CH_3}{|}}{C}HCH_3$$

(*b*) $$CH_3CH_2CH_3 \xrightarrow[uv]{Cl_2} CH_3CH_2CH_2Cl + (CH_3)_2CHCl \quad \textit{(separate the mixture)}$$

$$(CH_3)_2CHCl \xrightarrow[ether]{Li} (CH_3)_2CHLi$$

$$(CH_3)_2CHLi \xrightarrow[\text{2. } CH_3CH_2CH_2Cl]{\text{1. CuI}} (CH_3)_2CHCH_2CH_2CH_3$$

(*c*) $$^{14}CH_3Cl \xrightarrow[\text{2. CuI}\quad \text{3. } ^{14}CH_3Cl]{\text{1. Li}} {}^{14}CH_3{}^{14}CH_3 \xrightarrow[uv]{Cl_2} {}^{14}CH_3{}^{14}CH_2Cl \xrightarrow[\text{2. CuI}\quad \text{3. } ^{14}CH_3{}^{14}CH_2Cl]{\text{1. Li}} {}^{14}CH_3({}^{14}CH_2)_2{}^{14}CH_3$$

Problem 4.32 Synthesize the following deuterated compounds: (*a*) CH_3CH_2D, (*b*) CH_2DCH_2D.

(*a*) $$CH_3CH_2Br \xrightarrow[ether]{Mg} CH_3CH_2MgBr \xrightarrow{D_2O} CH_3CH_2D$$

(*b*) $$H_2C{=}CH_2 + D_2 \xrightarrow{Pt} H_2CDCH_2D$$

Problem 4.33 In the dark at 150°C, tetraethyl lead, $Pb(C_2H_5)_4$, catalyzes the chlorination of CH_4. Explain in terms of the mechanism.

$Pb(C_2H_5)_4$ readily undergoes thermal homolysis of the Pb—C bond.

$$Pb(C_2H_5)_4 \longrightarrow \cdot\dot{\underset{\cdot}{P}}b\cdot + 4CH_3CH_2\cdot$$

The $CH_3CH_2\cdot$ then generates the Cl· that initiates the propagation steps.

$$CH_3CH_2\cdot + Cl{:}Cl \longrightarrow CH_3CH_2Cl + Cl\cdot$$

$$\left.\begin{aligned} CH_4 + Cl\cdot &\longrightarrow H_3C\cdot + HCl \\ H_3C\cdot + Cl{:}Cl &\longrightarrow H_3CCl + Cl\cdot \end{aligned}\right\} \text{(propagation steps)}$$

Problem 4.34 Hydrocarbons are monochlorinated with *tert*-butyl hypochlorite, *t*-BuOCl.

$$t\text{-BuOCl} + RH \longrightarrow RCl + t\text{-BuOH}$$

Write the propagating steps for this reaction if the initiating step is

$$t\text{-BuOCl} \longrightarrow t\text{-BuO}\cdot + Cl\cdot$$

The propagating steps must give the products and also form chain-carrying free radicals. The formation of *t*-BuOH suggests H-abstraction from RH by *t*-BuO·, not by Cl·. The steps are:

$$RH + t\text{-BuO}\cdot \longrightarrow R\cdot + t\text{-BuOH}$$

$$R\cdot + t\text{-BuOCl} \longrightarrow RCl + t\text{-BuO}\cdot$$

R· and *t*-BuO· are the chain-carrying radicals.

Problem 4.35 Calculate the heat of combustion of methane at 25°C. The bond energies for C—H, O═O, C═O, and O—H are, respectively, 413.0, 498.3, 803.3, and 462.8 kJ/mol.

First, write the balanced equation for the reaction.

$$CH_4 + 2O_2 \longrightarrow CO_2 + 2H_2O$$

The energies for the bonds broken are calculated. These are endothermic processes and ΔH is positive.

$$CH_4 \longrightarrow C + 4H \quad \Delta H = 4(+431.0) = +1652.0\ \text{kJ/mol}$$

$$2O_2 \longrightarrow 4O \quad \Delta H = 2(+498.3) = +996.6\ \text{kJ/mol}$$

Next, the energies are calculated for the bonds formed. Bond formation is exothermic, so the ΔH values are made negative.

$$C + 2O \longrightarrow O{=}C{=}O \quad \Delta H = 2(-803.3) = -1606.6\ \text{kJ/mol}$$

$$4H + 2O \longrightarrow 2H{-}O{-}H \quad \Delta H = 4(-462.8) = -1851.2\ \text{kJ/mol}$$

The enthalpy for the reaction is the sum of these values:

$$+1652.0 + 996.6 - 1606.6 - 1851.2 = -809.2\ \text{kJ/mol} \quad \text{(Reaction is exothermic)}$$

Problem 4.36 (*a*) Deduce structural formulas and give IUPAC names for the nine isomers of C_7H_{16}. (*b*) Why is 2-ethylpentane not among the nine?

(*a*) **seven-C chain** 1. $CH_3CH_2CH_2CH_2CH_2CH_2CH_3$ Heptane

six-C chain 2. $CH_3\overset{CH_3}{\overset{|}{C}}HCH_2CH_2CH_2CH_3$ 2-Methylhexane

3. $CH_3CH_2\overset{CH_3}{\overset{|}{C}}HCH_2CH_2CH_3$ 3-Methylhexane

five-C chain 4. $CH_3\overset{CH_3}{\overset{|}{C}}H\underset{CH_3}{\underset{|}{C}}HCH_2CH_3$ 2,3-Dimethylpentane

5. $CH_3\overset{CH_3}{\overset{|}{C}}HCH_2\overset{CH_3}{\overset{|}{C}}HCH_3$ 2,4-Dimethylpentane

6. $CH_3\underset{CH_3}{\underset{|}{\overset{CH_3}{\overset{|}{C}}}}CH_2CH_2CH_3$ 2,2-Dimethylpentane

$$\begin{array}{c} CH_3 \\ | \\ \text{7. } CH_3CH_2CCH_2CH_3 \\ | \\ CH_3 \end{array}$$ 3,3-Dimethylpentane

$$\begin{array}{c} CH_3 \\ | \\ CH_2 \\ | \\ \text{8. } CH_3CH_2CHCH_2CH_3 \end{array}$$ 3-Ethylpentane

four-C chain $$\begin{array}{c} CH_3\ CH_3 \\ |\quad\ | \\ \text{9. } CH_3C{-}CHCH_3 \\ | \\ CH_3 \end{array}$$ 2,2,3-Trimethylbutane

(*b*) Because the longest chain has six C's, and it is 3-methylhexane.

Problem 4.37 Singlet methylene, :CH_2 (Section 3.2), may be generated from diazomethane, CH_2N_2; the other product is N_2. It can insert between C—H bonds of alkanes:

$$:\mathbf{CH_2} + -\overset{|}{\underset{|}{C}}-H \longrightarrow -\overset{|}{\underset{|}{C}}-\mathbf{CH_2}-H$$

Determine the selectivity and reactivity of :CH_2 from the yields of products from methylene insertion in pentane:

$$\overset{1}{C}H_3\overset{2}{C}H_2\overset{3}{C}H_2\overset{2}{C}H_2\overset{1}{C}H_3 + :\mathbf{CH_2} \longrightarrow \mathbf{HCH_2}CH_2CH_2CH_2CH_3 + CH_3\underset{\mathbf{HCH_2}}{\underset{|}{C}H}CH_2CH_2CH_3 + CH_3CH_2\underset{\mathbf{HCH_2}}{\underset{|}{C}H}CH_2CH_3$$

Name, Kind of CH insertion	Hexane, CH^1	2-Methylpentane, CH^2	3-Methylpentane, CH^3
Yield	48%	35%	17%

Calculate the theoretical % yield based only on the probability factor and then compare the theoretical and observed % yields.

	H's OF PENTANE			
PRODUCT	KIND	NUMBER	PROPORTION × 100% = CALC. % YIELD	OBSERVED % YIELD
Hexane	*1*	6	6/12 = 50	48
2-Methylpentane	2	4	4/12 = 33.3	35
3-Methylpentane	*3*	2	2/12 = 16.7	17
Total		12		

The almost identical agreement validates the assumption that methylene is one of the most reactive and least selective species in organic chemistry.

Problem 4.38 How would the energy-conformation diagrams of ethane and propane differ?

They would both have the same general appearance showing minimum energy in the staggered form and maximum energy in the eclipsed form. But the difference in energy between these forms would be greater in propane than in ethane (13.8 versus 12.5 kJ/mol). The reason is that the eclipsing strain of the Me group and H is greater than that of two H's.

Problem 4.39 1,2-Dibromoethane has a zero dipole moment, whereas ethylene glycol, CH_2OHCH_2OH, has a measurable dipole moment. Explain.

1,2-Dibromoethane exists in the *anti* form, so that the C⟼Br dipoles cancel and the net dipole moment is zero. When the glycol exists in the *gauche* form, intramolecular H-bonding occurs. Intramolecular H-bonding is a stabilizing effect which cannot occur in the *anti* conformer.

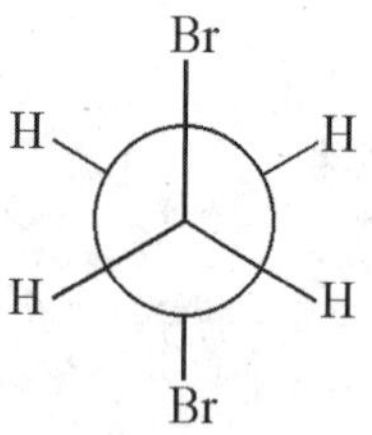

anti CH_2BrCH_2Br 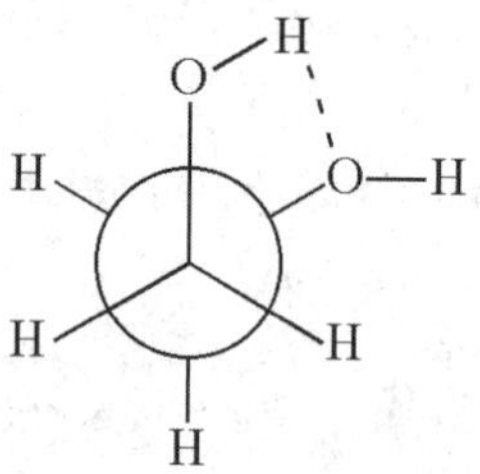 *gauche* CH_2OHCH_2OH

CHAPTER 5

Stereochemistry

5.1 Stereoisomerism

Stereoisomers (stereomers) have the same bonding order of atoms but differ in the way these atoms are arranged in space. They are classified by their symmetry properties in terms of certain symmetry elements. The two most important elements are as follows:

1. A **symmetry plane** divides a molecule into equivalent halves. It is like a mirror placed so that half the molecule is a mirror image of the other half.
2. A **center (point) of symmetry** is a point in the *center* of a molecule to which a line can be drawn from any atom such that, when extended an equal distance past the center, the line meets another atom of the same kind.

A **chiral** stereoisomer is *not* superimposable on its mirror image. It does *not* possess a plane or center of symmetry. The nonsuperimposable mirror images are called **enantiomers**. A mixture of equal numbers of molecules of each enantiomer is a **racemic form (racemate)**. The conversion of an enantiomer into a racemic form is called **racemization. Resolution** is the separation of a racemic form into individual enantiomers. Stereomers which are not mirror images are called **diastereomers**.

Molecules with a plane or center of symmetry have superimposable mirror images; they are **achiral**.

Problem 5.1 Classify as *chiral* or *achiral*: (*a*) a hand, (*b*) a screw, (*c*) an Erlenmeyer flask, (*d*) a clock face, (*e*) a rectangle, (*f*) a parallelogram, (*g*) a helix, (*h*) the letter A, (*i*) the letter R.

The achiral items are (*c*) and (*h*), with planes of symmetry; (*e*), with a center of symmetry and two planes of symmetry; and (*f*), with a center of symmetry. This leaves (*a*), (*b*), (*d*), (*g*), and (*i*) as chiral. [*Chiral* derives from the Greek for hand.]

Problem 5.2 Indicate the planes and centers of symmetry in (*a*) methane (CH_4), (*b*) chloroform ($CHCl_3$), and (*c*) ethene ($H_2C{=}CH_2$).

In molecular structures, planes of symmetry can cut through atoms or through bonds between atoms. (*a*) CH_4 has no center but has two planes, as shown in Fig. 5.1(*a*). Each plane cuts the C and a pair of H's, while bisecting the other H—C—H bond angle. (*b*) $CHCl_3$ has no center and has one plane that cuts the C, H, and one of the Cl's while bisecting the Cl—C—Cl bond angle, as shown in Fig. 5.1(*b*). (*c*) In ethene, all six atoms lie in one plane, which, in Fig. 5.1(*c*), is the plane of the paper. It has a center of symmetry between the C's. It also has three symmetry planes, one of which is the aforementioned plane of the paper. Another plane bisects the C═C bond and is perpendicular to the plane of the paper. The third plane cuts each C while bisecting each H—C—H bond.

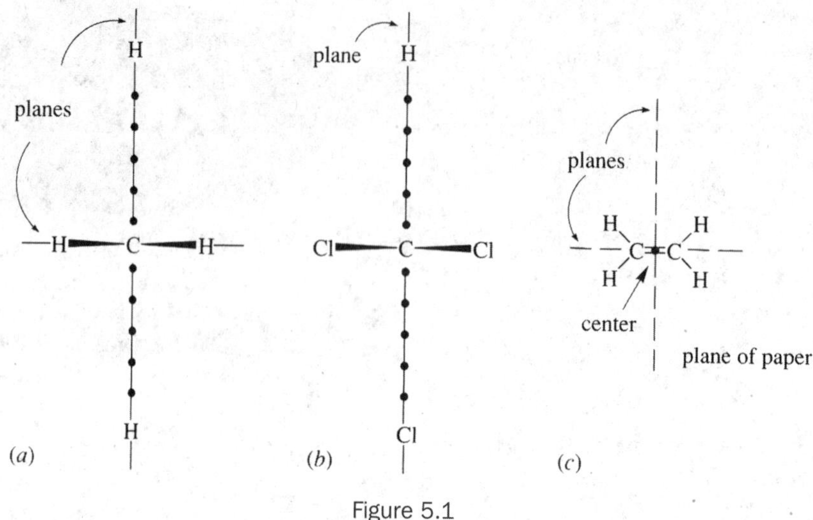

Figure 5.1

5.2 Optical Isomerism

Plane-polarized light (light vibrating in only one plane) passed through a chiral substance emerges vibrating in a different plane. The enantiomer that rotates the plane of polarized light clockwise (to the right) as seen by an observer is **dextrorotatory**; the enantiomer rotating to the left is **levorotatory**. The symbols (+) and (−) designate rotation to the right and left, respectively. Because of this optical activity, enantiomers are called **optical isomers**. The racemic form (±) is optically inactive, since it does not rotate the plane of polarized light.

The **specific rotation** $[\alpha]_\lambda^T$ is an inherent physical property of an enantiomer which, however, varies with the solvent used, temperature (T in °C), and wavelength of light used (λ). It is defined as the observed rotation per unit length of light path, per unit concentration (for a solution) or density (for a pure liquid) of the enantiomer; thus:

$$[\alpha]_\lambda^T = \frac{\alpha}{lc}$$

where $\alpha \equiv$ observed rotation, in degrees
$l \equiv$ length of path, in decimeters (dm)
$c \equiv$ concentration or density, in $\text{g/mL} \equiv \text{kg/dm}^3$

and where, by convention, the units $\text{deg}\cdot\text{dm}^2/\text{kg}$ of $[\alpha]_\lambda^T$ are abbreviated to degrees (°).

Problem 5.3 A 1.5-g sample of an enantiomer is dissolved in ethanol to make 50 mL of solution. Find the specific rotation at 20°C for sodium light (λ = 589.3 nm, the D line), if the solution has an observed rotation of +2.79° in a 10-cm polarimeter tube.

First, change the data to the appropriate units: 10 cm = 1 dm and 1.5 g/50 mL = 0.03 g/mL. The specific rotation is then

$$[\alpha]_D^{20^\circ} = \frac{\alpha}{lc} = \frac{+2.79^\circ}{(1)(0.03)} = +93^\circ \text{ (in ethanol)}$$

Problem 5.4 A 0.5-g sample of cholesterol isolated from an atherosclerotic arterial specimen yields an observed rotation of −0.76° when dissolved in 20 mL of chloroform and placed in a 1-dm cell. Determine the specific rotation [α] of cholesterol for: (*a*) the entities as defined above; (*b*) an increase in solute from 0.5 g to 1.0 g; (*c*) an increase in solvent from 20 mL to 40 mL; and (*d*) an increase in path length from 1 dm to 2 dm.

From the equation employed in Problem 5.3, simple substitution of the defined values yields: (*a*) $-30.4°$; (*b*) $-15.2°$; (*c*) $-60.8°$; and (*d*) $-15.2°$.

Problem 5.5 How could it be decided whether an observed dextrorotation of $+60°$ is not actually a levorotation of $-300°$?

Halving the concentration or the tube length would halve the number of optically active molecules, and the new rotation would be $+30°$ if the substance was dextrorotatory or $-150°$ if levorotatory.

Many chiral organic molecules have at least one carbon atom *bonded to four different atoms or groups*, called **ligands**. Such a carbon atom is called a **chiral center** and is indicated as C*. A chiral center is a particular **stereocenter**, defined as an atom for which interchange of a pair of ligands gives a different stereomer. In the case of a chiral center, the new stereomer is the enantiomer. Enantiomers can be depicted by planar projection formulas, as shown for lactic acid, $H_3CCH(OH)COOH$, in Fig. 5.2 (see also Fig. 1.2). In the **Fischer projection**, [Fig. 5.2(*c*)], the chiral C is at the intersection, horizontal groups project toward the viewer, and vertical groups project away from the viewer.

Problem 5.6 Draw (*a*) Newman and (*b*) Fischer projections for enantiomers of $CH_3CHIC_2H_5$.

See Fig. 5.3.

Problem 5.7 Determine if the configuration of the left-hand enantiomer of Fig. 5.3(*b*) is changed by a rotation in the plane of the paper of (i) 90° and (ii) 180°.

Rotations of 90° and 180° give the following Fischer projections:

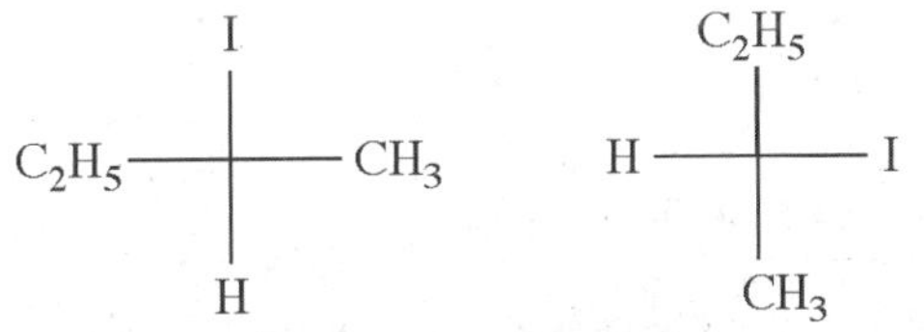

(*i*): top I; left C_2H_5; right CH_3; bottom H

(*ii*): top C_2H_5; left H; right I; bottom CH_3

The best way to see if the configuration has changed is to determine how many swaps of groups must be made to go from the initial to the final Fischer projection. An even number of swaps leaves the configuration unchanged, while an odd number results in a changed configuration. Convince yourself that to get (*i*) takes three swaps, while to get (*ii*) takes two swaps (the horizontal pairs and the vertical pairs).

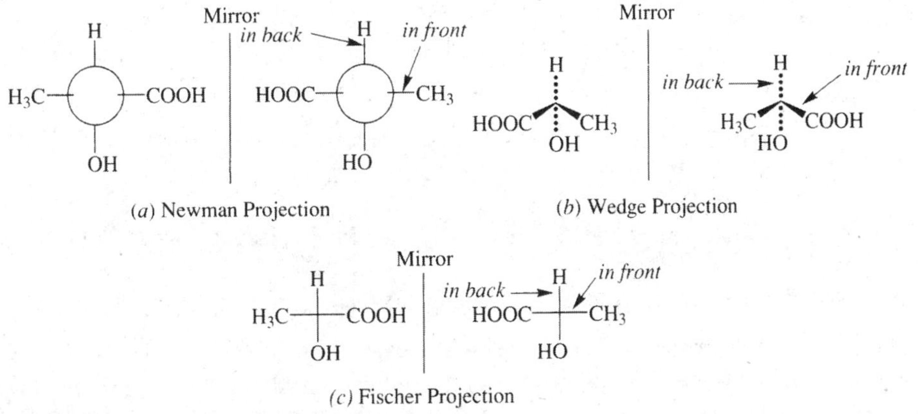

Mirror
H; H_3C; COOH; OH | in back; H; in front; HOOC; CH_3; HO

(*a*) Newman Projection

Mirror
H; HOOC; CH_3; OH | in back; H; in front; H_3C; COOH; HO

(*b*) Wedge Projection

Mirror
H; H_3C; COOH; OH | in back; H; in front; HOOC; CH_3; HO

(*c*) Fischer Projection

Figure 5.2

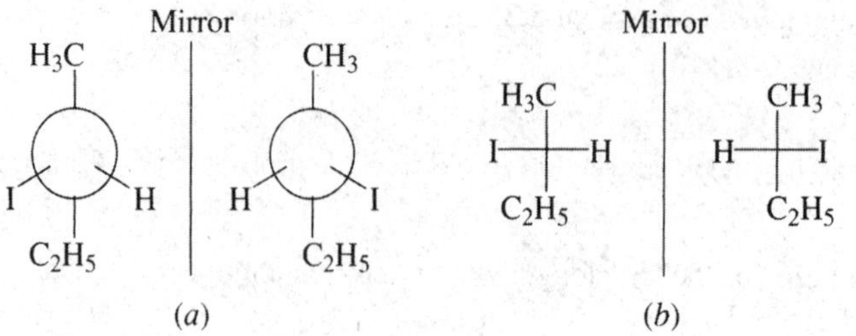

Mirror | Mirror

H_3C | CH_3 | H_3C | CH_3

I — H | H — I | I — H | H — I

C_2H_5 | C_2H_5 | C_2H_5 | C_2H_5

(*a*) (*b*)

Figure 5.3

Problem 5.8 Write structural formulas for the monochloroisopentanes. Place an asterisk on any chiral C and indicate the four different groups about the C*.

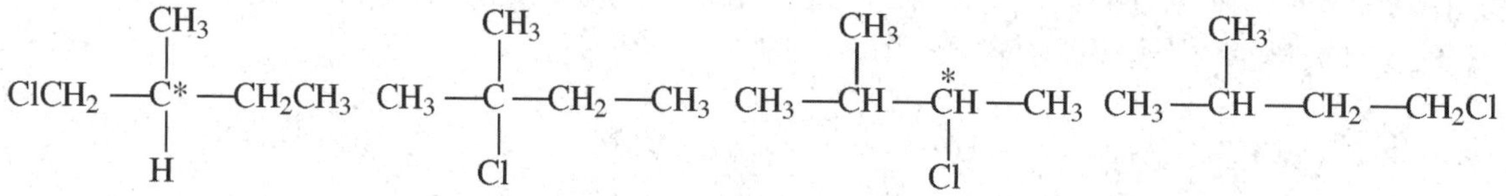

$ClCH_2—C^*(CH_3)(H)—CH_2CH_3$ $CH_3—C(CH_3)(Cl)—CH_2—CH_3$ $CH_3—CH(CH_3)—\overset{*}{C}H(Cl)—CH_3$ $CH_3—CH(CH_3)—CH_2—CH_2Cl$

1-Chloro-2-methylbutane (I)
(H, CH_3, CH_2CH_3, $ClCH_2$)

2-Chloro-2-methylbutane (II)

2-Chloro-3-methylbutane (III)
(H, Cl, CH_3, $(CH_3)_2CH$)

1-Chloro-3-methylbutane (IV)

In looking for chirality, one considers the entire group—for example, CH_2CH_3—attached to the C* and not just the attached atom.

5.3 Relative and Absolute Configuration

Configuration is the spatial arrangement of ligands in a stereoisomer. Enantiomers have opposite configurations. For enantiomers with a single chiral site, to pass from one configuration to the other (**inversion**) requires the breaking and interchanging of two bonds. A second interchange of bonds causes a return to the original configuration. Configurations may change as a result of chemical reactions. To understand the mechanism of reactions, it is necessary to assign configurations to enantiomers. For this purpose, the sign of rotation cannot be used because *there is no relationship between configuration and sign of rotation.*

Problem 5.9 Esterification of (+)-lactic acid with methyl alcohol gives (−)-methyl lactate. Has the configuration changed?

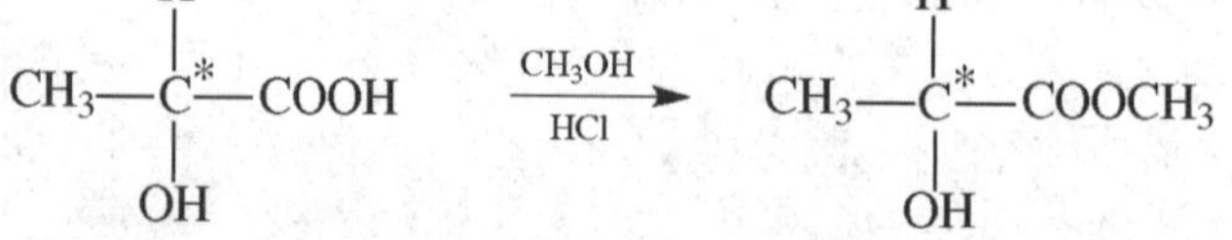

$$CH_3—\overset{*}{C}(H)(OH)—COOH \xrightarrow[HCl]{CH_3OH} CH_3—\overset{*}{C}(H)(OH)—COOCH_3$$

(+)-Lactic acid = +3.3° (–)-Methyl lactate = –8.2°

No; even though the sign of rotation changes, there is no breaking of bonds to the chiral C*.

The **Cahn-Ingold-Prelog rules** (1956) are used to designate the configuration of each chiral C in a molecule in terms of the symbols ***R*** and ***S***. These symbols come from the Latin, *R* from *rectus* (right) and *S* from *sinister* (left). Once told that the configuration of a chiral C is *R* or *S*, a chemist can write the correct projection or Fischer structural formulas. In our statement of the three rules, the numerals 1, 2, 3, and 4 are used; some chemists use letters a, b, c, and d in their place.

Rule 1

Ligands to the chiral C are assigned *increasing priorities* based on the *increasing atomic number* of the atom bonded directly to the C. (Recall that the atomic number, which is the number of protons in the nucleus, may be indicated by a presubscript on the chemical symbol, e.g., $_8O$. For isotopes, the one with higher mass has the higher priority, e.g., deuterium over hydrogen.) The priorities of the ligands will be given by numbers in parentheses, using (1) for the highest priority (the heaviest ligand) and (4) for the lowest priority (the lightest ligand). The lowest-priority ligand, (4), must project away from the viewer, behind the paper, leaving the other three ligands projecting forward. In the Fischer projection, the priority-(4) ligand must be in a vertical position (if necessary, make two interchanges of ligands to achieve this configuration). Then, for the remaining three ligands, if the sequence of decreasing priority, (1) to (2) to (3), is *counterclockwise*, the configuration is designated *S*; if it is *clockwise*, the configuration is designated *R*. The rule is illustrated for 1-chloro-1-bromoethane in Fig. 5.4. Both configuration and sign of optical rotation are included in the complete name of a species, for example, (*S*)-(+)-1-chloro-1-bromoethane.

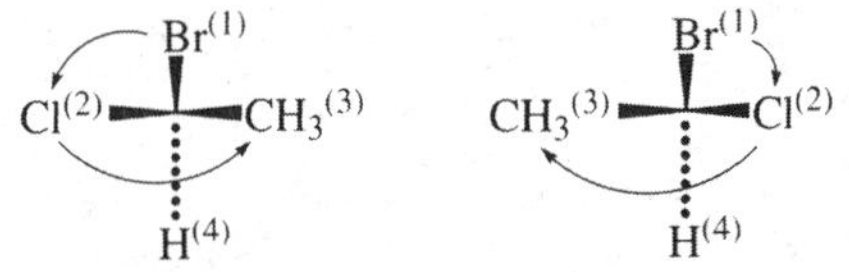

$Br^{(1)}$, $Cl^{(2)}$, $CH_3^{(3)}$, $H^{(4)}$ — Counterclockwise, *S*

$Br^{(1)}$, $CH_3^{(3)}$, $Cl^{(2)}$, $H^{(4)}$ — Clockwise, *R*

Figure 5.4

Rule 2

If the first bonded atom is the same in at least two ligands, the priority is determined by comparing the *next* atoms in each of these ligands. Thus, ethyl (H_3CCH_2—), with one C and two H's on the first bonded C, has priority over methyl (—CH_3), with three H's on the C. For butyl groups, the order of decreasing priority (or increasing ligand number) is

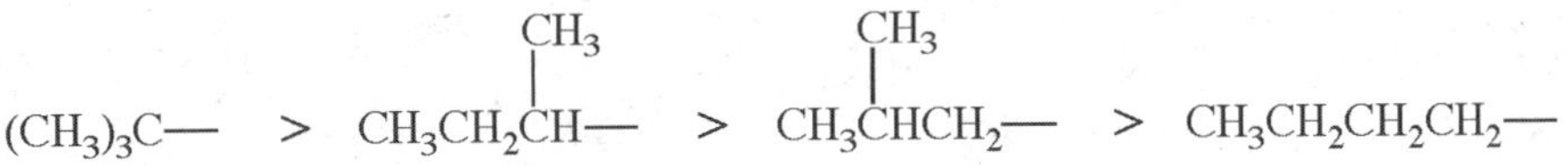

$$(CH_3)_3C— > CH_3CH_2\underset{}{CH}(CH_3)— > CH_3\underset{}{CH}(CH_3)CH_2— > CH_3CH_2CH_2CH_2—$$

Rule 3

For purposes of assigning priorities, replace

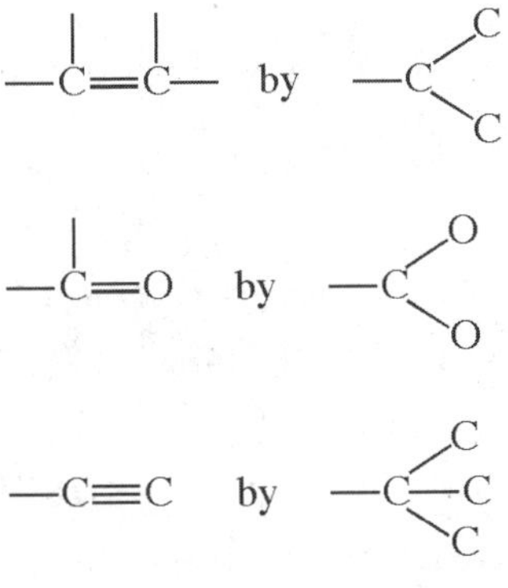

$$—\overset{|}{C}=\overset{|}{C}— \text{ by } —C\begin{smallmatrix} C \\ C \end{smallmatrix}$$

$$—\overset{|}{C}=O \text{ by } —C\begin{smallmatrix} O \\ O \end{smallmatrix}$$

$$—C\equiv C \text{ by } —C\begin{smallmatrix} C \\ C \\ C \end{smallmatrix}$$

Problem 5.10 Structures of CHClBrF are written below in seven Fischer projection formulas. Relate structures (*b*) through (*g*) to structure (*a*).

(*a*) H (top), F—|—Cl, Br (bottom) (*b*) Br (top), F—|—Cl, H (bottom) (*c*) F (top), H—|—Cl, Br (bottom) (*d*) F (top), H—|—Br, Cl (bottom)

(*e*) Br (top), Cl—|—F, H (bottom) (*f*) H (top), Cl—|—Br, F (bottom) (*g*) Cl (top), H—|—Br, F (bottom)

If two structural formulas differ by an odd number of interchanges, they are enantiomers; if by an even number, they are identical. See Table 5.1.

TABLE 5.1

	SEQUENCE OF GROUP INTERCHANGES	NUMBER OF INTERCHANGES	RELATIONSHIP TO (*a*)
(*b*)	H, Br	1 (odd)	Enantiomer
(*c*)	H, F	1 (odd)	Enantiomer
(*d*)	H, F; Br, Cl	2 (even)	Same
(*e*)	H, Br; Cl, F	2 (even)	Same
(*f*)	F, Br; F, Cl	2 (even)	Same
(*g*)	F, Br; Br, Cl; H, Cl	3 (odd)	Enantiomer

Problem 5.11 Put the following groups in decreasing order of priority.

(*a*) —C$_6$H$_5$ (phenyl) (*b*) $—CH{=}CH_2$ (*c*) $—C{\equiv}N$ (*d*) $—CH_2I$

(*e*) $—C{=}O$ with H (*f*) $—C{=}O$ with OH (*g*) $—CH_2NH_2$ (*h*) $—C(=O)—NH_2$ (*i*) $—C{=}O$ with CH_3

In each case, the first bonded atom is a C. Therefore, the second bonded atom determines the priority. In decreasing order of priority, these are ${}_{53}I > {}_{8}O > {}_{7}N < {}_{6}C$. The equivalencies are

(*a*) —C(C, C, C) (*b*) —C(H, C, C) (*c*) —C(N, N, N) (*d*) —C(H, H, I) (*e*) —C(H, O, O)

(*f*) —C(O, O, O) (*g*) —C(H, H, N) (*h*) —C(O, O, N) (*i*) —C(C, O, O)

The order of decreasing priority is (*d*) > (*f*) > (*h*) > (*i*) > (*e*) > (*c*) > (*g*) > (*a*) > (*b*). Note that in (*d*), one I has a greater priority than three O's in (*f*).

Problem 5.12 Designate as *R* or *S* the configuration of

(*a*) Cl on top, $ClCH_2$ on the left, $CH(CH_3)_2$ on the right, CH_3 on the bottom

(*b*) H on top, $H_2C{=}CH$ on the left, CH_2CH_3 on the right, Br on the bottom

(*c*) NH_2 on top, H on the left, COOH on the right, $COOC_2H_5$ on the bottom

(*a*) The decreasing order of priorities is Cl (1), CH_2Cl (2), $CH(CH_3)_2$ (3), and CH_3 (4). CH_3, with the lowest priority, is projected in back of the plane of the paper and is not considered in the sequence. The sequence of decreasing priority of the other groups is counterclockwise and the configuration is *S*.

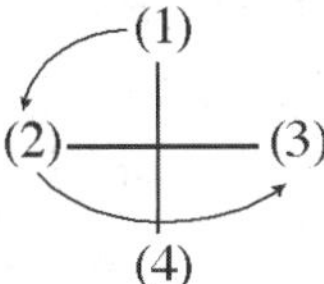

The compound is (*S*)-1,2-dichloro-2,3-dimethylbutane.

(*b*) The sequence of priorities is Br (1), $H_2C{=}CH—$ (2), $CH_3CH_2—$ (3) and H (4). The name is (*R*)-3-bromo-1-pentene.

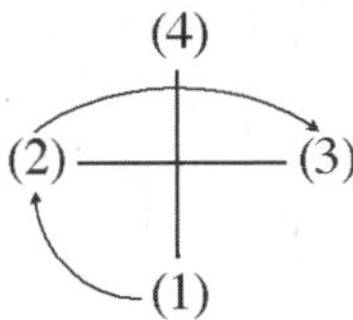

(*c*) H is exchanged with NH_2 to put H in the vertical position. Then the other two ligands are swapped, so that, with two exchanges, the original configuration is kept. Now the other three groups can be projected forward with no change in sequence. A possible identical structure is

H(4) on top, (1) H_2N on the left, (2) $COOC_2H_5$ on the right, (3) COOH on the bottom

The sequence is clockwise or *R*.

An alternative approach would leave H horizontal, thus generating an answer known to be wrong. If the wrong answer is *R* or *S*, the right answer is *S* or *R*.

Problem 5.13 Draw and specify as *R* and *S* the enantiomers, if any, of all the monochloropentanes.

n-Pentane has 3 monochloro-substituted products, 1-chloro, 2-chloro, and 3-chloropentane

$$ClCH_2CH_2CH_2CH_2CH_3 \qquad CH_3\overset{*}{C}HClCH_2CH_2CH_3 \qquad CH_3CH_2CHClCH_2CH_3$$

Only 2-chloropentane has a chiral C, whose ligands in order of decreasing priority are Cl (1), CH_3CH_2 (2), CH_3 (3), and H (4). The configurations are

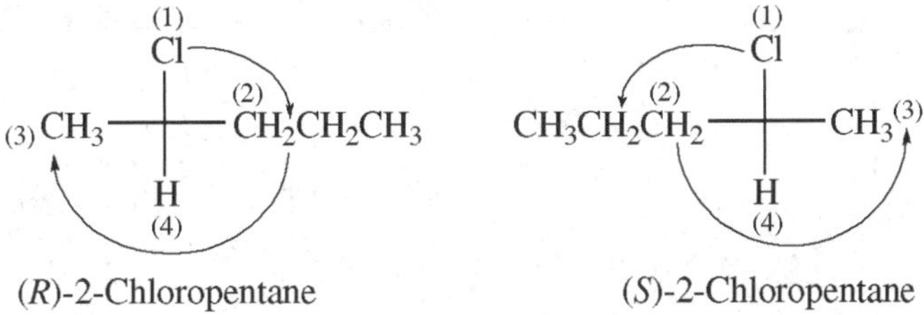

(*R*)-2-Chloropentane (*S*)-2-Chloropentane

The structures of the monochloroisopentanes are given in Problem 5.8. The sequence of decreasing priority of ligands for structure I is $ClCH_2 > CH_2CH_3 > CH_3 > H$, while for III is $Cl > (CH_3)_2CH > CH_3 > H$. The configurations are

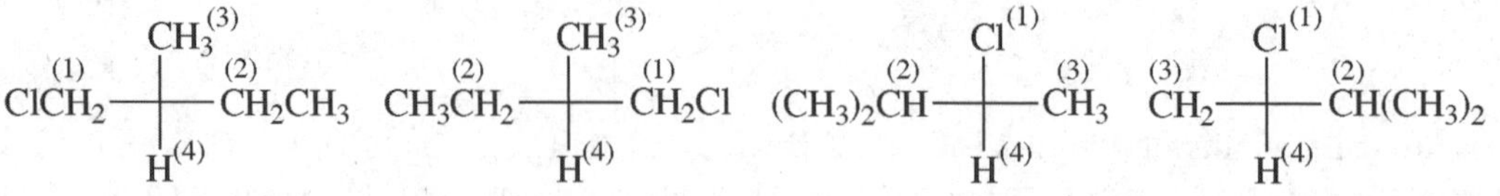

$CH_3^{(3)}$ / $ClCH_2^{(1)}$—C—$CH_2CH_3^{(2)}$ / $H^{(4)}$ (S)-(I)

$CH_3^{(3)}$ / $CH_3CH_2^{(2)}$—C—$CH_2Cl^{(1)}$ / $H^{(4)}$ (R)-(I)

$Cl^{(1)}$ / $(CH_3)_2CH^{(2)}$—C—$CH_3^{(3)}$ / $H^{(4)}$ (S)-(III)

$Cl^{(1)}$ / $CH_3^{(3)}$—C—$CH(CH_3)_2^{(2)}$ / $H^{(4)}$ (R)-(III)

Neopentyl chloride, $(CH_3)_3CCH_2Cl$, has no chiral C and therefore no enantiomers.

Relative configuration is the experimentally determined relationship between the configuration of a given chiral molecule and the arbitrarily assigned configuration of a reference chiral molecule. The **D,L system** (1906) uses glyceraldehyde, $HOCH_2C^*H(OH)CHO$, as the reference molecule. The (+)- and (−)- rotating enantiomers were arbitrarily assigned the configurations shown below in Fischer projections, and were designated as D and L, respectively.

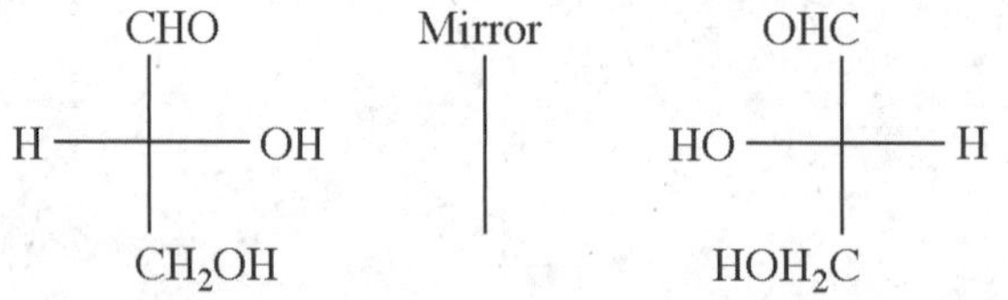

CHO / H—C—OH / CH_2OH D-(+)-Glyceraldehyde | Mirror | OHC / HO—C—H / HOH_2C L-(−)-Glyceraldehyde

Relative configurations about chiral C's of other compounds were then established by synthesis of these compounds from glyceraldehyde. It is easy to see that D-(+)-glyceraldehyde has the *R* configuration and the L-(−) isomer is *S*. This arbitrary assignment was shown to be correct by Bijvoet (1951), using X-ray diffraction. Consequently, we can now say that *R* and *S* denote **absolute configuration,** which is the actual spatial arrangement about a chiral center.

Problem 5.14 D-(+)-Glyceraldehyde is oxidized to (−)-glyceric acid, $HOCH_2CH(OH)COOH$. Give the D,L designation of the acid.

Oxidation of the D-(+)-aldehyde does not affect any of the bonds to the chiral C. The acid has the same D configuration even though the sign of rotation is changed.

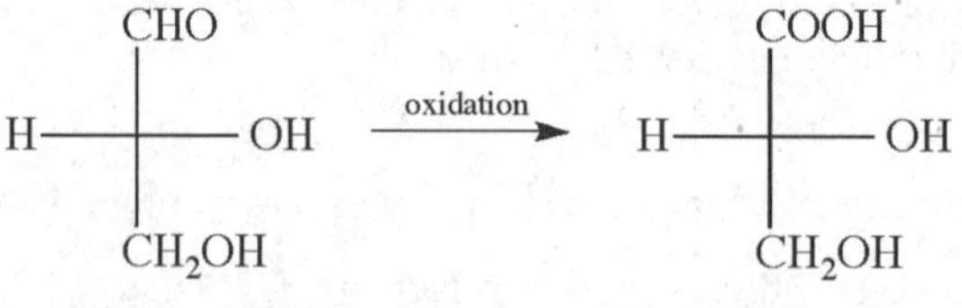

CHO / H—C—OH / CH_2OH D-(+)-Glyceraldehyde —oxidation→ COOH / H—C—OH / CH_2OH D-(−)-Glyceric acid

Problem 5.15 Does configuration change in the following reactions? Designate products (D,L) and (*R*,*S*).

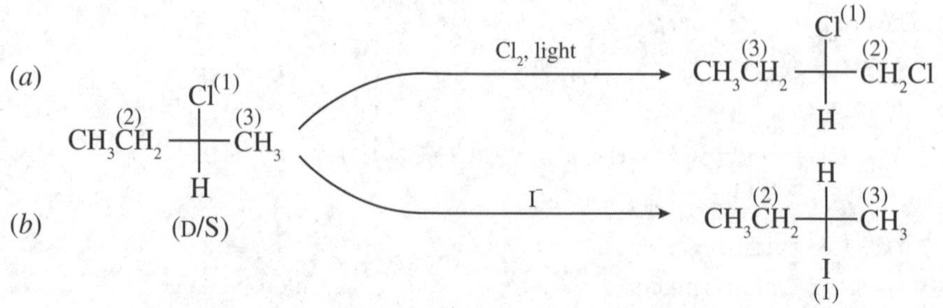

$Cl^{(1)}$ / $CH_3CH_2^{(2)}$—C—$CH_3^{(3)}$ / H (D/S)

(*a*) —Cl_2, light→ $Cl^{(1)}$ / $CH_3CH_2^{(3)}$—C—$CH_2Cl^{(2)}$ / H

(*b*) —I^-→ H / $CH_3CH_2^{(2)}$—C—$CH_3^{(3)}$ / $I^{(1)}$

(*c*) $\underset{(3)}{CH_3}$—C($Cl^{(1)}$)(H)—$\underset{(2)}{COOCH_3}$ (D/*R*) $\xrightarrow{CN^-}$ $\underset{(3)}{CH_3}$—C(H)($\underset{(2)}{CN}$)—$\underset{(1)}{COOCH_3}$

(*a*) No bond to the chiral C is broken, and the configuration is unchanged. Therefore, the configuration of both reactant and product is D. The product is *R*: *S* converts to *R* because there is a priority change. Thus, a change from *R* to *S* does not necessarily signal an inversion of configuration; it does only if the order of priority is unchanged.

(*b*) A bond to the chiral C is broken when I^- displaces Cl^-. An inversion of configuration has taken place, and there is a change from D to L. The product is *R*. This change from *S* to *R* also shows inversion, because there is no change in priorities.

(*c*) Inversion has occurred and there is a change from D to L. The product is *R*. Even though an inversion of configuration occurred, the reactant and product are both *R*. This is so because there is a change in the order of priority. The displaced Cl has priority (1), but the incoming CN^- has priority (2).

In general, the D,L convention signals a configuration change, if any, as follows: D → D or L → L means no change (retention), and D → L or L → D means change (inversion). The *RS* conversion can also be used, but it requires working out the priority sequences.

Problem 5.16 Draw three-dimensional and Fischer projections, using superscripts to show priorities, for (*a*) (*S*)–2- bromopropanaldehyde, (*b*) (*R*)-*3-iodo-2-chloropropanol.*

(*a*) Three-dimensional: $CH{=}O^{(2)}$ (top), $H_3C^{(3)}$, $Br^{(1)}$, $H^{(4)}$. Fischer: CH=O (top), CH_3 (left), Br (right), H (bottom).

(*b*) Three-dimensional: $Cl^{(1)}$ (top), $HOCH_2^{(3)}$, $CH_2I^{(2)}$, $H^{(4)}$. Fischer: Cl (top), $HOCH_2$ (left), CH_2I (right), H (bottom).

5.4 Molecules with More Than One Chiral Center

With *n* dissimilar chiral atoms, the number of stereoisomers is 2^n and the number of racemic forms is 2^{n-1}, as illustrated below for 2-chloro-3-bromobutane ($n = 2$). The *R, S* configuration is shown next to the C's.

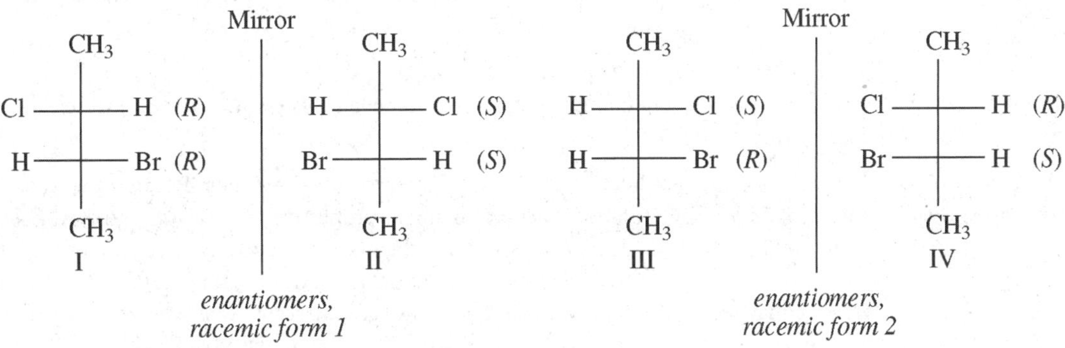

If $n = 2$ and the two chiral atoms are identical in that each holds the same four different groups, there are only three stereoisomers, as illustrated for 2,3-dichlorobutane.

	V	VI	VII	VIII
top	CH_3	CH_3	CH_3	CH_3
	Cl—C—H (*R*)	H—C—Cl (*S*)	H—C—Cl (*S*)	Cl—C—H (*R*)
	H—C—Cl (*R*)	Cl—C—H (*S*)	H—C—Cl (*R*)	Cl—C—H (*S*)
bottom	CH_3	CH_3	CH_3	CH_3
	2(*R*)-3(*R*)-Dichlorobutane	2(*S*)-3(*S*)-Dichlorobutane	2(*S*)-3(*R*)-Dichlorobutane	2(*R*)-3(*S*)-Dichlorobutane

Mirror — V and VI: *enantiomers*; Mirror — VII and VIII: *meso* (Symmetry Plane)

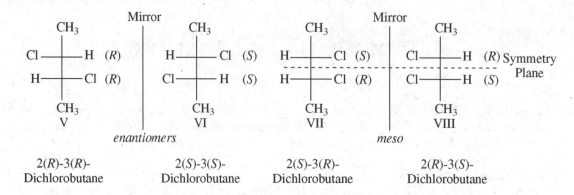

Structures VII and VIII are identical because rotating either one 180° in the plane of the paper makes it superimposable with the other one. VII possesses a symmetry plane and is achiral. Achiral stereoisomers which have chiral centers are called *meso*. The *meso* structure is a diastereomer of either of the enantiomers. The *meso* structure with two chiral sites always has the (*RS*) configuration.

Problem 5.17 For the stereoisomers of 3-iodo–2-butanol, (*a*) assign *R* and *S* configurations to C^2 and C^3. (*b*) Indicate which are (i) enantiomers and (ii) diastereomers. (*c*) Will rotation about the C—C bond alter the configurations?

	I	II	III	IV
top	H	CH_3	CH_3	I
C^3	I, CH_3	H, I	I, H	H, CH_3
C^2	CH_3, OH	CH_3, H	H, CH_3	H, OH
bottom	H	OH	OH	CH_3

(*a*)

	I	II	III	IV
C^2	*S*	*R*	*S*	*R*
C^3	*S*	*S*	*R*	*R*

(*b*) (i) I and IV; II and III. (ii) I and IV diastereomeric with II and III; II and III diastereomeric with I and IV.
(*c*) No.

Problem 5.18 Compare physical and chemical properties of (*a*) enantiomers, (*b*) an enantiomer and its racemic form, and (*c*) diastereomers.

(*a*) With the exception of rotation of plane-polarized light, enantiomers have identical physical properties, for example, boiling point, melting point, solubility. Their chemical properties are the *same* toward *achiral* reagents, solvents, and conditions. Toward *chiral* reagents, solvents, and catalysts, enantiomers react at *different* rates. The transition states produced from the chiral reactant and the individual enantiomers are diastereomeric and, hence, have different energies; the $\Delta H^{\ddagger}$ values are different, as are the rates of reaction.

(*b*) Enantiomers are optically active; the racemic form is optically inactive. Other physical properties of an enantiomer and its racemic form may differ depending on the racemic form. The chemical properties are the *same* toward *achiral* reagents, but *chiral* reagents at *different* rates.

(*c*) Diastereomers have different physical properties, and have different chemical properties with both achiral and chiral reagents. The rates are different and the products may be different.

Problem 5.19 How can differences in the solubilities of diastereomers be used to resolve a racemic form into individual enantiomers?

The reaction of a racemic form with a chiral reagent—for example, a racemic (±) acid with a (−) base, yields two diastereomeric salts (+)(−) and (−)(−) with different solubilities. These salts can be separated by fractional crystallization, and then each salt is treated with a strong acid (HCl) which liberates the enantiomeric organic acid. This is shown schematically:

Racemic Form		**Chiral Base**		**Diastereomeric Salts**
(±)RCOOH	+	(–)B	⟶	$(+)RCOO^-(-)BH^+ + (-)RCOO^-(-)BH^+$

separated

HCl ↓ (each salt)

$BH^+Cl^- + (+)RCOOH \quad (-)RCOOH + BH^+Cl^-$

separated enantiomeric acids

The most frequently used chiral bases are the naturally occurring, optically active alkaloids, such as strychnine, brucine, and quinine. Similarly, racemic organic bases are resolved with naturally occurring, optically active, organic acids, such as tartaric acid.

5.5 Synthesis and Optical Activity

1. Optically inactive reactants with achiral catalysts or solvents yield optically inactive products. With a chiral catalyst, for instance, an enzyme, any chiral product will be optically active.
2. A second chiral center generated in a chiral compound may not have an equal chance for *R* and *S* configurations; a 50:50 mixture of diastereomers is *not* usually obtained.
3. Replacement of a group or atom on a chiral center can occur with retention or inversion of configuration or with a mixture of the two (complete or partial racemization), depending on the mechanism of the reaction.

Problem 5.20 (*a*) What two products are obtained when C^3 of (*R*)-2-chlorobutane is chlorinated? (*b*) Are these diastereomers formed in equal amounts? (*c*) In terms of mechanism, account for the fact that

$$rac(\pm)\text{-}ClCH_2C(Cl)(CH_3)CH_2CH_3$$

is obtained when (*R*)-$ClCH_2\text{-}\overset{*}{C}H(CH_3)CH_2CH_3$ is chlorinated.

(*a*) $\overset{1}{C}H_3\text{-}\overset{2}{C}(Cl)(H)\text{-}\overset{3}{C}H_2\overset{4}{C}H_3$ (*R*) ⟶ $\overset{1}{C}H_3\text{-}\overset{2}{C}(Cl)(H)\text{-}\overset{3}{C}(H)(Cl)\text{-}\overset{4}{C}H_3$ (*RR*) optically active enantiomer + $\overset{1}{C}H_3\text{-}\overset{2}{C}(Cl)(H)\text{-}\overset{3}{C}(Cl)(H)\text{-}\overset{4}{C}H_3$ (*RS*) *meso*

In the products, C^2 retains the *R* configuration, since none of its bonds were broken and there was no change in priority. The configuration at C^3, the newly created chiral center, can be either *R* or *S*. As a result, two diastereomers are formed, the optically active *RR* enantiomer and the optically inactive *RS meso* compound.

(*b*) No. The numbers of molecules with *S* and *R* configurations at C^3 are not equal. This is so because the presence of the C^2-stereocenter causes an unequal likelihood of attack at the faces of C^3. Faces which give rise to diastereomers when attacked by a fourth ligand are **diastereotopic faces**.

(*c*) Removal of the H from the chiral C leaves the achiral free radical $ClCH_2\dot{C}(CH_3)CH_2CH_3$. Like the radical in Problem 5.21, it reacts with Cl_2 to give a racemic form.

The C^3 of 2-chlorobutane, of the general type RCH_2R', which becomes chiral when one of its H's is replaced by another ligand, is said to be **prochiral**.

Problem 5.21 Answer True or False to each of the following statements and explain your choice. (*a*) There are two broad classes of stereoisomers. (*b*) Achiral molecules cannot possess chiral centers. (*c*) A reaction catalyzed by an enzyme always gives an optically active product. (*d*) Racemization of an enantiomer must result in the breaking of at least one bond to the chiral center. (*e*) An attempted resolution can distinguish a racemate from a *meso* compound.

(*a*) True. The two classes are enantiomers and diastereomers.
(*b*) False. *Meso* compounds are achiral, yet they possess chiral centers.
(*c*) False. The product could be achiral.
(*d*) True. Only by breaking a bond could the configuration be changed.
(*e*) True. A racemate can be resolved, but a *meso* compound cannot be, because it does not consist of enantiomers.

Problem 5.22 In Fig. 4.4, give the stereochemical relationship between (*a*) the *two gauche* conformers, (*b*) the *anti* and either *gauche* conformer.

(*a*) They are stereomers, because they have the same structural formulas but different spatial arrangements. However, since they readily interconvert by rotation about a σ bond, they are not typical, isolatable, *configurational* stereomers; rather, they are **conformational stereomers**. The two *gauche* forms are non-superimposable mirror images; they are **conformational enantiomers**.
(*b*) They are **conformational diastereomers**, because they are stereomers but not mirror images.

Configurational stereomers differ from conformational stereomers in that they are interconverted only by breaking and making chemical bonds. The energy needed for such changes is of the order of 200–600 kJ/mol, which is large enough to permit their isolation, and is much larger than the energy required for interconversion of conformers.

SUPPLEMENTARY PROBLEMS

Problem 5.23 (*a*) What is the necessary and sufficient condition for the existence of enantiomers? (*b*) What is the necessary and sufficient condition for measurement of optical activity? (*c*) Are all substances with chiral atoms optically active and resolvable? (*d*) Are enantiomers possible in molecules that do not have chiral carbon atoms? (*e*) Can a prochiral carbon ever be primary or tertiary? (*f*) Can conformational enantiomers ever be resolved?

(*a*) Chirality in molecules having nonsuperimposable mirror images. (*b*) An excess of one enantiomer and a specific rotation large enough to be measured. (*c*) No. Racemic forms are not optically active but are resolvable. *Meso* compounds are inactive and not resolvable. (*d*) Yes. The presence of a chiral atom is a sufficient but not necessary condition for enantiomerism. For example, properly disubstituted allenes have no plane or center of symmetry and are chiral molecules even though they have no chiral C's:

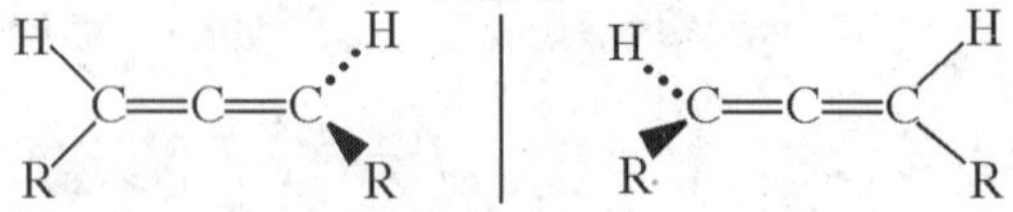

Mirror

H(R)C=C=C(H)(R) | (H)(R)C=C=C(H)(R)

a chiral allene
(nonsuperimposable enantiomers)

(*e*) No. Replacing one H of σ 1° CH_3 group by an X group would leave an achiral —CH_2X group. In order for a 3° CH group to be chiral when the H is replaced by X, it would already have to be chiral when bonded to the H.

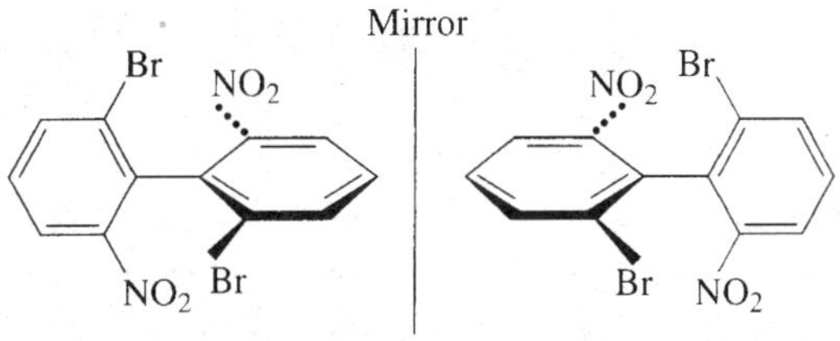

Mirror

Figure 5.5

(*f*) Yes. There are molecules that have a large enthalpy of activation for rotating about a σ bond because of severe steric hindrance. Examples are properly substituted biphenyls, for example, 2,2′-dibromo-6,6′-dinitrobiphenyl (Fig. 5.5). The four bulky substituents prevent the two flat rings from being in the same plane, a requirement for free rotation.

Problem 5.24 Select the chiral atoms in each of the following compounds:

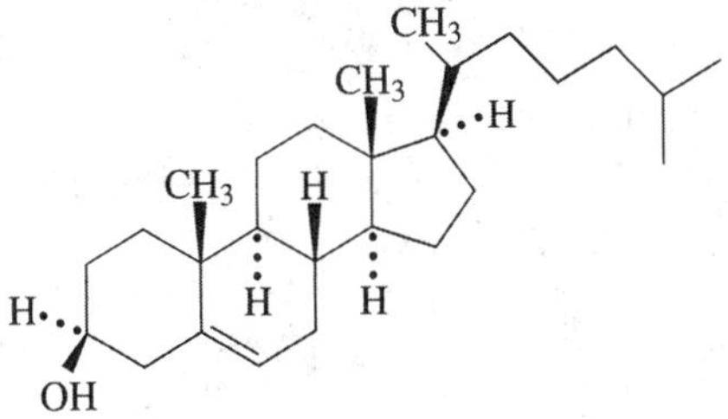

Cholesterol (*a*)

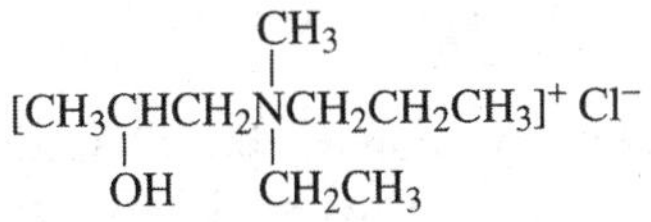

$[CH_3CHCH_2NCH_2CH_2CH_3]^+ Cl^-$ (with OH on the second C, CH_3 and CH_2CH_3 on N)

quaternary ammonium salt (*b*)

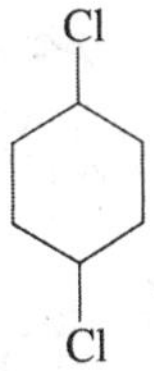

1,4-Dichlorocyclohexane (*c*)

(*a*) There are eight chiral C's: three C's attached to a CH_3 group, four C's attached to lone H's, and one C attached to OH. (*b*) Since N is bonded to four different groups, it is a chiral center, as is the C bonded to the OH. (*c*) There are no chiral atoms in this molecule. The two sides of the ring—CH_2CH_2—joining the C's bonded to Cl's are the same. Hence, neither of the C's bonded to a Cl is chiral.

Problem 5.25 Draw examples of (*a*) a *meso* alkane having the molecular formula C_8H_{18} and (*b*) the simplest alkane with a chiral quaternary C. Name each compound.

(*a*)

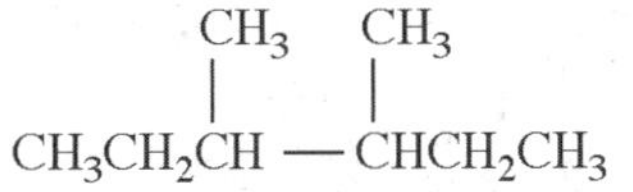

$CH_3CH_2CH(CH_3)—CH(CH_3)CH_2CH_3$

3,4-*Dimethylhexane*

(*b*) The chiral C in this alkane must be attached to the four simplest alkyl groups. These are CH_3—, CH_3CH_2—, $CH_3CH_2CH_2$—, and $(CH_3)_2CH$—, and the compound is

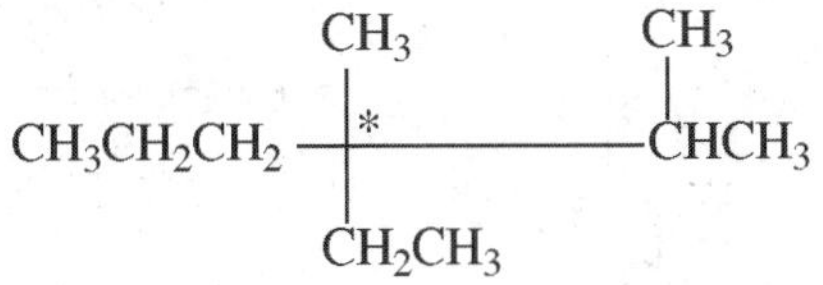

$CH_3CH_2CH_2—C^*(CH_3)(CH_2CH_3)—CH(CH_3)CH_3$

2,3-Dimethyl-3-ethylhexane

Problem 5.26 Relative configurations of chiral atoms are sometimes established by using reactions in which there is no change in configuration because no bonds to the chiral atom are broken. Which of the following reactions can be used to establish relative configurations?

(*a*) (S)-$CH_3CHClCH_2CH_3 + Na^+OCH^-_3 \longrightarrow CH_3CH(OCH_3)CH_2CH_3 + Na^+Cl^-$

(*b*) (S)-$CH_3CH_2\overset{\overset{\Large CH_3}{|}}{C}HO^-Na^+ + CH_3Br \longrightarrow CH_3CH_2\overset{\overset{\Large CH_3}{|}}{C}HOCH_3 + Na^+\ Br^-$

(*c*) (R)-$CH_3CH_2\overset{\overset{\Large CH_3}{|}}{C}OHCH_2Cl + PCl_5 \longrightarrow CH_3CH_2\overset{\overset{\Large CH_3}{|}}{C}ClCH_2Cl + POCl_3 + HCl$

(*d*) (S)-$(CH_3)_2C(OH)CHBrCH_3 + Na^+CN^- \longrightarrow (CH_3)_2C(OH)CH(CN)CH_3 + Na^+\ Br^-$

(*e*) (R)-$(CH_3)CH_2\underset{\underset{\Large OH}{|}}{C}HCH_3 + Na \longrightarrow CH_3CH_2\underset{\underset{\Large O^-Na^+}{|}}{C}HCH_3 + \frac{1}{2}H_2$

(*b*) and (*e*). The others involve breaking bonds to the chiral C.

Problem 5.27 Account for the disappearance of optical activity observed when (R)-2-butanol is allowed to stand in aqueous H_2SO_4 and when (S)-2-iodooctane is treated with aqueous KI solution.

Optically active compounds become inactive if they lose their chirality because the chiral center no longer has four different groups, or if they undergo racemization. In the two reactions cited, C remains chiral, and it must be concluded that in both reactions, racemization occurs.

Problem 5.28 For the following compounds, draw projection formulas for all stereoisomers and point out their R,S specifications, optical activity (where present), and *meso* compounds: (*a*) 1,2,3,4-tetrahydroxybutane, (*b*) 1-chloro–2,3-dibromobutane, (*c*) 2,4-diiodopentane, (*d*) 2,3,4-tribromohexane, (*e*) 2,3,4-tribromopentane.

(*a*) $HOCH_2\overset{*}{C}HOH\overset{*}{C}HOHCH_2OH$, with two similar chiral C's, has one *meso* form and two optically active enantiomers.

1 CH_2OH H—*2*—OH H—*3*—OH *4*CH_2OH	*1* CH_2OH H—*2*—OH HO—*3*—H *4*CH_2OH	*1* CH_2OH HO—*2*—H H—*3*—OH *4*CH_2OH
(2S, 3R) *meso*	(2S, 3S)	(2R, 3R)
	Racemic form	

(*b*) $ClCH_2\overset{*}{C}HBr\overset{*}{C}HBrCH_3$ has two different chiral C's. There are four (2^2) optically active enantiomers.

CH_2Cl H—Br H—Br CH_3	CH_2Cl Br—H Br—H CH_3	CH_2Cl H—Br Br—H CH_3	CH_2Cl Br—H H—Br CH_3
(2S, 3R)	(2R, 3S)	(2S, 3S)	(2R, 3R)
erythro enantiomers		*threo* enantiomers	

The two sets of diastereomers are differentiated by the prefix *erythro* for the set in which at least two identical or similar substituents on chiral C's are eclipsed. The other set is called *threo*.

(*c*) $CH_3\overset{*}{C}HICH_2\overset{*}{C}HICH_3$ has two similar chiral C's, C^2 and C^4, separated by a CH_2 group. There are two enantiomers comprising a (±) pair and one *meso* compound.

$^{1}CH_3$	CH_3	CH_3
H—^{2}—I	H——I	I——H
$^{3}CH_2$	CH_2	CH_2
H—^{4}—I	I——H	H——I
$^{5}CH_3$	CH_3	CH_3
(2*S*, 4*R*) *meso*	(2*S*, 4*S*)	(2*R*, 4*R*)

(2*S*, 4*S*) and (2*R*, 4*R*): Racemate

(*d*) With three different chiral C's in $CH_3\overset{*}{C}HBr\overset{*}{C}HBr\overset{*}{C}HBrCH_2CH_3$, there are eight ($2^3$) enantiomers and four racemic forms.

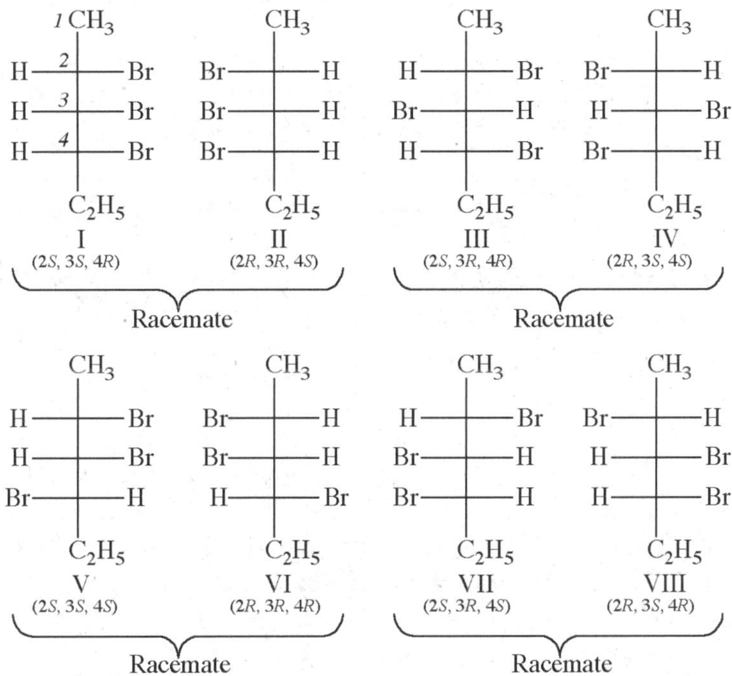

(*e*) $\overset{1}{C}H_3\overset{\overset{2}{*}}{C}HBr\overset{3}{C}HBr\overset{\overset{4}{*}}{C}HBr\overset{5}{C}H_3$ has two similar chiral atoms (C^2 and C^4). There are two enantiomers in which the configurations of C^2 and C^4 are the same, *RR* or *SS*. When C^2 and C^4 have different configurations, one *R* and one *S*, C^3 becomes a stereocenter and there are two *meso* forms.

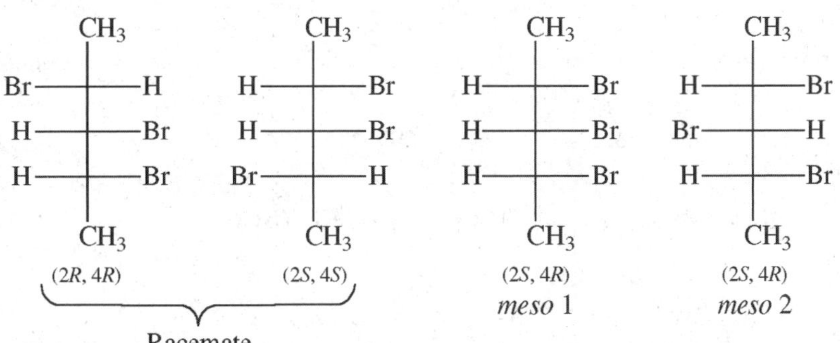

Problem 5.29 The specific rotation of (*R*)-(−)-2-bromooctane is −36°. What is the percentage composition of a mixture of enantiomers of 2-bromooctane whose rotation is +18°?

Let x = mole fraction of R, $1 - x$ = mole fraction of S.

$$x(-36^\circ) + (1 - x)(36^\circ) = 18^\circ \quad \text{or} \quad x = \tfrac{1}{4}$$

The mixture has 25% R and 75% S; it is 50% racemic and 50% S.

Problem 5.30 Predict the yield of stereoisomeric products, and the optical activity of the mixture of products, formed from chlorination of a racemic mixture of 2-chlorobutane to give 2,3-dichlorobutane.

The (S)−2-chlorobutane comprising 50% of the racemic mixture gives 35.5% of the *meso* (SR) product and 14.5% of the RR enantiomer. The R enantiomer gives 35.5% *meso* and 14.5% RR products. The total yield of *meso* product is 71%, and the combination of 14.5% RR and 14.5% SS gives 29% racemic product. The total reaction mixture is optically inactive. This result confirms the generalization that optically inactive starting materials, reagents, and solvents always lead to optically inactive products.

Problem 5.31 For the following reactions, give the number of stereoisomers that are isolated, their R,S configurations, and their optical activities. Use Fischer projections.

(*a*) *meso*-$HOCH_2CHOHCHOHCH_2OH \xrightarrow{\text{oxidation}} HOCH_2CHOHCHOHCOOH$

(*b*) (R)-$ClCH_2CH(CH_3)CH_2CH_3 \xrightarrow[\text{cuprate}]{\text{via}} CH_3CH_2CH(CH_3)CH_2CH_2CH(CH_3)CH_2CH_3$

(*c*) *rac* (±)-$CH_3—\overset{\overset{\large O}{\|}}{C}—CHOH—CH_3 \xrightarrow{H_2/Ni} CH_3CH(OH)CH(OH)CH_3$

(*d*) (S)-$CH_3CH_2—\overset{\overset{\large CH_2}{\|}}{C}—CH(CH_3)CH_2CH_3 \xrightarrow{H_2/Ni} CH_3CH_2CH(CH_3)CH(CH_3)CH_2CH_3$

(*a*) This *meso* alcohol is oxidized at either terminal CH_2OH to give an optically inactive racemic form. The chiral C next to the oxidized C undergoes a change in priority order; CH_2OH (3) goes to COOH (2). Therefore, if this C is R in the reactant, it becomes S in the product; if S, it goes to R.

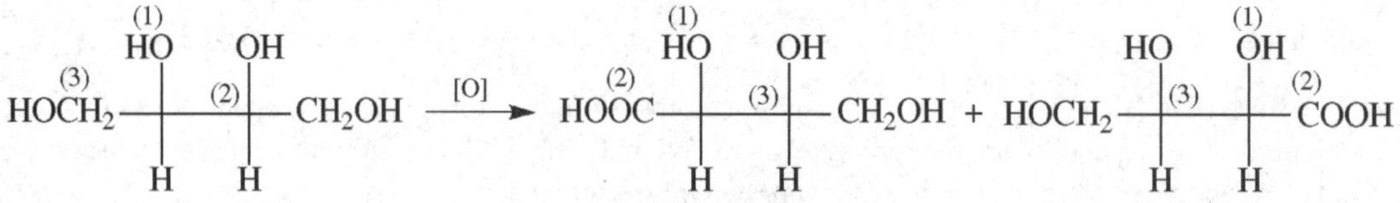

$$\underset{meso;\ (2R,\ 3S)}{HOCH_2^{(3)}—\underset{H}{\overset{HO^{(1)}}{C}}—^{(2)}\underset{H}{\overset{OH}{C}}—CH_2OH} \xrightarrow{[O]} \underset{(2S,\ 3S)\ (50\%)}{HOOC^{(2)}—\underset{H}{\overset{HO^{(1)}}{C}}—^{(3)}\underset{H}{\overset{OH}{C}}—CH_2OH} + \underset{(2R,\ 3R)\ (50\%)}{HOCH_2—\underset{H}{\overset{HO}{C}}—^{(3)}\underset{H}{\overset{OH^{(1)}}{C}}—COOH^{(2)}}$$

(*b*) Replacement of Cl by the isopentyl group does not change the priorities of the groups on the chiral C. There is one optically active product, whose two chiral C's have R configurations.

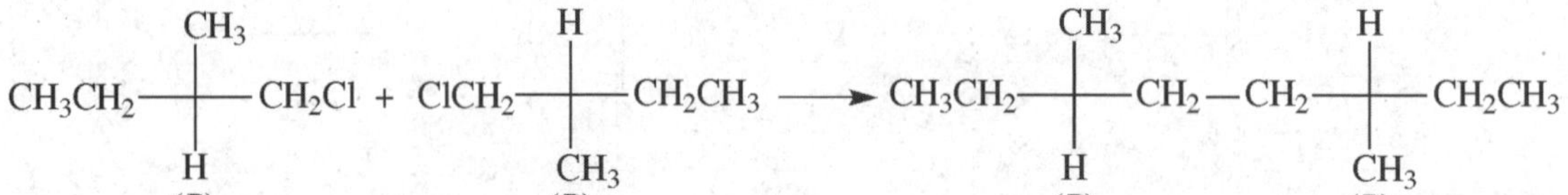

$$\underset{(R)}{CH_3CH_2—\underset{H}{\overset{CH_3}{C}}—CH_2Cl} + \underset{(R)}{ClCH_2—\underset{CH_3}{\overset{H}{C}}—CH_2CH_3} \longrightarrow \underset{(R)}{CH_3CH_2—\underset{H}{\overset{CH_3}{C}}}—CH_2—CH_2—\underset{(R)}{\underset{CH_3}{\overset{H}{C}}—CH_2CH_3}$$

(*c*) This reduction generates a second chiral center.

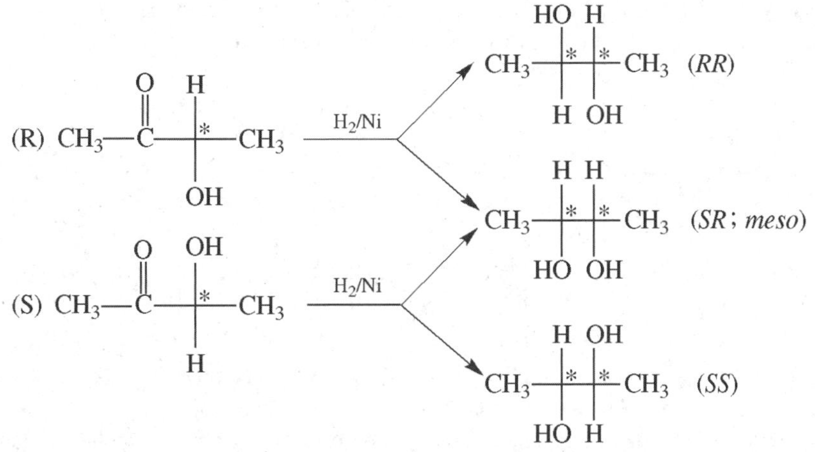

$$\text{(R) } CH_3-\overset{O}{\overset{\|}{C}}-\overset{H}{\underset{OH}{C^*}}-CH_3 \xrightarrow{H_2/Ni} CH_3-\overset{HO}{\underset{H}{C^*}}-\overset{H}{\underset{OH}{C^*}}-CH_3 \ (RR) \quad + \quad CH_3-\overset{H}{\underset{HO}{C^*}}-\overset{H}{\underset{OH}{C^*}}-CH_3 \ (SR;\ meso)$$

$$\text{(S) } CH_3-\overset{O}{\overset{\|}{C}}-\overset{OH}{\underset{H}{C^*}}-CH_3 \xrightarrow{H_2/Ni} CH_3-\overset{H}{\underset{HO}{C^*}}-\overset{H}{\underset{OH}{C^*}}-CH_3 \ (SR;\ meso) \quad + \quad CH_3-\overset{H}{\underset{HO}{C^*}}-\overset{OH}{\underset{H}{C^*}}-CH_3 \ (SS)$$

RR and *SS* enantiomers are formed in equal amounts to give a racemic form. The *meso* and racemic forms are in unequal amounts.

(*d*) Reduction of the double bond makes C^3 chiral. Reduction occurs on either face of the planar π bond to form molecules with *R* and molecules with *S* configurations at C^3. These are in unequal amounts because of the adjacent chiral C that has an *S* configuration. Since both chiral atoms in the product are structurally identical, the products are a *meso* structure (*RS*) and an optically active diastereomer (*SS*).

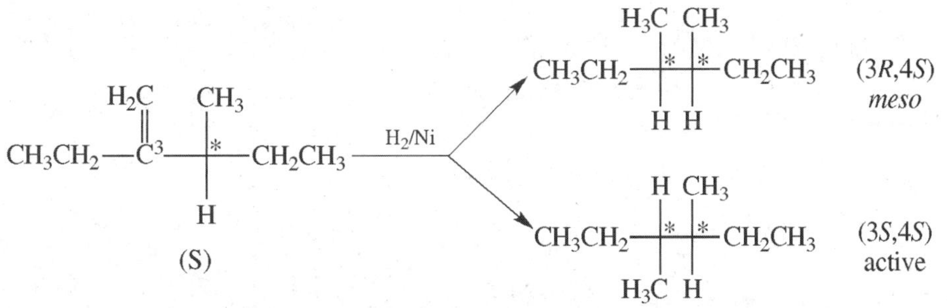

$$CH_3CH_2-\overset{H_2C}{\overset{\|}{C^3}}-\overset{CH_3}{\underset{H}{C^*}}-CH_2CH_3 \ \text{(S)} \xrightarrow{H_2/Ni} CH_3CH_2-\overset{H_3C}{\underset{H}{C^*}}-\overset{CH_3}{\underset{H}{C^*}}-CH_2CH_3 \ (3R,4S)\ meso \quad + \quad CH_3CH_2-\overset{H}{\underset{H_3C}{C^*}}-\overset{CH_3}{\underset{H}{C^*}}-CH_2CH_3 \ (3S,4S)\ \text{active}$$

Problem 5.32 Designate the following compounds as *erythro* or *threo* structures.

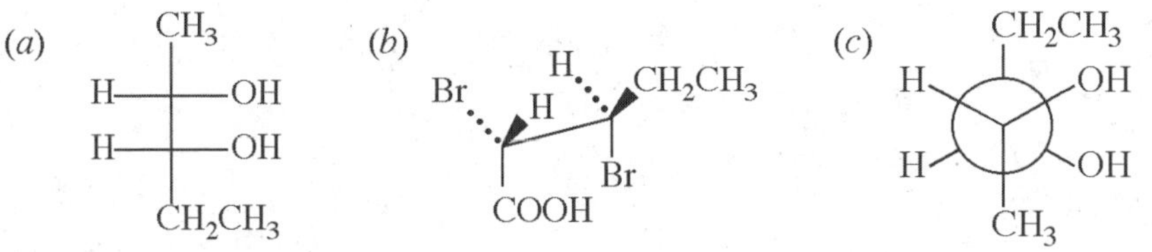

(*a*) Fischer projection: CH_3 (top); H–C–OH; H–C–OH; CH_2CH_3 (bottom)

(*b*) Br, H (wedge/dash) on C bearing COOH; H, CH_2CH_3 (dash/wedge) on C bearing Br

(*c*) Newman projection: front: CH_2CH_3, H, OH; back: H, OH, CH_3

(*a*) *Erythro* (see Problem 5.28).

(*b*) *Erythro*; it is best to examine eclipsed conformations. If either of the chiral C's is rotated 120° to an eclipsed conformation for the two Br's, the H's are also eclipsed.

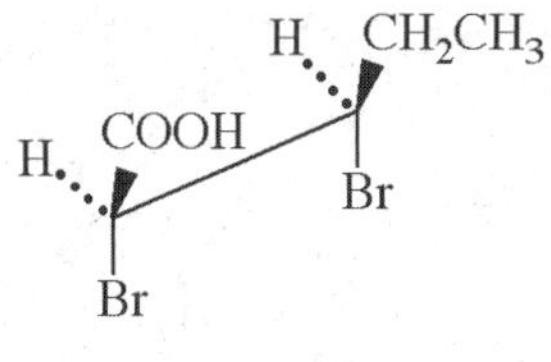

Eclipsed structure: H, COOH, Br on one C; H, CH_2CH_3, Br on the other C

(*c*) *Threo*; a 60° rotation of one of the chiral C's eclipses the H's but not the OH's.

Problem 5.33 Glyceraldehyde can be converted to lactic acid by the two routes shown below. These results reveal an ambiguity in the assignment of relative D,L configuration. Explain.

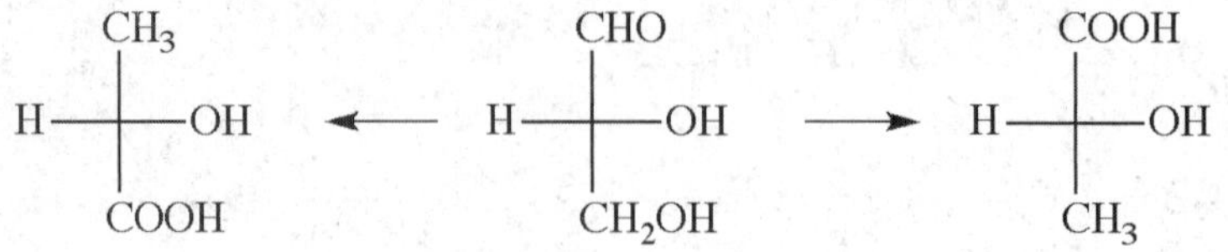

$$\begin{array}{ccccc} \mathrm{CH_3} & & \mathrm{CHO} & & \mathrm{COOH} \\ \mathrm{H}\!-\!\!+\!\!-\mathrm{OH} & \longleftarrow & \mathrm{H}\!-\!\!+\!\!-\mathrm{OH} & \longrightarrow & \mathrm{H}\!-\!\!+\!\!-\mathrm{OH} \\ \mathrm{COOH} & & \mathrm{CH_2OH} & & \mathrm{CH_3} \\ (R)\text{-}(-)\text{-Lactic acid} & & \text{D-}(+)\text{-Glyceraldehyde} & & (S)\text{-}(+)\text{-Lactic acid} \end{array}$$

In neither route is there a change in the bonds to the chiral C. Apparently, both lactic acids should have the D configuration, since the original glyceraldehyde was D. However, since the CH_3 and COOH groups are interchanged, the two lactic acids must be enantiomers. Indeed, one is (+) and the other is (−). This shows that, for an unambiguous assignment of D or L, it is necessary to specify the reactions in the chemical change. Because of such ambiguity, *R,S* is used. The (+) lactic acid is *S*, the (−) enantiomer is *R*.

Problem 5.34 Deduce the structural formula for an optically active alkene, C_6H_{12}, which reacts with H_2 to form an optically inactive alkane, C_6H_{14}.

The alkene has a group attached to the chiral C, which must react with H_2 to give a group identical to one already attached, resulting in loss of chirality.

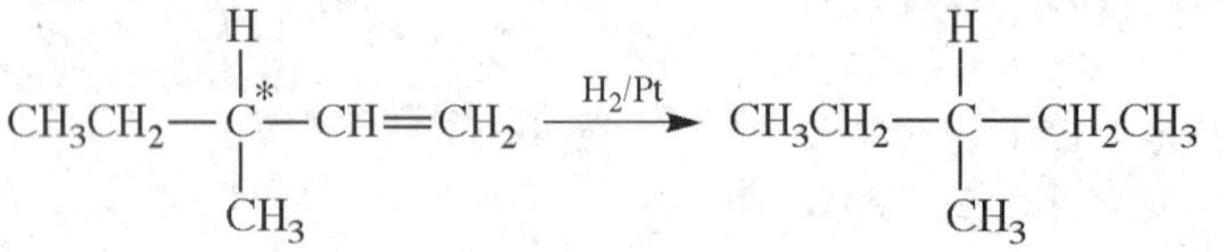

$$\mathrm{CH_3CH_2}-\overset{\mathrm{H}}{\underset{\mathrm{CH_3}}{\overset{|}{\underset{|}{\mathrm{C}^*}}}}-\mathrm{CH{=}CH_2} \xrightarrow{\mathrm{H_2/Pt}} \mathrm{CH_3CH_2}-\overset{\mathrm{H}}{\underset{\mathrm{CH_3}}{\overset{|}{\underset{|}{\mathrm{C}}}}}-\mathrm{CH_2CH_3}$$

CHAPTER 6

Alkenes

6.1 Nomenclature and Structure

Alkenes (olefins) contain the functional group

$$\rangle C{=}C\langle \quad \textit{a double bond}$$

and have the general formula C_nH_{2n}. These **unsaturated** hydrocarbons are isomeric with the saturated **cycloalkanes**.

$$\underset{\text{Propylene (Propene)}}{CH_3CH{=}CH_2} \qquad \boxed{C_3H_6} \qquad \underset{\text{Cyclopropane}}{H_2C{-}CH_2{-}CH_2 \text{ (ring)}}$$

In the IUPAC system, the longest continuous chain of C's *containing the double bond* is assigned the name of the corresponding alkane, with the suffix changed from *-ane* to *-ene*. The chain is numbered so that the position of the double bond is designated by assigning the lowest possible number to the first doubly bonded C.

$$\underset{\text{2-Hexene}}{CH_3CH{=}CHCH_2CH_2CH_3} \qquad \underset{\text{2-Methyl-2-butene}}{CH_3{-}\overset{CH_3}{\overset{|}{C}}{=}CHCH_3} \qquad \underset{\text{3-Ethyl-1-hexene}}{CH_3CH_2{-}\overset{\overset{^1CH_2}{\|}}{\overset{^2CH}{|}}\underset{3}{C}H\overset{4}{C}H_2\overset{5}{C}H_2\overset{6}{C}H_3}$$

A few important unsaturated groups that have trivial names are: $H_2C{=}CH{-}$ (Vinyl), $H_2C{=}CH{-}CH_2{-}$ (Allyl), and $CH_3CH{=}CH{-}$ (Propenyl).

Problem 6.1 Write structural formulas for (*a*) 3-bromo-2-pentene, (*b*) 2,4-dimethyl-3-hexene, (*c*) 2,4,4-trimethyl-2-pentene, (*d*) 3-ethylcyclohexene.

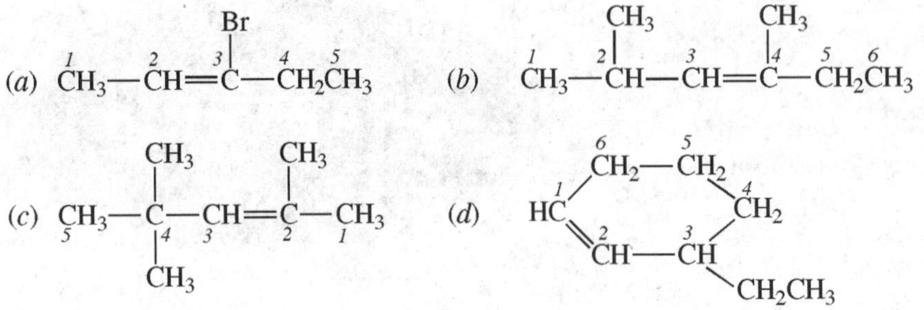

(a) $\overset{1}{C}H_3-\overset{2}{C}H=\overset{3}{C}(Br)-\overset{4}{C}H_2\overset{5}{C}H_3$ (b) $\overset{1}{C}H_3-\overset{2}{C}H(CH_3)-\overset{3}{C}H=\overset{4}{C}(CH_3)-\overset{5}{C}H_2\overset{6}{C}H_3$

(c) $\overset{5}{C}H_3-\overset{4}{C}(CH_3)_2-\overset{3}{C}H=\overset{2}{C}(CH_3)-\overset{1}{C}H_3$ (d) $\overset{1}{H}C=\overset{2}{C}H-\overset{3}{C}H(CH_2CH_3)-\overset{4}{C}H_2-\overset{5}{C}H_2-\overset{6}{C}H_2-$

Problem 6.2 Supply the structural formula and IUPAC name for (*a*) trichloroethylene, (*b*) *sec*-butylethylene, (*c*) *sym*-divinylethylene.

Alkenes are also named as derivatives of ethylene. The ethylene unit is shown in a box.

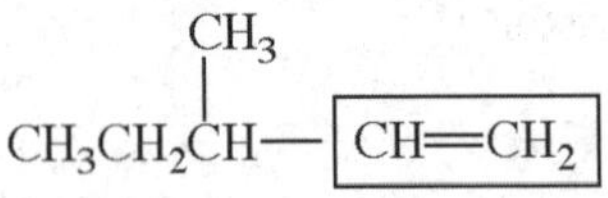

(*a*) $Cl-\boxed{C(Cl)=C(H)}-Cl$ Trichloroethene

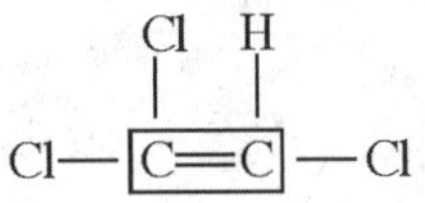

(*b*) $CH_3CH_2CH(CH_3)-\boxed{CH=CH_2}$ 3-Methyl-1-pentene

(*c*) $H_2C=CH-\boxed{CH=CH}-CH=CH_2$ 1,3,5-Hexatriene

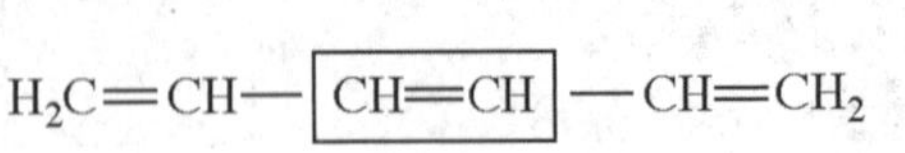

6.2 Geometric (*cis-trans*) Isomerism

The C=C consists of a σ bond and a π bond. The π bond is in a plane at right angles to the plane of the single bonds to each C (Fig. 6.1). The π bond is weaker and more reactive than the σ bond. The reactivity of the π bond imparts the property of unsaturation to alkenes; alkenes therefore readily undergo addition reactions. The π bond prevents free rotation about the C=C, and therefore, an alkene having two different substituents on each doubly bonded C has geometric isomers. For example, there are two 2-butenes:

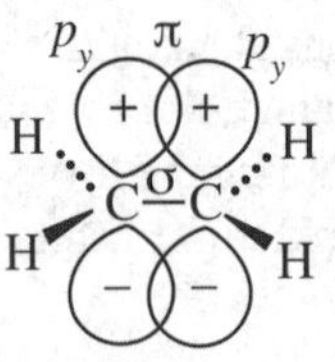

Figure 6.1

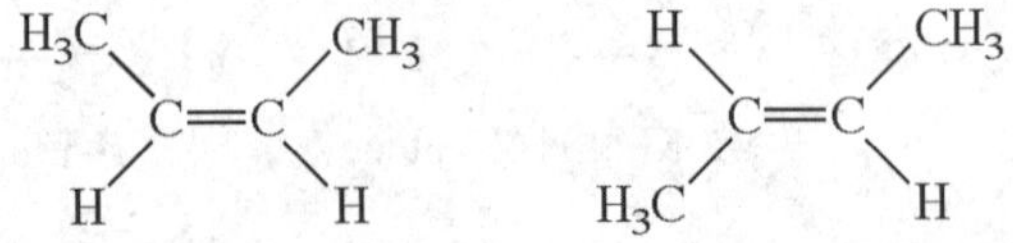

$(H_3C)(H)C=C(CH_3)(H)$ — CH_3's on same side; called *cis*-

$(H)(H_3C)C=C(CH_3)(H)$ — CH_3's on opposite sides; called *trans*-

Geometric (*cis-trans*) isomers are stereoisomers because they differ only in the spatial arrangement of the groups. They are diastereomers and have different physical properties (m.p., b.p., etc.).

In place of *cis-trans*, the letter *Z* is used if the higher-priority substituents (Section 5.3) on each C are on the same side of the double bond. The letter *E* is used if they are on opposite sides.

Problem 6.3 Predict (*a*) the geometry of ethylene, $H_2C{=}CH_2$; (*b*) the relative C-to-C bond lengths in ethylene and ethane; (*c*) the relative C—H bond lengths and bond strengths in ethylene and ethane; (*d*) the relative bond strengths of C—C and C=C.

(*a*) Each C in ethylene (ethene) uses sp^2 HO's (Fig. 2.8) to form three trigonal σ bonds. All five σ bonds (four C—H and one C—C) must lie in the same plane: ethylene is a **planar** molecule. All bond angles are approximately 120°.

(*b*) The C=C atoms, having four electrons between them, are closer to each other than the C—C atoms, which are separated by only two electrons. Hence, the C=C length (0.134 nm) is less than the C—C length (0.154 nm).

(*c*) The more *s* character in the hybrid orbital used by C to form a σ bond, the closer the electrons are to the nucleus and the shorter is the σ bond. Thus, the C—H bond length in ethylene (0.108 nm) is less than the length in ethane (0.110 nm). The shorter bond is also the stronger bond.

(*d*) Since it takes more energy to break two bonds than one bond, the bond energy of C=C in ethylene (611 kJ/mol) is greater than that of C—C in ethane (348 kJ/mol). However, note that the bond energy of the double bond is less than twice that of the single bond. This is so because it is easier to break a π bond than a σ bond.

Problem 6.4 Which of the following alkenes exhibit geometric isomerism? Supply structural formulas and names for the isomers.

(*a*) $CH_3CH_2C(CH_3){=}C(C_2H_5)CH_2CH_3$ (drawn with CH_3 above and C_2H_5 below the double-bonded carbons) (*b*) $H_2C{=}C(Cl)CH_3$ (*c*) $C_2H_5C(H){=}C(H){-}CH_2I$

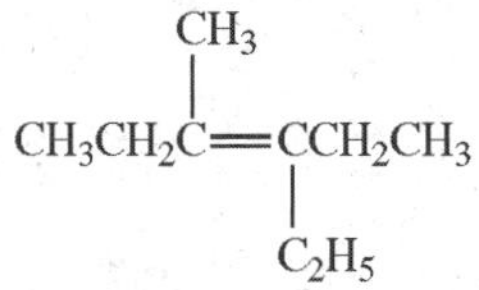

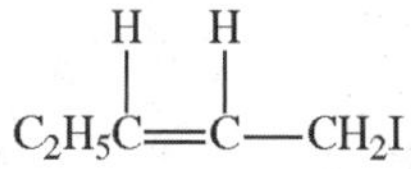

(*d*) $CH_3CH{=}CH{-}CH{=}CH_2$ (*e*) $CH_3CH{=}CH{-}CH{=}CHCH_2CH_3$ (*f*) $CH_3CH{=}CH{-}CH{=}CHCH_3$

(*a*) No geometric isomers because one double-bonded C has two C_2H_5's.

(*b*) No geometric isomers; one double-bonded C has two H's.

(*c*) Has geometric isomers because each double-bonded C has two different substituents:

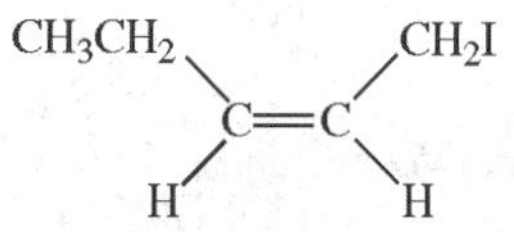

$CH_3CH_2(H)C{=}C(H)CH_2I$ (H's on same side)
cis- or (*Z*)-1-Iodo-2-pentene

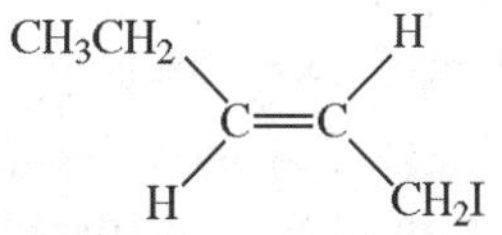

$CH_3CH_2(H)C{=}C(H)CH_2I$ (H's on opposite sides)
trans- or (*E*)-1-Iodo-2-pentene

(*d*) There are two geometric isomers because one of the double bonds has two different substituents:

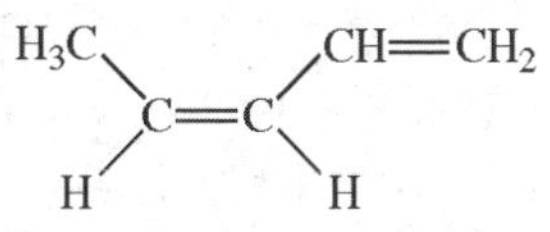

$H_3C(H)C{=}C(H)CH{=}CH_2$ (H's on same side)
(*Z*)-1,3-Pentadiene (*cis*)

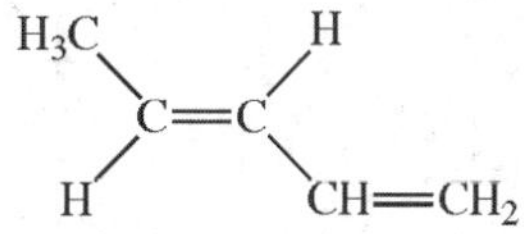

$H_3C(H)C{=}C(H)CH{=}CH_2$ (H's on opposite sides)
(*E*)-1,3-Pentadiene (*trans*)

(*e*) Both double bonds meet the conditions for geometric isomers, and there are four diastereomers of 2,4-heptadiene:

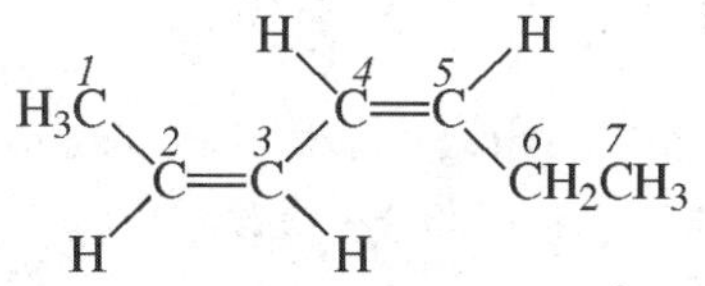

$\overset{1}{H_3C}{-}\overset{2}{C}H{=}\overset{3}{C}H{-}\overset{4}{C}H{=}\overset{5}{C}H{-}\overset{6,7}{CH_2CH_3}$
cis, cis or (*Z,Z*)

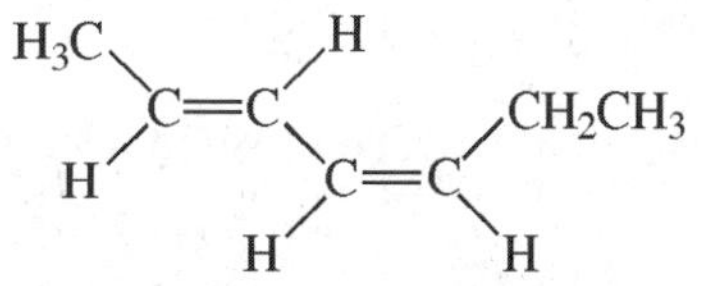

$H_3C{-}CH{=}CH{-}CH{=}CH{-}CH_2CH_3$
trans, cis or (*E,Z*)

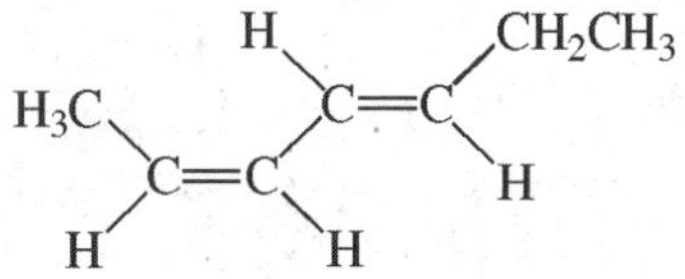

$H_3C{-}CH{=}CH{-}CH{=}CH{-}CH_2CH_3$
cis, trans or (*Z,E*)

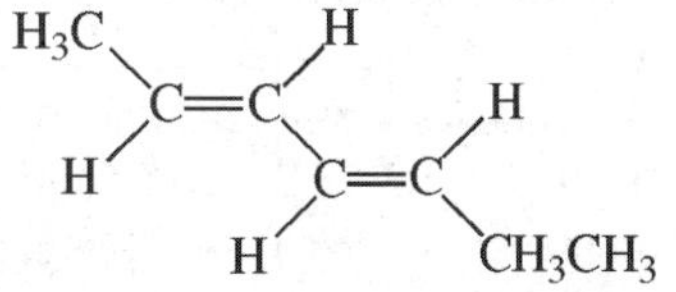

$H_3C{-}CH{=}CH{-}CH{=}CH{-}CH_3CH_3$
trans, trans or (*E,E*)

Note that *cis* and *trans* and *E* and *Z* are listed in the *same order* as the *bonds* are *numbered*.

(*f*) There are now only three isomers because *cis-trans* and *trans-cis* geometries are identical:

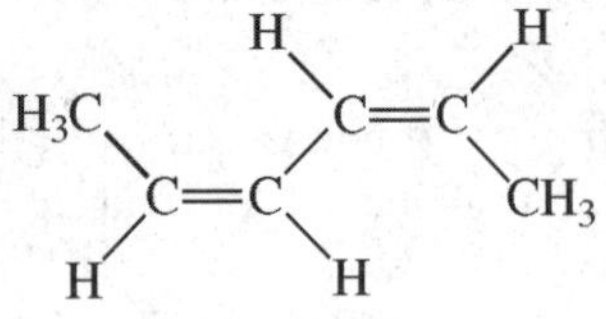

cis, cis-2,4-Hexadiene or (*Z, Z*)-2,4-Hexadine

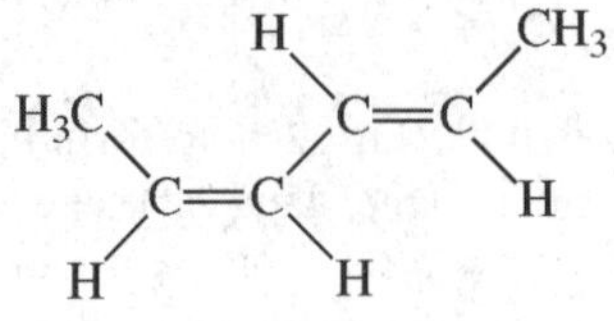

cis, trans-2,4-Hexadiene or (*Z, E*)-2,4-Hexadiene

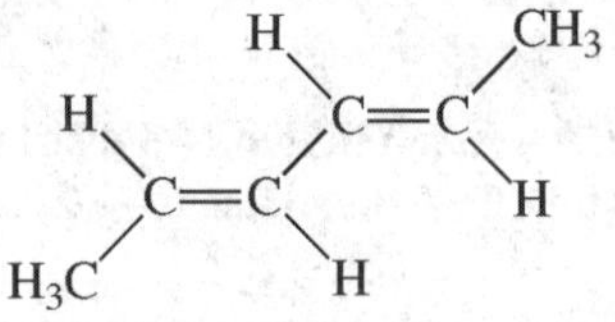

trans, trans-2,4-Hexadiene or (*E, E*)-2,4-Hexadiene

Problem 6.5 Write structural formulas for (*a*) (*E*)-2-methyl-3-hexene (*trans*), (*b*) (*S*)-3-chloro-1-pentene, (*c*) (*R*),(*Z*)-2-chloro-3-heptene (*cis*).

(*a*) (*b*) $H_2C{=}C$ CH_2CH_3 (*c*) H_3C $CH_2CH_2CH_3$

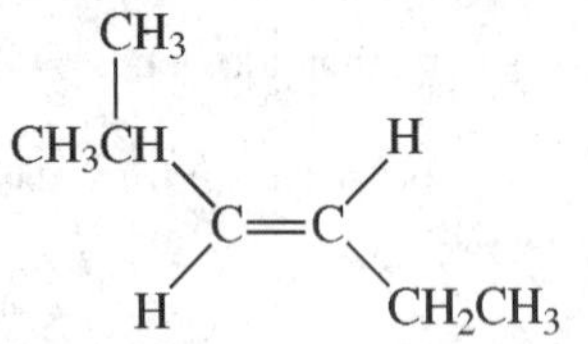

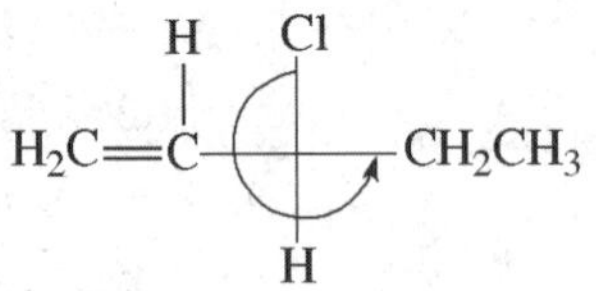

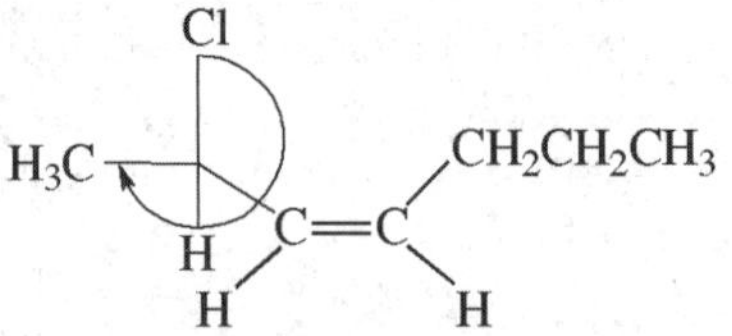

Problem 6.6 How do boiling points and solubilities of alkanes compare with those of corresponding alkanes?

Alkanes and alkenes are nonpolar compounds whose corresponding structures have almost identical molecular weights. Boiling points of alkenes are close to those of alkanes and similarly have 20° increments per C atom. Both are soluble in nonpolar solvents and insoluble in water, except that lower-molecular-weight alkenes are slightly more water-soluble because of attraction between the π bond and H_2O.

Problem 6.7 Show the directions of individual bond dipoles and net dipole of the molecule for (*a*) 1,1-dichloroethylene, (*b*) *cis*- and *trans*-1,2-dichloroethylene.

The individual dipoles are shown by the arrows on the bonds between C and Cl. The net dipole for the molecule is represented by an arrow that bisects the angle between the two Cl's. C—H dipoles are insignificant and are disregarded.

(*a*) (*b*) 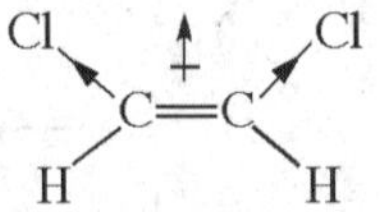 *cis* (*c*) *trans*

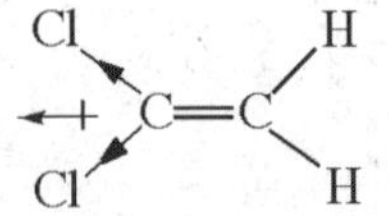

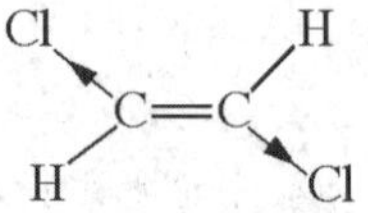

In the *trans* isomer, the C—Cl moments are equal but in opposite directions; they cancel and the *trans* isomer has a zero dipole moment.

Problem 6.8 How can heats of combustion be used to compare the differences in stability of the geometric isomers of alkenes?

The thermodynamic stability of isomeric hydrocarbons is determined by burning them to CO_2 and H_2O and comparing the heat evolved per mole ($-\Delta H$ combustion). The more stable isomer has the smaller ($-\Delta H$) value. *Trans* alkenes have the smaller values and hence are more stable than the *cis* isomers. This is supported by the exothermic (ΔH negative) conversion of *cis* to *trans* isomers by ultraviolet light and some chemical reagents.

The *cis* isomer has higher energy because there is greater repulsion between its alkyl groups on the same side of the double bond than between an alkyl group and an H in the *trans* isomer. These repulsions are greater with larger alkyl groups, which produce larger energy differences between geometric isomers.

6.3 Preparation of Alkenes

1. 1,2-Eliminations

Also called **β-eliminations**, these constitute the principal laboratory method whereby two atoms or groups are removed from adjacent bonded C's.

(*a*) Dehydrohalogenation

$$\mathrm{B\overset{-}{:}} + -\overset{\overset{\textstyle H}{|}}{\underset{|}{C}}-\overset{\overset{\textstyle X}{|}}{\underset{|}{C}}- \longrightarrow -\underset{|}{C}=\underset{|}{C}- + \mathrm{B{:}H} + X^-$$

KOH in ethanol is most often used as the source of the base, B:$^-$, which then is mainly $C_2H_5O^-$.

(*b*) Dehydration

$$-\overset{\overset{\textstyle H}{|}}{\underset{|}{C}}-\overset{\overset{\textstyle OH}{|}}{\underset{|}{C}}- \xrightarrow{\text{acid}} -\underset{|}{C}=\underset{|}{C}- + H_2O$$

(*c*) Dehalogenation

$$\text{Mg (or Zn)} + -\overset{\overset{\textstyle Br}{|}}{\underset{|}{C}}-\overset{\overset{\textstyle Br}{|}}{\underset{|}{C}}- \longrightarrow -\underset{|}{C}=\underset{|}{C}- + MgBr_2\ (\text{or } ZnBr_2)$$

(*d*) Dehydrogenation

$$-\overset{\overset{\textstyle H}{|}}{\underset{|}{C}}-\overset{\overset{\textstyle H}{|}}{\underset{|}{C}}- \xrightarrow{\text{Pt or Pd}} -\underset{|}{C}=\underset{|}{C}- + H_2 \quad \text{(mainly a special industrial process)}$$

Cracking (Section 4.4) of petroleum hydrocarbons is the source of commercial alkenes.

In dehydration and dehydrohalogenation, the preferential order for removal of an H is 3° > 2° > 1° (**Saytzeff rule**). We can say "the poor get poorer." This order holds true because the more R's on the C=C group, the more stable is the alkene. The stability of alkenes in decreasing order of substitution by R is

$$R_2C{=}CR_2 > R_2C{=}CRH > R_2C{=}CH_2, \qquad RCH{=}CHR > RCH{=}CH_2 > CH_2{=}CH_2$$

2. Partial Reduction of Alkynes

$$\underset{\textit{an alkyne}}{RC{\equiv}CR} \xrightarrow{H_2/Pd} \text{(R)(H)C=C(R)(H)} \quad \textit{cis}\text{-alkene}$$

$$\underset{\textit{an alkyne}}{RC{\equiv}CR} \xrightarrow{Na,\ NH_3} \text{(R)(H)C=C(H)(R)} \quad \textit{trans}\text{-alkene}$$

These reactions, which give only one of two possible stereomers, are called **stereoselective**. In this case, they are more specifically called **diastereoselective**, because the stereomers are diastereomers.

Problem 6.9 (*a*) How does the greater enthalpy of *cis*- vis-a-vis *trans*-2-butene affect the ratio of isomers formed during the dehydrohalogenation of 2-chlorobutane? (*b*) How does replacing the CH_3 groups of 2-chlorobutane by *t*-butyl groups to give $CH_3C(CH_3)_2CH_2CHClC(CH_3)_2CH_3$ alter the distribution of the alkene geometric isomers?

(*a*) The transition states for the formation of the geometric isomers reflect the relative stabilities of the isomers. The greater repulsion between the nearby CH_3's in the *cis*-like transition state (TS) causes this TS to have a higher enthalpy of activation ($\Delta H^\ddagger$) than the *trans*-like TS. Consequently, the *trans* isomer predominates.

(*b*) The repulsion of the bulkier *t*-butyl groups causes a substantial increase in the $\Delta H^\ddagger$ of the *cis*-like TS, and the *trans* isomer is practically the only product.

Problem 6.10 Give the structural formulas for the alkenes formed on dehydrobromination of the following alkyl bromides and underline the principal product in each reaction: (*a*) 1-bromobutane, (*b*) 2-bromobutane, (*c*) 3-bromopentane, (*d*) 2-bromo-2-methylpentane, (*e*) 3-bromo-2-methylpentane, (*f*) 3-bromo-2,3-dimethylpentane.

The Br is removed with an atom from an adjacent C.

(*a*) $H_2C{-}CHCH_2CH_3$ (Br on C1, H^1 on C2) $\longrightarrow$ $H_2C{=}CHCH_2CH_3$ (only 1 adjacent H; only 1 product)

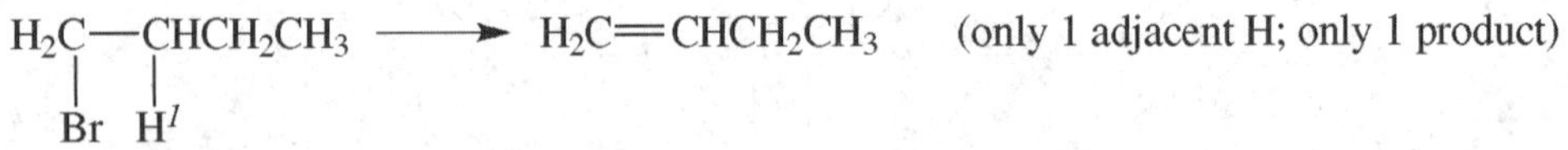

(*b*) $H_2C{-}CHCHCH_3$ (H^1 on C1, Br on C2, H^2 on C3) $\longrightarrow$ $H_2C{=}CHCH_2CH_3$ ($-H^1$) + *cis*- and $\underline{trans\text{-}CH_3CH{=}CHCH_3}$ ($-H^2$) di-R-substituted

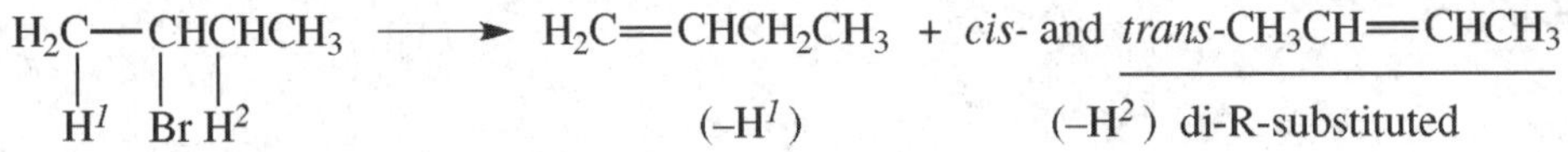

(*c*) $CH_3CH\,CH\,CHCH_3$ (H^1, Br, H^1) $\longrightarrow$ $\underline{cis\text{- and }trans\text{-}CH_3CH{=}CHCH_2CH_3}$ (adjacent H's are equivalent)

(*d*) $H_2C{-}C(CH_2H^1){-}CHCH_2CH_3$ (H^1 on C1, Br on C2, H^2 on C3) $\longrightarrow$ $H_2C{=}C(CH_3){-}CH_2CH_2CH_3$ ($-H^1$) and $\underline{CH_3C(CH_3){=}CHCH_2CH_3}$ ($-H^2$)

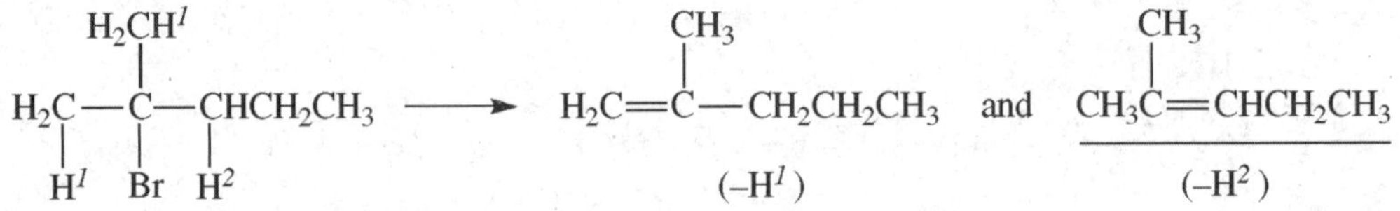

(*e*) $(CH_3)_2C{-}CHCHCH_3$ (H^1, Br, H^2) $\longrightarrow$ $\underline{(CH_3)_2C{=}CHCH_2CH_3}$ ($-H^1$) tri-R-substituted + *cis*- and *trans*-$(CH_3)_2CHCH{=}CHCH_3$ ($-H^2$) di-R-substituted

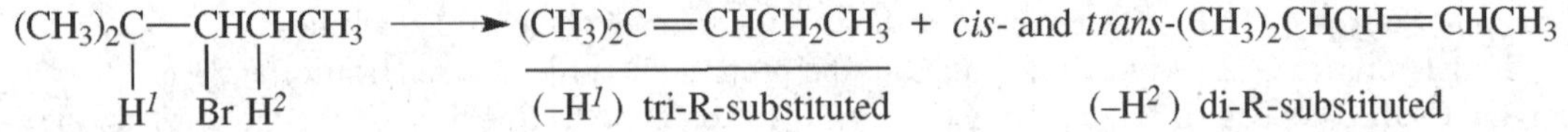

(*f*) $CH_3CH{-}C(CH_2H^2){-}C(CH_3)_2$ (H^1, Br, H^3) $\longrightarrow$ *cis* and *trans*-$CH_3CH{=}C(CH_3)CH(CH_3)_2$ ($-H^1$) tri-R-alkene

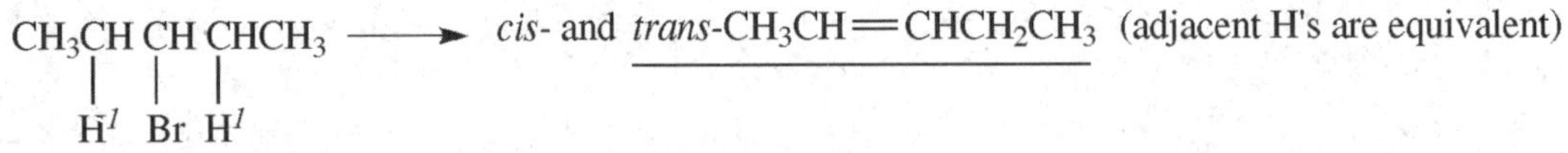

+ $\underline{CH_3CH_2C(CH_3){=}C(CH_3)_2}$ ($-H^3$) tetra-R-alkene + $CH_3CH_2{-}C({=}CH_2){-}CH(CH_3)_2$ ($-H^2$) di-R-alkene

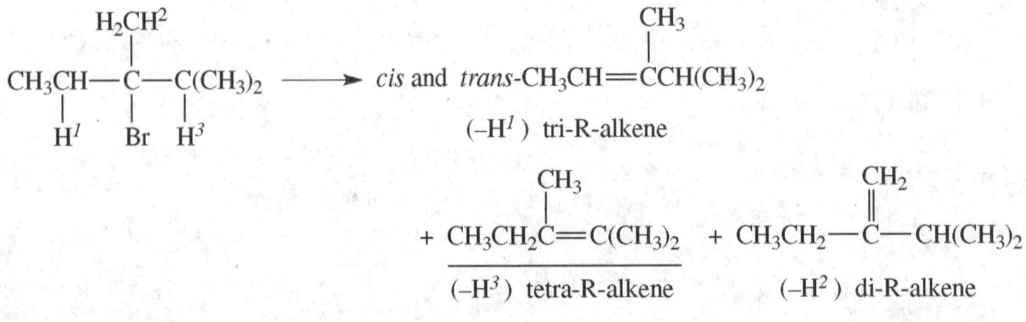

Problem 6.11 (*a*) Suggest a mechanism for the dehydration of $CH_3CHOHCH_3$ that proceeds through a carbocation intermediate. Assign a catalytic role to the acid and keep in mind that the O in ROH is a basic site like the O in H_2O. (*b*) Select the slow rate-determining step, and justify your choice. (*c*) Use transition states to explain the order of reactivity of ROH: 3° > 2° > 1°.

(*a*)

Step 1 $CH_3CH(OH)CH_2H + H_2SO_4 \overset{\text{fast}}{\rightleftharpoons} CH_3CH(\overset{+}{O}H_2)CH_2H + HSO_4^-$

(Reactant: CH_3CHCH_2 with H on CH_2 and $:\ddot{O}H$ on C; product: CH_3CHCH_2 with H on CH_2 and $H:\ddot{O}H^+$ on C)

base$_1$ acid$_2$ acid$_1$ *an onium ion* base$_2$

Step 2 $CH_3C(H_2O^+)—CH_2(H) \xrightarrow{\text{slow}} CH_3\overset{+}{C}HCH_2(H) + H_2O$

Isopropyl cation

Step 3 $CH_3\overset{+}{C}H—CH_2(\text{H}) + HSO_4^- \xrightarrow{\text{fast}} CH_3CH{=}CH_2 + H_2SO_4$

very strong acid$_1$ base$_2$ base$_1$ acid$_2$

Instead of HSO_4^-, a molecule of alcohol could act as the base in Step 3 to give ROH_2^+.

(*b*) Carbocation formation, Step 2, is the slow step, because it is a heterolysis leading to a very high-energy carbocation possessing an electron-deficient C.

(*c*) The order of reactivity of the alcohols reflects the order of stability of the incipient carbocation (3° > 2° > 1°) in the TS of Step 2, the rate-determining step. See Fig. 6.2.

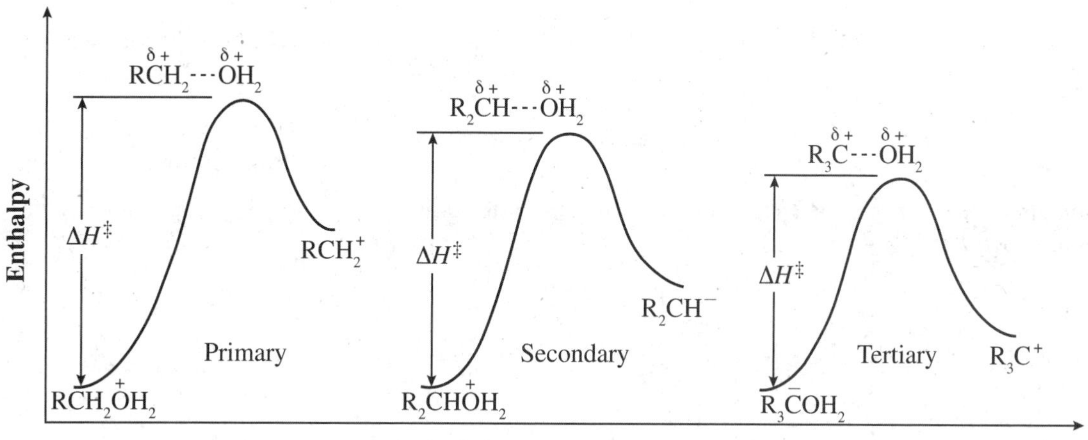

Figure 6.2

Problem 6.12 Account for the fact that dehydration of: (*a*) $CH_3CH_2CH_2CH_2OH$ yields mainly $CH_3CH{=}CHCH_3$ rather than $CH_3CH_2CH{=}CH_2$, (*b*) $(CH_3)_3CCHOHCH_3$ yields mainly $(CH_3)_2C{=}C(CH_3)_2$.

(*a*) The carbocation (R^+) formed in a reaction like Step 2 of Problem 6.11(*a*) is 1° and rearranges to a more stable 2° R_2CH^+ by a **hydride shift** (indicates as ~H:; the H migrates with its bonding pair of electrons).

$$CH_3CH_2C(H)(\ddot{H})—\overset{+}{C}H_2 \xrightarrow{\text{~H:}} CH_3CH_2\overset{+}{C}(H)—CH_2\ddot{H} \xrightarrow[-H^+]{+ROH} CH_3CH{=}CHCH_3 + R\overset{+}{O}H_2$$

1° RCH_2^+ (*n*-Butyl cation) 2° R_2CH^+ (*sec*-Butyl cation) (*b*) The 2° R_2CH^+ formed undergoes a methide shift (~:CH_3) to the more stable 3° R_3C^+.

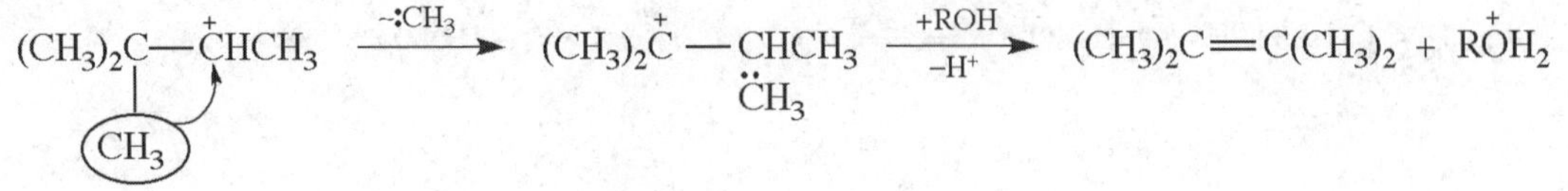

$$(CH_3)_2C(CH_3)-\overset{+}{C}HCH_3 \xrightarrow{\sim:CH_3} (CH_3)_2\overset{+}{C}-CH(CH_3)CH_3 \xrightarrow[-H^+]{+ROH} (CH_3)_2C{=}C(CH_3)_2 + R\overset{+}{O}H_2$$

2° R_2CH^+ (3,3-Dimethyl-2-butyl cation) → 3° R_3C^+ (2,3-Dimethyl-2-butyl cation)

Carbocations are always prone to rearrangement, especially when rearrangement leads to a more stable carbocation. The alkyl group may actually begin to migrate as the leaving group (e.g., H_2O) is departing—even before the carbocation is fully formed.

Problem 6.13 Assign numbers from 1 for LEAST to 3 for MOST to indicate the relative ease of dehydration, and justify your choices.

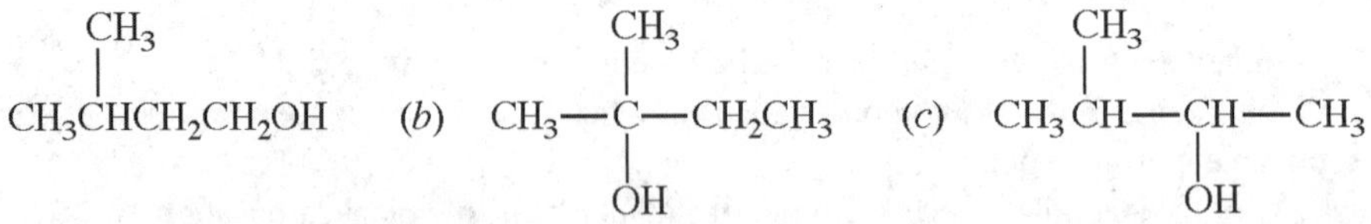

(*a*) $CH_3CH(CH_3)CH_2CH_2OH$ (*b*) $CH_3-C(CH_3)(OH)-CH_2CH_3$ (*c*) $CH_3CH(CH_3)-CH(OH)-CH_3$

(*a*) 1, (*b*) 3, (*c*) 2. The ease of dehydration depends on the relative ease of forming an R^+, which depends in turn on its relative stability. This is greatest for the 3° alcohol (*b*) and least for the 1° alcohol (*a*).

Problem 6.14 Give structural formulas for the reactants that form 2-butene when treated with the following reagents: (*a*) heating with conc. H_2SO_4, (*b*) alcoholic KOH, (*c*) zinc dust and alcohol, (*d*) hydrogen and a catalyst.

(*a*) $CH_3CHOHCH_2CH_3$ (*b*) $CH_3CHBrCH_2CH_3$ (*c*) $CH_3CHBrCHBrCH_3$ (*d*) $CH_3C{\equiv}CCH_3$.

Problem 6.15 Write the structural formula and name of the principal organic compound formed in the following reactions:

(*a*) $CH_3CCl(CH_3)CH_2CH_3$ + alc. KOH ⟶

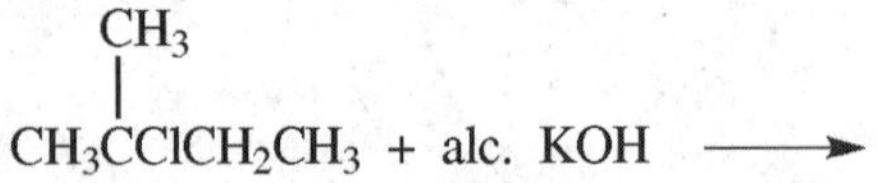

(*b*) $HOCH_2CH_2CH_2CH_2OH + BF_3$, heat ⟶

(*c*) *Trans*- H_2C ring: CH_2-CH_2, CH_2, CHBr-CHBr (trans-1,2-dibromocyclohexane) + Zn in alcohol ⟶

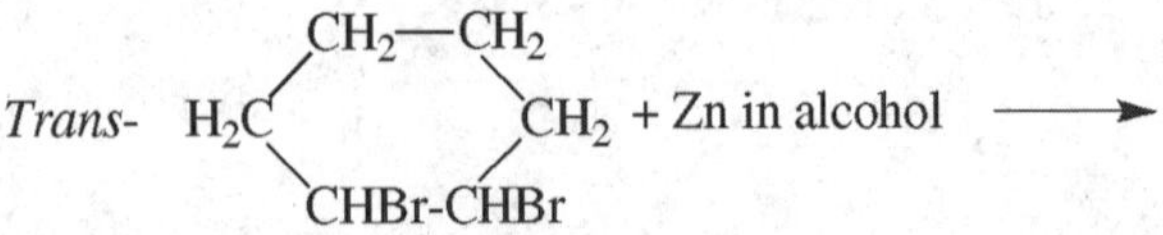

(*d*) $CH_3-C(CH_3)(Br)-CH_3 + CH_3COO^-$ ⟶

(*a*) $CH_3C(CH_3){=}CHCH_3$ 2-Methyl-2-butene (*b*) $H_2C{=}CH-CH{=}CH_2$ 1,3-Butadiene

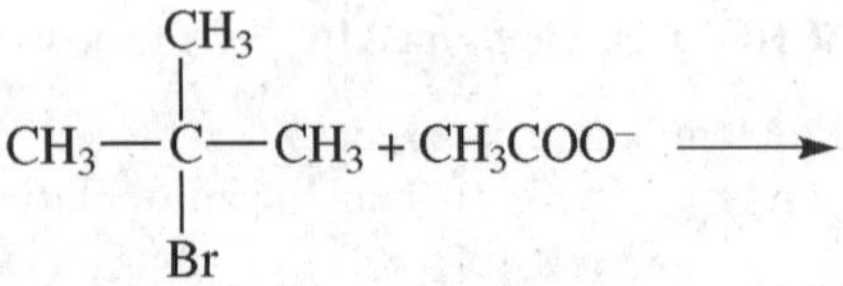

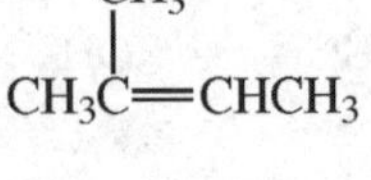

(*c*) H_2C ring: CH_2-CH_2, CH_2, $CH{=}CH$ Cyclohexene (*d*) $CH_3-C(CH_3){=}CH_2$ Isobutylene

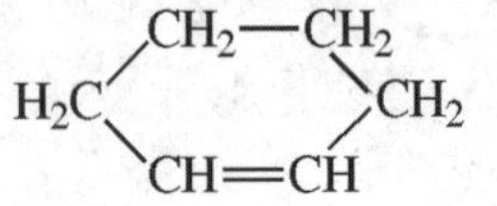

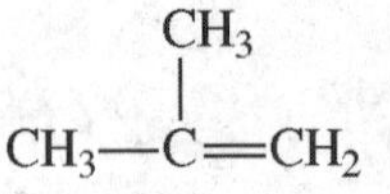

6.4 Chemical Properties of Alkenes

Alkenes undergo addition reactions at the double bond. The π electrons of alkenes are a nucleophilic site, and they react with electrophiles by three mechanisms (see Problem 3.37).

$$\mathrm{>C{=}C<} + \mathrm{E{:}Nu} \longrightarrow \mathrm{>C(E){-}C^{+}<} + \mathrm{{:}Nu^-} \longrightarrow \mathrm{>C(E){-}C(Nu)<} \quad \text{(Ionic)}$$

Intermediate R^+

$$\mathrm{E\cdot} + \mathrm{>C{=}C<} \longrightarrow \mathrm{>C(E){-}\dot{C}<} + \mathrm{E{:}E} \longrightarrow \mathrm{>C(E){-}C(E)<} + \mathrm{E\cdot} \quad \text{(Free radical)}$$

Intermediate $R\cdot$

$$\mathrm{RCH{=}CHR} + \mathrm{E{-}Nu} \longrightarrow \left[\begin{array}{c}\mathrm{RCH{\cdots}CHR} \\ \vdots \quad \vdots \\ \mathrm{E{\cdots}Nu}\end{array}\right]^{\ddagger} \longrightarrow \mathrm{RCH(E){-}CHR(Nu)} \quad \text{(Cyclic, one-step, rare)}$$

Transition State

Reduction to Alkanes

1. Addition of H_2

$$\mathrm{RCH{=}CHR} + H_2 \xrightarrow{\text{Pt, Pd or Ni}} \mathrm{RCH_2CH_2R} \quad \text{(heterogeneous catalysis)}$$

H_2 can also be added under homogeneous conditions in solution by using transition-metal coordination complexes such as the rhodium compound, $Rh[P(C_6H_5)_3Cl]$ (**Wilkinson's catalyst**). The relative rates of hydrogenation

$$H_2C{=}CH_2 > RCH{=}CH_2 > R_2C{=}CH_2, \quad RCH{=}CHR > R_2C{=}CHR > R_2C{=}CR_2$$

indicate that the rate is decreased by steric hindrance.

2. Reductive Hydroboration

$$\mathrm{RCH{=}CHR} \xrightarrow[\text{from } B_2H_6]{[BH_3]} \underset{\text{Alkylborane}}{\mathrm{RCH(H){-}CHR(BH_2)}} \xrightarrow{CH_3COOH} \mathrm{RCH_2CH_2R} \quad \text{(homogeneous catalysis)}$$

The compound BH_3 does not exist; the stable borohydride is diborane, B_2H_6. In syntheses, B_2H_6 is dissolved in tetrahydrofuran (THF), a cyclic ether, to give the complex THF:BH_3,

$$\text{(THF ring)}\mathrm{O{:}BH_3}$$

in which BH_3 is the active reagent.

Problem 6.16 Given the following heats of hydrogenation, $-\Delta H_h$, in kJ/mol: 1 -pentene, 125.9; *cis*-2-pentene, 119.7; *trans*-2-pentene, 115.5. (*a*) Use an enthalpy diagram to derive two generalizations about the relative stabilities of alkenes. (*b*) Would the ΔH_h of 2-methyl-2-butene be helpful in making your generalizations? (*c*) The corresponding heats of combustion, $-\Delta H_c$, are: 3376, 3369, and 3365 kJ/mol. Are these values consistent with your generalizations in part (*a*)? (*d*) Would the ΔH_c of 2-methyl-2-butene be helpful in your comparison? (*e*) Suggest a relative value for the ΔH_c of 2-methyl-2-betene.

(*a*) See Fig. 6.3. The lower ΔH_h, the more stable the alkene. (1) The alkene with more alkyl groups on the double bond is more stable; 2-pentene > 1-pentene. (2) The *trans* isomer is usually more stable than the *cis*. Bulky alkyl groups are *anti*-like in the *trans* isomer and eclipsed-like in the *cis* isomer.
(*b*) No. The alkenes being compared *must* give the same product on hydrogenation.
(*c*) Yes. Again the highest value indicates the least stable isomer.
(*d*) Yes. On combustion, all four isomers give the same products, H_2O and CO_2.
(*e*) Less than 3365 kJ/mol, since this isomer is a trisubstituted alkene and the 2-pentenes are disubstituted.

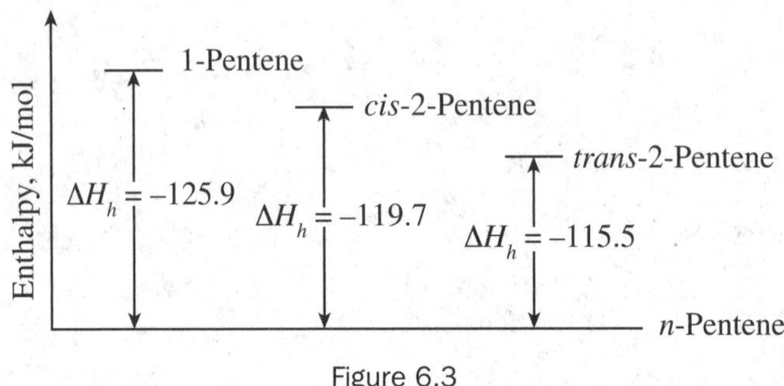

Figure 6.3

Problem 6.17 What is the stereochemistry of the catalytic addition of H_2 if *trans*- $CH_3CBr{=\!=}CBrCH_3$ gives *rac*-$CH_3CHBrCHBrCH_3$ and its *cis* isomer gives the *meso* product?

In hydrogenation reactions, two H atoms add stereoselectively *syn* (*cis*) to the π bond of the alkene.

from topside addition *from bottomside addition*

trans → (*SS*) + (*RR*) — *racemate*

from topside addition *from bottomside addition*

cis → (*identical*) — *meso*

Electrophilic Polar Addition Reactions

Table 6.1 shows the results of electrophilic addition of polar reagents to ethylene.

Problem 6.18 Unsymmetrical reagents like HX add to unsymmetrical alkenes such as propene according to **Markovnikov's rule**: The positive portion—for instance, H of HX—adds to the C that has more H's ("the rich get richer"). Explain by stability of the intermediate cation.

TABLE 6.1

REAGENT		PRODUCT	
NAME	STRUCTURE	NAME	STRUCTURE
Halogens (Cl_2, Br_2 only)	X:X	Ethylene dihalide	CH_2XCH_2X
Hydrohalic acids	$\overset{\delta+}{H}:\overset{\delta-}{X}$	Ethyl halide	CH_3CH_2X
Hypohalous acids	$\overset{\delta+}{X}:\overset{\delta-}{OH}$	Ethylene halohydrin	CH_3XCH_2OH
Sulfuric acid (cold)	$\overset{\delta+}{H}:\overset{\delta-}{OSO_3OH}$	Ethyl bisulfate	$CH_3CH_2OSO_3H$
Water (dil. H_3O^+)	$\overset{\delta+}{H}:\overset{\delta-}{OH}$	Ethyl alcohol	CH_3CH_2OH
Borane	$\overset{\delta+}{H_2B}:\overset{\delta-}{H}$	Ethyl borane	$[CH_3CH_2BH_2] \rightarrow (CH_3CH_2)_3B$
Peroxyformic acid	$\overset{\delta+}{H}:O-\overset{\delta-}{O}\underset{\underset{O}{\Vert}}{C}H$	Ethylene glycol	$[CH_2OHCH_2O\underset{\underset{O}{\Vert}}{C}H] \longrightarrow HOCH_2CH_2OH$
Mercuric acetate, H_2O	$\overset{\delta+}{Hg}(\overset{\delta-}{O_2CCH_3})_2$ H_2O	Ethanol	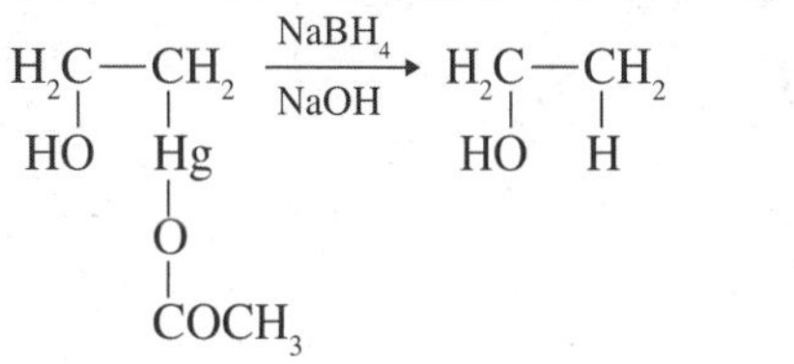 $\underset{HO}{H_2C}-\underset{\underset{\underset{COCH_3}{\vert}}{\underset{O}{\vert}}}{\underset{Hg}{CH_2}} \xrightarrow[NaOH]{NaBH_4} \underset{HO}{H_2C}-\underset{H}{CH_2}$

The more stable cation (3° > 2° > 1°) has a lower $\Delta H^{\ddagger}$ for the transition state and forms more rapidly (Fig. 6.4). Markovnikov additions are called **regioselective**, since they give mainly one of several possible structural isomers.

Problem 6.19 Give the structural formula of the major organic product formed from the reaction of $CH_3CH{=}CH_2$ with: (*a*) Br_2, (*b*) HI, (*c*) BrOH, (*d*) H_2O in acid, (*e*) cold H_2SO_4, (*f*) BH_3 from B_2H_6, (*g*) peroxyformic acid (H_2O_2 and HCOOH).

The positive (δ+) part of the addendum is an electrophile (E^+) which forms $CH_3\overset{+}{C}HCH_2E$ rather than $CH_3CHE\overset{+}{C}H_2$. The Nu: part then forms a bond with the carbocation.
The E^+ is in a box; the $Nu:^-$ is encircled.

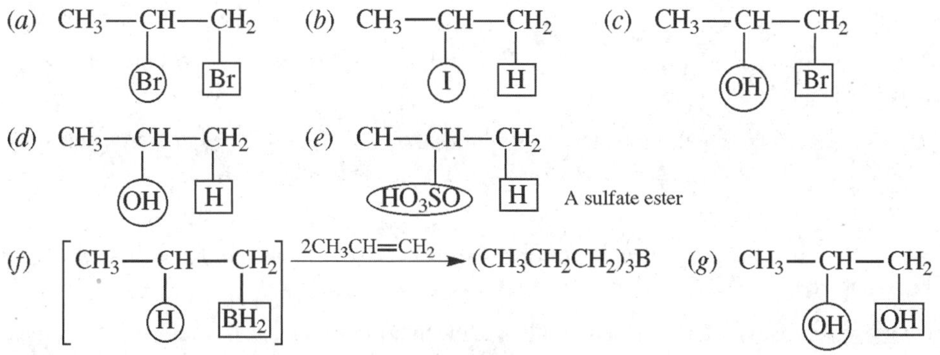

(*a*) $CH_3-\underset{(Br)}{CH}-\underset{\boxed{Br}}{CH_2}$ (*b*) $CH_3-\underset{(I)}{CH}-\underset{\boxed{H}}{CH_2}$ (*c*) $CH_3-\underset{(OH)}{CH}-\underset{\boxed{Br}}{CH_2}$

(*d*) $CH_3-\underset{(OH)}{CH}-\underset{\boxed{H}}{CH_2}$ (*e*) $CH-\underset{(HO_3SO)}{CH}-\underset{\boxed{H}}{CH_2}$ A sulfate ester

(*f*) $\left[CH_3-\underset{(H)}{CH}-\underset{\boxed{BH_2}}{CH_2}\right] \xrightarrow{2CH_3CH{=}CH_2} (CH_3CH_2CH_2)_3B$ (*g*) $CH_3-\underset{(OH)}{CH}-\underset{\boxed{OH}}{CH_2}$

(Anti-Markovnikov orientation; with nonbulky alkyl groups, all H's of BH_3 add to form a trialkylborane.)

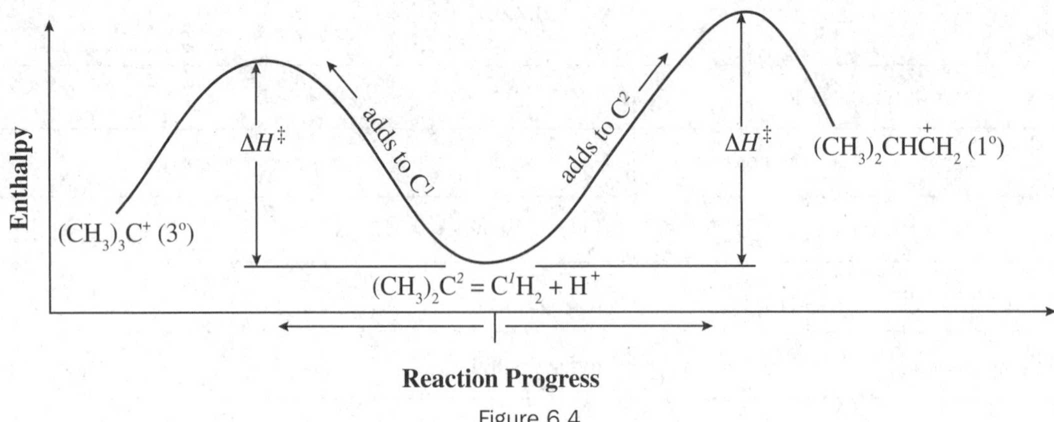

Figure 6.4

Problem 6.20 Account for the anti-Markovnikov orientation in Problem 6.19(*f*).

The electron-deficient B of BH_3, as an electrophilic site, reacts with the π electrons of C=C, as the nucleophilic site. In typical fashion, the bond is formed with the C having the greater number of H's—in this case, the terminal C. As this bond forms, one of the H's of BH_3 begins to break away from the B as it forms a bond to the other doubly bonded C atom, giving a four-center transition state shown in the equation. The product from this step, $CH_3CH_2CH_2BH_2$ (*n*-propyl borane), reacts stepwise in a similar fashion with two more molecules of propene, eventually giving $(CH_3CH_2CH_2)_3B$.

$$CH_3CH{=}CH_2 + BH_3 \longrightarrow CH_3\overset{\sigma+}{C}\text{---}CH_3 \;(\text{H---}\overset{\sigma-}{B}H_2\text{ bridging}) \longrightarrow CH_3CH_2CH_2BH_2 \xrightarrow{2\,CH_2{=}CHCH_3} (CH_3CH_2CH_2)_3B$$

This reaction is a stereoselective and regioselective *syn* addition.

Problem 6.21 (*a*) What principle is used to relate the mechanisms for dehydration of alcohols and hydration of alkenes? (*b*) What conditions favor dehydration rather than hydration reactions?

(*a*) The **principle of microscopic reversibility** states that every reaction is reversible, even if only to a microscopic extent. Furthermore, the reverse process proceeds through the same intermediates and transition states, but in the opposite order:

$$RCH_2CH_2OH \overset{H^+}{\rightleftharpoons} RCH{=}CH_2 + H_2O$$

(*b*) Low H_2O concentration and high temperature favor alkene formation by dehydration, because the volatile alkene distills out of the reaction mixture and shifts the equilibrium. Hydration of alkenes occurs at low temperature and with dilute acid, which provides a high concentration of H_2O as reactant.

Problem 6.22 Why are dry gaseous hydrogen halides (HX) acids and not their aqueous solutions used to prepare alkyl halides from alkenes?

Dry hydrogen halides are stronger acids and better electrophiles than the H_3O^+ formed in their water solutions. Furthermore, H_2O is a nucleophile that can react with R^+ to give an alcohol.

Problem 6.23 Arrange the following alkenes in order of increasing reactivity upon addition of hydrohalogen acids: (*a*) $H_2C{=}CH_2$, (*b*) $(CH_3)_2C{=}CH_2$, (*c*) $CH_3CH{=}CHCH_3$.

The relative reactivities are directly related to the stabilities of the intermediate R^+'s. Isobutylene, (*b*), is most reactive because it forms the 3° $(CH_3)_2\overset{+}{C}CH_3$. The next-most reactive compound is 2-butene, (*c*), which forms the 2° $CH_3\overset{+}{C}HCH_2CH_3$. Ethylene forms the 1° $CH_3\overset{+}{C}H_2$ and is least reactive. The order of increasing reactivity is (*a*) < (*c*) < (*b*).

Problem 6.24 The addition of HBr to some alkenes gives a mixture of the expected alkyl bromide and an isomer formed by rearrangement. Outline the mechanism of formation and structures of products from the reaction of HBr with (*a*) 3-methyl-1-butene, (*b*) 3,3-dimethyl-1-butene.

No matter how formed, an R^+ can undergo H: or :CH_3 (or other alkyl) shifts to form a more stable R'^+.

(*a*) $H_2C{=}CHCH(CH_3)CH_3 \xrightarrow{H^+} CH_3{-}\overset{+}{C}H{-}CH(CH_3){-}CH_3$ (2°)

→ (Br^-) $CH_3CHBrCH(CH_3)CH_3$ 2-Bromo-3-methylbutane

→ (~:H) $CH_3CH_2{-}\overset{+}{C}(CH_3){-}CH_3$ (3°) $\xrightarrow{Br^-}$ $CH_3CH_2{-}CBr(CH_3){-}CH_3$ 2-Bromo-2-methylbutane

(*b*) $H_2C{=}CH{-}C(CH_3)_2{-}CH_3 \xrightarrow{H^+} CH_3{-}\overset{+}{C}H{-}C(CH_3)_2{-}CH_3$ (2°, *less stable*)

→ (Br^-) $CH_3{-}CHBr{-}C(CH_3)_2{-}CH_3$ 3-Bromo-2,2-dimethylbutane

→ (~:CH_3) $CH_3{-}CH(CH_3){-}\overset{+}{C}(CH_3){-}CH_3$ (3°, *more stable*) $\xrightarrow{Br^-}$ $CH_3CH(CH_3){-}CBr(CH_3){-}CH_3$ 2-Bromo-2,3-dimethylbutane

Problem 6.25 Compare and explain the relative rates of addition to alkenes (reactivities) of HCl, HBr, and HI.

The relative reactivity depends on the ability of HX to donate an H^+ (acidity) to form an R^+ in the rate-controlling first step. The acidity and reactivity order is HI > HBr > HCl.

Problem 6.26 (*a*) What does each of the following observations tell you about the mechanism of the addition of Br_2 to an alkene? (i) In the presence of a Cl^- salt, in addition to the *vic*-dibromide, some *vic*-bromochloroalkane is isolated but no dichloride is obtained. (ii) With *cis*-2-butene, only *rac*-2,3-dibromobutane is formed. (iii) With *trans*-2-butene, only *meso*–2,3-dibromobutane is produced. (*b*) Give a mechanism compatible with these observations.

(*a*) (i) Br_2 adds in two steps. If Br_2 added in one step, no bromochloroalkane would be formed. Furthermore, the first step must be the addition of an electrophile (the Br^+ part of Br_2) followed by addition of a nucleophile, which could now be Br^- or Cl^-. This explains why the products must contain at least one Br. (ii) One Br adds from above the plane of the double bond; the second Br adds from below. This is an *anti* (*trans*) addition. Since a Br^+ can add from above to either C, a racemic form results.

cis $\xrightarrow{Br_2}$ (*RR*) + (*SS*) — *racemic*

(iii) This substantiates the *anti* addition.

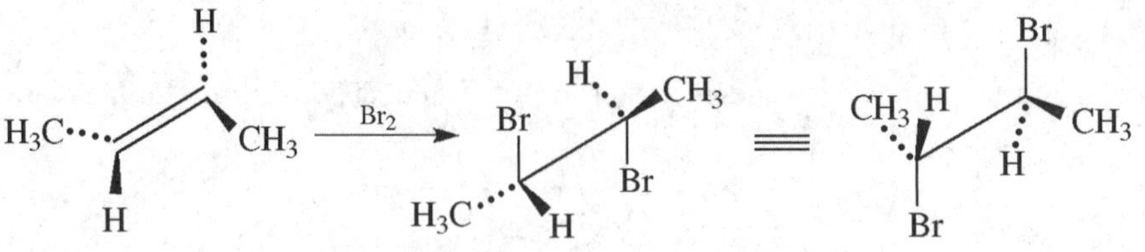

trans $\xrightarrow{Br_2}$ ≡ meso

The reaction is also stereospecific because different stereoisomers give stereochemically different products—for instance, *cis* ⟶ racemic and *trans* ⟶ *meso*. Because of this stereospecificity, the intermediate *cannot* be the free carbocation $CH_3CHBr\overset{+}{C}HCH_3$. The same carbocation would arise from either *cis*- or *trans*-2-butene, and the product distribution from both reactants would be identical.

(*b*) The open carbocation is replaced by a cyclic bridged ion having Br^+ partially bonded to each C (**bromonium ion**). In this way, the stereochemical differences of the starting materials are retained in the intermediate. In the second step, the nucleophile attacks the side *opposite* the bridging group to yield the *anti* addition product.

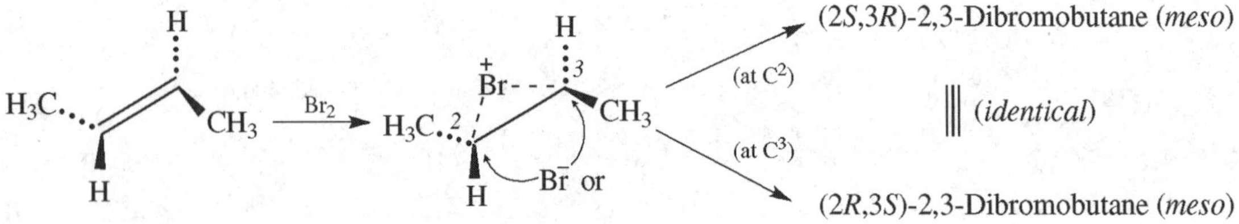

trans $\xrightarrow{Br_2}$ *a bromonium ion* (attacked by Br^- at C^2 or at C^3)

(at C^2) → (2*S*,3*R*)-2,3-Dibromobutane (*meso*)

‖ (*identical*)

(at C^3) → (2*R*,3*S*)-2,3-Dibromobutane (*meso*)

Br_2 does not break up into Br^+ and Br^-. More likely, the π electrons attack one of the Br's, displacing the other as an anion (Fig. 6.5).

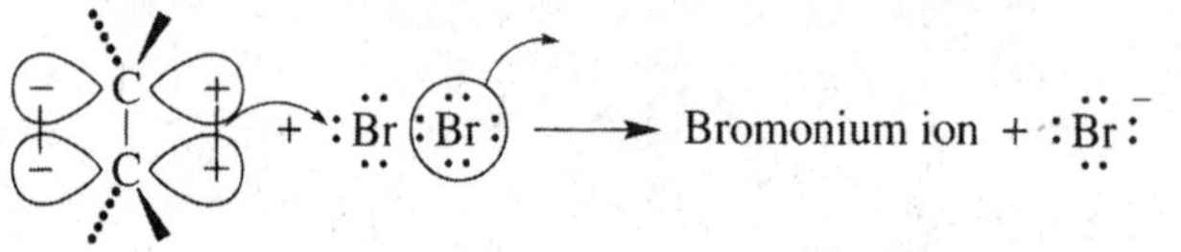

$+ \,:\!Br\!:\!Br\!: \longrightarrow$ Bromonium ion $+ \,:\!Br\!:^-$

Figure 6.5

Problem 6.27 Alkenes react with aqueous Cl_2 or Br_2 to yield *vic*-halohydrins, —CXCOH. Give a mechanism for this reaction that also explains how Br_2 and $(CH_3)_2C{=}CH_2$ give $(CH_3)_2C(OH)CH_2Br$.

The reaction proceeds through a bromonium ion [Problem 6.26(*b*)], which reacts with the nucleophilic H_2O to give

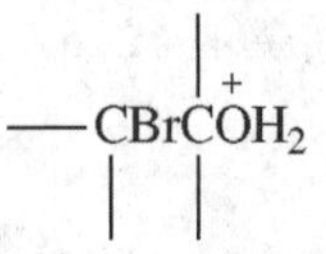

$$—CBrC\overset{+}{O}H_2$$

This protonated halohydrin then loses H^+ to the solvent, giving the halohydrin. The partial bonds between the C's and Br engender $\delta+$ charges on the C's. Since the bromonium ion of 2-methylpropene has more partial positive charge on the 3° carbon than on the 1° carbon, H_2O binds to the 3°C to give the observed product. In general, X appears on the C with the greater number of H's. The addition, like that of Br_2, is *anti* because H_2O binds to the C from the side away from the side where the Br is positioned.

Problem 6.28 (*a*) Describe the stereochemistry of glycol formation with peroxyformic acid (HCO_3H) if *cis*-2-butene gives a racemic glycol and *trans*-2-butene gives the *meso* form. (*b*) Give a mechanism for *cis*.

(*a*) The reaction is a stereospecific *anti* addition similar to that of addition of Br_2.
(*b*)

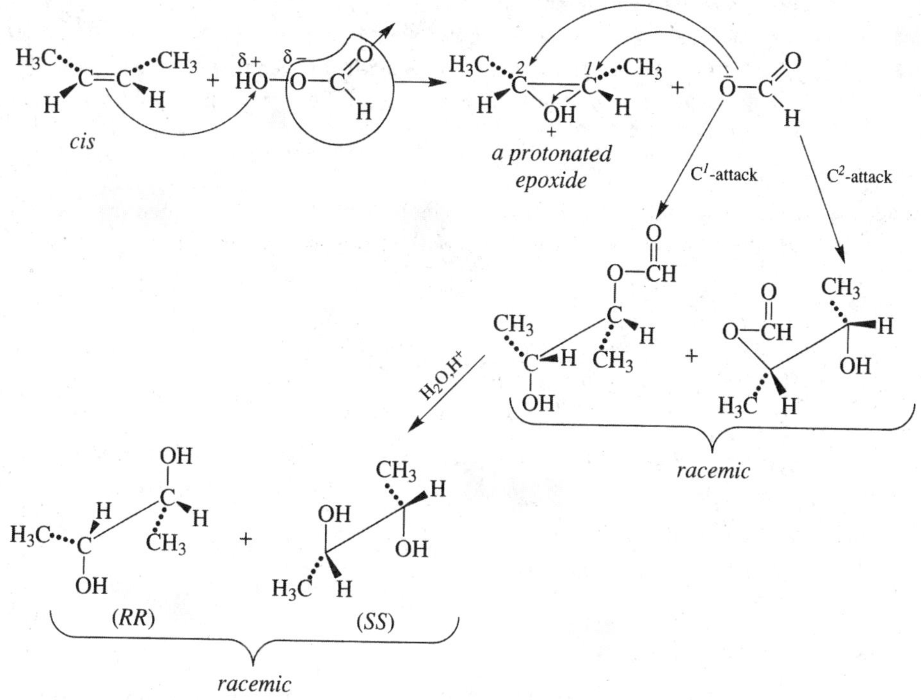

cis
a protonated epoxide
C^1-attack
C^2-attack
racemic
H_2O, H^+
(*RR*) (*SS*)
racemic

Dimerization and Polymerization

Under proper conditions, a carbocation (R^+), formed by adding an electrophile such as H^+ or BF_3 to an alkene, may add to the C=C bond of another alkene molecule to give a new dimeric R'^+; here, R^+ acts as an electrophile and the π bond of C=C acts as a nucleophilic site. R'^+ may then lose an H^+ to give an alkene **dimer**.

Problem 6.29 (*a*) Suggest a mechanism for the dimerization of isobutylene, $(CH_3)_2C{=}CH_2$. (*b*) Why does $(CH_3)_3C^+$ add to the "tail" carbon rather than to the "head" carbon? (*c*) Why are the Brönsted acids H_2SO_4 and HF typically used as catalysts, rather than HCl, HBr, or HI?

(*a*)

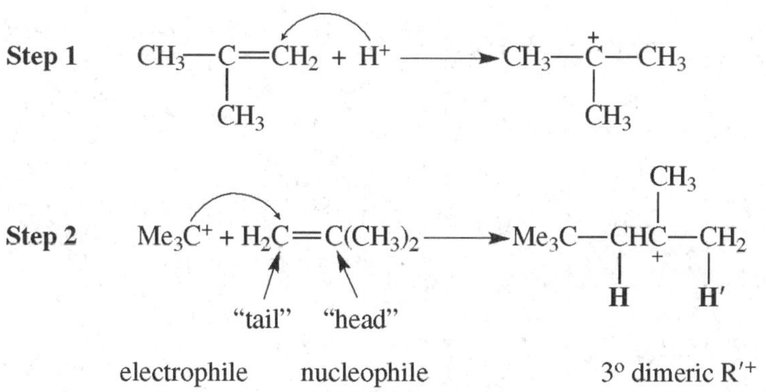

Step 1 $CH_3{-}C(CH_3){=}CH_2 + H^+ \longrightarrow CH_3{-}\overset{+}{C}(CH_3){-}CH_3$

Step 2 $Me_3C^+ + H_2C{=}C(CH_3)_2 \longrightarrow Me_3C{-}CH(H){-}\overset{+}{C}(CH_3){-}CH_2H'$

"tail" "head"

electrophile nucleophile 3° dimeric R'^+

Step 3 $R'^+{-}H^+ \longrightarrow Me_3C{-}CH{=}C(CH_3)_2$ (major, Saytzeff product)

$R'^+{-}H'^+ \longrightarrow Me_3C{-}CH_2C(CH_3){=}CH_2$ (minor, non-Saytzeff product)

(*b*) Step 2 is a Markovnikov addition. Attachment at the "tail" gives the 3° R'^+; attach at the "head" would give the much less stable, 1° carbocation $^+CH_2C(CH_3)_2CMe_3$.
(*c*) The catalytic acid must have a weakly nucleophilic conjugate base to avoid addition of HX to the C=C. The conjugate bases of HCl, HBr, and HI (Cl^-, Br^-, and I^-) are good nucleophiles that bind to R^+.

The newly formed R'^+ may also add to another alkene molecule to give a **trimer**. The process whereby simple molecules, or **monomers**, are merged can continue, eventually giving high-molecular-weight molecules called **polymers**. This reaction of alkenes is called **chain-growth (addition) polymerization**. The repeating unit in the polymer is called the **mer**. If a mixture of at least two different monomers polymerizes, a **copolymer** is obtained.

Problem 6.30 Write the structural formula for (*a*) the major trimeric alkene formed from $(CH_3)_2C{=}CH_2$, labeling the mer; (*b*) the dimeric alkene from $CH_3CH{=}CH_2$. [Indicate the dimeric R^+.]

(*a*) [$(CH_3)_3C$]—[CH_2—$C(CH_3)_2$—] [$CH{=}C(CH_3)_2$]
mer

The individual combining units are boxed.

(*b*) $(CH_3)_2CHCH{=}CHCH_3$; $[(CH_3)_2CHCH_2\overset{+}{C}HCH_3]$

Stereochemistry of Polymerization

The polymerization of propylene gives stereochemically different polypropylenes having different physical properties:

$$\text{H}\cdots\overset{*}{\text{C}}(\text{CH}_3) \quad \text{or} \quad \text{H}\cdots\overset{*}{\text{C}}(\text{R})$$

The C's of the mers are chiral, giving millions of stereoisomers, which are grouped into three classes depending on the arrangement of the branching Me(R) groups relative to the long "backbone" chain of the polymer (Fig. 6.6).

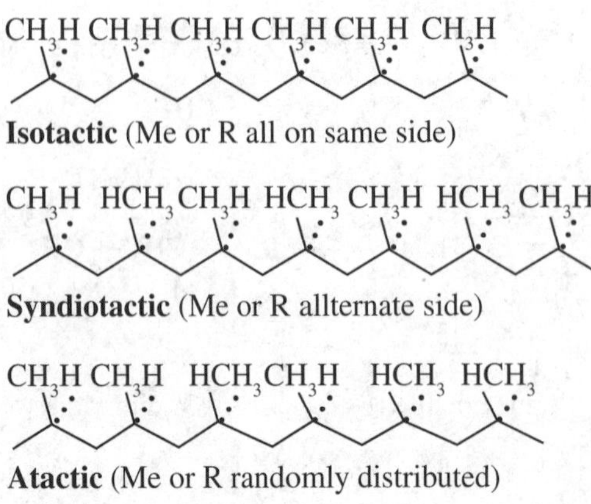

Figure 6.6

Addition of Alkanes

$$(CH_3)_2C{=}CH_2 + HC(CH_3)_3 \xrightarrow[0^\circ]{HF} \underset{\text{2,2,4-Trimethylpentane}}{(CH_3)_2CHCH_2C(CH_3)_3}$$

Problem 6.31 Suggest a mechanism for alkane addition where the key step is an intermolecular **hydride** (H:) **transfer**.

See Steps 1 and 2 in Problem 6.29 for formation of the dimeric R^+.

$$\underset{\text{dimeric } R^+}{(CH_3)_3CCH_2\overset{+}{C}(CH_3)_2} \longleftarrow \text{(H:)}C(CH_3)_3 \longrightarrow (CH_3)_3CCH_2C(CH_3)_2\text{(:H)} + \overset{+}{C}(CH_3)_3$$

This intermolecular H: transfer forms the $(CH_3)_3C^+$ ion, which adds to another molecule of $(CH_3)_2C{=}CH_2$ to continue the chain. A 3° H usually transfers to leave a 3° R^+.

Free-Radical Additions

$$RCH{=}CH_2 + HBr \xrightarrow[ROOR]{O_2\text{ or}} RCH_2CH_2Br \quad \text{(anti-Markovnikov; not with HF, HCl or HI)}$$

$$RCH{=}CH_2 + HSH \xrightarrow{ROOR} RCH_2CH_2SH \quad \text{(anti-Markovnikov)}$$

$$RCH{=}CH_2 + HCCl_3 \xrightarrow{ROOR} RCH(\boxed{H})CH_2\boxed{CCl_3} \quad \text{(anti-Markovnikov)}$$

$$RCH{=}CH_2 + BrCCl_3 \xrightarrow{ROOR} RCH\boxed{Br}CH_2\boxed{CCl_3}$$

Problem 6.32 Suggest a chain-propagating free-radical mechanism for addition of HBr in which Br· attacks the alkene to form the more stable carbon radical.

Initiation Steps

$$R{-}O{-}O{-}R \xrightarrow{heat} 2R{-}O\cdot \quad (-\ddot{O}-\ddot{O}- \text{ bond is weak})$$

$$RO\cdot + HBr \longrightarrow Br\cdot + R{-}O{-}H$$

Propagation Steps For Chain Reaction

$$\underset{(1^\circ\text{ radical})}{CH_3CHBr\dot{C}H_2} \not\longleftarrow \boxed{CH_3CH{=}CH_2} + Br\cdot \longrightarrow \underset{(2^\circ\text{ radical})}{CH_3\dot{C}HCH_2Br}$$

$$CH_3\dot{C}HCH_2Br + HBr \longrightarrow CH_3CH_2CH_2Br + Br\cdot$$

The Br· generated in the second propagation step continues the chain.

Carbene Addition

$$>C=C< + :CH_2 \longrightarrow >C{-}C< \text{ (cyclopropane ring with } CH_2\text{, C bearing H H)}$$

$$CH_2N_2 \xrightarrow[-N_2]{\text{light}} :CH_2$$

CH_2N_2
Diazomethane

Cleavage Reactions

Ozonolysis

$$>C=C< \xrightarrow{O_3} \text{Ozonide} \xrightarrow[Zn]{H_2O} >C=O + O=C<$$

Ozonide Carbonyl compounds

Problem 6.33 Give the products formed on ozonolysis of (*a*) $H_2C{=}CHCH_2CH_3$, (*b*) $CH_3CH{=}CHCH_3$, (*c*) $(CH_3)_2C{=}CHCH_2CH_3$, (*d*) cyclobutene, (*e*) $H_2C{=}CHCH_2CH{=}CHCH_3$.

To get the correct answers, erase the double bond and attach a $=O$ to each of the formerly double-bonded C's. The total numbers of C's in the carbonyl products and in the alkene reactant must be equal.

(*a*) $H_2C{=}O + O{=}CHCH_2CH_3$.
(*b*) $CH_3CH{=}O$; the alkene is symmetrical and only one carbonyl compound is formed.
(*c*) $(CH_3)_2C{=}O + O{=}CHCH_2CH_3$.
(*d*) $O{=}CHCH_2CH_2CH{=}O$; a cycloalkene gives only a dicarbonyl compound.
(*e*) $H_2C{=}O + O{=}CHCH_2CH{=}O + O{=}CHCH_3$. Noncyclic polyenes give a mixture of monocarbonyl compounds formed from the terminal C's and dicarbonyl compounds from the internal doubly bonded C's.

Problem 6.34 Deduce the structures of the following alkenes:

(*a*) An alkene $C_{10}H_{20}$ on ozonolysis yields only $CH_3{-}C({=}O){-}CH_2CH_2CH_3$.

(*b*) An alkene C_9H_{18} on ozonolysis gives $(CH_3)_3C(H)C{=}O$ and $CH_3{-}C({=}O){-}CH_2CH_3$.

(*c*) A compound C_8H_{14} adds one mole of H_2 and forms, upon ozonolysis, the dialdehyde

$$O{=}CH\underset{CH_3}{CH}CH_2CH_2\underset{CH_3}{CH}CH{=}O$$

(*d*) A compound C_8H_{12} adds two mole of H_2 and undergoes ozonolysis to give two moles of the dialdehyde $O{=}CHCH_2CH_2CH{=}O$.

(*a*) The formation of only one carbonyl compound indicates that the alkene is symmetrical about the double bond. Write the structure of the ketone twice so that the $C{=}O$ groups face each other. Replacement of the two O's by a double bond gives the alkene structure.

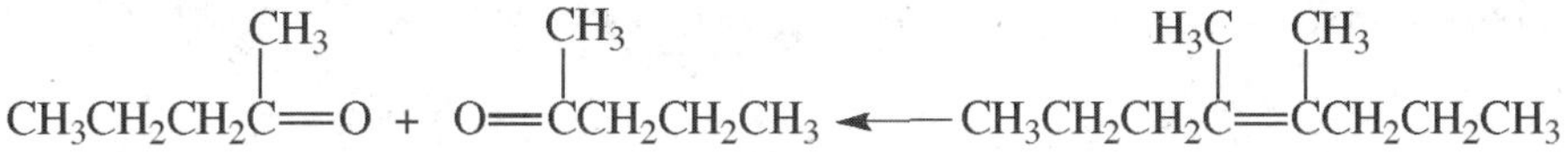

(*b*)

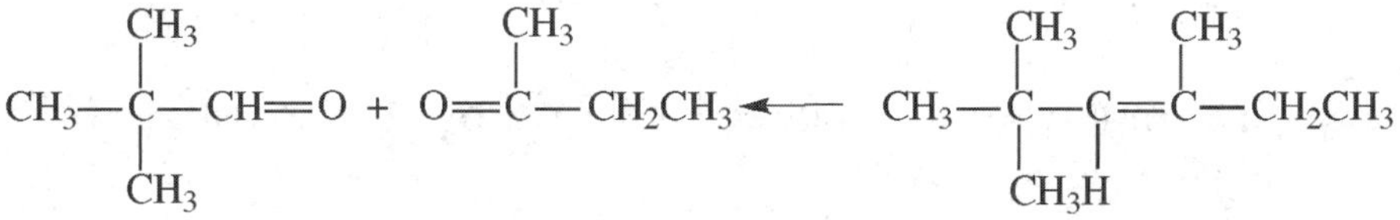

cis or trans

(*c*) C_8H_{14} has four fewer H's than the corresponding alkane, C_8H_{18}. There are two degrees of unsaturation. One of these is accounted for by a C$=$C because the alkene adds 1 mole of H_2. The second degree of unsaturation is a ring structure. The compound is a cycloalkene whose structure is found by writing the two terminal carbonyl groups facing each other:

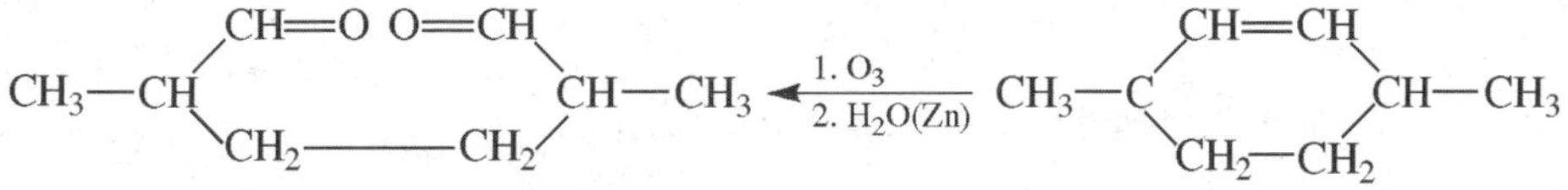

(*d*) The difference of six H's between C_8H_{12} and the alkane C_8H_{18} shows three degrees of unsaturation. The 2 mol of H_2 absorbed indicates two double bonds. The third degree of unsaturation is a ring structure. When two molecules of product are written with the pairs of C$=$O groups facing each other, the compound is seen to be a cyclic diene:

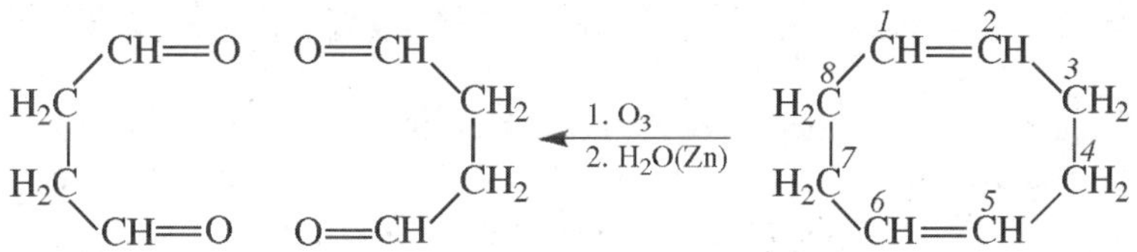

1,5-Cyclooctadiene

6.5 Substitution Reactions at the Allylic Position

Allylic carbons are the ones bonded to the doubly bonded C's; the H's attached to them are called allylic H's.

$$Cl_2 + H_2C{=}CHCH_3 \xrightarrow{\text{high temperature}} H_2C{=}CHCH_2Cl + HCl$$

$$Br_2 + H_2C{=}CHCH_3 \xrightarrow[\text{of } Br_2]{\text{low concentration}} H_2C{=}CHCH_2Br + HBr$$

The low concentration of Br_2 comes from *N*-bromosuccinimide (NBS):

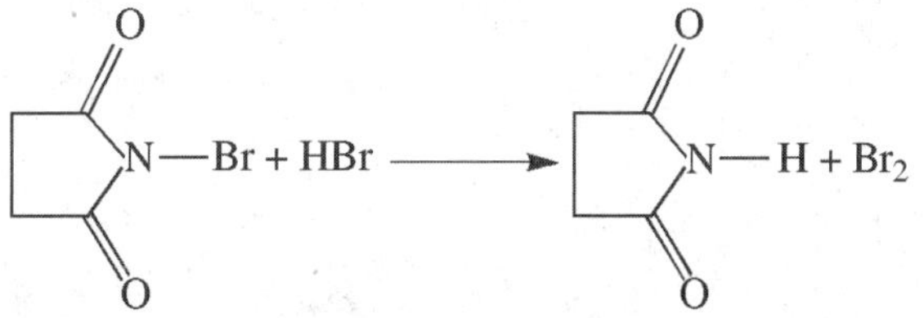

NBS *product of bromination* Succinimide

$$SO_2Cl_2 + H_2C{=}CHCH_3 \xrightarrow[\text{peroxide}]{\text{uv or}} H_2C{=}CHCH_2Cl + HCl + SO_2$$

Sulfuryl chloride

These halogenations are like free-radical substitutions of alkanes (see Section 4.4). The order of reactivity of H-abstraction is

$$\text{allyl} > 3° > 2° > 1° > \text{vinyl}$$

Problem 6.35 Use the concepts of (*a*) resonance and (*b*) extended π orbital overlap (delocalization) to account for the extraordinary stability of the allyl-type radical.

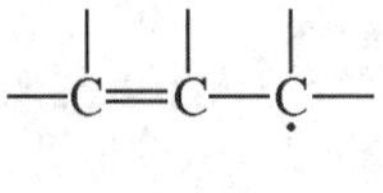

$$-\overset{|}{C}=\overset{|}{C}-\overset{|}{\underset{\cdot}{C}}-$$

(*a*) Two equivalent resonance structures can be written:

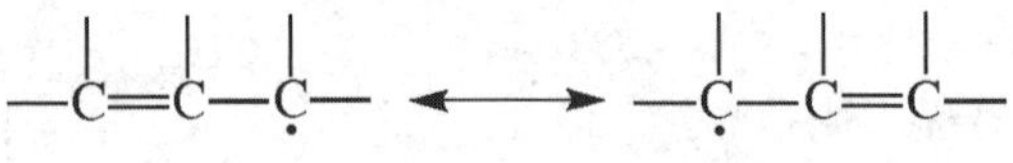

$$-\overset{|}{C}=\overset{|}{C}-\overset{|}{\underset{\cdot}{C}}- \longleftrightarrow -\overset{|}{\underset{\cdot}{C}}-\overset{|}{C}=\overset{|}{C}-$$

therefore, the allyl-type radical has considerable resonance energy (Section 2.7) and is relatively stable.

(*b*) The three C's in the allyl unit are sp^2-hybridized, and each has a p orbital lying in a common plane (Fig. 6.7). These three p orbitals overlap forming an extended π system, thereby delocalizing the odd electron. Such delocalization stabilizes the allyl-type free radical.

Problem 6.36 Designate the type of each set of H's in $CH_3CH{=}CHCH_2CH_2{-}CH(CH_3)_2$ (e.g. 3°, allylic, etc.) and show their relative reactivity toward a Br· atom, using (1) for the most reactive, (2) for the next, and so on.

Labeling the H's as

$$\overset{(a)}{CH_3}\,\overset{(b)}{CH}{=}\overset{(b)}{CH}\,\overset{(c)}{CH_2}\,\overset{(d)}{CH_2}\,\overset{(e)}{CH}(\overset{(f)}{CH_3})_2$$

we have (*a*) 1°, allylic (2); (*b*) vinylic (6); (*c*) 2°, allylic (1); (*d*) 2° (4); (*e*) 3° (3); (*f*) 1° (5).

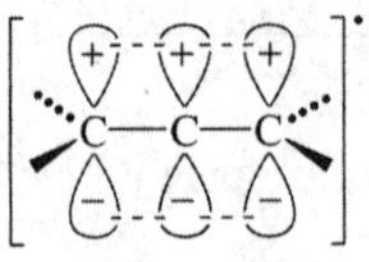

Figure 6.7

6.6 Summary of Alkene Chemistry

PREPARATION

1. **Dehydrohalogenation of RX**

   $RCHXCH_3$, RCH_2CH_2X + alc. KOH

2. **Dehydration of ROH**

   $RCHOHCH_3$, $RCH_2CH_2OH + H_2SO_4$, heat

3. **Dehalogenation of *vic* – Dihalide**

   $RCHXCH_2X$ + Zn

4. **Dehydrogenation of Alkanes**

   RCH_2CH_3, heat, Pt / Pd

5. **Reduction of RC≡CR**

   R—C≡CH + H_2 or Na / C_2H_5OH

(All of the above → $RCH{=}CH_2$)

PROPERTIES (of $RCH{=}CH_2$)

1. **Addition Reactions**

   (*a*) **Hydrogenation**

   Heterogeneous: H_2/Pt → RCH_2CH_3

   Homogeneous: $H_2 \xrightarrow{Rh[PPh_3]_3Cl} RCH_2CH_3$

   Chemical: (BH_3), CH_3CO_2H → RCH_2CH_3

   (*b*) **Polar Mechanism**

   $+ X_2 \rightarrow RCHXCH_2X$ (X = Cl, Br)

   $+ HX \rightarrow RCHXCH_3$

   $+ X_2, H_2O \rightarrow RCH(OH)CH_2X$

   $+ H_2O \xrightarrow{H^+} RCH(OH)CH_3$

   $+ H_2SO_4 \rightarrow RCH(OSO_3H)\ CH_3$

   $+ BH_3, H_2O_2 + NaOH \rightarrow RCH_2CH_2OH$

   + dil. cold $KMnO_4 \rightarrow RCH(OH)CH_2OH$

   + hot $KMnO_4 \rightarrow RCOOH + CO_2$

   $+ R'CO_3H \xrightarrow{-H^+}$ RCH—CH₂ (epoxide, bridged by O)

   $+ RCO_3H, H_3O^+ \rightarrow RCH(OH)CH_2OH$

   $+ HF, HCMe_3 \xrightarrow{0°C} RCH_2CH_2CMe_3$

   $+ O_3, Zn, H_2O \rightarrow RCH{=}O + CH_2{=}O$

   (*c*) **Free-Radical Mechanism**

   $+ HBr \rightarrow RCH_2CH_2Br$

   $+ CHCl_3 \rightarrow RCH_2CH_2CCl_3$

   $+ H^+$ or $BF_3 \rightarrow$ polymer

2. **Allylic Substitution Reactions**

   $$R{-}CH_2{-}CH{=}CH_2 + X_2 \xrightarrow[\text{or uv}]{\Delta} R{-}CHX{-}CH{=}CH_2$$

SUPPLEMENTARY PROBLEMS

Problem 6.37 Write structures for the following:

(*a*) 2,3-dimethyl-2-pentene
(*b*) 4-chloro-2,4-dimethyl-2-pentene
(*c*) allyl bromide

(*d*) 2,3-dimethylcyclohexene
(*e*) 3-isopropyl-1-hexene
(*f*) 3-isopropyl-2,6-dimethyl-3-heptene

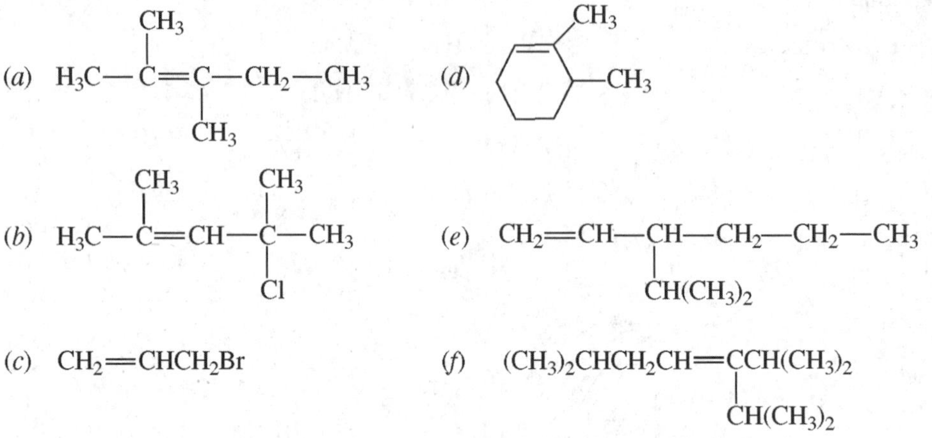

(*a*) $H_3C—C(CH_3)=C(CH_3)—CH_2—CH_3$

(*d*) CH_3 CH_3

(*b*) $H_3C—C(CH_3)=CH—C(CH_3)(Cl)—CH_3$

(*e*) $CH_2=CH—CH(CH(CH_3)_2)—CH_2—CH_2—CH_3$

(*c*) $CH_2=CHCH_2Br$

(*f*) $(CH_3)_2CHCH_2CH=C(CH(CH_3)_2)CH(CH_3)_2$

Problem 6.38 (*a*) Give structural formulas and systematic names for all alkenes with the molecular formula C_6H_{12} that exist as stereomers. (*b*) Which stereomer has the lowest heat of combustion (is the most stable), and which is the least stable?

(*a*) For a molecule to possess stereomers, it must be chiral, exhibit geometric isomerism, or both. There are five such constitutional isomers of C_6H_{12}. Since double-bonded C's cannot be chiral, one of the remaining four C's must be chiral. This means that a *sec*-butyl group must be attached to the C=C group, and the enantiomers are as follows:

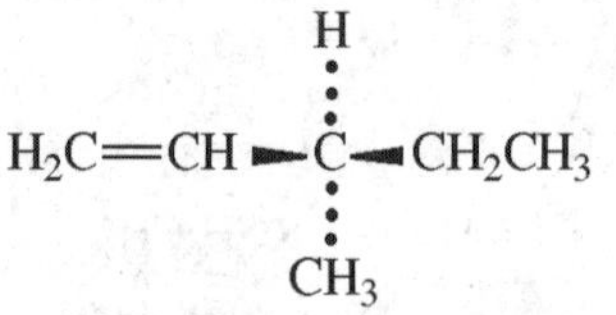

$H_2C=CH$ ▶ C(H)(CH₃) ◀ CH_2CH_3
(*R*)-3-Methyl-1-pentene

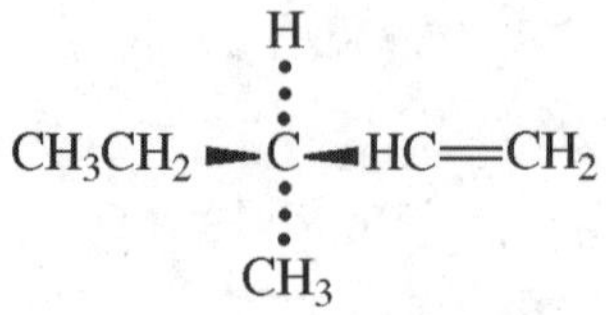

CH_3CH_2 ▶ C(H)(CH₃) ◀ $HC=CH_2$
(*S*)-3-Methyl-1-pentene

A molecule with a terminal $=CH_2$ cannot have geometric isomers. The double bond must be internal and is first placed between C^2 and C^3 to give three more sets of stereomers:

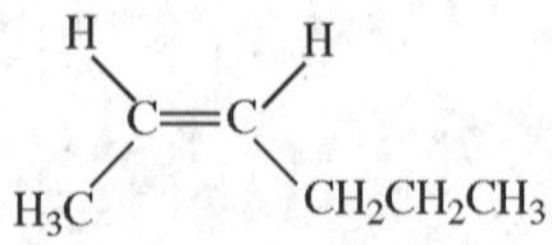

(Z)-2-Hexene: H, H_3C on one C; H, $CH_2CH_2CH_3$ on the other (H's cis)

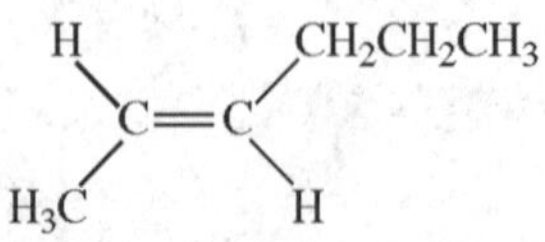

(*E*)-2-Hexene: H, H_3C on one C; $CH_2CH_2CH_3$, H on the other

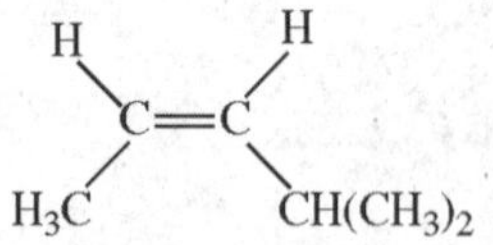

(Z)-4-Methyl-2-pentene: H, H_3C on one C; H, $CH(CH_3)_2$ on the other (H's cis)

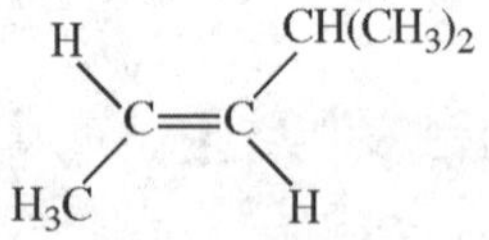

(*E*)-4-Methyl-2-pentene: H, H_3C on one C; $CH(CH_3)_2$, H on the other

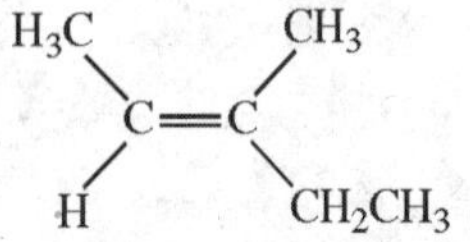

(*E*)-3-Methyl-2-pentene: H_3C, H on one C; CH_3, CH_2CH_3 on the other (H_3C cis to CH_3)

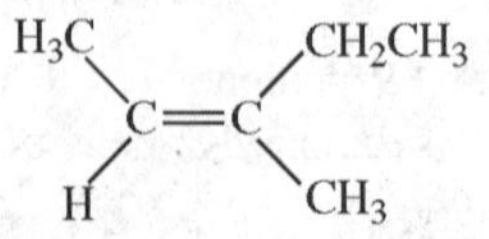

(Z)-3-Methyl-2-pentene: H_3C, H on one C; CH_2CH_3, CH_3 on the other (H_3C cis to CH_2CH_3)

The double bond is now placed between C^3 and C^4 to give the isomers:

$$\underset{\text{(Z)-3-Hexene}}{\begin{matrix} CH_3CH_2 & & CH_2CH_3 \\ & C{=}C & \\ H & & H \end{matrix}} \qquad \underset{\text{(E)-3-Hexene}}{\begin{matrix} CH_3CH_2 & & H \\ & C{=}C & \\ H & & CH_2CH_3 \end{matrix}}$$

(*Z*)-3-Hexene (*E*)-3-Hexene

(*b*) The 3-methyl-2-pentenes each have three R groups on the C=C group and are more stable than one or the other isomers, which have only two R's on C=C. The *E* isomer is the more stable because it has the smaller CH_3's *cis* to each other, whereas the *Z* isomer has the larger CH_2CH_3 group *cis* to a CH_3 group. 3-Methyl-1-pentene is the least stable, because it has the fewest R groups, only one, on the C=C group. Among the remaining isomers, the *trans* are more stable than the *cis* because the R groups are on opposite sides of the C=C and are thus more separated.

Problem 6.39 Write structural formulas for the organic compounds designated by a ?, and show the stereochemistry where requested.

(*a*) $CH_3{-}\overset{\displaystyle CH_3}{\underset{\displaystyle Br}{C}}{-}CH_2CH_2CH_3$ + alcoholic KOH ⟶ ? (major) + ? (minor)

(*b*) $\begin{matrix} H & & H \\ & C{=}C & \\ CH_3 & & CH_3 \end{matrix}$ + Br_2 ⟶ ? + ? (Stereochemistry)

(*c*) ? (alkene) + ? (reagent) ⟶ $CH_3{-}\overset{\displaystyle CH_3}{\underset{\displaystyle H}{C}}{-}CH_2{-}\overset{\displaystyle CH_3}{\underset{\displaystyle OH}{C}}{-}CH_3$

(*d*) $CH_3CH_2CH_2CH_2CH{=}CH_2 + CHBr_3$ + peroxide ⟶ ?

(*e*) $CH_2\begin{matrix} CH_2CH_2 \\ \\ CH{=}C \end{matrix}CH_2$ (ring), with CH_3 on the C= ; + HOBr ⟶ ? (Stereochemistry)

(*f*) $CH_3CH_2\overset{\displaystyle CH_3}{C}HCH_2OH + H_2SO_4$, heat ⟶ ? (major) + ? (minor)

(*a*) $\boxed{CH_3}\overset{\displaystyle \boxed{CH_3}}{C}{=}CH\boxed{CH_2CH_3}$ $\qquad H_2C{=}\overset{\displaystyle \boxed{CH_3}}{C}{-}\boxed{CH_2CH_2CH_3}$

major: has 3 R's on the C=C. minor: has only 2 R's.

(*b*) *Anti* addition to a *cis* diastereomer gives a racemic mixture:

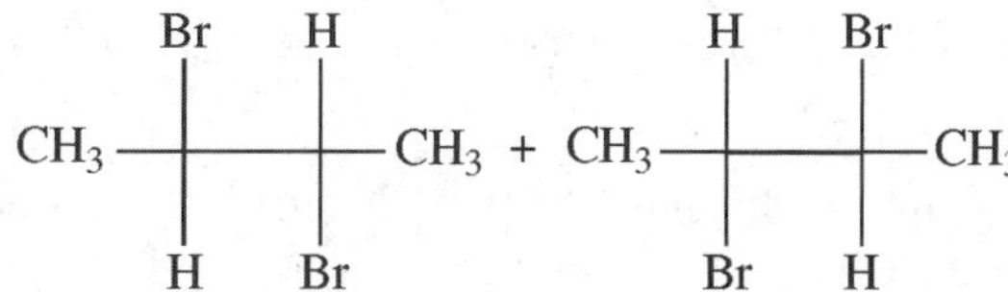

(*c*) The 3° alcohol is formed by acid-catalyzed hydration of

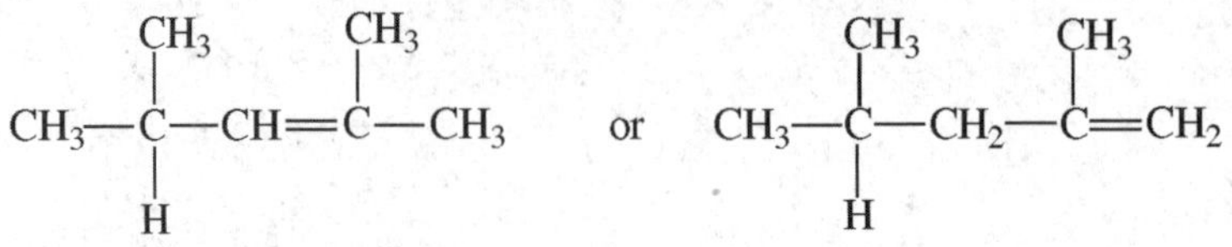
CH3
CH3
CH3—C—CH=C—CH3
H
or
CH3
CH3
CH3—C—CH2—C=CH2
H

The reagent is dilute aq. H_2SO_4.

(*d*) $CH_3CH_2CH_2CH_2CH_2CH_2CBr_3$.

(*e*) In this polar Markovnikov addition, the positive Br adds to the C having the H. The addition is *anti*, so the Br will be *trans* to OH but *cis* to CH_3. The product is racemic:

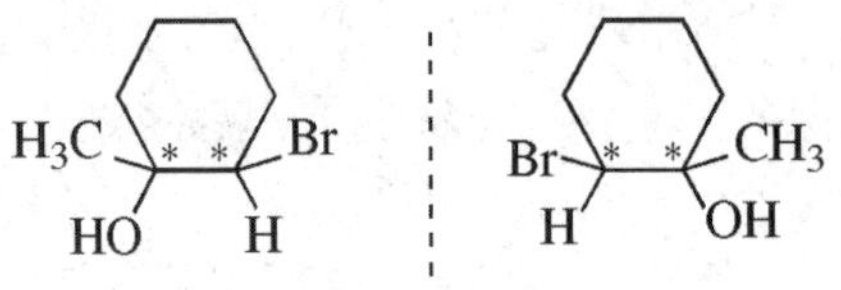
H3C
Br
HO
H
Br
CH3
H
OH

(*f*) The formation of products is shown:

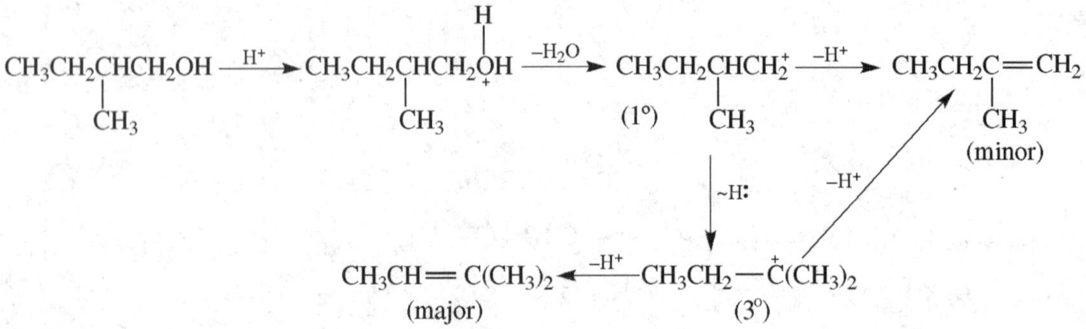
CH3CH2CHCH2OH
CH3
H+
CH3CH2CHCH2OH
H
CH3
–H2O
CH3CH2CHCH2+
(1°)
CH3
–H+
CH3CH2C=CH2
CH3
(minor)
~H:
–H+
CH3CH=C(CH3)2
(major)
–H+
CH3CH2—C(CH3)2
(3°)

Problem 6.40 Draw an enthalpy-reaction progress diagram for addition of Br_2 to an alkene.

See Fig. 6.8.

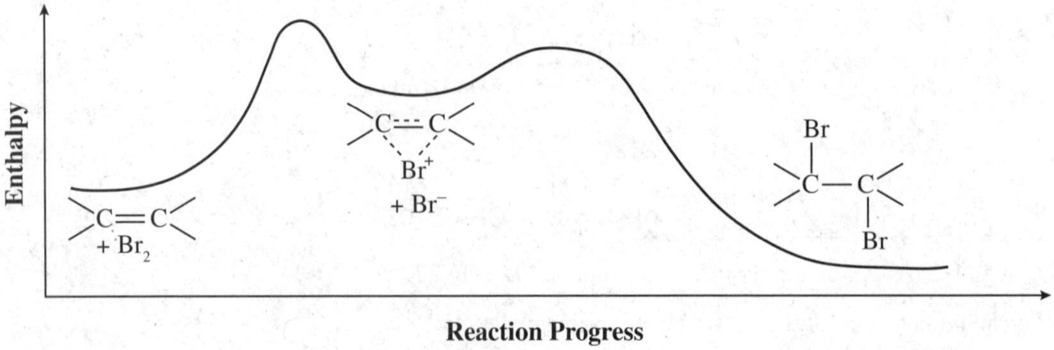
Enthalpy
C=C
+ Br2
C---C
Br+
+ Br−
Br
C—C
Br
Reaction Progress

Figure 6.8

Problem 6.41 Write the initiation and the propagation steps for a free-radical-catalyzed (RO·) addition of $CH_3CH{=}O$ to 1-hexene to form methyl *n*-hexyl ketone:

$$CH_3CH_2CH_2CH_2CH_2CH_2-\overset{\overset{O}{\|}}{C}-CH_3$$

The initiation step is

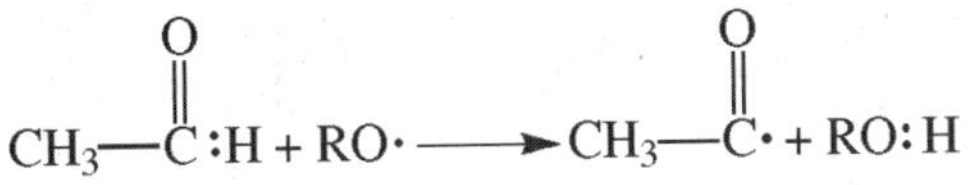

$$CH_3-\overset{\overset{O}{\|}}{C}:H + RO\cdot \longrightarrow CH_3-\overset{\overset{O}{\|}}{C}\cdot + RO:H$$

and the propagation steps are

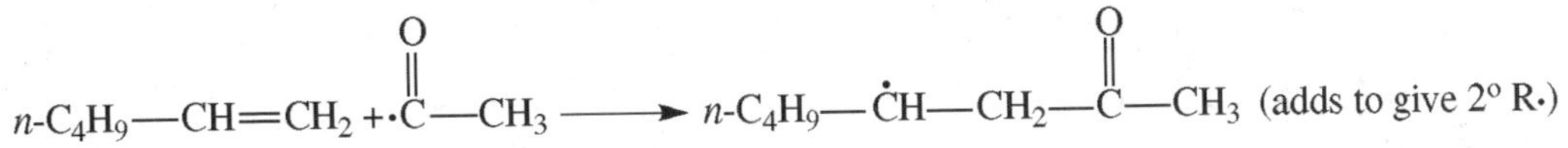

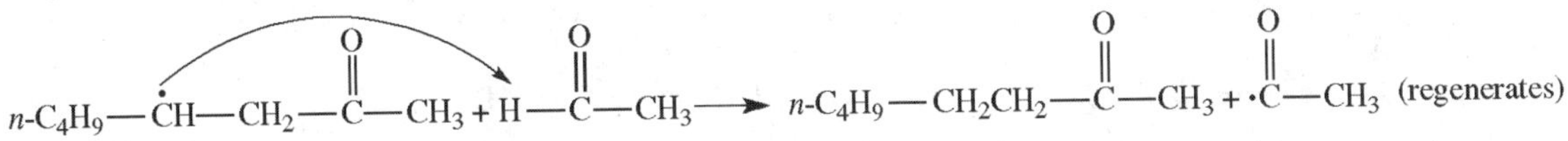

$$n\text{-}C_4H_9-CH=CH_2 + \cdot\overset{\overset{O}{\|}}{C}-CH_3 \longrightarrow n\text{-}C_4H_9-\dot{C}H-CH_2-\overset{\overset{O}{\|}}{C}-CH_3 \text{ (adds to give 2° R}\cdot\text{)}$$

$$n\text{-}C_4H_9-\dot{C}H-CH_2-\overset{\overset{O}{\|}}{C}-CH_3 + H-\overset{\overset{O}{\|}}{C}-CH_3 \longrightarrow n\text{-}C_4H_9-CH_2CH_2-\overset{\overset{O}{\|}}{C}-CH_3 + \cdot\overset{\overset{O}{\|}}{C}-CH_3 \text{ (regenerates)}$$

Problem 6.42 Suggest a radical mechanism to account for the interconversion of *cis* and *trans* isomers by heating with I_2.

I_2 has a low bond dissociation energy (151 kJ/mol) and forms 2I· on heating. I· adds to the C=C to form a carbon radical which rotates about its sigma bond and assumes a different conformation. However, the C—I bond is also weak (235 kJ/mol), and the radical loses I· under these conditions. The double bond is reformed, and the two conformations produce a mixture of *cis* and *trans* isomers.

$$\underset{trans}{RCH=CHR} \underset{-I\cdot}{\overset{+I\cdot}{\rightleftharpoons}} R\overset{I}{C}H-\dot{C}HR \rightleftharpoons R\overset{I}{C}H-\dot{C}HR \underset{+I\cdot}{\overset{-I\cdot}{\rightleftharpoons}} \underset{cis}{RCH=CHR}$$

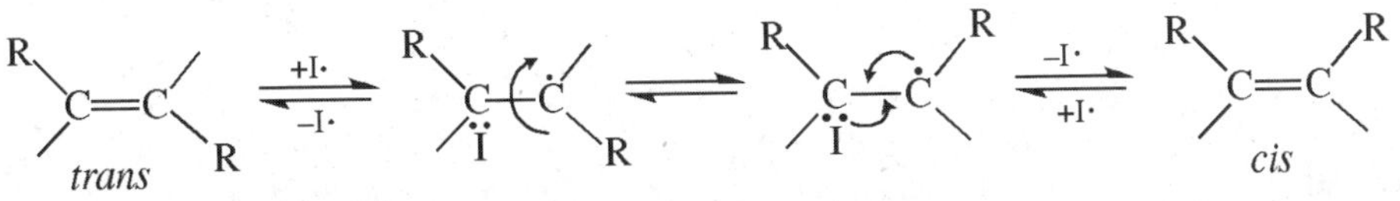

Problem 6.43 Write structures for the products of the following polar addition reactions:

(*a*) $(CH_3)_2C=CHCH_3 + I-Cl \longrightarrow ?$ (*b*) $(CH_3)_2C=CH_2 + HSCH_3 \longrightarrow ?$

(*c*) $(CH_3)_3\overset{+}{N}-CH=CH_2 + HI \longrightarrow ?$ (*d*) $H_2C=CHCF_3 + HCl \longrightarrow ?$

(*a*) $(CH_3)_2C\text{(Cl)}CH\boxed{I}CH_3$. I is less electronegative than Cl in $\overset{\delta+}{I}-\overset{\delta-}{Cl}$ and adds to the C with more H's.

(*b*)

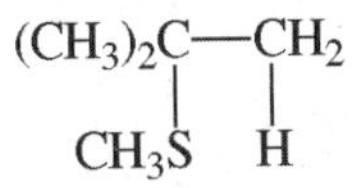

$$\underset{CH_3S}{\overset{|}{(CH_3)_2C}}-\underset{H}{\overset{|}{CH_2}}$$

H is less electronegative than S; $\overset{\delta+}{H}-\overset{\delta-}{S}CH_3$.

(*c*) $(CH_3)_3\overset{+}{N}-CH_2-CH_2-I$

The + charge on N destabilizes an adjacent + charge. $(CH_3)_3\overset{+}{N}CH_2\overset{+}{C}H_2$ is more stable than $(CH_3)_3\overset{+}{N}\overset{+}{C}HCH_3$. Addition is anti-Markovnikov.

(*d*) $ClCH_2CH_2CF_3$. The strong electron-attracting CF_3 group destabilizes an adjacent + charge so that $\overset{+}{C}H_2CH_2CF_3$ is the intermediate rather than $CH_3\overset{+}{C}HCF_3$.

Problem 6.44 Explain the following observations: (*a*) Br_2 and propene in C_2H_5OH gives not only $BrCH_2CHBrCH_3$ but also $BrCH_2CH(OC_2H_5)CH_3$. (*b*) Isobutylene is more reactive than 1-butene toward peroxide-catalyzed addition of CCl_4. (*c*) The presence of Ag^+ salts enhance the solubility of alkenes in H_2O.

(*a*) The intermediate bromonium ion reacts with both Br^- and $C_2H_5\ddot{O}H$ as nucleophiles to give the two products. (*b*) The more stable the intermediate free radical, the more reactive the alkene. $H_2C{=}CHCH_2CH_3$ adds $\cdot CCl_3$ to give the less stable 2° radical $Cl_3CCH_2\dot{C}HCH_2CH_3$, whereas $H_2C{=}C(CH_3)_2$ reacts to give the more stable 3° radical $Cl_3CCH_2\dot{C}(CH_3)_2$. (*c*) Ag^+ coordinates with the alkene by *p-d* π bonding to give an ion similar to bromonium ion, but more stable:

$$\overset{+}{Ag} \quad \text{(bridging)} \quad \diagdown C{\overset{\cdots}{=}}C \diagup$$

Problem 6.45 Supply the structural formulas of the alkenes and the reagents which react to form: (*a*) $(CH_3)_3CI$, (*b*) CH_3CHBr_2, (*c*) $BrCH_2CHClCH_3$, (*d*) $BrCH_2CHOHCH_2Cl$.

(*a*) $CH_3{-}\overset{\overset{\displaystyle CH_3}{|}}{C}{=}CH_2 + HI$

(*b*) $H_2C{=}CHBr + HBr$

(*c*) $H_2C{=}CCl{-}CH_3 + HBr + \text{peroxide}$
or $H_2C{=}CHCH_3 + BrCl$

(*d*) $BrCH_2{-}CH{=}CH_2 + HOCl$
or $H_2C{=}CH{-}CH_2Cl + HOBr$

Problem 6.46 Outline the steps needed for the following syntheses in reasonable yield. Inorganic reagents and solvents may also be used. (*a*) 1-Chloropentane to 1,2-dichloropentane. (*b*) 1-Chloropentane to 2-chloropentane. (*c*) 1-Chloropentane to 1-bromopentane. (*d*) 1-Bromobutane to 1,2-dihydroxybutane. (*e*) Isobutyl chloride to

$$CH_3{-}\underset{\underset{\displaystyle CH_3}{|}}{\overset{\overset{\displaystyle CH_3}{|}}{C}}{-}CH_2{-}\underset{\underset{\displaystyle I}{|}}{\overset{\overset{\displaystyle CH_3}{|}}{C}}{-}CH_3$$

Syntheses are best done by working backwards, keeping in mind your starting material.

(*a*) The desired product is a *vic*-dichloride made by adding Cl_2 to the appropriate alkene, which in turn is made by dehydrochlorinating the starting material.

$$ClCH_2CH_2CH_2CH_2CH_3 \xrightarrow[\text{KOH}]{\text{alc.}} H_2C{=}CHCH_2CH_2CH_3 \xrightarrow{Cl_2} ClCH_2CHClCH_2CH_2CH_3$$

(*b*) To get a pure product, add HCl to 1-pentene as made in part (*a*).

$$H_2C{=}CHCH_2CH_2CH_3 + HCl \longrightarrow H_3CCHClCH_2CH_2CH_3$$

(*c*) An anti-Markovnikov addition of HBr to 1-pentene [part (*a*)].

$$H_2C{=}CHCH_2CH_2CH_3 + HBr \xrightarrow{\text{peroxide}} BrCH_2CH_2CH_2CH_2CH_3$$

(*d*) Glycols are made by mild oxidation of alkenes.

$$BrCH_2CH_2CH_2CH_3 \xrightarrow[\text{KOH}]{\text{alc.}} H_2C{=}CHCH_2CH_3 \xrightarrow[\text{RT}]{KMnO_4} HOCH_2CHOHCH_2CH_3$$

(*e*) The product has twice as many C's as does the starting material. The skeleton of C's in the product corresponds to that of the dimer of $(CH_3)_2C{=}CH_2$.

$$(CH_3)_2CHCH_2Cl \xrightarrow[\text{KOH}]{\text{alc.}} (CH_3)_2C{=}CH_2 \xrightarrow{H_2SO_4} (CH_3)_3CCH{=}C(CH_3)_2 + (CH_3)_3CCH_2\overset{\overset{\large CH_3}{|}}{C}{=}CH_2 \xrightarrow{HI} (CH_3)_3CCH_2CI(CH_3)_2$$

Problem 6.47 Show how propene can be converted to (*a*) 1,5-hexadiene, (*b*) 1-bromopropene, (*c*) 4-methyl-1-pentene.

(*a*) $CH_3CH{=}CH_2 \xrightarrow{Cl_2,\ 500\ ^\circ C} ClCH_2CH{=}CH_2 \xrightarrow[3.\ ClCH_2CH=CH_2]{1.\ Li.\ 2.\ CuI} H_2C{=}CHCH_2CH_2CH{=}CH_2$

(*b*) $CH_3CH{=}CH_2 \xrightarrow{Br_2(CCl_4)} CH_3CHBrCH_2Br \xrightarrow[\text{KOH}]{\text{alc.}} CH_3CH{=}CHBr$

(Little $CH_3CBr{=}CH_2$ is formed because the 2° H of—CH_2Br is more acidic than the 3° H of—$\overset{|}{C}HBr$.)

(*a*) $CH_3CH{=}CH_2 \xrightarrow{HBr} CH_3{-}\underset{\underset{\large Br}{|}}{CH}{-}CH_3 \xrightarrow{Mg} CH_3{-}\underset{\underset{\large MgBr}{|}}{CH}{-}CH_3$

$CH_3CH{=}CH_2 \xrightarrow{Cl_2,\ 500\ ^\circ C} ClCH_2CH{=}CH_2$

$CH_3{-}\underset{\underset{\large MgBr}{|}}{CH}{-}CH_3 + ClCH_2CH{=}CH_2 \xrightarrow{CuBr} CH_3{-}\overset{\overset{\large CH_3}{|}}{CH}{-}CH_2CH{=}CH_2$

Problem 6.48 (*a*) Br_2 is added to (S)—$\overset{1}{H_2C}{=}\overset{2}{C}H\overset{3}{C}HBr\overset{4}{C}H_3$. Give Fischer projections and *R*, *S* designations for the products. Are the products optically active? (*b*) Repeat (*a*) with HBr.

(*a*) C^2 becomes chiral and the configuration of C^3 is unchanged. There are two optically active diastereomers of 1,2,3-tribromobutane. It is best to draw formulas with H's on vertical lines.

$$\underset{(S)}{H_2\overset{1}{C}{=}\overset{2}{C}H{-}\overset{3}{C}(Br)(H){-}\overset{4}{C}H_3} \longrightarrow \underset{\substack{(2R,\,3S)\\ \textit{optically active}}}{Br\overset{1}{C}H_2{-}\overset{2}{C}(Br)(H){-}\overset{3}{C}(Br)(H){-}\overset{4}{C}H_3} + \underset{\substack{(2S,\,3S)\\ \textit{optically active}}}{Br\overset{1}{C}H_2{-}\overset{2}{C}(H)(Br){-}\overset{3}{C}(Br)(H){-}\overset{4}{C}H_3}$$

(Fischer projections: in each, the top/bottom groups are — starting material: C3 Br top, H bottom; first product: C2 Br/H, C3 Br/H; second product: C2 H/Br, C3 Br/H.)

(*b*) There are two diastereomers of 2,3-dibromobutane:

$$\underset{(S)}{H_2\overset{1}{C}{=}\overset{2}{C}H{-}\overset{3}{C}(Br)(H){-}\overset{4}{C}H_3} \longrightarrow \underset{\substack{(2R,\,3S)\text{-}meso\\ \textit{optically active}}}{H_3C{-}C(Br)(H){-}C(Br)(H){-}CH_3} + \underset{\substack{(2R,\,3S)\\ \textit{optically inactive}}}{H_3C{-}C(H)(Br){-}C(Br)(H){-}CH_3}$$

(Fischer projections: first product: C2 Br top/H bottom, C3 Br top/H bottom; second product: C2 H top/Br bottom, C3 Br top/H bottom.)

Problem 6.49 Polypropylene can be synthesized by the acid-catalyzed polymerization of propylene. (*a*) Show the first three steps. (*b*) Indicate the repeating unit (mer).

(*a*) $CH_3CH{=}CH_2 \xrightarrow{H^+} (CH_3)_2\overset{+}{C}H \xrightarrow{H_2C=CHCH_3} (CH_3)_2CH{-}CH_2{-}\overset{+}{C}H{-}CH_3 \xrightarrow{H_2C=CHCH_3} (CH_3)_2CH{-}CH_2{-}CH(CH_3){-}CH_2{-}\overset{+}{C}H{-}CH_3$

monomer — monomeric carbocation — dimeric carbocation — trimeric carbocation

(*b*) $—CH_2—CH(CH_3)—$

Problem 6.50 List the five kinds of reactions of carbocations, and give an example of each.

(*a*) They combine with nucleophiles.

$$(CH_3)_3C^+ + :\ddot{O}H_2 \xrightarrow{-H^+} (CH_3)_3C{-}OH$$

(*b*) As strong acids, they lose a vicinal H [deprotonate], to give an alkene.

$$(CH_3)_3C^+ \xrightarrow{-H^+} CH_3{-}C(CH_3){=}CH_2$$

(*c*) They rearrange to give a more stable carbocation.

$$\underset{1^\circ}{CH_3{-}C(CH_3)(H){-}\overset{+}{C}H_2} \longrightarrow \underset{3^\circ}{CH_3{-}\overset{+}{C}(CH_3){-}CH_3}$$

(*d*) They add to an alkene to give a carbocation with higher molecular weight.

$$\underset{\text{electrophile}}{(CH_3)_3C^+} + \underset{\text{nucleophile}}{H_2C{=}C(CH_3){-}CH_3} \longrightarrow CH_3{-}C(CH_3)_2{-}CH_2{-}\overset{+}{C}(CH_3){-}CH_3$$

(*e*) They may remove :H (a hybride transfer) from a tertiary position in an alkane.

$$(CH_3)_3C(:H) + \overset{+}{C}H_2CH_3 \longrightarrow CH_3{-}CH_3 + (CH_3)_3C^+$$

Problem 6.51 From propene, prepare (*a*) CH_3CHDCH_3, (*b*) $CH_3CH_2CH_2D$.

(*a*) $CH_3CH{=}CH_2 + HCl \longrightarrow CH_3CHClCH_3 \xrightarrow{Mg} CH_3CHMgClCH_3 \xrightarrow{D_2O} CH_3CHDCH_3$

or propene + $B_2D_6 \longrightarrow (CH_3CHDCH_2)_3B \xrightarrow{CH_3COOH}$ product

(*b*) $CH_3CH{=}CH_2 \xrightarrow{B_2H_6} (CH_3CH_2CH_2)_3B \xrightarrow{CH_3COOD} CH_3CH_2CH_2D$

Problem 6.52 Give four simple chemical tests to distinguish an alkene from an alkane.

A positive simple chemical test is indicated by one or more detectable events, such as a change in color, formation of a precipitate, evolution of a gas, uptake of a gas, and evolution of heat.

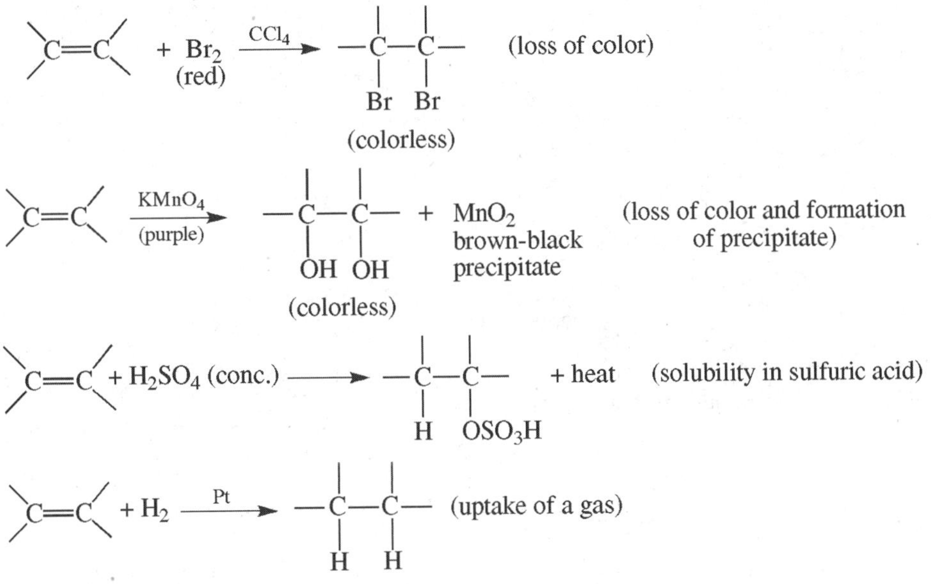

$$\text{>C=C<} + \underset{\text{(red)}}{Br_2} \xrightarrow{CCl_4} \underset{\text{(colorless)}}{-\overset{|}{\underset{Br}{\overset{}{C}}}-\overset{|}{\underset{Br}{C}}-} \quad \text{(loss of color)}$$

$$\text{>C=C<} \xrightarrow[\text{(purple)}]{KMnO_4} \underset{\text{(colorless)}}{-\underset{OH}{\overset{|}{C}}-\underset{OH}{\overset{|}{C}}-} + \underset{\text{brown-black precipitate}}{MnO_2} \quad \text{(loss of color and formation of precipitate)}$$

$$\text{>C=C<} + H_2SO_4\,\text{(conc.)} \longrightarrow -\underset{H}{\overset{|}{C}}-\underset{OSO_3H}{\overset{|}{C}}- \quad + \text{heat} \quad \text{(solubility in sulfuric acid)}$$

$$\text{>C=C<} + H_2 \xrightarrow{Pt} -\underset{H}{\overset{|}{C}}-\underset{H}{\overset{|}{C}}- \quad \text{(uptake of a gas)}$$

Alkanes give none of these tests.

Problem 6.53 Give the configuration, stereochemical designation, and *R,S* specification for the indicated tetrahydroxy products.

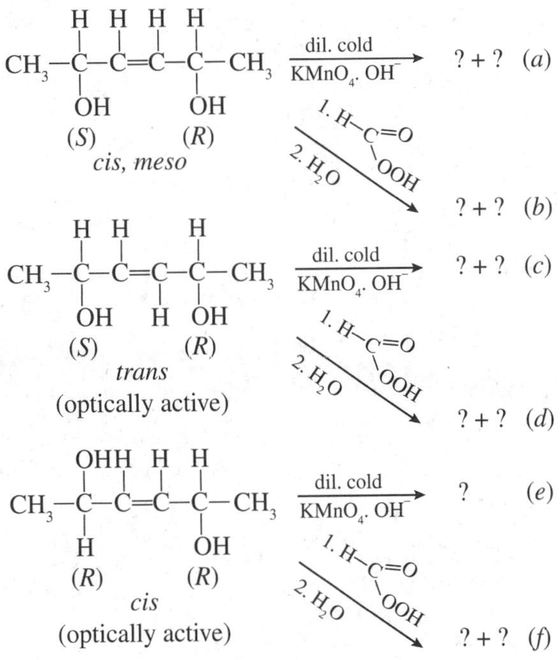

$CH_3-CH(OH)-CH{=}CH-CH(OH)-CH_3$ (H, H on the double-bond carbons on the same side); (*S*) (*R*); *cis, meso*

$\xrightarrow{\text{dil. cold } KMnO_4,\ OH^-}$? + ? (*a*)

$\xrightarrow{\text{1. } H-C(=O)-OOH \quad \text{2. } H_2O}$? + ? (*b*)

$CH_3-CH(OH)-CH{=}CH-CH(OH)-CH_3$ (H on the double-bond carbons on opposite sides); (*S*) (*R*); *trans* (optically active)

$\xrightarrow{\text{dil. cold } KMnO_4,\ OH^-}$? + ? (*c*)

$\xrightarrow{\text{1. } H-C(=O)-OOH \quad \text{2. } H_2O}$? + ? (*d*)

$CH_3-CH(OH)-CH{=}CH-CH(OH)-CH_3$ (OH up, H down at C2; H, H on the double-bond carbons on the same side); (*R*) (*R*); *cis* (optically active)

$\xrightarrow{\text{dil. cold } KMnO_4,\ OH^-}$? (*e*)

$\xrightarrow{\text{1. } H-C(=O)-OOH \quad \text{2. } H_2O}$? + ? (*f*)

(*a*) *syn* Addition of encircled OH's:

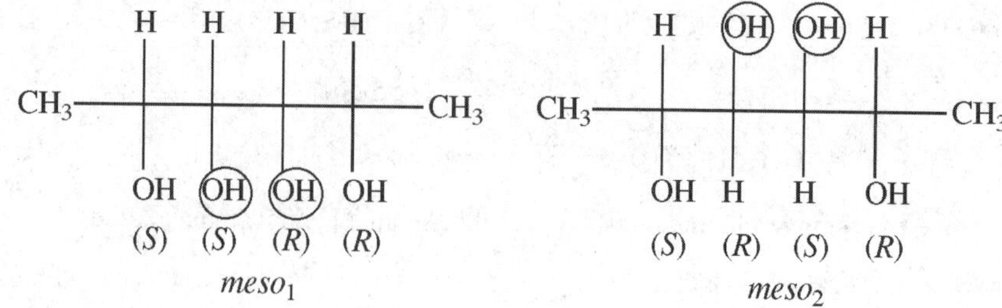

(*b*) *anti* Addition of encircled OH's:

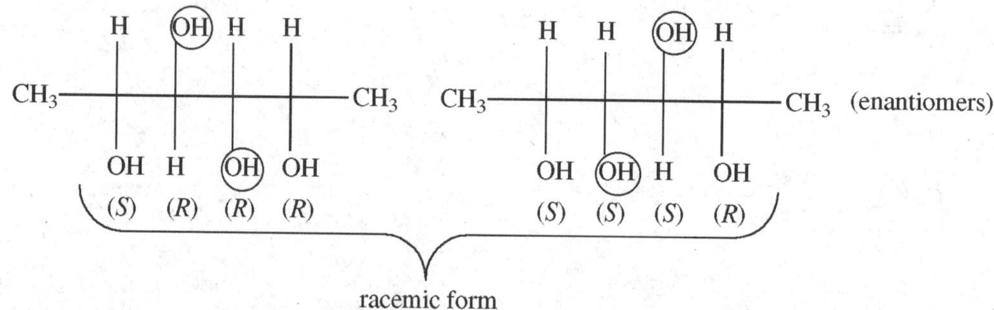

(*c*) *syn* Addition; same products as part (*b*).
(*d*) *syn* Addition; same products as part (*a*).
(*e*) *syn* Addition:

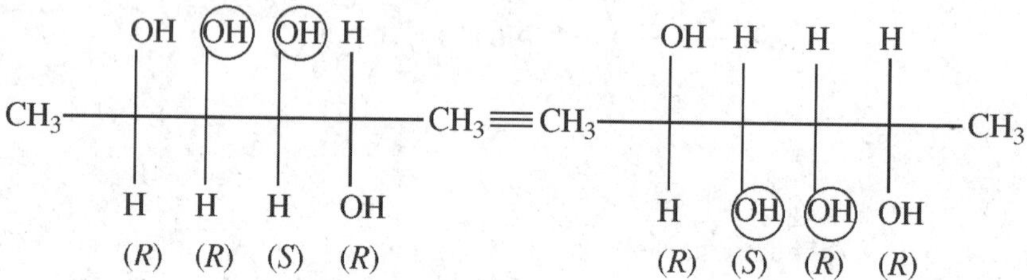

One optically active stereoisomer is formed.
(*f*) *anti* Addition:

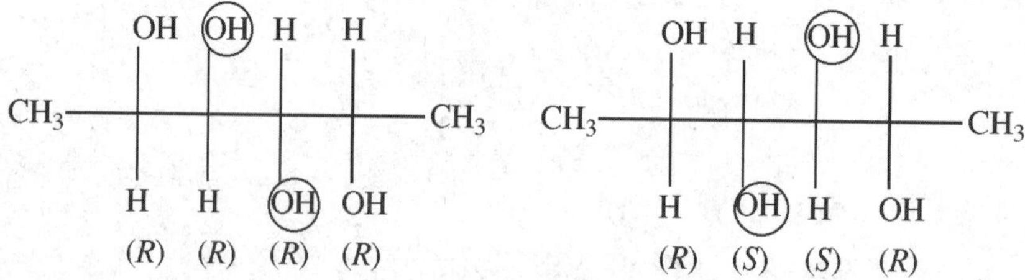

Two optically active diasteriomers.

Problem 6.54 Describe (*a*) **radical-induced** and (*b*) **anion-induced** polymerization of alkenes. (*c*) What kind of alkenes undergo anion-induced polymerization?

(*a*) See Problem 6.32 for formation of a free-radical initiator, RO·, which adds according to the Markovnikov rule.

$$RO\cdot + H_2C{=}CHCH_3 \longrightarrow \underset{RO}{H_2C}{-}\overset{\cdot}{\underset{H\ CH_3}{C}} \xrightarrow{H_2C{=}CHCH_3} \underset{RO}{H_2C}{-}\underset{H\ CH_3}{C}{-}CH_2\overset{\cdot}{\underset{H\ CH_3}{C}}$$

The polymerization can terminate when the free-radical terminal C of a long chain forms a bond with the terminal C of another long chain (**combination**), RC· + ·CR′ ⟶ RC—CR′. Termination may also occur when the terminal free-radical C's of two long chains **disproportionate**, in a sort of auto-redox reaction. One C picks off an H from the C of the other chain, to give an alkane at one chain end and an alkene group at the other chain end:

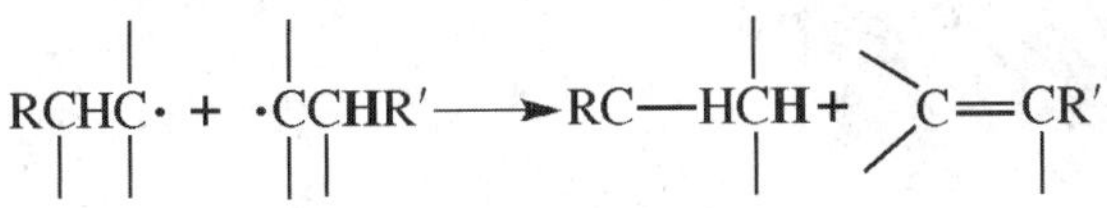

$$RCHC\cdot + \cdot CCHR' \longrightarrow RC{-}HCH + \;\rangle C{=}CR'$$

(*b*) Typical anions are carbanions, R:$^-$, generated from lithium or Grignard organometallics.

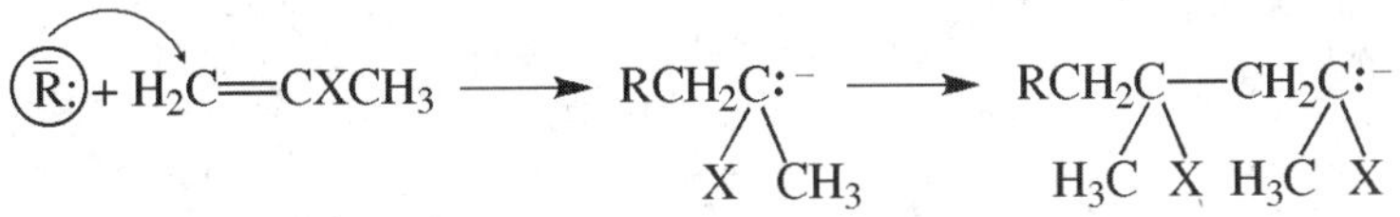

$$R:^- + H_2C{=}CXCH_3 \longrightarrow RCH_2C(X)(CH_3):^- \longrightarrow RCH_2C(H_3C)(X){-}CH_2C(H_3C)(X):^-$$

These types of polymerizations also have stereochemical consequences.

(*c*) Since alkenes do not readily undergo anionic additions, the alkene must have a functional group X (such as —CN or O=COR′) on the C=C that can stabilize the negative charge:

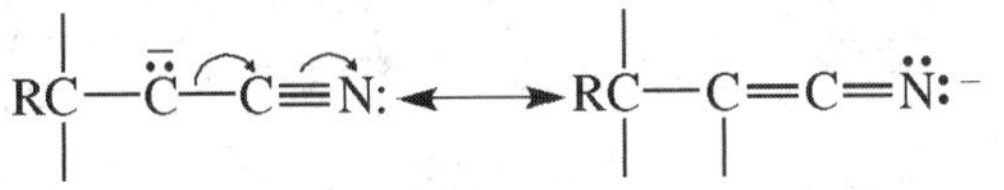

$$RC{-}\ddot{C}{-}C{\equiv}N: \longleftrightarrow RC{-}C{=}C{=}\ddot{N}:^-$$

Problem 6.55 Alkenes undergo **oxidative cleavage** with acidic $KMnO_4$ and, as a result, each C of the C=C ends up in a molecule in its highest oxidation state. Give the products resulting from the oxidative cleavage of (*a*) $H_2C{=}CHCH_2CH_3$, (*b*) (*E*)- or (*Z*)-$CH_3CH{=}CHCH_3$, (*c*) $(CH_3)_2C{=}CHCH_2CH_3$.

(*a*) $H_2C{=}CHCH_2CH_3 \xrightarrow{KMnO_4} CO_2 + HOOCCH_2CH_3$ ($H_2C{=}$ gives CO_2)

(*b*) $CH_3CH{=}CHCH_3 \xrightarrow{KMnO_4} CH_3COOH + HOOCCH_3$ ($RCH{=}$ gives RCOOH)

(*c*) $CH_3{-}C(CH_3){=}CHCH_2CH_2CH_3 \xrightarrow{KMnO_4} CH_3{-}C(CH_3){=}O + HOOCCH_2CH_2CH_3$ ($R_2C{=}$ gives $R_2C{=}O$)

CHAPTER 7

Alkyl Halides

7.1 Introduction

Alkyl halides have the general formula RX, where R is an alkyl or substituted alkyl group and X is any halogen atom (F, Cl, Br, or I).

Problem 7.1 Write structural formulas and IUPAC names for all isomers of: (*a*) $C_5H_{11}Br$, and classify the isomers as to whether they are tertiary (3°), secondary (2°), or primary (1°); and (*b*) $C_4H_8Cl_2$, and classify the isomers that are *gem*-dichlorides and *vic*-dichlorides.

Take each isomer of the parent hydrocarbon and replace one of each type of equivalent H by X. The correct IUPAC name is written to avoid duplication.

(*a*) The parent hydrocarbons are the isomeric pentanes. From pentane, $CH_3CH_2CH_2CH_2CH_3$, we get three monobromo products, shown with their classification.

$BrCH_2CH_2CH_2CH_2CH_3$	$CH_3CHBrCH_2CH_2CH_3$	$CH_3CH_2CHBrCH_2CH_3$
1-Bromopentane (1°)	2-Bromopentane (2°)	3-Bromopentane (2°)

Classification is based on the structural features: RCH_2Br is 1°, R_2CHBr is 2°, and R_3CBr is 3°.

From isopentane, $(CH_3)_2CHCH_2CH_3$, we get four isomers:

$BrCH_2CH(CH_3)CH_2CH_3$	$CH_3CBr(CH_3)CH_2CH_3$	$CH_3CH(CH_3)CHBrCH_3$	$CH_3CH(CH_3)CH_2CH_2Br$
1-Bromo-2-methyl-butane (1°)	2-Bromo-2-methyl-butane (3°)	2-Bromo-3-methyl-butane (2°)	1-Bromo-3-methyl-butane (1°)

$CH_3CH(CH_2Br)CH_2CH_3$ is also 1-bromo-2-methylbutane; the two CH_3's on C^2 are equivalent.

Neopentane has 12 equivalent H's and has only one monobromo substitution product: $(CH_3)_3CCH_2Br$(1°), 1-bromo-2,2-dimethylpropane.

(*b*) For the dichlorobutanes, the two Cl's are first placed on one C of the straight chain. These are geminal or *gem*-dichlorides.

$Cl_2CHCH_2CH_2CH_3$	$CH_3CCl_2CH_2CH_3$
1,1-Dichlorobutane	2,2-Dichlorobutane

Then the Cl's are placed on different C's. The isomers with the Cl's on adjacent C's are vicinal or *vic*-dichlorides.

$ClCH_2CH(Cl)CH_2CH_3$	$ClCH_2CH_2CH(Cl)CH_3$	$ClCH_2CH_2CH_2CH_2Cl$	$CH_3CHClCHClCH_3$
1,2-Dichloro-butane (*vic*)	1,3-Dichloro-butane	1,4-Dichloro-butane	2,3-Dichloro-butane (*vic*)

From isobutane we get

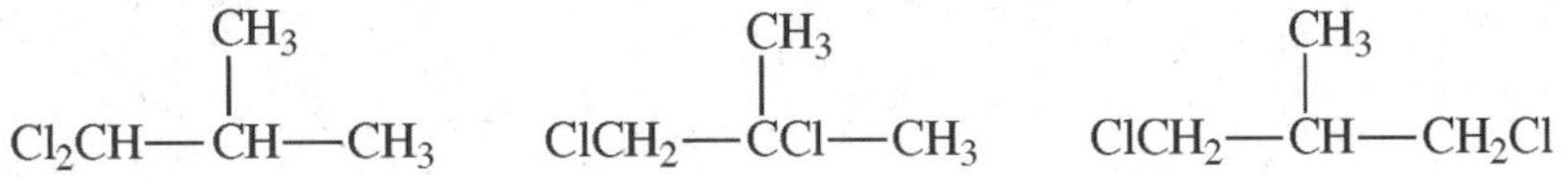

$Cl_2CH—CH(CH_3)—CH_3$	$ClCH_2—CCl(CH_3)—CH_3$	$ClCH_2—CH(CH_3)—CH_2Cl$
1,1-Dichloro-2-methyl-propane (*gem*)	1,2-Dichloro-2-methyl-propane (*vic*)	1,3-Dichloro-2-methyl-propane

Problem 7.2 Compare and account for differences in the (*a*) dipole moment, (*b*) boiling point, (*c*) density, and (*d*) solubility in water of an alkyl halide RX and its parent alkane RH.

(*a*) RX has a larger dipole moment because the C—X bond is polar. (*b*) RX has a higher boiling point, since it has a larger molecular weight and also is more polar. (*c*) RX is more dense, since it has a heavy X atom; the order of decreasing density is RI > RBr > RCl > RF. (*d*) RX, like RH, is insoluble in H_2O, but RX is somewhat more soluble because some H-bonding can occur:

$$\overset{\delta+}{R}—\overset{\delta-}{X}\!:\!\text{---}\overset{\delta+}{H}—\overset{\delta-}{OH}$$

This effect is greatest for RF.

7.2 Synthesis of RX

1. Halogenation of alkanes with Cl_2 or Br_2 (Section 4.4).
2. From alcohols (ROH) with HX or PX_3 (X = I, Br, Cl); $SOCl_2$ (*major method*).
3. Addition of HX to alkenes (Section 6.4).
4. X_2 (X = Br, Cl) + alkenes give *vic*-dihalides (Section 6.4).
5. $RX + X'^- \rightarrow RX' + X^-$ (halogen exchange).

Problem 7.3 Give the products of the following reactions:

(*a*) $CH_3CH_2CH_2OH + HI \longrightarrow$
(*b*) $n\text{-}C_4H_9OH + NaBr + H_2SO_4 \longrightarrow$
(*c*) $CH_3CH_2OH + PI_3(P + I_2) \longrightarrow$
(*d*) $(CH_3)_2CHCH_2OH + SOCl_2 \longrightarrow$
(*e*) $H_2C{=}CH_2 + Br_2 \longrightarrow$
(*f*) $CH_3CH{=}C(CH_3)_2 + HI \longrightarrow$
(*g*) $CH_3CH_2CH_2Br + I^- \longrightarrow$

(*a*) $CH_3CH_2CH_2I + H_2O$
(*b*) $n\text{-}C_4H_9Br + NaHSO_4 + H_2O$
(*c*) $CH_3CH_2I + H_3PHO_3$ (phosphorous acid)
(*d*) $(CH_3)_2CHCH_2Cl + HCl(g) + SO_2(g)$
(*e*) H_2CBrCH_2Br
(*f*) $CH_3CH_2CI(CH_3)_2$
(*g*) $CH_3CH_2CH_2I + Br^-$

Problem 7.4 Which of the following chlorides can be made in good yield by light-catalyzed monochlorination of the corresponding hydrocarbon?

(*a*) CH_3CH_2Cl (*b*) $CH_3CH_2CH_2CH_2Cl$ (*c*) $(CH_3)_3CCH_2Cl$ (*d*) $(CH_3)_3CCl$ (*e*) H, H, H—, —H, H, Cl (*f*) $H_2C{=}CHCH_2Cl$

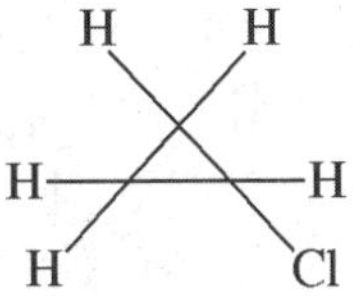

To get good yields, all the reactive H's of the parent hydrocarbon must be equivalent. This is true for

(*a*) CH_3CH_3 (*c*) $(CH_3)_3CCH_3$ (*e*) H_2C—CH_2 (with CH_2 bridging, cyclopropane) (*f*) $H_2C=CHCH_3$

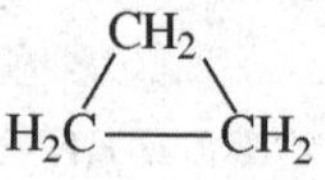

Although the H's of (*f*), propene, are not equivalent, the three allylic H's of CH_3 are much more reactive than the inert vinylic H's on the double-bonded C's. The precursors for (*b*) and (*d*), which are $CH_3CH_2CH_2CH_3$ and $(CH_3)_3CH$, respectively, both have more than one type of equivalent H and would give mixtures.

Problem 7.5 Prepare

(*a*) $CH_3CHBrCH_3$ (*b*) $CH_3CH_2CH_2CH_2I$

(*c*) $CH_3-C(CH_3)(Cl)-CH_3$ (*d*) $ClCH_2C(CH_3)=C(CH_3)-CH_3$ (*e*) $CH_3CH(Cl)CH_2Cl$

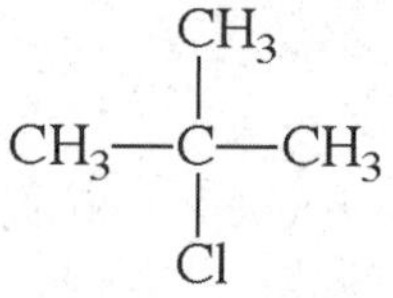

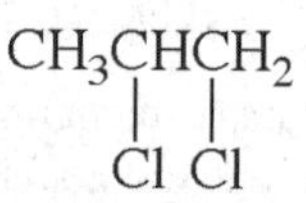

from a hydrocarbon or an alcohol.

(*a*) Two ways:

$$CH_3CH=CH_2 + HBr \longrightarrow \boxed{CH_3CHBrCH_3} \xleftarrow{PBr_3} CH_3CHOHCH_3$$

(*b*) Two ways:

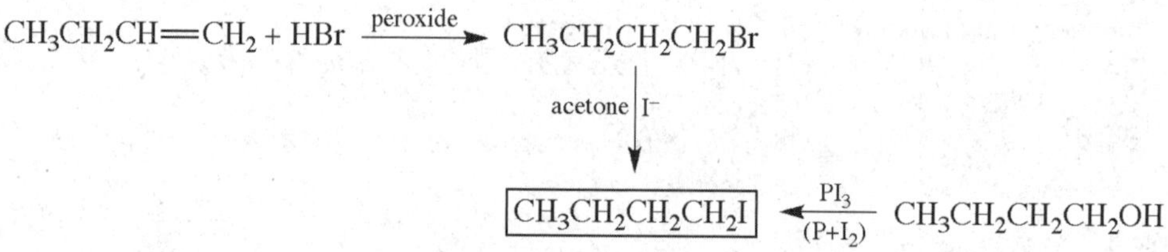

$$CH_3CH_2CH=CH_2 + HBr \xrightarrow{\text{peroxide}} CH_3CH_2CH_2CH_2Br$$

$$CH_3CH_2CH_2CH_2Br \xrightarrow{\text{acetone } I^-} \boxed{CH_3CH_2CH_2CH_2I} \xleftarrow[(P+I_2)]{PI_3} CH_3CH_2CH_2CH_2OH$$

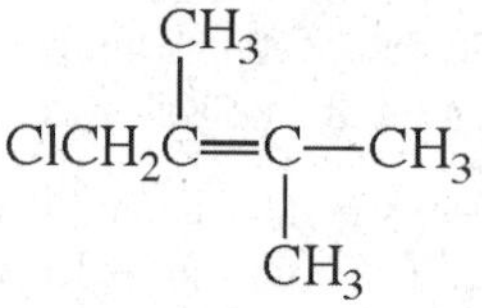

HI does not undergo an anti-Markovnikov radical addition.

(*c*) Two ways:

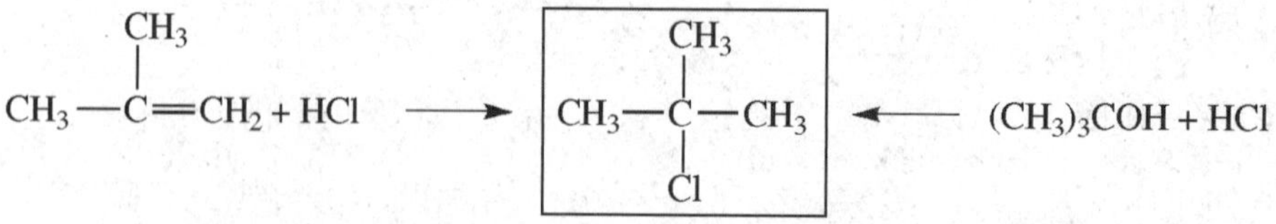

$$CH_3-C(CH_3)=CH_2 + HCl \longrightarrow \boxed{CH_3-C(CH_3)(Cl)-CH_3} \longleftarrow (CH_3)_3COH + HCl$$

(*d*)

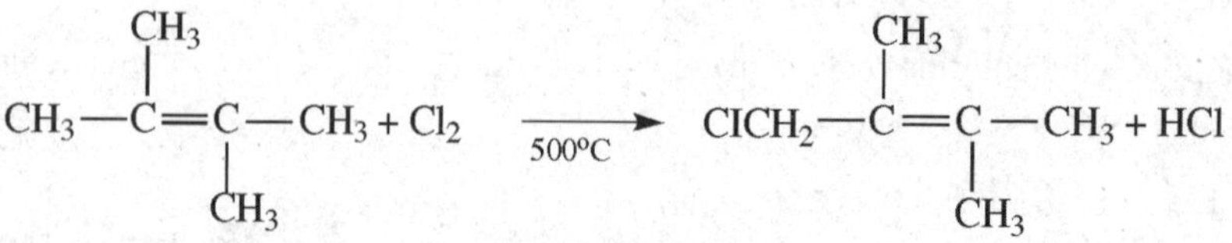

$$CH_3-C(CH_3)=C(CH_3)-CH_3 + Cl_2 \xrightarrow[500^\circ C]{} ClCH_2-C(CH_3)=C(CH_3)-CH_3 + HCl$$

(*e*)

$$CH_3CH=CH_2 + Cl_2 \longrightarrow CH_3CH(Cl)CH_2Cl$$

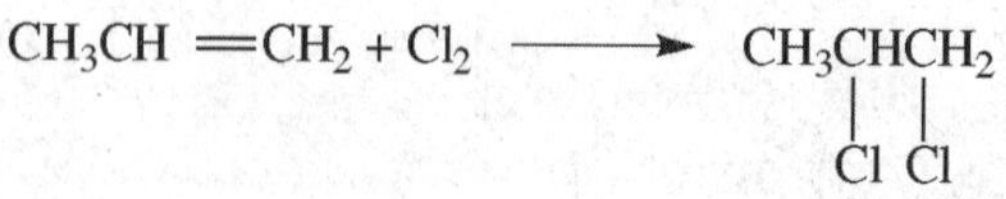

7.3 Chemical Properties

Alkyl halides react mainly by heterolysis of the polar $\overset{\delta+}{C}—\overset{\delta-}{X}$ bond.

Nucleophilic Displacement

$$\left.\begin{array}{c}Nu:^- \\ \text{or} \\ Nu:\end{array}\right\} + \overset{\delta+}{R}\overset{\delta-}{(:X)} \longrightarrow \left\{\begin{array}{c}R:Nu \\ \text{or} \\ [R:Nu]^+\end{array}\right. + :X^-$$

Nucleophile + Substrate ⟶ Product + Leaving Group

The weaker the Brönsted basicity of X^-, the better leaving group is X^- and the more reactive is RX. Since the order of basicities of the halide ions X^- is $I^- < Br^- < Cl^- < F^-$, the order of reactivity of RX is RI > RBr > RCl > RF.

The *equilibrium* of nucleophilic displacements favors the side with the weaker Brönsted base; the stronger Brönsted base displaces the weaker Brönsted base. The *rate* of the displacement reaction on the C of a given substrate depends on the **nucleophilicity** of the attacking base. Basicity and nucleophilicity differ as shown:

basicity $B:^- + H(:X) \longrightarrow B\text{–}H + :X^-$

nucleophilicity $(B:^-) + —\overset{|}{\underset{|}{C}}(:X) \longrightarrow B—\overset{|}{\underset{|}{C}}— + :X^-$

Problem 7.6 What generalizations about the relationship of basicity and nucleophilicity can be made from the following relative rates of nucleophilic displacements:

(*a*) $OH^- \gg H_2O$ and $NH_2^- \gg NH_3$ (*b*) $H_3C:^- > :\ddot{O}H^- > :\ddot{F}:^-$

(*c*) $:\ddot{I}:^- > :\ddot{Br}:^- > :\ddot{Cl}:^- > :\ddot{F}:^-$ (*d*) $CH_3O^- > OH^- > CH_3COO^-$

(*a*) Bases are better nucleophiles than their conjugate acids.
(*b*) In going from left to right in the periodic table, basicity and nucleophilicity are directly related—they both decrease.
(*c*) In going down a group in the periodic table, they are inversely related, in that nucleophilicity increases and basicity decreases.
(*d*) When the nucleophilic and basic sites are the same atom (here an O), nucleophilicity parallels basicity.

The order in Problem 7.6(*c*) may occur because the valence electrons of a larger atom could be more available for bonding with the C, being farther away from the nucleus and less firmly held. Alternatively, the greater ease of distortion of the valence shell (induced polarity) makes easier the approach of the larger atom to the C atom. This property is called **polarizability**. The larger, more polarizable species (e.g., I, Br, S, and P) exhibit enhanced nucleophilicity; they are called **soft bases**. The smaller, more weakly polarizable bases (e.g., N, O, and F) have diminished nucleophilicity; they are called **hard bases**.

Problem 7.7 Explain why the order of reactivity of Problem 7.6(*c*) is observed in nonpolar, weakly polar aprotic, and polar protic solvents, but is reversed in polar aprotic solvents.

In nonpolar and weakly polar aprotic solvents, the salts of $:Nu^-$ are present as ion pairs (or ion clusters) in which the nearby cations diminish the reactivity of the anion. Since, with a given cation, ion pairing is strongest with the smallest ion, F^-, and weakest with the largest ion, I^-, the reactivity of X^- decreases as the size of the anion decreases. In polar protic solvents, hydrogen-bonding, which also lessens the reactivity of X^-, is weakest with the largest ion, again making the largest ion more reactive. Polar aprotic solvents solvate only the cations, leaving free, unencumbered anions. The reactivities of all anions are enhanced, but the effect is more pronounced the smaller the anion. Hence, the order of Problem 7.6(*c*) is reversed.

Problem 7.8 Write equations for the reaction of RCH_2X with

(*a*) $:\ddot{I}:^-$ (*b*) $:\ddot{O}H^-$ (*c*) $:\ddot{O}R'^-$ (*d*) $R':^-$ (*e*) $RC(=O)\ddot{O}:^-$ (*f*) $H_3N:$ (*g*) $:CN^-$

and classify the functional group in each product.

	Reaction	Product
(*a*)	$:\ddot{I}:^- + RCH_2X \longrightarrow RCH_2I + :\ddot{X}:^-$	Iodide
(*b*)	$^-:\ddot{O}H + RCH_2X \longrightarrow RCH_2OH + :\ddot{X}:^-$	Alcohol
(*c*)	$^-:\ddot{O}R' + RCH_2X \longrightarrow RCH_2OR' + :\ddot{X}:^-$	Ether
(*d*)	$^-:R' + RCH_2X \longrightarrow RCH_2R' + :\ddot{X}:^-$	Alkane (coupling)
(*e*)	$^-:\ddot{O}OCR' + RCH_2X \longrightarrow RCH_2OOCR' + :\ddot{X}:^-$	Ester
(*f*)	$:NH_3 + RCH_2X \longrightarrow RCH_2NH_3^+ + :\ddot{X}:^-$	Ammonium salt
(*g*)	$^-:CN + RCH_2X \longrightarrow RCH_2CN + :\ddot{X}:^-$	Nitrile (or Cyanide)

Problem 7.9 Compare the effectiveness of acetate (CH_3COO^-), phenoxide ($C_6H_5O^-$), and benzenesulfonate ($C_6H_5SO_3^-$) anions as leaving groups if the acid strengths of their conjugate acids are given by the pK_a values 4.5, 10.0, and 2.6, respectively.

The best leaving group is the weakest base, $C_6H_5SO_3^-$; the poorest is $C_6H_5O^-$, which is the strongest base.

Sulfonates are excellent leaving groups—much better than the halides. One of the best leaving groups (10^8 times better than Br^-) is $CF_3SO_3^-$, called **triflate.**

1. S_N1 and S_N2 Mechanisms

The two major mechanisms of nucleophilic displacement are outlined in Table 7.1.

Problem 7.10 Give the three steps for the mechanism of the S_N1 hydrolysis of 3° Rx, Me_3CBr.

(*1*) $Me_3C{:}\ddot{B}r: \longrightarrow Me_3C^+ \; :\ddot{B}r:^-$

(*2*) $Me_3C^+ + :\ddot{O}H_2 \longrightarrow Me_3C{:}\ddot{O}H_2^+$

(*3*) $Me_3C\ddot{O}H_2^+ + H_2O \longrightarrow Me_3C\ddot{O}H + H_3O^+$

Problem 7.11 Give examples of the four charge-types of S_N2 reactions, as shown in the first line of Table 7.1.

$$CH_3CH_2Br + :\ddot{O}H^- \longrightarrow CH_3CH_2\ddot{O}H + :\ddot{B}r:^-$$

$$CH_3CH_2Br + :NH_3 \longrightarrow CH_3CH_2NH_3^+ + :\ddot{B}r:^-$$

$$(CH_3)_2\overset{+}{\ddot{S}}{:}CH_3 + :\ddot{O}H^- \longrightarrow CH_3\ddot{O}H + (CH_3)_2\ddot{S}: \text{ (leaving group)}$$

a sulfonium cation Dimethyl sulfide

$$(CH_3)_2\overset{+}{\ddot{S}}—CH_3 + :N(CH_3)_3 \longrightarrow CH_3\overset{+}{N}(CH_3)_3 + (CH_3)_2\ddot{S}:$$

TABLE 7.1

	S_N1	S_N2
Steps	Two: (1) $R{:}X \xrightarrow{\text{slow}} \underset{\text{carbocation}}{R^+} + {:}X^-$ (2) $R^+ + {:}NuH \xrightarrow{\text{fast}} RNuH^+$	One: $R{:}X + {:}Nu^- \longrightarrow RNu + {:}X^-$ or $R{:}X + {:}Nu \longrightarrow RNu^+{:}X^-$ or $R{:}X^+ + Nu^- \longrightarrow RNu + X{:}$ or $R{:}X^+ + Nu \longrightarrow RNu^+ + X{:}$
Rate	= k[RX] (1st-order)	= k[RX][:Nu^-] (2nd-order)
TS of slow step	$\overset{\delta+}{C}$- - -:$\overset{\delta-}{X}$	$\overset{\delta-}{:Nu}$- -C- -$\overset{\delta-}{X}$ (with :Nu^-)
Molecularity	Unimolecular	Bimolecular
Stereochemistry	Inversion and racemization	Inversion (backside attack)
Reactivity Structure of R Determining factor Nature of X Solvent effect on rate	 3° > 2° > 1° > CH_3 Stability of R^+ RI > RBr > RCl > RF Rate increase in polar solvents	 CH_3 > 1° > 2° > 3° Steric hindrance in R group RI > RBr > RCl > RF With Nu^-, there is a large rate increase in polar aprotic solvents
Effect of nucleophile	R^+ reacts with nucleophilic solvents rather than with :Nu^- (**solvolysis**), except when R^+ is relatively stable	Rate depends on nucleophilicity $I^- > Br^- > Cl^-$; $RS{:}^- > RO{:}^-$ Equilibrium lies toward weaker Brönsted base
Catalysis	Lewis acid, e.g., Ag^+, $AlCl_3$, $ZnCl_2$	(1) Aprotic polar solvent (2) Phase-transfer
Competition, reaction	Elimination, rearrangement	Elimination, especially with 3° RX in strong Brönsted base

Problem 7.12 (*a*) Give an orbital representation for an S_N2 reaction with (*S*)- RCHDX and :Nu^-, if in the transition state, the C on which displacement occurs uses sp^2 hybrid orbitals. (*b*) How does this representation explain (i) inversion and (ii) the order of reactivity 3° > 2° > 1°?

(*a*) See Fig. 7.1.

(*b*) (i) The reaction is initiated by the nucleophile beginning to overlap with the tail of the sp^3 hybrid orbital holding X. In order for the tail to become the head, the configuration must change; inversion occurs. (ii) As H's on the attacked C are replaced by R's, the TS becomes more crowded and has a higher enthalpy. With a 3° RX, there is a higher $\Delta H^{\ddagger}$ and a lower rate.

Problem 7.13 (*a*) Give a representation of an S_N1 TS which assigns a role to the nucleophilic protic solvent molecules (HS:) needed to solvate the ion. (*b*) In view of this representation, explain why (i) the reaction is first-order;

(ii) R^+ reacts with solvent rather than with stronger nucleophiles that may be present; (iii) catalysis by Ag^+ takes place; (iv) the more stable the R^+, the less inversion and the more racemization occurs.

(*a*)

H-bond

$\delta+ \ \delta-$

HS:---C---X---HS:

Solvent-assisted S_N1 TS

(*b*) (i) Although the solvent HS: appears in the TS, solvents do not appear in the rate expression. (ii) HS: is already partially bonded, via solvation, with the incipient R^+. (iii) Ag^+ has a stronger affinity for X^- than has a solvent molecule; the dissociation of X^- is accelerated. (iv) The HS: molecule solvating an unstable R^+ is more apt to form a bond, causing inversion. When R^+ is stable, the TS gives an intermediate that reacts with another HS: molecule to give a symmetrically solvated cation:

$[HS--C--SH]^+$

which collapses to a racemic product:

$H\overset{+}{S}—C— \ + \ —C—\overset{+}{S}H$

The more stable is R^+, the more selective it is and the more it can react with the nucleophilic anion Nu^-.

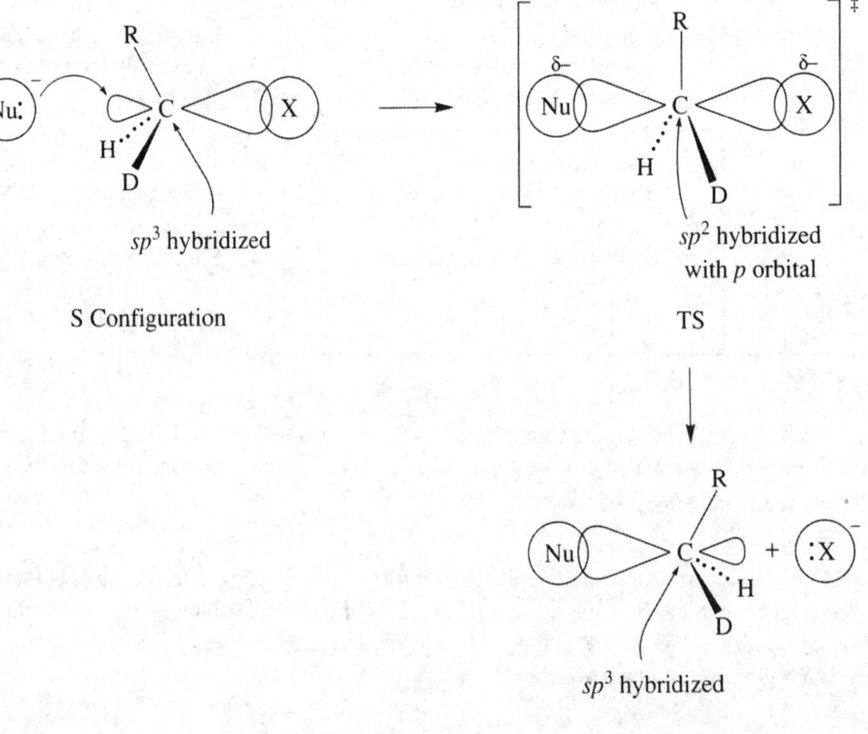

Figure 7.1

Problem 7.14 Give differences between S_N1 and S_N2 transition states.

1. In the S_N1 TS, *there is considerable positive charge on* C; there is much weaker bonding between the attacking group and leaving groups with C. There is little or no charge on C in the S_N2 TS.
2. The S_N1 TS is approached by separation of the leaving group; the S_N2 TS by attack of :Nu^- or :Nu.
3. The $\Delta H^\ddagger$ of the S_N1 TS (and the rate of the reaction) depends on the stability of the incipient R^+. When R^+ is more stable, $\Delta H^\ddagger$ is lower and the rate is greater. The $\Delta H^\ddagger$ of the S_N2 TS depends on the steric effects. When there are more R's on the attacked C or when the attacking :Nu^- is bulkier, $\Delta H^\ddagger$ is greater and the rate is less.

Problem 7.15 How can the stability of an intermediate R^+ in an S_N1 reaction be assessed from its enthalpy-reaction diagram?

The intermediate R^+ is a trough between two transition-state peaks. More stable R^+'s have deeper troughs and differ less in energy from the reactants and products.

Problem 7.16 (*a*) Formulate $(CH_3)_3COH + HCl \rightarrow (CH_3)_3CCl + H_2O$ as an S_N1 reaction. (*b*) Formulate the reaction $CH_3OH + HI \rightarrow CH_3I + H_2O$ as an S_N2 reaction.

(*a*) **Step 1** $\underset{\text{base}_1}{(CH_3)_3COH} + \underset{\text{acid}_2}{HCl} \xrightleftharpoons{\text{fast}} \underset{\substack{\text{acid}_1 \\ \textit{oxonium ion}}}{(CH_3)_3C\overset{+}{O}H_2} + \underset{\text{base}_2}{Cl^-}$

Step 2 $(CH_3)_3C\overset{+}{O}H_2 \xrightleftharpoons{\text{slow}} (CH_3)_3C^+ + H_2O$

Step 3 $(CH_3)_3C^+ + Cl^- \xrightarrow{\text{fast}} (CH_3)_3CCl$

(*b*) **Step 1** $CH_3OH + HI \xrightleftharpoons{\text{fast}} CH_3\overset{+}{O}H_2 + I^-$

Step 2 $:\ddot{\underset{..}{I}}:^- + H_3C\text{—}\overset{+}{O}H_2 \xrightarrow{\text{slow}} ICH_3 + H_2O$

Problem 7.17 ROH does not react with NaBr, but adding H_2SO_4 forms RBr. Explain.

Br^-, an extremely weak Brönsted base, cannot displace the strong base OH^-. In acid, $R\overset{+}{O}H_2$ is first formed. Now, Br^- displaces H_2O, which is a very weak base and a good leaving group.

Problem 7.18 Optically pure (*S*)-(+)-CH_3CHBr-*n*-C_6H_{13} has $[\alpha]_D^{25} = +36.0°$. A partially racemized sample having a specific rotation of +30° is reacted with dilute NaOH to form (*R*)-(−)-$CH_3CH(OH)$-*n*-C_6H_{13} ($[\alpha]_D^{25} = -5.97°$), whose specific rotation is −10.3° when optically pure. (*a*) Write an equation for the reaction using projection formulas. (*b*) Calculate the percent optical purity of reactant and product. (*c*) Calculate percentages of racemization and inversion. (*d*) Calculate percentages of front-side and back-side attack. (*e*) Draw a conclusion concerning the reactions of 2° alkyl halides. (*f*) What change in conditions would increase inversion?

(*a*) $H\ddot{\underset{..}{O}}:^- + CH_3\overset{C_6H_{13}}{\underset{\underset{(S)}{H}}{+}}\ddot{\underset{..}{Br}}: \longrightarrow HO\overset{C_6H_{13}}{\underset{\underset{(R)}{H}}{+}}CH_3 + :\ddot{\underset{..}{Br}}:^-$

(*b*) The percentage of optically active enantiomer (optical purity) is calculated by dividing the observed specific rotation by that of pure enantiomer and multiplying the quotient by 100%. The optical purities are as follows:

$$\text{Bromide} = \frac{+30°}{+36°}(100\%) = 83\% \qquad \text{Alcohol} = \frac{-5.97°}{-10.3°}(100\%) = 58\%$$

(*c*) The percentage of inversion is calculated by dividing the percentage of optically active alcohol of opposite configuration by that of reacting bromide. The percentage of racemization is the difference between this percentage and 100%.

$$\text{Percentage inversion} = \frac{58\%}{83\%}(100\%) = 70\%$$

$$\text{Percentage racemization} = 100\% - 70\% = 30\%$$

(*d*) Inversion involves only backside attack, while racemization results from equal backside and frontside attack. The percentage of backside reaction is the sum of the inversion and one-half of the racemization; the percentage of frontside attack is the remaining half of the percentage of racemization.

$$\text{Percentage backside reaction} = 70\% + \tfrac{1}{2}(30\%) = 85\%$$

$$\text{Percentage frontside reaction} = \tfrac{1}{2}(30\%) = 15\%$$

(*e*) The large percentage of inversion indicates chiefly S_N2 reaction, while the smaller percentage of racemization indicates some S_N1 pathway. This duality of reaction mechanism is typical of 2° alkyl halides.

(*f*) The S_N2 rate is increased by raising the concentration of the nucleophile—in this case, OH^-.

Problem 7.19 Account for the following stereochemical results:

$$C_6H_5-\underset{\large H}{\overset{\large CH_3}{\overset{|}{\underset{|}{C}}}}-Br \xrightarrow{H_2O} HO-\underset{\large H}{\overset{\large CH_3}{\overset{|}{\underset{|}{C}}}}-C_6H_5 + HBr \qquad \begin{matrix}2\%\text{ inversion}\\ 98\%\text{ racemization}\end{matrix}$$

$$C_6H_5-\underset{\large H}{\overset{\large CH_3}{\overset{|}{\underset{|}{C}}}}-Br \xrightarrow{CH_3OH} CH_3O-\underset{\large H}{\overset{\large CH_3}{\overset{|}{\underset{|}{C}}}}-C_6H_5 + HBr \qquad \begin{matrix}27\%\text{ inversion}\\ 73\%\text{ racemization}\end{matrix}$$

H_2O is more nucleophilic and polar than CH_3OH. It is better able to react to give $HS\!:\!---\overset{\delta+}{R}---\!:\!SH$ (see Problem 7.13), leading to racemization.

Problem 7.20 NH_3 reacts with RCH_2X to form an ammonium salt, $RCH_2NH_3^+X^-$. Show the transition state, indicating the partial charges.

$$\overset{\delta+}{H_3N}\!:\!---\underset{H\quad R}{\overset{H}{\overset{|}{C}}}---\!:\!\overset{\delta-}{X}$$

N gains $\delta+$ as it begins to form a bond.

Problem 7.21 $H_2C{=}CHCH_2Cl$ is solvolyzed faster than $(CH_3)_2CHCl$. Explain.

Solvolyses go by an S_N1 mechanism. Relative rates of different reactants in S_N1 reactions depend on the stabilities of intermediate carbocations. $H_2C{=}CHCH_2Cl$ is more reactive because

$$[H_2C \overset{\cdots}{=} CH \overset{\cdots}{=} CH_2]^+$$

is more stable than $(CH_3)_2CH^+$. See Problem 6.35 for a corresponding explanation of stability of an allyl radical.

Problem 7.22 In terms of (*a*) the inductive effect and (*b*) steric factors, account for the decreasing stability of R^+:

$$Me_3\overset{+}{C} > Me_2\overset{+}{C}H > Me\overset{+}{C}H_2 > \overset{+}{C}H_3$$

(*a*) Compared to H, R has an electron-releasing inductive effect. Replacing H's on the positive C by CH_3's disperses the positive charge and thereby stabilizes R^+.

(*b*) Steric acceleration also contributes to this order of R^+ stability. Some steric strain of the three Me's in Me_3C—Br separated by a 109° angle (sp^3) is relieved upon going to a 120° separation in R^+ with a C using sp^2 hybrid orbitals.

Carbocations have been prepared as long-lived species by the reaction $RF + SbF_5 \rightarrow R^+ + SbF_6^-$. SbF_5, a covalent liquid, is called a **superacid** because it is a stronger Lewis acid than R^+.

Problem 7.23 How does the S_N1 mechanism for the hydrolysis of 2-bromo-3-methylbutane, a 2° alkyl bromide, to give exclusively the 3° alcohol 2-methyl-2-butanol, establish a carbocation intermediate?

The initial slow step is dissociation to the 2° 1,2-dimethylpropyl cation. A hydride shift yields the more stable 3° carbocation, which reacts with H_2O to form the 3° alcohol.

$$CH_3—\underset{\textstyle Br}{CH}—\overset{\textstyle CH_3}{\underset{\textstyle H}{C}}—CH_3 \xrightarrow{\text{slow}} Br^- + CH_3—\overset{+}{CH}—\overset{\textstyle CH_3}{\underset{\textstyle \ddot{H}}{C}}—CH_3 \quad (2°)$$

$$\downarrow \sim:H^-$$

$$CH_3—\underset{\textstyle H}{CH}—\overset{\textstyle CH_3}{\underset{+}{C}}—CH_3 \; (3°) \xrightarrow[\text{fast}]{2H_2O} CH_3—\underset{\textstyle H}{CH}—\overset{\textstyle CH_3}{\underset{\textstyle OH}{C}}—CH_3 + H_3\overset{+}{O}$$

Problem 7.24 RBr reacts with $AgNO_2$ to give RNO_2 and RONO. Explain.

The nitrite ion

$$\left[:\ddot{N} \begin{matrix} \diagup \ddot{O}: \\ \diagdown\!\!\diagdown \ddot{O}: \end{matrix} \right]^-$$

has two different nucleophilic sites: the N and either O. Reaction with the unshared pair on N gives RNO_2, while RONO is formed by reaction at O. [Anions with two nucleophilic sites are called **ambident** anions.]

2. Role of the Solvent

Polar solvents stabilize and lower the enthalpies of charged reactants and charged transition states. The more diffuse the charge on the species, the less effective the stabilization by the polar solvent.

Problem 7.25 In terms of transition-state theory, account for the following solvent effects: (*a*) The rate of solvolysis of a 3° RX increases as the polarity of the protic nucleophilic solvent (:SH) increases; for example:

$$H_2O > HCOOH > CH_3OH > CH_3COOH$$

(*b*) The rate of the S_N2 reaction $:Nu^- + RX \rightarrow RNu + :X^-$ decreases slightly as the polarity of protic solvent increases. (*c*) The rate of the S_N2 reaction $:Nu + RX \rightarrow RNu^+ + :X^-$ increases as the polarity of the solvent increases. (*d*) The rate of reaction in (*b*) is greatly increased in a polar aprotic solvent. (*e*) The rate of the reaction in (*b*) is less in nonpolar solvents than in aprotic polar solvents.

See Table 7.2.

TABLE 7.2

GROUND STATE (GS)	TS	RELATIVE CHANGE	EFFECT OF SOLVENT CHANGE	$\Delta H^{\ddagger}$	RATE
(*a*) RX + HS:	$\overset{\delta+}{\text{HS---R}}\overset{\delta-}{\text{---X}}\text{---HS:}$	None in GS charge in TS	A lower *H* of TS	Decreases	Increases
(*b*) RX + Nu$^-$	$\overset{\delta-}{\text{Nu}}\text{---R---}\overset{\delta-}{\text{X}}$	Full in GS diffuse in TS	A lower *H* of GS A less lower *H* of TS	Increases	Decreases
(*c*) RX + Nu	$\overset{\delta+}{\text{Nu}}\text{---R---}\overset{\delta-}{\text{X}}$	None in GS charge in TS	A lower *H* of TS	Decreases	Increases
(*d*) Same as (*b*)			A big rise in *H* of GS* A rise in *H* of TS	Decreases	Increases

*Aprotic solvents do not solvate anions.

(*e*) In nonpolar solvents, Nu$^-$ is less reactive because it is ion-paired with its countercation, M$^+$.

S_N2 reactions with Nu$^-$'s are typical of reactions between water-soluble salts and organic substrates that are soluble only in nonpolar solvents. These incompatible reactants can be made to mix by adding small amounts of **phase-transfer catalysts**, such as quaternary ammonium salts, Q^+A^-. Q^+ has a water-soluble ionic part and nonpolar R groups that tend to be soluble in the nonpolar solvent. Hence, Q^+ shuttles between the two immiscible solvents while transporting Nu$^-$, as Q^+Nu^-, into the nonpolar solvent; there Nu$^-$ quickly reacts with the organic substrate. Q^+ then moves back to the water while it transports the leaving group, X^-, as Q^+X^-. Since the positive charge on the N of Q^+ is surrounded by the R groups, ion pairing between Nu$^-$ and Q^+ is loose, and Nu$^-$ is quite free and very reactive.

Problem 7.26 Write equations for and explain the use of tetrabutyl ammonium chloride, $Bu_4N^+Cl^-$, to facilitate the reaction between 1-heptyl chloride and cyanide ion.

$$Bu_4N^+Cl^- + Na^+CN^- \longrightarrow Bu_4N^+CN^- + Na^+Cl^- \quad \text{(in water)}$$
$$n\text{-}C_7H_{15}Cl + Bu_4N^+CN^- \longrightarrow n\text{-}C_7H_{15}CN + Bu_4N^+Cl^- \quad \text{(in nonolar phase)}$$

The phase-transfer catalyst, $Bu_4N^+Cl^-$, reacts with CN$^-$ to form a quaternary cyanide salt that is slightly soluble in the organic phase because of the bulky, nonpolar, butyl groups. Reaction with CN$^-$ to form the nitrile is rapid because it is not solvated or ion-paired in the organic phase; it is a free, strong nucleophile. The phase-transfer catalyst, regenerated in the organic phase, returns to the aqueous phase, and the chain process is propagated.

Elimination Reactions

In a *β*-elimination (dehydrohalogenation) reaction, a halogen and a hydrogen atom are removed from adjacent carbon atoms to form a double bond between the two C's. The reagent commonly used to remove HX is the strong base KOH in ethanol (cf. Section 6.2).

$$\underset{\text{ethyl halide}}{\text{H—}\overset{\beta}{\text{C}}\text{H}_2\text{—}\overset{\alpha}{\text{C}}\text{HX—H}} \xrightarrow{\text{alc. KOH}} \underset{\text{ethylene}}{\text{H}_2\text{C=CH}_2} + \text{HX (as KX and H}_2\text{O)}$$

TABLE 7.3

	E1	E2
Steps	Two: 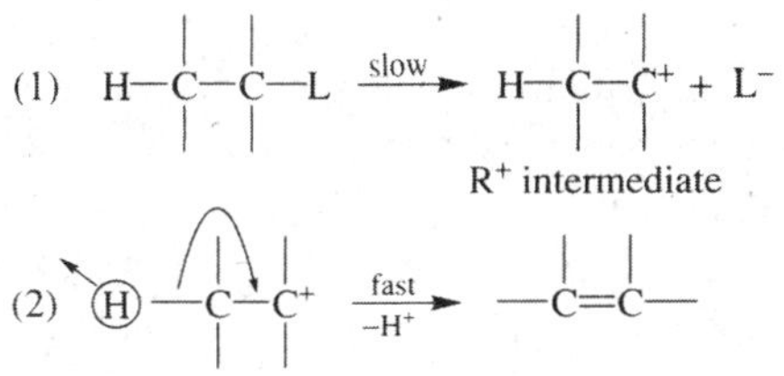 (1) H—C—C—L $\xrightarrow{\text{slow}}$ H—C—C$^+$ + L$^-$ (R$^+$ intermediate) (2) (H)—C—C$^+$ $\xrightarrow[-H^+]{\text{fast}}$ —C=C—	One: B: + H—C—C—L ↓ B:H + >C=C< + :L$^-$
Transition states	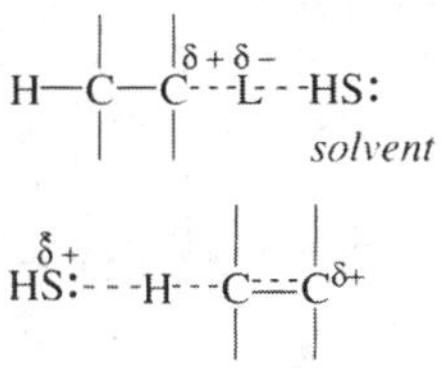 H—C—C$^{\delta+}$- - -L$^{\delta-}$- -HS: *solvent* $^{\delta+}$HS:- - -H- - -C==C$^{\delta+}$	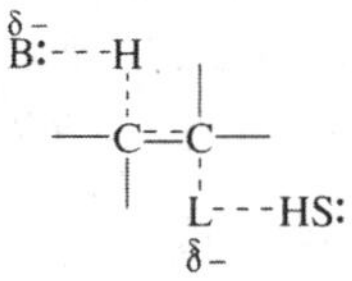 $^{\delta-}$B:- - -H —C==C— L- - -HS: $\delta-$
	INDICATES E1	**INDICATES E2**
Kinetics	First-order Rate = k[RL] Ionization determines rate Unimolecular	Second-order Rate = k[RL] [:B$^-$] Bimolecular
Stereochemistry	Nonstereospecific	*anti* Elimination (*syn* when *anti* impossible)
Reactivity order factor	3° > 2° > 1° RX Stability of R$^+$	3° > 2° > 1° RX Stability of alkenes (Saytzeff rule)
Rearrangements	Common	None
Deuterium isotope effect	None	Observed
Competing reaction	S_N1, S_N2	S_N2
Regioselectivity	Saytzeff	(see Problem 7.33)
	FAVORS E1	**FAVORS E2**
Alkyl group	3° > 2° > 1°	3° > 2° > 1°
Loss of H	No effect	Increased acidity
Base Strength Concentration	Weak Low	Strong High
Leaving group	Weak base I$^-$ > Br$^-$ > Cl$^-$ > F$^-$	Weak base I$^-$ > Br$^-$ > Cl$^-$ > F$^-$
Catalysis	Ag$^+$	Phase-transfer
Solvent	Polar protic	Polar aprotic

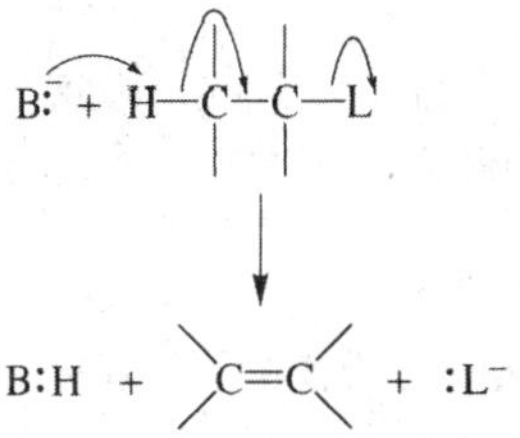

n-Propyl bromide $CH_3CH_2CH_2Br \xrightarrow{\text{alc. KOH}} CH_3CH{=}CH_2$ Propene

sec-Butyl chloride $CH_3CHClCH_2CH_3 \xrightarrow{\text{alc. KOH}} CH_3CH{=}CHCH_3$ 2-Butene (mainly *trans*)

1. E1 and E2 Mechanisms

The two major mechanisms for *β*-eliminations involving the removal of an H and an adjacent functional group are the **E2** and **E1**. Their features are compared and summarized in Table 7.3. A third mechanism, $\mathbf{E_{1cb}}$, is occasionally observed. Table 7.4 compares **E2** and $\mathbf{S_N2}$.

Problem 7.27 (*a*) Why do alkyl halides rarely undergo the E1 reaction? (*b*) How can the E1 mechanism be promoted?

(*a*) RX reacts by the E1 mechanism only when the base is weak and has a very low concentration; as the base gets stronger and more concentrated, the E2 mechanism begins to prevail. On the other hand, if the base is too weak or too dilute, either the R^+ reacts with the nucleophilic solvent to give the S_N1 product or, in nonpolar solvents, RX fails to react.
(*b*) By the use of catalysts, such as Ag^+, which help pull away the leaving group X^-.

2. H-D Isotope Effect

The C—H bond is broken at a faster rate than is the stronger C—D bond. The ratio of the rate constants, k_H/k_D, measures this **H-D isotope effect**. The observation of an isotope effect indicates that C—H bond-breaking occurs in the rate-controlling step.

Problem 7.28 Why does CH_3CH_2I undergo loss of HI with strong base faster than CD_3CH_2I loses DI?

These are both E2 reactions in which the C—H or C—D bonds are broken in the rate-controlling step. Therefore, the H-D isotope effect accounts for the faster rate of reaction of CH_3CH_2I.

Problem 7.29 Explain the fact that whereas 2-bromopentane undergoes dehydrohalogenation with $C_2H_5O^-K^+$ to give mainly 2-pentene (the Sayzteff product), with $Me_3CO^-K^+$ it gives mainly 1-pentene (the anti-Sayzteff, **Hofmann**, product).

Since Me_3CO^- is a bulky base, its attack is more sterically hindered at the 2° H than at the 1° H. With Me_3CO^-,

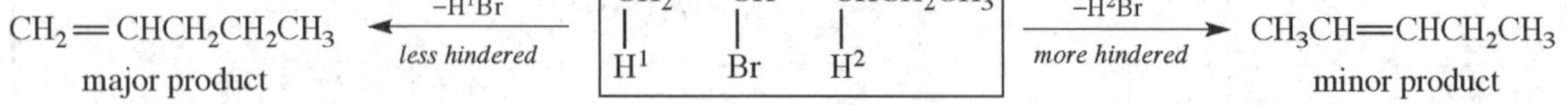

$CH_2{=}CHCH_2CH_2CH_3$ (major product) $\xleftarrow[\text{less hindered}]{-H^1Br}$ $\boxed{CH_2(H^1)—CH(Br)—CH(H^2)CH_2CH_3}$ $\xrightarrow[\text{more hindered}]{-H^2Br}$ $CH_3CH{=}CHCH_2CH_3$ (minor product)

Problem 7.30 Assuming that *anti* elimination is favored, illustrate the stereospecificity of the E2 dehydrohalogenation by predicting the products formed from (*a*) *meso-* and (*b*) either of the enantiomers of 2,3-dibromobutane. Use the wedge-sawhorse and Newman projections.

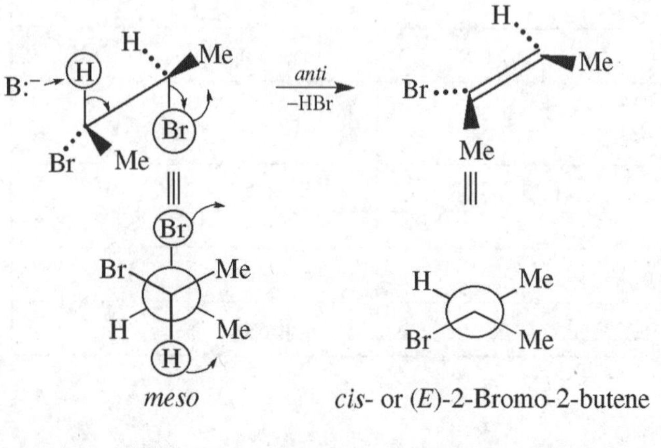

B:⁻ → (H), H, Me, (Br), Br, Me $\xrightarrow[-HBr]{anti}$ Br, H, Me, Me

Newman: (Br), Br, Me, H, Me, (H) — H, Me, Br, Me

meso *cis-* or (*E*)-2-Bromo-2-butene

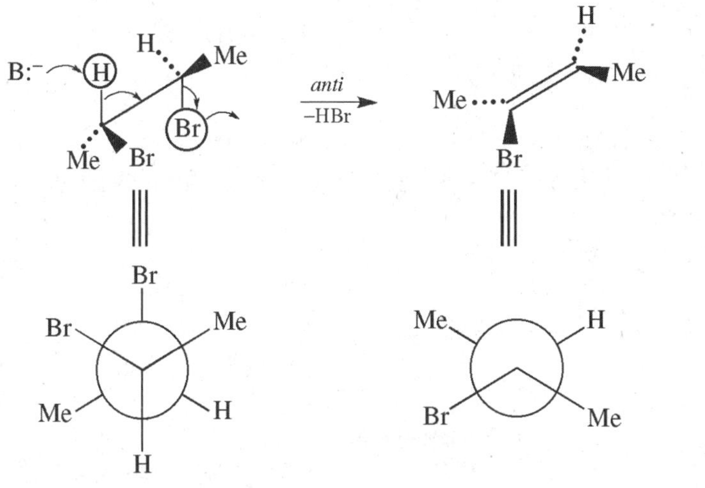

enantiomer (RR) *trans*- or (*Z*)-2-Bromo-2-butene

Problem 7.31 Account for the percentages of the products, (i) $(CH_3)_2CHOC_2H_5$ and (ii) $CH_3CH{=}CH_2$, of the reaction of $CH_3CHBrCH_3$ with

(*a*) $C_2H_5ONa/C_2H_5OH \rightarrow$ 79% (ii) +21% (i) (*b*) $C_2H_5OH \rightarrow$ 3% (ii) +97% (i)

(*a*) $C_2H_5O^-$ is a strong base and E2 predominates. (*b*) C_2H_5OH is weakly basic but nucleophilic, and S_N1 is favored.

Problem 7.32 Account for the following observations: (*a*) In a polar solvent such as water, the S_N1 and El reactions of a 3° RX have the same rate. (*b*) $(CH_3)_3CI + H_2O \rightarrow (CH_3)_3COH + HI$ but $(CH_3)_3CI + OH^- \rightarrow (CH_3)_2C{=}CH_2 + H_2O + I^-$.

TABLE 7.4

	FAVORS E2	FAVORS S_N2
Structure of R	3° > 2° > 1°	1° > 2° > 3°
Reagent	Strong bulky Brönsted base (e.g., Me_3CO^-)	Strong nucleophile
Temperature	High	Low
Low-polarity solvent	Yes	No
Structure of L	I > Br > Cl > F	I > Br > Cl > F
	FAVORS E2	**FAVORS S_N1**
Structure of R	3° > 2° > 1°	3° >> 2° > 1°
Base Strength Connection	 Strong High	 Very weak Low
Structure of L	I > Br > Cl >> F	I > Br > Cl >> F

(*a*) The rate-controlling step for both El and S_N1 reactions is the same:

$$\overset{\delta+}{R}\text{---}\overset{\delta-}{X} \xrightarrow{\text{slow}} R^+ + X^-$$

and therefore the rates are the same. (*b*) In a nucleophilic solvent in the absence of a strong base, a 3° RX undergoes an S_N1 solvolysis. In the presence of a strong base (OH^-), a 3° RX undergoes mainly E2 elimination.

7.4 Summary of Alkyl Halide Chemistry

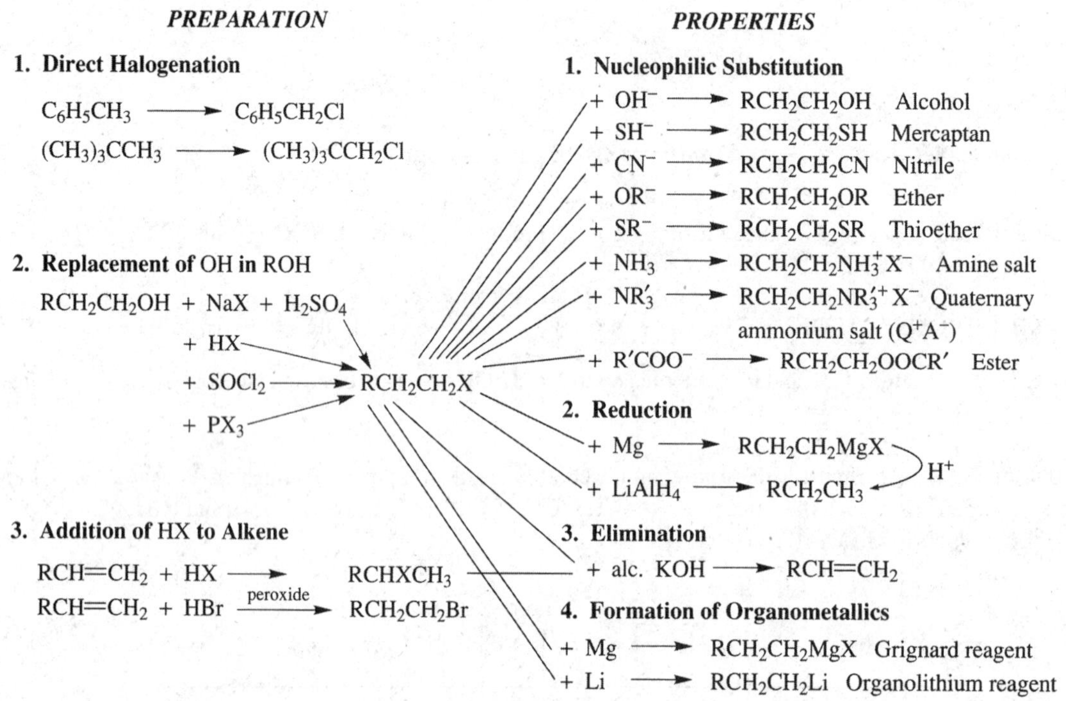

SUPPLEMENTARY PROBLEMS

Problem 7.33 Write the structure of the only tertiary halide having the formula $C_5H_{11}Br$.

In order for the halide to be *tertiary*, Br must be attached to a C that is attached to three other C's—in other words, to no H's. This gives the skeletal arrangement:

```
    C
    |
 C—C—C
    |
    Br
```

involving four C's. A fifth C must be added, which must be attached to one of the C's on the central C, giving:

```
    C
    |
 C—C—C—C
    |
    Br
```

and with the H's.

$$\begin{array}{c} CH_3 \\ | \\ CH_3—C—CH_2—CH_3 \\ | \\ Br \end{array}$$

2-Bromo-2-methylbutane

Problem 7.34 On substitution of one H by a Cl in the isomers of C_5H_{12}, (*a*) which isomer gives only a primary halide? (*b*) Which isomers give secondary halides? (*c*) Which isomer gives a tertiary halide?

(*a*) 2,2-Dimethylpropane. (*b*) $CH_3CH_2CH_2CH_2CH_3$ gives 2-chloropentane and 3-chloropentane. 2-Methylbutane gives 2-chloro–3-methylbutane. (*c*) 2-Methylbutane gives 2-chloro–2-methylbutane.

Problem 7.35 Complete the following table:

	Substance	*Treated with*	*Yields*
1.	CH_3CH_3	(*a*)	CH_3CH_2Cl
2.	(*b*)	$SOCl_2$	$CH_3\underset{\displaystyle Cl}{\underset{\vert}{C}}HCH_3$
3.	(*c*)	Br_2	$CH_3\underset{\displaystyle Br}{\underset{\vert}{C}}HCH_2Br$
4.	$CH_3CH{=\!=}CH_2$	HBr, peroxides	(*d*)

(*a*) Cl_2, UV (*b*) $CH_3\underset{\displaystyle OH}{\underset{\vert}{C}}HCH_3$ (*c*) $CH_3CH{=\!=}CH_2$ (*d*) $CH_3CH_2CH_2Br$

Problem 7.36 Give the organic product in the following substitution reactions. The solvent is given above the arrow.

(*a*) $HS^- + CH_3CH_2CHBrCH_3 \xrightarrow{CH_3OH}$

(*b*) $I^- + (CH_3)_3CBr \xrightarrow{HCOOH}$

(*c*) $CH_3CH_2Br + AgCN \longrightarrow$ 2 products

(*d*) $CH_3CHBrCH_3 + CH_3\ddot{N}H_2 \longrightarrow$

(*e*) $CH_3CHBrCH_3 + (CH_3)_2\ddot{S}\!: \longrightarrow$

(*f*) $CH_3CH_2Br + :P(C_6H_5)_3 \longrightarrow$

(*g*) $CH_3CH_2Br + \left[S—\underset{\displaystyle O}{\underset{|}{\overset{\displaystyle O}{\overset{|}{S}}}}—O \right]^{2-}$ (thiosulfate ion) $\longrightarrow$

(*a*) $CH_3CH_2CHSHCH_3$ (a mercaptan).

(*b*) $(CH_3)_3CO\overset{\displaystyle O}{\overset{\|}{C}}H$; 3° RX undergoes S_N1 solvolysis.

(*c*) $CH_3CH_2CN + CH_3CH_2NC$; $:C{\equiv}N:^-$ is an ambident anion (Problem 7.24).

(*d*) $\left[(CH_3)_2CH\overset{H}{\underset{H}{N}}CH_3\right]^+$ Br^-; an ammonium salt.

(*e*) $[(CH_3)_2CHS(CH_3)_2]^+$ Br^-; a sulfonium salt.

(*f*) $[CH_3CH_2P(C_6H_5)_3]^+$ Br^-; a phosphonium salt.

(*g*) $CH_3CH_2\text{—S—}\overset{O}{\underset{O}{S}}\text{—}O^-$; S is a more nucleophilic site than is O.

Problem 7.37 Account for the observation that catalytic amounts of KI enhance the rate of reaction of RCH_2Cl with OH^- to give the alcohol RCH_2OH.

Since I^- is a better nucleophile than OH^-, it reacts rapidly with RCH_2Cl to give RCH_2I. But since I^- is a much better leaving group than Cl^-, RCH_2I reacts faster with OH^- than does RCH_2Cl. Only a catalytic amount of I^- is needed, because the regenerated I^- is recycled in the reaction.

$$RCH_2Cl + OH^- \xrightarrow{\text{slower}} RCH_2OH + Cl^-$$

$$RCH_2Cl \xrightarrow{+I^-\text{(fast)}} RCH_2I \xrightarrow[-I^-]{+OH^-\text{(fast)}} RCH_2OH + Cl^-$$

Problem 7.38 Account for the following products from the reaction of $CH_3CHOHC(CH_3)_3$ with HBr: (*a*) $CH_3CHBrC(CH_3)_3$, (*b*) $H_2C{=}CHC(CH_3)_3$, (*c*) $(CH_3)_2CHCHBr(CH_3)_2$, (*d*) $(CH_3)_2CH{=}CH(CH_3)_2$, (*e*) $(CH_3)_2CH\underset{CH_3}{C}{=}CH_2$

For this 2° alcohol, oxonium-ion formation is followed by loss of H_2O to give a 2° carbocation that rearranges to a 3° carbocation. Both carbocations react by two pathways: they form bonds to Br^- to give alkyl bromides, or they lose H^+ to yield alkenes.

$$CH_3CHOHC(CH_3)_3 \xrightarrow{HBr} CH_3\underset{\overset{+}{O}H_2}{C}HC(CH_3)_3 \xrightarrow{-H_2O} CH_3\overset{+}{C}HC(CH_3)_3 \text{ (2° carbocation)}$$

$$CH_3\overset{+}{C}HC(CH_3)_3 \xrightarrow{+Br^-} CH_3CHBrC(CH_3)_3 \ (a)$$

$$CH_3\overset{+}{C}HC(CH_3)_3 \xrightarrow{-H^+} H_2C{=}CHC(CH_3)_2 \ (b)$$

$$CH_3\overset{+}{C}HC(CH_3)_3 \xrightarrow{\sim:CH_3} (CH_3)_2CH\overset{+}{C}(CH_3)_2 \text{ (3° carbocation)}$$

$$(CH_3)_2CH\overset{+}{C}(CH_3)_2 \xrightarrow{+Br^-} (CH_3)_2CHCBr(CH_3)_2 \ (c)$$

$$(CH_3)_2CH\overset{+}{C}(CH_3)_2 \xrightarrow{-H^+} (CH_3)_2C{=}C(CH_3)_2 \ (d) \quad + \quad (CH_3)_2CH\overset{CH_3}{C}{=}CH_2 \ (e)$$

Problem 7.39 Show steps for the following conversions:

(*a*) $BrCH_2CH_2CH_2CH_3 \rightarrow CH_3CHBrCH_2CH_3$ (*b*) $CH_3CHBrCH_2CH_3 \rightarrow BrCH_2CH_2CH_2CH_3$
(*c*) $CH_3CH_2CH_3 \rightarrow CH_2ClCHClCH_2Cl$

To do syntheses, it is best to work backward while keeping in mind your starting material. As you do this, keep asking what is needed to make what you want. Always try to use the fewest steps.

(*a*) The precursors of the possible product are the corresponding alcohol, 1-butene, and 2-butene. The alcohol is a poor choice, because it would have to be made from either of the alkenes and an extra step would be needed. 2-Butene cannot be made directly from the 1° halide, but 1-butene can.

$$BrCH_2CH_2CH_2CH_3 \xrightarrow{\text{alc, KOH}} CH_2{=}CHCH_2CH_3 \xrightarrow{HBr} CH_3CHBrCH_2CH_3$$

(*b*) To ensure getting 1-butene, the needed precursor, use a bulky base for the dehydrohalogenation of the starting material.

$$CH_3CHBrCH_2CH_3 \xrightarrow{Me_3CO^-K^+} CH_2{=}CHCH_2CH_3 \xrightarrow[\text{peroxides}]{HBr} BrCH_2CH_2CH_2CH_3$$

(*c*) The precursor for *vic* dichlorides is the corresponding alkene—in this case, $H_2C{—}CHCH_2Cl$, which is made by allylic chlorination of propene. Although free-radical chlorination of propane gives a mixture of isomeric propyl chlorides, the mixture can be dehydrohalogenated to the same alkene, making this particular initial chlorination a useful reaction.

$$CH_3CH_2CH_3 \xrightarrow[uv]{Cl_2} ClCH_2CH_2CH_3 + CH_3CHClCH_3 \xrightarrow[KOH]{alc.} CH_2{=}CHCH_3 \xrightarrow[uv]{Cl_2} CH_2{=}CHCH_2Cl \xrightarrow{Cl_2} ClCH_2CHClCH_2Cl$$

Problem 7.40 Indicate the products of the following reactions, and point out the mechanism as S_N1, S_N2, E1, or E2.

(*a*) $CH_3CH_2CH_2Br + LiAlH_4$ (source of :H^-)
(*b*) $(CH_3)_3CBr + C_2H_5OH$, heat at 60°C
(*c*) $CH_3CH{=}CHCl + NaNH_2$
(*d*) $BrCH_2CH_2Br$ + Mg (ether)
(*e*) $BrCH_2CH_2CH_2Br$ + Mg (ether)
(*f*) $CH_3CHBrCH_3 + NaOCH_3$ in CH_3OH

(*a*) $CH_3CH_2CH_3$; an S_N2 reaction, :H^- of AlH_4^- replaces Br^-.

(*b*)

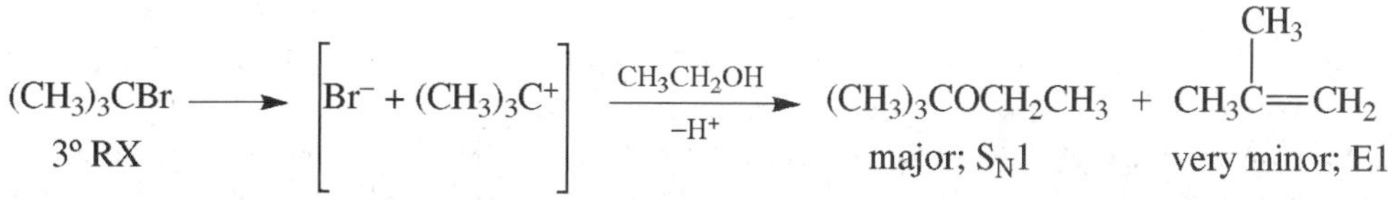

$$\underset{3^\circ\ RX}{(CH_3)_3CBr} \longrightarrow \left[Br^- + (CH_3)_3C^+\right] \xrightarrow[-H^+]{CH_3CH_2OH} \underset{\text{major; } S_N1}{(CH_3)_3COCH_2CH_3} + \underset{\text{very minor; E1}}{CH_3\overset{\overset{\displaystyle CH_3}{|}}{C}{=}CH_2}$$

(*c*) $CH_3CH{=}CHCl + NaNH_2 \longrightarrow CH_3C{\equiv}CH + NH_3 + NaCl$ (E2)
Vinyl halides are quite inert toward S_N2 reactions.

(*d*) $BrCH_2CH_2Br + Mg \longrightarrow H_2C{=}CH_2 + MgBr_2$
This is an E2 type of β–elimination via an alkyl magnesium iodide.

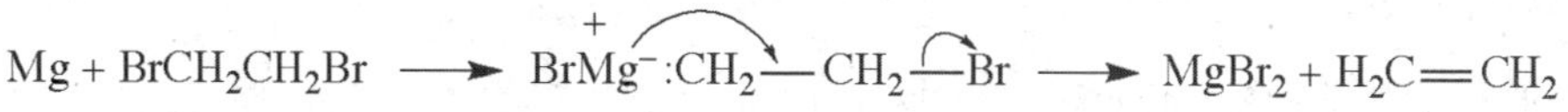

$$Mg + BrCH_2CH_2Br \longrightarrow \overset{+}{Br}Mg^-{:}CH_2{—}CH_2{—}Br \longrightarrow MgBr_2 + H_2C{=}CH_2$$

(*e*) This reaction resembles that in (*d*) and is an internal S_N2 reaction.

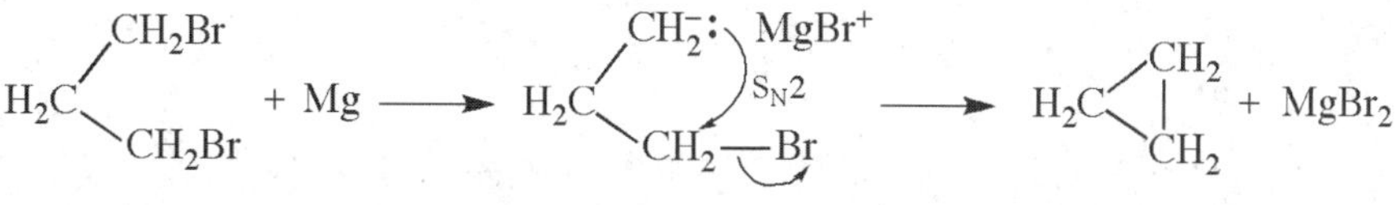

$$H_2C(CH_2Br)_2 + Mg \longrightarrow H_2C\begin{matrix}CH_2^-{:}\ MgBr^+ \\ CH_2{—}Br\end{matrix}\ (S_N2) \longrightarrow \text{cyclo-}H_2C(CH_2)_2 + MgBr_2$$

(*f*) This 2° RBr undergoes both E2 and S_N2 reactions to form propylene and isopropyl methyl ether.

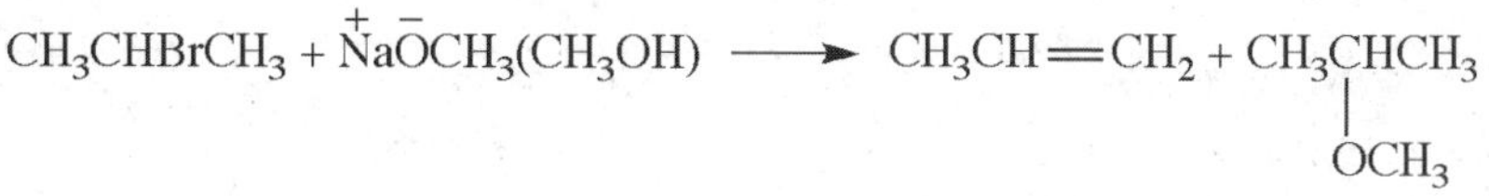

$$CH_3CHBrCH_3 + \overset{+}{Na}\overset{-}{O}CH_3(CH_3OH) \longrightarrow CH_3CH{=}CH_2 + CH_3\underset{\underset{\displaystyle OCH_3}{|}}{C}HCH_3$$

Problem 7.41 Give structures of the organic products of the following reactions and account for their formation:

(*a*) $ClCH_2CH_2CH_2CH_2Br + NaCN \xrightarrow{C_2H_5OH}$

(*b*) $CH_3CHBrCH_3 + NaI \xrightarrow{\text{acetone}}$

(*c*) $ClCH_2CH{=}CH_2 + NaI \xrightarrow{\text{acetone}}$

(*a*) $ClCH_2CH_2CH_2CH_2CN$. Br^- is a better leaving group than Cl^-.
(*b*) CH_3CHICH_3. Equilibrium is shifted to the right because NaI is soluble in acetone, while NaBr is not and precipitates.
(*c*) $ICH_2CH{=}CH_2$. NaI is soluble in acetone and NaCl is insoluble.

Problem 7.42 Account for the following observations when (*S*)-$CH_3CH_2CH_2CHID$ is heated in acetone solution with NaI: (*a*) The enantiomer is racemized. (*b*) If radioactive $*I^-$ is present in excess, the rate of racemization is twice the rate at which the radioactive $*I^-$ is incorporated into the compound.

(*a*) Since enantiomers have identical energy, reaction proceeds in both directions until a racemic equilibrium mixture is formed.

$$*I^- + CH_3CH_2CH_2-\underset{H}{\overset{D}{C}}-I \underset{\text{inversion}}{\overset{\text{inversion}}{\rightleftharpoons}} *I-\underset{H}{\overset{D}{C}}-CH_2CH_2CH_3 + I^- \quad (S_N2 \text{ reaction})$$

(*R*) (*S*)

(*b*) Each radioactive $*I^-$ incorporated into the compound forms one molecule of enantiomer. Now one unreacted molecule and one molecule of its enantiomer, resulting from reaction with $*I^-$, form a racemic modification. Since two molecules are racemized when one $*I^-$ reacts, the rate of racemization will be twice that at which $*I^-$ reacts.

Problem 7.43 Indicate the effect on the rate of S_N1 and S_N2 reactions of the following: (*a*) Doubling the concentration of substrate (RL) or Nu^-. (*b*) Using a mixture of ethanol and H_2O or only acetone as solvent. (*c*) Increasing the number of R groups on the C bonded to the leaving group, L. (*d*) Using a strong Nu^-.

(*a*) Doubling either [RL] or $[Nu^-]$ doubles the rate of the S_N2 reaction. For S_N1 reactions, the rate is doubled only by doubling [RL] and is not affected by any change in $[Nu^-]$.
(*b*) A mixture of ethanol and H_2O has a high dielectric constant and therefore enhances the rate of S_N1 reactions. This usually has little effect on S_N2 reactions. Acetone has a low dielectric constant and is aprotic and favors S_N2 reactions.
(*c*) Increasing the number of R's on the reaction site enhances S_N1 reactivity through electron release and stabilization of R^+. The effect is opposite in S_N2 reactions because bulky R's sterically hinder formation of, and raise $\Delta H^\ddagger$ for, the transition state.
(*d*) Strong nucleophiles favor S_N2 reactions and do not affect S_N1 reactions.

Problem 7.44 List the following alkyl bromides in order of decreasing reactivity in the indicated reactions.

(I)	(II)	(III)
$CH_3\underset{Br}{\overset{CH_3}{C}}-CH_2CH_3$	$CH_3CH_2CH_2CH_2CH_2Br$	$CH_3CH_2\underset{Br}{C}HCH_2CH_3$

(*a*) S_N1 reactivity, (*b*) S_N2 reactivity, (*c*) reactivity with alcoholic $AgNO_3$.

(*a*) Reactivity for the S_N1 mechanism is 3°(I) > 2°(III) > 1°(II).
(*b*) The reverse reactivity for S_N2 reactions gives 1°(II) > 2°(III) > 3°(I).
(*c*) Ag^+ catalyzes S_N1 reactions and the reactivities are 3°(I) > 2°(III) > 1°(II).

Problem 7.45 Potassium *tert*-butoxide, $K^+\bar{O}CMe_3$, is used as a base in E2 reactions. (*a*) How does it compare in effectiveness with ethylamine, $CH_3CH_2NH_2$? (*b*) Compare its effectiveness in the solvents *tert*-butyl alcohol and dimethylsulfoxide (DMSO). (*c*) Give the major alkene product when it reacts with $(CH_3)_2CClCH_2CH_3$.

(*a*) K^+OCMe_3 is more effective because it is more basic. Its larger size also precludes S_N2 reactions.
(*b*) Its reactivity is greater in aprotic DMSO because its basic anion is not solvated. Me_3COH reduces the effectiveness of Me_3CO^- by H-bonding.
(*c*) Me_3CO^- is a bulky base and gives the anti-Saytzeff (Hofmann) product $CH_2{=}C(CH_3)CH_2CH_3$.

Problem 7.46 Give structures of all alkenes formed, and underline the major product expected from E2 elimination of: (*a*) 1-chloropentane, (*b*) 2-chloropentane.

(*a*) $CH_3CH_2CH_2CH_2CH_2Cl \longrightarrow CH_3CH_2CH_2CH{=}CH_2$
A 1° alkyl halide, therefore one alkene.

(*b*) $CH_3CH_2CH_2\underset{\displaystyle Cl}{\underset{|}{C}}H{-}CH_3 \longrightarrow CH_3CH_2CH_2CH{=}CH_2 + \underline{CH_3CH_2CH{=}CHCH_3}$

A 2° alkyl halide flanked by two R's; therefore, two alkenes are formed. The more substituted alkene is the major product because of its greater stability.

Problem 7.47 How is conformational analysis used to explain the 6 : 1 ratio of *trans*- to *cis*-2-butene formed on dehydrochlorination of 2-chlorobutane?

For either enantiomer, there are two conformers in which the H and Cl eliminated are *anti* to each other:

(I) Newman projection (front: Cl, H, CH_3; back: H_3C, H, H) $\xrightarrow{B:}$ [transition state with B:--H anti to Cl]‡ $\longrightarrow$ *trans*-2-Butene

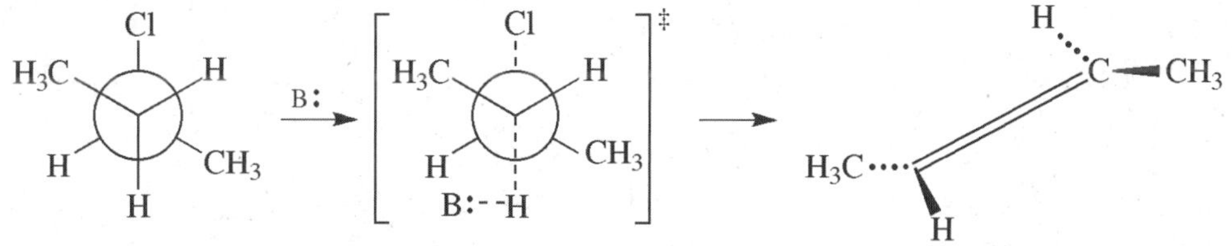

(II) Newman projection (front: Cl, H, CH_3; back: H, CH_3, H) $\xrightarrow{B:}$ [transition state with B:--H anti to Cl]‡ $\longrightarrow$ *cis*-2-Butene

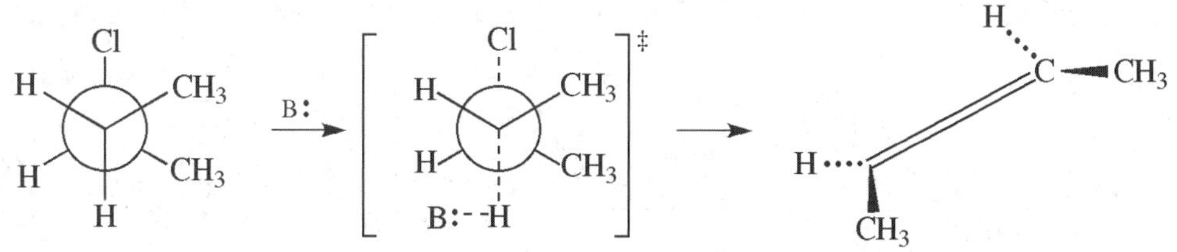

Conformer I has a less crowded, lower-enthalpy transition state than conformer II. Its $\Delta H^\ddagger$ is less and reaction rate greater; this accounts for the greater amount of *trans* isomer obtained from conformer I and the smaller amount of *cis* isomer from conformer II.

Problem 7.48 State whether each of the following R^+'s is stabilized or destabilized by the attached atom or group:

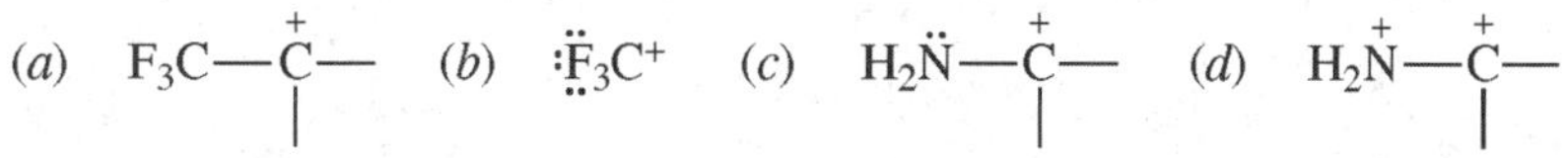

(*a*) $F_3C{-}\overset{+}{C}{-}$ (*b*) $:\ddot{F}_3C^+$ (*c*) $H_2\ddot{N}{-}\overset{+}{C}{-}$ (*d*) $H_2\overset{+}{N}{-}\overset{+}{C}{-}$

If an electron-withdrawing group is adjacent to the positive C, it will tend to destabilize the carbocation. Electron-donating groups, on the other hand, delocalize the + charge and serve to stabilize the carbocation.

(*a*) Destabilized. The strongly electron-withdrawing F's place a $\delta+$ on the atom adjacent to C^+:

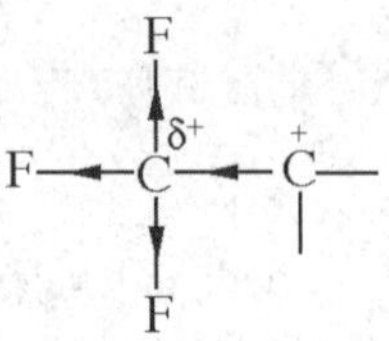

$$F\leftarrow \overset{\overset{F}{\uparrow}\ \delta+}{\underset{\underset{F}{\downarrow}}{C}} \leftarrow \overset{+}{\underset{|}{C}}-$$

(Arrows indicate withdrawn electron density.)

(*b*) Stabilized. Each F has an unshared pair of electrons in a *p* orbital which can be shifted to $-\overset{+}{\underset{|}{C}}-$ via *p-p* orbital overlap.

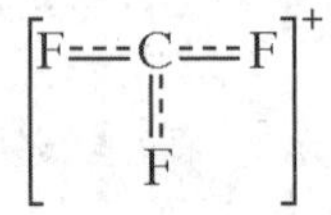

$$\left[F \text{---} \underset{\underset{F}{\vdots}}{C} \text{---} F\right]^+$$

(*c*) Stabilized. The unshared pair of electrons on N can be contributed to C^+.

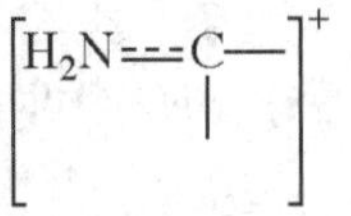

$$\left[H_2N \text{---} \underset{|}{C}-\right]^+$$

(*d*) Destabilized. The adjacent N has a + charge.

Problem 7.49 Account for the formation of

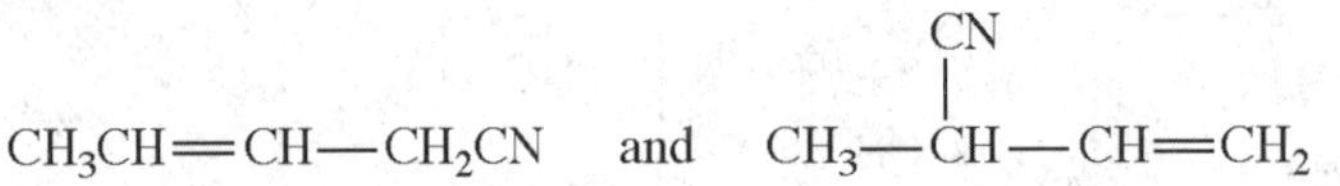

$$CH_3CH{=}CH{-}CH_2CN \quad \text{and} \quad CH_3{-}\overset{\overset{CN}{|}}{C}H{-}CH{=}CH_2$$

from the reaction with CN^- of 1-chloro-2-butene, $CH_3CH{=}CH{-}CH_2Cl$.

Formation of 1-cyano–2-butene results from S_N2 reaction at the terminal C.

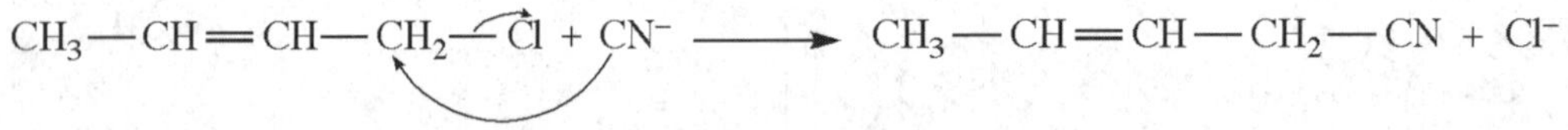

$$CH_3{-}CH{=}CH{-}CH_2{-}\ddot{C}l + CN^- \longrightarrow CH_3{-}CH{=}CH{-}CH_2{-}CN + Cl^-$$

Attack by CN^- can also occur at C^3 with the π electrons of the double bond acting as nucleophile to displace Cl^- in an allylic rearrangement:

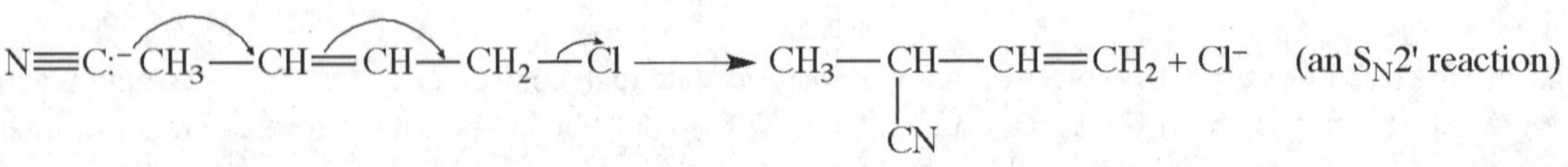

$$N{\equiv}C{:}^- \ CH_3{-}CH{=}CH{-}CH_2{-}Cl \longrightarrow CH_3{-}\underset{\underset{CN}{|}}{C}H{-}CH{=}CH_2 + Cl^- \quad (\text{an } S_N2' \text{ reaction})$$

Problem 7.50 Calculate the rate for the S_N2 reaction of 0.1-M C_2H_5I with 0.1-M CN^- if the reaction rate for 0.01 M concentration is 5.44×10^{-9} mol/L· s.

The rates are proportional to the products of the concentrations:

$$\frac{\text{Rate}}{5.44 \times 10^{-9}\ \text{mol/L}\cdot\text{s}} = \frac{[0.1][0.1]}{[0.01][0.01]}$$

$$\text{Rate} = 100 \times 5.44 \times 10^{-9}\ \text{mol/L}\cdot\text{s} = 5.44 \times 10^{-7}\ \text{mol/L}\cdot\text{s}$$

Problem 7.51 Give reactions for tests that can be carried out rapidly in a test tube to differentiate the following compounds: hexane, $CH_3CH{=}CHCl$, $H_2C{=}CHCH_2Cl$, and $CH_3CH_2CH_2Cl$.

Hexane is readily distinguished from the other three compounds because there is a negative test for Cl^- after Na fusion and treatment with acidic $AgNO_3$. The remaining three compounds are differentiated by their reactivity with alcoholic $AgNO_3$ solution. $CH_3CH{=}CHCl$ is a vinylic chloride and does not react even on heating. $H_2C{=}CHCH_2Cl$ is most reactive (allylic) and precipitates AgCl in the cold, while $CH_3CH_2CH_2Cl$ gives a precipitate of AgCl on warming with the reagent.

Problem 7.52 Will the following reactions be primarily displacement or elimination?

(*a*) $CH_3CH_2CH_2Cl + I^- \longrightarrow$ (*b*) $(CH_3)_3CBr + CN^-$ (ethanol) $\longrightarrow$

(*c*) $CH_3CHBrCH_3 + OH^-$ (H_2O) $\longrightarrow$ (*d*) $CH_3CHBrCH_3 + OH^-$ (ethanol) $\longrightarrow$

(*e*) $(CH_3)_3CBr + H_2O \longrightarrow$

(*a*) S_N2 displacement. I^- is a good nucleophile, and a poor base.
(*b*) E2 elimination. A 3° halide and a fairly strong base.
(*c*) Mainly S_N2 displacement.
(*d*) Mainly E2 elimination. A less polar solvent than that in (*c*) favors E2.
(*e*) S_N1 displacement. H_2O is not basic enough to remove a proton to give elimination.

Problem 7.53 Depending on the solvent, ROH reacts with $SOCl_2$ to give RCl by two pathways, each of which involves formation of a chlorosulfite ester:

$$\mathrm{RO\overset{\overset{\displaystyle O}{\|}}{S}Cl}$$

along with HCl. Use the following stereochemical results to suggest mechanisms for the two pathways: In pyridine, a 3° amine base:

$$(R)\text{-}CH_3CH(OH)CH_2CH_3 \longrightarrow (S)\text{-}CH_3CHClCH_2CH_3$$

and, in either, the same (R)-ROH $\rightarrow$ (R)-RCl.

Pyridine (Py) reacts with the initially formed HCl to give PyH^+Cl^-, and the free nucleophilic Cl^- attacks the chiral C with inversion, displacing the $OSOCl^-$ as SO_2 and Cl^-. With no change in priority, inversion gives the (*S*)-RCl. Ether is too weakly basic to cause enough dissociation of HCl. In the absence of Cl^-, the Cl of the —OSOCl attacks the chiral C from the side to which the group is attached. This **internal nucleophilic substitution (S_Ni)** reaction proceeds through an ion pair and leads to **retention** of configuration.

(H)(CH₃)(C₂H₅)C—OH (*R*) $\xrightarrow[\text{Py}]{SOCl_2}$ PyH^+Cl^- + (H)(CH₃)(C₂H₅)C—O—S(=O)—Cl (*R*) $\xrightarrow{S_N2}$ Cl—C(CH₃)(H)(C₂H₅) (*S*) + SO_2 + Cl^-

Alkyl chlorosulfite

(H)(CH₃)(C₂H₅)C—OH (*R*) $\xrightarrow[\text{Et}_2\text{O}]{SOCl_2}$ HCl + (H)(CH₃)(C₂H₅)C(—O—S(=O)—Cl) (*R*) $\longrightarrow$ (H)(CH₃)(C₂H₅)C^+ Cl^- (*ion pair*) + SO_2 $\xrightarrow{S_Ni}$ (H)(CH₃)(C₂H₅)C—Cl (*R*) + SO_2

Alkynes and Dienes

8.1 Alkynes

Nomenclature and Structure

Alkynes or **acetylenes** (C_nH_{2n-2}) have a —C≡C— and are isomeric with **alkadienes**, which have two double bonds. In IUPAC, a —C≡C— is indicated by the suffix **-yne**.

Acetylene, C_2H_2, is a linear molecule in which each C uses two *sp* HO's to form two σ bonds with a 180° angle. The unhybridized *p* orbitals form two π bonds.

Problem 8.1 Name the structures below by the IUPAC system:

(*a*) $CH_3C{\equiv}CCH_3$

(*b*) $CH_3C{\equiv}CCH_2CH_3$

(*c*) $CH_3—C(CH_3)(H)—C{\equiv}C—C(CH_3)_2—CH_3$

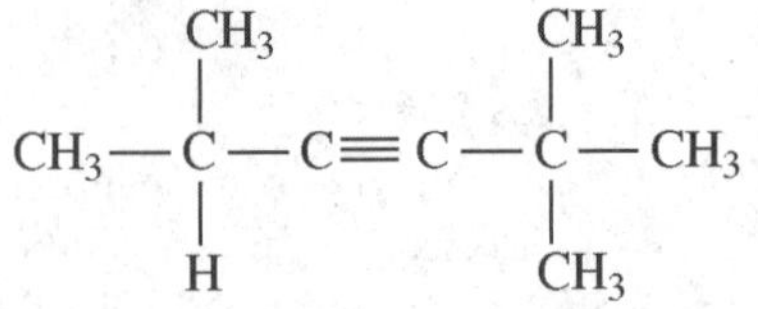

(*d*) $HC{\equiv}C—CH_2CH{=}CH_2$

(*e*) $HC{\equiv}C—CH_2CH_2Cl$

(*f*) $CH_3CH{=}CH—C{\equiv}C—C{\equiv}CH$

(*a*) 2-Butyne
(*b*) 2-Pentyne
(*c*) 2,2,5-Trimethyl-3-hexyne
(*d*) 1-Penten-4-yne
C=C has priority over C≡C and gets the smaller number.
(*e*) 4-Chloro-1-butyne
(*f*) 5-Hepten-1,3-diyne

Problem 8.2 Supply structural formulas and IUPAC names for all alkynes with the molecular formula (*a*) C_5H_8, (*b*) C_6H_{10}.

(*a*) Insert a triple bond where possible in *n*-pentane, isopentane, and neopentane. Placing a triple bond in an *n*-pentane chain gives $H—C{\equiv}C—CH_2CH_2CH_3$ (1-pentyne) and $CH_3—C{\equiv}C—CH_2CH_3$ (2-pentyne). Isopentane gives one compound,

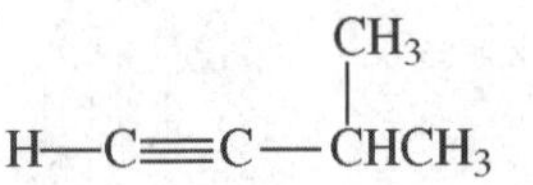

$$H—C{\equiv}C—CH(CH_3)CH_3 \qquad \text{3-Methyl-1-butyne}$$

because a triple bond cannot be placed on a 3° C. No alkyne is obtainable from neopentane, $(CH_3)_2C(CH_3)_2$.

(*b*) Inserting a triple bond in *n*-hexane gives

$H{-}C{\equiv}C{-}CH_2CH_2CH_2CH_3$	$CH_3{-}C{\equiv}C{-}CH_2CH_2CH_3$	$CH_3CH_2{-}C{\equiv}C{-}CH_2CH_3$
1-Hexyne	2-Hexyne	3-Hexyne

Isohexane yields two alkynes, and 3-methylpentane and 2,2-dimethylbutane one alkyne each.

$CH_3\overset{\overset{CH_3}{\vert}}{C}HCH_2C{\equiv}CH$	$CH_3\overset{\overset{CH_3}{\vert}}{C}H{-}C{\equiv}C{-}CH_3$	$HC{\equiv}C{-}\overset{\overset{CH_3}{\vert}}{C}HCH_2CH_3$	$CH_3{-}\overset{\overset{CH_3}{\vert}}{\underset{\underset{CH_3}{\vert}}{C}}{-}C{\equiv}C{-}H$
4-Methyl-1-pentyne	4-Methyl-2-pentyne	3-Methyl-1-pentyne	3,3-Dimethyl-1-butyne

Problem 8.3 Draw models of (*a*) *sp*-hybridized C and (*b*) C_2H_2 to show bonds formed by orbital overlap.

(*a*) See Fig. 8.1(*a*). Only one of three *p* orbitals of C is hybridized. The two unhybridized *p* orbitals (p_z and p_y) are at right angles to each other and also to the axis of the *sp* hybrid orbitals.
(*b*) See Fig. 8.1(*b*). Sidewise overlap of the p_y and p_z orbitals on each C forms the π_y and π_z bonds, respectively.

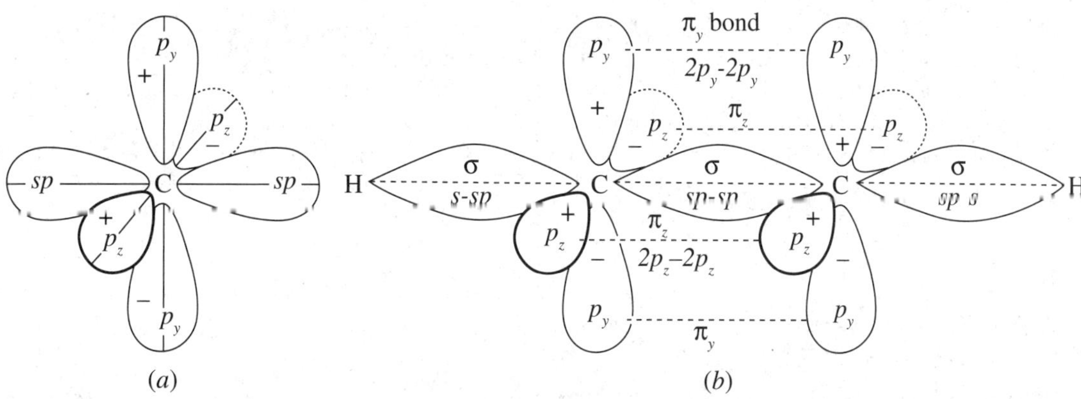

Figure 8.1

Problem 8.4 Why is the C≡C distance (0.120 nm) shorter than the C═C (0.133 nm) and C—C (0.154 nm)?

The carbon nuclei in C≡C are shielded by six electrons (from three bonds) rather than by four or two electrons as in C═C or C—C, respectively. With more shielding electrons present, the C's of —C≡C— can get closer, thereby affording more orbital overlap and stronger bonds.

Problem 8.5 Explain how the orbital picture of —C≡C— accounts for (*a*) the absence of geometric isomers in $CH_3C{\equiv}CC_2H_5$; (*b*) the acidity of an acetylenic H, for example:

$$HC{\equiv}CH + NH_2^- \rightarrow HC{\equiv}C{:}^- + NH_3 \quad (pK_a = 25)$$

(*a*) The *sp*-hybridized bonds are linear, ruling out *cis-trans* isomers in which substituents must be on different sides of the multiple bond.
(*b*) We apply the principle: "The more *s* character in the orbital used by the C of the C—H bond, the more acidic is the H." Therefore, the order of acidity of hydrocarbons is

$$\underset{sp}{\equiv C{-}H} > \underset{sp^2}{-\overset{\vert}{C}{-}H} > \underset{sp^3}{-\overset{\vert}{\underset{\vert}{C}}{-}H}$$

Problem 8.6 (*a*) Relate the observed C—H and C—C bond lengths and bond energies given in Table 8.1 in terms of the hybrid orbitals used by the C's involved. (*b*) Predict the relative C—C bond lengths in CH_3CH_3, $CH_2{=}CH{-}CH{=}CH_2$, and $H{-}C{\equiv}C{-}C{\equiv}C{-}H$.

TABLE 8.1

COMPOUND	BOND	BOND LENGTH, nm	BOND ENERGY, kJ/mol
(1) $CH_3{-}CH_3$	—C—H	0.110	410
(2) $CH_2{=}CH_2$	=C—H	0.108	423
(3) $H{-}C{\equiv}C{-}H$	≡C—H	0.106	460
(4) $CH_3{-}CH_3$	C—C—	0.154	356
(5) $CH_3{-}CH{=}CH_2$	C—C=	0.151	377
(6) $CH_3{-}C{\equiv}C{-}H$	C—C≡	0.146	423

Bond energy increases as bond length decreases; the shorter bond length makes for greater orbital overlap and a stronger bond.

(*a*) The hybrid nature of C is: (1) C_{sp^3}-H_s, (2) C_{sp^2}-H_s, (3) C_{sp}-H_s, (4) C_{sp^3}-C_{sp^3}, (5) C_{sp^3}-C_{sp^2}, and (6) C_{sp^3}-C_{sp}. In going from (1) to (3), the C—H bond length decreases as the *s* character of the hybrid orbital used by C increases. The same situation prevails for the C—C bond in going from (4) to (6). Bonds to C therefore become shorter as the *s* character of the hybridized orbital used by C increases.

(*b*) The hybrid character of the C's in the C—C bond is: for $CH_3{-}CH_3$, C_{sp^3}-C_{sp^3}; $H_2C{=}CH{-}CH{=}CH_2$, C_{sp^2}-C_{sp^2}; and $H{-}C{\equiv}C{-}C{\equiv}C{-}H$, C_{sp}-C_{sp}. Bond length becomes shorter as *s* character increases and hence relative C—C bond lengths should decrease in the order

$$CH_3CH_3 > CH_2{=}CH{-}CH{=}CH_2 > HC{\equiv}C{-}C{\equiv}CH$$

The observed bond lengths are, respectively, 0.154 nm, 0.149 nm, and 0.138 nm.

Laboratory Methods of Preparation

1. Dehydrohalogenation of *vic-Dihalides* or *gem*-Dihalides

The vinyl (alkenyl) halide requires the stronger base sodamide ($NaNH_2$).

$$\underset{\text{a } gem\text{-dihalide}}{-\overset{H}{\underset{H}{C}}-\overset{X}{\underset{X}{C}}-} \text{ or } \underset{\text{a } vic\text{-dihalide}}{-\overset{H}{\underset{X}{C}}-\overset{H}{\underset{X}{C}}-} \xrightarrow{\text{alc. KOH}} KX + H_2O + \underset{\text{a vinly halide}}{-\overset{H}{C}{=}\underset{X}{C}-} \xrightarrow{NaNH_2} NaX + NH_3 + \underset{\text{an acteylene}}{-C{\equiv}C-}$$

(dihalide directly to acetylene: $NaNH_2$, liq. NH_3(b.p. –33 °C))

2. Primary Alkyl Substitution in Acetylene; Acidity of ≡C—H [see Problem 8.5(*b*)]

$$R{-}C{\equiv}C{-}H + \left\{\begin{matrix} NaNH_2 \\ \text{or} \\ Na \end{matrix}\right. \longrightarrow \underset{\text{an acetylide anion}}{R{-}C{\equiv}C{:}^- \; Na^+} + \left\{\begin{matrix} NH_3 \\ \text{or} \\ \tfrac{1}{2}H_2 \end{matrix}\right.$$

$$R{-}C{\equiv}C{:}^- + 1^\circ\ R'{-}CH_2(:X) \longrightarrow R{-}C{\equiv}C{-}CH_2{-}R' + X^- \quad (S_N2 \text{ mechanism})$$

Problem 8.7 Explain why $CH_3CHBrCH_2Br$ *does not* react with KOH to give $CH_2{=}CHCH_2Br$.

In E2 eliminations, the more acidic H is removed preferably. The inductive effect of the Br's increases the acidities of the H's on the C's to which the Br's are bonded. To get this product, the less acidic H (one of the CH_3 group) must be removed.

Problem 8.8 Outline a synthesis of propyne from isopropyl or propyl bromide.

The needed *vic*-dihalide is formed from propene, which is prepared from either of the alkyl halides.

$$\left.\begin{array}{c}\underset{n\text{-Propyl bromide}}{CH_3CH_2CH_2Br}\\ \text{or}\\ \underset{\text{Isopropyl bromide}}{CH_3CHBrCH_3}\end{array}\right\}\xrightarrow[\text{KOH}]{\text{alc.}}\underset{\text{Propene}}{CH_3{-}CH{=}CH_2}\xrightarrow{Br_2}\underset{\text{Propylene bromide}}{CH_3CHBrCH_2Br}\xrightarrow[\text{KOH}]{\text{alc.}}\underset{\text{Propenyl bromide}}{CH_3CH{=}CHBr}\xrightarrow{NaNH_2}\underset{\text{Propyne}}{CH_3C{\equiv}CH}$$

(Propylene bromide $\xrightarrow{NaNH_2}$ Propyne directly.)

Problem 8.9 Synthesize the following compounds from $HC{\equiv}CH$ and any other organic and inorganic reagents (do not repeat steps): (*a*) 1-pentyne, (*b*) 2-hexyne.

$$(a)\quad H{-}C{\equiv}C{-}H\xrightarrow{NaNH_2}H{-}C{\equiv}\bar{C}{:}\overset{+}{Na}\xrightarrow{CH_3CH_2CH_2I}H{-}C{\equiv}C{-}CH_2CH_2CH_3$$

$$(b)\quad \overset{+}{Na}\,{:}\bar{C}{\equiv}C{-}H\xrightarrow{CH_3I}CH_3{-}C{\equiv}C{-}H\xrightarrow{NaNH_2}CH_3{-}C{\equiv}\bar{C}{:}\overset{+}{Na}\xrightarrow{CH_3CH_2CH_2I}CH_3{-}C{\equiv}C{-}CH_2CH_2CH_3$$

Problem 8.10 Industrially, acetylene is made from calcium carbide, $CaC_2 + 2H_2O \rightarrow HC{\equiv}CH + Ca(OH)_2$. Formulate the reaction as a Brönsted acid-base reaction.

The carbide anion C_2^{2-} is the base formed when $HC{\equiv}CH$ loses two H^+'s.

$$\underset{\text{base}_1}{[{:}C{\equiv}C{:}]^{2-}}+\underset{\text{acid}_2}{2HOH}\longrightarrow\underset{\text{acid}_1}{H{-}C{\equiv}C{-}H}+\underset{\text{base}_2}{2OH^-}\quad(Ca^{2+}\text{ precipitates as }Ca(OH)_2)$$

8.2 Chemical Properties of Acetylenes

Addition Reactions at the Triple Bond

Nucleophilic π electrons of alkynes add H_2 and electrophiles in reactions similar to additions to alkenes. Alkynes can add one or two moles of reagent but are less reactive (except to H_2) than alkenes.

1. Hydrogen

$$(a)\quad CH_3{-}C{\equiv}C{-}CH_2CH_3+2H_2\xrightarrow{Pt}CH_3CH_2CH_2CH_2CH_3$$

$$(b)\quad \underset{cis\text{-2-Pentene}}{\overset{CH_3\quad C_2H_5}{\underset{H\qquad H}{C{=}C}}}\xleftarrow[\substack{cis\ (syn)\\ \text{addition}}]{H_2/Pd(Pb)\ (\text{Lindlar's catalyst})}\underset{\substack{\text{2-Pentyne}\\ stereospecific\ reductions}}{\boxed{CH_3{-}C{\equiv}C{-}C_2H_5}}\xrightarrow[\substack{trans\ (anti)\\ \text{addition}}]{\text{Na, liq. }NH_3}\underset{trans\text{-2-Pentene}}{\overset{CH_3\qquad H}{\underset{H\quad C_2H_5}{C{=}C}}}$$

2. HX (HCl, HBr, HI)—an *anti* addition for first mole

$$CH_3{-}C{\equiv}C{-}H\xrightarrow{HBr}CH_3{-}CBr{=}CH_2\xrightarrow{HBr}\underset{\text{a }gem\text{-dihalide}}{CH_3{-}CBr_2{-}CH_3}\quad(\text{Markovnikov addition})$$

$$CH_3{-}C{\equiv}C{-}H+HBr\xrightarrow{\text{peroxide}}CH_3{-}CH{=}CHBr\quad(\text{anti-Markovnikov})$$

3. **Halogen** (Br_2, Cl_2)— an *anti* addition for first mole

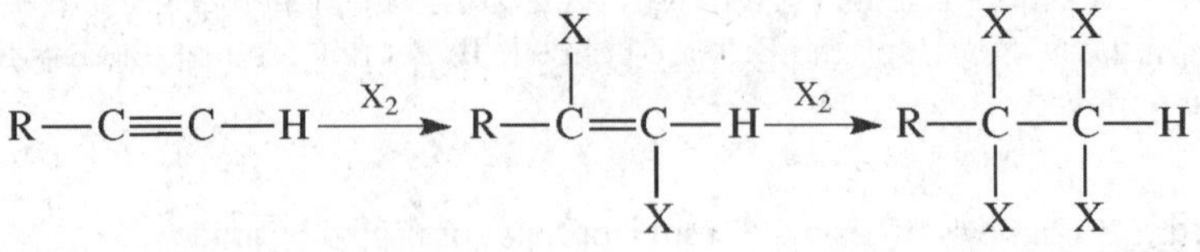

$$R-C\equiv C-H \xrightarrow{X_2} R-\underset{X}{\overset{X}{\underset{|}{\overset{|}{C}}}}=C-H \xrightarrow{X_2} R-\underset{X}{\overset{X}{C}}-\underset{X}{\overset{X}{C}}-H$$

4. **H_2O (Hydration to Carbonyl Compounds)**

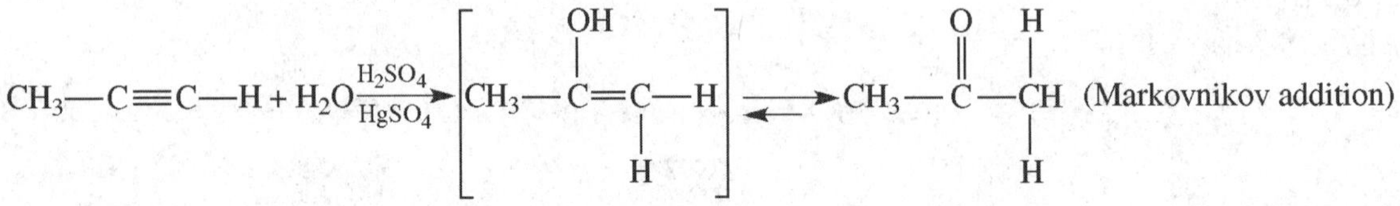

$$CH_3-C\equiv C-H + H_2O \xrightarrow[HgSO_4]{H_2SO_4} \left[CH_3-\overset{OH}{C}=\underset{H}{C}-H\right] \rightleftharpoons CH_3-\overset{O}{\overset{\|}{C}}-\underset{H}{\overset{H}{CH}} \text{ (Markovnikov addition)}$$

Propyne a *vinyl alcohol* (*enol*) (*unstable*) Acetone

5. **Boron Hydride**

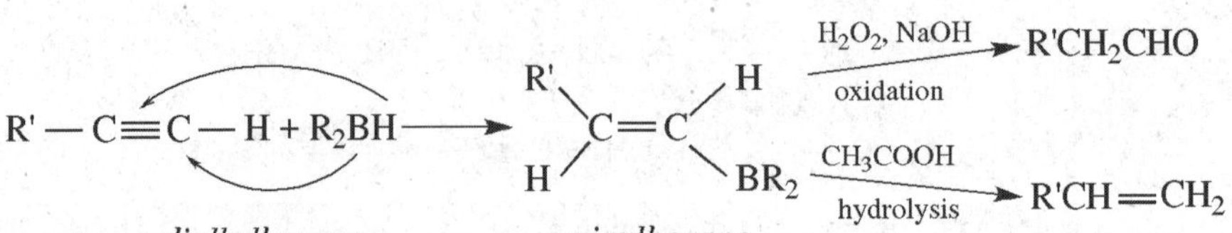

$$R'-C\equiv C-H + R_2BH \longrightarrow \underset{H}{\overset{R'}{}}C=C\underset{BR_2}{\overset{H}{}} \quad \begin{cases} \xrightarrow[\text{oxidation}]{H_2O_2,\ NaOH} R'CH_2CHO \\ \xrightarrow[\text{hydrolysis}]{CH_3COOH} R'CH=CH_2 \end{cases}$$

a *dialkylborane* a *vinylborane*

With dialkylacetylenes, the products of hydrolysis and oxidation are *cis*-alkenes and ketones, respectively.

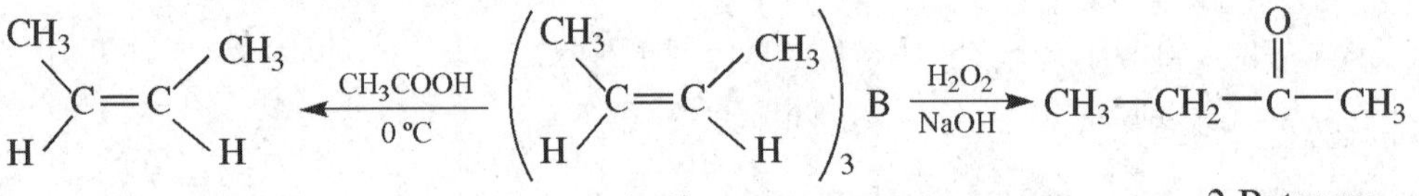

$$\underset{H}{\overset{CH_3}{}}C=C\underset{H}{\overset{CH_3}{}} \xleftarrow[0\,^\circ C]{CH_3COOH} \left(\underset{H}{\overset{CH_3}{}}C=C\underset{H}{\overset{CH_3}{}}\right)_3 B \xrightarrow[NaOH]{H_2O_2} CH_3-CH_2-\overset{O}{\overset{\|}{C}}-CH_3$$

cis-2-Butene a *vinylborane* 2-Butanone *a ketone*

6. **Dimerization**

$$2H-C\equiv C-H \xrightarrow[H_2O]{\overset{+}{Cu(NH_3)_2}\,Cl^-} H_2C=CH-C\equiv C-H$$

Vinylacetylene

7. **Nucleophiles**

$$CH_3C\equiv CCH_3 + CN^- \quad HCN \longrightarrow CH_3CH=C(CN)CH_3$$

Problem 8.11 In terms of the mechanism, explain why alkynes are less reactive than alkenes toward electrophilic addition of, for example, HX or BR_2.

The mechanism of electrophilic addition is similar for alkenes and alkynes. When HX adds to a triple bond, the intermediate is a carbocation having a positive charge on an *sp*-hybridized C atom:

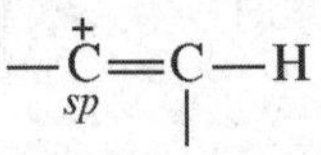

$$-\underset{sp}{\overset{+}{C}}=\underset{|}{C}-H$$

This vinyl-type carbocation is less stable than its analog formed from an alkene, which has the positive charge on an sp^2-hybridized C atom:

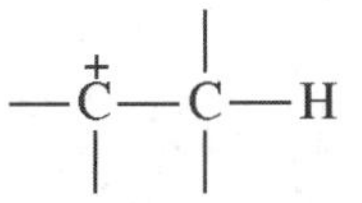

$$-\overset{+}{\underset{|}{C}}-\overset{|}{\underset{|}{C}}-H$$

An addendum such as Br_2 forms an intermediate bromonium-type ion:

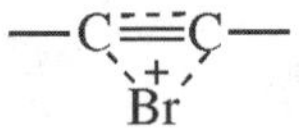

$$-C\equiv C- \text{ (bridged by } \overset{+}{Br}\text{)}$$

In this ion, some positive charge is dispersed to the C's, which, because of their *sp*-like hybrid character, are less able to bear the positive charge than the sp^2 C's in the alkene's bromonium ion. Such situations cause alkynes to be less reactive than alkenes toward Br_2.

Problem 8.12 Alkynes differ from alkenes in adding nucleophiles such as CN^-. Explain.

The intermediate carbanion from addition of CN^- to an alkyne has the unshared electron pair on an sp^2-hybridized C. It is more stable and is formed more readily than the sp^3-hybridized carbanion formed from a nucleophile and an alkene.

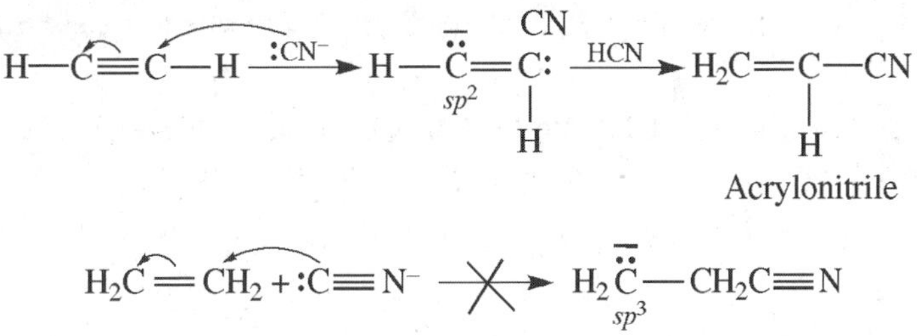

$$H-C\equiv C-H \xrightarrow{:CN^-} \underset{sp^2}{H-\ddot{C}^-}=\underset{H}{\overset{CN}{C}}: \xrightarrow{HCN} H_2C=\underset{H}{C}-CN$$

Acrylonitrile

$$H_2C=CH_2 + :C\equiv N^- \not\longrightarrow H_2\underset{sp^3}{\ddot{C}^-}-CH_2C\equiv N$$

Problem 8.13 Dehydrohalogenation of 3-bromohexane gives a mixture of *cis*-2-hexene and *trans*-2-hexene. How can this mixture be converted to pure (*a*) *cis*-2-hexene? (*b*) *trans*-2-hexene?

Relatively pure alkene geometric isomers are prepared by stereoselective reduction of alkynes.

(*a*) Hydrogenation of 2-hexyne with Lindlar's catalyst gives 98% *cis*-2-hexene.

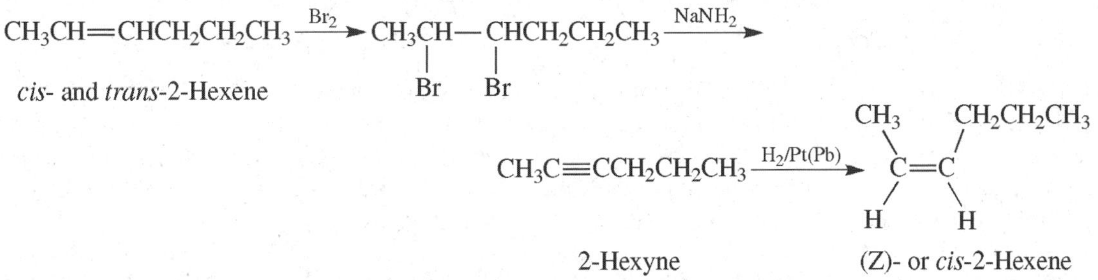

$$\underset{cis\text{- and }trans\text{-2-Hexene}}{CH_3CH=CHCH_2CH_2CH_3} \xrightarrow{Br_2} CH_3\underset{Br}{CH}-\underset{Br}{CH}CH_2CH_2CH_3 \xrightarrow{NaNH_2}$$

$$\underset{\text{2-Hexyne}}{CH_3C\equiv CCH_2CH_2CH_3} \xrightarrow{H_2/Pt(Pb)} \underset{(Z)\text{- or } cis\text{-2-Hexene}}{(CH_3)(H)C=C(CH_2CH_2CH_3)(H)}$$

(*b*) Reduction of 2-hexyne with Na in liquid NH_3 gives the *trans* product.

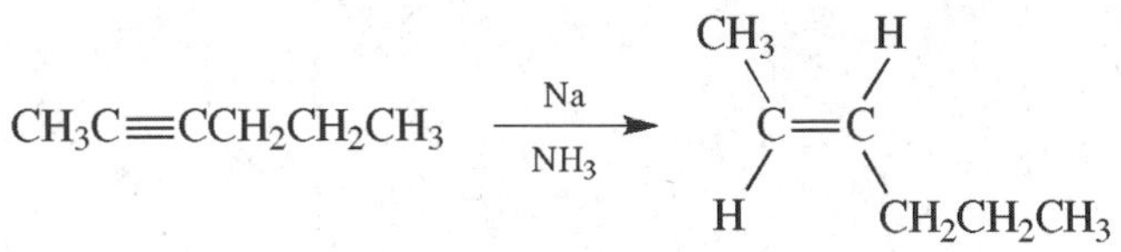

$$CH_3C\equiv CCH_2CH_2CH_3 \xrightarrow[NH_3]{Na} \underset{(E)\text{- or } trans\text{-2-Hexene}}{(CH_3)(H)C=C(H)(CH_2CH_2CH_3)}$$

Problem 8.14 Outline steps for the conversion of $CH_3CH_2CH_2Br$ to (*a*) $CH_3CBr=CH_2$, (*b*) $CH_3CCl_2CH_3$, (*c*) $CH_3CH=CHBr$.

As usual, we think backward (the **retrosynthetic** approach). Each product is made from $CH_3C{\equiv}CH$, which in turn is synthesized from $CH_3CH{=}CH_2$.

$$CH_3CH_2CH_2Br \xrightarrow{\text{alc. KOH}} CH_3CH{=}CH_2 \xrightarrow{Br_2} CH_3CHBrCH_2Br \xrightarrow{NaNH_2} CH_3C{\equiv}CH \begin{cases} \xrightarrow{2HCl} (b)\ CH_3CCl_2CH_3 \\ \xrightarrow{HBr} (a)\ CH_3CBr{=}CH_2 \\ \xrightarrow[\text{peroxides}]{HBr} (c)\ CH_3CH{=}CHBr \end{cases}$$

Acidity and Salts of 1-Alkynes [see Problem 8.5(b)]

$$CH_3C{\equiv}CH + Ag^+ \xrightarrow[-H^+]{NH_3} CH_3C{\equiv}CAg(s) \xrightarrow{HNO_3} CH_3C{\equiv}CH + Ag^+$$

Problem 8.15 Will the following compounds react? Give any products and the reason for their formation.

(*a*) $CH_3{-}C{\equiv}C{-}H + \text{aq. } Na^+\,OH^- \longrightarrow$
(*b*) $CH_3CH_2C{\equiv}C{-}MgI + CH_3OH \longrightarrow$
(*c*) $CH_3C{\equiv}C{:}^-\,Na^+ + NH_4^+ \longrightarrow$

(*a*) No. The products would be the stronger acid H_2O and the stronger base $CH_3C{\equiv}C{:}^-$.
(*b*) Yes. The products are the weaker acid $CH_3CH_2C{\equiv}C{:}H$ and the weaker base $MgI(OCH_3)$.
(*c*) Yes. The products are the weaker acid propyne and the weaker base NH_3.

Problem 8.16 Deduce the structure of a C_5H_8 compound which forms a precipitate with Ag^+ and is reduced to 2-methylbutane.

The precipitate shows an acetylene bond at the end of a chain with an acidic H. With $-C{\equiv}CH$, the other three carbons must be present, as a $(CH_3)_2CH-$ group, because of reduction of $(CH_3)_2CH{-}C{\equiv}CH$ to $(CH_3)_2CHCH_2CH_3$.

8.3 Alkadienes

Problem 8.17 Name by the IUPAC method and classify as *cumulated, conjugated,* or *isolated*:

(*a*) $H_2C{=}CH{-}CH{=}CHCH_3$
(*b*) $H_2C{=}CHCH_2\overset{\displaystyle CH_2CH_3\atop|}{C}{=}CHCH_2CH_3$
(*c*) $H_2C{=}C{=}CH_2$
(*d*) $H_2C{=}CH{-}CH{=}CHCH{=}CH_2$

(*a*) 1,3-Pentadiene. *Conjugated* diene, since it has alternating double and single bonds—that is, $-C{=}C{-}C{=}C-$. (*b*) 4-Ethyl-1,4-heptadiene. *Isolated* dience, since the double bonds are separated by at least one sp^3-hybridized C—in other words, $-C{=}C{-}(CH_2)_n{-}C{=}C-$. (*c*) 1,2-Propadiene (allene). *Cumulated* diene, since two double bonds are on the same C—in other words, $-C{=}C{=}C-$. (*d*) 1,3,5-Hexatriene. *Conjugated diene,* since it has alternating single and double bonds.

Problem 8.18 Compare the stabilities of the three types of dienes from the following heats of hydrogenation, ΔH_h (in kJ/mol). (For comparison, ΔH_h for 1-pentene is -126.)

Conjugated	$H_2C{=}\overset{\displaystyle H\atop|}{C}{-}CH{=}CH{-}CH_3$ 1,3-Pentadiene	–230
Isolated	$H_2C{=}CH{-}CH_2{-}CH{=}CH_2$ 1,4-Pentadiene	–252
Cumulated	$H_2C{=}C{=}CH{-}CH_2CH_3$ 1,2-Pentadiene	–297

The calculated ΔH_h, assuming no interaction between the double bonds, is $2(-126) = -252$. The more negative the observed value of ΔH_h compared to -252, the less stable the diene; the less negative the observed value, the more stable the diene. Conjugated dienes are most stable, and cumulated dienes are least stable; under the proper conditions, allenes tend to rearrange to conjugated dienes.

Problem 8.19 Give steps for the conversion $HC{\equiv}CCH_2CH_2CH_3 \rightarrow H_2C{=}CH{-}CH{=}CHCH_3$.

$$HC{\equiv}CCH_2CH_2CH_3 \xrightarrow{H_2/Pt(Pb)} H_2C{=}CHCH_2CH_2CH_3 \xrightarrow[500^\circ C]{Cl_2}$$

(allylic substitution)

$$H_2C{=}CHCHClCH_2CH_3 \xrightarrow{\text{alc. KOH}} H_2C{=}CHCH{=}CHCH_3$$

Problem 8.20 Account for the stability of conjugated dienes by (*a*) extended π bonding, (*b*) resonance theory.

(*a*) The four *p* orbitals of conjugated dienes are adjacent and parallel (Fig. 8.2) and overlap to form an extended π system involving all four C's. This results in *greater stability* and decreased energy.

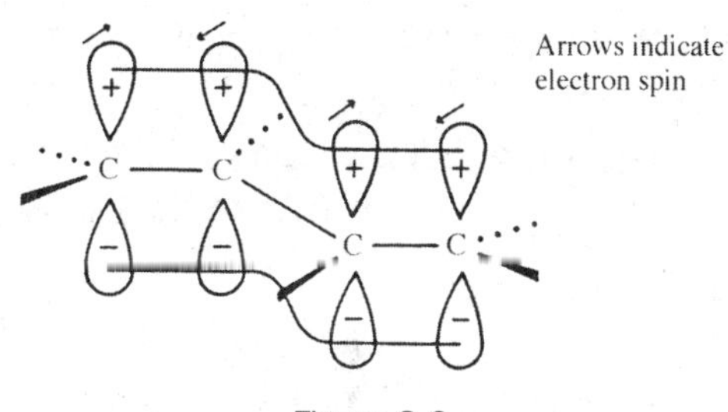
Arrows indicate electron spin

Figure 8.2

(*b*) A conjugated diene is a resonance hybrid:

$$-\overset{|}{C}{=}\overset{|}{C}-\overset{|}{C}{=}\overset{|}{C}- \longleftrightarrow -\overset{..}{\underset{|}{C}}-\overset{|}{C}{=}\overset{|}{C}-\overset{+}{\underset{|}{C}}- \longleftrightarrow -\overset{+}{\underset{|}{C}}-\overset{|}{C}{=}\overset{|}{C}-\overset{..}{\underset{|}{C}}-$$

(i)

Structure (i) has 11 bonds and makes a more significant contribution than the other two structures, which have only 10 bonds. Since the contributing structures are not equivalent, the resonance energy is small.

8.4 MO Theory and Delocalized π Systems

Review Section 2.2. The MO theory focuses attention on the interacting *p* AO's of the delocalized π systems, such as conjugated polyenes. The theory states that the number of interacting *p* AO's is the same as the number of π molecular orbitals formed. The molecular orbitals are considered to be stationary waves, and their relative energies increase as the numbers of nodal points in the corresponding waves increase. Nodes may appear at a C atom, as indicated with a 0 rather than a + or a − sign. In a linear system with an even number of molecular orbitals, half are bonding MO's and half are antibonding MO*'s. With an odd number of molecular orbitals, the middle-energy molecular orbital is nonbonding (MO^n). The electrons in the delocalized π system are placed first in the bonding, then in the nonbonding (if present), and then, if necessary, into the antibonding, molecular orbitals—with no more than two electrons in each molecular orbital. Electrons in MO's add to bonding strength; those in

MO*'s diminish bonding strength; those in MO^n's have no effect. We often simplify our representations of molecular orbitals by showing only the signs of the upper lobes of the *p* AO's, and not the entire orbital.

Problem 8.21 Apply the MO theory to the π system of ethene.

Each of the doubly bonded C's, C═C, has a *p* AO. These two *p* AO's provide two molecular orbitals: a lower-energy bonding MO and a higher-energy antibonding MO*. Each *p* AO has one electron, giving two electrons for placement in the π molecular orbitals. Molecular orbitals receive electrons in the order of their increasing energies, with no more than two of opposite spins in any given molecular orbital. For ethene, the two *p* electrons, shown as ↑ and ↓, are placed in the bonding MO (π); the antibonding MO*(π*) is devoid of electrons. Note the simplification of showing only the signs of the upper lobes of the interacting *p* orbitals. The stationary waves, with any nodes, are shown superimposed on the energy levels.

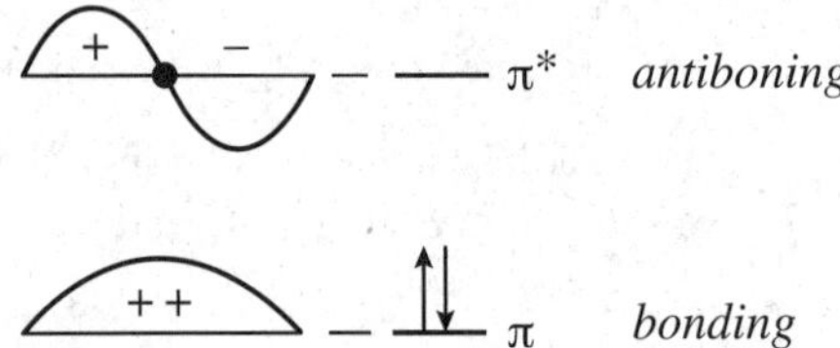

Problem 8.22 Apply the MO theory to 1,3-butadiene, and compare the relative energies of its molecular orbitals with those of ethene (Problem 8.21).

Four *p* AO's (see Fig. 8.2) give four molecular orbitals, as shown in Fig. 8.3. Wherever there is a switch from + to −, there is a node, as indicated by a heavy dot. Note that π_1 of the diene has a lower energy than π of ethene.

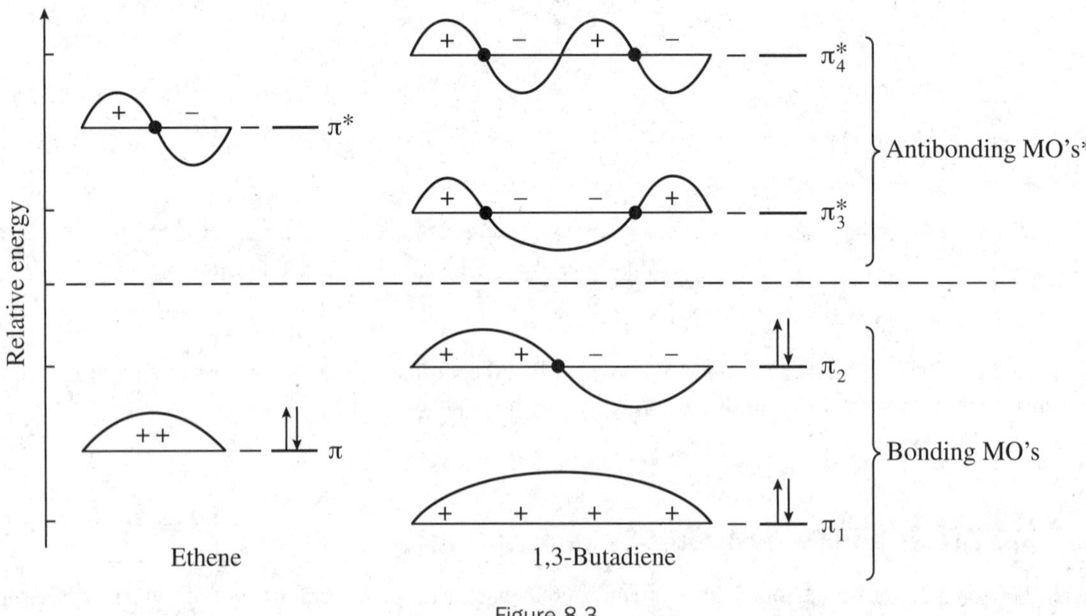

Figure 8.3

In a linear π system, the relative energies of the molecular orbitals are determined by the pairwise overlaps of adjacent *p* orbitals along the chain. An excess of bonding interactions, + with + or − with −, denotes a bonding MO; an excess of antibonding interactions, + with −, denotes an antibonding MO*.

Problem 8.23 Explain how the energies shown in Fig. 8.3 are consistent with the fact that a conjugated diene is more stable than an isolated diene.

The energy of $\pi_1 + \pi_2$ of the conjugated diene is less than twice the energy of an ethene π bond. Two ethene π bonds correspond to an isolated diene.

Problem 8.24 (*a*) Apply the MO theory to the allyl system. Indicate the relative energies of the molecular orbitals, and state if they are bonding, nonbonding, or antibonding. (*b*) Insert the electrons for the carbocation $C_3H_5^+$, the free radical $C_3H_5\cdot$, and the carbanion $C_3H_5^-$, and compare the relative energies of these three species.

(*a*) Three *p* AO's give three molecular orbitals, as indicated in Fig. 8.4. Since there are an odd number of *p* AO's in this linear system, the middle-energy molecular orbital is nonbonding (π_2^n). Note that the node of this MO^n is at a C, indicated by a 0. An MO^n can be recognized if the number of bonding pairs equals the number of antibonding pairs or if there is no overlap.

(*b*) R^+ $\quad \pi_1 \uparrow\downarrow \qquad R\cdot \quad \begin{matrix}\pi_2^n \uparrow \\ \pi_1 \uparrow\downarrow\end{matrix} \qquad R:^- \quad \begin{matrix}\pi_2^n \uparrow\downarrow \\ \pi_1 \uparrow\downarrow\end{matrix}$

The electrons in the π_2^n orbital do not appreciably affect the stability of the species. Therefore, all three species are more stable than the corresponding alkyl systems $C_3H_7^+$, $C_3H_7\cdot$, and $C_3H_7^-:$. The extra electrons do increase the repulsive forces between electrons slightly, so the order of stability is $C_3H_5^+ > C_3H_5\cdot > C_3H_5^-$.

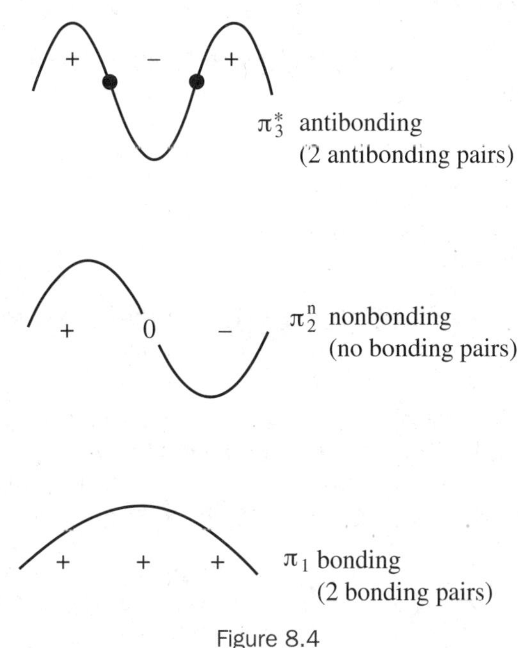

Figure 8.4

8.5 Addition Reactions of Conjugated Dienes

1,2- and 1,4-Additions

Typical of conjugated dienes, 1,3-butadiene undergoes both 1,2- and 1,4-addition, as illustrated with HBr:

$$\underset{\displaystyle H\qquad\qquad\quad Br}{H_2C\text{—}CH\text{=}CHCH_2} \xleftarrow{\text{1,4-addition}} \boxed{H_2C\text{=}CHCH\text{=}CH_2 + HBr} \xrightarrow{\text{1,2-addition}} \underset{\displaystyle H\quad Br}{H_2C\text{—}CH\text{—}CH\text{=}CH}$$

Problem 8.25 Explain 1,4-addition in terms of the mechanism of electrophilic addition.

The electrophile (H^+) adds to form an allylic carbocation with positive charge delocalized at C^2 and C^4 (resonance forms II and III). This cation adds the nucleophile at C^2 to form the 1,2-addition product or at C^4 to form the 1,4-addition product.

$$\underset{(\mathrm{I})}{H_2C{=}CH{-}CH{=}CH_2} \xrightarrow[-Br^-]{H{:}Br} \left\{ \begin{array}{c} \overset{1}{H_2C}(H){-}\overset{2}{\overset{+}{C}H}{-}\overset{3}{C}H{=}\overset{4}{C}H_2 \ (\mathrm{II}) \\ \updownarrow \\ \overset{1}{H_2C}(H){-}\overset{2}{C}H{=}\overset{3}{C}H{-}\overset{4}{\overset{+}{C}H_2} \ (\mathrm{III}) \end{array} \right\} \xrightarrow{+Br^-} \left\{ \begin{array}{c} H_2C(H){-}CH(Br){-}CH{=}CH_2 \\ \text{1,2-adduct} \\ H_2C(H){-}CH{=}CH{-}CH_2(Br) \\ \text{1,4-adduct} \end{array} \right.$$

The relative rates of formation of carbocations are: 3°, 2° allyl > 1° allyl, 2° > 1° > CH_3^+.

Controlling Factors

A **rate-controlled** reaction is one whose major product is formed through the transition state with the lowest $\Delta H^\ddagger$. A **thermodynamic-controlled** reaction is one whose major product has the lower (more negative) ΔH of reaction. Reactions may shift from rate to thermodynamic control with increasing temperature, especially when the formation of the rate-controlled product is reversible.

Problem 8.26 Use an enthalpy-reaction diagram to explain the following observations. Start from the allylic carbocation, the common intermediate.

$$HBr + H_2C{=}CH{-}CH{=}CH_2 \xrightarrow{-80^\circ C} 80\%\ \underset{\text{1,2-adduct}}{H_2C(H){-}CH(Br){-}CH{=}CH_2} + 20\%\ \underset{\text{1,4-adduct}}{H_2C(H){-}CH{=}CH{-}CH_2(Br)}$$

$$\downarrow +40^\circ C$$

$$HBr + H_2C{=}CH{-}CH{=}CH_2 \xrightarrow{+40^\circ C} 20\%\ H_2C(H){-}CH(Br){-}CH{=}CH_2 + 80\%\ H_2C(H){-}CH{=}CH{-}CH_2(Br)$$

The different products arise from enthalpy differences in the second step, the reaction of Br^- and the allyl R^+. See Fig. 8.5. At −80°C, the 1,2-adduct, the rate-controlled product, is favored because its formation has the lower $\Delta H^\ddagger$. 1,2-Adduct formation can reverse to refurnish the intermediate allylic carbocation, R^+. At 40°C, R^+ goes through the higher-energy transition state for formation of the more stable 1,4-adduct, the thermodynamic-controlled

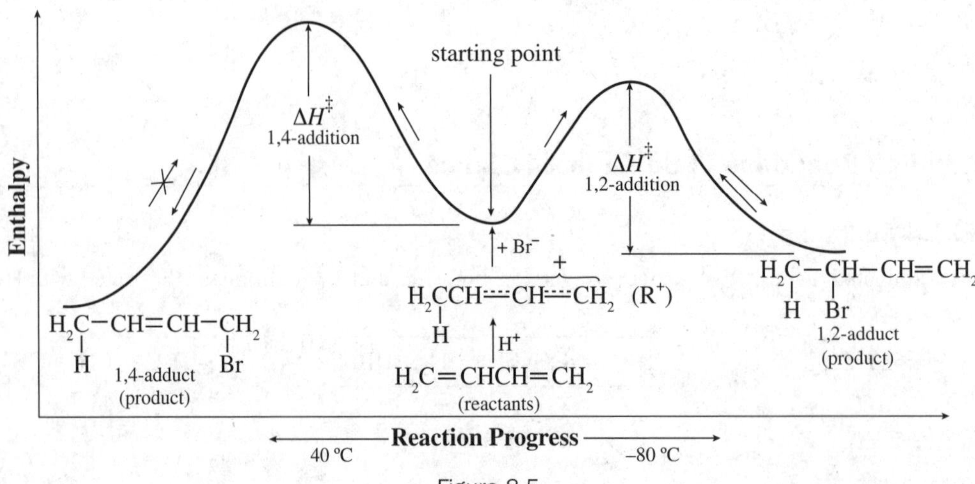

Figure 8.5

product. The 1,4-adduct accumulates because the addition, having a greater $\Delta H^{\ddagger}$, is more difficult to reserve than that for the 1,2-adduct. The 1,4-adduct has a lower enthalpy because it has more R groups on the C$=$C.

Problem 8.27 Explain why the conjugated 1,3-pentadiene reacts with one mole of Br_2 at a faster rate than does the isolated 1,4-pentadiene.

The reaction products are shown:

$$\underset{\text{conjugated}}{H_2C{=}CH{-}CH{=}CH{-}CH_3} \xrightarrow{+Br_2} \left[\begin{array}{c} BrCH_2{-}\overset{+}{C}H{-}CH{=}CH{-}CH_3 \\ \updownarrow \\ BrCH_2{-}CH{=}CH{-}\overset{+}{C}H{-}CH_3 \end{array}\right] \xrightarrow{+Br^-} \begin{array}{l} BrCH_2{-}CH(Br){-}CH{=}CH{-}CH_3 \\ BrCH_2{-}CH{=}CH{-}CH(Br){-}CH_3 \end{array}$$

an allylic R^+

$$\underset{\text{isolated}}{H_2C{=}CH{-}CH{-}CH{=}CH_2} + Br_2 \longrightarrow \underset{\text{an isolated } R^+}{BrCH_2{-}\overset{+}{C}H{-}CH_2{-}CH{=}CH_2} + Br^- \longrightarrow BrCH_2{-}CH(Br){-}CH_2{-}CH{=}CH_2$$

The intermediate carbocation formed from the conjugated diene is allylic and is more stable than the isolated carbocation from the isolated diene. Since the transition state for the rate-controlling first step leading to the lower-enthalpy allylic R^+ also has a lower enthalpy, $\Delta H^{\ddagger}$ for this reaction is smaller and the reaction is faster. It is noteworthy that although the conjugated diene is more stable, it nevertheless reacts faster.

Problem 8.28 (*a*) Which additional C's in the following R^+ bear some + charge?

$$-\underset{7}{C}{=}\underset{6}{C}{-}\underset{5}{C}{=}\underset{4}{C}{-}\underset{3}{C}{=}\underset{2}{C}{-}\underset{1}{\overset{+}{C}}-$$

(*b*) With which of these C's will $:Nu^-$ react to give the thermodynamic-controlled product?

(*a*) C^3, C^5, and C^7. These are alternating sites.

$$\underset{7}{C}{=}\underset{6}{C}{-}\underset{5}{C}{=}\underset{4}{C}{-}\underset{3}{\overset{+}{C}}{-}\underset{2}{C}{=}\underset{1}{C} \longleftrightarrow \underset{7}{C}{=}\underset{6}{C}{-}\underset{5}{\overset{+}{C}}{-}\underset{4}{C}{=}\underset{3}{C}{-}\underset{2}{C}{=}\underset{1}{C} \longleftrightarrow \underset{7}{\overset{+}{C}}{-}\underset{6}{C}{=}\underset{5}{C}{-}\underset{4}{C}{=}\underset{3}{C}{-}\underset{2}{C}{=}\underset{1}{C}$$

(*b*) $:Nu^-$ adds to equivalent C^1 or C^7 to give the conjugated triene

$$-C{=}C{-}C{=}C{-}C{=}C{-}C{-}Nu \quad \text{(most stable)}$$

Addition at C^3 (or C^5) gives a triene with only two conjugated C$=$C's:

$$-C{=}C{-}\underset{Nu}{C}{-}C{=}C{-}C{=}C- \quad \text{(less stable)}$$

Problem 8.29 Write structural formulas for major and minor products from acid-catalyzed dehydration of $H_2C{=}CHCH_2CH(OH)CH_3$.

Dehydration can occur by removal of H from either C^3 or C^5.

$$\underset{\substack{\text{4-Hydroxy-1-pentene}\\ \text{(1-Penten-4-ol)}}}{\overset{5\;\;4\;\;3\;\;2\;\;1}{H_2C(H^5)\text{—}CH(OH)\text{—}CH(H^3)\text{=}CH\text{—}CH_2}} \longrightarrow \underset{\substack{\text{H (from }-H^3)\\ \text{1,3-Pentadiene (major)}\\ \text{conjugated diene}}}{H_2C\text{—}CH\text{=}CH\text{—}CH\text{=}CH_2} + \underset{\substack{\text{H (from }-H^5)\\ \text{1,4-Pentadiene (minor)}\\ \text{isolated diene}}}{H_2C\text{=}CH\text{—}CH\text{—}CH\text{=}CH_2}$$

Problem 8.30 Write the structures of the intermediate R^+'s and the two products obtained from the reaction of $H_2C{=}C(CH_3)CH{=}CH_2$ with (*a*) HBr, (*b*) Cl_2.

(*a*) H^+ adds to C^1 to form the more stable allylic 3° R^+, rather than to C^2 or C^3 to form the nonallylic 1° R^+'s $H_2\overset{+}{C}\text{—}CH(CH_3)CH{=}CH_2$ and $H_2C{=}C(CH_3)CH_2\text{—}\overset{+}{C}H_2$, respectively; or to C^4 to yield the 2° allylic R^+ $H_2C{=}C(CH_3)\overset{+}{C}HCH_3$.

$$\underset{1\;\;\;2\;\;\;3\;\;\;4}{H_2C{=}C(CH_3)\text{—}CH{=}CH_2} + \overset{\delta+\,\delta-}{HBr} \longrightarrow CH_3\text{—}\overset{+}{C}(CH_3)\text{—}CH{=}CH_2 \longleftrightarrow CH_3\text{—}C(CH_3){=}CH\text{—}\overset{+}{C}H_2$$

$$\xrightarrow{Br^-} \underset{\substack{\text{3-Methyl-3-bromo-}\\ \text{1-butene}\\ \text{(minor)}}}{CH_3\text{—}CBr(CH_3)\text{—}CH{=}CH_2} + \underset{\substack{\text{1-Bromo-3-methyl-}\\ \text{2-butene}\\ \text{(major)}}}{CH_3\text{—}C(CH_3){=}CH\text{—}CH_2Br}$$

(*b*) Cl^+ also adds to C^1 to form a hybrid allylic R^+.

$$H_2C{=}C(CH_3)\text{—}CH{=}CH_2 + Cl_2 \longrightarrow ClH_2C\text{—}C(CH_3)\overset{\cdots}{=}CH\overset{\cdots}{=}CH_2\;(+) + Cl^- \longrightarrow \underset{\substack{\text{3-Methyl-3,4-}\\ \text{dichloro-1-butene}\\ \text{(minor)}}}{ClH_2C\text{—}CCl(CH_3)\text{—}CH{=}CH_2} + \underset{\substack{\text{2-Methyl-1,4-}\\ \text{dichloro-2-butene}\\ \text{(major)}}}{ClH_2C\text{—}C(CH_3){=}CH\text{—}CH_2Cl}$$

Problem 8.31 Write initiation and propagation steps in radical-catalyzed addition of $BrCCl_3$ to 1,3-butadiene, and show how the structure of the intermediate accounts for: (*a*) greater reactivity of conjugated dienes than alkenes, (*b*) orientation in addition.

Initiation

$$\text{Rad}\cdot + Br{:}CCl_3 \longrightarrow \text{Rad}{:}Br + \cdot CCl_3$$

Propagation

$$Cl_3C\cdot + \text{—}C{=}C\text{—}C{=}C\text{—} \longrightarrow Cl_3C\text{—}C\text{—}\dot{C}\text{—}C{=}C \longleftrightarrow Cl_3C\text{—}C\text{—}C{=}C\text{—}\dot{C}\text{—} \text{ or } Cl_3C\text{—}C\text{—}C\overset{\cdot}{\cdots}C\cdots C\text{—}$$

$$Cl_3C\text{—}C\text{—}C\overset{\cdot}{\cdots}C\cdots C\text{—} + Br{:}CCl_3 \longrightarrow Cl_3C\cdot + \underset{\substack{\text{1,4-Adduct}\\ \text{(major)}}}{Cl_3C\text{—}C\text{—}C{=}C\text{—}C(Br)\text{—}} + \underset{\substack{\text{1,2-Adduct}\\ \text{(minor)}}}{Cl_3C\text{—}C\text{—}C(Br)\text{—}C{=}C\text{—}}$$

(*a*) The allyl radical formed in the first propagation step is more stable and requires a lower $\Delta H^{\ddagger}$ than the alkyl free radical from alkenes. The order of free-radical stability is allyl > 3° > 2° > 1°.
(*b*) The 1,4-orientation is similar to ionic addition because of the relative stabilities of the two products.

8.6 Polymerization of Dienes

Electrophilic Catalysis

$$E^+ + \overset{1}{H_2C}=\overset{2}{C}H-\overset{3}{C}H=\overset{4}{C}H_2 \longrightarrow \overset{1}{E}CH_2-\overset{2}{C}H=\overset{3}{C}H-\overset{+}{\underset{4}{C}H_2} \xrightarrow{\overset{5}{H_2C}=\overset{6}{C}H\overset{7}{C}H=\overset{8}{C}H_2}$$

monomer

$$\overset{1}{E}CH_2-\overset{2}{C}H=\overset{3}{C}H\overset{4}{C}H_2\overset{5}{C}H_2\overset{6}{C}H=\overset{7}{C}H\overset{+}{\underset{8}{C}H_2} \xrightarrow{n(CH_2=CHCH=CH_2)} (E[-CH_2CH=CHCH_2-]_{n+2})^+$$

mer of polymer

Nucleophilic or Anionic Polymerization

$$Nu:^- + H_2C=\underset{|}{\overset{CH_3}{C}}-CH=CH_2 \longrightarrow Nu:CH_2-\overset{CH_3}{C}=CH-\ddot{C}H_2 \xrightarrow{H_2C=\overset{CH_3}{C}-CH=CH_2}$$

$$Nu:CH_2\overset{CH_3}{C}=CH-CH_2:CH_2-\overset{CH_3}{C}=CH-\ddot{C}H_2 \xrightarrow{n(CH_2=\overset{CH_3}{C}-CH=CH_2)} Nu\left[-CH_2\overset{CH_3}{C}=CHCH_2-\right]_{n+2}^-$$

dimeric anion — mer of polymer

The reaction is stereospecific in yielding a polymer with an all-*cis* configuration.

Conjugated dienes undergo nucleophilic attack more easily than simple alkenes because they form more stable allyl carbanions,

$$Nu-C-C\!\overset{\dots}{=}\!C\!\overset{\dots}{=}\!C-$$

Like the allyl cation, the allylic anion is stabilized by charge delocalization through extended π bonding.

Radical Polymerization

$$Z\cdot + -C=C-C=C- \longrightarrow Z:C-C=C-C\cdot \xrightarrow[\text{Monomer}]{-C=C-C=C-} Z:C-C=C-C-C-C=C-C\cdot$$

monomer — monomeric free radical — dimeric free radical

Conjugated diene polymers are modified and improved by copolymerizing them with other unsaturated compounds, such as acrylonitrile, $H_2C=CH-C\equiv N$.

$$H_2C=CH-CH=CH_2 + H_2C=\underset{C\equiv N}{CH} + H_2C=CH-CH=CH_2 + H_2C=CH-CH=CH_2 \longrightarrow$$

Acrylonitrile

$$\left[-H_2C-CH=CH-CH_2-CH_2-\underset{C\equiv N}{CH}-CH_2-CH=CH-CH_2-CH_2-CH=CH-CH_2-\right]_n$$

Buna N rubber unit

8.7 Cycloaddition

A useful synthetic reaction is cycloaddition of an alkene, called a **dienophile**, to a conjugated diene by 1,4- addition.

$$\underset{\text{1,3-Butadiene}}{H_2C{=}CH{-}CH{=}CH_2} + \underset{\text{Ethylene}}{H_2C{=}CH_2} \xrightarrow{200\ ^\circ C} \underset{\text{Cyclohexene}}{\text{cyclo-}(HC{=}CH{-}CH_2{-}CH_2{-}CH_2{-}CH_2)}$$

Diels – Alder reaction

8.8 Summary of Alkyne Chemistry

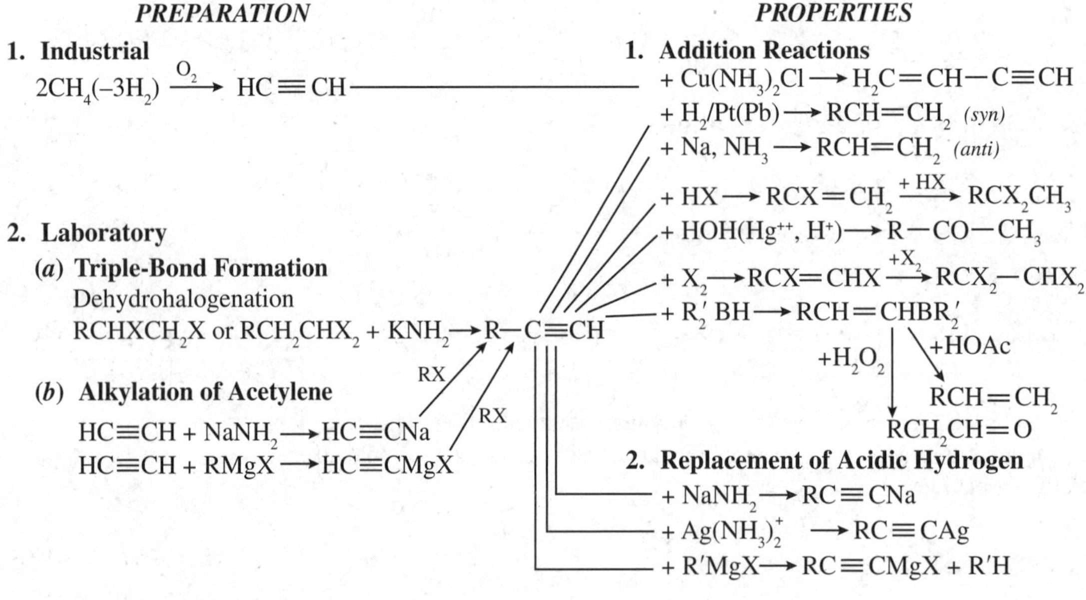

8.9 Summary of Diene Chemistry

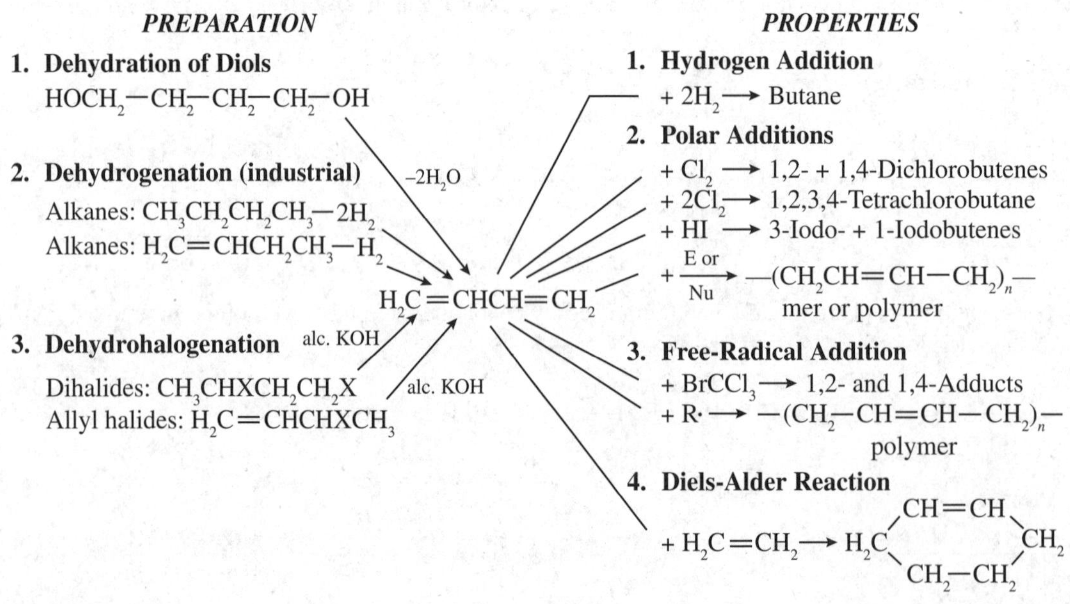

SUPPLEMENTARY PROBLEMS

Problem 8.32 For the conjugated and isolated dienes of molecular formula C_6H_{10}, tabulate (*a*) structural formula and IUPAC name, (*b*) possible geometric isomers, (*c*) ozonolysis products.

In Table 8.2, a box is placed about the C=C associated with geometric isomers.

Problem 8.33 Show reagents and reactions needed to prepare the following compounds from the indicated starting compounds. (*a*) Acetylene to ethylidene iodide (1,1-diiodoethane). (*b*) Propyne to isopropyl bromide. (*c*) 2-Butyne to racemic 2,3-dibromobutane. (*d*) 2-Bromobutane to *trans*-2-butene. (*e*) *n*-Propyl bromide to 2-hexyne. (*f*) 1-Pentene to 2-pentyne.

(*a*) $H{-}C{\equiv}C{-}H \xrightarrow{HI} H_2C{=}CHI \xrightarrow{HI} CH_3CHI_2$

(*b*) $CH_3C{\equiv}C{-}H \xrightarrow{H_2/Pt} CH_3{-}CH{=}CH_2 \xrightarrow{HBr} CH_3CHBrCH_3$

Add H_2 first; the reaction can be stopped after 1 mol is added.

(*c*) $CH_3C{\equiv}CCH_3 \xrightarrow{H_2/Pt(Pb)}$ *cis*-$CH_3CH{=}CHCH_3 \xrightarrow{Br_2}$ *rac*-(±)-$CH_3CHBrCHBrCH_3$ (*trans* addition)

(*d*)* $CH_3CHBrCH_2CH_3 \xrightarrow{\text{alc. KOH}}$ *cis* + *trans*-$CH_3CH{=}CHCH_3 \xrightarrow{Br_2}$

$\underset{\textit{meso}\text{ and }\textit{rac}}{CH_3CHBrCHBrCH_3} \xrightarrow{KNH_2} CH_3C{\equiv}CCH_3 \xrightarrow{Na,\ NH_3}$ *trans*-$CH_3CH{=}CHCH_3$

(*e*) $CH_3CH_2CH_2Br \xrightarrow{\text{alc. KOH}} CH_3CH{=}CH_2 \xrightarrow{Br_2} CH_3CHBrCH_2Br \xrightarrow{KNH_2}$

$CH_3C{\equiv}CH \xrightarrow{Na} CH_3C{\equiv}C^{-}{:}Na^{+} \xrightarrow{n\text{-}C_3H_7Br} CH_3C{\equiv}C{-}CH_2CH_2CH_3$

(*f*) $H_2C{=}CHCH_2CH_2CH_3 \xrightarrow{HBr} CH_3CHBrCH_2CH_2CH_3 \xrightarrow{\text{alc. KOH}}$

$CH_3CH{=}CHCH_2CH_3 \xrightarrow{Br_2} CH_3CHBrCHBrCH_2CH_3 \xrightarrow{KNH_2}$
(Saytzeff product)

$CH_3C{\equiv}CCH_2CH_3$ (not the less stable allene, $CH_3CH{=}CH{=}CHCH_3$)

Problem 8.34 Write a structural formula for organic compounds (A) through (N):

(*a*) $HC{\equiv}CCH_2CH_2CH_3 \xrightarrow{Ag(NH_3)_2^+} (A) \xrightarrow{HNO_3} (B)$

(*b*) $CH_3C{\equiv}CH \xrightarrow{CH_3MgBr}$ (C, a gas) + (D) $\xrightarrow{CH_3I}$ (E)

(*c*) $CH_3CH_2C{\equiv}CH + Na^+NH_2^- \longrightarrow (F) \xrightarrow{C_2H_5I} (G) \xrightarrow{H_3O^+,\ Hg^{++}} (H)$

(*d*) $CH_3C{\equiv}CH + BH_3 \longrightarrow (I) \xrightarrow{CH_3COOH} (J) \xrightarrow[KMnO_4]{\text{dil. aq.}} (K)$

(*e*) $ClCH_2{-}\overset{\overset{\displaystyle CH_3}{|}}{C}HCHClCH_3 + \text{alc. KOH} \longrightarrow (L) \xrightarrow[\text{peroxide}]{BrCCl_3} (M) + (N)$

(*a*) $AgC{\equiv}CCH_2CH_2CH_3$ (A) $HC{\equiv}CCH_2CH_2CH_3$ (B)

(*b*) CH_4 (C) + $CH_3C{\equiv}CMgBr$ (D) $CH_3C{\equiv}CCH_3$ (E)

(*c*) $CH_3CH_2C{\equiv}C{:}^{-}\,Na^+$ (F) $CH_3CH_2C{\equiv}CCH_2CH_3$ (G) $CH_3CH_2{-}\overset{\overset{\displaystyle O}{||}}{C}{-}CH_2CH_2CH_3$ (H)

(*d*) $(CH_3CH{=}CH)_3B$ (I) $CH_3CH{=}CH_2$ (J) $CH_3CHOHCH_2OH$ (K)

---

* *Trans*- and *cis*-alkenes are made by stereospecific reductions of corresponding alkynes.

TABLE 8.2

(A) FORMULA AND NAME	(B) GEOMETRIC ISOMERS	(C) OZONOLYSIS PRODUCTS
(1) $H_2C{=}CH{-}\boxed{CH{=}CH}{-}CH_2{-}CH_3$ 1,3-Hexadiene	2	$H_2C{=}O$, $O{=}CH{-}CH{=}O$, $O{=}CHCH_2CH_3$
(2) $H_2C{=}CH{-}CH_2{-}\boxed{CH{=}CH}{-}CH_3$ 1,4-Hexadiene	2	$H_2C{=}O$, $O{=}CHCH_2CH{=}O$, $O{=}CHCH_3$
(3) $H_2C{=}CH{-}CH_2{-}CH_2{-}CH{=}CH_2$ 1,5-Hexadiene	None	$H_2C{=}O$, $O{=}CHCH_2CH_2CH{=}O$, $O{=}CH_2$
(4) $H_2C{=}C(CH_3){-}\boxed{CH{=}CH}{-}CH_3$ 2-Methyl-1,3-pentadiene	2	$H_2C{=}O$, $O{=}C(CH_3){-}CH{=}O$, $O{=}CHCH_3$
(5) $H_2C{=}C(CH_3){-}CH_2{-}CH{=}CH_2$ 2-Methyl-1,4-pentadiene	None	$H_2C{=}O$, $O{=}C(CH_3){-}CH_2CH{=}O$, $O{=}CH_2$
(6) $H_2C{=}CH{-}\boxed{C(CH_3){=}CH}{-}CH_3$ 3-Methyl-1,3-pentadiene	2	$H_2C{=}O$, $O{=}CH{-}C(CH_3){=}O$, $O{=}CHCH_3$
(7) $H_2C{=}CH{-}CH{=}C(CH_3){-}CH_3$ 4-Methyl-1,3-pentadiene	None	$H_2C{=}O$, $O{=}CH{-}CH{=}O$, $O{=}C(CH_3){-}CH_3$
(8) $CH_3{-}\boxed{CH{=}CH}{-}\boxed{CH{=}CH}{-}CH_3$ 2,4-Hexadiene	2 *cis, cis;* *cis, trans;* *trans, trans*	$CH_3CH{=}O$, $O{=}CH{-}CH{=}O$, $O{=}CHCH_3$
(9) $H_2C{=}C(CH_3){-}C(CH_3){=}CH_2$ 2,3-Dimethyl-1.3-butadiene	None	$H_2C{=}O$, $O{=}C(CH_2CH_3){-}CH{=}O$, $O{=}CH_2$
(10) $H_2C{=}C(CH_2CH_3){-}CH{=}CH_2$ 2-Ethyl-1,3-butadiene	None	$H_2C{=}O$, $O{=}C(CH_2CH_3){-}CH{=}O$, $O{=}CH_2$
(11) $H_2C{=}CH{-}CH(CH_3)CH{=}CH_2$ 3-Methyl-1,4-pentadiene	None	$H_2C{=}O$, $O{=}CHCH(CH_3){-}CH{=}O$, $O{=}CH_2$

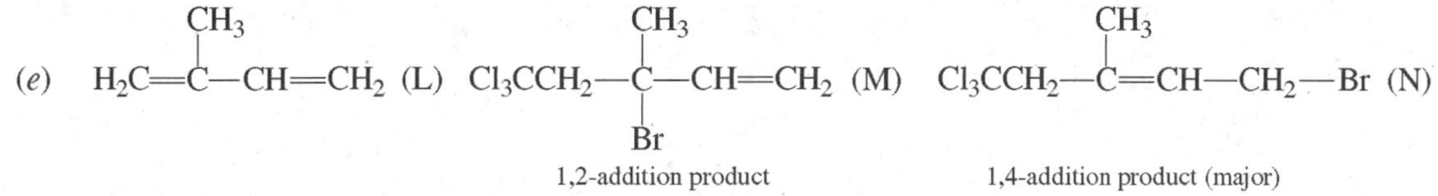

(*e*) $H_2C{=}C(CH_3){-}CH{=}CH_2$ (L) $Cl_3CCH_2{-}C(CH_3)(Br){-}CH{=}CH_2$ (M) $Cl_3CCH_2{-}C(CH_3){=}CH{-}CH_2{-}Br$ (N)

1,2-addition product (M); 1,4-addition product (major) (N)

Problem 8.35 Assign numbers from 1 for LEAST to 5 for MOST to indicate the relative reactivity on HBr addition to the following compounds:

(*a*) $H_2C{=}CH{-}CH_2CH_3$ (*b*) $CH_3{-}CH{=}CH{-}CH_3$ (*c*) $H_2C{=}CH{-}CH{=}CH_2$

(*d*) $CH_3{-}CH{=}CH{-}CH{=}CH_2$ (*e*) $H_2C{=}C(CH_3){-}C(CH_3){=}CH_2$

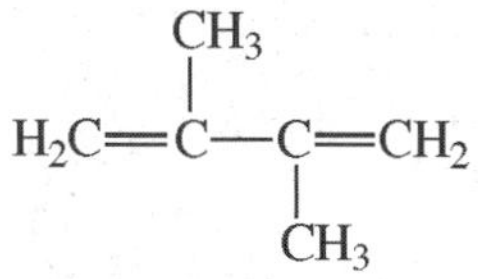

Conjugated dienes form the more stable allyl R^+'s and therefore are more reactive than alkenes. Alkyl groups on the unsaturated C's increase reactivity. Relative reactivities are (*a*) 1, (*b*) 2, (*c*) 3, (*d*) 4, (*e*) 5.

Problem 8.36 For the reaction of propyne with Br_2 + NaOH, give the structures of the products and the mechanisms of their formation.

Propyne reacts with strong bases to form a nucleophilic carbanion which displaces $:\ddot{Br}:^-$ from Br_2 by attacking $\overset{\delta+}{Br}$ to form 1-bromopropyne.

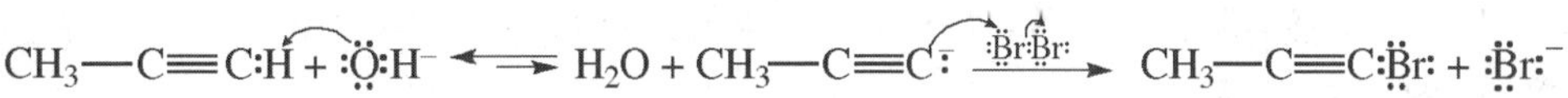

$$CH_3{-}C{\equiv}C{:}H + :\ddot{O}{:}H^- \rightleftharpoons H_2O + CH_3{-}C{\equiv}C:^- \xrightarrow{:\ddot{Br}:\ddot{Br}:} CH_3{-}C{\equiv}C{:}\ddot{Br}: + :\ddot{Br}:^-$$

1-Bromopropyne

Problem 8.37 $^{14}CH_3CH{=}CH_2$ is subjected to allylic free-radical bromination. Will the reaction product be exclusively labeled $H_2C{=}CH^{14}CH_2Br$? Explain.

No. The product consists of an equal number of $H_2C{=}CH^{14}CH_2Br$ and $^{14}CH_2{=}CHCH_2Br$ molecules. H-abstraction produces a resonance hybrid of two contributing structures having both ^{12}C and ^{14}C as equally reactive, free-radical sites that attack Br_2.

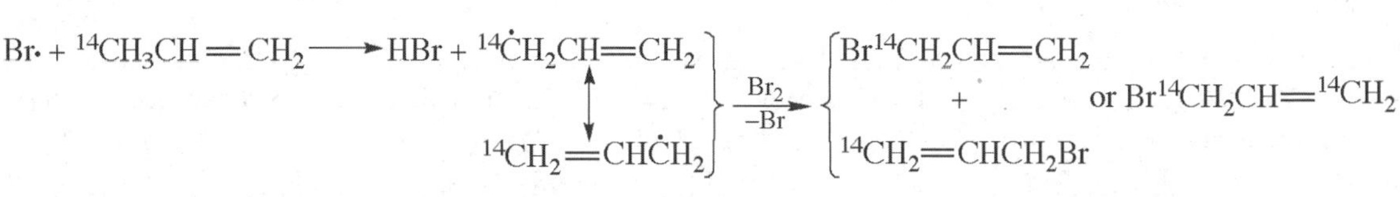

$$Br\cdot + {}^{14}CH_3CH{=}CH_2 \longrightarrow HBr + \left\{ {}^{14}\dot{C}H_2CH{=}CH_2 \updownarrow {}^{14}CH_2{=}CH\dot{C}H_2 \right\} \xrightarrow[-Br]{Br_2} \left\{ Br^{14}CH_2CH{=}CH_2 + {}^{14}CH_2{=}CHCH_2Br \right\} \text{ or } Br^{14}CH_2CH{=}^{14}CH_2$$

Problem 8.38 (*a*) Write a schematic structure for the mer of the polymer from head-to-tail reaction of 2-methyl–1,3-butadiene. (*b*) Account for this orientation in polymerization. (*c*) Show how the structure is deduced from the product

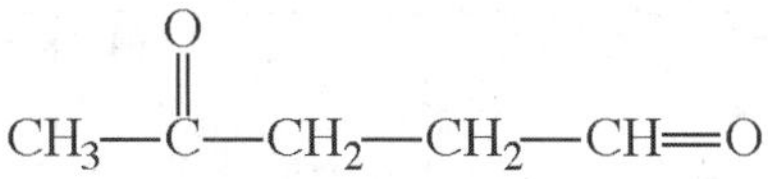

$$CH_3{-}\overset{O}{\overset{\|}{C}}{-}CH_2{-}CH_2{-}CH{=}O$$

obtained from ozonolysis of the polymer.

(*a*) 1,4-Addition with regular head-to-tail orientation produces a polymer with the following repeating unit (mer):

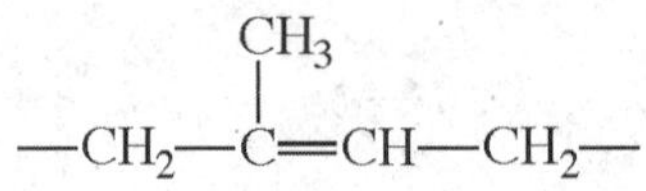

$$\text{—CH}_2\text{—}\underset{}{\overset{\text{CH}_3}{\overset{|}{\text{C}}}}\text{=CH—CH}_2\text{—}$$

(*b*) This orientation results from more rapid formation of the more stable intermediate free radical.

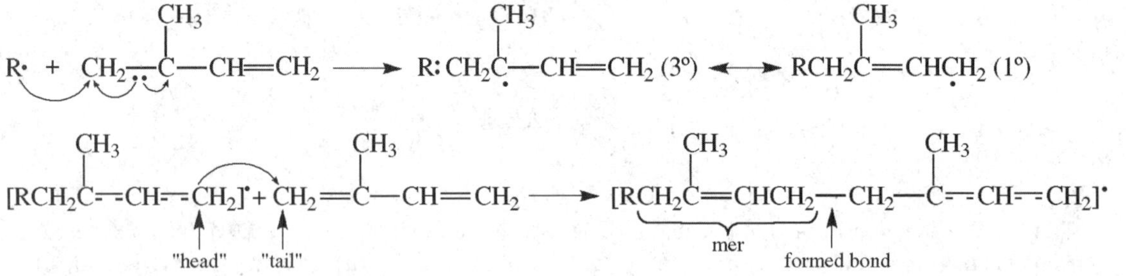

$$\text{R}\cdot + \text{CH}_2\text{=}\overset{\text{CH}_3}{\overset{|}{\text{C}}}\text{—CH=CH}_2 \longrightarrow \text{R:CH}_2\overset{\text{CH}_3}{\overset{|}{\dot{\text{C}}}}\text{—CH=CH}_2\ (3°) \longleftrightarrow \text{RCH}_2\overset{\text{CH}_3}{\overset{|}{\text{C}}}\text{=CH}\dot{\text{C}}\text{H}_2\ (1°)$$

$$[\text{RCH}_2\overset{\text{CH}_3}{\overset{|}{\text{C}}}\text{=-CH=-CH}_2]^\cdot + \text{CH}_2\text{=}\overset{\text{CH}_3}{\overset{|}{\text{C}}}\text{—CH=CH}_2 \longrightarrow [\underbrace{\text{RCH}_2\overset{\text{CH}_3}{\overset{|}{\text{C}}}\text{=CHCH}_2}_{\text{mer}}\text{—CH}_2\text{—}\overset{\text{CH}_3}{\overset{|}{\text{C}}}\text{=-CH=-CH}_2]^\cdot$$

"head" "tail" formed bond

The 1° allylic site is more reactive than the 3° allylic site. Attack at the other terminal $=CH_2$ gives the less stable free radical.

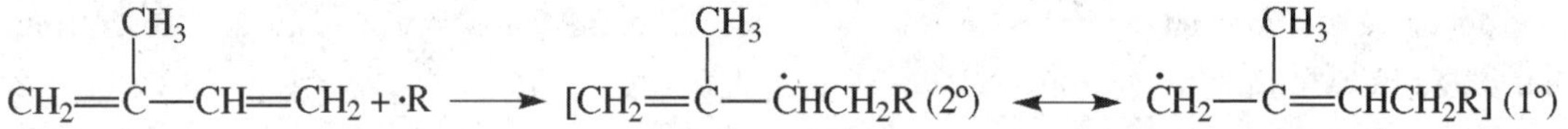

$$\text{CH}_2\text{=}\overset{\text{CH}_3}{\overset{|}{\text{C}}}\text{—CH=CH}_2 + \cdot\text{R} \longrightarrow [\text{CH}_2\text{=}\overset{\text{CH}_3}{\overset{|}{\text{C}}}\text{—}\dot{\text{C}}\text{HCH}_2\text{R}\ (2°) \longleftrightarrow \dot{\text{C}}\text{H}_2\text{—}\overset{\text{CH}_3}{\overset{|}{\text{C}}}\text{=CHCH}_2\text{R]}\ (1°)$$

(*c*) Write the ozonolysis products with the O's pointing at each other. Now erase the O's and join the C's by a double bond.

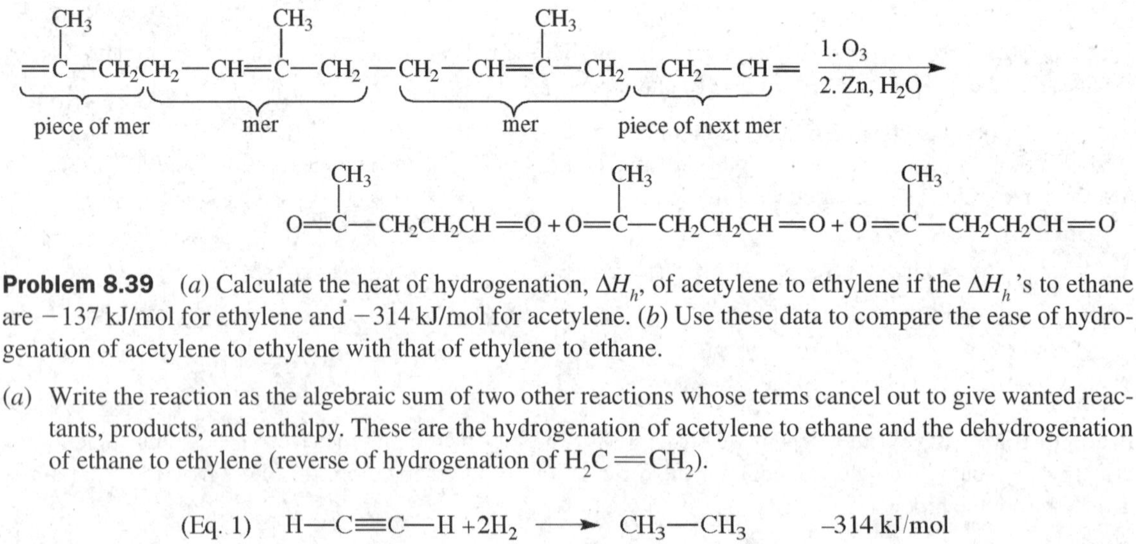

$$\underbrace{\text{=}\overset{\text{CH}_3}{\overset{|}{\text{C}}}\text{—CH}_2\text{CH}_2}_{\text{piece of mer}}\underbrace{\text{—CH=}\overset{\text{CH}_3}{\overset{|}{\text{C}}}\text{—CH}_2}_{\text{mer}}\underbrace{\text{—CH}_2\text{—CH=}\overset{\text{CH}_3}{\overset{|}{\text{C}}}\text{—CH}_2}_{\text{mer}}\underbrace{\text{—CH}_2\text{—CH=}}_{\text{piece of next mer}} \xrightarrow[\text{2. Zn, H}_2\text{O}]{\text{1. O}_3}$$

$$\text{O=}\overset{\text{CH}_3}{\overset{|}{\text{C}}}\text{—CH}_2\text{CH}_2\text{CH=O} + \text{O=}\overset{\text{CH}_3}{\overset{|}{\text{C}}}\text{—CH}_2\text{CH}_2\text{CH=O} + \text{O=}\overset{\text{CH}_3}{\overset{|}{\text{C}}}\text{—CH}_2\text{CH}_2\text{CH=O}$$

Problem 8.39 (*a*) Calculate the heat of hydrogenation, ΔH_h, of acetylene to ethylene if the ΔH_h's to ethane are -137 kJ/mol for ethylene and -314 kJ/mol for acetylene. (*b*) Use these data to compare the ease of hydrogenation of acetylene to ethylene with that of ethylene to ethane.

(*a*) Write the reaction as the algebraic sum of two other reactions whose terms cancel out to give wanted reactants, products, and enthalpy. These are the hydrogenation of acetylene to ethane and the dehydrogenation of ethane to ethylene (reverse of hydrogenation of $H_2C{=}CH_2$).

(Eq. 1)	$\text{H—C≡C—H} + 2\text{H}_2$	$\longrightarrow$	$\text{CH}_3\text{—CH}_3$	–314 kJ/mol
(Eq. 2)	$\text{CH}_3\text{—CH}_3$	$\longrightarrow$	$\text{H}_2\text{C=CH}_2 + \text{H}_2$	+137 kJ/mol
	$\text{H—C≡C—H} + \text{H}_2$	$\longrightarrow$	$\text{H}_2\text{C=CH}_2$	–177 kJ/mol

Equation 2 (dehydrogenation) is the reverse of hydrogenation ($\Delta H_h = -137$ kJ/mol). Hence, the ΔH_h of Eq. 2 has a + value.

(*b*) Acetylene is less stable thermodynamically relative to ethylene than ethylene is to ethane because ΔH_h for acetylene → ethylene is -177 kJ/mol, while for ethylene → ethane, it is -137 kJ/mol. Therefore, acetylene is more easily hydrogenated and the process can be stopped at the ethylene stage. In general, hydrogenation of alkynes can be stopped at the alkene stage.

Problem 8.40 Deduce the structural formula of a compound of molecular formula C_6H_{10} which adds 2 mol of H_2 to form 2-methylpentane, forms a carbonyl compound in aqueous H_2SO_4-$HgSO_4$ solution, and does not react with ammoniacal $AgNO_3$ solution, $[Ag(NH_3)_2]^+NO_3^-$.

There are two degrees of unsaturation, since the compound C_6H_{10} lacks four H's from being an alkane. The addition of 2 mol of H_2 excludes a cyclic compound. It may be either a diene or an alkyne, and the latter functional group is established by hydration to a carbonyl compound. The skeleton must be

```
    C
    |
C—C—C—C—C
```

as established by the reduction product. The two possible alkynes with this skeleton are

$$(CH_3)_2CHCH_2C{\equiv}CH \quad \text{and} \quad (CH_3)_2CH{-}C{\equiv}C{-}CH_3$$

The negative test for a 1-alkyne with Ag^+ establishes the second structure, 4-methyl-2-pentyne.

Problem 8.41 The allene 2,3-pentadiene ($\overset{1}{C}H_3\overset{2}{C}H{=}\overset{3}{C}{=}\overset{4}{C}H\overset{5}{C}H_3$) does not have a chiral C but is resolved into enantiomers. (*a*) Draw an orbital picture that accounts for the chirality [see Problem 5.23(*d*)]. (*b*) What structural features must a chiral allene have?

(*a*) C^3 is *sp* hybridized and forms two σ bonds by *sp-sp*^{2} overlap with the orbitals of C^2 and C^4. The two remaining *p* orbitals of C^3 form two π bonds, one with C^2 and one with C^4. These π bonds are at right angles to each other. The H and CH_3 on C^2 are in a plane at right angles to the plane of the H and CH_3 on C^4. See Fig. 8.6.

Because there is no free rotation about the two π bonds, the two H's and two CH_3's have a fixed spatial relationship. Whenever the two substituents on C^2 are different and the two substituents on C^4 are different, the molecule lacks symmetry and is chiral.

(*b*) Individually, the terminal C's of the allenic system must have two different attached groups, for example, $RHC{=}C{=}CHR'(R)$. The groups could be other than H's. $H_2C{=}C{=}CHR$ is not chiral.

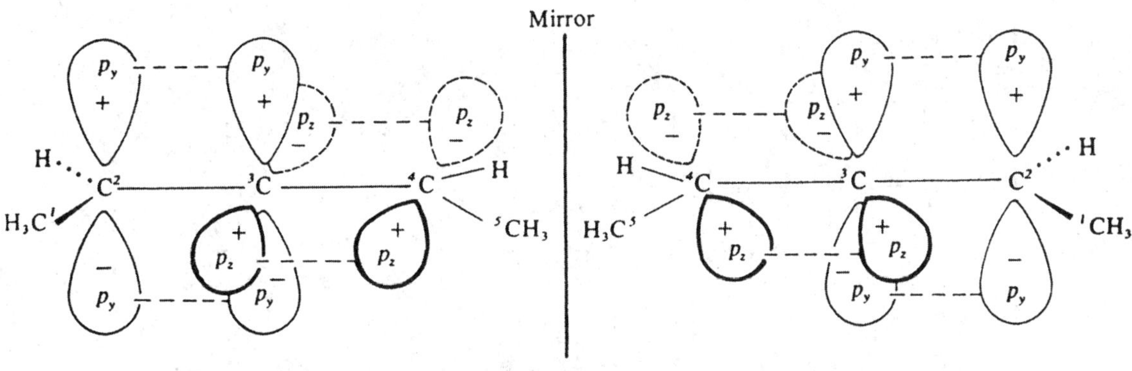

Figure 8.6

Problem 8.42 Heating C_4H_9Br (A) with alcoholic KOH forms an alkene, C_4H_8 (B), which reacts with bromine to give $C_4H_8Br_2$ (C). (C) is transformed by KNH_2 to a gas, C_4H_6 (D), which forms a precipitate when passed through ammoniacal CuCl. Give the structures of compounds (A) through (D).

The precipitate with ammoniacal CuCl indicates that (D) is a 1-alkyne, which can only be 1-butyne. The reactions and compounds are:

$$\underset{\text{(D)}}{H{-}C{\equiv}CCH_2CH_3} \xleftarrow{KNH_2} \underset{\text{(C)}}{BrCH_2CHBrCH_2CH_3} \xleftarrow{Br_2} \underset{\text{(B)}}{H_2C{=}CHCH_2CH_3} \xleftarrow[\text{KOH}]{\text{alc.}} \underset{\text{(A)}}{BrCH_2CH_2CH_2CH_3}$$

(A) cannot be $CH_3CHBrCH_2CH_3$, which would give mainly $H_3CCH{=}CHCH_3$ and finally $CH_3C{\equiv}CCH_3$.

Problem 8.43 Is the fact that conjugated dienes are more stable and more reactive than isolated dienes an incongruity?

No. Reactivity depends on the relative $\Delta H^\ddagger$ values. Although the ground-state enthalpy for the conjugated diene is lower than that of the isolated diene, the transition-state enthalpy for the conjugated system is lower by a greater amount (see Fig. 8.7).

$$\Delta H^\ddagger \text{ Conjugated} < \Delta H^\ddagger \text{ isolated and rate}_{\text{conjugated}} > \text{rate}_{\text{isolated}}$$

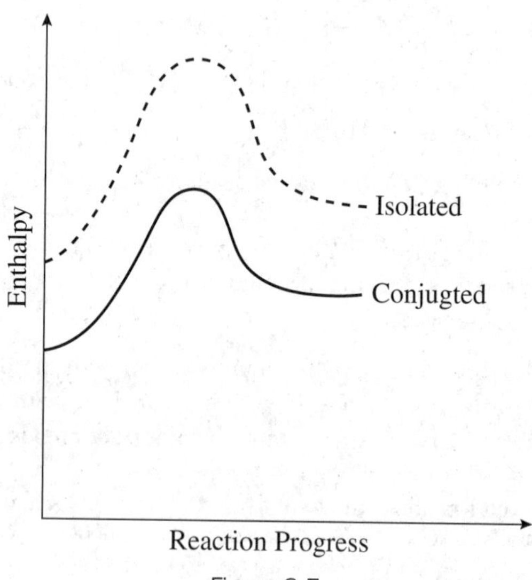

Figure 8.7

Problem 8.44 Explain why 1,3-butadiene and O_2 do not react unless irradiated by UV light to give the 1,4-adduct.

O—O

Ordinary ground-state O_2 is a diradical,

:Ö—Ö:
↑ ↑

One bond could form, but the intermediate has two electrons with the same spin and a second bond cannot form.

$$\underset{\uparrow}{\text{O}}\text{—}\underset{\uparrow}{\text{O}} + \text{H}_2\text{C}\overset{\downarrow\uparrow}{\text{—}}\text{CH—CH}\overset{\uparrow\downarrow}{\text{—}}\text{CH}_2 \longrightarrow \underset{\uparrow}{\text{O}}\text{—O—CH}_2\text{—CH=CH—}\underset{\uparrow}{\text{CH}_2}$$

When irradiated, O_2 is excited to the singlet spin-paired state

:Ö—Ö:
↑↓

Singlet O_2 reacts by a concerted mechanism to give the product.

Problem 8.45 Is there any inconsistency between the facts that the C—H bond in acetylene has the greatest bond energy of all C—H bonds and that it is also the most acidic?

No. Bond energy is a measure of homolytic cleavage, $\equiv$C:H → $\equiv$C· + ·H. Acidity is due to a heterolytic cleavage, $\equiv$C:H + Base → $\equiv$C:$^-$ + H^+ (Base).

Cyclic Hydrocarbons

9.1 Nomenclature and Structure

Cyclic hydrocarbons are called **cycloalkanes**; they are examples of **alicyclic** (*ali*phatic *cyclic*) **compounds**. Cycloalkanes, having the general formula C_nH_{2n}, are isomeric with alkenes but, unlike alkenes, they are saturated compounds. They are named by combining the prefix **cyclo-** with the name of the alkane having the same number of C's as are in the ring. Two or more substituents are listed alphabetically and are assigned the lowest possible numbers. Alicyclic compounds are usually symbolized by the appropriate-sized rings, without the C's of the ring and with or without the attached H's.

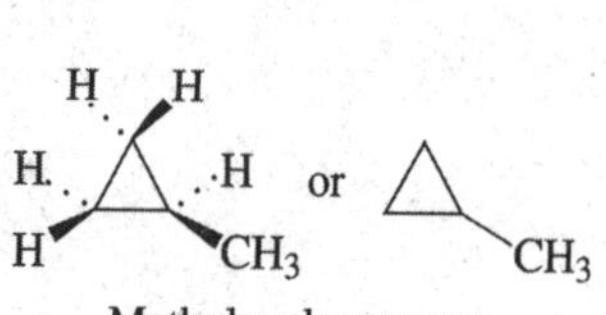

Methylcyclopropane

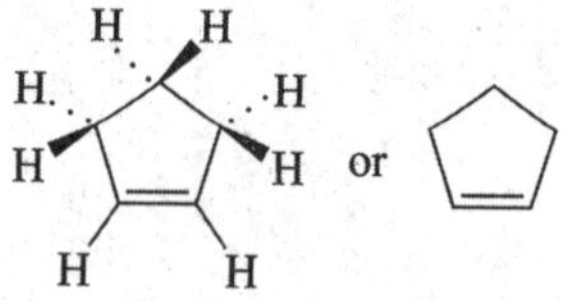

Cyclopentene

Polycyclic compounds have more than one ring; those with exactly two rings are **bicyclic**. The rings may merely be bonded to each other, as in Problem 9.1(*d*). Those that share a C—C bond are said to be **fused**, as exemplified by decalin (bicyclo[4.4.0]decane). The C's that are common to the two rings, shown encircled in decalin that follows, are called **bridgehead** C's.

a zero-carbon bridge

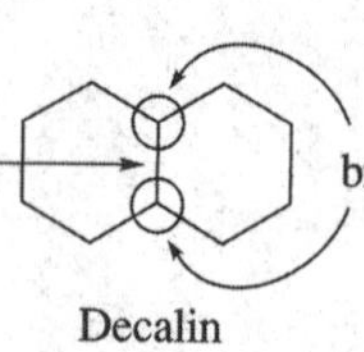

bridgehead C's

Decalin

Unlike fused bicyclics, **bridged bicyclics** have one or more C's separating the bridgehead C's, as in Problem 9.2(*f*).

For bicyclic compounds, the prefix **bicyclo-** is combined with a pair of brackets enclosing numbers separated by periods, which is followed by a name indicating the total number of atoms in the bridged rings. The bracketed numbers show how many C's are in each bridge joining the bridgehead C's and are cited in the order of decreasing size.

Problem 9.1 Draw structural formulas for (*a*) bromocycloheptane, (*b*) 1-ethylcyclopentene, (*c*) bicyclo-[3.1.0]hexane, (*d*) cyclobutylcyclohexane.

(*a*) 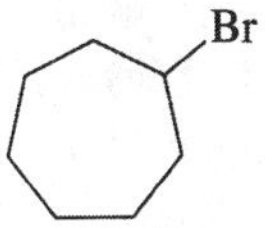 Br (*b*) 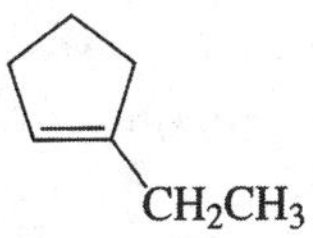 CH_2CH_3 (*c*) 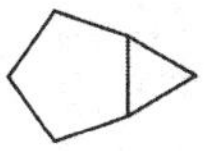 (*d*)

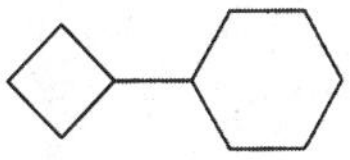

Problem 9.2 Name the following compounds:

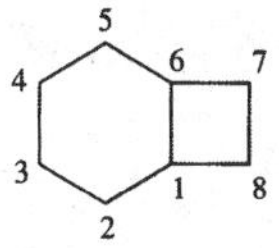

(*a*) 1 2 3 4 5 6 7 8

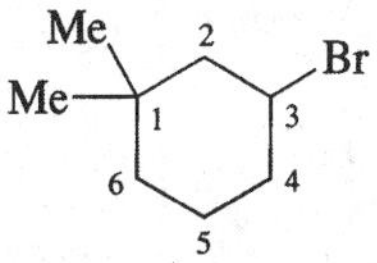

(*b*) $\triangleright$—CH=CH—CH(CH_3)—CH_2CH_3

(*c*) Me, Me, Br; 1 2 3 4 5 6

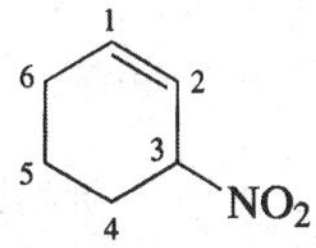

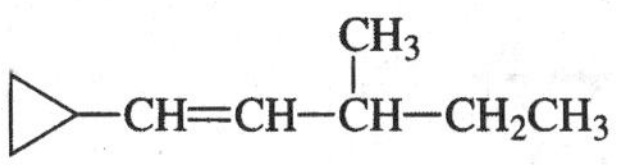

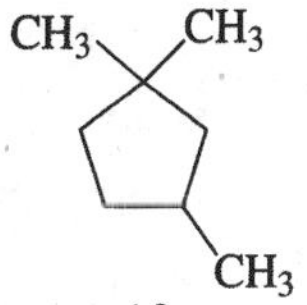

(*d*) CH_3 CH_3 CH_3

(*e*) 1 2 3 4 5 6 NO_2

(*f*) Me Me 7 ← *a one-carbon bridge*; Me Me 1 2 3 4 5 6; bridgehead C's (common to 3 rings)

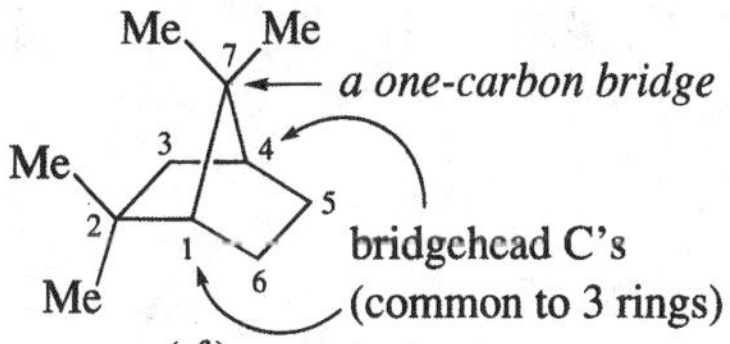

In (*f*), C^1, C^2, C^3, C^4, C^5, C^6 constitute one ring; C^1, C^7, C^4, C^5, C^6 another; C^1, C^2, C^3, C^4, C^7 a third.

(*a*) This fused-ring bicyclic compound has a four-carbon bridge made up of C^2, C^3, C^4, and C^5, and a two-carbon bridge of C^7 and C^8; the bridge between the bridgehead C's has 0 carbon atoms. There are eight C's in the compound. The name is bicyclo[4.2.0]octane. (*b*) Consider the cyclopropane ring to be a substituent on the longest carbon chain. The name is 1-cyclopropyl-3-methyl-1-pentene. (*c*) The substituents, written alphabetically, are numbered so that the C's have the lowest possible numbers. The name is 3-bromo-1,1-dimethylcyclohexane. (*d*) 1,1,3-Trimethylcyclopentane. (*e*) 3-Nitrocyclohexene (*not* 6-nitrocyclohexene). The doubly bonded C's are numbered *1* and *2*, to give the smaller number to the substituent. (*f*) The numbering of C's of bicyclic compounds starts with the bridgehead C closest to a substituent. Substituents on the largest bridge get the smallest numbers. The name is 2,2,7,7-tetramethylbicyclo[2.2.1]heptane.

9.2 Geometric Isomerism and Chirality

Review Chapter 5. The inability of atoms in rings to rotate completely about their σ bonds leads to *cis-trans* (geometric) isomers in cycloalkanes.

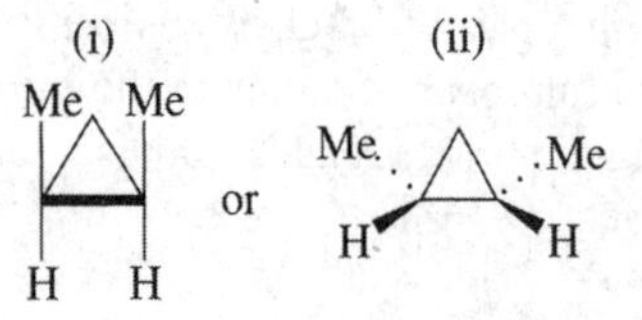

cis-1,2-Dimethylcyclopropane

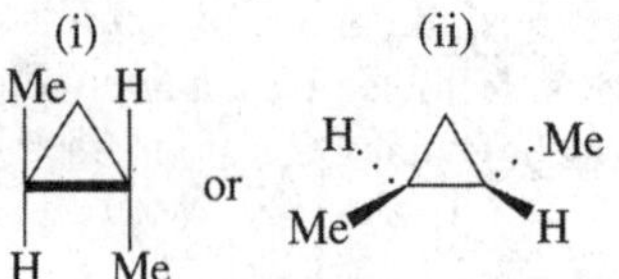

trans-1,2-Dimethylcyclopropane

In such diagrams, either (i) the flat ring is perpendicular to the plane of the paper, with the bond(s) facing the viewer drawn heavy and with the substituents in the plane of the paper and projecting up and down, or (ii) the flat ring is in the plane of the paper, with "wedges" projecting toward the viewer and "dots" away from the viewer.

Since the ring C's are sp^3-hybridized, they may be chiral centers. Therefore, substituted cycloalkanes may be geometric isomers, as well as being enantiomers or *meso* compounds.

Problem 9.3 Give the names, structural formulas, and stereochemical designations of the isomers of (*a*) bromochlorocyclobutane, (*b*) dichlorocyclobutane, (*c*) bromochlorocyclopentane, (*d*) diiodocyclopentane, (*e*) dimethylcyclohexane. Indicate chiral C's.

(*a*) There is only one structure for 1-bromo-1-chlorocyclobutane: 

With 1-bromo–2-chlorocyclobutane, there are *cis* and *trans* isomers and both substituted C's are chiral. Both geometric isomers form racemic mixtures.

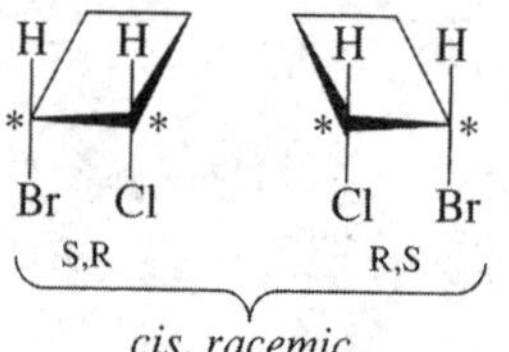

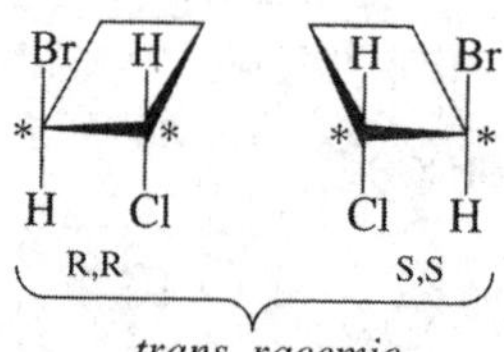

cis, racemic *trans, racemic*

1-Bromo-2-chlorocyclobutane

In 1-bromo-3-chlorocyclobutane, there are *cis* and *trans* isomers, but no enantiomers; C^1 and C^3 are not chiral, because a plane perpendicular to the ring bisects them and their four substituents. The sequence of atoms is identical going around the ring clockwise or counterclockwise from C^1 to C^3.

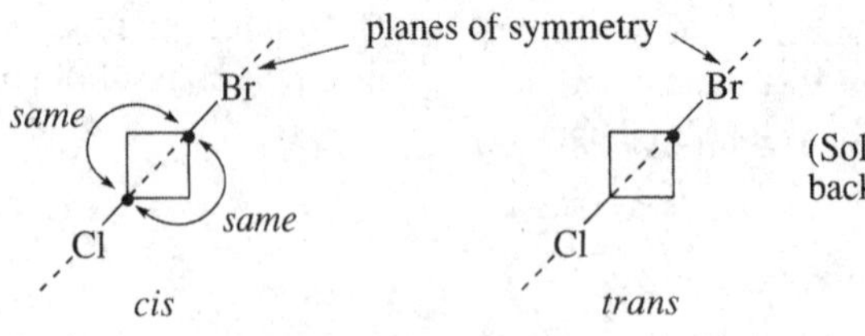

(Solid • indicates an H is in the back, no dot for an H in front.)

1-Bromo-3-chlorocyclobutane

In these structural formulas, the other atoms on C^1 and C^3 are directly in back of those shown and are bisected by the indicated plane.

(*b*) Same as (*a*) except that the *cis*-1,2-dichlorocyclobutane has a plane of symmetry (dashed line below) and is *meso*.

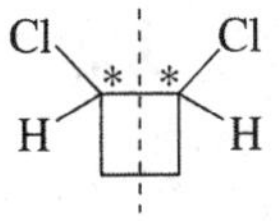

Cl Cl H H *cis, meso*

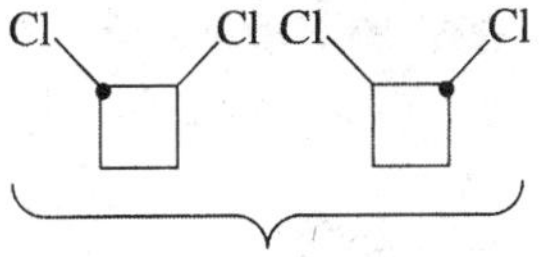

Cl Cl Cl Cl *trans, racemic*

(*c*) There are nine isomers because both 1,2- and 1,3-isomers have *cis* and *trans* geometric isomers, and these have enantiomers.

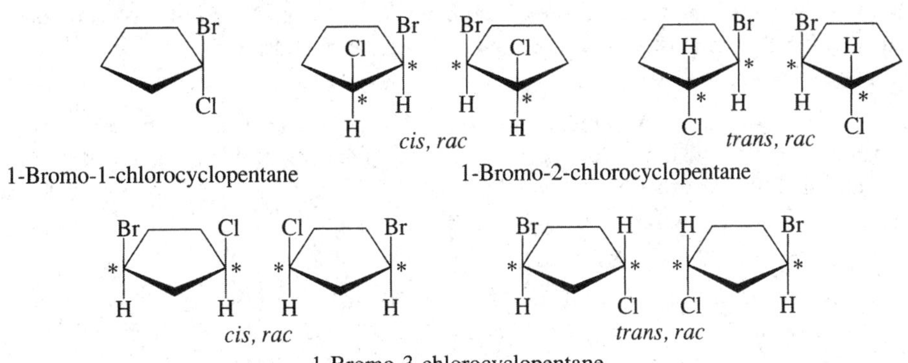

1-Bromo-1-chlorocyclopentane

cis, rac *trans, rac*

1-Bromo-2-chlorocyclopentane

cis, rac *trans, rac*

1-Bromo-3-chlorocyclopentane

(*d*) The diiodocyclopentanes are similar to the bromochloro derivative, except that both the *cis*-1,2- and the *cis*-1,3-diiodo derivatives are *meso*. They both have planes of symmetry.

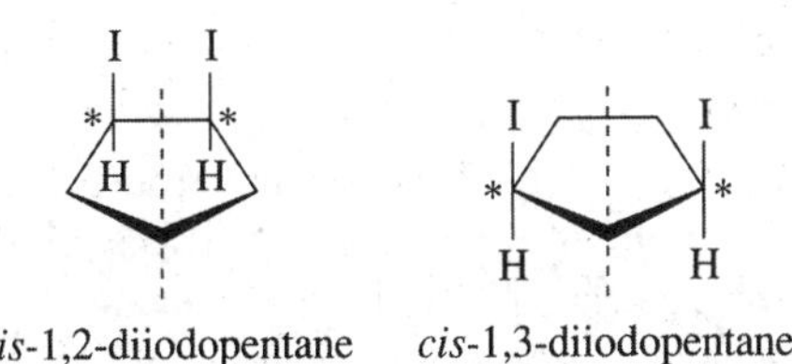

cis-1,2-diiodopentane *cis*-1,3-diiodopentane

(*e*) There are nine isomeric dimethylcyclohexanes.

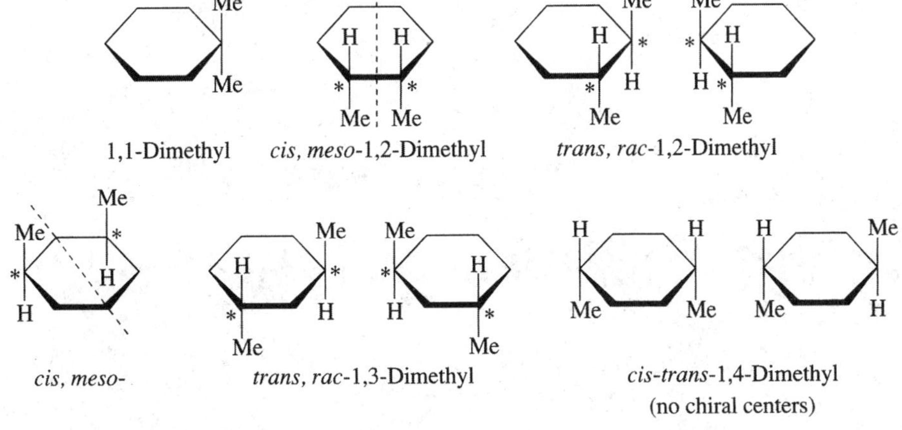

1,1-Dimethyl *cis, meso*-1,2-Dimethyl *trans, rac*-1,2-Dimethyl

cis, meso- *trans, rac*-1,3-Dimethyl *cis-trans*-1,4-Dimethyl (no chiral centers)

9.3 Conformations of Cycloalkanes

Ring Strain

The relative stabilities of cycloalkanes can be determined by comparing their ΔH's of combustion (Problem 4.35) on a per-CH_2-unit basis. Rings have different ΔH's of combustion per CH_2 unit because they have different amounts of **ring strain**.

Problem 9.4 (*a*) Calculate ΔH of combustion per CH_2 unit for the first four cycloalkanes, given the following ΔH's of combustion, in kJ/mol: cyclopropane, −2091; cyclobutane, −2744; cyclopentane, −3320; cyclohexane, −3952. (*b*) Write (i) the thermochemical equation for the combustion of cyclopropane and (ii) the theoretical equation for the combustion of a CH_2 unit of any given ring. (*c*) How do ring stability and ring size correlate for the first four cycloalkanes?

(*a*) Divide the given ΔH values by the number of CH_2 units in the ring (3, 4, 5, and 6, respectively), to obtain: cyclopropane, −697; cyclobutane, −686; cyclopentane, −664; cyclohexane, −659. Observe that these *per-unit* ΔH's are in the reverse order of the *total* ΔH's.

(*b*) (i) $C_3H_6 + \frac{9}{2}O_2 \rightarrow 3CO_2 + 3H_2O \qquad \Delta H = -2091$ kJ/mol

(ii) $—CH_2— + \frac{3}{2}O_2 \rightarrow CO_2 + H_2O \qquad \Delta H < 0$

(*c*) In (ii) of (*b*), $\Delta H \equiv H(CO_2) + H(H_2O) - H(\frac{3}{2}O_2) - H(\text{unit})$. Thus, for the four different ring-memberships under consideration:

$$H(\text{unit}) = \text{constant} - \Delta H$$

Now, from (*a*), ΔH increases (becomes less negative) with increasing size of the ring. Thus, H(unit) decreases with increasing size, which implies that H(ring) also decreases with increasing size. But a *decreasing H*(ring) means an *increasing* ring stability. In short, *stability increases with ring size*.

Problem 9.5 Account for the ring strain in cyclopropane in terms of geometry and orbital overlap.

The C's of cyclopropane form an equilateral triangle with C—C—C bond angles of 60°—a significant deviation from the tetrahedral bond angle of 109.5°. This deviation from the "normal" bond angle constitutes **angle strain**, a major component of the ring strain of cyclopropane. In terms of orbital overlap, the strongest chemical bonds are formed by the greatest overlap of atomic orbitals. For sigma bonding, maximum overlap is achieved when the orbitals meet head-to-head along the bond axis, as in Fig. 9.1(*a*). This type of overlap in cyclopropane could not lead to ring closure for sp^3-hybridized C's because it would demand bond angles of 109.5°. Hence, the overlap must be off the bond axis to give a *bent* bond, as shown in Fig. 9.1(*b*).

In order to minimize the angle strain, the C's assume more *p* character in the orbitals forming the ring and more *s* character in the external bonds, in this case, the C—H bonds. Additional *p* character narrows the expected angle, while more *s* character expands the angle. The observed H—C—H bond angle of 114° confirms this suggestion. Clearly, there are deviations from pure *p*, *sp*, sp^3, and sp^2 hybridizations.

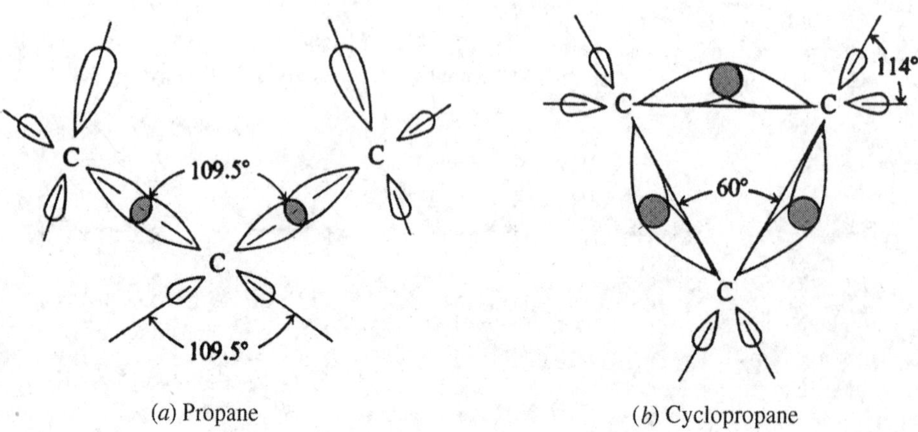

(*a*) Propane (*b*) Cyclopropane

Figure 9.1

Problem 9.6 What factor besides angle strain contributes to the ring strain of cyclopropane?

Because the cyclopropane molecule is a planar ring, the three pairs of H's eclipse each other as in *n*-butane (see Fig. 4.4) to introduce an **eclipsing (torsional) strain**.

Problem 9.7 (*a*) Why is the ΔH of *cis*-1,2-dimethylcyclopropane greater than that of its *trans* isomer? (*b*) Which isomer is more stable?

(*a*) There is more eclipsing in the *cis* isomer because the methyl groups are closer. (*b*) The *trans* is more stable.

Conformations of Cyclobutane and Cyclopentane

Problem 9.8 Explain why the ring strain of cyclobutane is only slightly less than that of cyclopropane.

If the C's of the cyclobutane ring were coplanar, they would form a rigid square with internal bond angles of 90°. The deviation from 109.5° would not be as great as that for cyclopropane, and there would be less angle strain in cyclopropane. However, this is somewhat offset by the fact that the eclipsing strain involves four pairs of H's, one pair more than in cyclopropane.

Actually, eclipsing strain is reduced because cyclobutane is not a rigid, flat molecule. Rather, there is an equilibrium mixture of two flexible puckered conformations that rapidly flip back and forth (Fig. 9.2), thereby relieving eclipsing strain. Puckering more than offsets the slight increase in angle strain (angle is now 88°). The boxed H's in Fig. 9.2 alternate between up-and-down and outward-projecting positions relative to the ring.

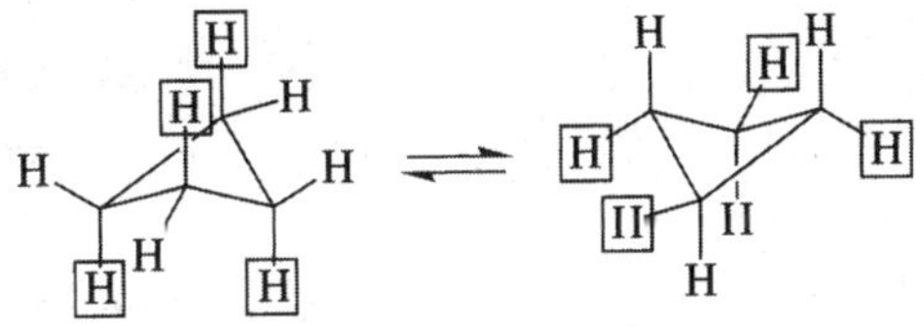

Figure 9.2

Problem 9.9 Depict the flexible puckered conformation of cyclobutane in a Newman projection.

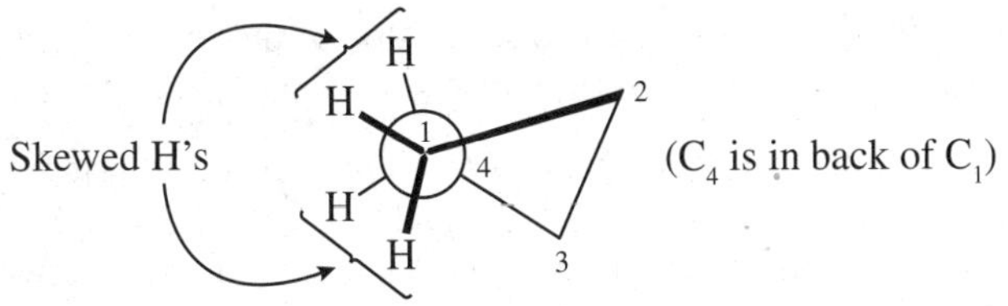

Skewed H's

(C_4 is in back of C_1)

Figure 9.3

See Fig. 9.3. The circle represents the C's of any given ring C—C bond. The other C's of the ring bridge these two C's, one to the front C (heavy line) and the other to the back C (ordinary line). This Newman projection formula reveals that the H's on adjacent C's are skewed in the puckered conformation.

Problem 9.10 Are the following compounds stable?

(*a*) No. A *trans*-cyclohexene is too strained. The *trans* unit C—C=C—C cannot be bridged by two more C's. *trans*-Cycloalkenes are stable for eight-membered, (*b*), and larger rings. (*c*) No. Cycloalkynes of fewer than eight C's are too strained. The triple bond imposes linearity on four C's, C—C≡C—C, which cannot be bridged by two more C's but can be bridged by four C's, as in (*d*). (*e*) No. A bridgehead bicyclic cannot have a double bond at a bridgehead position (unless one of the rings has at least eight C's). Such a bridgehead C and the three atoms bonded to it cannot assume the flat, planar structure required of an sp^2-hybridized C. This is known as **Bredt's rule**. (*f*) Yes. This exists because one of the bridges has no C, and these bridgehead C's can easily use sp^2-hybridized orbitals to form triangular, planar sigma bonds. (*g*) Yes. Compounds (**spiranes**) having a single C, which is a junction for two separate rings, are known for all size rings. However, the rings must be at right angles.

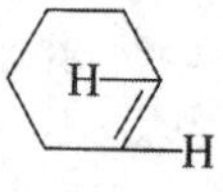

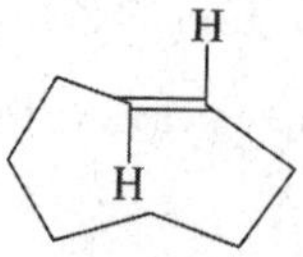

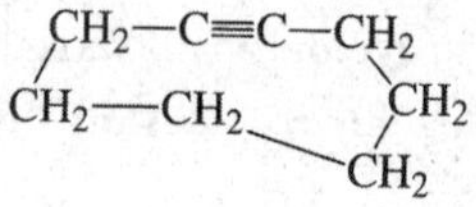

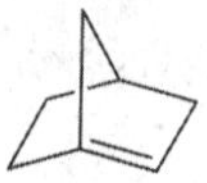

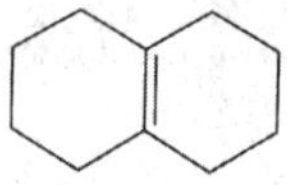

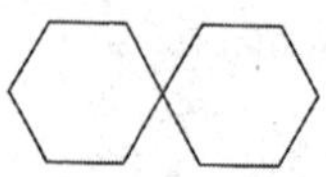

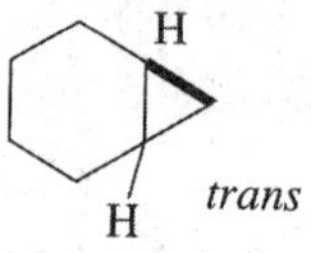

(*h*) No. Three- and six-membered rings cannot be fused *trans*, since there is too much strain.

Conformations of Cyclohexane

Problem 9.11 Among the cycloalkanes with up to 13 ring C's, cyclohexane has the least ring strain. Would one expect this if cyclohexane had a flat hexagonal ring?

No. A flat hexagon would have considerable ring strain. It would have six pairs of eclipsed H's and C—C—C bond angles of 120°, which deviate by 10.5° from the tetrahedral angle, 109.5°.

Cyclohexane minimizes its ring strain by being puckered rather than flat. The two extreme conformations are the more stable **chair** and the less stable **boat**. The **twist-boat** conformer is less stable than the chair by about 23 kJ/mol, but is more stable than the boat. It is formed from the boat by moving one "flagpole" to the left and the other to the right. See Fig. 9.4.

Problem 9.12 In terms of eclipsing interactions, explain why (*a*) the chair conformation is more stable than the boat conformation and (*b*) the twist-boat conformation is more stable than the boat conformation.

(*a*) In the boat form, Fig. 9.4(*b*), the following pairs of C—C bonds are eclipsed: C^1—C^2 with C^3—C^4, and C^1—C^6 with C^4—C^5. Furthermore, the H's on C^2—C^3 and C^5—C^6 are also eclipsed. Additional strain arises from the crowding of the "flagpole" H's on C^1 and C^4, which point toward each other. This is called **steric strain** because the H's tend to occupy the same space. In the chair form, Fig. 9.4(*a*), all C—C bonds are skew and one pair of H's on adjacent C's is *anti* and the other pair is *gauche* (see the Newman projection).

(*b*) Twisting C^1 and C^4 of the boat form away from each other gives the flexible twist-boat conformation in which the steric and all the eclipsed interactions are reduced [Fig. 9.4(*c*)].

Substituted Cyclohexanes

1. Axial and Equatorial Bonds

Six of the twelve H's of cyclohexane are **equatorial** (e); they project from the ring, forming a belt around the ring perimeter as in Fig. 9.5(*a*). The other 6 H's, shown in Fig. 9.5(*b*), are **axial** (*a*); they are perpendicular

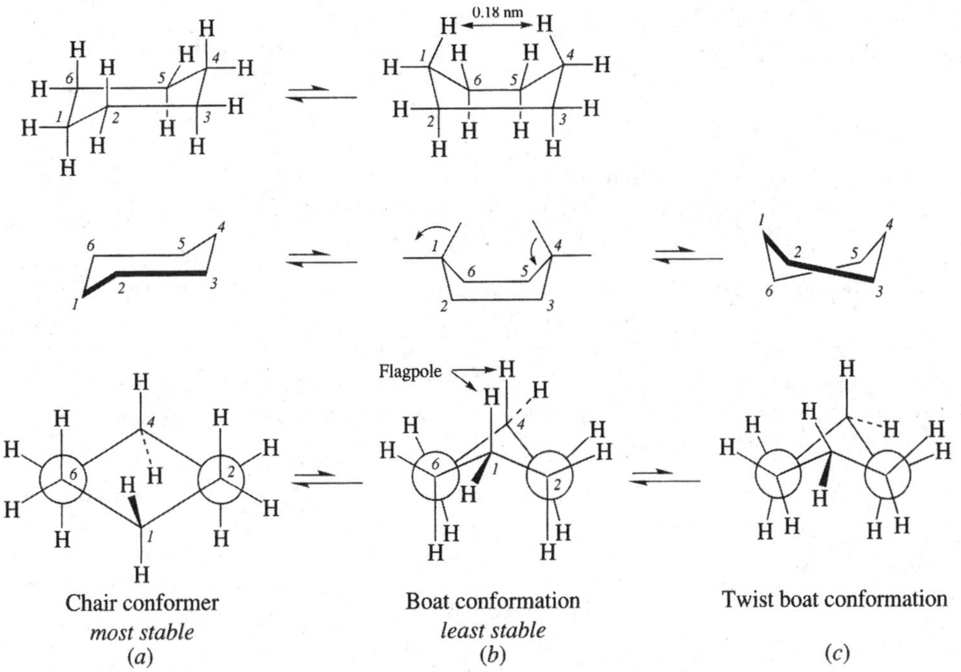

0.18 nm

Flagpole

Chair conformer
most stable
(*a*)

Boat conformation
least stable
(*b*)

Twist boat conformation
(*c*)

Figure 9.4

to the plane of the ring and parallel to each other. Three of these axial H's on alternate C's extend up, and the other three point down.

Converting one chair conformer to the other also changes the axial bonds, shown as heavy lines in Fig. 9.5(*c*), to equatorial bonds in Fig. 9.5(*d*). The equatorial bonds of Fig. 9.5(*c*) similarly become axial bonds in Fig. 9.5(*d*).

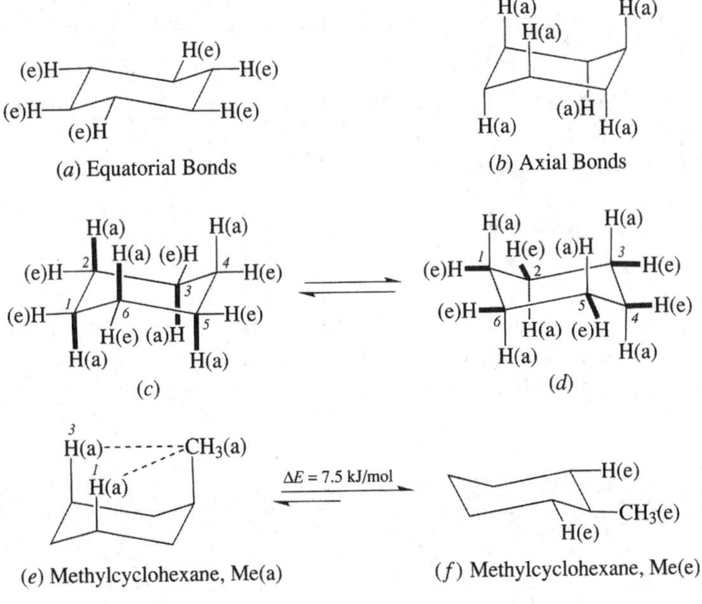

(*a*) Equatorial Bonds

(*b*) Axial Bonds

(*c*)

(*d*)

(*e*) Methylcyclohexane, Me(a)

(*f*) Methylcyclohexane, Me(e)

Figure 9.5

Problem 9.13 Give the conformational designations of the boxed H's in Fig. 9.2.

Axial on the left; *equatorial* on the right.

2. Monosubstituted Cyclohexanes

Replacing H by CH_3 gives two different chair conformations; in Fig. 9.5(*e*), CH_3 is axial, in Fig. 9.5(*f*), CH_3 is equatorial. For methylcyclohexane, the conformer with the axial CH_3 is less stable and has 7.5 kJ/mol more energy. This difference in energy can be analyzed in either of two ways:

1,3-Diaxial interactions (transannular effect). In Fig. 9.5(*e*), the axial CH_3 is closer to the two axial H's than is the equatorial CH_3 to the adjacent equatorial H's in Fig. 9.5(*f*). The steric strain for each CH_3—H 1,3-diaxial interaction is 3.75 kJ/mol, and the total is 7.5 kJ/mol for both.
Gauche interaction. In Fig. 9.6, an axial CH_3 on C^1 has a gauche interaction with the C^2—C^3 bond of the ring. One gauche interaction is also 3.75 kJ/mol; for the two, the difference in energy is 7.5 kJ/mol. The equatorial CH_3 indicated as (CH_3) is *anti* to the C^2—C^3 ring bond.

In general, a *given substituent prefers the less crowded equatorial position to the more crowded axial position.*

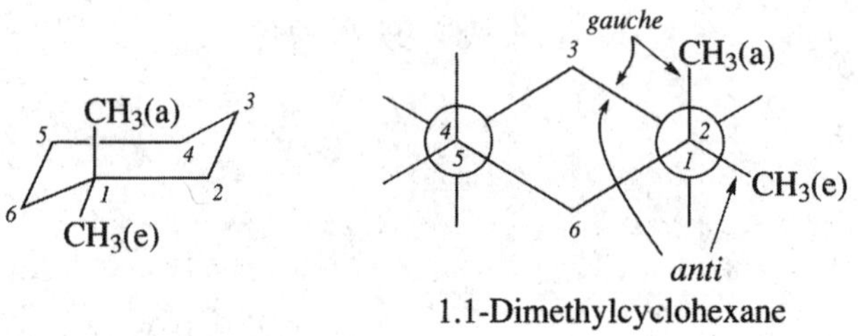

CH3(a) 5 3 4 6 1 2 CH3(e)

gauche 3 CH3(a) 4 5 2 1 CH3(e) 6 anti

1.1-Dimethylcyclohexane

Figure 9.6

Problem 9.14 (*a*) Draw the possible chair conformational structures for the following pairs of dimethylcyclohexanes: (i) *cis*- and *trans*-1,2-; (ii) *cis*- and *trans*-1,3-; (iii) *cis*- and *trans*-1,4-. (*b*) Compare the stabilities of the more stable conformers for each pair of geometric isomers. (*c*) Determine which of the isomers of dimethylcyclohexane are chiral.

With chair conformers, the better way to determine whether substituents are *cis* or *trans* is to look at the axial rather than the equatorial groups. If one axial bond is up and the other is down, the isomer is *trans*; if both axial bonds are up (or down), the geometric isomer is *cis*.

(*a*) (i) In the 1,2-isomer, since one axial bond is up and one is down, they are *trans* (Fig. 9.7). The equatorial bonds are also *trans*, although this is not obvious from the structure. In the *cis*-1,2-isomer, an H and CH_3 are *trans* to each other (Fig. 9.8).
(ii) In the 1,3-isomer, both axial bonds are up (or down) and *cis* (Fig. 9.9). In the more stable conformer [Fig. 9.9(*a*)], both CH_3's are equatorial. In the *trans* isomer, one CH_3 is axial and one equatorial (Fig. 9.10).
(iii) In the 1,4-isomer, the axial bonds are in opposite directions and are *trans* (Fig. 9.11).

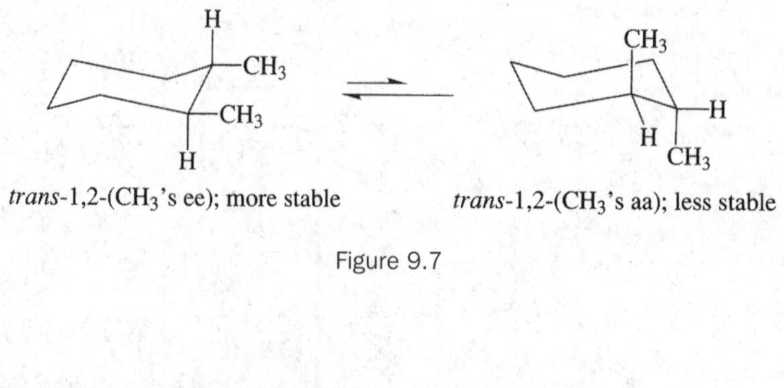

H CH3 CH3 H ⇌ CH3 H H CH3

trans-1,2-(CH_3's ee); more stable *trans*-1,2-(CH_3's aa); less stable

Figure 9.7

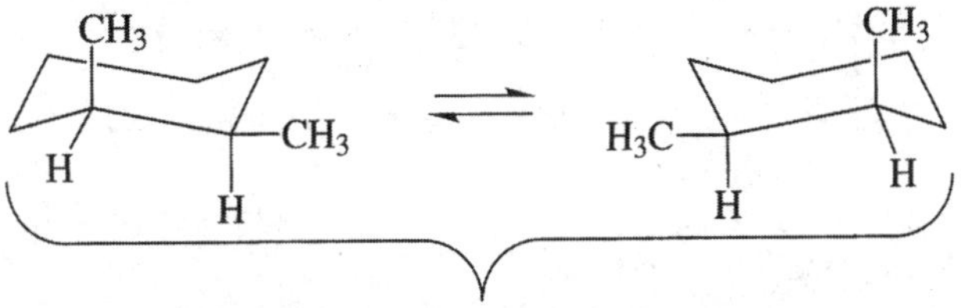

cis-1,2-(CH_3's ea); conformational enantiomers

Figure 9.8

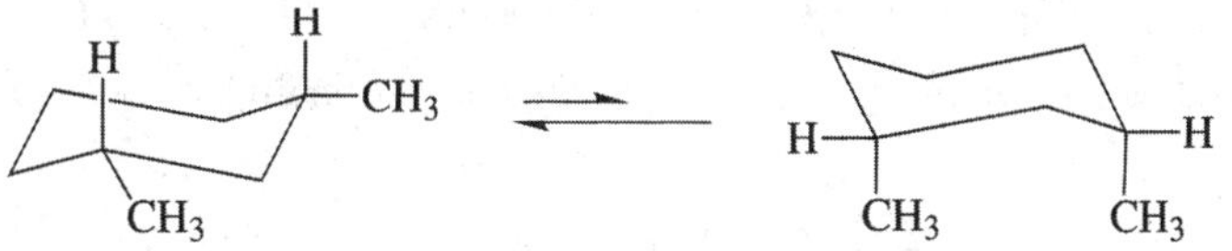

(*a*) *cis*-1,3-(CH_3's ee); more stable (*b*) *cis*-1,3-(CH_3's aa); less stable

Figure 9.9

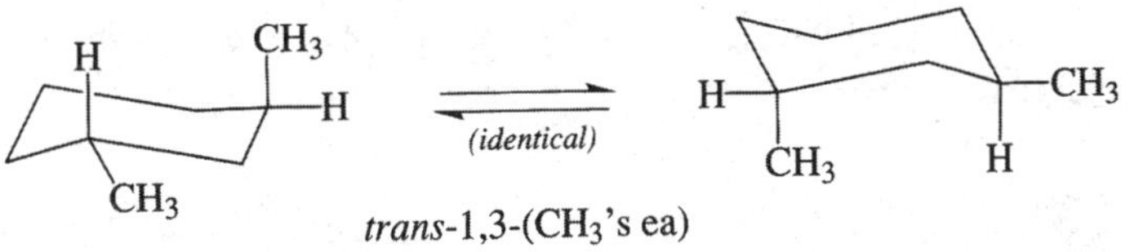

(identical)

trans-1,3-(CH_3's ea)

Figure 9.10

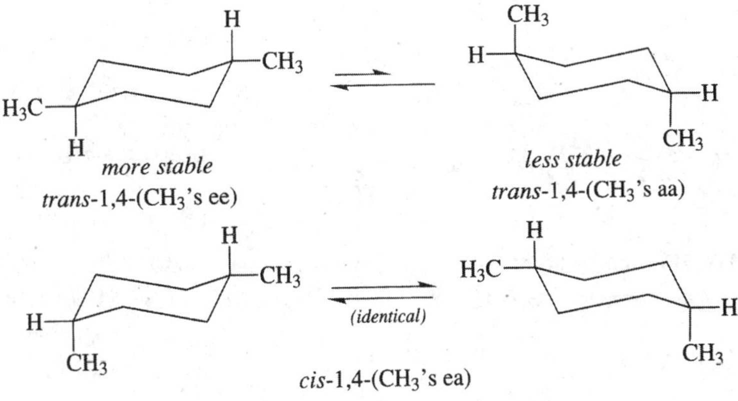

Figure 9.11

(*b*) Since an (e) substituent is more stable than an (a) substituent, in each case the (CH_3's ee) isomer is more stable than the (CH_3's ea) isomer.

(i) *trans* > *cis* (ii) *cis* > *trans* (iii) *trans* > *cis*

(*c*) The best way to detect chirality in cyclic compounds is to examine the flat structures as in Problem 9.3 (*e*); *trans*-1,2- and *trans*-1,3- are the chiral isomers.

Problem 9.15 Give your reasons for selecting the isomers of dimethylcyclohexane shown in Figs. 9.7 to 9.11 that exist as: (*a*) a pair of configurational enantiomers, each of which exists in one conformation; (*b*) a pair of conformational diastereomers; (*c*) a pair of configurational enantiomers, each of which exists as a pair of conformational diastereomers; (*d*) a single conformation; (*e*) a pair of conformational enantiomers.

(*a*) *trans*-1,3-Dimethylcyclohexane is chiral and exists as two enantiomers. Each enantiomer is (ae) and has only one conformer.
(*b*) Both *cis*-1,3- and *trans*-1,4-dimethylcyclohexane have conformational diastereomers, the stable (ee) and unstable (aa). Neither has configurational isomers.
(*c*) *trans*-1,2-Dimethylcyclohexane is a racemic form of a pair of configurational enantiomers. Each enantiomer has (ee) and (aa) conformational diastereomers.
(*d*) *cis*-1,4-Dimethylcyclohexane has no chiral C's and has only a single (ae) conformation.
(*e*) *cis*-1,2-Dimethylcyclohexane has two (ae) conformers that are nonsuperimposable mirror images.

Problem 9.16 Use 1,3-interactions and *gauche* interactions, when needed, to find the difference in energy between (*a*) *cis*- and *trans*-1,3-dimethylcyclohexane; (*b*) (ee) *trans*-1,2- and (aa) *trans*-1,2-dimethylcyclohexane.

Each CH_3/H 1,3-interaction and each CH_3/CH_3 *gauche* interaction imparts 3.75 kJ/mol of instability to the molecule.

(*a*) In the *cis*-1,3-isomer (Fig. 9.9), the more stable conformer has (ee) CH_3's and thus has no 1,3-interactions. The *trans* isomer has (ea) CH_3's. The axial CH_3 has two CH_3/H 1,3-interactions, accounting for 2(3.75) = 7.5 kJ/mol of instability. The *cis* isomer is *more* stable than the *trans* isomer by 7.5 kJ/mol.
(*b*) See Fig. 9.12; (ee) is more stable than (aa) by 15.0 − 3.75 = 11.25 kJ/mol.

Problem 9.17 Write the structure of the preferred conformation of (*a*) *trans*-1-ethyl-3 isopropylcyclohexane, (*b*) *cis*-2-chloro-*cis*-4-chlorocyclohexyl chloride.

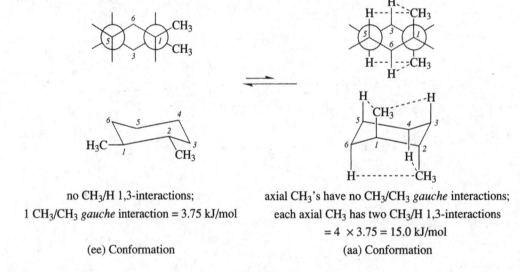

no CH_3/H 1,3-interactions;
1 CH_3/CH_3 *gauche* interaction = 3.75 kJ/mol

(ee) Conformation

axial CH_3's have no CH_3/CH_3 *gauche* interactions;
each axial CH_3 has two CH_3/H 1,3-interactions
= 4 × 3.75 = 15.0 kJ/mol

(aa) Conformation

Figure 9.12

(*a*) The *trans*-1,3-isomer is (ea); the bulkier group, in this case *i*-propyl, is equatorial, and the smaller group, in this case ethyl, is axial. See Fig. 9.13(*a*).
(*b*) See Fig. 9.13(*b*).

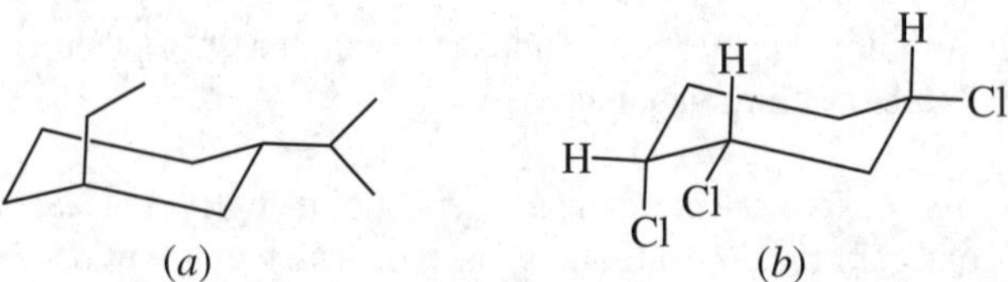

Figure 9.13

Problem 9.18 You wish to determine the relative rates of reaction of an axial and an equatorial Br in an S_N2 displacement. Can you compare (*a*) *cis*- and *trans*-1-methyl-4-bromocyclohexane? (*b*) *cis*- and *trans*-1-*t*-butyl-4-bromocyclohexane? (*c*) *cis*-3,5-dimethyl-*cis*-1-bromocyclohexane and *cis*-3,5-dimethyl-*trans*-1-bromocyclohexane?

(*a*) The *trans* substituents are (ee). The *cis* substituents are (ea). Although CH_3 is bulkier and has a greater (e) preference than has Br, the difference in preference is small and an appreciable number of molecules exist with the Br (e) and the CH_3 (a). At no time are there conformers with Br only in an (a) position. These isomers, therefore, cannot be used for this purpose.

(*b*) The bulky *t*-butyl group can only be (e). In practically all molecules of the *cis* isomer, Br is forced to be (a). All molecules of the *trans* isomer have an (e) Br. Because *t*-butyl "freezes" the conformation and prevents interconversion, these isomers *can* be used.

(*c*) The *cis*-3,5-dimethyl groups are almost exclusively (ee) to avoid severe CH_3/CH_3 1,3-interactions were they to be (aa). These *cis*-CH_3's freeze the conformation. When Br at C^1 is *cis*, it has an (e) position; when it is *trans*, it has an (a) position. These isomers can be used.

9.4 Synthesis

Intramolecular Cyclization

This technique applies to many open-chain compounds, as discussed in later chapters. Pertinent here is the intramolecular cyclization of polyenes (an **electrocyclic reaction**).

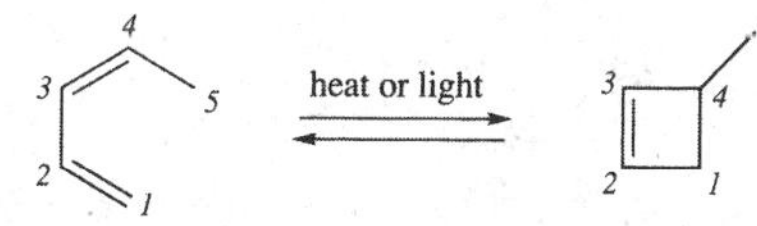

1,3-Pentadiene (a diene) $\xrightleftharpoons{\text{heat or light}}$ 3-Methylcyclobutene

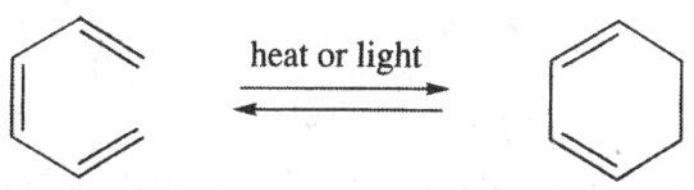

1,3,5-Hexatriene (a triene) $\xrightleftharpoons{\text{heat or light}}$ 1,3-Cyclohexadiene

Intermolecular Cyclization

In this method, two or occasionally more, open-chain compounds are merged into a ring. Examples are the common syntheses of cyclopropanes by the addition of carbene (CH_2) or substituted carbenes to alkenes (Section 6.4).

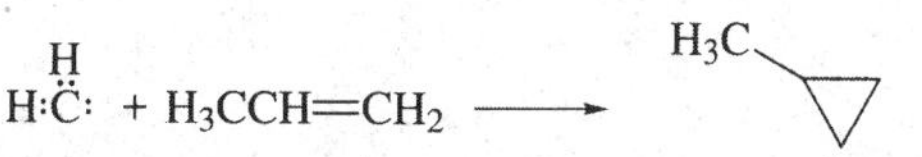

$$\underset{\text{Carbene}}{\text{H:}\ddot{\text{C}}\text{H:}} + \underset{\text{Propylene}}{H_3CCH{=}CH_2} \longrightarrow \underset{\text{Methylcyclopropane}}{H_3C\text{–}\triangledown}$$

Methylene can be transferred directly from the reagent mixture, CH_2I_2 + Zn-Cu alloy, to the alkene without being generated as an intermediate (**Simmons–Smith reaction**).

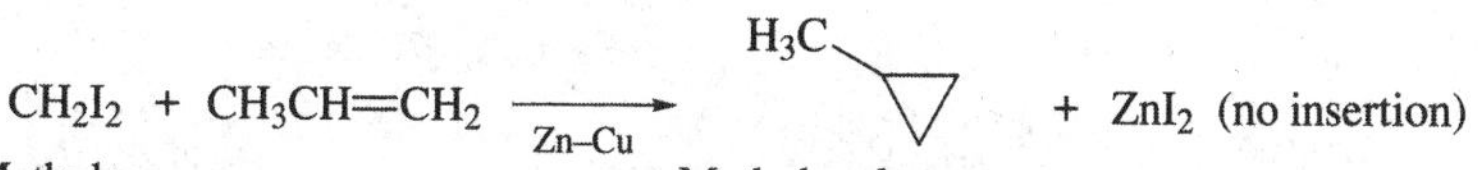

$$\underset{\text{Methylene diiodide}}{CH_2I_2} + CH_3CH{=}CH_2 \xrightarrow[\text{Zn–Cu}]{} \underset{\text{Methylcyclopropane}}{H_3C\text{–}\triangledown} + ZnI_2 \text{ (no insertion)}$$

Problem 9.19 The carbene $:CCl_2$ generated from chloroform, $CHCl_3$, and KOH in the presence of alkenes gives substituted cyclopropanes. Write the equation for the reaction of $:CCl_2$ and propene.

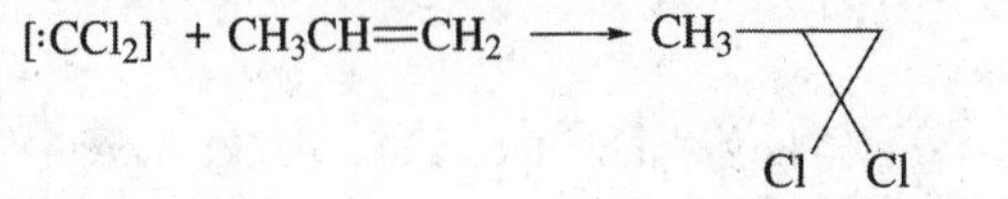

$$[:CCl_2] + CH_3CH{=}CH_2 \longrightarrow CH_3\text{-cyclopropane(Cl, Cl)}$$

Dichlorocarbene [see Problem 7.56(*c*)] — 1,1-Dichloro-2-methylcyclopropane

Cycloaddition Reactions of Alkenes and Alkynes

(*a*) [2 + 2]. Ultraviolet-light-catalyzed dimerization of alkenes yields cyclobutanes in one step.

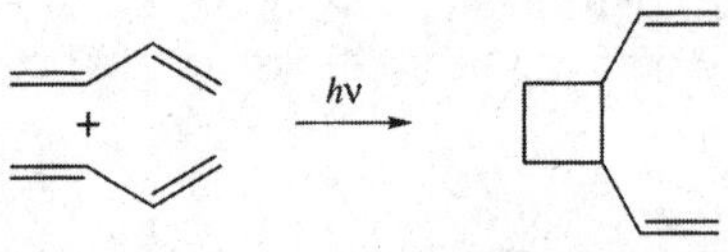

1,3-Butadiene + 1,3-Butadiene $\xrightarrow{h\nu}$ 1,2-Divinylcyclobutane (one of several products)

(*b*) [2 + 4] (the Diels-Alder reaction; Section 8.7). A conjugated diene and an alkene form a cyclohexene. Reactive alkenes (dienophiles) have electron-attracting groups on their unsaturated C's.

1,3-Butadiene + Ethylene $\xrightarrow{\text{heat}}$ Cyclohexene

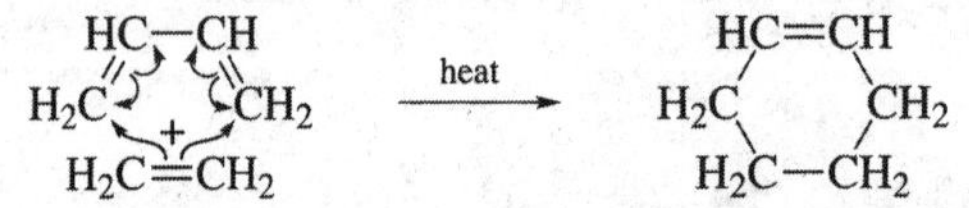

1,3-Butadiene + Acrylic aldehyde (*a dienophile*) $\xrightarrow{\text{heat}}$ 1,2,3,6-Tetrahydrobenzaldehyde

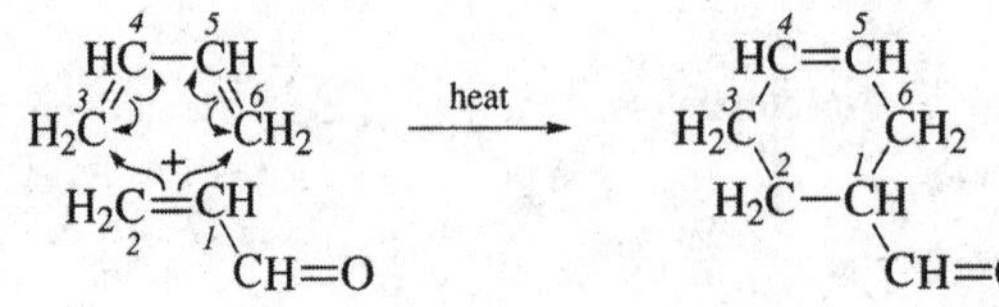

(*c*) [2 + 2 + 2 + 2] $4H{-}C{\equiv}C{-}H \xrightarrow[50°C]{Ni(CN)_2}$ Cyclooctatetraene (access to cyclooctane)

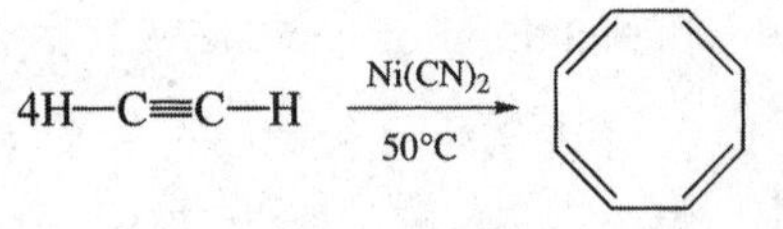

Cyclohexanes may be formed by hydrogenating compounds with benzene rings, many of which are isolated from coal, as illustrated with toluene:

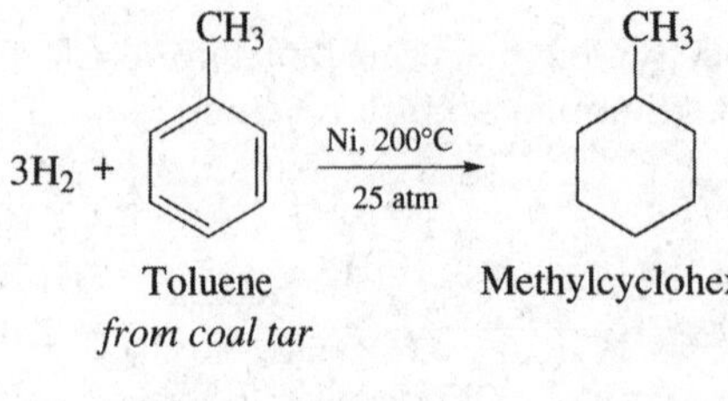

$$3H_2 + \text{Toluene} \xrightarrow[25\ atm]{Ni,\ 200°C} \text{Methylcyclohexane}$$

Toluene *from coal tar* — Methylcyclohexane

9.5 Chemistry

The chemistry of cyclic hydrocarbons and their corresponding open-chain analogs is similar. Exceptions are the cyclopropanes, whose strained rings open easily, and the cyclobutanes, whose rings open with difficulty. The larger rings are stable [see Problem 9.4(*c*)].

Problem 9.20 Although cyclopropanes are less reactive than alkenes, they undergo similar addition reactions. (*a*) Account for this by geometry and orbital overlap. (*b*) How does HBr addition to 1,1-dimethylcyclopropane resemble Markovnikov addition?

(*a*) Because of the ring strain in cyclopropane (Problem 9.5), there is less orbital overlap (Fig. 9.1) and the sigma electrons are accessible to attack by electrophiles.

(*b*) The proton of HBr is attacked by an electron pair of a bent cyclopropane sigma bond to form a carbocation that adds Br^- to give a 1,3-Markovnikov-addition product.

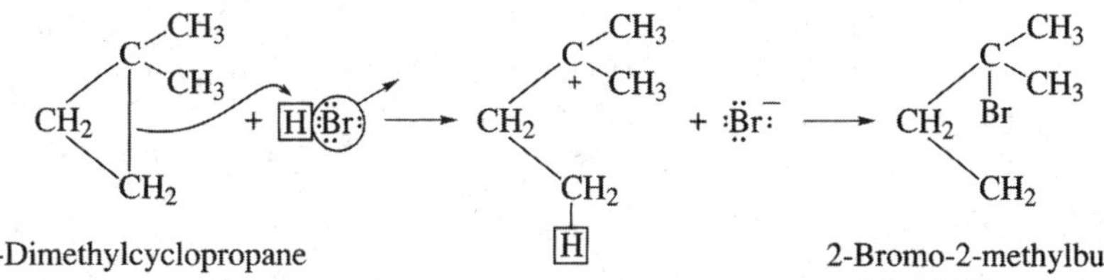

1,1-Dimethylcyclopropane + H:Br: ⟶ 3° R^+ + :Br:$^-$ ⟶ 2-Bromo-2-methylbutane, 1,3-*addition product*

Problem 9.21 Which conformation of 1,3-butadiene participates in the Diels-Alder reaction with, for example, ethene?

The two conformations of 1,3-butadiene are *s-cis* (**cisoid**) and *s-trans* (**transoid**):

s-cis *s-trans*

Although *s-trans* is the more favorable conformer, reaction occurs with *s-cis* because this conformation has its double bonds on the same side of the single bond connecting them; hence, the stable form of cyclohexene with a *cis* double bond is formed. Reaction of the *s-trans* conformer with ethene would give the impossibly strained *trans*-cyclohexene [Problem 9.10(*a*)]. As the *s-cis* conformer reacts, the equilibrium between the two conformers shifts toward the *s-cis* side, and in this way, all the unreactive *s-trans* reverts to the reactive *s-cis* conformer.

Problem 9.22 Outline a synthesis of the following alicyclic compounds from acyclic compounds.

(*a*) 1,1-Dimethylcyclopropane (*b*) cyclohexenyl—CH=CH$_2$ (*c*) Cyclooctane

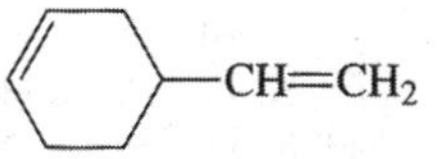

(*a*) $CH_3—C(CH_3)=CH_2 + CH_2I_2 \xrightarrow{\text{Zn–Cu}}$ 1,1-dimethylcyclopropane

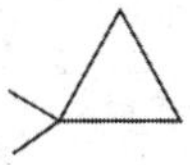

(*b*) $CH_2=CH—CH=CH_2 + CH_2=CH—CH=CH_2 \xrightarrow{\text{heat}}$ 4-vinylcyclohexene (—CH=CH$_2$)

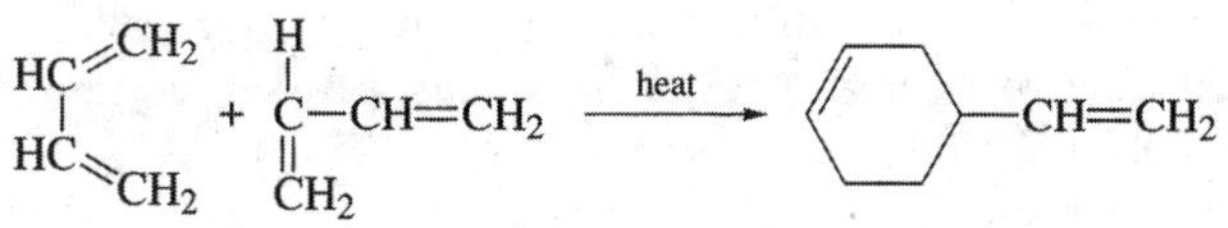

(*c*) Acetylene $\xrightarrow[50°C]{Ni(CN)_2}$ Cyclooctatetraene $\xrightarrow{H_2/Pd}$ Cyclooctane

Problem 9.23 Starting with cyclopentanol, show the reactions and reagents needed to prepare (*a*) cyclopentene, (*b*) 3-bromocyclopentane, (*c*) 1,3-cyclopentadiene, (*d*) *trans*-1,2-dibromocyclopentane, (*e*) cyclopentane.

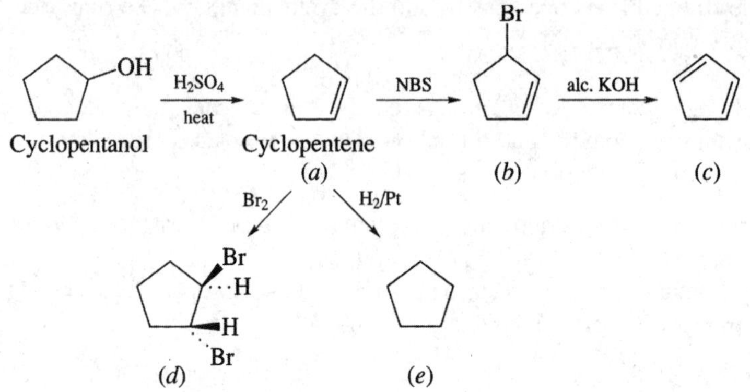

Cyclopentanol —(H_2SO_4, heat)→ Cyclopentene (*a*) —(NBS)→ 3-bromocyclopentene (*b*) —(alc. KOH)→ 1,3-cyclopentadiene (*c*)

Cyclopentene (*a*) —(Br_2)→ *trans*-1,2-dibromocyclopentane (*d*)

Cyclopentene (*a*) —(H_2/Pt)→ cyclopentane (*e*)

Problem 9.24 Complete the following reactions:

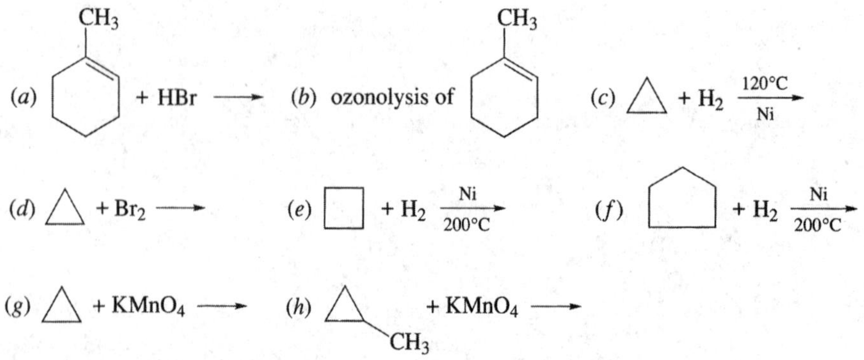

(*a*) 1-methylcyclohexene + HBr ⟶ (*b*) ozonolysis of 1-methylcyclohexene (*c*) cyclopropane + H_2 —(120°C, Ni)→

(*d*) cyclopropane + Br_2 ⟶ (*e*) cyclobutane + H_2 —(Ni, 200°C)→ (*f*) cyclopentane + H_2 —(Ni, 200°C)→

(*g*) cyclopropane + $KMnO_4$ ⟶ (*h*) methylcyclopropane (CH_3) + $KMnO_4$ ⟶

Cycloalkenes behave chemically like alkenes.

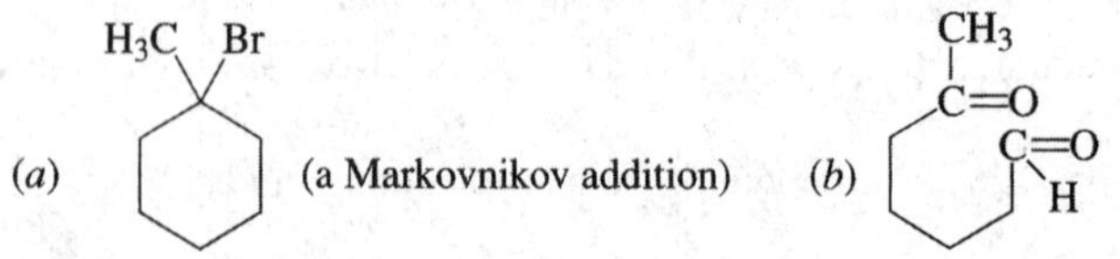

(*a*) 1-bromo-1-methylcyclohexane (H_3C, Br on the same carbon) (a Markovnikov addition) (*b*) ring-opened product bearing CH_3–C=O and C(=O)H groups

(*c*) $CH_3CH_2CH_3$. Under these conditions, the strained three-membered ring opens. (*d*) $BrCH_2CH_2CH_2Br$. Again, the three-membered ring opens. (*e*) $CH_3CH_2CH_2CH_3$. The strained four-membered ring opens, but a higher temperature is needed than in part (*c*). (*f*) No reaction. The five-membered ring has no ring strain. (*g*) No reaction. Even the strained rings are stable toward oxidation. (*h*) ▽–COOH, cyclopropanecarboxylic acid.

Problem 9.25 Cycloalkanes with more than six C's are difficult to synthesize by intramolecular ring closures, yet they are stable. On the other hand, cyclopropanes are synthesized this way, yet they are the least stable cycloalkanes. Are these facts incompatible? Explain.

No. The relative ease of synthesis of cycloalkanes by intramolecular cyclization depends on both ring stability and the probability of bringing the two ends of the chain together to form a C-to-C bond, thereby closing the ring. This probability is greatest for the smallest rings and decreases with increasing ring size. The interplay of ring stability and this probability factor are as follows (numbers represent ring sizes):

Probability of ring closure	$3 > 4 > 5 > 6 > 7 > 8 > 9$
Thermal stability	$6 > 7, 5 > 8, 9 \gg 4 > 3$
Ease of synthesis	$5 > 3, 6 > 4, 7, 8, 9$

The high yield of cyclopropane indicates that a favorable probability factor outweighs the ring instability. For rings with more than six C's the ring stability effect is outweighed by the highly unfavorable probability factor.

Problem 9.26 Account for the fact that intramolecular cyclizations to rings with more than six C's are effected at extremely low concentrations (**Ziegler method**).

Chains can also react intermolecularly to form longer chains. Although intramolecular reactions are ordinarily faster than intermolecular reactions, the opposite is true in the reaction of chains leading to rings with more than six C's. This side reaction from collisions between different chains is minimized by carrying out the reaction in extremely dilute solutions.

Very Dilute Solution **Concentrated Solution**

$$(CH_2)_n\begin{matrix} CH_2 \\ | \\ CH_2 \end{matrix} \xleftarrow{-AB} ACH_2(CH_2)_nCH_2B \xrightarrow{-AB} ACH_2(CH_2)_nCH_2{-}CH_2(CH_2)_nCH_2B$$

Ring closure Chain lengthening

9.6 MO Theory of Pericyclic Reactions

The formation of alicyclics by electrocyclic and cycloaddition reactions (Section 9.4) proceeds by one-step cyclic transition states having little or no ionic or free-radical character. Such **pericyclic** (ring closure) reactions are interpreted by the **Woodward-Hoffmann rules**; in the reactions, the new σ bonds of the ring are formed from the "head-to-head" overlap of p orbitals of the unsaturated reactants.

Intermolecular Reactions

The rules state the following:

1. Reaction occurs when the lowest unoccupied molecular orbital (LUMO) of one reactant overlaps with the highest occupied molecular orbital (HOMO) of the other reactant. If different molecules react, either can furnish the HOMO and the other the LUMO.
2. The reaction is possible only when the overlapping lobes of the p orbitals of the LUMO and the HOMO have the same sign (or shading).
3. Only the terminal p AO's of the interacting molecular orbitals are considered, as it is their overlap that produces the two new σ bonds to close the ring.

1. Ethene Dimerization [2 + 2] to Cyclobutane

The bracketed numbers indicate that the cycloaddition involves two species each having two π electrons. Without ultraviolet light we have the situation indicated in Fig. 9.14(*a*). Irradiation with UV causes a $\pi \rightarrow \pi^*$ transition (Fig. 8.3), and now the proper orbital symmetry for overlap prevails [Fig. 9.14(*b*)].

2. Diels-Alder Reaction [2 + 4]

See Fig. 9.15.

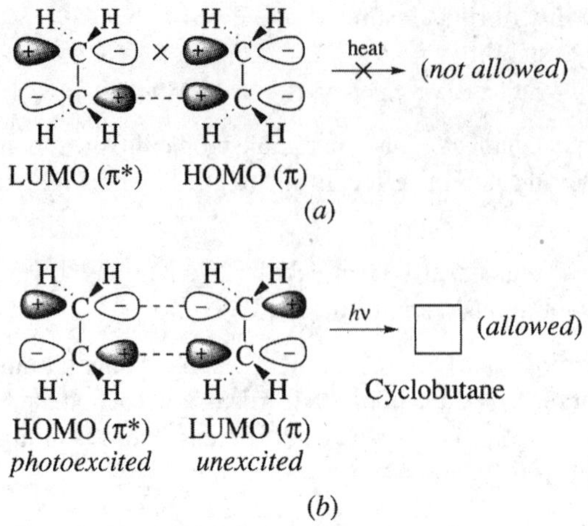

LUMO (π^*) × HOMO (π) $\xrightarrow[\times]{\text{heat}}$ (*not allowed*)

(*a*)

HOMO (π^*) *photoexcited* + LUMO (π) *unexcited* $\xrightarrow{h\nu}$ Cyclobutane (*allowed*)

(*b*)

Figure 9.14

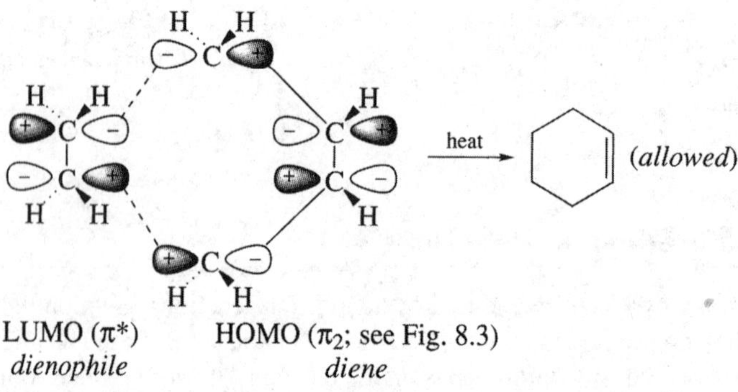

Figure 9.15

Electrocyclic (Intramolecular) Reactions

In electrocyclic reactions of conjugated polyenes, one double bond is lost and a single bond is formed between the terminal C's to give a ring. The reaction is reversible.

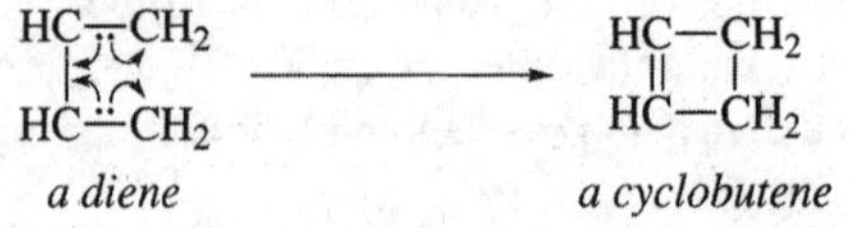

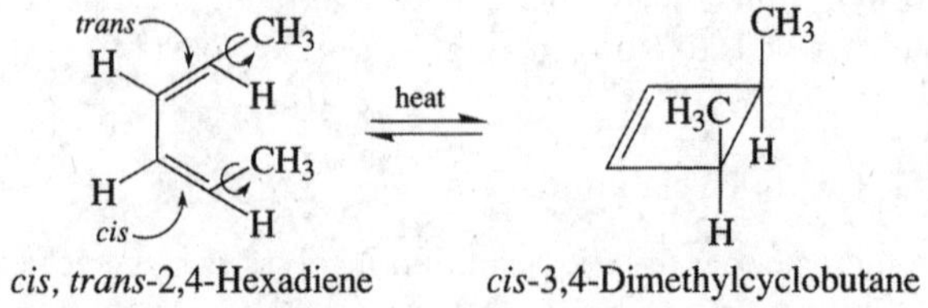

$HC{=}CH_2$ / $HC{=}CH_2$ (*a diene*) $\longrightarrow$ $HC{-}CH_2$ / $HC{-}CH_2$ (*a cyclobutene*)

cis, trans-2,4-Hexadiene $\underset{}{\overset{\text{heat}}{\rightleftharpoons}}$ *cis*-3,4-Dimethylcyclobutane

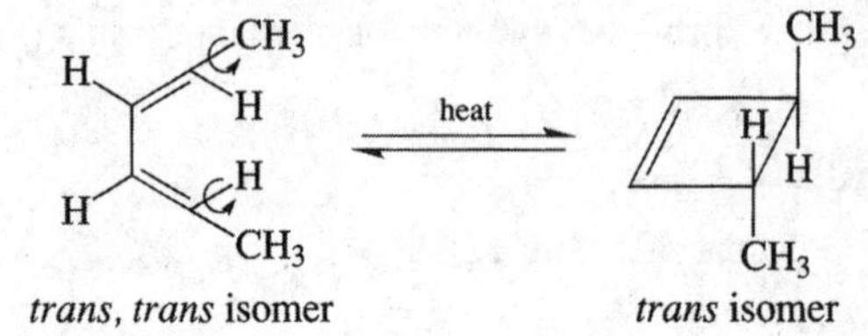

trans, trans isomer $\underset{}{\overset{\text{heat}}{\rightleftharpoons}}$ *trans* isomer

To achieve this stereospecificity, both terminal C's rotate 90° in the *same* direction, called a **conrotatory motion**. Movement of these C's in opposite directions (one clockwise and one counterclockwise) is termed **disrotatory**.

The Woodward-Hoffmann rule that permits the proper analysis of the stereochemistry is: **The orbital symmetry of the HOMO must be considered**, and rotation occurs to permit overlap of two like-signed lobes of the p orbitals to form the σ bond after rehybridization.

The HOMO for the thermal reaction then requires a conrotatory motion [Fig. 9.16(*a*)]. Irradiation causes a disrotatory motion by exciting an electron from $\pi_2 \rightarrow \pi_3^*$, which now becomes the HOMO [Fig. 9.16(*b*)].

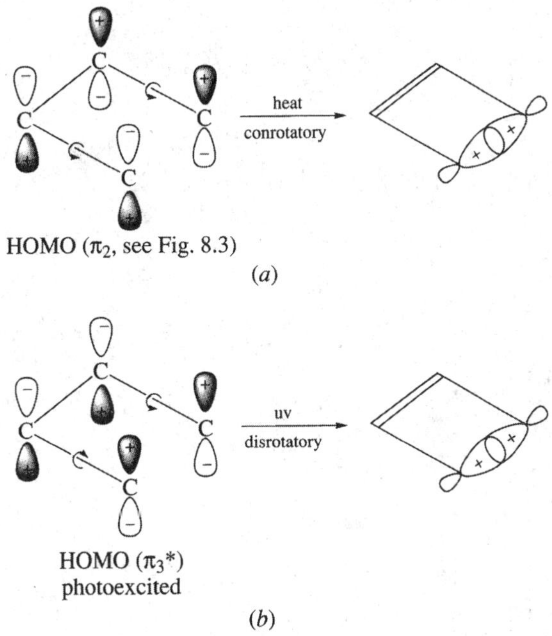

Figure 9.16

Problem 9.27 When applying the Woodward-Hoffmann rules to the Diels-Alder reaction, (*a*) would the same conclusion be drawn if the LUMO of the dienophile interacts with the HOMO of the diene? (*b*) Would the reaction be light-catalyzed?

(*a*) Yes; see Fig. 9.17(*a*). (*b*) No; see Fig. 9.17(*b*).

Problem 9.28 Use Woodward-Hoffmann rules to predict whether the following reaction would be expected to occur thermally or photochemically.

$^{1}CH_2{=}{}^{2}CH_2$ + $^{-3}$:CH_2–$C^4H{=}{}^{5}CH_2$ ⟶ cyclopentyl carbanion (ring positions 1, 2, 3, 4:−, 5)

allyl carbanion

The MO energy levels of the allyl carbanion π system showing the distribution of the four π electrons (two from π double bond and two unshared) are indicated in Fig. 9.18. The 0 is used whenever a node point is at an atom. The allowed reaction occurs thermally as shown in Fig. 9.19(*a*). The photoreaction is forbidden [Fig. 9.19(*b*)].

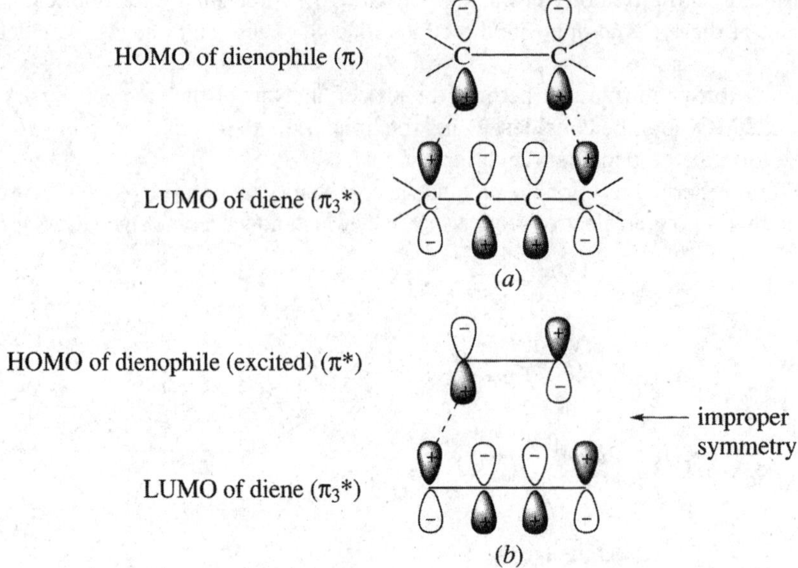
HOMO of dienophile (π)
LUMO of diene (π3*)
(a)
HOMO of dienophile (excited) (π*)
improper symmetry
LUMO of diene (π3*)
(b)

Figure 9.17

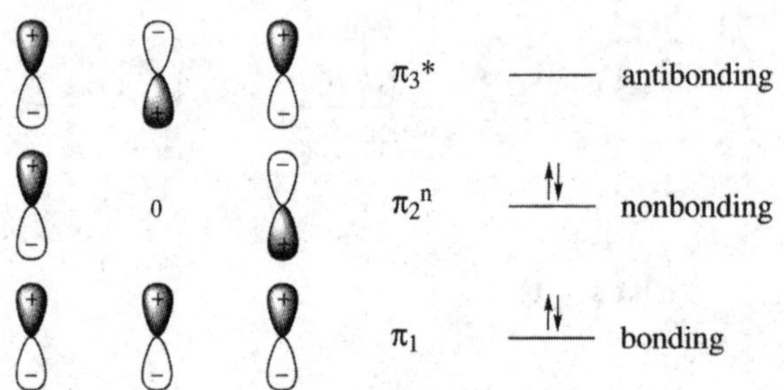
π3*
antibonding
π2n
nonbonding
π1
bonding

Figure 9.18

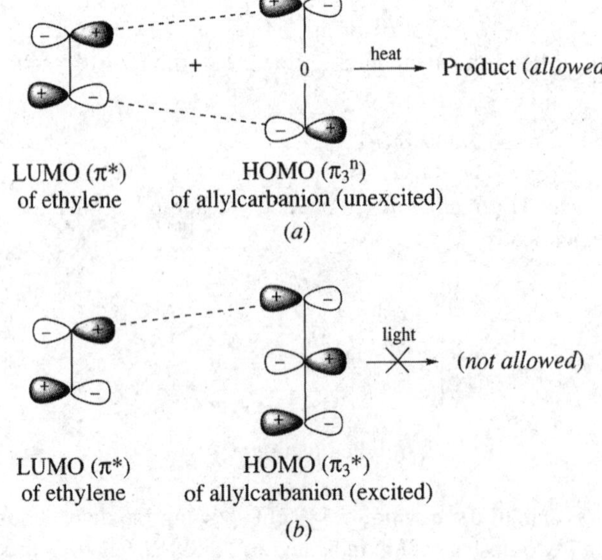
heat
Product (allowed)
LUMO (π*) of ethylene
HOMO (π3n) of allylcarbanion (unexcited)
(a)
light
(not allowed)
LUMO (π*) of ethylene
HOMO (π3*) of allylcarbanion (excited)
(b)

Figure 9.19

9.7 Terpenes and The Isoprene Rule

The carbon skeleton

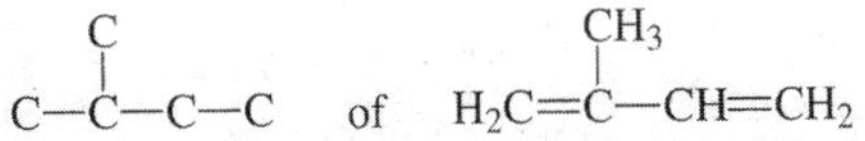

$$\underset{}{\overset{\displaystyle C}{\overset{|}{\text{C—C—C—C}}}} \quad \text{of} \quad \underset{\text{Isoprene}}{H_2C{=}\overset{\displaystyle CH_3}{\overset{|}{C}}{-}CH{=}CH_2}$$

is the structural unit of many naturally occurring compounds, among which are the **terpenes**, whose generic formula is $(C_5H_8)_n$.

Problem 9.29 Pick out the isoprene units in the terpenes limonene, myrcene, and α-phellandrene, and in vitamin A, shown below.

In the structures that follow, dashed lines separate the isoprene units.

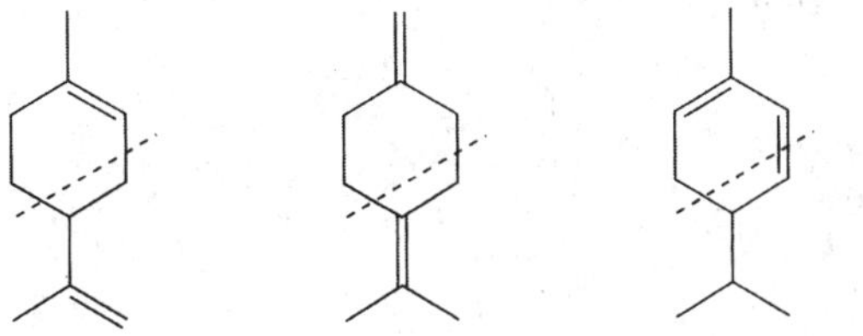

Limonene   Myrcene   α-Phellandrene

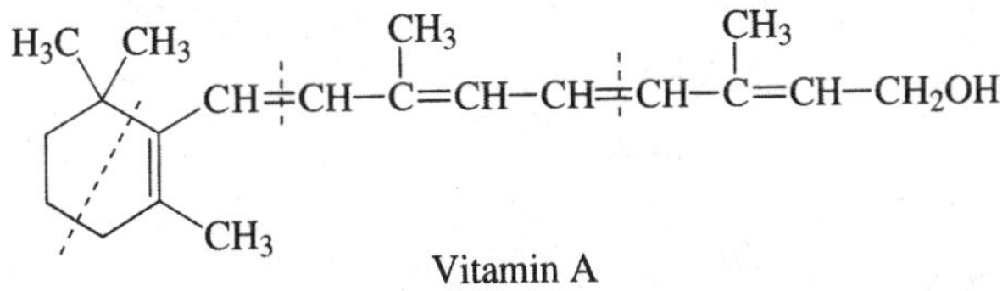

Vitamin A

SUPPLEMENTARY PROBLEMS

Problem 9.30 Draw formulas for (*a*) isopropylcyclopentane; (*b*) *cis*-1,3-dimethylcyclooctane; (*c*) bicyclo-[4.4.1]undecane; (*d*) *trans*-1-propyl-4-butylcyclohexane.

(*a*) or (*b*) or

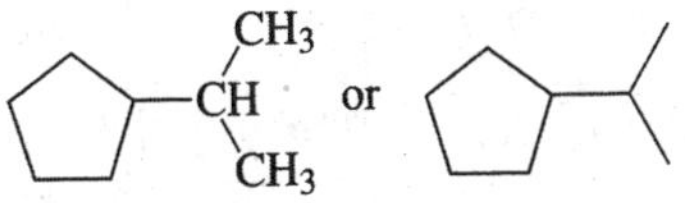

(*c*) (*d*) or

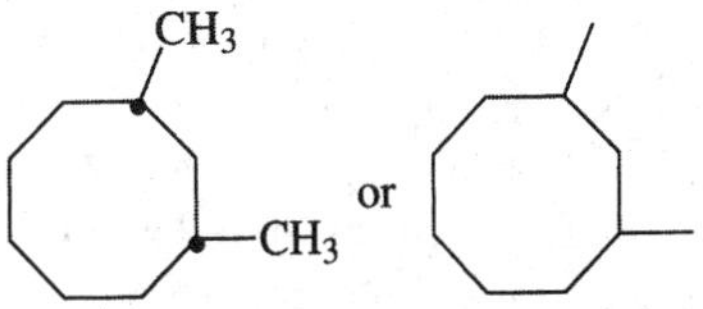

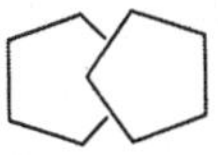

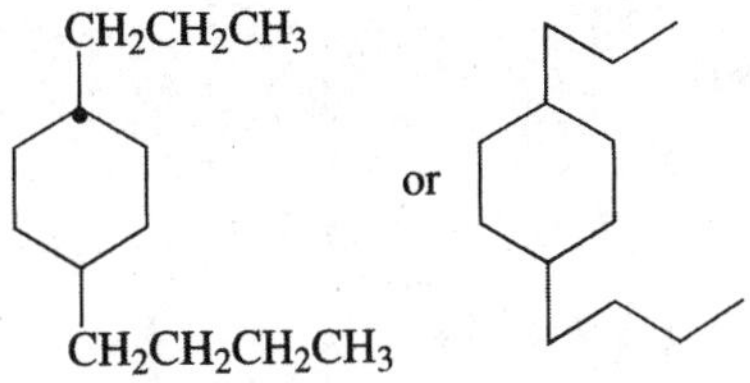

Problem 9.31 Name each of the following compounds and indicate which, if any, is chiral.

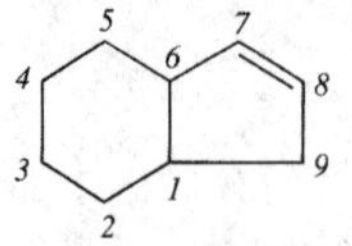

(*a*) 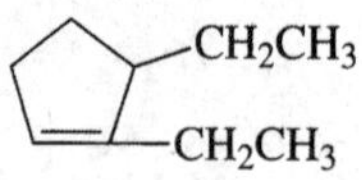 (*b*) 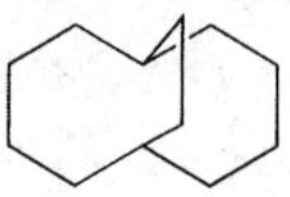 (*c*)

(*a*) The rings are numbered as shown, starting at a bridgehead C and going around the larger ring first, in such a way that the first doubly bonded C attains the lowest possible number. The name is bicyclo[4.3.0]non-7-ene. The molecule is chiral; C^1 and C^6 are chiral centers.
(*b*) 2,3-Diethylcyclopentene. The molecule is chiral; C^3 is a chiral center.
(*c*) Bicyclo[4.4.2]dodecane. The molecule is achiral.

Problem 9.32 Draw the structural formulas and give the stereochemical designation (*meso, rac, cis, trans*, achiral) of all the isomers of trichlorcyclobutane.

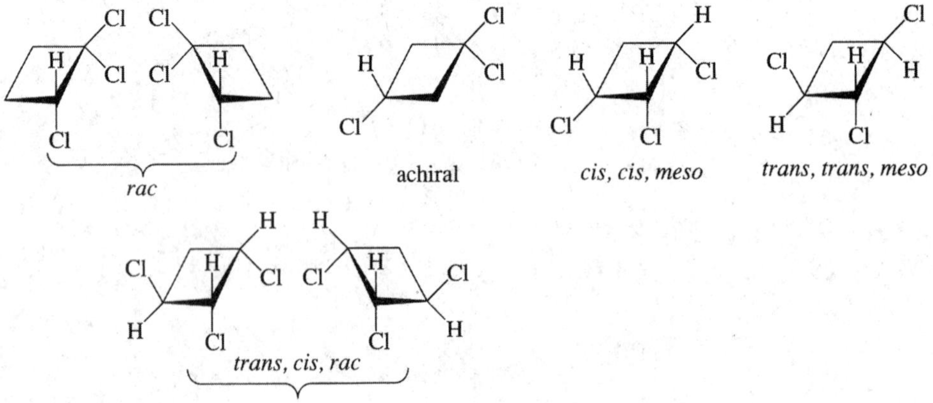

rac achiral *cis, cis, meso* *trans, trans, meso*

trans, cis, rac

Problem 9.33 Show steps in the synthesis of cyclohexane from phenol, C_6H_5OH.

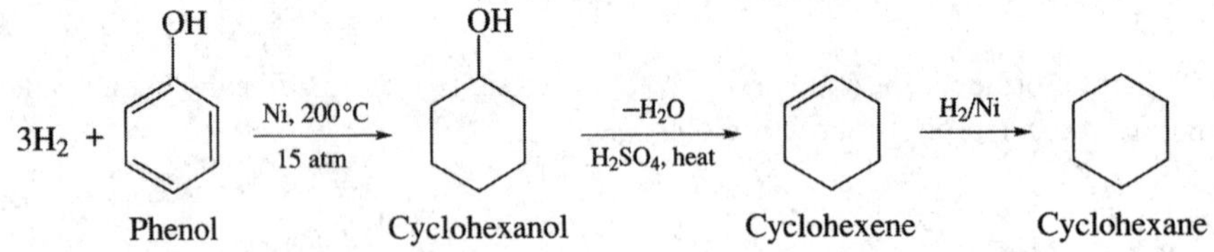

$$3H_2 + \text{Phenol} \xrightarrow[15\ \text{atm}]{\text{Ni, 200°C}} \text{Cyclohexanol} \xrightarrow[H_2SO_4,\ \text{heat}]{-H_2O} \text{Cyclohexene} \xrightarrow{H_2/Ni} \text{Cyclohexane}$$

Problem 9.34 Write structural formulas for the organic compounds designated by a ?. Indicate the stereochemistry where necessary, and account for the products.

(*a*) Methylcyclopropane $+ Br_2(CCl_4) \longrightarrow ? + ?$

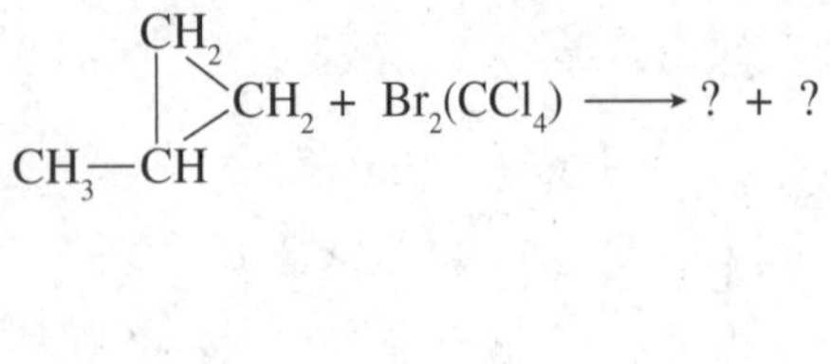

(*b*) $\begin{matrix} CH_2-CH_2 \\ | \quad\quad | \\ CH_2-CH_2 \end{matrix} + Br_2 \xrightarrow{uv} ?$

(*c*) $\begin{matrix} CH_3 & & H \\ & C=C & \\ H & & CH_3 \end{matrix} + CH_2N_2$ (liquid phase) $\xrightarrow{uv} ?$

(*d*) $\begin{matrix} CH_3 & & CH_3 \\ & C=C & \\ H & & H \end{matrix} + CH_2N_2$ (in presence of an inert gas, argon) $\xrightarrow{uv} ? + ?$

(*e*) Same as (*d*) but with added $O_2 \longrightarrow ?$

(*f*) $CH_2 \begin{matrix} / CH(CH_3)-CH \\ \quad\quad\quad \| \\ \backslash CH_2 \text{———} CH \end{matrix} + CHCl_3 \xrightarrow{(CH_3)_3CO^-K^+} ? + ?$

(*g*) $CH_2 \begin{matrix} / CH_2-C-CH_3 \\ \quad\quad \| \\ \backslash CH_2-CH \end{matrix} + CHClBr_2 \xrightarrow{(CH_3)_3CO^-K^+} ? + ?$

(*h*) $\begin{matrix} CH_2 \\ CH_2 \quad CH_2 \\ CH_3-C=CH \end{matrix} + Br_2 \xrightarrow{CCl_4} ?$

(*i*) $CH_2 \begin{matrix} / CH_2-CH_2 \backslash \\ \quad\quad\quad\quad CH_2 \\ \backslash CH_2-CHOH / \end{matrix} \xrightarrow[heat]{H_2SO_4} ? \xrightarrow[KMnO_4]{dil.\ aq.} ?$

(*a*) Two Br's add to each of the two nonequivalent single bonds I and II of

$\overset{b}{\ } \text{II}$, I, c, a I (methylcyclopropane ring with bonds labeled)

to form two products, (±)-$\overset{c}{C}H_2Br-\overset{b}{C}H_2-\overset{a}{C}HBr-CH_3$ by breaking (I) and

$\overset{b}{C}H_2BR-\overset{a}{C}H(CH_3)-\overset{c}{C}H_2BR$ by breaking (II)

(*b*) $\begin{matrix} CH_2-CHBr \\ | \quad\quad | \\ CH_2-CH_2 \end{matrix}$ by radical substitution; unlike the three-membered ring, the four-membered ring is stable.

(*c*) In the liquid phase, we get singlet CH_2 which adds *cis; trans*-2-butene forms *trans*-1,2-dimethylcyclopropane:

$\begin{matrix} CH_3 \ H \\ \triangledown \\ H \quad CH_3 \end{matrix}$

(*d*) Some initially formed singlet CH_2 collides with inert-gas molecules and changes to triplet

$\overset{\uparrow\ \uparrow}{CH_2}$

which adds nonstereospecifically. *cis*-2-Butene yields a mixture of *cis*- and *trans*-1,2-dimethylcyclopropane:

CH_3 CH_3 / H H and CH_3 H / H CH_3

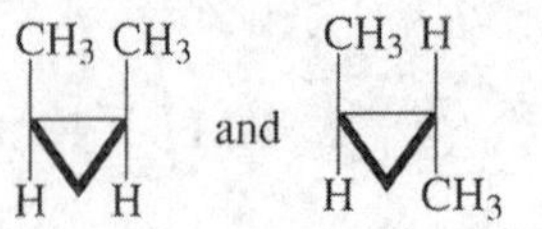

(*e*) O_2 is a diradical which combines with triplet carbenes, leaving the singlet species to react with *cis*-2-butene to give *cis*-1,2-dimethylcyclopropane.

(*f*) Dichlorocarbene adds *cis* to C=C, but either *cis* or *trans* to the Me.

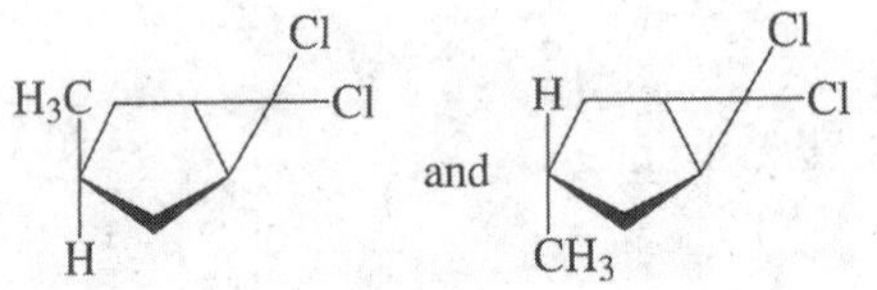

H_3C, H, Cl, Cl and H, CH_3, Cl, Cl

(*g*) $CHClBr_2$ loses Br^-, the better leaving group, rather than Cl^- to give ClBrC:, which then adds so that either Cl or Br can be *cis* to CH_3.

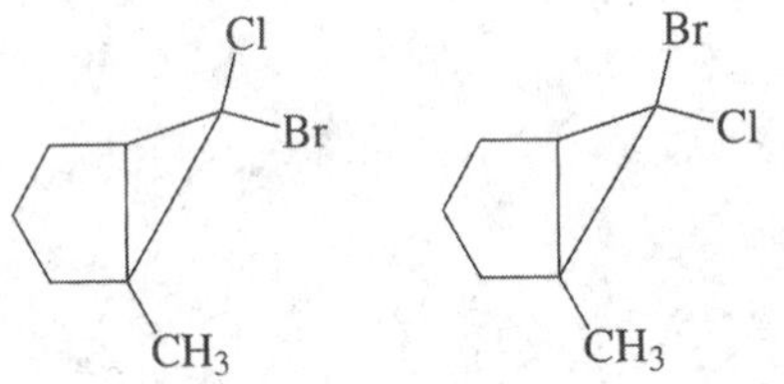

Cl, Br, CH_3 Br, Cl, CH_3

(*h*) H Br / Br CH_3 + Br H / H_3C Br

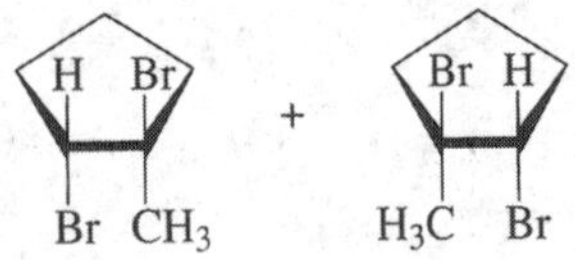

trans addition; 2 enantiomers (*rac*)

(*i*) → H H / HO OH

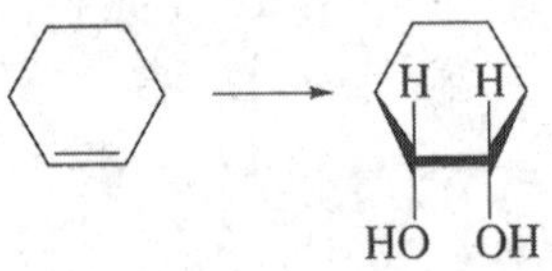

Cyclohexanol is dehydrated to cyclohexene, which forms a *meso* glycol by *cis* addition of two OH's.

Problem 9.35 Explain why (*a*) a carbene is formed by dehydrohalogenation of $CHCl_3$ but not from methyl, ethyl, or *n*-propyl chlorides; (*b*) *cis*-1,3- and *trans*-1,4-di-*tert*-butylcyclohexane exist in chair conformations, but their geometric isomers, *trans*-1,3- and *cis*-1,4-, do not.

(*a*) Carbene is formed from $CHCl_3$ because the three strongly electronegative Cl's make this compound sufficiently acidic to have its proton abstracted by a base. CH_3Cl has only one Cl and is considerably less acidic. Carbene formation is an α-elimination of HCl from the same C; it does not occur with ethyl or propyl chlorides because protons are more readily eliminated from the β C's to form alkenes.

(*b*) Both *cis*-1,3 and *trans*-1,4 compounds exist in the chair form because of the stability of their (ee) conformers. *Trans*-1,3- and *cis*-1,4- are (ea). An axial *t*-butyl group is very unstable, so that a twist-boat with a quasi-(ee) conformation (Fig. 9.20) is more stable than the chair.

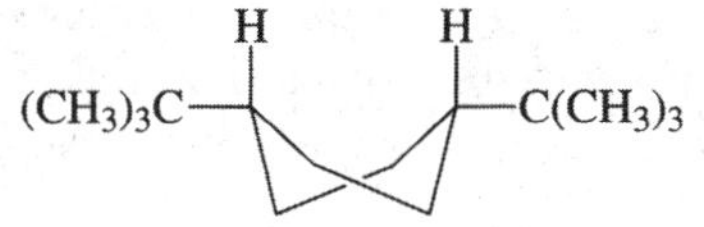

H H
$(CH_3)_3C$ — — $C(CH_3)_3$

Twist-boat
Twisting reduces eclipsed and "flagpole" interactions

Figure 9.20

Problem 9.36 Assign structures or configurations for A through D. (*a*) Two isomers, A and B, with formula C_8H_{14} differ in that one adds 1 mol and the other 2 mol of H_2. Ozonolysis of A gives only one product, $O{=}CH(CH_2)_6CH{=}O$, while the same reaction with 1 mol of B produces 2 mol of $CH_2{=}O$ and 1 mol of $O{=}CH(CH_2)_6CH{=}O$. (*b*) Two stereoisomers, C and D, of 3,4-dibromocyclopentane-1,1-dicarboxylic acid undergo decarboxylation as shown:

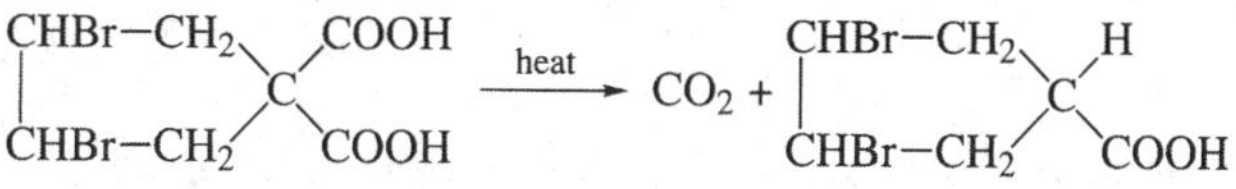

CHBr—CH₂ C(COOH)(COOH) / CHBr—CH₂ —heat→ CO_2 + CHBr—CH₂ C(H)(COOH) / CHBr—CH₂

a *gem*-dicarboxylic acid — a monocarboxylic acid

C gives one, while D yields two, monocarboxylic acids.

(*a*) Both compounds have two degrees of unsaturation (see Problem 6.34). B absorbs two moles of H_2 and has two multiple bonds. A absorbs one mole of H_2 and has a ring and a double bond; it is a cycloalkene. As a cycloalkene, A can form only a single product, a dicarbonyl compound, on ozonolysis.

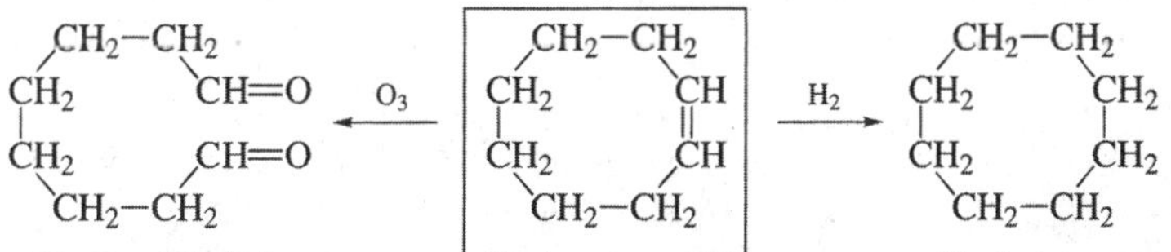

Octane-1,8-dial ←O_3— Cyclooctene (A) —H_2→ Cyclooctane

Since one molecule of B gives three carbonyl molecules, it must be a diene and not an alkyne.

$$H_2C{=}O + O{=}HC(CH_2)_4CH{=}O + O{=}CH_2 \xleftarrow{O_3} H_2C{=}CH(CH_2)_4CH{=}CH_2$$

1,6-Hexanedial — 1,7-Octadiene (B)

$$\xrightarrow{2H_2} CH_3CH_2(CH_2)_4CH_2CH_3$$

Octane

(*b*) The Br's of the dicarboxylic acid may be *cis* or *trans*. Decarboxylation of the *cis* isomer yields two isomeric products in which both Br's are *cis* (E) or *trans* (F) with respect to COOH. The *cis* isomer is D. In the monocarboxylic acid G formed from C (*trans* isomer), one Br is *cis* and the other *trans* with respect to COOH and there is only one isomer.

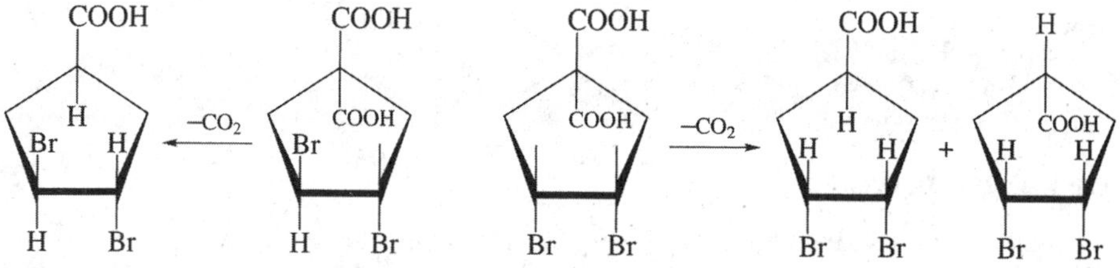

(G) ←$-CO_2$— (C) *trans*   (D) *cis* —$-CO_2$→ (F) + (E)

Problem 9.37 Outline the reactions and reagents needed to synthesize the following from any acyclic compounds having up to four C's and any needed inorganic reagents: (*a*) *cis*-1-methyl-2-ethylcyclopropane; (*b*) *trans*-1,1-dichloro-2-ethyl-3-*n*-propylcyclopropane; (*c*) 4-cyanocyclohexene; (*d*) bromocyclobutane.

(*a*) The *cis*-disubstituted cyclopropane is prepared by stereospecific additions of singlet carbene to *cis*-2-pentene.

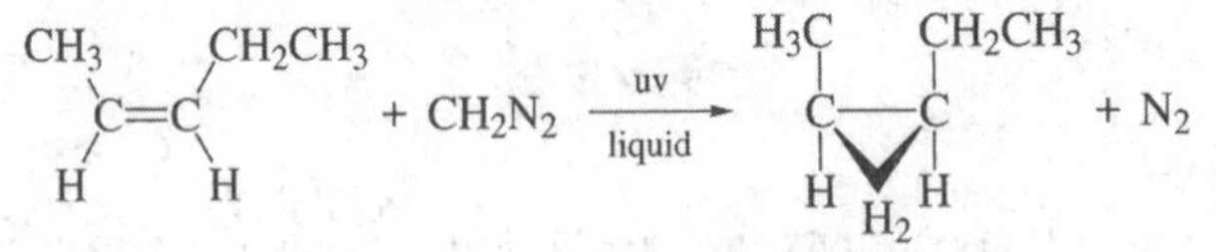

$$\text{(CH}_3\text{)(H)C=C(H)(CH}_2\text{CH}_3\text{)} + CH_2N_2 \xrightarrow[\text{liquid}]{\text{uv}} \text{H}_3\text{C(H)C—C(H)CH}_2\text{CH}_3\ (\text{ring }H_2) + N_2$$

The alkene, having five C's, is best formed from 1-butyne, a compound with four C's.

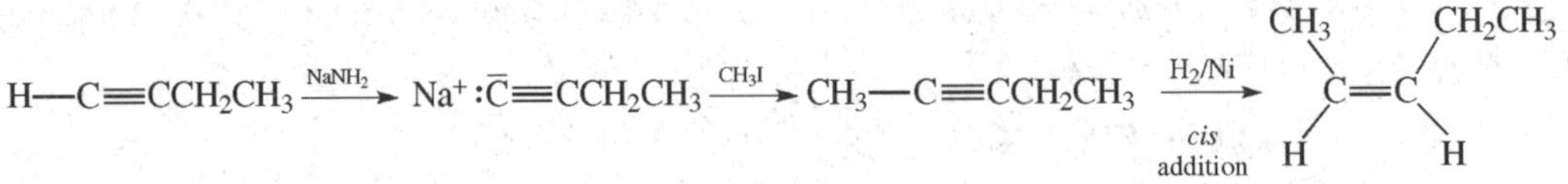

$$H—C\equiv CCH_2CH_3 \xrightarrow{NaNH_2} Na^+ :\bar{C}\equiv CCH_2CH_3 \xrightarrow{CH_3I} CH_3—C\equiv CCH_2CH_3 \xrightarrow[\text{cis addition}]{H_2/Ni} \text{(CH}_3\text{)(H)C=C(H)(CH}_2\text{CH}_3\text{)}$$

(*b*) Add dichlorocarbene to *trans*-3-heptene, which is formed from 1-butyne.

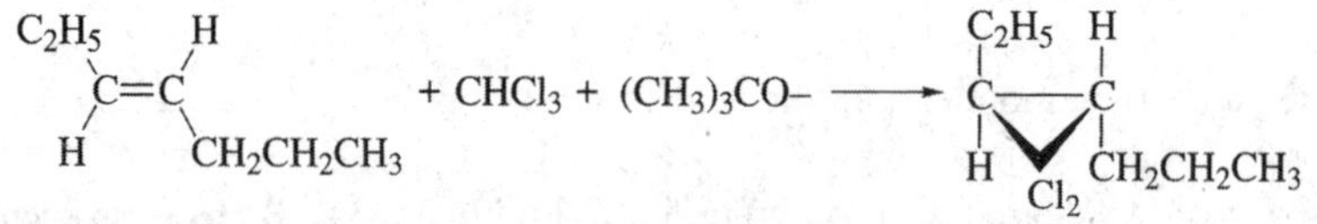

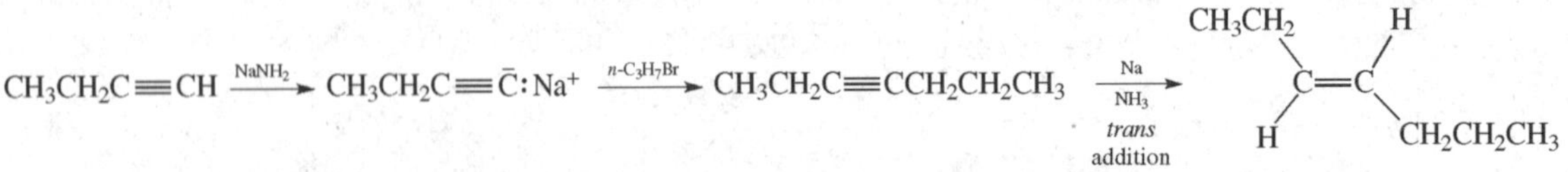

$$\text{(C}_2\text{H}_5\text{)(H)C=C(H)(CH}_2\text{CH}_2\text{CH}_3\text{)} + CHCl_3 + (CH_3)_3CO^- \longrightarrow \text{(C}_2\text{H}_5\text{)(H)C—C(H)(CH}_2\text{CH}_2\text{CH}_3\text{)}\ (\text{ring }Cl_2)$$

$$CH_3CH_2C\equiv CH \xrightarrow{NaNH_2} CH_3CH_2C\equiv \bar{C}{:}Na^+ \xrightarrow{n\text{-}C_3H_7Br} CH_3CH_2C\equiv CCH_2CH_2CH_3 \xrightarrow[\text{trans addition}]{Na/NH_3} \text{(CH}_3\text{CH}_2\text{)(H)C=C(H)(CH}_2\text{CH}_2\text{CH}_3\text{)}$$

(*c*) Cyclohexenes are best made by Diels-Alder reactions. The CN group is strongly electron-withdrawing and when attached to the C=C engenders a good dienophile.

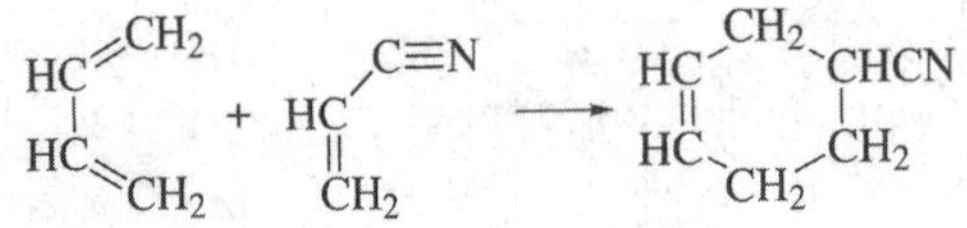

$$\underset{\text{1,3-Butadiene}}{CH_2=CH—CH=CH_2} + \underset{\text{Acrylonitrile}}{CH_2=CH—C\equiv N} \longrightarrow \text{cyclo-}(HC=CH—CH_2—CHCN—CH_2—CH_2) \quad \text{(a Diels–Alder reaction)}$$

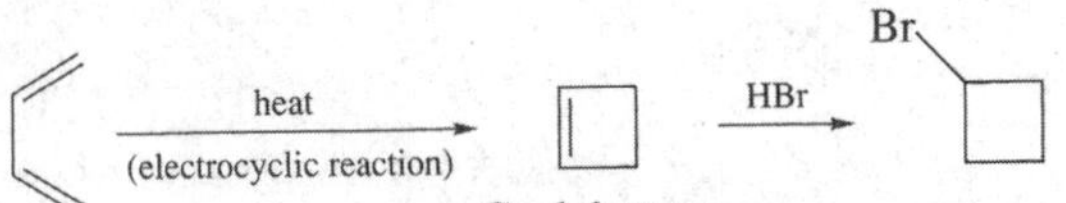

(*d*)

$$\text{1,3-butadiene} \xrightarrow[\text{(electrocyclic reaction)}]{\text{heat}} \underset{\text{Cyclobutene}}{\square} \xrightarrow{HBr} \text{Br—cyclobutane}$$

Problem 9.38 Write planar structures for the cyclic derivatives formed in the following reactions, and give their stereochemical labels.

(*a*) 3-Cyclohexenol + dil. aq. $KMnO_4$ ⟶

(*b*) 3-Cyclohexenol + HCO_3H and then H_2O ⟶

(*c*) (+)-*trans*-1,2-Dibromocyclopropane + Br_2 $\xrightarrow{\text{light}}$

(*d*) *meso-cis*-1,2-Dibromocyclopropane + Br_2 $\xrightarrow{\text{light}}$

(*e*) 1-Methylcyclohexene + HBr (peroxide) ⟶ (an *anti* addition)

(*a*) HO OH, OH, H H, H + HO H, H, H HO, OH
meso *rac*

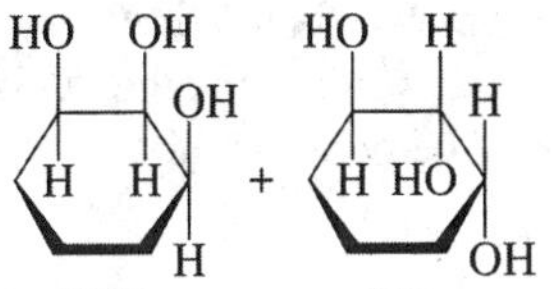

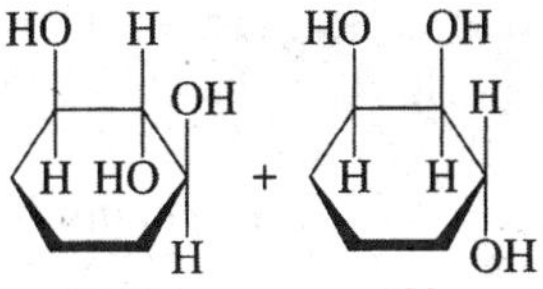

(*b*) HO H, OH, H HO, H + HO OH, H, H H, OH
meso *rac*

(*c*) Br H, H, Br Br, H + Br H, Br, H Br, H
Optically active
cis, trans
Optically inactive
(*meso*)

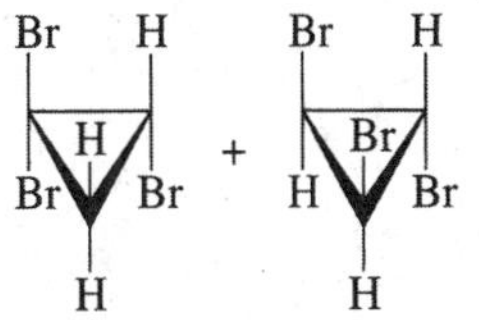

(*d*) Br Br, *, Br H + Br Br, *, H Br + Br Br, Br, H H, H + Br Br, *, H, *, H H, Br
racemate
all *cis*
no chiral C's
cis, trans-meso

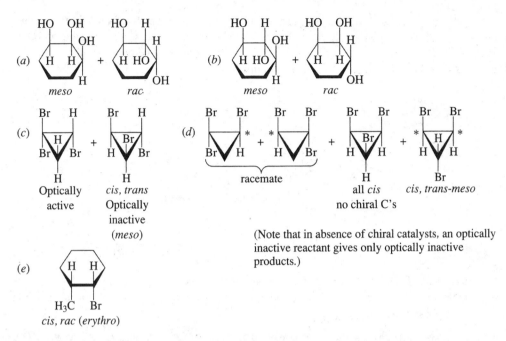

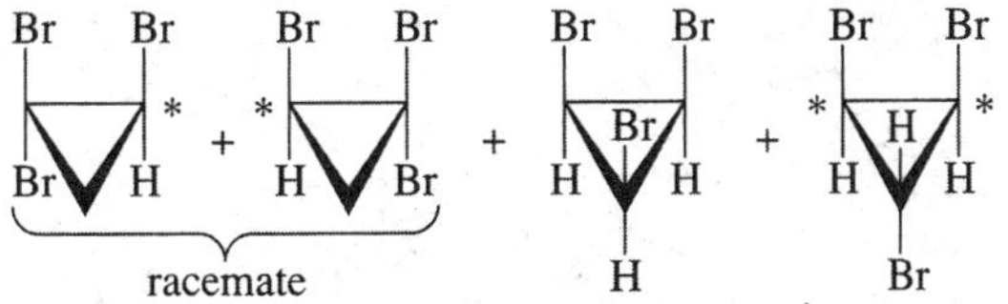

(Note that in absence of chiral catalysts, an optically inactive reactant gives only optically inactive products.)

(*e*) H H, H_3C Br
cis, rac (*erythro*)

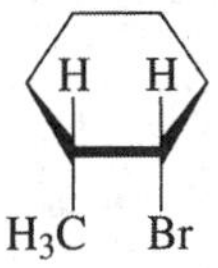

Problem 9.39 Use cyclohexanol and any inorganic reagents to synthesize (*a*) *trans*-1,2 -dibromocyclohexane; (*b*) *cis*-1,2-dibromocyclohexane; (*c*) *trans*-1,2-cyclohexanediol.

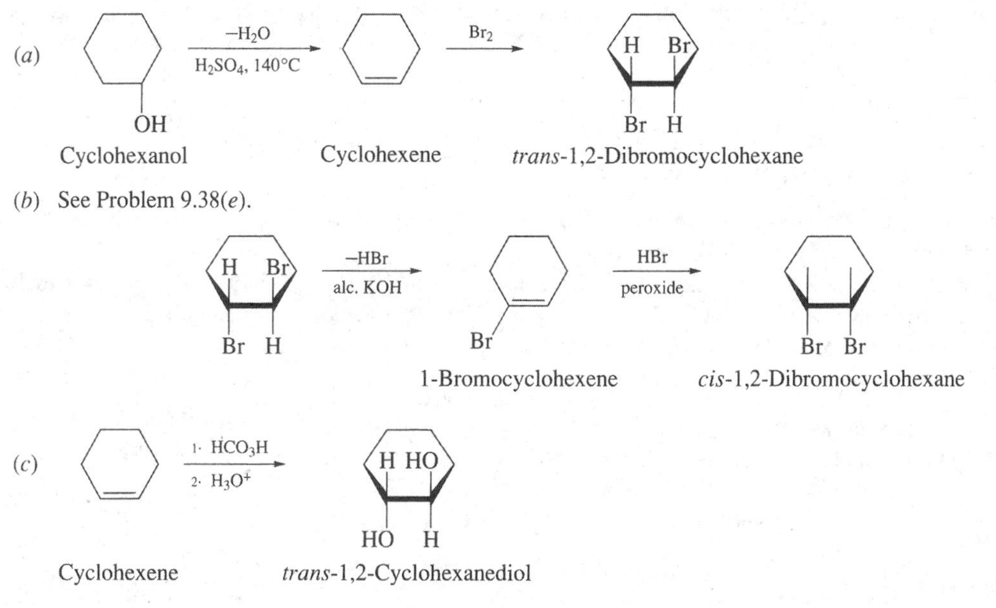

Problem 9.40 Decalin, $C_{10}H_{18}$, has *cis* and *trans* isomers that differ in the configurations about the two shared C's as shown below. Draw their conformational structural formulas.

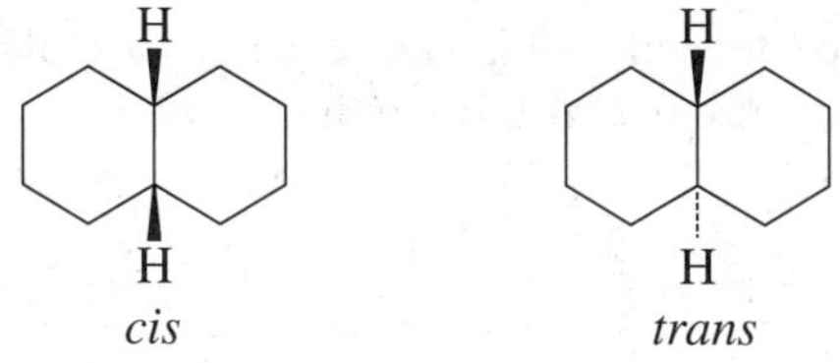

For each ring, the other ring can be viewed as 1,2-substituents. For the *trans* isomer, only the rigid (ee) conformation is possible structurally. As shown in Fig. 9.21, diaxial bonds point 180° away from each other and cannot be bridged by only four C's to complete the second ring. *Cis* fusion is (ea) and the bonds can be twisted to reverse the (a) and (e) positions, yielding conformation enantiomers.

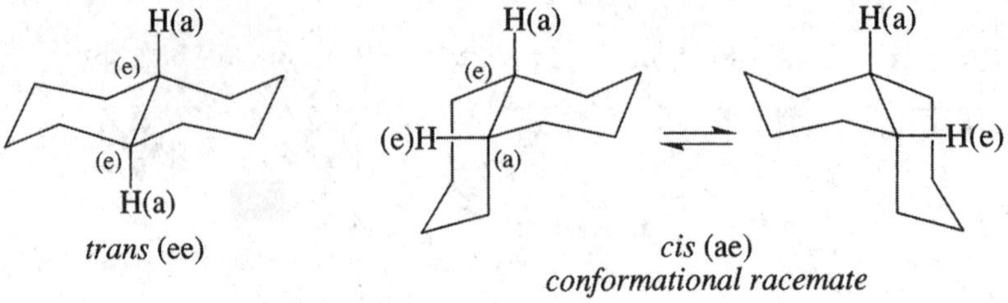

H(a) (e) (e) H(a)

trans (ee)

H(a) (e) (e)H (a) ⇌ H(a) H(e)

cis (ae)
conformational racemate

Figure 9.21

Problem 9.41 Explain the following facts in terms of the structure of cyclopropane. (*a*) The H's of cyclopropane are more acidic than those of propane. (*b*) The Cl of chlorocyclopropane is less reactive toward S_N2 and S_N1 displacements than the Cl in $CH_3CHClCH_3$.

(*a*) The external C—H bonds of cyclopropane have more *s* character than those of an alkane (Fig. 9.1). The more *s* character in the C—H bond, the more acidic the H.

(*b*) The C—Cl bond of chlorocyclopropane also has more *s* character, which diminishes the reactivity of the Cl. Remember that vinyl chlorides are inert in S_N2 and S_N1 reactions. The R^+ formed during the S_N1 reaction would have very high energy, since the C would have to use sp^2-hybrid orbitals needing a bond angle of 120°. The angle strain of the R^+ is much more severe (120°–60°) than in cyclopropane itself (109°–60°).

Problem 9.42 Use quantitative and qualitative tests to distinguish between (*a*) cyclohexane, cyclohexene, and 1,3-cyclohexadiene; (*b*) cyclopropane and propene.

(*a*) Cyclohexane does not decolorize Br_2 in CCl_4. The uptake of H_2, measured quantitatively, is 2 mol for 1 mol of the diene, but 1 mol for 1 mol of the cycloalkene.

(*b*) Cyclopropane resembles alkenes and alkynes, and differs from other cycloalkanes in decolorizing Br_2 slowly, adding H_2, and reacting readily with H_2SO_4. However, it is like other cycloalkanes and differs from multiple-bonded compounds in not decolorizing aqueous $KMnO_4$.

Problem 9.43 (*a*) Give the structure of the major product, A, whose formula is C_5H_8, resulting from the dehydration of cyclobutylmethanol. On hydrogenation, A yields cyclopentane. (*b*) Give a mechanism for this reaction.

(*a*) Compound A is cyclopentene, which gives cyclopentane on hydrogenation.

(*b*) CH_2OH $\xrightarrow[-H_2O]{+H^+}$ $\overset{+}{C}H_2$ ⟶ + $\xrightarrow{-H^+}$

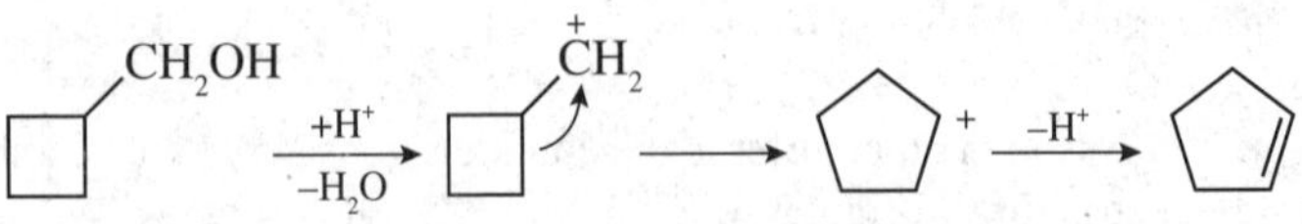

The side of a ring migrates, thereby converting a $R\overset{+}{C}H_2$ having a strained four-membered ring to a much more stable R_2CH^+ with a strain-free five-membered ring.

Benzene and Polynuclear Aromatic Compounds

10.1 Introduction

Benzene, C_6H_6, is the prototype of **aromatic** compounds, which are unsaturated compounds showing a low degree of reactivity. The **Kekulé structure** (1865) for benzene has only one monosubstituted product:

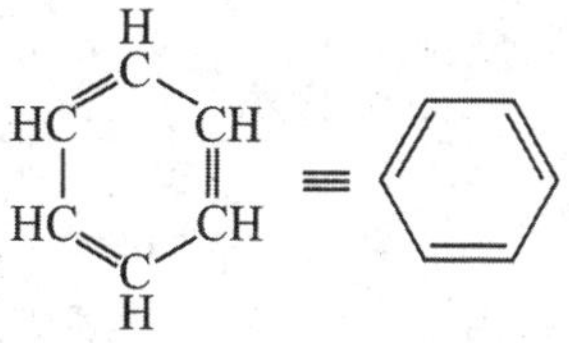

(C_6H_5Y), since all six H's are equivalent. There are three disubstituted benzenes—the 1,2-, 1,3-, and 1,4-position isomers—designated as *ortho, meta,* and *para*, respectively.

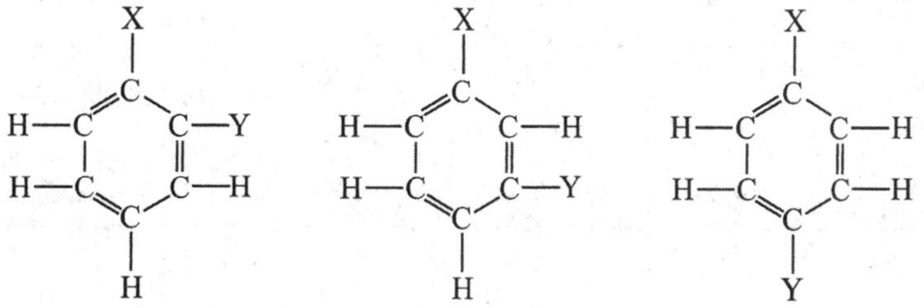

1,2- or *ortho* (*o*-) 1,3- or *meta* (*m*-) 1,4- or *para* (*p*-)

Problem 10.1 Benzene is a planar molecule with bond angles of 120°. All six C-to-C bonds have the identical length, 0.139 nm. Is benzene the same as 1,3,5-cyclohexatriene?

No. The bond lengths in 1,3,5-cyclohexatriene would alternate between 0.153 nm for the single bond and 0.132 nm for the double bond. The C-to-C bonds in benzene are intermediate between single and double bonds.

Problem 10.2 (*a*) How do the following heats of hydrogenation (ΔH_h, kJ/mol) show that benzene is not the ordinary triene 1,3,5-cyclohexatriene? Cyclohexene, −119.7; 1,4-cyclohexadiene, −239.3; 1,3-cyclohexadiene, −231.8; and benzene, −208.4. (*b*) Calculate the delocalization energy of benzene. (*c*) How does the delocalization energy of benzene compare to that of 1,3,5-hexatriene ($\Delta H_h = -336.8$ kJ/mol)? Draw a conclusion about the relative reactivities of the two compounds.

In computing the first column of Table 10.1, we assume that in the absence of any orbital interactions, each double bond should contribute −119.7 kJ/mol to the total ΔH_h of the compound, since this is the ΔH_h of an isolated C═C (in cyclohexane). Any difference between such a calculated ΔH_h value and the observed value is the delocalization energy. Since ΔH_h for 1,4-cyclohexadiene is 7.5 kJ/mol less than that for 1,3-cyclohexadiene, conjugation stabilizes the 1,3-isomer. [Remember that the smaller (more negative) the energy, the more stable the structure.]

(*a*) 1,3,5-Cyclohexatriene should behave as a typical triene and have $\Delta H_h = -359.1$ kJ/mol. The observed ΔH_h for benzene is −208.4 kJ/mol. Benzene is *not* 1,3,5-cyclohexatriene; in fact, the latter does not exist.
(*b*) See Table 10.1.
(*c*) The delocalization energy of benzene (−150.7 kJ/mol) is much smaller than that of 1,3,5-hexatriene (−22.3 kJ/mol). Three conjugated double bonds engender a large negative delocalization energy only when

TABLE 10.1

	CALCULATED ΔH_h, kJ/mol	OBSERVED ΔH_h, kJ/mol	DELOCALIZATION ENERGY
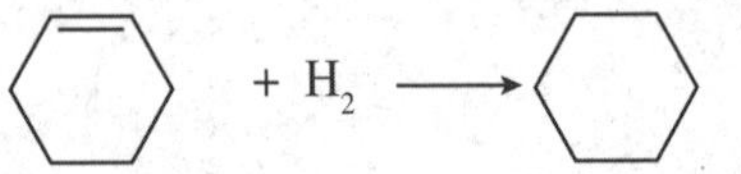 Cyclohexene + H_2 ⟶ Cyclohexane		−119.7	
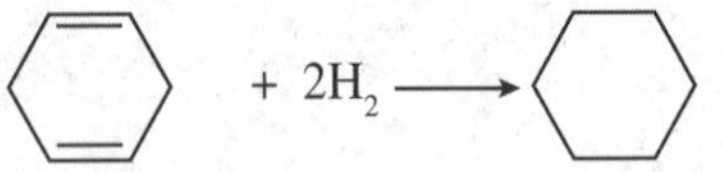 1,4-Cyclohexadiene + $2H_2$ ⟶ cyclohexane	2(−119.7) = −239.4	−239.3	0.0
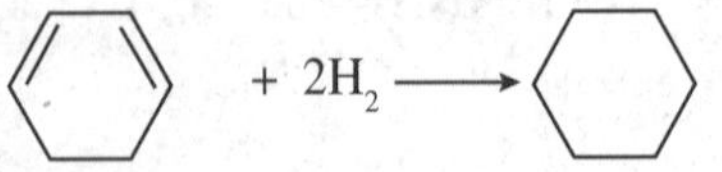 1,3-Cyclohexadiene + $2H_2$ ⟶ cyclohexane	2(−119.7) = −239.4	−231.8	−7.6
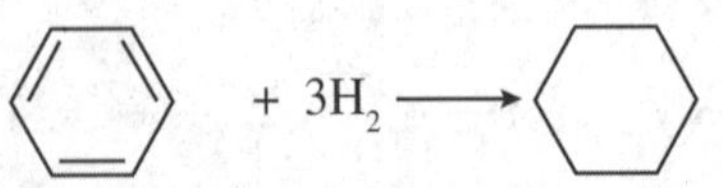 Benzene + $3H_2$ ⟶ cyclohexane	3(−119.7) = −359.1	−208.4	−150.7
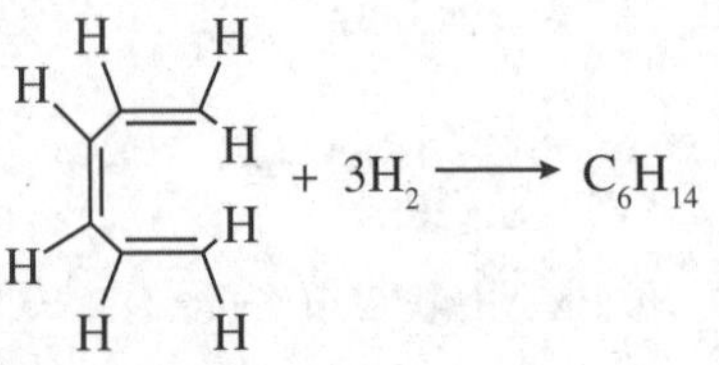 *cis*-1,3,5-Hexatriene + $3H_2$ ⟶ C_6H_{14}	3(−119.7) = −359.1	−336.8	−22.3

they are in a ring. Since the ground-state enthalpy of benzene is much smaller in absolute value than that of the triene, the $\Delta H^{\ddagger}$ for addition of H_2 to benzene is much greater, and benzene reacts much slower. Benzene is less reactive than open-chain trienes toward all electrophilic addition reactions.

Problem 10.3 (*a*) Use the ΔH_h's for complete hydrogenation of cyclohexene, 1,3-cyclohexadiene, and benzene, as given in Problem 10.2, to calculate ΔH_h for the addition of 1 mol of H_2 to (i) 1,3-cyclohexadiene, (ii) benzene. (*b*) What conclusion can you draw from these values about the rate of adding 1 mol of H_2 to these three compounds? (The ΔH of a reaction step is not necessarily related to $\Delta H^{\ddagger}$ of the step. However, in the cases being considered in this problem, $\Delta H_{\text{reaction}}$ is directly related to $\Delta H^{\ddagger}$.) (*c*) Can cyclohexadiene and cyclohexene be isolated on controlled hydrogenation of benzene?

Equations are written for the reactions so that their algebraic sum gives the desired reactant, products, and enthalpy.

(*a*) (i) Add reactions (1) and (2):

(1) Cyclohexane $- H_2 \longrightarrow$ Cyclohexene $\quad \Delta H = +119.7$ kJ/mol

(2) 1,3-Cyclohexadiene $+ 2H_2 \longrightarrow$ Cyclohexane $\quad \Delta H_h = -231.8$

(1) + (2) = (3) 1,3-Cyclohexadiene $+ H_2 \longrightarrow$ Cyclohexene $\quad \Delta H_h = -112.1$ kJ/mol

Note that reaction (1) is a dehydrogenation (reverse of hydrogenation) and that its ΔH is positive. (ii) Add the following two reactions:

Cyclohexane $- 2H_2 \longrightarrow$ 1,3-Cyclohexadiene $\quad \Delta H = +231.8$ kJ/mol

Benzene $+ 3H_2 \longrightarrow$ Cyclohexane $\quad \Delta H_h = -208.4$

Benzene $+ H_2 \longrightarrow$ 1,3-Cyclohexadiene $\quad \Delta H_h = +23.4$ kJ/mol

(*b*) The reaction with the largest negative ΔH_h value is the most exothermic and, in this case, also has the fastest rate. The ease of addition of 1 mol of H_2 is:

$$\text{cyclohexene } (-119.7) > \text{1,3-cyclohexadiene } (-112.1) \gg \text{benzene } (+23.4)$$

(*c*) No. When one molecule of benzene is converted to the diene, the diene is reduced all the way to cyclohexane by two more molecules of H_2 before more molecules of benzene react. If 1 mol each of benzene and H_2 are reacted, the product is $\frac{1}{3}$ mol of cyclohexane and $\frac{2}{3}$ mol of unreacted benzene.

Problem 10.4 The observed heat of combustion (ΔH_c) of C_6H_6 is -3301.6 kJ/mol.* Theoretical values are calculated for C_6H_6 by adding the contributions from each bond obtained experimentally from other compounds; these are (in kJ/mol) -492.4 for C=C, -206.3 for C—C and -225.9 for C—H. Use these data to calculate the heat of combustion for C_6H_6 and the difference between this and the experimental value. Compare the difference with that from heats of hydrogenation.

The contribution is calculated for each bond, and these are totaled for the molecule:

$$
\begin{aligned}
\text{Six C—H bonds} &= 6(-225.9) = -1355.4 \text{ kJ/mol} \\
\text{Three C—C bonds} &= 3(-206.3) = -618.9 \\
\text{Three C=C bonds} &= 3(-492.4) = \underline{-1477.2} \\
\text{TOTAL} &= -3451.5 \text{ (calculated } \Delta H_c \text{ for } C_6H_6) \\
\text{Experimental} &= \underline{-3301.6} \\
\text{DIFFERENCE} &= -149.9 \text{ kJ/mol}
\end{aligned}
$$

This difference is the delocalization energy of C_6H_6; essentially the same value is obtained from ΔH_h (Table 10.1).

Problem 10.5 How is the structure of benzene explained by (*a*) resonance, (*b*) the orbital picture, and (*c*) molecular orbital theory?

(*a*) Benzene is a hybrid of two equal-energy (Kekulé) structures differing only in the location of the double bonds:

(*b*) Each C is sp^2-hybridized and is σ bonded to two other C's and one H (Fig. 10.1). These σ bonds compose the skeleton of the molecule. Each C also has one electron in a *p* orbital at right angles to the plane of the ring. These *p* orbitals overlap *equally* with each of the two adjacent *p* orbitals to form a π system parallel to and above and below the plane of the ring (Fig. 10.2). The six *p* electrons in the π system are associated with all six C's. They are therefore more *delocalized*, and this accounts for the great stability and large resonance energy of aromatic rings.

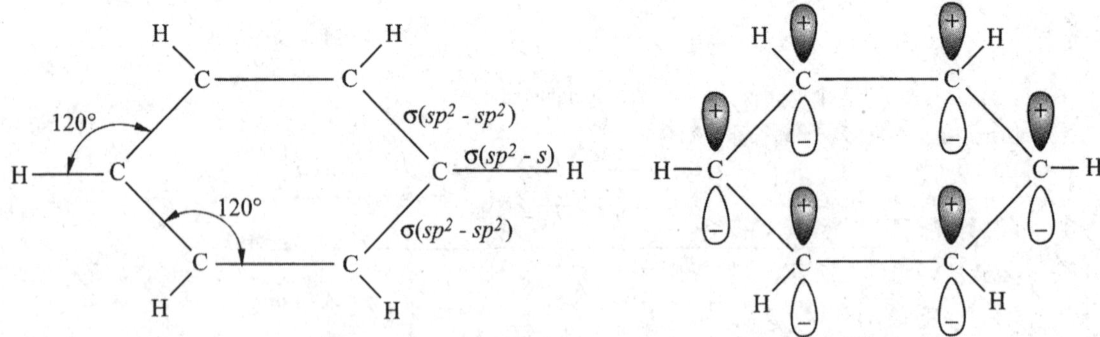

Figure 10.1

(*c*) The six *p* AO's discussed in part (*b*) interact to form six π MO's. These are indicated in Fig. 10.3, which gives the signs of the upper lobes (cf. Fig. 8.3 for butadiene). Since benzene is cyclic, the stationary waves representing the electron clouds are cyclic and have nodal planes, shown as lines, instead of nodal points. See Problem 9.28 for the significance of a 0 sign. The six *p* electrons fill the three bonding MO's, thereby accounting for the stability of C_6H_6.

* Some books define heat of combustion as $-\Delta H_c$, and values are given as positive numbers.

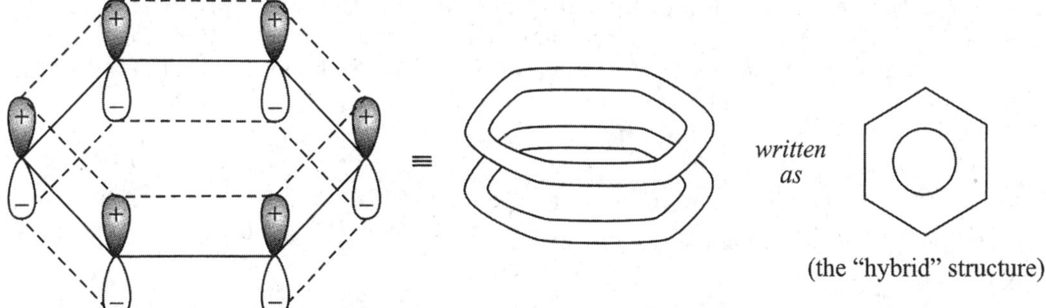

Figure 10.2

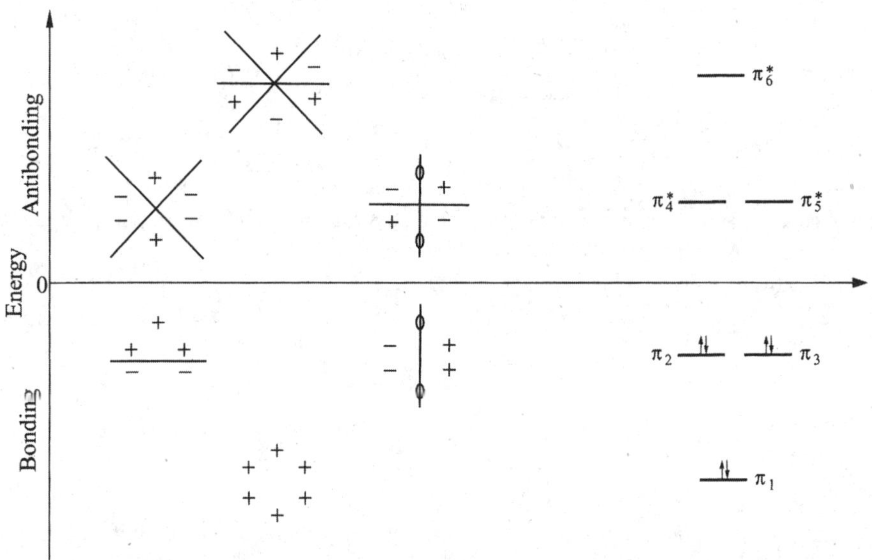

Figure 10.3

The unusual benzene properties collectively known as **aromatic character** are as follows:

1. **Thermal stability**.
2. **Substitution rather than addition reactions** with polar reagents such as HNO_3, H_2SO_4, and Br_2. In these reactions, the aromatic unsaturated ring is preserved.
3. **Resistance to oxidation** by aq. $KMnO_4$, HNO_3 and all but the most vigorous oxidants.
4. **Unique nuclear magnetic resonance spectra**. (See Table 12.4.)

10.2 Aromaticity and Hückel's Rule

Hückel's rule (1931) for **planar species** states that **if the number of π electrons is equal to $2 + 4n$, where n equals zero or a whole number, the species is aromatic**. The rule was first applied to carbon-containing monocyclics in which each C is capable of being sp^2-hybridized to provide a p orbital for extended π bonding; it has been extended to unsaturated heterocyclic compounds and fused-ring compounds. Note that benzene corresponds to $n = 1$.

Problem 10.6 Account for aromaticity observed in: (*a*) 1,3-cyclopentadienyl anion but not 1,3 cyclopentadiene; (*b*) 1,3,5-cycloheptatrienyl cation but not 1,3,5-cycloheptatriene; (*c*) cyclopropenyl cation; (*d*) the heterocycles pyrrole, furan, and pyridine.

(*a*) 1,3-Cyclopentadiene has an sp^3-hybridized C, making cyclic *p* orbital overlap impossible. Removal of H^+ from this C leaves a carbanion whose C is now sp^2-hybridized and has a *p* orbital capable of overlapping to give a cyclic π system. The four π electrons from two double bonds plus the two unshared electrons total six π electrons; the anion is aromatic ($n = 1$).

+ B: ⟶ B:H⁺ + or

cyclopentadienyl anion

(*b*) Although the triene has six *p* electrons in three C═C bonds, the lone sp^3-hybridized C prevents cyclic overlap of *p* orbitals.

⟶ Br^- + or

cycloheptatrienyl cation

Generation of a carbocation by ionization permits cyclic overlap of *p* orbitals on each C. With six π electrons, the cation is aromatic ($n = 1$).

(*c*) Cyclopropenyl cation has two π electrons and $n = 0$.

X^- ≡ X^-

a salt

The ions in parts (*a*), (*b*), and (*c*) are reactive, but they are much more stable than the corresponding open-chain ions.

(*d*) Hückel's rule is extended to heterocyclic compounds as follows:

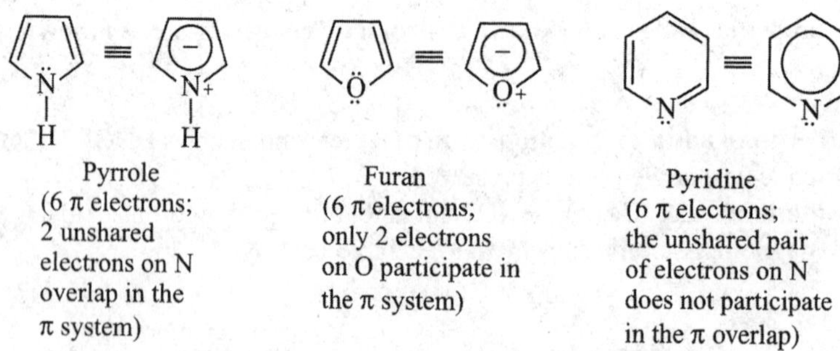

Pyrrole
(6 π electrons;
2 unshared
electrons on N
overlap in the
π system)
Furan
(6 π electrons;
only 2 electrons
on O participate in
the π system)
Pyridine
(6 π electrons;
the unshared pair
of electrons on N
does not participate
in the π overlap)

Note that dipoles are generated in pyrrole and furan because of delocalization of electrons from the heteroatoms.

10.3 Antiaromaticity

Planar cyclic conjugated species less stable than corresponding acyclic unsaturated species are called **antiaromatic**. They have $4n$ π electrons. 1,3-Cyclobutadiene ($n = 1$), for which one can write two equivalent contributing structures, is an extremely unstable antiaromatic molecule. This shows that the ability to write equivalent contributing structures is not sufficient to predict stability.

Problem 10.7 Cyclooctatetraene (C_8H_8), unlike benzene, is not aromatic; it decolorizes both dil. aq. $KMnO_4$ and Br_2 in CCl_4. Its experimentally determined heat of combustion is -4581 kJ/mol. (*a*) Use the Hückel rule to account for the differences in chemical properties of C_8H_8 from those of benzene. (*b*) Use thermochemical data of Problem 10.4 to calculate the resonance energy. (*c*) Why is this compound not antiaromatic? (*d*) Styrene, $C_6H_5CH{=}CH_2$, with heat of combustion -4393 kJ/mol, is an isomer of cyclooctatetraene. Is styrene aromatic?

(*a*) C_8H_8 has eight rather than six *p* electrons. Since it is not aromatic, it undergoes addition reactions.
(*b*) The calculated heat of combustion is

$$\begin{aligned} 8\ \text{C—H bonds} &= 8(-225.9) = -1807\ \text{kJ/mol} \\ 4\ \text{C—C bonds} &= 4(-206.3) = -825 \\ 4\ \text{C=C bonds} &= 4(-492.4) = \underline{-1970} \\ \text{TOTAL} &= -4602\ \text{kJ/mol} \end{aligned}$$

The difference $-4602 - (-4581) = -21$ kJ/mol shows small (negative) resonance energy and no aromaticity.
(*c*) Although the molecule has ($n = 2$) π electrons, it is *not* antiaromatic, because it is not planar. It exists chiefly in a "tub" conformation (Fig. 10.4).

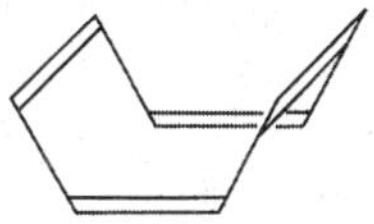

Figure 10.4

(*d*) Styrene is aromatic; its delocalization energy is $-4602 - (-4393) = -209$ kJ/mol. This is attributable to the presence of the benzene ring. Styrene has a more negative energy than does benzene [-150 kJ/mol] because its ring is conjugated to the C=C bond, thereby extending the delocalization of the electron cloud.

Problem 10.8 Deduce the structure and account for the stability of the following substances which are insoluble in nonpolar but soluble in polar solvents. (*a*) A red compound formed by reaction of 2 mol of $AgBF_4$ with 1 mol of 1,2,3,4-tetraphenyl-3,4-dibromocyclobut-1-ene. (*b*) A stable compound from the reaction of 2 mol of K with 1 mol of 1,3,5,7-cyclooctatetraene with no liberation of H_2.

The solubility properties suggest that these compounds are salts. The stability of the organic ions formed indicates that they conform to the Hückel rule and are aromatic.

(*a*) Two Br^-'s are abstracted by two Ag^+'s to form two AgBr and a tetraphenylcyclobutenyl dication.

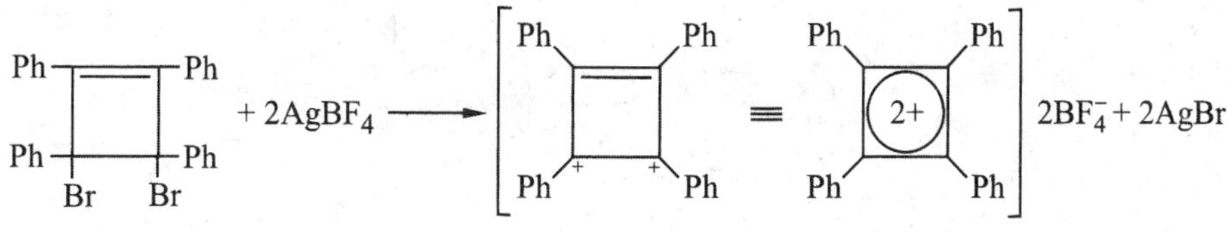

Ph, Ph, Ph, Ph, Br, Br + $2AgBF_4$ ⟶ [Ph, Ph, Ph, Ph ≡ Ph, Ph, 2+, Ph, Ph] $2BF_4^- + 2AgBr$

an aromatic cation ($n = 0$)

(*b*) Since K· is a strong reductant and no H_2 is evolved, two K's supply two electrons to form a cyclooctatetraenyl dianion (Fig. 10.5). This planar conjugated unsaturated monocycle has 10 electrons, conforms to the Hückel rule ($n = 2$), and is aromatic.

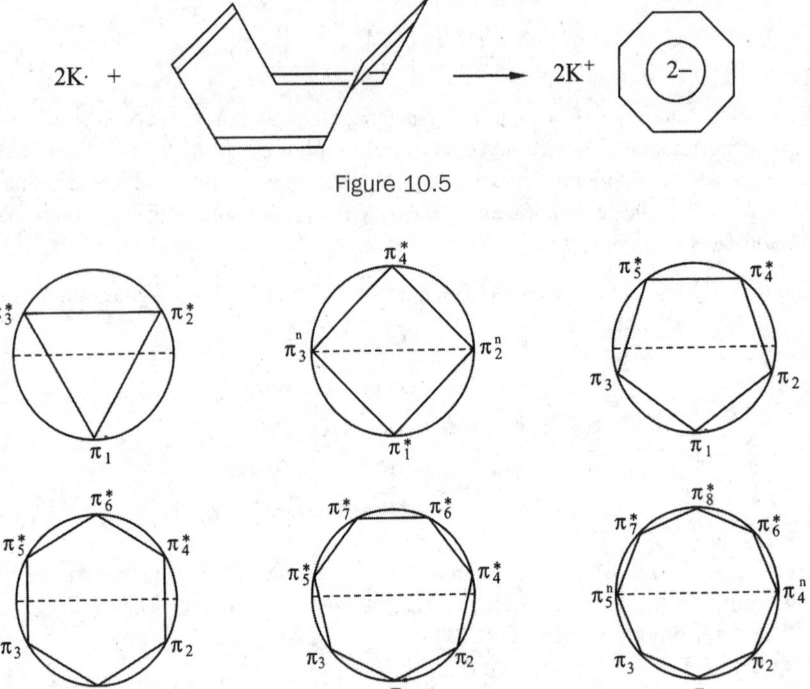

Figure 10.5

Figure 10.6

Problem 10.9 Sondheimer synthesized a series of interesting conjugated cyclopolyalkenes that he designated the [*n*]-**annulenes**, where *n* is the number of C's in the ring.

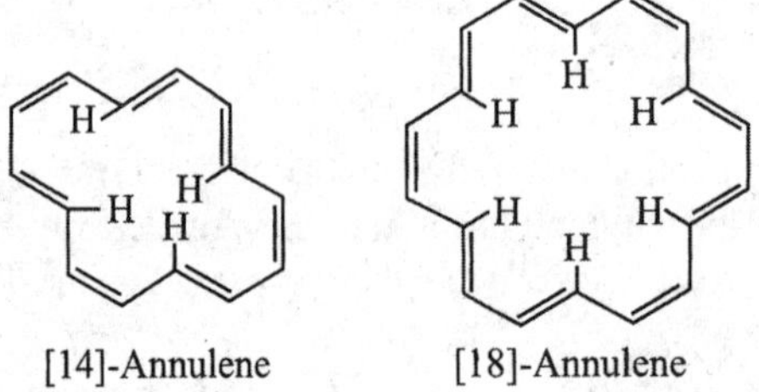

[14]-Annulene [18]-Annulene

Account for his observation that (*a*) [18]-annulene is somewhat aromatic, [16]- and [20]-annulene are not; (*b*) [18]-annulene is more stable than [14]-annulene.

(*a*) The somewhat aromatic [18]-annulene has $4n + 2$ ($n = 4$) π electrons; there are $4n$ π electrons in the nonaromatic, nonplanar [16]- and [20]-annulenes.

(*b*) [14]-Annulene is somewhat strained because the H's in the center of the ring are crowded. This steric strain prevents a planar conformation, which diminishes aromaticity.

Problem 10.10 Use the Hückel rule to indicate whether the following planar species are aromatic or antiaromatic:

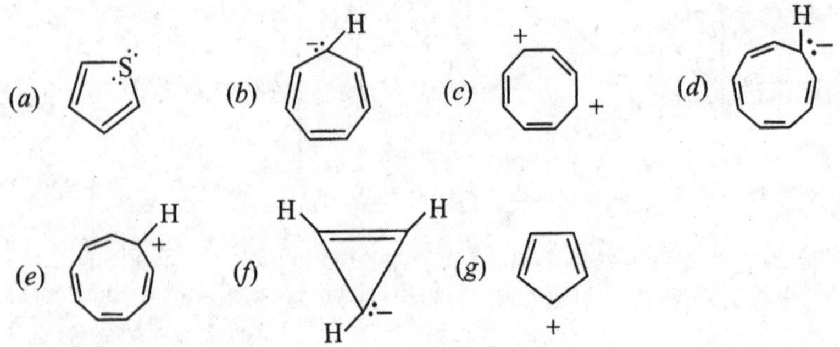

(*a*) Aromatic. There are 2 π electrons from each C=C and 2 from an electron pair on S to make an aromatic sextet. (*b*) Antiaromatic. There are $4n$ ($n = 2$) π electrons. (*c*) Aromatic. There are 6 π electrons. (*d*) Aromatic. There are 10 π electrons and this anion conforms to the $(4n + 2)$ rule ($n = 2$). (*e*) Antiaromatic. The cation has $4n$ ($n = 2$) π electrons. (*f*) and (*g*) Antiaromatic. They have $4n$ ($n = 1$) π electrons.

Problem 10.11 The relative energies of the MO's of conjugated cyclic polyenes can be determined by the following simple **polygon rule** instead of using nodal planes as in Problem 10.5(*c*). Inscribe a regular polygon in a circle, with one vertex at the bottom of the circle and with the total number of vertices equal to the number of MO's. Then the height of a vertex is proportional to the energy of the associated MO. Vertices below the horizontal diameter are bonding π, those above are antibonding π^*, and those on the diameter are nonbonding π^n. Apply the method to 3-, 4-, 5-, 6-, 7-, and 8-carbon systems and indicate the character of the MO's.

See Fig. 10.6.

10.4 Polynuclear Aromatic Compounds

Most of these compounds have fused benzene rings. The prototype is naphthalene, $C_{10}H_8$.

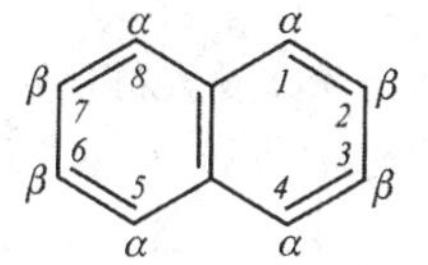

Naphthalene

Although the Hückel $4n + 2$ rule is rigorously derived for monocyclic systems, it is also applied in an approximate way to fused-ring compounds. Since two fused rings must share a pair of π electrons, the aromaticity and the delocalization energy per ring is less than that of benzene itself. Decreased aromaticity of polynuclear aromatics is also revealed by the different C—C bond lengths.

Problem 10.12 Draw a conclusion about the stability and aromaticity of naphthalene from the fact that the experimentally determined heat of combustion is 255 kJ/mol smaller in absolute value than that calculated from the structural formula.

The difference, −255 kJ/mol, is naphthalene's resonance energy. Naphthalene is less aromatic than benzene because a per-ring resonance energy of $\frac{1}{2}(-255) = -127.5$ kJ/mol is smaller in absolute value than that of benzene (−150 kJ/mol).

Problem 10.13 Deduce an orbital picture (like Fig. 10.2) for naphthalene, a planar molecule with bond angles of 120°.

See Fig. 10.7. The C's use sp^2 hybrid atomic orbitals to form σ bonds with each other and with the H's. The remaining p orbitals at right angles to the plane of the C's overlap laterally to form a π electron cloud.

Problem 10.14 (*a*) Draw three resonance structures for naphthalene. (*b*) Which structure makes the major contribution to the structure of the hybrid in that it has the smallest energy? (*c*) There are four kinds of C-to-C bonds in naphthalene: C^1-C^2, C^2-C^3, C^1-C^9, and C^9-C^{10}. Select the shortest bond, and account for your choice.

(*a*)

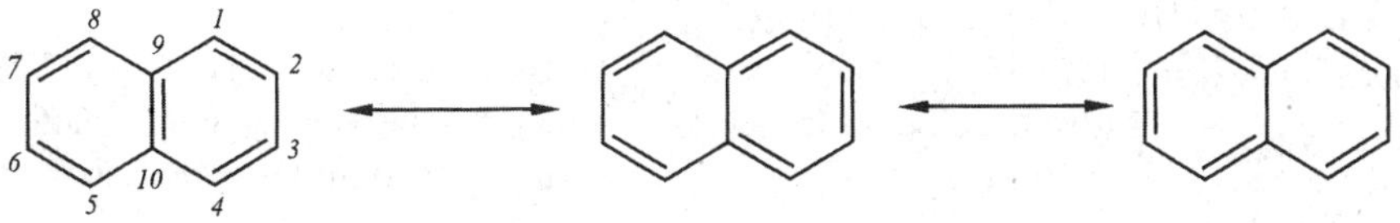

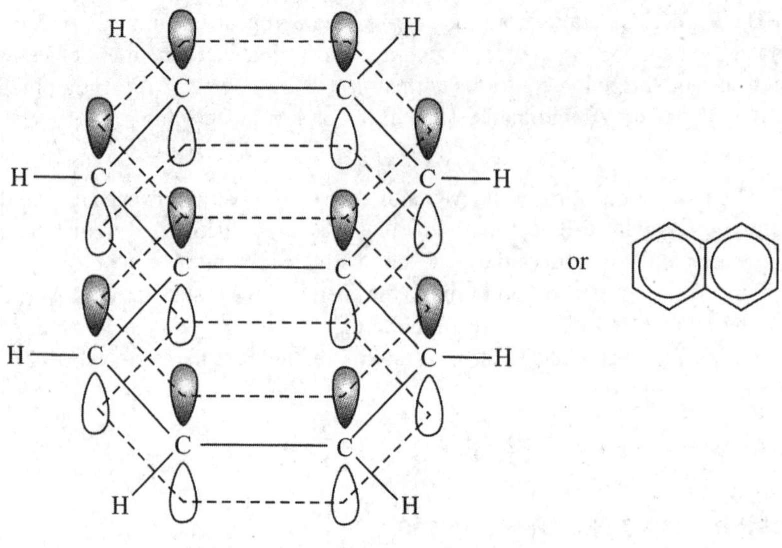

Figure 10.7

(*b*) Structure I has the smallest energy because only it has two intact benzene rings.
(*c*) The bond, in all of its positions, that most often has double-bond character in the three resonance structures is the shortest. This is true for C^1-C^2 (also for C^3-C^4, C^5-C^6, and C^7-C^8).

Anthracene and phenanthrene are isomers ($C_{14}H_{10}$) having three fused benzene rings.

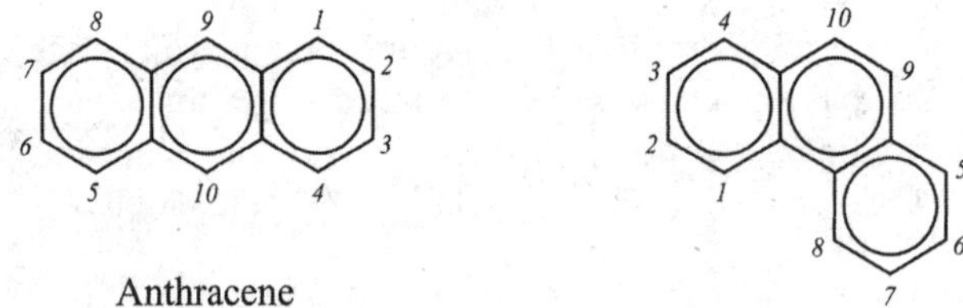

Anthracene

Phenanthrene

As the number of fused rings increases, the delocalization energy per ring continues to decrease in absolute value and compounds become more reactive, especially toward addition. The delocalization energies per ring for anthracene and phenanthrene are -117.2 and -126.8 kJ/mol, respectively.

10.5 Nomenclature

Some common names of benzene derivatives are **toluene** ($C_6H_5CH_3$), **xylene** ($C_6H_4(CH_3)_2$), **phenol** (C_6H_5OH), **aniline** ($C_6H_5NH_2$), **benzaldehyde** (C_6H_5CHO), **benzoic acid** (C_6H_5COOH), **benzene-sulfonic acid** ($C_6H_5SO_3H$), **styrene** ($C_6H_5CH{=}CH_2$), **mesitylene** (1,3,5-$(CH_3)_3C_6H_3$), and **anisole** ($C_6H_5OCH_3$).

Derived names combine the name of the substituent as a prefix with the word *benzene*. Examples are **nitrobenzene** ($C_6H_5NO_2$), **ethylbenzene** ($C_6H_5\,CH_2CH_3$), and **fluorobenzene** (C_6H_5F).

Some **aryl** (Ar—) groups are C_6H_5— (**phenyl**), C_6H_5—C_6H_4— (**biphenyl**), *p*-$CH_3C_6H_4$— (*p*-**tolyl**) and $(CH_3)_2C_6H_3$— (**xylyl**). Some **arylalkyl** groups are $C_6H_5CH_2$— (**benzyl**), $C_6H_5\overset{}{C}H$— (**benzal**), $C_6H_5C{\equiv}$ (**benzo**), $(C_6H_5)_2CH$— (**benzhydryl**), and $(C_6H_5)_3C$— (**trityl**).

The order of decreasing priorities of common substituents is as follows: COOH, SO_3H, CHO, CN, C$=$O, OH, NH_2, R, NO_2, X. For disubstituted benzenes with a group giving the ring a common name, *o*-, *p*-, or *m*- is used to designate the position of the second group. Otherwise, positions of groups are designated by the lowest combination of numbers.

Problem 10.15 Name the compounds:

(*a*)  (*b*) 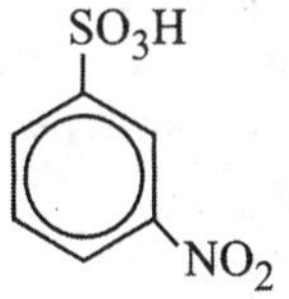 (*c*)

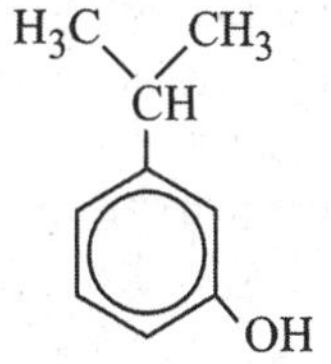

(*d*) 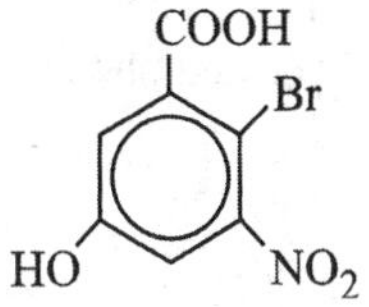 (*e*)

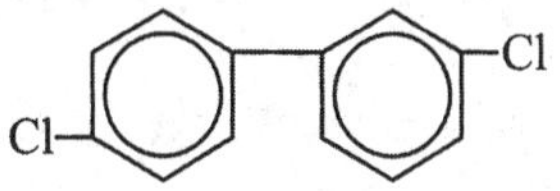

(*a*) *p*-Aminobenzoic acid. (*b*) *m*-Nitrobenzenesulfonic acid. (*c*) *m*-Isopropylphenol. (*d*) 2-Bromo-3-nitro-5-hydroxybenzoic acid. (Named as a benzoic acid rather than a phenol because COOH has priority over OH.) (*e*) 3,4′-Dichlorobiphenyl; the notational system in biphenyl is

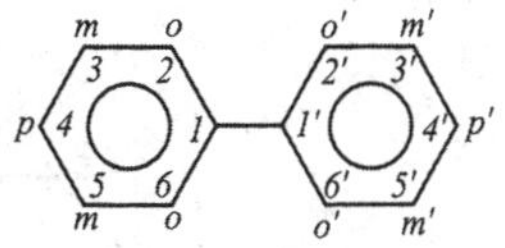

Problem 10.16 Give the structural formulas for (*a*) 2,4,6-tribromoaniline, (*b*) *m*-toluenesulfonic acid, (*c*) *p*-bromobenzalbromide, (*d*) di-*o*-tolylmethane, (*e*) trityl chloride.

(*a*) 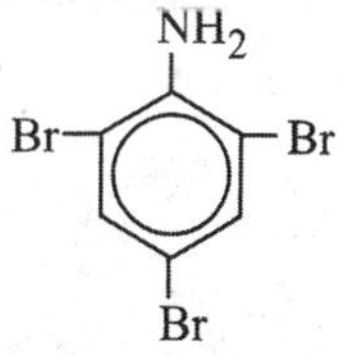 (*b*) 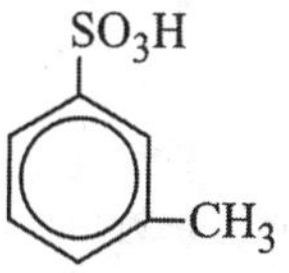 (*c*)

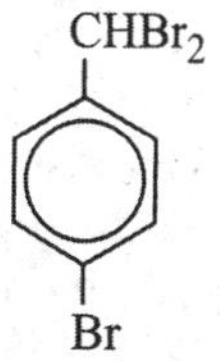

(*d*) 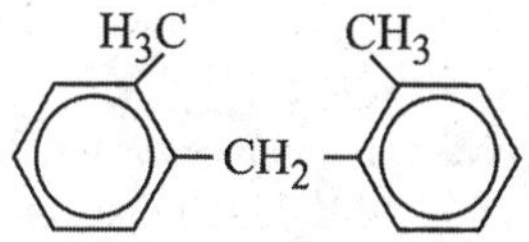 (*e*)

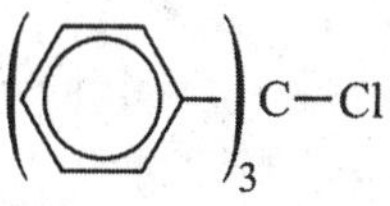

Problem 10.17 Name the following compounds:

(*a*) 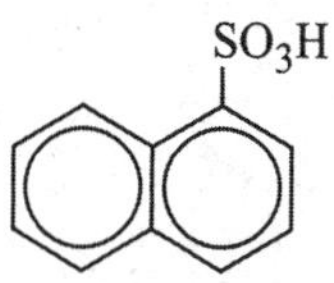 (*b*) 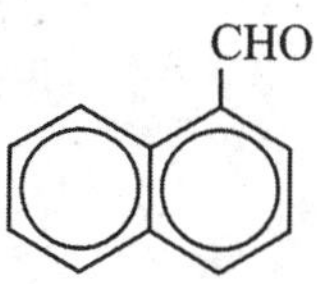 (*c*)

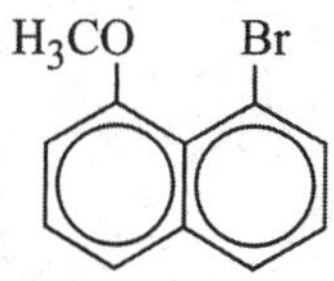

(*a*) 1-naphthalenesulfonic acid or α-naphthalenesulfonic acid, (*b*) 1-naphthaldehyde or α-naphthaldehyde, (*c*) 8-bromo–1-methoxynaphthalene.

10.6 Chemical Reactions

The unusual stability of the benzene ring dominates the chemical reactions of benzene and naphthalene. Both compounds resist addition reactions which lead to destruction of the aromatic ring. Rather, they undergo substitution reactions, discussed in detail in Chapter 11, in which a group or atom replaces an H from the ring, thereby preserving the stable aromatic ring. Atoms or groups other than H may also be replaced.

Reduction

Benzene is resistant to catalytic hydrogenation (high temperatures and high pressures of H_2 are needed) and to reduction with Na in alcohol. With Na in refluxing ethanol, naphthalene gains two H's, giving the unconjugated 1,4-dihydronaphthalene. In higher-boiling alcohols, two more H's are gained, at C^1 and C^4, to give the tetrahydro derivative (tetralin). Anthracene and phenanthrene each pick up H's at C^9 and C^{10}, and acquire no more.

Problem 10.18 (*a*) Write equations for the reductions of (i) naphthalene, (ii) anthracene, and (iii) phenanthrene. (*b*) Explain why naphthalene is reduced more easily than benzene. (*c*) Explain why anthracene and phenanthrene react at the C^9-C^{10} double bond and go no further.

(*a*)

(i) Naphthalene $\xrightarrow[132°C]{Na,\ C_5H_{11}OH}$ 1,2,3,4-Tetrahydronaphthalene (Tetralin)

Naphthalene $\xrightarrow[78°C]{Na,\ C_2H_5OH}$ 1,4-Dihydronaphthalene $\xrightarrow{H_2/Pt}$ 1,2,3,4-Tetrahydronaphthalene (Tetralin)

(ii) Anthracene $\xrightarrow[ROH]{Na}$ 9,10-Dihydroanthracene

(iii) Phenanthrene $\xrightarrow[ROH]{Na}$ 9,10-Dihydrophenanthrene

(*b*) Each ring of naphthalene is less aromatic than the ring of benzene and therefore is more reactive.

(*c*) These reactions leave two benzene rings having a combined resonance energy of $2(-150) = -300$ kJ/mol. If an attack were to occur in an end ring, a naphthalene derivative having a resonance energy of -255 kJ/mol would remain. Two phenyls are less energetic (more stable) than one naphthyl.

Problem 10.19 In the **Birch reduction,** benzene is reduced with an active metal (Na or Li) in alcohol and liquid NH_3(-33°C) to a cyclohexadiene that gives only $OCHCH_2CHO$ on ozonolysis. What is the reduction product?

Since the diene gives only a single product on ozonolysis, it must be symmetrical. The reduction product is 1,4-cyclohexadiene.

1,4-Cyclohexadiene $\xrightarrow[2.\ Zn,\ H^+]{1.\ O_3}$ $OHC-CH_2-CHO$ + $OHC-CH_2-CHO$

Malonic aldehyde

Problem 10.20 Typical of mechanisms for reductions with active metals in protic solvents, two electrons are transferred from the metal atoms to the substrate to give the most stable dicarbanion, which then accepts two H^+'s from the protic solvent molecules to give the product. (*a*) Give the structural formula for the dicarbanion formed from C_6H_6, and (*b*) explain why it is preferentially formed.

(*a*)

(*b*) The repulsion between the negative charges results in their maximum separation (*para* orientation). Product formation is controlled by the stability of the intermediate dicarbanion and not by the stability of the product. In this case, the more stable product would be the conjugated 1,3-cyclohexadiene.

Oxidation

Benzene is very stable to oxidation except under very vigorous conditions. In fact, when an alkylbenzene is oxidized, the alkyl group is oxidized to a COOH group, while the benzene ring remains intact. For this reaction to proceed, there must be at least one H atom on the C attached to the ring.

$$C_6H_5CHRR \rightarrow C_6H_5COOH \quad \text{(poor yields)}$$

However, in naphthalene, one benzene ring undergoes oxidation under mild conditions—for instance, $Cr_2O_7^{2-}$ in H^+. Under more vigorous conditions, a benzene dicarboxylic acid is formed.

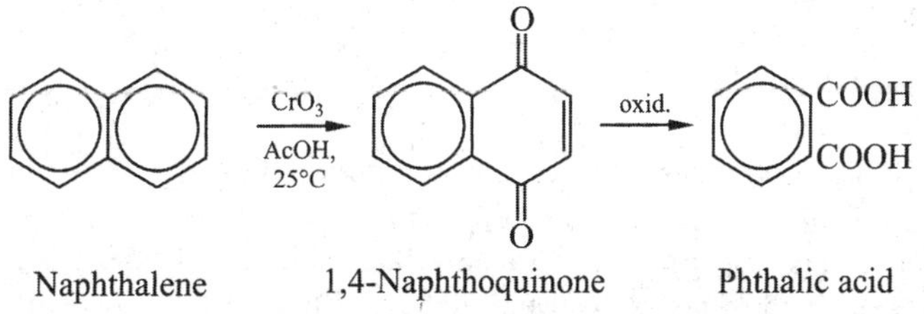

CrO_3, AcOH, 25°C; oxid.

Naphthalene 1,4-Naphthoquinone Phthalic acid

Problem 10.21 Oxidation of 1-nitronaphthalene yields 3-nitrophthalic acid. However, if 1-nitronaphthalene is reduced to α-naphthylamine and if this amine is oxidized, the product is phthalic acid.

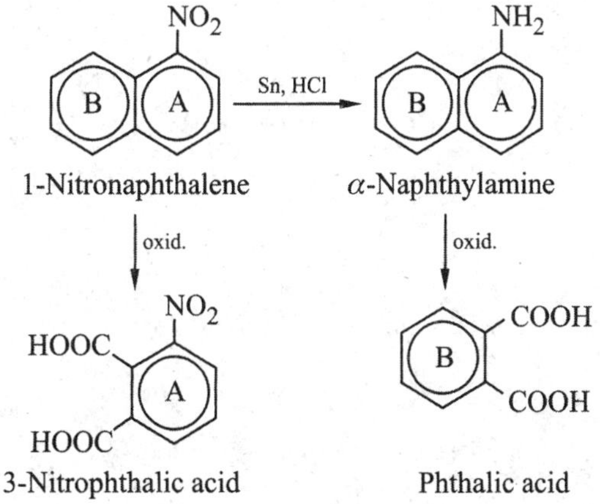

Sn, HCl; oxid.; oxid.

1-Nitronaphthalene α-Naphthylamine

3-Nitrophthalic acid Phthalic acid

How do these reactions establish the gross structure of naphthalene?

The electron-attracting —NO_2 stabilizes ring A of 1-nitronaphthalene to oxidation, and ring B is oxidized to form 3-nitrophthalic acid. By orbital overlap, —$\ddot{N}H_2$ releases electron density, making ring A more susceptible to oxidation, and α-naphthylamine is oxidized to phthalic acid. The NO_2 labels one ring and establishes the presence of two fused benzene rings in naphthalene.

Problem 10.22 Like naphthalene, anthracene and phenanthrene are readily oxidized to a quinone. Suggest the products and account for your choice.

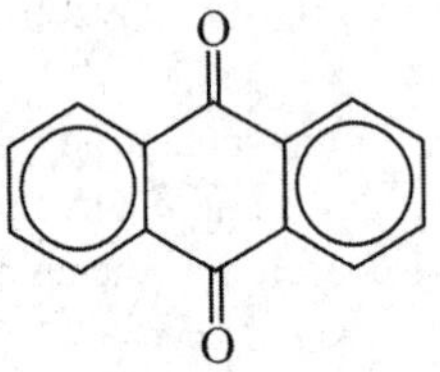

9,10-Anthraquinone

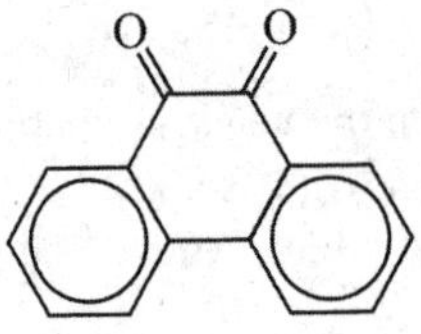

9,10-Phenanthraquinone

Oxidation at C^9 and C^{10} leaves two stable intact benzene rings. (See Problem 10.18(*c*).)

Halogenation

Under typical polar conditions, benzene and napthalene undergo halogen substitution and not addition. However, in the presence of UV light, benzene adds Cl_2 to give 1,2,3,4,5,6-hexachlorocyclohexane, an insecticide.

10.7 Synthesis

Benzene, naphthalene, toluene, and the xylenes are naturally occurring compounds obtained from coal tar. Industrial synthetic methods, called **catalytic reforming**, utilize alkanes and cycloalkanes isolated from petroleum. Thus, cyclohexane is dehydrogenated (**aromatization**), and *n*-hexane (**cyclization**) and methylcyclopentane (**isomerization**) are converted to benzene. Aromatization is the reverse of catalytic hydrogenation, and in the laboratory, the same catalysts—Pt, Pd, and Ni—can be used. The stability of the aromatic ring favors dehydrogenation.

Problem 10.23 Use the Diels-Alder reaction to synthesize benzoic acid, C_6H_5COOH.

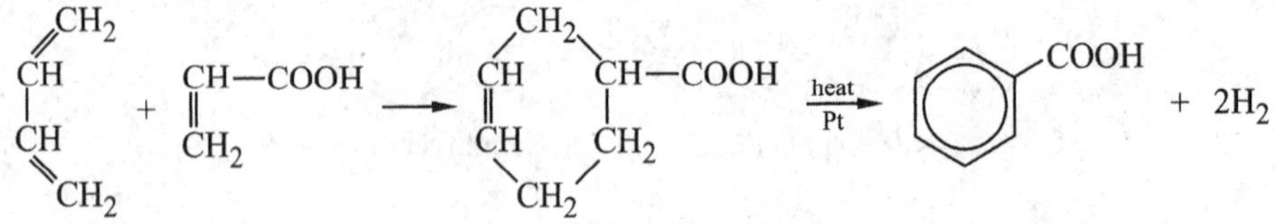

$$CH_2{=}CH{-}CH{=}CH_2 + CH_2{=}CH{-}COOH \longrightarrow \text{(cyclohexene-4-carboxylic acid)} \xrightarrow[\text{Pt}]{\text{heat}} C_6H_5COOH + 2H_2$$

S and Se can be used in place of Pt, and H_2S and H_2Se are then the respective products.

SUPPLEMENTARY PROBLEMS

Problem 10.24 (*a*) Draw two Kekulè structures for 1,2-dimethylbenzene (*o*-xylene). (*b*) Why are these structures not isomers? What are they? (*c*) Give the carbonyl products formed on ozonolysis.

(*a*) [structures of two Kekulé forms of o-xylene, each with two CH_3 groups]

(*b*) These structures differ only in the position of the π electrons and therefore are contributing (resonance) structures, not isomers.

(*c*) Although neither contributing structure exists, the isolated ozonolysis products arising from the resonance hybrid are those expected from either one:

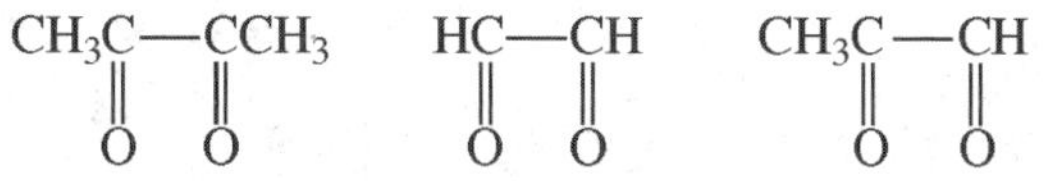

$$CH_3\underset{\underset{O}{\|}}{C}-\underset{\underset{O}{\|}}{C}CH_3 \qquad H\underset{\underset{O}{\|}}{C}-\underset{\underset{O}{\|}}{C}H \qquad CH_3\underset{\underset{O}{\|}}{C}-\underset{\underset{O}{\|}}{C}H$$

Problem 10.25 What are the necessary conditions for (*a*) aromaticity and (*b*) antiaromaticity?

(*a*) (1) A planar cyclic molecule or ion. (2) Each atom in the ring must have a *p* AO. (3) These *p* AO's must be parallel, so that they can overlap side by side. (4) The overlapping π system must have $(4n + 2)$ π electrons (Hückel).

(*b*) In (*a*), change $(4n + 2)$ to $4n$.

Problem 10.26 Design a table showing the structure, number of π electrons, energy levels of π MO's and electron distribution, and state of aromaticity of: (*a*) cyclopropenyl cation, (*b*) cyclopropenyl anion, (*c*) cyclobutadiene, (*d*) cyclobutadienyl dication, (*e*) cyclopentadienyl anion, (*f*) cyclopentadienyl cation, (*g*) benzene, (*h*) cycloheptatrienyl anion, (*i*) cyclooctatetraene, (*j*) cyclooctatetraenyl dianion.

See Table 10.2. (H's are understood to be attached to each doubly bonded C.)

Problem 10.27 Explain aromaticity and antiaromaticity in terms of the MO's of Problem 10.26.

Aromaticity is observed when all bonding MO's are filled and nonbonding MO's, if present, are empty or completely filled. Hückel's rule arises from this requirement. A species is antiaromatic if it has electrons in antibonding MO's or if it has half-filled bonding or nonbonding MO's, provided it is planar.

Problem 10.28 Name the monobromo derivatives of (*a*) anthracene, (*b*) phenanthrene.

(*a*) There are 3 isomers: 1-bromo-, 2-bromo-, and 9-bromoanthracene.

(*b*) There are 5 isomers: 1-bromo-, 2-bromo-, 3-bromo-, 4-bromo-, and 9-bromophenanthrene.

Problem 10.29 What is the Diels-Alder addition product of anthracene and ethene?

Reaction occurs at the (most reactive) C^9 and C^{10} positions.

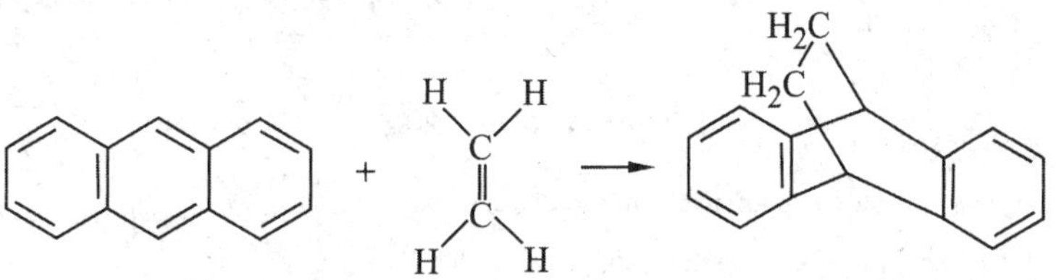

[anthracene + $H_2C{=}CH_2$ (H, H / C=C / H, H) → bicyclic adduct with H_2C–H_2C bridge across C9 and C10]

TABLE 10.2

	STRUCTURE	NUMBER OF π ELECTRONS	π MO's	AROMATICITY
(*a*)	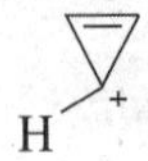 H +	2	π_3^*— —π_2^* ⇅ π_1	Aromatic
(*b*)	 H :–	4	π_3^*↑ ↑π_2^* ⇅ π_1	Antiaromatic
(*c*)	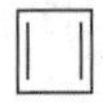	4	—π_4^* π_3^n↑ ↑π_2^n ⇅ π_1	Antiaromatic
(*d*)	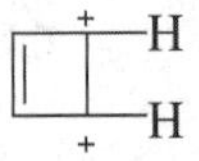 + H + H	2	—π_4^* π_3^n— —π_2^n ⇅ π_1	Aromatic
(*e*)	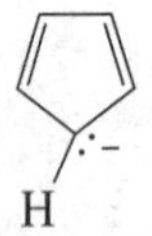 :– H	6	π_5^*— —π_4^* π_3 ⇅ ⇅ π_2 ⇅ π_1	Aromatic
(*f*)	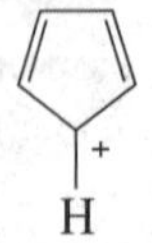 + H	4	π_5^*— —π_4^* π_3↑ ↑π_2 ⇅ π_1	Antiaromatic
(*g*)		6	—π_6^* π_5^*— —π_4^* π_3 ⇅ ⇅ π_2 ⇅ π_1	Aromatic
(*h*)	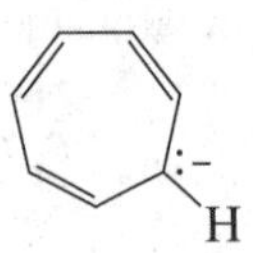 :– H	8	π_7^*— —π_6^* π_5^*↑ ↑π_4^* π_3 ⇅ ⇅ π_2 ⇅ π_1	Antiaromatic
(*i*)	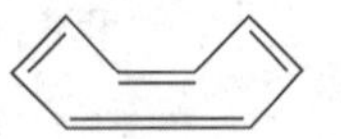	8	—π_8^* π_7^*— —π_6^* π_5^n↑ ↑π_4^n π_3 ⇅ ⇅ π_2 ⇅ π_1	Nonaromatic† (nonplanar)
(*j*)	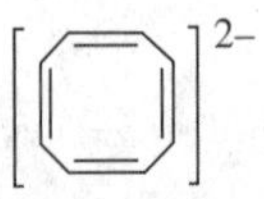 2–	10	—π_8^* π_7^*— —π_6^* π_5^n ⇅ ⇅ π_4^n π_3 ⇅ ⇅ π_2 ⇅ π_1	Aromatic

† If it were planar, it would be antiaromatic; to avoid this, (i) is nonplanar.

Aromatic Substitution; Arenes

11.1 Aromatic Substitution by Electrophiles (Lewis Acids, E^+ or E)

Mechanism

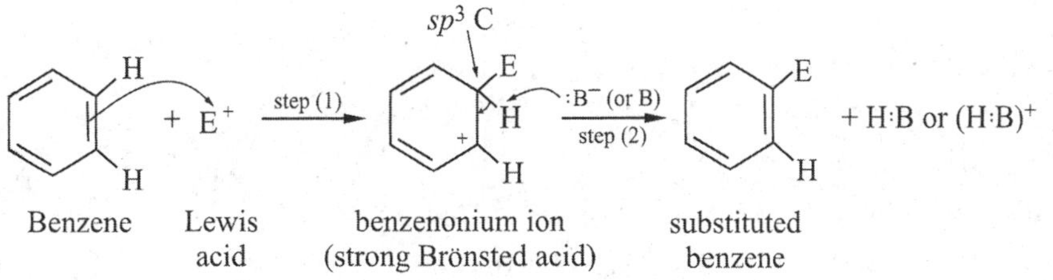

sp^3 C

Benzene + E^+ (Lewis acid) $\xrightarrow{\text{step (1)}}$ benzenonium ion (strong Brönsted acid) $\xrightarrow[\text{step (2)}]{:B^- \text{ (or B)}}$ substituted benzene + H:B or $(H{:}B)^+$

or

$$C_6H_6 + E^+ \longrightarrow \left[C_6H_5 \begin{matrix} E \\ H \end{matrix} \right]^+ \xrightarrow[\text{or B}]{:B^-} C_6H_5E + H{:}B \text{ or } (H{:}B)^+$$

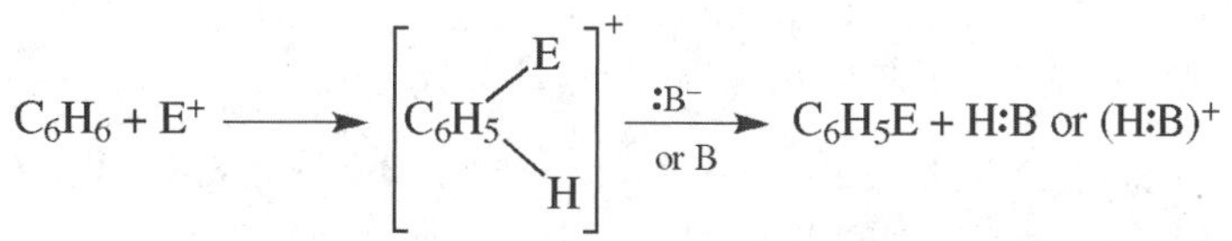

Step (1) is reminiscent of electrophilic addition to an alkene. Aromatic substitution differs in that the intermediate carbocation (a **benzenonium** ion) loses a cation (most often H^+) to give the substitution product, rather than adding a nucleophile to give the addition product. The benzenonium ion is a specific example of an **arenonium** ion, formed by an electrophilic attack on an **arene** (Section 11.4). It is also called a **sigma complex**, because it arises by formation of a σ bond between E and the ring. See Fig. 11.1 for a typical enthalpy-reaction curve for the nitration of an arene.

Problem 11.1 Account for the relative stability of the benzenonium ion by (*a*) resonance theory and (*b*) charge delocalization.

(*a*) [resonance structures of benzenonium ion with H and E on the sp^3 carbon, + at ortho, para, ortho positions]

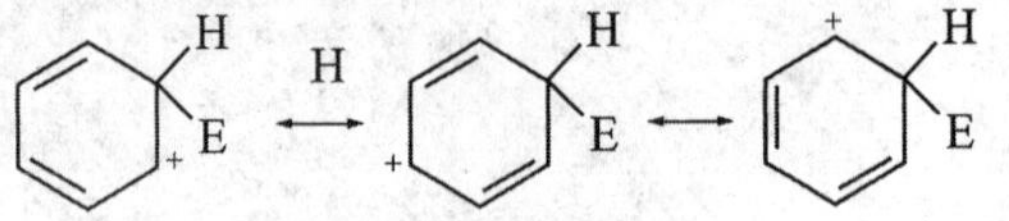

contributing structures

Note that + is at C's *ortho* and *para* to sp^3-hybridized C, which is the one bonded to E^+.

(*b*) The benzenonium ion is a type of allylic cation [see Problem 8.24(*b*)]. The five remaining C's using sp^2-hybridized orbitals each have a *p* orbital capable of overlapping laterally to give a delocalized π structure, or σ complex:

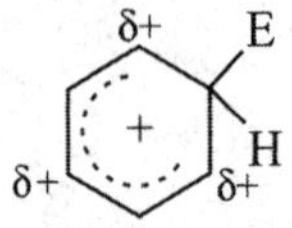

[hybrid structure with $\delta+$ at three positions, + in ring, E and H on the sp^3 carbon]

delocalized (hybrid) structure

The $\delta+$ indicates positions where + charge exists.

Problem 11.2 For each electrophilic aromatic substitution in Table 11.1, give equations for formation of E^+ and indicate what is B^- or B (several bases may be involved). In reaction (*c*), the electrophile is a molecule, E.

(*a*) $X_2 + FeX_3 \longrightarrow X^+(E^+) + FeX_4^-\ (B^-)$ (forms $HX + FeX_3$)

(*b*) $H_2SO_4 + HONO_2 \longrightarrow HSO_4^-\ (B^-) + H_2\overset{+}{O}NO_2 \longrightarrow H_2O + NO_2^+\ (E^+)$

(*c*) $2H_2SO_4 \longrightarrow H_3O^+ + HSO_4^-\ (B^-) + SO_3\ (E)$

$C_6H_6 + SO_3 \longrightarrow \overset{+}{C_6H_5}(H)(SO_3^-)\ (B^-) \longrightarrow C_6H_5SO_3H$

(*d*) $RX + AlX_3 \longrightarrow R^+\ (E^+) + AlX_4^-\ (B^-)$ (forms $HAlX_4$)

$ROH + HF \longrightarrow R^+\ (E^+) + H_2O\ (B) + F^-\ (B^-)$

$-\overset{|}{C}=\overset{|}{C}- + H_3PO_4 \longrightarrow -\overset{|}{\underset{+}{C}}-\overset{|}{C}H\ (E^+) + H_2PO_4^-\ (B^-)$

(*e*) $RCOCl + AlCl_3 \longrightarrow \underbrace{R\overset{+}{C}{=}\ddot{O}: \longleftrightarrow RC{\equiv}\overset{+}{O}:}_{\text{Oxycarbonium ion}}\ (E^+) + AlCl_4^-\ (B^-)$

Problem 11.3 How does the absence of a primary isotope effect prove experimentally that the first step in aromatic electrophilic substitution is rate-determining?

A C—H bond is broken faster than is a C—D bond. This rate difference (isotope effect, k_H/k_D) is observed only if the C—H (or C—D) bond is broken in the rate-determining step. If no differene is observed, as is the case for most aromatic electrophilic substitutions, C—H bond-breaking must occur in a fast step (in this case, the second step). Therefore, the first step, involving no C—H bond-breaking, is rate-determining. This slow step requires the loss of aromaticity, the fast second step restores the aromaticity.

Problem 11.4 How is E^+ generated, and what is the base, in the following reactions? (*a*) Nitration of reactive aromatics with HNO_3 alone. (*b*) Chlorination with HOCl using HCl as catalyst. (*c*) Nitrosation (introduction of a NO group) of reactive aromatics with HONO in strong acid. (*d*) Deuteration with DCl.

TABLE 11.1

REACTION	REAGENT	CATALYST	PRODUCT	E^+ OR E
(*a*) Halogenation	X_2 (X = Cl, Br)	FeX_3 (from Fe + X_2)	ArCl, ArBr	X^+
(*b*) Nitration	HNO_3	H_2SO_4	$ArNO_2$	$^+NO_2$
(*c*) Sulfonation	H_2SO_4 or $H_2S_2O_7$	none	$ArSO_3H$	SO_3
(*d*) Friedel-Crafts alkylation	RX, $ArCH_2X$ ROH RCH=CH (H H)	$AlCl_3$ HF, H_2SO_4, or BF_3 H_3PO_4 or HF	Ar—R, Ar—CH_2Ar Ar—R Ar—CH(CH_3)—R	R^+
(*e*) Friedel–Crafts acylation	RCOCl	$AlCl_3$	Ar—C(=O)—R	$R\overset{+}{C}=O$

(*a*) $HNO_3 + H—O—NO_2 \longrightarrow NO_3^- + \left[H—\overset{+}{O}(H)—NO_2\right] \longrightarrow H_2O\ (Base) + \overset{+}{N}O_2\ (E^+)$

unstable   Nitronium ion

(*b*) $H^+ + H—O—Cl \longrightarrow H—\overset{+}{O}(H)—Cl \longrightarrow H_2O\ (Base) + Cl^+\ (E^+)$

(*c*) $H—O—N=O + H^+ \longrightarrow H—\overset{+}{O}(H)—N=O \longrightarrow H_2O\ (Base) + NO^+\ (E^+)$

Nitrosonium ion

(*d*) D^+, transferred by DCl to benzene. Base is Cl^-.

$$C_6H_6 + DCl \longrightarrow \left[C_6H_5\langle^{H}_{D}\right]^+ + Cl^- \longrightarrow C_6H_5D + HCl$$

base$_1$ acid$_2$ acid$_1$ base$_2$

Problem 11.5 Since the initial step of aromatic electrophilic substitution is identical with that of alkene addition, explain why (*a*) aromatic substitution is slower than alkene addition; (*b*) catalysts are needed for aromatic substitution; (*c*) the intermediate carbocation eliminates a proton instead of adding a nucleophile.

(*a*) The intermediate benzenonium ion is less stable than benzene; hence, its formation has a high $\Delta H^{\ddagger}$ and the reaction is slowed. Loss of aromaticity is energetically more unfavorable than loss of π bond.
(*b*) The catalysts are acids which polarize the reagent and make it more electrophilic.
(*c*) The addition reaction would be endothermic and would produce the less stable cyclohexadiene. Loss of a proton, on the other hand, produces a stable aromatic ring.

Problem 11.6 Sulfonation resembles nitration and halogenation in being an electrophilic substitution, but differs in being *reversible* and in having a *moderate primary kinetic* isotope effect. Illustrate with diagrams of enthalpy (H) versus reaction coordinate.

In nitration (and other irreversible electrophilic substitutions), the transition state (TS) for the reaction wherein

$$\left[Ar\langle^{H}_{NO_2} \right]^+$$

loses H^+, has a considerably smaller $\Delta H^\ddagger$ than does the TS for the reaction in which NO_2^+ is lost. In sulfonation, the $\Delta H^\ddagger$ for loss of SO_3 from

$$\left[\overset{+}{Ar}\langle^{H}_{SO_3^-} \right]$$

is only slightly more than that for loss of H^+.

In terms of the specific rate constants

$$\underset{(1)}{ArH + SO_3} \underset{k_{-1}}{\overset{k_1}{\rightleftharpoons}} \overset{+}{Ar}\langle^{H}_{SO_3^-} \xrightarrow{k_2} \underset{(2)}{ArSO_3^- + H^+}$$

k_2 is about equal to k_{-1}. (For nitration, $k_2 \gg k_{-1}$.) Therefore, in sulfonation, the intermediate can go almost equally well in either direction, and sulfonation is reversible. Furthermore, since the rate of Step 2 affects the overall rate, the substitution of D for H decreases the rate because $\Delta H^\ddagger$ for loss of D^+ from

$$\overset{+}{Ar}\langle^{D}_{SO_3^-}$$

is greater than $\Delta H^\ddagger$ for loss of H^+ from the protonated intermediate. Hence, there is a modest primary isotope effect.

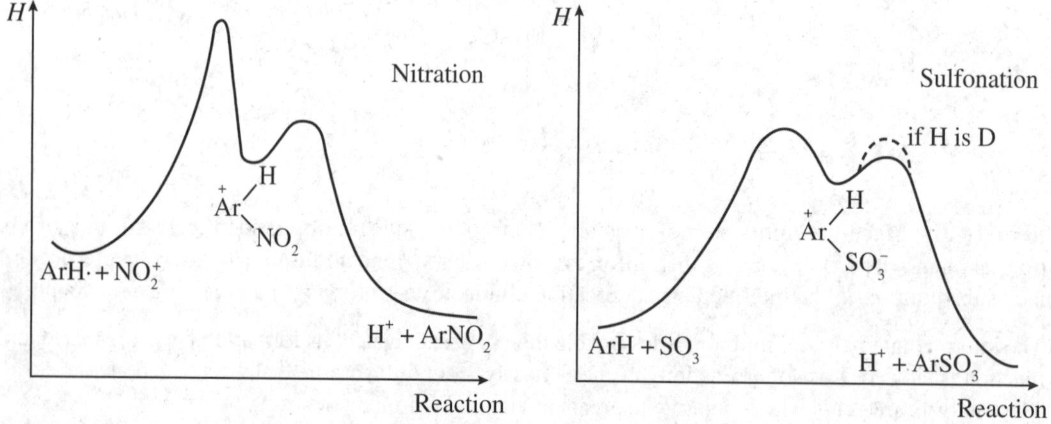

Figure 11.1

Groups other than H^+ can be displaced during electrophilic aromatic attack. The acid-catalyzed reversal of sulfonation (**desulfonation**) exemplifies such a reaction; here, H^+ displaces SO_3H as SO_3 and H^+:

$$ArSO_3H + H_2O \underset{\text{fum. } H_2SO_4}{\overset{50\%\text{ aq. } H_2SO_4,\ 150°C}{\rightleftharpoons}} ArH + H_2SO_4$$

Problem 11.7 Use the principle of microscopic reversibility (Problem 6.21) to write a mechanism for desulfonation.

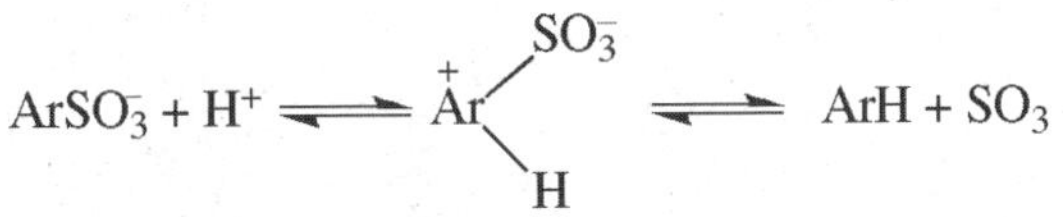

$$ArSO_3^- + H^+ \rightleftharpoons \overset{+}{Ar}(SO_3^-)(H) \rightleftharpoons ArH + SO_3$$

Orientation and Activation of Substituents

The 5 ring H's of monosubstituted benzenes, C_6H_5G, are not equally reactive. Introduction of E into C_6H_5G rarely gives the statistical distribution of 40% *ortho*, 40% *meta*, and 20% *para* disubstituted benzenes. The ring substituent(s) determine(s) (*a*) the orientation of E (*meta* or a mixture of *ortho* and *para*) and (*b*) the reactivity of the ring toward substitution.

Problem 11.8 (*a*) Give the delocalized structure (Problem 11.1) for the 3 benzenonium ions resulting from the common ground state for electrophilic substitution, $C_6H_5G + E^+$. (*b*) Give resonance structures for the *para*-benzenonium ion when G is OH. (*c*) Which ions have G attached to a positively charged C? (*d*) If the products from this reaction are usually determined by rate control (Section 8.5), how can the **Hammond principle** be used to predict the relative yields of *op* (i.e., the mixture of *ortho* and *para*) as against *m* (*meta*) products? (*e*) In terms of electronic effects, what kind of G is a (i) *op*-director, (ii) *m*-director? (*f*) Classify G in terms of its structure and its electronic effect.

(*a*) G, δ+, H, E, δ+, δ+ — *ortho*; G, δ+, δ+, δ+, H, E — *para*; G, δ+, δ+, E, H, δ+ — *meta*

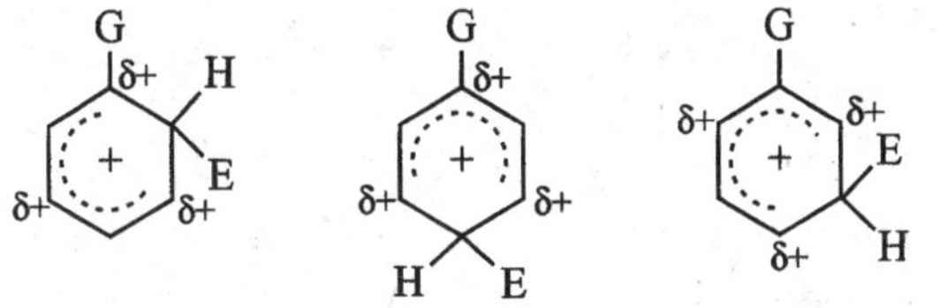

(*b*) (i) OH, H, E ⟷ OH, H, E ⟷ OH, H, E ⟷ $^+$OH, H, E (major contributor because all atoms obey octet rule)

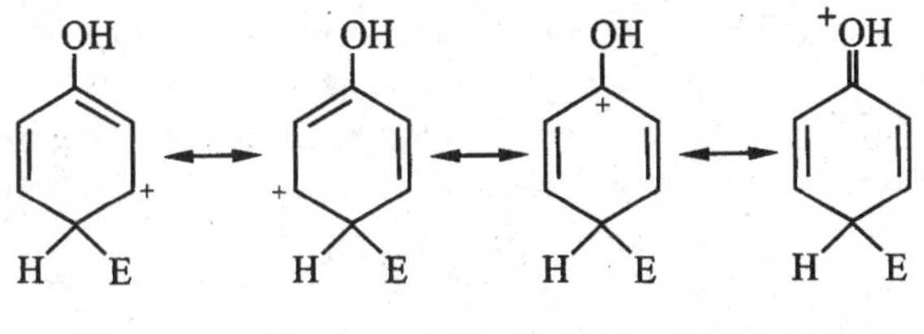

(*c*) The *ortho* and *para*. This is why G is either an *op*- or an *m*-director.

(*d*) Because of kinetic control, the intermediate with the lowest-enthalpy transition state (TS) is formed in the greatest amount. Since this step is endothermic, the Hammond principle says that the intermediate resembles the TS. We then evaluate the relative energies of the intermediates (*op* vs. *m*) and predict that the one with the lowest enthalpy has the lowest $\Delta H^{\ddagger}$ and is formed in the greatest yield.

(*e*) (i) An electron-donating G can better stabilize the intermediate when it is attached directly to positively charged (*op*) C's. Such G's are *op*-directing. (ii) An electron-withdrawing G destabilizes the ion to a greater extent when attached directly to positively charged (*op*) C's. They destabilize less when attached to *meta* and are thus *m*-directors.

(*f*) **Electron-donating** (*op*-directors): (i) Those that have an unshared pair of electrons on the atom bonded to the ring, which can be delocalized to the ring by extended π bonding.

Other examples are —Ö—, —Ẍ: (halogen), and —S̈—.
(ii) Those with an attached atom participating in an electron-rich π bond, for example:

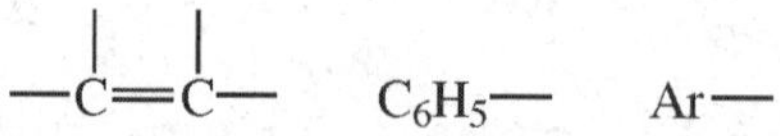

$$-\overset{|}{C}=\overset{|}{C}- \quad C_6H_5- \quad Ar-$$

(iii) Those without an unshared pair, which are electron-donating by induction or by **hyperconjugation** (absence of bond resonance), for example, alkyl groups.

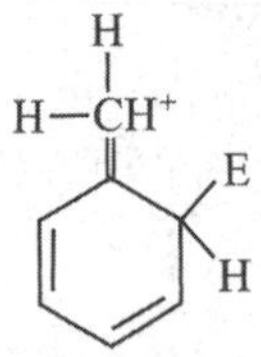

hyperconjugated structure

Electron-withdrawing (*m*-**directors**): The attached atom has no unshared pair of electrons and has some positive charge, for example:

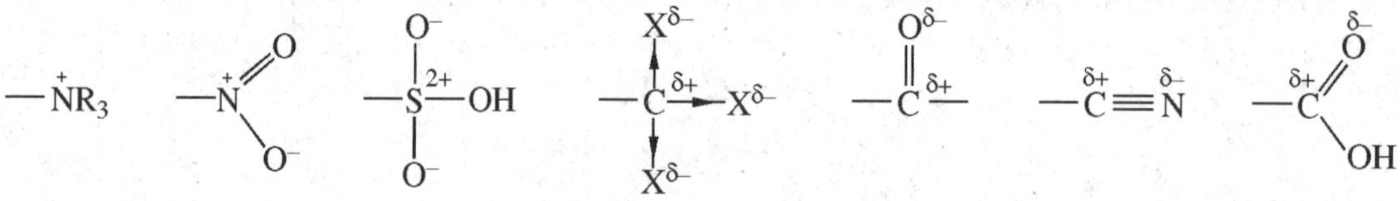

X = F, Cl

Problem 11.9 Explain: (*a*) All *m*-directors are deactivating. (*b*) Most *op*-directing substituents make the ring more reactive than benzene itself—they are activating. (*c*) As exceptions, the halogens are *op*-directors but are deactivating.

(*a*) All *m*-directors are electron-attracting and destabilize the incipient benzenonium ion in the TS. They therefore diminish the rate of reaction as compared to the rate of reaction of benzene.

(*b*) Most *op*-directors are, on balance, electron-donating. They stabilize the incipient benzenonium ion in the TS, thereby increasing the rate of reaction as compared to the rate of reaction of benzene. For example, the ability of the —ÖH group to donate electrons by extended *p* orbital overlap (resonance) far outweighs the ability of the ÖH group to withdraw electrons by its inductive effect.

(*c*) In the halogens, unlike the OH group, the electron-withdrawing inductive effect predominates and, consequently, the halogens are deactivating. The *o*-, *p*-, and *m*-benzenonium ions each have a higher $\Delta H^\ddagger$ than does the cation from benzene itself. However, on demand, the halogens contribute electron density by extended π bonding.

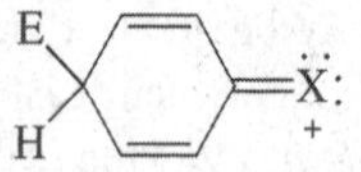

(showing delocalization of + to X)

and thereby lower the $\Delta H^\ddagger$ of the *ortho* and *para* intermediates but not the *meta* cation. Hence, the halogens are *op*-directors, but deactivating.

Problem 11.10 Compare the activating effects of the following *op*-directors:

(*a*) $—\ddot{\underset{..}{O}}H$, $—\ddot{\underset{..}{O}}:^-$ and $—\ddot{\underset{..}{O}}\underset{\underset{O}{\|}}{C}—CH_3$ (*b*) $—\ddot{N}H_2$ and $—\ddot{N}H—\underset{\underset{O}{\|}}{C}—CH_3$

Explain your order.

(*a*) The order of activation is $—O^- > —OH > —OCOCH_3$. The $—O^-$, with a full negative charge, is best able to donate electrons, thereby giving the very stable uncharged intermediate

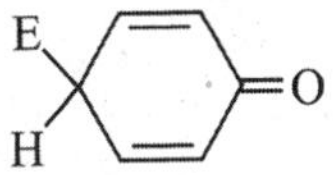

E, H, =O

In $—OCOCH_3$, the C of the $\overset{\delta+}{C}=\overset{\delta-}{O}$ group has + charge and makes demands on the $—\ddot{\underset{..}{O}}—$ for electron density, thereby diminishing the ability of this $—\ddot{\underset{..}{O}}—$ to donate electrons to the benzenonium ion.

(*b*) The order is $—NH_2 > —NHCOCH_3$ for the same reason that OH is a better activator than $—OCOCH_3$.

Table 11.2 extends the results of Problem 11.10.

Problem 11.11 (*a*) Draw enthalpy-reaction diagrams for the first step of electrophilic attack on benzene, toluene (*meta* and *para*), and nitrobenzene (*meta* and *para*). Assume all ground states have the same energy. (*b*) Where would the *para* and *meta* substitution curves for C_6H_5Cl lie on this diagram?

(*a*) Since CH_3 is an activating group, the intermediates and TS's from $PhCH_3$ have less enthalpy than those from benzene. The *para* intermediate has less enthalpy than the *meta* intermediate. The TS and intermediates for $PhNO_2$ have higher enthalpies than those for C_6H_6, with the *meta* at a lower enthalpy than the *para*. See Fig. 11.2.

(*b*) They would both lie between those for benzene and *p*-nitrobenzene, with the *para* lower than the *meta*.

Problem 11.12 (*a*) Explain in terms of the reactivity-selectivity principle (Section 4.4) the following yields of *meta* substitution observed with toluene: Br_2 in CH_3COOH, 0.5%; HNO_3 in CH_3COOH, 3.5%; CH_3CH_2Br in $GaBr_3$,

TABLE 11.2

op-DIRECTORS		*m*-DIRECTORS
ACTIVATING	WEAKLY DEACTIVATING	DEACTIVATING
$—\ddot{\underset{..}{O}}:^-$	$—\ddot{\underset{..}{F}}:$ $—\ddot{\underset{..}{Cl}}:$ $—\ddot{\underset{..}{Br}}:$ $—\ddot{\underset{..}{I}}:$ $—\ddot{N}=\ddot{O}:$	$—NR_3^+$
$—\ddot{\underset{..}{O}}H$ $—\ddot{N}H_2$ $—\ddot{N}R_2$		$—NO_2$ $—CF_3$ $—CCl_3$
$—\ddot{\underset{..}{O}}R$ $—\ddot{N}H\underset{\underset{O}{\|}}{C}R$		$—CN$ $—SO_3H$
$—R$ $C_6H_5—$ $—\overset{\vert}{C}=\overset{\vert}{C}—$		$—\underset{\underset{O}{\|}}{C}OH$ $—\underset{\underset{O}{\|}}{C}OR$ $—\underset{\underset{O}{\|}}{C}H$ $—\underset{\underset{O}{\|}}{C}R$

increase ↑ (activating, left) increase ↑ (deactivating, right)

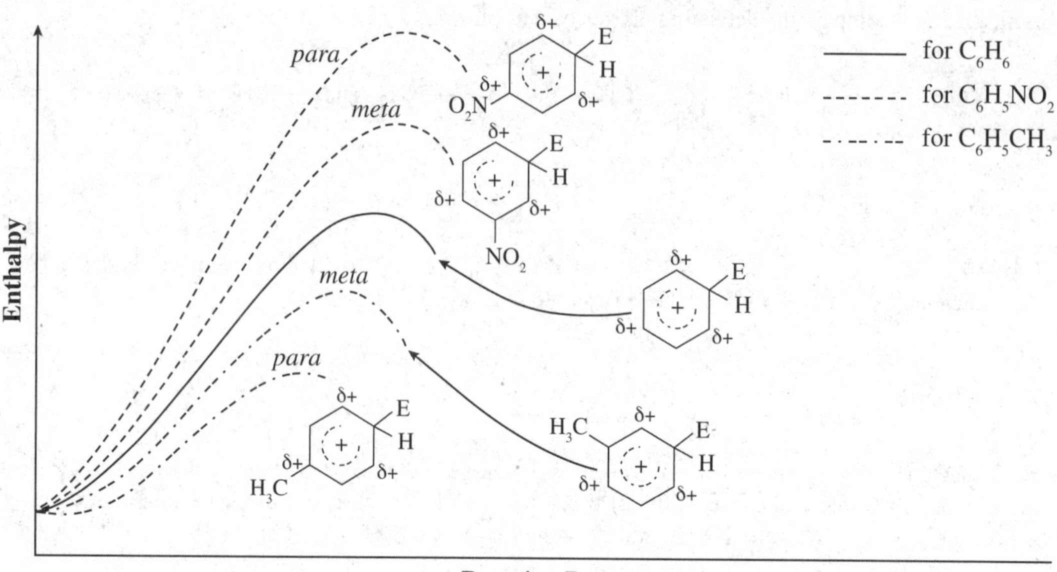

Figure 11.2

21%. (*b*) In terms of kinetic versus thermodynamic control, explain the following effect of temperature on isomer distribution in sulfonation of toluene: at 0°C, 43% *o*- and 53% *p*-; at 100°C, 13% *o*- and 79% *p*-.

(*a*) The most reactive electrophile is least selective and gives the most *meta* isomer. The order of reactivity is

$$CH_3CH_2^+ > NO_2^+ > Br_2(Br^+)$$

(*b*) Sulfonation is one of the few reversible electrophilic substitutions, and, therefore, kinetic and thermodynamic products can result. At 100°C, the thermodynamic product predominates; this is the *para* isomer. The *ortho* isomer is somewhat more favored by kinetic control at 0°C.

Problem 11.13 $PhNO_2$, but not C_6H_6, is used as a solvent for the Friedel-Crafts alkylation of PhBr. Explain.

C_6H_6 is more reactive than PhBr and would preferentially undergo alkylation. —NO_2 is so strongly deactivating that $PhNO_2$ does not undergo Friedel-Crafts alkylation or acylations.

Problem 11.14 Account for the percentages of *m*-orientation in the following compounds: (*a*) $C_6H_5CH_3$ (4.4%), $C_6H_5CH_2Cl$ (15.5%), $C_6H_5CHCl_2$ (33.8%), $C_6H_5CCl_3$ (64.6%); (*b*) $C_6H_5N^+(CH_3)_3$ (100%), $C_6C_5CH_2N^+(CH_3)_3$ (88%), $C_6H_5(CH_2)_2N^+(CH_3)_3$ (19%).

(*a*) Substitution of the CH_3 H's by Cl's causes a change from electron-release ($\leftarrow CH_3$) to electron-attraction ($\rightarrow CCl_3$) and *m*-orientation increases.

(*b*) $^+NMe_3$ has a strong electron-attracting inductive effect and is *m*-orienting. When CH_2 groups are placed between this N^+ and the ring, this inductive effect falls off rapidly, as does the *m*-orientation. When two CH_2's intercede, the electron-releasing effect of the CH_2 bonded directly to the ring prevails, and chiefly *op*-orientation is observed.

Problem 11.15 Predict and explain the reaction, if any, of (*a*) phenol (PhOH), (*b*) PhH, and (*c*) benzenesulfonic acid with D_2SO_4 in D_2O.

(*a*) D_2SO_4 transfers D^+, an electrophile, to form 2,4,6-trideuterophenol. Reaction is rapid because of the activating —OH group. The *meta* positions are deactivated. (*b*) PhH reacts slowly to give hexadeuterobenzene. (*c*) The sulfonic acid does not react, because —SO_3H is too deactivating.

Rules for Predicting Orientation in Disubstituted Benzenes

1. If the groups *reinforce* each other, the orientation can be inferred from either group.
2. If an *op*-director and an *m*-director are *not reinforcing*, the *op*-director controls the orientation. (The incoming group goes mainly *ortho* to the *m*-director.)
3. A *strongly activating* group, competing with a *weakly activating* group, controls the orientation.
4. When *two weakly activating or deactivating* groups or *two strongly activating or deactivating* groups compete, substantial amounts of both isomers are obtained; there is little preference.
5. Very little substitution occurs in the *sterically hindered* position between *meta* substituents.
6. Very little substitution occurs *ortho* to a bulky *op*-directing group such as *t*-butyl.

Problem 11.16 Indicate by an arrow the position(s) most likely to undergo electrophilic substitution in each of the following compounds. List the number of the above rule(s) used in making your prediction. (*a*) *m*-xylene, (*b*) *p*-nitrotoluene, (*c*) *m*-chloronitrobenzene, (*d*) *p*-methoxytoluene, (*e*) *p*-chlorotoluene, (*f*) *m*-nitrotoluene, (*g*) *o*-methylphenol (*o*-cresol).

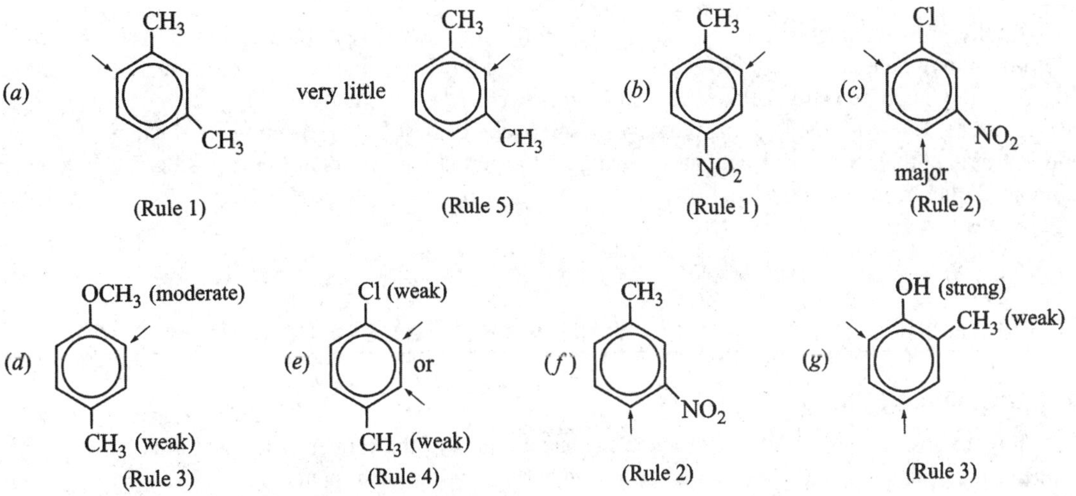

(*a*) CH_3, CH_3 (Rule 1); very little CH_3, CH_3 (Rule 5)
(*b*) CH_3, NO_2 (Rule 1)
(*c*) Cl, NO_2, major (Rule 2)

(*d*) OCH_3 (moderate), CH_3 (weak) (Rule 3)
(*e*) Cl (weak), or, CH_3 (weak) (Rule 4)
(*f*) CH_3, NO_2 (Rule 2)
(*g*) OH (strong), CH_3 (weak) (Rule 3)

Electrophilic Substitution of Naphthalene

1. Electrophilic substitution of naphthalene occurs preferentially at α position.
2. Examples of β-substitution are (*a*) sulfonation at high temperatures (at low temperatures, α- substitution occurs); (*b*) acylation with RCOCl and $AlCl_3$, in $C_6H_5NO_2$ as solvent (in CS_2 or CH_2ClCH_2Cl, α-substitution occurs).
3. Substitution occurs in a ring holding an activating (electron-releasing) group: (*a*) *para* to an α- substituent; (*b*) *ortho* to an α-substituent if the *para* position is blocked; (*c*) to α *ortho* position if the activating group is a β-substituent.
4. A deactivating group (electron-withdrawing) directs electrophiles into the other ring, usually at α positions.

Problem 11.17 Account for (*a*) formation of the α-isomer in nitration and halogenation of naphthalene, (*b*) formation of α-naphthalenesulfonic acid at 80°C and β-naphthalenesulfonic acid at 160°C.

(*a*) The mechanism of electrophilic substitution is the same as that for benzene. Attack at the α position has a lower $\Delta H^{\ddagger}$ because intermediate I, an allylic R^+ with an intact benzene ring, is more stable than intermediate II from β-attack.

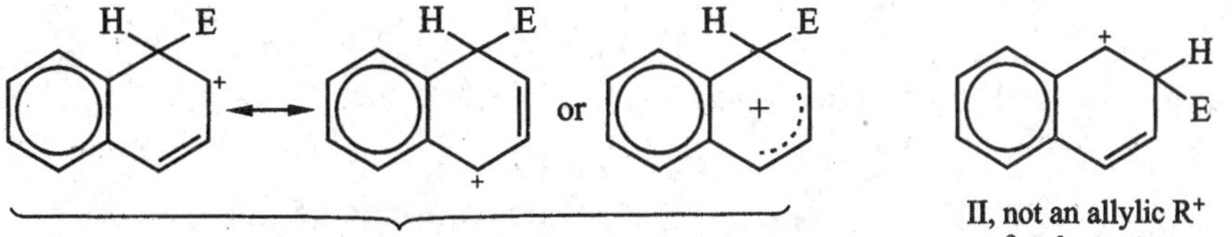

H E, H E, or, H E — I, an allylic R^+ *α-substitution*

H, E — II, not an allylic R^+ *β-substitution*

In II, the + charge is isolated from the remaining double bond, and hence, there is no direct delocalization of charge to the double bond without involvement of the stable benzene ring. In both I and II, the remaining aromatic ring has the same effect on stabilizing the + charge. Since I is more stable than II, α-substitution predominates.

(*b*) α-Naphthalenesulfonic acid is the kinetic-controlled product [see part (*a*)]. However, sulfonation is a reversible reaction, and at 160°C, the *thermodynamic*-controlled product, β-naphthalenesulfonic acid, is formed.

Problem 11.18 Name the product and account for the orientation in the following electrophilic substitution reactions:

(*a*) 1-Methylnaphthalene + Br_2, Fe
(*b*) 2-Ethylnaphthalene + Cl_2, Fe
(*c*) 2-Ethylnaphthalene + C_2H_5COCl + $AlCl_3$
(*d*) 1-Methylnaphthalene + CH_3COCl, $AlCl_3(CS_2)$
(*e*) 2-Methoxynaphthalene + HNO_3 + H_2SO_4
(*f*) 2-Nitronapththalene + Br_2, Fe

(*a*) 1-Methyl-4-bromonaphthalene. Br substitutes in the more reactive α position of the activated ring. (*b*) 1-Chloro-2-ethylnaphthalene. C^1 (*ortho* and α) is activated by C_2H_5, since C^4, which is also α, is *meta* to C_2H_5. (*c*) 1-(2-Ethylnaphthyl) ethyl ketone. Same reason as in (*b*). (*d*) 4-(1-Methylnaphtyl) methyl ketone. Same reason as in (*a*). (*e*) 1-Nitro-2-methoxynaphthalene. Same reason as in (*b*). (*f*) 1-Bromo-6-nitronaphthalene and 1-bromo-7-nitronaphthalene. NO_2 deactivates its ring and bromination occurs at the α positions of the other ring, more at C^5 that bears no delocalized δ^+.

11.2 Electrophilic Substitutions in Syntheses of Benzene Derivatives

Order of Introducing Groups

To do such syntheses, it is essential to introduce the substituents in the proper order, which is based on the knowledge of the orientation and activation of both ring and incoming substituents.

Problem 11.19 From C_6H_6 (PhH) or $PhCH_3$, synthesize: (*a*) *p*-$ClC_6H_4NO_2$, (*b*) *m*-$ClC_6H_4NO_2$, (*c*) *p*-$O_2NC_6H_4COOH$, (*d*) *m*-$O_2NC_6H_4COOH$.

In the synthesis of disubstituted benzenes, the first substituent present determines the position of the incoming second. Therefore, the order of introducing substituents must be carefully planned to yield the desired isomer.

(*a*) Since the two substituents are *para*, it is necessary to introduce the *op*-directing Cl first:

$$\text{PhH} \xrightarrow[\text{Cl}_2]{\text{Fe}} \text{PhCl} \xrightarrow[\text{H}_2\text{SO}_4]{\text{HNO}_3} p\text{-ClC}_6\text{H}_4\text{NO}_2$$

(*b*) Since the substituents are *meta*, the *m*-directing NO_2 is introduced first:

$$\text{PhH} \xrightarrow[\text{H}_2\text{SO}_4]{\text{HNO}_3} \text{PhNO}_2 \xrightarrow[\text{Fe}]{\text{Cl}_2} m\text{-ClC}_6\text{H}_4\text{NO}_2$$

(*c*) The COOH group is formed by oxidation of CH_3. Since *p*-$O_2NC_6H_4COOH$ has two *m*-directing groups, the NO_2 must be added while the *op*-directing CH_3 is still present.

$$\text{PhCH}_3 \xrightarrow[\text{H}_2\text{SO}_4]{\text{HNO}_3} p\text{-CH}_3\text{C}_6\text{H}_4\text{NO}_2 + o\text{-CH}_3\text{C}_6\text{H}_4\text{NO}_2$$

The *para* isomer is usually easily separated from the *op* mixture:

$$p\text{-}O_2NC_6H_4CH_3 \xrightarrow[H^+]{KMnO_4} p\text{-}O_2NC_6H_4COOH \quad \text{(use phase transfer catalysts, Prob. 7.26.)}$$

(*d*) Now the substituents are *meta*, and NO_2 is introduced when the *m*-directing COOH is present:

$$PhCH_3 \xrightarrow[H^+]{KMnO_4} PhCOOH \xrightarrow[H_2SO_4]{HNO_3} m\text{-}O_2NC_6H_4COOH$$

Use of Blocking Groups

If, in most cases, electrophilic substitution of a phenyl derivative with an *op*-directing group yields mainly the *para* isomer, how are good yields of the *ortho* isomer obtained? Answer: First introduce an easily removed **blocking group** into the *para* position; then introduce the *ortho* substituent; and, finally, remove the blocking group. Two good blocking groups are —SO_3H and —$C(CH_3)_3$.

Problem 11.20 Show steps in the synthesis of (*a*) *o*-chlorotoluene and (*b*) 1,3-dimethyl-2-ethylbenzene.

(*a*) In this synthesis, —SO_3H is the blocking group. In the second step, the *op*-directing CH_3 and the *m*-directing SO_3H reinforce each other.

$$C_6H_5CH_3 \xrightarrow{H_2SO_4} p\text{-}HO_3SC_6H_4CH_3 \xrightarrow[Cl_2]{Fe} \text{4-}HO_3S\text{-2-Cl-}C_6H_3CH_3 \xrightarrow[heat]{H_2SO_4} o\text{-}ClC_6H_4CH_3$$

(*b*) In this synthesis, —$C(CH_3)_3$ is the blocking group. In the second step, although C_2H_5 and $C(CH_3)_3$ are competing *op*-directors, the bulkiness of the latter group inhibits attack *ortho* to itself.

$$C_6H_5CH_3 \xrightarrow[HF]{(CH_3)_2C{=}CH_2} p\text{-}(CH_3)_3CC_6H_4CH_3 \xrightarrow[AlCl_3]{2CH_3Cl} 1,2,3\text{-}(CH_3)_3\text{-5-}C(CH_3)_3C_6H_2 \xrightarrow{HF} 1,2,3\text{-}(CH_3)_3C_6H_3 + (CH_3)_2C{=}CH_2$$

In the reaction with HF, the electrophile H^+ replaces $C(CH_3)_3^+$, which forms $(CH_3)_2C{=}CH_2$.

11.3 Nucleophilic Aromatic Substitutions

Addition–Elimination Reactions

Nucleophilic aromatic substitutions of H are rare. The intermediate benzenanion in aromatic nucleophilic substitution is analogous to the intermediate benzenonium ion in aromatic electrophilic substitution, except that negative charge is dispersed to the *op*-positions.

Benzenanion (ring carbon bearing H and Nu; $\delta-$ at the two *ortho* positions and the *para* position)

$$:Nu^- + ArH \longrightarrow \left[Ar\begin{matrix}H\\Nu\end{matrix}\right]^- \xrightarrow[K_3Fe(CN)_6]{O_2 \text{ or}} ArNu + H_2O$$

$$\left[Ar\begin{matrix}H\\Nu\end{matrix}\right]^- \not\longrightarrow ArNu + H{:}^-$$

Oxidants such as O_2 and $K_3Fe(CN)_6$ facilitate the second step, which may be rate-controlling, by oxidizing the ejected $:H^-$, a powerful base and a very poor leaving group, to H_2O.

Electron-withdrawing groups positioned at the *op*-positions bearing the negative charges greatly stabilize the formation of the intermediate benzenanion. Thus, groups (such as NO_2, CN, and halogen) which deactivate the ring toward electrophilic attack, encourage nucleophilic attack. These groups are *op*-directors toward nucleophilic aromatic substitution.

Problem 11.21 Account for the product in the reaction.

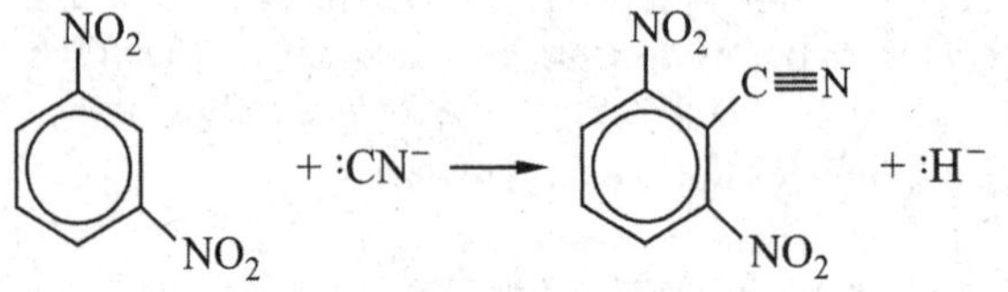

$$m\text{-}C_6H_4(NO_2)_2 + :CN^- \longrightarrow 2,6\text{-}(NO_2)_2C_6H_3C{\equiv}N + :H^-$$

CN^- is a nucleophile. NO_2's activate the ring toward nucleophilic substitution at *op*-positions by withdrawing the electron density and placing charge on the O's of NO_2:

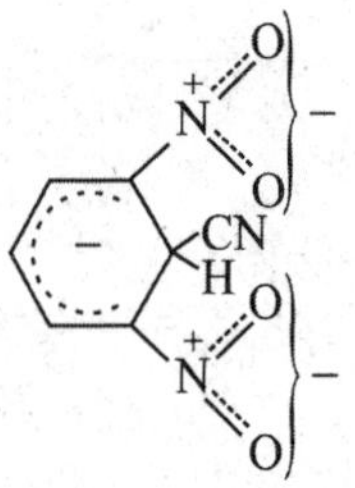

When not sterically hindered, the *ortho* position may be more reactive. CN^- is a "thin" nucleophile, and its insertion *ortho* to each NO_2 is not hindered.

A good leaving group, such as halide ion (X^-), is more easily displaced than H^- from a benzene ring by nucleophiles. Electron-attracting substituents, such as NO_2 and CN, in *ortho* and *para* positions facilitate the nucleophilic displacement of X of aryl halides. The greater the number of such *ortho* and *para* substituents, the more rapid the reaction and the less vigorous the conditions needed.

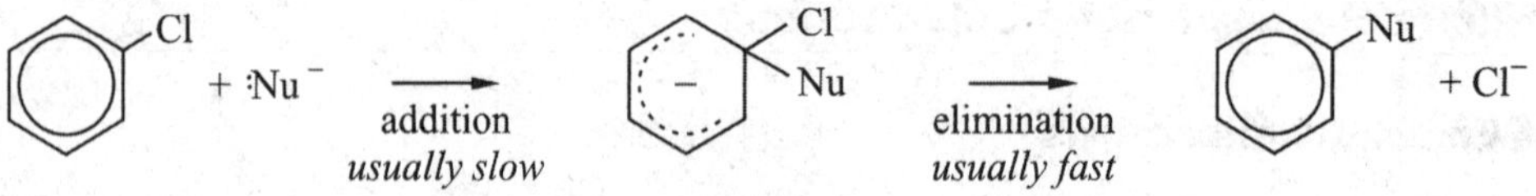

$$C_6H_5Cl + :Nu^- \xrightarrow[\textit{usually slow}]{\text{addition}} [C_6H_5(Cl)(Nu)]^- \xrightarrow[\textit{usually fast}]{\text{elimination}} C_6H_5Nu + Cl^-$$

$$C_6H_5Cl \xrightarrow{\text{NaOH, 300°C}} C_6H_5OH$$

$$p\text{-}O_2NC_6H_4Cl \xrightarrow[\text{160°C}]{\text{15\% NaOH}} p\text{-}O_2NC_6H_4OH$$

$$2,4\text{-}(O_2N)_2C_6H_3Cl \xrightarrow[\text{130°C}]{Na_2CO_3} 2,4\text{-}(O_2N)_2C_6H_3OH$$

$$2,4,6\text{-}(O_2N)_3C_6H_2Cl \xrightarrow[\text{warm}]{H_2O} 2,4,6\text{-}(O_2N)_3C_6H_2OH$$

Problem 11.22 Write resonance structures to account for activation in addition-elimination aromatic nucleophilic substitution from delocalization of the charge of the intermediate carbanion by the following *para* substituent groups: (*a*) —NO_2, (*b*) —CN, (*c*) —N=O, (*d*) CH=O.

Only the resonance structures with the negative charge on the *para* C are written to show delocalization of charge from ring C to the *para* substituent.

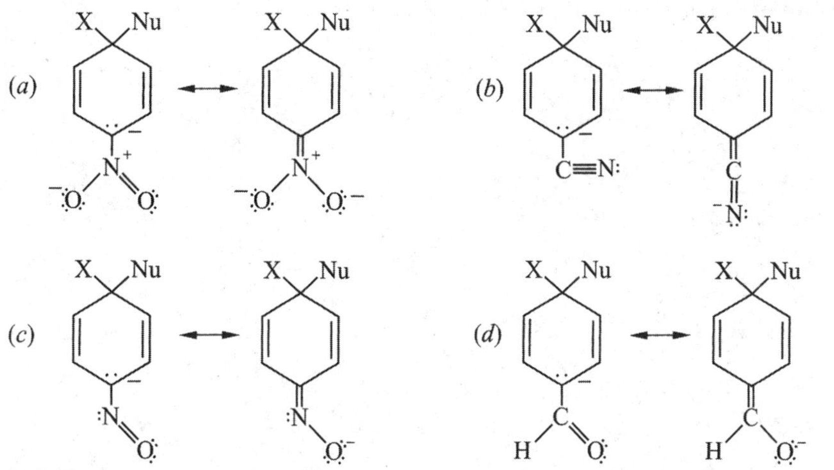

(*a*) (*b*) (*c*) (*d*)

Problem 11.23 Compare addition-elimination aromatic nucleophilic and electrophilic substitution reactions with aliphatic S_N2 reactions in terms of (*a*) number of steps and transition states and (*b*) character of intermediates.

(*a*) Nucleophilic and electrophilic aromatic substitutions are *two-step* reactions, having a first slow and rate-determining step followed by a rapid second step. Aliphatic S_N2 reactions have only *one step*. There are two transition states for the *aromatic* and, one for the *aliphatic* substitution. (*b*) S_N2 reactions have no intermediate. In *electrophilic* aromatic substitution, the intermediate is a carbocation, while that in *nucleophilic* substitution is a *carbanion*.

Problem 11.24 Why do the typical S_N2 and S_N1 mechanisms not occur in nucleophilic aromatic substitution?

The S_N2 back-side attack cannot occur, because of the high electron density of the delocalized π cloud of the benzene ring. Furthermore, inversion at the attacked C is sterically impossible. The S_N1 mechanism does not occur, because the intermediate $C_6H_5^+$, with a + charge on an *sp*-hybridized C, would have a very high energy. The ring would also have a large ring strain.

Elimination–Addition Reactions

With very strong bases, such as amide ion, NH_2^-, unactivated aryl halides undergo substitution by an elimination-addition (**benzyne**) mechanism.

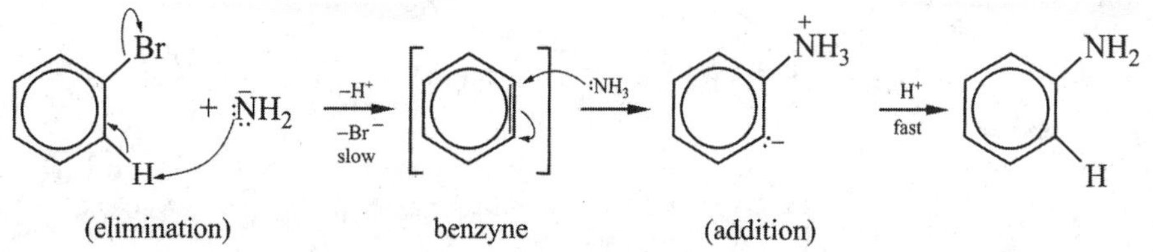

(elimination) benzyne (addition)

Problem 11.25 How do the following observations support the benzyne mechanism? (*a*) Compounds lacking *ortho* H's, such as 2,6-dimethylchlorobenzene, do not react. (*b*) 2,6-Dideuterobromobenzene reacts more slowly than bromobenzene. (*c*) *o*-Bromoanisole, *o*-$CH_3OC_6H_4Br$, reacts with $NaNH_2/NH_3$ to form *m*-$CH_3OC_6H_4NH_2$. (*d*) Chlorobenzene with Cl bonded to ^{14}C gives almost 50% aniline having NH_2 bonded to ^{14}C and 50% aniline with NH_2 bonded to an *ortho* C.

(*a*) With no H *ortho* to Cl, vicinal elimination cannot occur.
(*b*) This primary isotope effect (Problem 7.28) indicates that a bond to H is broken in the rate-determining step, which is consistent with the first step in the benzyne mechanism being rate-determining.
(*c*) NH_2^- need not attack the C^2 from which the Br^- left; it can add at C^3.

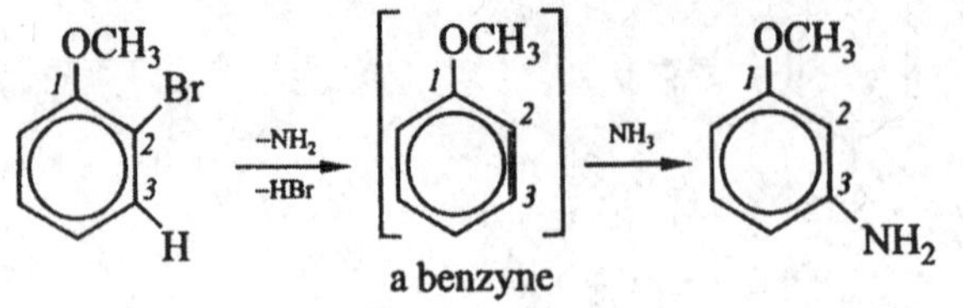

a benzyne

(*d*)

(50%)

or

(50%)

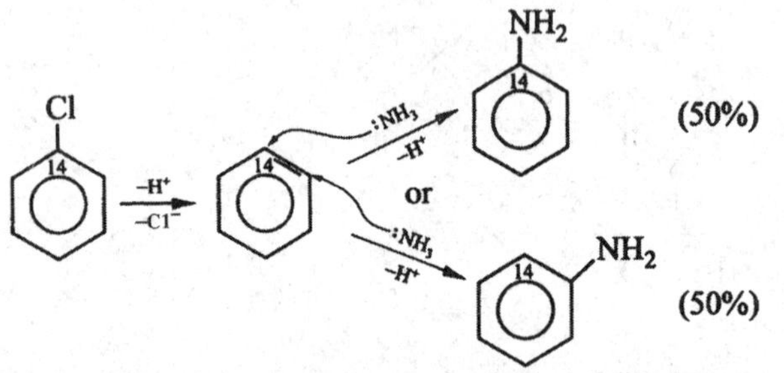

Problem 11.26 Account for the observation that NaOH reacts at 300°C with *p*-bromotoluene to give *m*- and *p*-cresols, while *m*-bromotoluene yields the three isomeric cresols.

The benzyne intermediate from *p*-bromotoluene has a triple bond between C^3 and C^4; both C's are independently attacked by OH^-, giving a mixture of *m*- and *p*-cresols ($HOC_6H_4CH_3$). Two isomeric benzynes are formed from *m*-bromotoluene, one with a C^2-to-C^3 triple bond and the other with a C^3-to-C^4 triple bond. Hence, this mixture of benzynes reacts with OH^- at all three C's, giving the mixture of three isomeric cresols.

11.4 Arenes

Nomenclature and Properties

Benzene derivatives with saturated or unsaturated C-containing side chains are **arenes**. Examples are cumene or isopropylbenzene, $C_6H_5CH(CH_3)_2$, and styrene or phenylethene, $C_6H_5CH{=}CH_2$.

Problem 11.27 Supply systematic and, where possible, common names for:

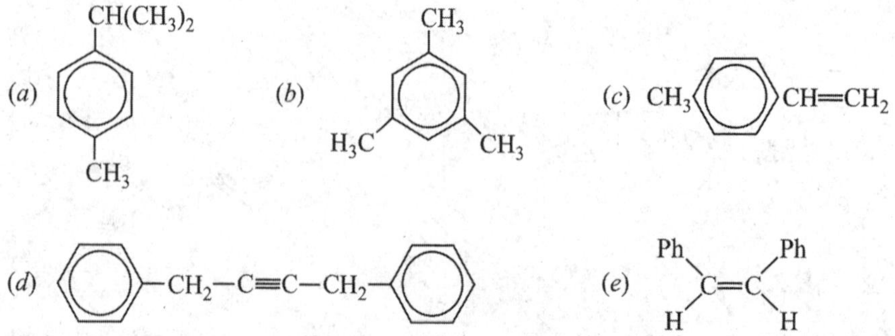

(*a*) (*b*) (*c*) $CH_3C_6H_4CH{=}CH_2$

(*d*) $C_6H_5-CH_2-C{\equiv}C-CH_2-C_6H_5$ (*e*)

(*a*) *p*-Isopropyltoluene (*p*-cymene). (*b*) 1,3,5-Trimethylbenzene (mesitylene). (*c*) *p*-Methylstyrene. (*d*) 1,4-Diphenyl-2-butyne (dibenzylacetylene). (*e*) (*Z*)-1,2-Diphenylethene (*cis*-stilbene).

Problem 11.28 Arrange the isomeric tetramethylbenzenes, prehnitene (1,2,3,4-) and durene (1,2,4,5-) in order of decreasing melting point and verify this order from tables of melting points.

The more symmetrical the isomer, the closer the molecules are packed in the crystal and the higher is the melting point. The order of decreasing symmetry, durene > prehnitene, corresponds to that of their respective melting points, +80°C > −6.5°C.

Syntheses

Friedel–Crafts alkylations and acylations, followed by reduction of the C═O group, are most frequently used to synthesize arenes. Coupling reactions can also be employed.

$$Ar_2CuLi + \underset{1^\circ}{RX} \longrightarrow ArR \quad \text{or} \quad R_2CuLi + ArX \longrightarrow ArR$$

Problem 11.29 Explain the following observations about the Friedel-Crafts alkylation reaction. (*a*) In monoalkylating C_6H_6 with RX in AlX_3, an excess of C_6H_6 is used. (*b*) The alkylation of PhOH and $PhNH_2$ gives poor yields. (*c*) Ph—Ph *cannot* be prepared by the reaction

$$PhH + PhCl \xrightarrow[\times]{AlCl_3} Ph{-}Ph + HCl$$

(*d*) At 0°C

$$PhH + 3CH_3Cl \xrightarrow{AlCl_3} \text{1,2,4-trimethylbenzene}$$

but at 100°C one gets 1,3,5-trimethylbenzene (mesitylene). (*e*) The reaction

$$PhH + CH_3CH_2CH_2Cl \xrightarrow{AlCl_3} PhCH_2CH_2CH_3 + HCl$$

gives poor yield, whereas

$$PhH + CH_3CHClCH_3 \xrightarrow{AlCl_3} PhCH(CH_3)_2 + HCl$$

gives very good yield.

(*a*) The monoalkylated product, C_6H_5R, which is more reactive than C_6H_6 itself since R is an activating group, will react to give $C_6H_4R_2$ and some $C_6H_3R_3$. To prevent polyalkylation, an excess of C_6H_6 is used to increase the chance for collision between R^+ and C_6H_6 and to minimize collision between R^+ and C_6H_5R.
(*b*) OH and NH_2 groups react with and inactivate the catalyst.
(*c*) $PhCl + AlCl_3 \not\longrightarrow Ph^+ + AlCl_4^-$, Ph^+ has a very high enthalpy and doesn't form.
(*d*) The alkylation reaction is *reversible* and therefore gives the kinetic-controlled product at 0°C and the thermodynamic-controlled product at 100°C.
(*e*) The R^+ intermediates, especially the 1° RCH_2^+ can undergo rearrangements. With $CH_3CH_2CH_2Cl$, we get

$$CH_3CH_2CH_2^+ \xrightarrow{\sim H:} CH_3\overset{+}{C}HCH_3$$

and the major product is $PhCH(CH_3)_2$.

Problem 11.30 Prepare $PhCH_2CH_2CH_3$ from PhH and any open-chain compound.

$$PhH + ClCH_2CH{=}CH_2 \xrightarrow{AlCl_3} PhCH_2CH{=}CH_2 \xrightarrow{H_2/Pt} PhCH_2CH_2CH_3$$

or

$$PhH + ClCOCH_2CH_3 \xrightarrow{AlCl_3} PhCOCH_2CH_3 \xrightarrow[\text{(Clemmensen reduction)}]{Zn/Hg,\ HCl} PhCH_2CH_2CH_3$$

$CH_3CH_2CH_2Cl$ cannot be used because the intermediate 1° $CH_3CH_2CH_2^+$ rearranges to the more stable 2° $(CH_3)_2CH^+$, to give $C_6H_5CH(CH_3)_2$ as the major product. The latter method is for synthesizing $PhCH_2R$.

Problem 11.31 Give the structural formula and the name for the major alkylation product:

(*a*) $C_6H_6 + (CH_3)_2CHCH_2Cl \xrightarrow{AlCl_3}$ (*b*) $C_6H_5CH_3 + (CH_3)_3CCH_2OH \xrightarrow{BF_3}$

(*c*) $C_6H_6 + CH_3CH_2CH_2CH_2Cl \xrightarrow[100\ °C]{AlCl_3}$ (*d*) *m*-xylene + $(CH_3)_3CCl \xrightarrow[100\ °C]{AlCl_3}$

(*a*) $(CH_3)_2CHCH_2Cl$ (Isobutyl chloride) $\xrightarrow{AlCl_3}$ $(CH_3)_2CHCH_2^+$ (Isobutyl cation (1°)) $\xrightarrow{\sim H:}$ $(CH_3)_3C^+$ (*tert*-Butyl cation (3°)) $\xrightarrow{C_6H_6}$ $(CH_3)_3CC_6H_5$ (*tert*-Butylbenzene)

(*b*) $(CH_3)_3CCH_2OH$ (Neopentyl alcohol) $\xrightarrow{BF_3}$ $(CH_3)_3CCH_2^+$ (Neopentyl cation (1°)) $\xrightarrow{\sim :CH_3}$ $(CH_3)_2\overset{+}{C}CH_2CH_3$ (*tert*-Pentyl cation (3°)) $\xrightarrow{C_6H_5CH_3}$ *p*-$CH_3C_6H_4—C(CH_3)_2CH_2CH_3$ (*p*-*tert*-Pentyltoluene)

(*c*) $CH_3CH_2CH_2CH_2Cl$ (*n*-Butyl chloride) $\xrightarrow{AlCl_3}$ $[CH_3CH_2CH_2CH_2^+ \xrightarrow{\sim H:} CH_3CH_2\overset{+}{C}HCH_3]$ $\xrightarrow{C_6H_6}$ 34% $PhCH_2CH_2CH_2CH_3$ (*n*-Butylbenzene) + 66% $C_6H_5—CH(CH_3)CH_2CH_3$ (*s*-Butylbenzene)

(*d*) CH_3 H_3C $C(CH_3)_3$

Thermodyanamic product; it has less steric strain and is more stable than the kinetic-controlled isomer having a bulky *t*-butyl group ***ortho*** to a CH_3.

Chemistry; Reactivity of the Benzylic (PhCH) *H* and *C*

The chemistries of the benzylic and allylic positions are very similar. Intermediate carbocations, free radicals, and carbanions formed at these positions are stabilized by delocalization with the adjacent π system, the benzene ring in the case of the benzylic position. Another aspect of arene chemistry is the enhanced stability of unsaturated arenes having double bonds conjugated with the benzene ring. This property is akin to the stability of conjugated di- and polyenes.

Problem 11.32 $PhCH_3$ reacts with Br_2 and Fe to give a mixture of three monobromo products. With Br_2 in light, only one compound, a fourth monobromo isomer, is isolated. What are the four products? Explain the formation of the light-catalyzed product.

With Fe, the products are *o*-, *p*-, and some *m*-$BrC_6H_4CH_3$. In light, the product is benzyl bromide, $PhCH_2Br$. Like allylic halogenation (Section 6.5), the latter reaction is a free-radical substitution:

(1) $Br_2 \xrightarrow{uv} 2Br\cdot$

(2) $Br\cdot + PhCH_3 \longrightarrow Ph\dot{C}H_2 + HBr$

(3) $Ph\dot{C}H_2 + Br_2 \longrightarrow PhCH_2Br + Br\cdot$

Steps (2) and (3) are the propagating steps.

Problem 11.33 Which is more reactive to radical halogenation, $PhCH_3$ or *p*-xylene? Explain.

p-Xylene reactivity depends on the rate of formation of the benzyl-type radical. Electron-releasing groups such as CH_3 stabilize the transition state, producing the benzyl radical on the other CH_3 and thereby lowering the $\Delta H^{\ddagger}$ and increasing the reaction rate.

Problem 11.34 Outline a synthesis of 2,3-dimethyl-2,3-diphenylbutane from benzene, propylene, and any needed inorganic reagents.

The symmetry of this hydrocarbon makes possible a self-coupling reaction with 2-bromo-2-phenylpropane.

$$C_6H_6 + CH_3CH{=}CH_2 \xrightarrow{HF} CH_3{-}\underset{CH_3}{\overset{C_6H_5}{C}}{-}H \xrightarrow[uv]{Br_2} CH_3{-}\underset{CH_3}{\overset{C_6H_5}{C}}{-}Br \xrightarrow[2.\ CuBr]{1.\ Li} CH_3{-}\underset{CH_3}{\overset{C_6H_5}{C}}{-}\underset{CH_3}{\overset{C_6H_5}{C}}{-}CH_3$$

Problem 11.35 Give all possible products of the following reactions, and underline the major product.

(*a*) $PhCH_2CHOHCH(CH_3)_2 \xrightarrow{H_2SO_4}$ (*b*) $PhCH_2CHBrCH(CH_3)_2 \xrightarrow[KOH]{alc.}$

(*c*) $PhCH{=}CHCH_3 + HBr \longrightarrow$ (*d*) $PhCH{=}CHCH_3 + HBr \xrightarrow{peroxide}$

(*e*) $PhCH{=}CHCH{=}CH_2 + Br_2$ (equimolar amounts) $\longrightarrow$

(*a*) $PhCH_2CH{=}C(CH_3)_2 + \underline{PhCH{=}CHCH(CH_3)_2}$. The major product has the C=C conjugated with the benzene ring, and therefore, even though it is a disubstituted alkene, it is more stable than the minor product, which is a trisubstituted nonconjugated alkene.

(*b*) Same as part (*a*) and for the same reason.

(*c*) $PhCH_2CHBrCH_3 + \underline{PhCHBrCH_2CH_3}$. H^+ adds to C=C to give the more stable benzyl-type $Ph\overset{+}{C}HCH_2CH_3$. Reaction with Br^- gives the major product. The benzyl-type cation $Ph\overset{+}{C}HR$ (like $CH_2{=}CHCH_2^+$) can be stabilized by delocalizing the + to the *op*-positions of the ring:

δ+ δ+ δ+ (ring with +) =C(H)(CH₂CH₃)

(*d*) $\underline{PhCH_2CHBrCH_3} + PhCHBrCH_2CH_3$. Br· adds to give the more stable benzyl-type $Ph\dot{C}HCHBrCH_3$, rather than $PhCHBr\dot{C}HCH_3$. We have already discussed the stability of benzylic free radicals.

(*e*) $PhCHBrCHBrCH{=}CH_2 + \underline{PhCH{=}CHCHBrCHBr} + PhCHBrCH{-}CHCH_2Br$. The major product is the conjugated alkene, which is more stable than the other two products [see part (*a*)].

Problem 11.36 Explain the following observations. (*a*) A yellow color is obtained with Ph_3COH (trityl alcohol) is reacted with concentrated H_2SO_4, or when Ph_3CCl is treated with $AlCl_3$. On adding H_2O, the color disappears and a white solid is formed. (*b*) Ph_3CCl is prepared by the Friedel-Crafts reaction of benzene and CCl_4. It does not react with more benzene to form Ph_4C. (*c*) A deep-red solution appears when Ph_3CH is added to a solution of $NaNH_2$ in liquid NH_3. The color disappears on adding water. (*d*) A red color appears when Ph_3CCl reacts with Zn in C_6H_6. O_2 decolorizes the solution.

(*a*) The yellow color is attributed to the stable Ph_3C^+, whose + is delocalized to the *op*-positions of the three rings.

$$Ph_3COH + H_2SO_4 \longrightarrow Ph_3C^+ + H_3O^+ + HSO_4^-$$

$$Ph_3CCl + AlCl_3 \longrightarrow Ph_3C^+ + AlCl_4^-$$

$$\underset{\text{Lewis acid}}{Ph_3C^+} + \underset{\text{Lewis base}}{2H_2O} \longrightarrow \underset{\textit{white solid}}{Ph_3COH} + H_3O^+$$

(*b*) With $AlCl_3$, Ph_3CCl forms a salt, $Ph_3C^+AlCl_4^-$, whose carbocation is too stable to react with benzene. Ph_3C^+ may also be too sterically hindered to react further.

(*c*) The strong base $:\ddot{N}H_2^-$ removes H^+ from Ph_3CH to form the stable, deep red-purple carbanion $Ph_3C:^-$, which is then decolorized on accepting H^+ from the feeble acid H_2O.

$$\underset{\text{acid}_1}{Ph_3CH} + \underset{\text{base}_2}{:\ddot{N}H_2^-} \longrightarrow \underset{\text{acid}_2}{H:\ddot{N}H_2} + \underset{\text{base}_1\text{ (deep red)}}{Ph_3C\bar{:}}$$

$$\underset{\text{base}_1}{Ph_3C\bar{:}} + \underset{\text{acid}_2}{H_2O} \longrightarrow \underset{\text{acid}_1}{Ph_3CH} + \underset{\text{base}_2}{OH^-}$$

The $Ph_3C:^-$ is stabilized because the – can be delocalized to the *op*-positions of the three rings (as in the corresponding carbocation and free radicals).

(*d*) Cl· is removed from Ph_3CCl by Zn to give the colored radical $Ph_3C\cdot$, which decolorizes as it forms the peroxide in the presence of O_2.

$$2Ph_3CCl + Zn \longrightarrow 2Ph_3C\cdot + ZnCl_2$$

$$2Ph_3C\cdot + \cdot\ddot{\underset{..}{O}}-\ddot{\underset{..}{O}}\cdot \longrightarrow Ph_3C:\ddot{\underset{..}{O}}:\ddot{\underset{..}{O}}:CPh_3$$

Problem 11.37 In the following hydrocarbons, the alkyl H's are designated by the Greek letters α, β, γ, and so on. Assign each letter an Arabic number, beginning with 1 for LEAST, in order of *increasing* ease of abstraction by a Br·.

(*a*) $\underset{\alpha}{CH_3}-\underset{\beta}{CH_2}-\underset{\gamma}{CH_2}-C_6H_4-\underset{\delta}{CH_2}-C_6H_4-\underset{\varepsilon}{CH_3}$

(*b*) $\underset{\alpha}{CH_3}-\underset{\beta}{CH_2}-\underset{\gamma}{CH_2}-\underset{\delta}{CH}=\underset{\delta}{CH}-\underset{\varepsilon}{CH_2}-C_6H_5$

(*c*) $\underset{\alpha}{CH_3}$ $\underset{\beta}{CH_3}$ $\underset{\gamma}{CH_3}$

See Table 11.3.

TABLE 11.3

	α	β	γ	δ	ε
(*a*)	1 (1°)	2 (2°)	4 (2°, benzylic)	5 (2°, dibenzylic)	3 (1°, benzylic)
(*b*)	2 (1°)	3 (2°)	4 (allylic)	1 (vinylic)	5 (allylic, benzylic)
(*c*)	3 (*op* to other CH_3's)	1 (*mo* to two other CH_3's; more hindered)	2 (less sterically hindered than β; *mp* to two other CH_3's)		

Problem 11.38 Use + and – signs for positive and negative tests in tabulating rapid chemical reactions that can be used to distinguish among the following compounds: (*a*) chlorobenzene, benzyl chloride, and cyclohexyl chloride; (*b*) ethylbenzene, styrene, and phenylacetylene.

See Tables 11.4(*a*) and 11.4(*b*).

TABLE 11.4(a)

REACTIONS	CHLOROBENZENE	BENZYL CHLORIDE	CYCLOHEXYL CHLORIDE
Ring sulfonation is exothermic	+	+	–
Alc. $AgNO_3$ (forms AgCl, a white precipitate)	–	+ (very fast)*	+ (much slower)*

* Ag^+ induces an S_N1 reaction; $Ph\overset{+}{C}H_2 > C_6H_{11}^+$.

TABLE 11.4(b)

REACTIONS	ETHYLBENZENE	STYRENE	PHENYLACETYLENE
Br_2 in CCl_4 (is decolorized)	–	+	+
$Ag(NH_3)_2^+$ (forms a precipitate)	–	–	+

11.5 Summary of Arene and Aryl Halide Chemistry

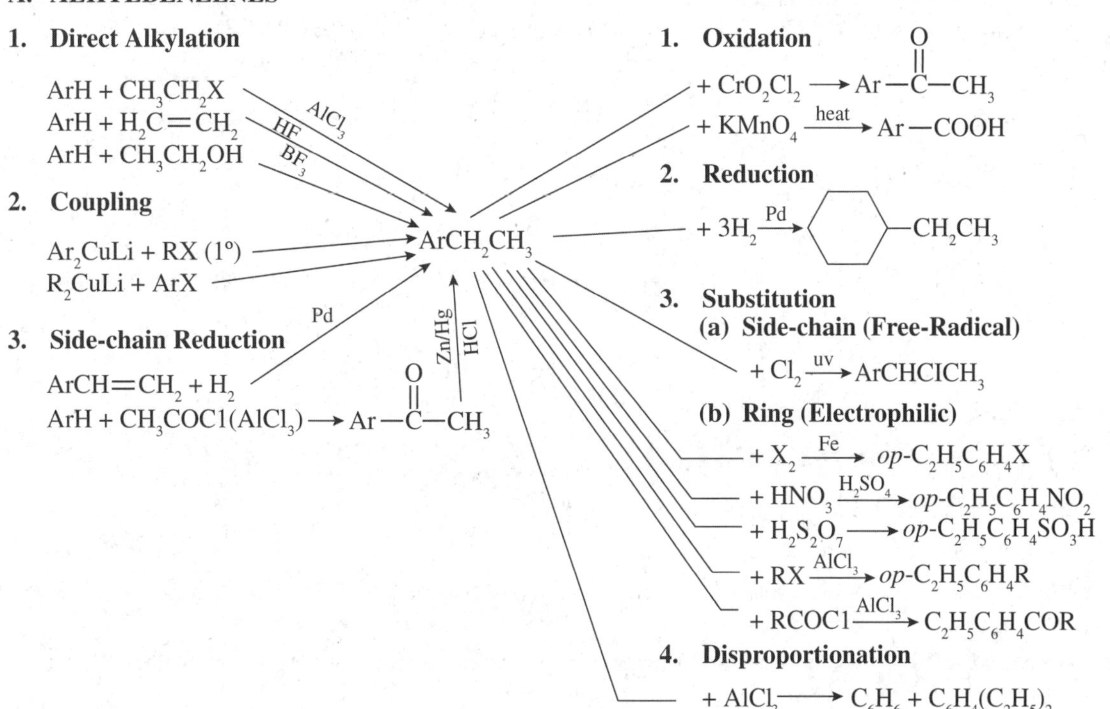

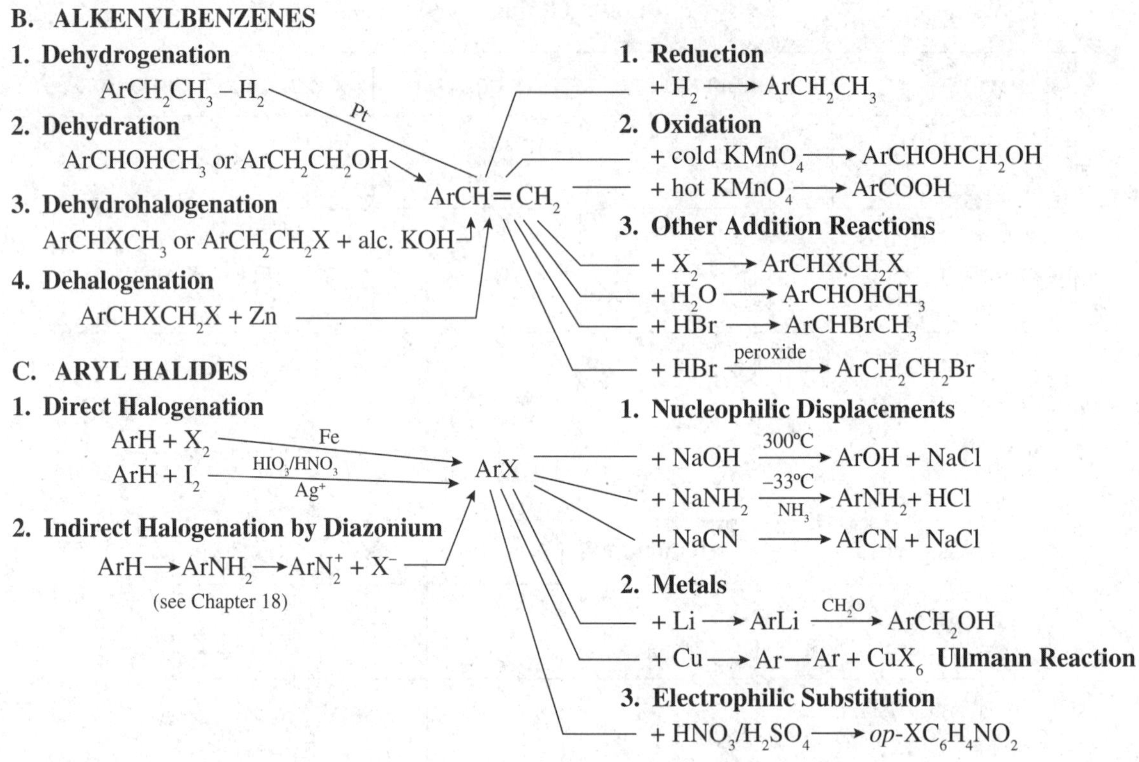

SUPPLEMENTARY PROBLEMS

Problem 11.39 Write structural formulas for the principal monosubstitution products of the indicated reactions from the following monosubstituted benzenes. For each, write an **S** or **F** to show whether reaction is SLOWER or FASTER than with benzene. (*a*) Monobromination, $C_6H_5CF_3$. (*b*) Mononitration, $C_6H_5COOCH_3$. (*c*) Monochlorination, $C_6H_5OCH_3$. (*d*) Monosulfonation, C_6H_5I. (*e*) Mononitration, $C_6H_5C_6H_5$. (*f*) Monochlorination, C_6H_5CN. (*g*) Mononitration, $C_6H_5NHCOCH_3$. (*h*) Monosulfonation, $C_6H_5CH(CH_3)CH_2CH_3$.

(*a*) F_3C–C_6H_4–Br (*meta*) **S** (*b*) $CH_3OC(=O)$–C_6H_4–NO_2 (*meta*) **S** (*c*) CH_3O–C_6H_4–Cl (*para*) **F**

(*d*) I–C_6H_4–SO_3H (*para*) **S** (*e*) C_6H_5–C_6H_4–NO_2 (*para*) **F** (*f*) $\overset{\delta-}{N}\equiv\overset{\delta+}{C}$–$C_6H_4$–Cl (*meta*) **S**

(*g*) CH_3CONH–C_6H_4–NO_2 (*para*) **F** (*h*) $CH_3CH_2CH(CH_3)$–C_6H_4–SO_3H (*para*) **F**

Problem 11.40 Which xylene is most easily sulfonated?

m-Xylene is most reactive and sulfonates at C^4 because its CH_3's reinforce each other [Rule 1; Problem 11.16(*a*)].

Problem 11.41 Write structures for the principal mononitration products of (*a*) *o*-cresol (*o*-methylphenol), (*b*) *p*-$CH_3CONHC_6H_4SO_3H$, (*c*) *m*-cyanotoluene (*m*-toluonitrile).

(*a*) H_3C (weak), OH (strong), NO_2 (*b*) $CH_3C(=O)NH$, NO_2, SO_3H (*c*) H_3C, CN, NO_2

Problem 11.42 Assign numbers from 1 for LEAST to 5 for MOST to designate relative reactivity to ring monobromination of the following groups.

(*a*) (I) $PhNH_2$, (II) $PhNH_3^+Cl^-$, (III) $PhNHCOCH_3$, (IV) PhCl, (V) $PhCOCH_3$.
(*b*) (I) $PhCH_3$, (II) PhCOOH, (III) PhH, (IV) PhBr, (V) $PhNO_2$.
(*c*) (I) *p*-xylene, (II) *p*-$C_6H_4(COOH)_2$, (III) PhMe, (IV) *p*-$CH_3C_6H_4COOH$, (V) *m*-xylene.

See Table 11.5.

TABLE 11.5

	(I)	(II)	(III)	(IV)	(V)
(*a*)	5	1	4	3	2
(*b*)	5	2	4	3	1
(*c*)	4	1	3	2	5

Problem 11.43 Use PhH, PhMe, and any aliphatic or inorganic reagents to prepare the following compounds in reasonable yields: (*a*) *m*-bromobenzenesulfonic acid; (*b*) 3-nitro-4-bromobenzoic acid; (*c*) 3,4-dibromonitrobenzene; (*d*) 2,6-dibromo-4-nitrotoluene.

(*a*) PhH $\xrightarrow{H_2S_2O_7}$ $PhSO_3H$ $\xrightarrow[Fe]{Br_2}$ *m*-$BrC_6H_4SO_3H$ (*m*-director is added first)

(*b*) $PhCH_3$ $\xrightarrow[Fe]{Br_2}$ *p*-$BrC_6H_4CH_3$ $\xrightarrow[H_2SO_4]{K_2Cr_2O_7}$ *p*-BrC_6H_4COOH $\xrightarrow[H_2SO_4]{HNO_3}$ 3-nitro-4-bromobenzoic acid (COOH, Br, NO_2)

Nitration of *p*-$BrC_6H_4CH_3$ would have given about a 50–50 mixture of two products; 2-nitro-4-bromotoluene would be unwanted. When oxidation precedes nitration, an excellent yield of the desired product is obtained.

(*c*) PhH $\xrightarrow[Fe]{Br_2}$ PhBr $\xrightarrow[H_2SO_4]{HNO_3}$ *p*-$BrC_6H_4NO_2$ $\xrightarrow[AlBr_3]{Br_2}$ 3,4-dibromonitrobenzene (only product)

Nitration followed by dibromination would give as the major product 2,5-dibromonitrobenzene (see Rule 2, preceding Problem 11.16).

(*d*) $PhCH_3$ $\xrightarrow[H_2SO_4]{HNO_3}$ *p*-$O_2NC_6H_4CH_3$ $\xrightarrow[Fe]{2Br_2}$ 2,6-dibromo-4-nitrotoluene (O_2N, CH_3, Br, Br)

Problem 11.44 Supply structures for organic compounds (A) through (O).

(*a*) $PhCH_2CH_2CH_3 + Br_2 \xrightarrow{uv} (A) \xrightarrow[KOH]{alc.} (B) \xrightarrow[KMnO_4]{cold\ dil.} (C) \xrightarrow[KMnO_4]{hot} (D)$

(*b*) $PhBr + Mg \xrightarrow{Et_2O} (E) \xrightarrow{H_2C{=}CHCH_2Br} (F) \xrightarrow[heat]{KOH} (G) \xrightarrow{NBS} (H)$

(*c*) $Ph{-}C{\equiv}CH + CH_3MgX \longrightarrow (I) \xrightarrow{ArCH_2Cl} (J) \xrightarrow{Li,\ NH_3} (K)$

(*d*) $p\text{-}CH_3C_6H_4C{\equiv}CPh + H_2/Pt \longrightarrow (L) \xrightarrow{HBr} (M)$

(*e*) Ph(H)C=C(H)(CH_3) [Ph and H on one carbon; H and CH_3 on the other] $+ Br_2(Fe) \longrightarrow (N) \xrightarrow[peroxide]{HBr} (O)$

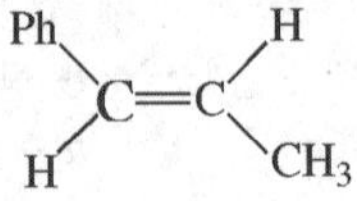

(*a*) (A) $PhCHBrCH_2CH_3$ (B) $PhCH{=}CHCH_3$ (C) $PhCHOHCHOHCH_3$

(D) $PhCOOH + CH_3COOH$

(*b*) (E) $PhMgBr$ (F) $PhCH_2CH{=}CH_2$ (G) $PhCH{=}CHCH_3$ (conjugated alkene is more stable) (H) $PhCH{=}CHCH_2Br$

(*c*) (I) $PhC{\equiv}CMgX(+CH_4)$ (J) $PhC{\equiv}CCH_2Ar$ (K) 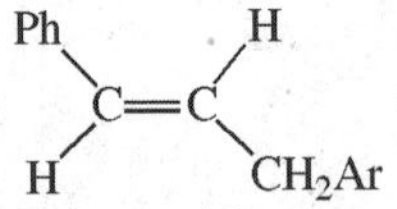 Ph(H)C=C(H)(CH_2Ar) (*trans*)

(*d*) (L) 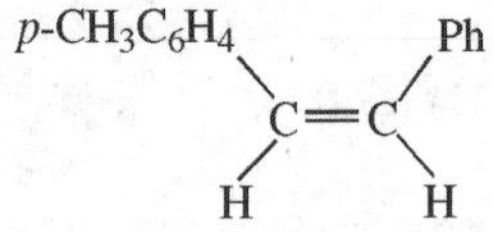 $p\text{-}CH_3C_6H_4$(H)C=C(Ph)(H) (*cis*) (M) $p\text{-}CH_3C_6H_4CH(Br){-}CH_2Ph$

In (M), H^+ adds to give more stable R^+, which is $p\text{-}CH_3C_6H_4\overset{+}{C}HCH_2Ph$ rather than $p\text{-}CH_3C_6H_4CH_2\overset{+}{C}HPh$ because of electron release by $p\text{-}CH_3$.

(*e*) (N) 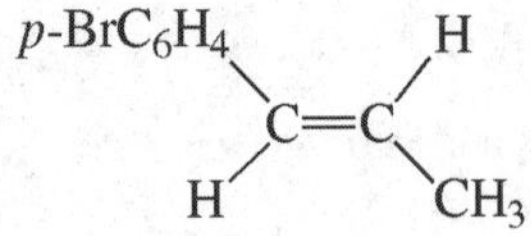 $p\text{-}BrC_6H_4$(H)C=C(H)(CH_3) (O) $p\text{-}BrC_6H_4{-}CH_2{-}CH(Br)CH_3$

In (O), Br· adds to give more stable R·, $p\text{-}BrC_6H_4\dot{C}HCHBrCH_3$, which is benzylic.

Problem 11.45 Assign numbers from 1 for LEAST to 3 for MOST to the Roman numerals for the indicated compounds to show their relative reactivities in the designated reactions.

(*a*) HBr addition to (I) $PhCH{=}CH_2$, (II) $p\text{-}CH_3C_6H_4CH{=}CH_2$, (III) $p\text{-}O_2NC_6H_4CH{=}CH_2$.

(*b*) Dehydration of (I) $p\text{-}O_2NC_6H_4CHOHCH_3$, (II) $p\text{-}H_2NC_6H_4CHOHCH_3$, (III) $C_6H_5CHOHCH_3$.

(*c*) Dehydration of (I) $Ph{-}C(CH_3)(OH){-}CH_2CH_3$, (II) $Ph{-}CH(CH_3){-}CH_2OH$, (III) $PhCHOHCH_2CH_2CH_3$.

(*d*) Solvolysis of (I) $C_6H_5CH_2Cl$, (II) $p\text{-}O_2NC_6H_4CH_2Cl$, (III) $p\text{-}CH_3OC_6H_4CH_2Cl$.

See Table 11.6.

Problem 11.46 Show the syntheses of the following compounds from benzene, toluene, and any inorganic reagents or aliphatic compounds having up to three C's:

(*a*) $p\text{-}BrC_6H_4CH_2Cl$ (*b*) $p\text{-}BrC_6H_4CH{=}CH_2$ (*c*) $Ph{-}C(CH_3)(OH){-}CH_3$

(*d*) $p\text{-}O_2NC_6H_4CH_2Ph$ (*e*) Ph_2CHCH_3

TABLE 11.6

	I	II	III
(*a*)	2	3 (*p*-Me stabilizes the benzylic R^+)	1 (*p*-NO_2 destabilizes the benzylic R^+)
(*b*)	1 (*p*-NO_2 destabilizes benzylic R^+)	3 (*p*-NH_2 stabilizes the benzylic R^+)	2
(*c*)	3 (get 3° benzylic R^+)	1 (get 1°, non-benzylic R^+)	2 (get 2°, benzylic R^+)
(*d*)	2	1 (*p*-NO_2 destabilizes benzylic R^+)	3 (*p*-CH_3O stabilizes benzylic R^+)

(*a*) $C_6H_5CH_3 \xrightarrow{Br_2,\ Fe} p\text{-}BrC_6H_4CH_3 \xrightarrow{Cl_2,\ uv} p\text{-}BrC_6H_4CH_2Cl$

(*b*) $C_6H_6 \xrightarrow[AlCl_3]{C_2H_5Cl} C_6H_5CH_2CH_3 \xrightarrow[Fe]{Br_2} p\text{-}BrC_6H_4CH_2CH_3 \xrightarrow[uv]{Cl_2}$

$p\text{-}BrC_6H_4CHClCH_3 \xrightarrow[KOH]{alc.} p\text{-}BrC_6H_4CH{=}CH_2$

The C═C must be introduced by a base-induced reaction. If acid is used, such as in the dehydration of an alcohol, the product would undergo polymerization.

(*c*) $PhH + (CH_3)_2CHCl \xrightarrow{AlCl_3} Ph{-}CH(CH_3)_2 \xrightarrow{NBS} Ph{-}C(Br)(CH_3)_2 \xrightarrow[(S_N1)]{H_2O} Ph{-}C(OH)(CH_3)_2$

(*d*) $C_6H_5CH_3 \xrightarrow[H_2SO_4]{HNO_3} p\text{-}O_2NC_6H_4CH_3 \xrightarrow[uv]{Cl_2} p\text{-}O_2NC_6H_4CH_2Cl \xrightarrow[AlCl_3]{C_6H_6} p\text{-}O_2NC_6H_4CH_2C_6H_5$

The deactivated $C_6H_5NO_2$ cannot be alkylated with $PhCH_2Cl$.

(*e*) $PhH \xrightarrow[AlCl_3]{C_2H_5Cl} PhCH_2CH_3 \xrightarrow{NBS} PhCHBrCH_3 \xrightarrow[KOH]{Alc.} PhCH{=}CH_2 \xrightarrow[HF]{PhH} Ph_2CHCH_3$

Problem 11.47 Deduce the structural formulas of the following arenes. (*a*) (i) Compound A ($C_{16}H_{16}$) decolorizes both Br_2 in CCl_4 and cold aqueous $KMnO_4$. It adds an equimolar amount of H_2. Oxidation with hot $KMnO_4$ gives a dicarboxylic acid, $C_6H_4(COOH)_2$, having only one monobromo substitution product. (ii) What structural feature is uncertain? (*b*) Arene B ($C_{10}H_{14}$) has five possible monobromo derivatives ($C_{10}H_{13}Br$). Vigorous oxidation of B yields an acidic compound, $C_8H_6O_4$, having only one mononitro substitution product, $C_8H_5O_4NO_2$.

(*a*) (i) Compound A has one C═C, since it adds one H_2. The other 8 degrees of unsaturation mean the presence of two benzene rings. Since oxidative cleavage gives a dicarboxylic acid, $C_6H_4(COOH)_2$, each benzene ring must be disubstituted. Since $C_6H_4(COOH)_2$ has only one monobromo derivative, the COOH's must be *para* to each other.

$p\text{-}CH_3C_6H_4CH{=}CHC_6H_4CH_3\text{-}p$ (A)

$\xrightarrow{Br_2\ (CCl_4)} p\text{-}CH_3C_6H_4CH(Br){-}CH(Br)C_6H_4CH_3\text{-}p$

$\xrightarrow{H_2/Pt} p\text{-}CH_3C_6H_4CH_2CH_2C_6H_4CH_3\text{-}p$

$\xrightarrow{KMnO_4} p\text{-}CH_3C_6H_4CH(OH){-}CH(OH)C_6H_4CH_3\text{-}p$

$\xrightarrow{oxid.} p\text{-}HOOCC_6H_4COOH \xrightarrow[Fe]{Br_2} HOOCC_6H_3(Br)COOH$

(ii) It may be *cis* or *trans*.

(*b*) $C_8H_6O_4$ must be a dicarboxylic acid, $C_6H_4(COOH)_2$, and, as in part (*a*), has *para* COOH's. Compound B must therefore be a *p*-dialkylbenzene.

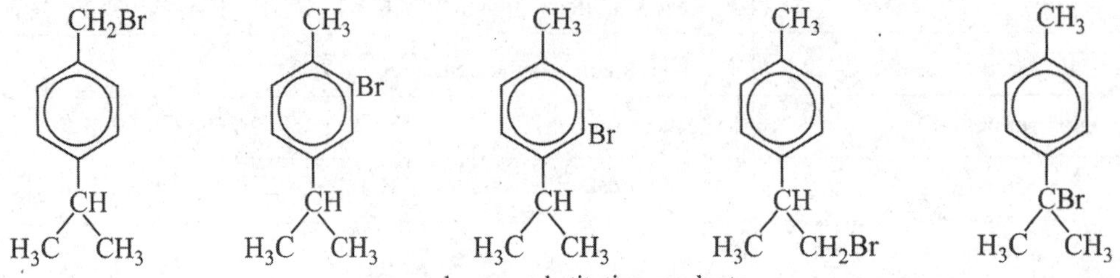

monobromo substitution products

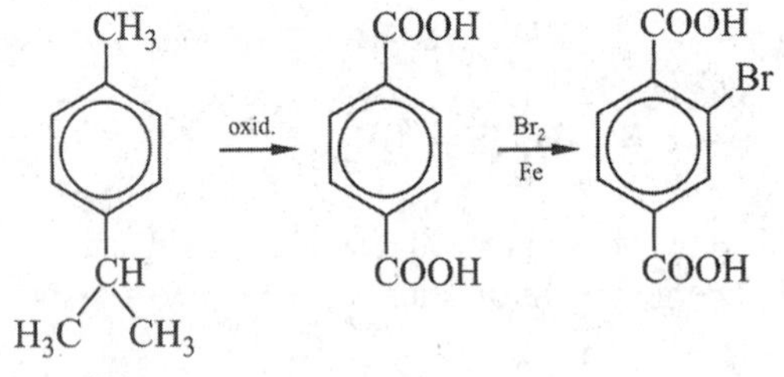

(B)

Problem 11.48 Outline practical laboratory syntheses from benzene or toluene and any needed inorganic reagents of: (*a*) *p*-chlorobenzal chloride, (*b*) 2,4-dinitroaniline, (*c*) *m*-chlorobenzotrichloride, (*d*) 2,5-dibromonitrobenzene.

(a) Toluene $\xrightarrow{Cl_2/Fe}$ *p*-chlorotoluene $\xrightarrow[hv]{Cl_2}$ *p*-$ClC_6H_4CHCl_2$

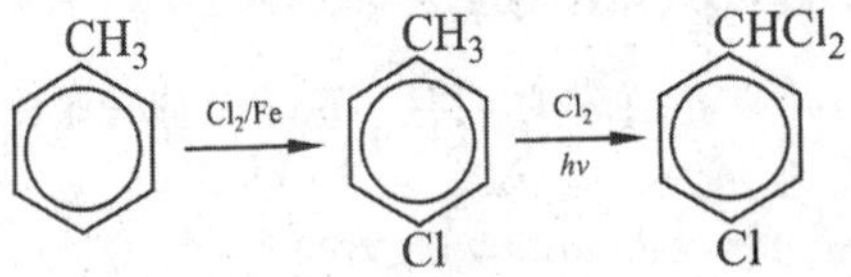

(*b*) $C_6H_6 \xrightarrow{Cl_2/Fe}$ chlorobenzene $\xrightarrow[H_2SO_4]{HNO_3}$ 2,4-dinitrochlorobenzene $\xrightarrow{NH_3}$ 2,4-dinitroaniline

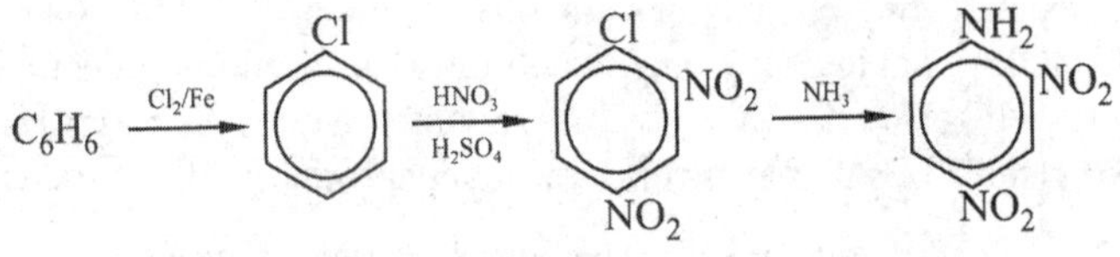

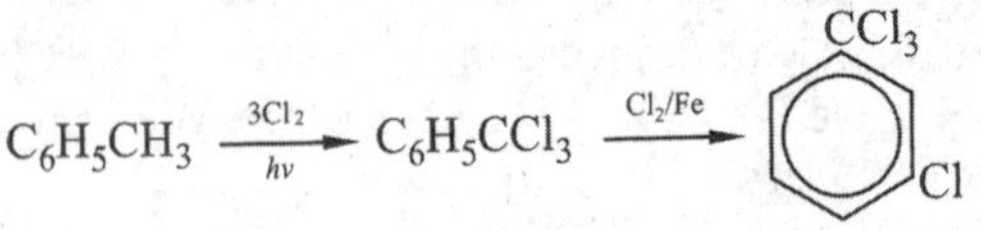

(*c*) $C_6H_5CH_3 \xrightarrow[hv]{3Cl_2} C_6H_5CCl_3 \xrightarrow{Cl_2/Fe}$ *m*-$ClC_6H_4CCl_3$

(*d*) $C_6H_6 \xrightarrow{Br_2/Fe}$ *p*-dibromobenzene $\xrightarrow[H_2SO_4]{HNO_3}$ 2,5-dibromonitrobenzene

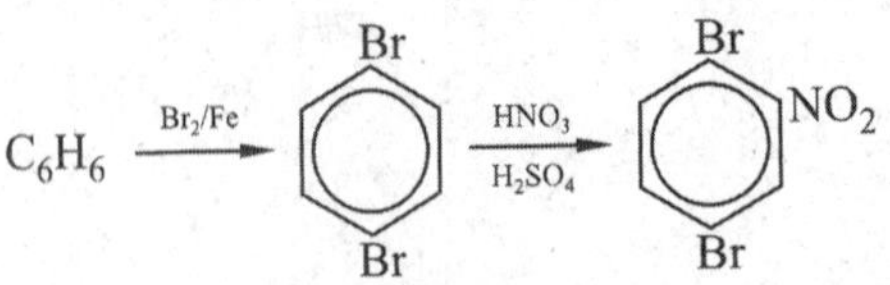

Problem 11.49 Assign numbers from 1 for LEAST to 3 for MOST to show the relative reactivities of the compounds with the indicated reagents: (*a*) $C_6H_5CH_2CH_2Br$ (I), C_6H_5 $CHBrCH_3$ (II) and $C_6H_5CH{=}CHBr$ (III) with alcoholic $AgNO_3$; (*b*) CH_3CH_2Cl (I), $C_6H_5CH_2Cl$ (II) and C_6H_5Cl (III) with KCN; (*c*) *m*-nitrochlorobenzene (I), 2,4-dinitrochlorobenzene (II) and *p*-nitrochlorobenzene (III) with sodium methoxide.

See Table 11.7.

TABLE 11.7

	I	II	III
(*a*) An S_N1 reaction	2 (primary)	3 (benzylic)	1 (vinylic)
(*b*) An S_N2 reaction	2 (primary)	3 (benzylic)	1 (aryl)
(*c*) An aromatic nucleophilic displacement	1 (*meta*)	3 (*o* and *p*)	2 (*para*)

Problem 11.50 Which Cl in 1,2,4-trichlorobenzene reacts with $^-OCH_2COO^-$ to form the herbicide "2,4-D"? Give the structure of "2,4-D."

Cl's are electron-withdrawing and activate the ring to nucleophilic attack. The Cl at C^1 is displaced because it is *ortho* and *para* to the other Cl's.

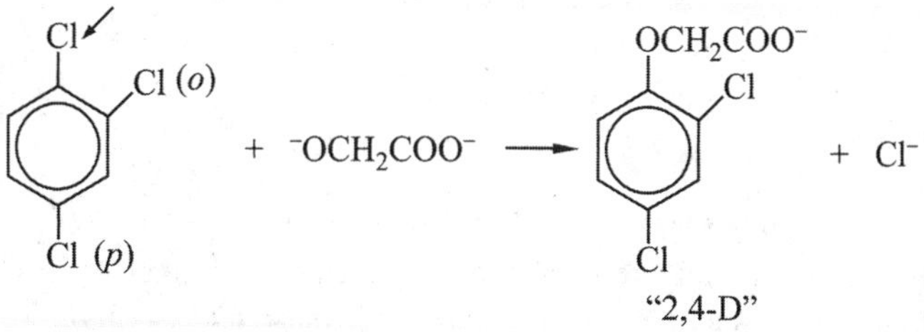

1,2,4-trichlorobenzene (Cl↙, Cl (*o*), Cl (*p*)) + $^-OCH_2COO^-$ ⟶ 2,4-dichlorophenoxyacetate (OCH_2COO^-, Cl, Cl) + Cl^-

"2,4-D"

Problem 11.51 Explain these observations: (*a*) *p*-Nitrobenzenesulfonic acid is formed from the reaction of *p* nitrochlorobenzene with $NaHSO_3$, but benzenesulfonic acid cannot be formed from chlorobenzene by this reaction. (*b*) 2,4,6-Trinitroanisole with $NaOC_2H_5$ gives the same product as 2,4,6-trinitrophenetole with $NaOCH_3$.

(*a*) Nucleophilic aromatic substitution occurs with *p*-nitrochlorobenzene, but not with chlorobenzene, because NO_2 stabilizes the carbanion [Problem 11.22(*a*)].

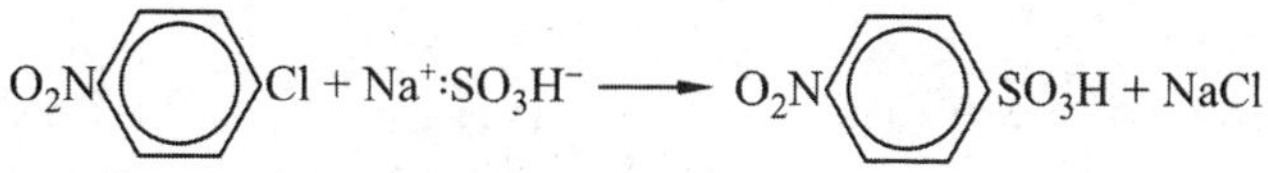

$$O_2N\text{-}C_6H_4\text{-}Cl + Na^+{:}SO_3H^- \longrightarrow O_2N\text{-}C_6H_4\text{-}SO_3H + NaCl$$

(*b*) The product is a sodium salt formed by addition of alkoxide.

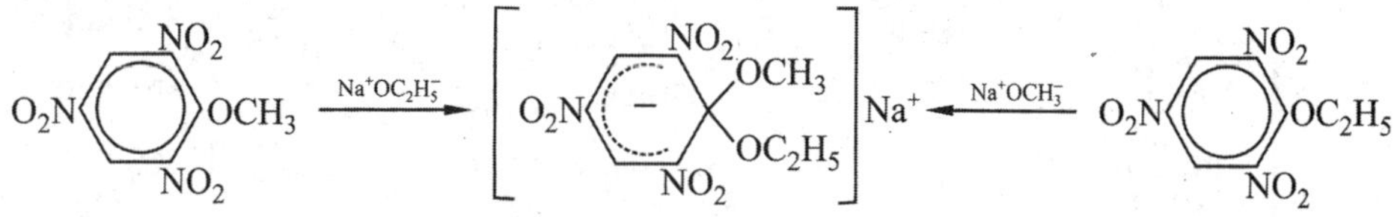

2,4,6-$(NO_2)_3C_6H_2OCH_3$ $\xrightarrow{Na^+OC_2H_5^-}$ $[O_2N(NO_2)_2C_6H_2^-(OCH_3)(OC_2H_5)]Na^+$ $\xleftarrow{Na^+OCH_3^-}$ 2,4,6-$(NO_2)_3C_6H_2OC_2H_5$

Spectroscopy and Structure

12.1 Introduction

Spectral properties are used to determine the structure of molecules and ions. Of special importance are ultraviolet (uv), infrared (ir), nuclear magnetic resonance (nmr), and mass spectra (ms). Free radicals are studied by electron spin resonance (esr).

The various types of molecular energy, such as electronic, vibrational, and nuclear spin, are quantized. That is, only certain energy states are permitted. The molecule can be raised from its lowest energy state (**ground state**) to a higher energy state (**excited state**) by a photon (quantum of energy) of electromagnetic radiation of the correct wavelength.

REGION OF ELECTROMAGNETIC SPECTRUM (↑ increasing energy)	TYPE OF EXCITATION	WAVELENGTH OF PHOTON
Far ultraviolet (uv)	Electronic	100–200 nm
Near ultraviolet (uv)	Electronic	200–350 nm
Visible	Electronic	350–800 nm
Infrared (ir)	Molecular vibration	1–300 μm
Radio	Spin (electronic of nuclear)	1 m

Wavelengths (λ) for ultraviolet spectra are expressed in **nanometers** (1 nm = 10^{-9} m); for the infrared, **micrometers** are used (1 μm = 10^{-6}m). Frequencies (ν) in the infrared are often specified by the **wave number**, $\bar{\nu}$, where $\bar{\nu} = 1/\lambda$; a common unit for $\bar{\nu}$ is the **reciprocal centimeter** (1 cm^{-1} = 100 m^{-1}). The basic SI units for frequency and energy are the **hertz** (Hz) and the **joule** (J), respectively.

Problem 12.1 (*a*) Calculate the frequencies of violet and red light if their wavelengths are 400 and 750 nm, respectively. (*b*) Calculate and compare the energies of their photons.

(*a*) The wavelengths are substituted into the equation $\nu = c/\lambda$, where c = speed of light = 3.0×10^8 m/s. Thus:

$$\text{Violet:}\quad \nu = \frac{3.0\times10^{8}\ \text{m/s}}{400\times10^{-9}\,\text{m}} = 7.5\times10^{14}\ \text{s}^{-1} = 750\ \text{THz}$$

$$\text{Red:}\quad \nu = \frac{3.0\times10^{8}\ \text{m/s}}{750\times10^{-9}\,\text{m}} = 4.0\times10^{14}\ \text{s}^{-1} = 400\ \text{THz}$$

where 1 THz = 10^{12} Hz = 10^{12} s^{-1}. Violet light has the shorter wavelength and higher frequency.

(*b*) The frequencies from part (*a*) are substituted into the equation $E = h\nu$, where $h = 6.624 \times 10^{-34}$ J·s (**Planck's constant**). Thus:

$$\text{Violet:}\quad E = (6.624\times10^{-34}\ \text{J}\cdot\text{s})(7.5\times10^{14}\ \text{s}^{-1}) = 5.0\times10^{-19}\ \text{J}$$

$$\text{Red:}\quad E = (6.624\times10^{-34}\ \text{J}\cdot\text{s})(4.0\times10^{14}\ \text{s}^{-1}) = 2.7\times10^{-19}\ \text{J}$$

Photons of violet light have more energy than those of red light.

Problem 12.2 Express 10 micrometers (*a*) in centimeters, (*b*) in angstroms (1 Å = 10^{-10} m), (*c*) in nanometers, (*d*) as a wave number.

$$(a)\quad 10\ \mu\text{m} = (10\times10^{-6}\ \text{m})\left(\frac{100\ \text{cm}}{1\ \text{m}}\right) = 10^{-3}\ \text{cm}$$

$$(b)\quad 10\ \mu\text{m} = (10\times10^{-6}\ \text{m})\left(\frac{1\ \text{Å}}{10^{-10}\ \text{m}}\right) = 10^{5}\ \text{Å}$$

$$(c)\quad 10\ \mu\text{m} = (10\times10^{-6}\ \text{m})\left(\frac{10^{9}\ \text{nm}}{1\ \text{m}}\right) = 10^{4}\ \text{nm}$$

$$(d)\quad \bar{\nu} = \frac{1}{10\times10^{-6}\text{m}} = 10^{5}\text{m}^{-1} = 10^{5}\text{m}^{-1}\left(\frac{1\ \text{cm}^{-1}}{100\ \text{m}^{-1}}\right) = 10^{3}\text{cm}^{-1}$$

In a typical spectrophotometer, a dissolved compound is exposed to electromagnetic radiation with a continuous spread in wavelength. The radiation passing through or absorbed is recorded on a chart against the wavelength or wave number. Absorption peaks are plotted as *minima* in *infrared*, and usually as *maxima* in *ultraviolet* spectroscopy.

At a given wavelength, absorption follows an exponential law of the form

$$A = \varepsilon Cl$$

where $A \equiv$ **absorbance** $\equiv -\log_{10}$ (fraction of incident radiation transmitted)
$\varepsilon \equiv$ **molar extinction coefficient**, cm^2/mol
$C \equiv$ concentration of solution, mol/L (= mol/cm^3)
$l \equiv$ thickness of solution presented to radiation, cm

The wavelength of maximum absorption, λ_{max}, and the corresponding ε_{max} are identifying properties of a compound. Units are normally omitted from specifications of ε.

12.2 Ultraviolet and Visible Spectroscopy

Ultraviolet and visible light cause an electron to be excited from a lower-energy HOMO to a high-energy LUMO. There are three kinds of electrons: those in σ bonds, those in π bonds, and unshared electrons, which are designated by the letter n for nonbonding. These are illustrated in formaldehyde:

H, σ, C, π, σ, n, Ö:n, σ, H

On absorbing energy, any of these electrons can enter excited states, which are either antibonding σ^* or π^*. All molecules have σ and σ^* orbitals, but only those with π orbitals have π^* orbitals.

Only the $n \rightarrow \pi^*$, $\pi \rightarrow \pi^*$, and more rarely the $n \rightarrow \sigma^*$ excitations occur in the near ultraviolet and visible regions, which are the available regions for ordinary spectrophotometers. Species which absorb in the visible region are colored, and black is observed when all visible light is absorbed.

Problem 12.3 The relative energy for various electronic states (MO's) is

Energy ↑	
σ^*	
π^*	
n	(nonbonding)
π	
σ	

List the three electronic transitions detectable by uv spectrophotometers in order of increasing ΔE.

$$n \rightarrow \pi^* < \pi \rightarrow \pi^* < \mathrm{n} \rightarrow \sigma^*$$

Problem 12.4 List all the electronic transitions possible for (*a*) CH_4, (*b*) CH_3Cl, (*c*) $H_2C{=}O$.

(*a*) $\sigma \rightarrow \sigma^*$. (*b*) $\sigma \rightarrow \sigma^*$ and $n \rightarrow \sigma^*$ (there are no π or π^* MO's). (*c*) $\sigma \rightarrow \sigma^*$, $\sigma \rightarrow \pi^*$, $\pi \rightarrow \sigma^*$, $n \rightarrow \sigma^*$, $\pi \rightarrow \pi^*$, and $n \rightarrow \pi^*$.

Problem 12.5 The uv spectrum of acetone shows two peaks of λ_{max} = 280 nm, ε_{max} = 15 and λ_{max} = 190 nm, ε_{max} = 100. (*a*) Identify the electronic transition for each. (*b*) Which is more intense?

(*a*) The longer wavelength (280 nm) is associated with the smaller-energy ($n \rightarrow \pi^*$) transition. $\pi \rightarrow \pi^*$ occurs at 190 nm.

(*b*) $\pi \rightarrow \pi^*$ has the larger ε_{max} and is the more intense peak.

Problem 12.6 Draw conclusions about the relationship of λ_{max} to the structure of the absorbing molecule from the following λ_{max} values (in nm): ethylene (170), 1,3-butadiene (217), 2,3-dimethyl-1,3-butadiene (226), 1,3-cyclohexadiene (256), and 1,3,5-hexatriene (274).

1. Conjugation of π bonds causes molecules to absorb at longer wavelengths.
2. As the number of conjugated π bonds increases, λ_{max} increases.
3. Cyclic polyenes absorb at higher wavelengths than do acyclic polyenes.
4. Substitution of alkyl groups on C=C causes a shift to longer wavelength (**red shift**).

Problem 12.7 Account for the following variations in λ_{max} (nm) of CH_3X: X = Cl(173), Br(204), and I(258).

The transition must be $n \rightarrow \sigma^*$ [Problem 12.4(*b*)]. On going from Cl to Br to I, the *n* electrons (*a*) are found in higher principal energy levels (the principal quantum numbers are 3, 4, 5, respectively), (*b*) are farther away from the attractive force of the nucleus, and (*c*) are more easily excited. Hence, absorption occurs at progressively higher λ_{max}, since less energy is required.

Problem 12.8 Identify the two geometric isomers of stilbene, $C_6H_5CH{=}CHC_6H_5$, from their λ_{max} values, 294 nm and 278 nm.

The higher-energy *cis* isomer has the shorter wavelength. Steric strain prevents full coplanarity of the *cis* phenyl groups, and the conjugative effect is attenuated.

Problem 12.9 The complementary color pairs are violet–yellow, blue–orange, and green–red. Given a red, an orange, and a yellow polyene, which is most and which is least conjugated?

The orange polyene absorbs *blue*, the red absorbs *green*, and the yellow absorbs *violet*. The most conjugated polyene—in this case, the red one—absorbs the color of longest wavelength, in this case green. Violet has the shortest wavelength, and therefore, the yellow polyene is the least conjugated.

12.3 Infrared Spectroscopy

Infrared radiation causes excitation of the quantized molecular vibration states. Atoms in a diatomic molecule, for example, H—H and H—Cl, vibrate in only one way; they move, as though attached by a coiled spring, toward and away from each other. This mode is called **bond stretching**. Triatomic molecules, such as CO_2 (O=C=O), possess two different stretching modes. In the **symmetrical stretch**, each O moves away from the C at the same time. In the **antisymmetrical stretch**, one O moves toward the C while the other O moves away.

Molecules with more than two atoms have, in addition, continuously changing bond angles. These **bending modes** are indicated in Fig. 12.1.

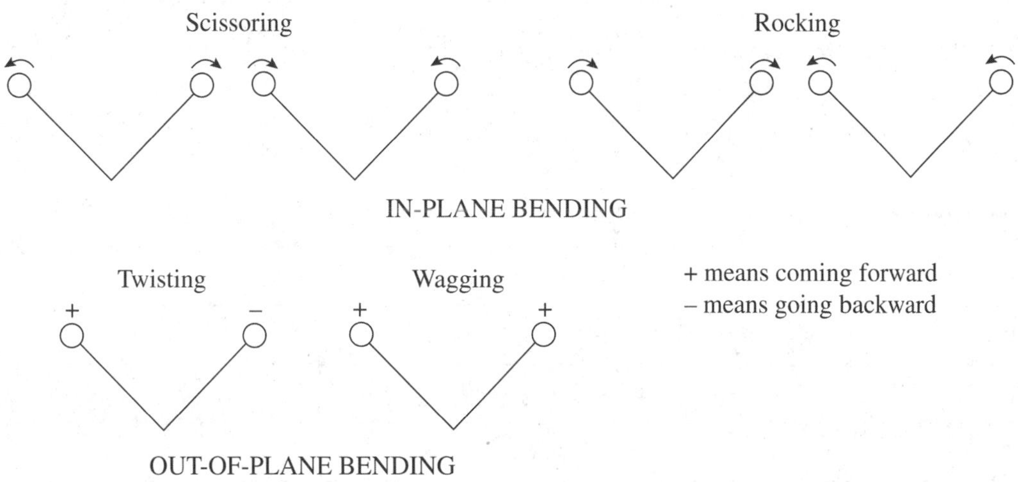

Figure 12.1

In a molecule, each bond, such as O—H, and each group of three or more atoms, such as NH_2 and CH_3, absorbs ir radiation at certain frequencies to give quantized excited stretching and bending vibrational states. See Tables 12.1 and 12.2. Only vibrations that cause a change in dipole moment give rise to an absorption band. Although absorption is affected only slightly by the molecular environment of the bond or group, it is possible to determine from small variations in the band frequencies such factors as the size of the ring containing a C=O group or whether C=O is part of a ketone, COCl, or COOH. An observed absorption band at a specific wavelength proves the identity of a particular bond or group of bonds in a molecule. Conversely, the absence of a certain band in the spectrum usually rules out the presence of the bond that would produce it.

Between 42 and 24 THz (1400 and 800 cm^{-1}), there are many peaks which are difficult to interpret. However, this range, called the **fingerprint region**, is useful for determining whether compounds are identical. (It is virtually impossible for two different organic compounds to have the same ir spectrum, because of the large number of peaks in the spectrum.)

Problem 12.10 How do the following factors affect absorption frequencies? Use data in Tables 12.1 and 12.2. (*a*) For C—H stretch, the hybrid orbitals used by C. (*b*) Bond strength—in other words, change in bond

TABLE 12.1 Infrared Absorption Peaks (mostly stretching)

$\bar{\upsilon}$, cm^{-1}	INTENSITY*	STRUCTURE
1050–1400	(s)	C—O (in ethers, alcohols, and esters)
1150–1360	(s)	SO_2 (in sulfonic acid derivatives)
1315–1475	(m-s)	C—H (in alkanes)
1340, 1500	(s)	NO_2
1450–1600	(s)	C═C bond in aromatic ring (usually shows several peaks)
1620–1680	(m)	C═C
1630–1690	(s)	C═O (in amides O═C—N)
1690–1750	(s)	C═O (in carbonyl compounds and esters)
1700–1725	(s)	OH C═O (in carboxylic acids)
1770–1820	(s)	Cl C═O (in acid chlorides)
2100–2200	(m)	C≡C
2210–2260	(m)	C≡N
2500	(w)	S—H
2700–2800	(w)	C—H (of aldehyde group)
2500–3000	(s)(vb)†	O—H in COOH
3000–3100	(m)	C—H (C is part of aromatic ring)
3300	(s)	C—H (C is acetylenic)
3020–3080	(m)	C—H (C is ethylenic)
2800–3000	(m-s)	C—H (in alkanes)
3300–3500	(m)	N—H (in amines and amides)
3200–3600	(s, b)†	O—H (in H-bonded ROH and ArOH)
3600–3650	(s)	O—H
2100	(s)	O—D

* Intensities: (s) = strong, (m) = medium, (w) = weak, (b) = broad, (vb) = very broad.

† Intermolecular H-bonded peaks are sharp, not broad.

TABLE 12.2 Bending Frequencies (cm^{-1}) of Hydrocarbons*

Alkanes	CH_3 1420–1470 1375	$=CH_2$ 1430–1470	$CH(CH_3)_2$ Doublet of equal intensities at 1370, 1385; also 1170	$C(CH_3)_3$ Doublet at 1370 (strong) 1395 (moderate)
Alkanes out-of-plane	$RCH{=}CH_2$ 910–920 990–1000	$R_2C{=}CH_2$ 880–900	RCH=CHR *cis* 675–730 (variable) *trans* 965–975	
Aromatic C—H out-of-plane	Monosubstituted 690–710 730–770		Disubstituted *ortho* 735–770 *meta* 690–710, 750–810	*para* 810–840

* Most peaks are medium-to-strong or strong.

multiplicity. (*c*) Change in mass of one of the bonded atoms; e.g., O—H versus O—D. (*d*) Stretching versus bending. (*e*) H-bonding of OH.

(*a*) The more *s* character in the C—H bond, the stiffer the bond and the higher the frequency:

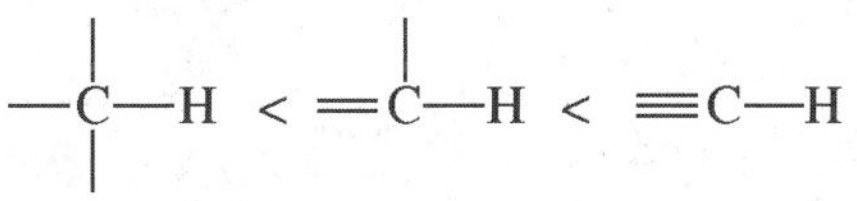

$$-\overset{|}{\underset{|}{C}}-H \; < \; =\overset{|}{C}-H \; < \; \equiv C-H$$

(*b*) Stretching frequencies parallel bond strengths. Because bond strength increases with the number of bonds between two given atoms, absorption frequencies increase with bond multiplicity:

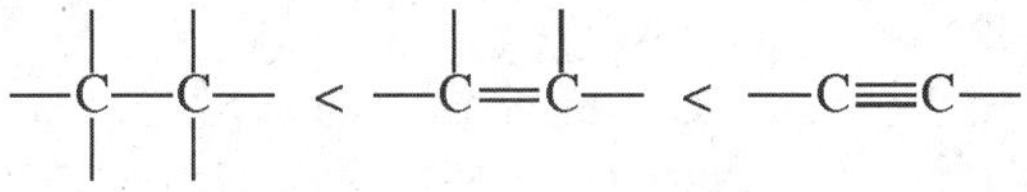

$$-\overset{|}{\underset{|}{C}}-\overset{|}{\underset{|}{C}}- \; < \; -\overset{|}{C}=\overset{|}{C}- \; < \; -C\equiv C-$$

(*c*) Frequencies are inversely related to the masses of the bonded atoms. Therefore, changing the lighter H to the heavier D causes a decrease in the stretching frequency.
(*d*) Most of the stretching frequencies in Table 12.1 are higher than the bending frequencies in Table 12.2.
(*e*) H-bonding causes a shift to lower frequencies ($3600\ cm^{-1} \rightarrow 3300\ cm^{-1}$). The band also becomes broader and less intense.

Problem 12.11 Identify the peaks marked by Roman numerals in Fig. 12.2, the ir spectrum of ethyl acetate, $CH_3\underset{\underset{O}{\|}}{C}OCH_2CH_3$.

The deep valleys represent transmission minima and are therefore absorption "peaks" or bands. At about 2800 cm^{-1}, peak I is due to $H—C_{sp^3}$ stretching. The peak at 1700 cm^{-1}, II, is due to stretching of the group.

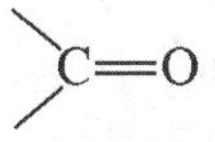

$$>C=O$$

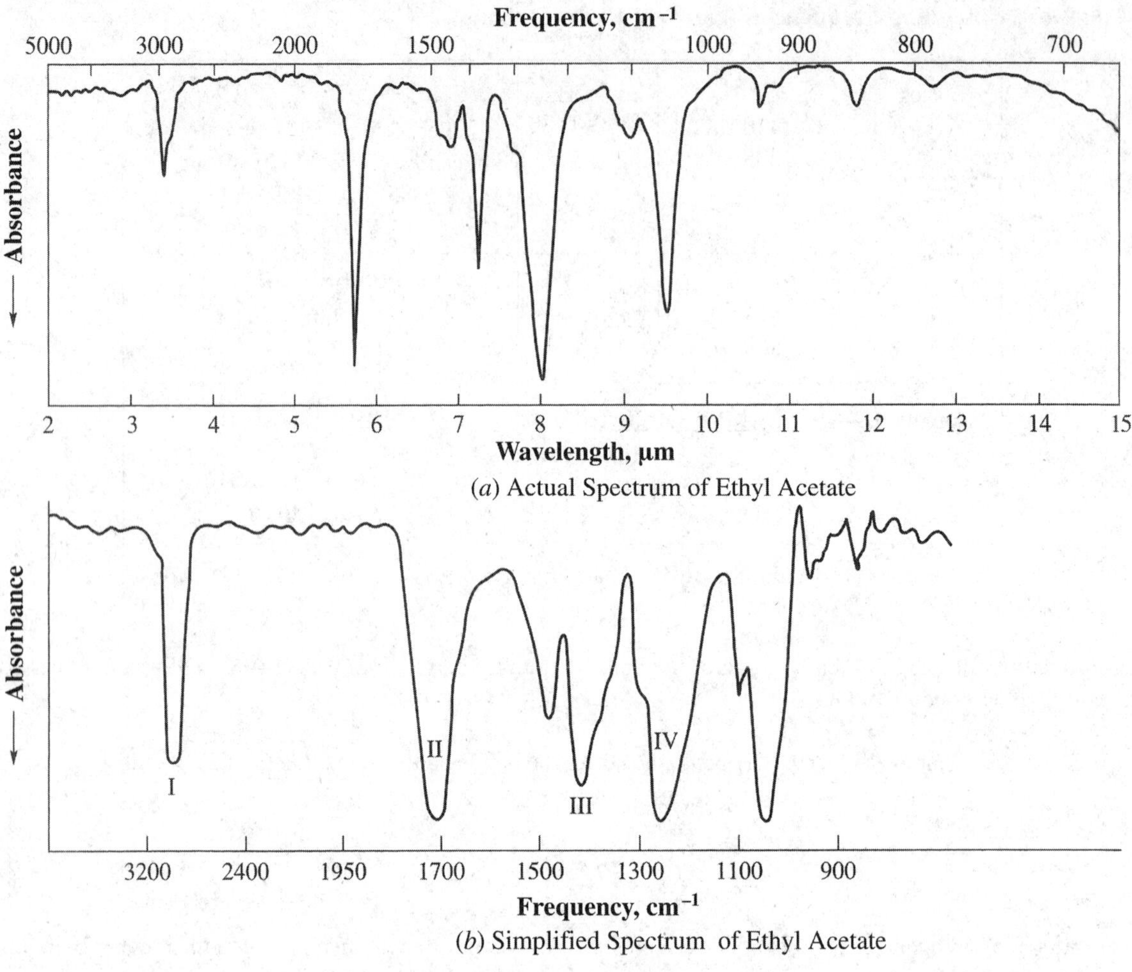

(*a*) Actual Spectrum of Ethyl Acetate

(*b*) Simplified Spectrum of Ethyl Acetate

Figure 12.2

The two bands at 1400–1500 cm^{-1}, III, are again due to C—H bonds. The one at 1250 cm^{-1}, IV, is due to the C—O stretch. (It is extremely difficult and impractical to attempt an interpretation of each band in the spectrum.)

Problem 12.12 Which of the following vibrational modes show no ir absorption bands? (*a*) Symmetrical CO_2 stretch, (*b*) antisymmetrical CO_2 stretch, (*c*) symmetrical O═C═S stretch, (*d*) C═C stretch in *o*-xylene, (*e*) C═C stretch in *p*-xylene, and (*f*) C═C stretch in *p*-bromotoluene.

Those vibrations that do not result in a change in dipole moment show no band. These are (*a*) and (*e*), which are symmetrical about the axis of the stretched bonds.

12.4 Nuclear Magnetic Resonance (Proton, PMR)

Origin of Spectra

Nuclei with an odd number of protons or of neutrons have permanent magnetic moments and quantized nuclear spin states. For example, an H in a molecule has two equal-energy nuclear spin states, which are assigned the quantum numbers $+\frac{1}{2}(\uparrow)$ and $-\frac{1}{2}(\downarrow)$ [Fig. 12.3(*a*)]. When a compound is placed in a magnetic field, its H's align their own fields either *with* or *against* the applied magnetic field, H_0, giving rise to two separated energy states, as shown in Fig. 12.3(*b*). In the higher energy state, E_2, the fields are aligned *against* each other; in the lower energy state, E_1, they are aligned *with* each other. The difference in energy between the two states, which is directly proportional to H_0, corresponds to a frequency of radiowaves. For this reason, radio-frequency photons can "flip" H nuclei from lower to higher energy states, reversing their spin in the process. When such spin-flip occurs, the nucleus (a proton, in the case of H) is said to be in **resonance** with the applied radiation; hence the name **nuclear magnetic resonance** (**nmr**) spectroscopy. When used for medical diagnosis, this technique is called **magnetic resonance imaging** (**mri**).

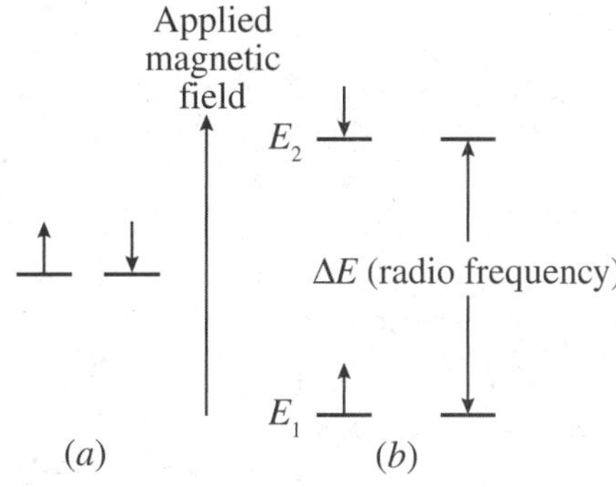

Figure 12.3

In practice, it is easier and cheaper to fix the radio frequency and slowly to vary the magnetic field strength. When the H_0 value is reached enabling the proton to absorb the radio photon and to spin-flip, a **signal** (peak) is traced on calibrated chart paper. When the radiowaves are removed, the excited nuclei quickly return to the lower-energy spin state. The same sample can then be used for obtaining repeated spectra. The sample compound is dissolved in a proton-free solvent such as CCl_4 or a deuterated solvent such as $DCCl_3$. (Although D is nmr-active, it does not absorb in the frequency range that H does and, therefore, does not interfere with proton spectra.) The solution is placed in a long, thin tube, which is spun in the magnetic field so that all molecules of the compound feel the same magnetic field strength at any given instant.

Problem 12.13 Which of the following atoms do not exhibit nuclear magnetic resonance? ^{12}C, ^{16}O, ^{14}N, ^{15}N, ^{2}H, ^{19}F, ^{31}P, ^{13}C, and ^{32}S.

Atoms with odd numbers of protons and/or neutrons are nmr-active. The inactive atoms are $^{12}C(6p, 6n)$, $^{16}O(8p, 8n)$, and $^{32}S(16p, 16n)$. To detect the nmr activity of atoms other than ^{1}H requires alteration of the nmr spectrometer. The ordinary spectrometer selects the range of radiowave frequency that excites only ^{1}H.

Chemical Shift

Nmr spectroscopy is useful because not all H's change spin at the same applied magnetic field, for the energy absorbed depends on the bonding environment of the H. The magnetic field experienced by an H is not necessarily that which is applied by the magnet, because the electrons in the bond to the H and the electrons in nearby π bonds *induce* their own magnetic fields. This induced field, H^*, partially shields the proton from the applied H_0. The field "felt" by the proton, the **effective field**, is $H_0 - H^*$.

The typical nmr spectrogram, Fig. 12.4, is produced by the application of an external field H_0 that *very slowly decreases* in time. First to appear are the signals of the most **upfield** (most shielded) protons; the greater the shielding, the smaller the effective field at the proton and, hence, the lower the signal's frequency. As we move **downfield**, H_0 remains essentially constant; so the other proton signals come in, in the order of decreasing shielding (increasing effective field, increasing frequency).

The displacement of a signal from the hypothetical position of maximum shielding is called its **chemical shift**, notated as δ (delta) and measured in parts per million (ppm). As indicated on Fig. 12.4, the zero of the δ scale is conventionally located at the signal produced by the H's of tetramethylsilane (TMS), $(CH_3)_4Si$. This compound serves because its H-signal is usually isolated in the extreme upfield region. Clues to the structure of an unknown compound can be obtained by comparing the chemical shifts of its spectrum to the δ values in such tabulations as Table 12.3.

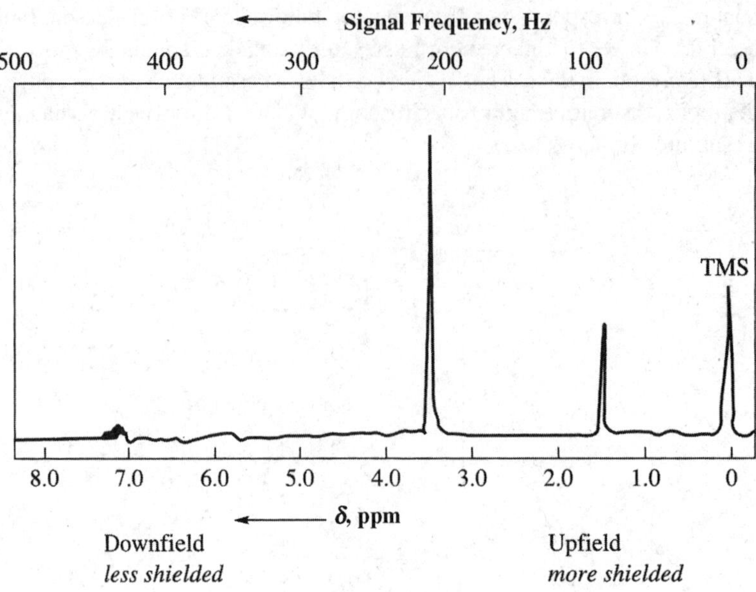

Figure 12.4

Some generalizations about molecular structure and proton chemical shift in ^{1}H nmr (pmr) are as follows:

1. Electronegative atoms, such as N, O, and X, lessen the shielding of H's and cause downfield shifts. The extent of the downfield shift is directly proportional to the electronegativity of the atom and its proximity to the H. The influence of electronegativity is illustrated with the methyl halides, MeX.

δ, ppm	4.3	3.1	2.7	2.2
X	F	Cl	Br	I

2. H's attached to π-bonded C's are less shielded than those in alkanes. The order of δ values is

$$\text{H—}\overset{|}{\text{C}}\text{=O} > \text{Ar—H} > \text{H—}\overset{|}{\text{C}}\text{=}\overset{|}{\text{C}}\text{—} > \text{H—C}\equiv\text{C—} > \text{H—}\underset{|}{\overset{|}{\text{C}}}\text{—}$$

3. Ar, C═O, C═C, and C≡C are electron-withdrawing by induction and cause a downfield shift of an H on an adjacent C, as in

$$\underset{\alpha\text{-carbonyl}}{\text{O=C—}\underset{|}{\overset{|}{\text{C}}}\text{—H}} \quad \underset{\text{benzylic-type}}{\text{Ar—}\underset{|}{\overset{|}{\text{C}}}\text{—H}} \quad \underset{\text{allylic}}{\text{H—}\underset{|}{\overset{|}{\text{C}}}\text{—C=C}} \quad \underset{\text{propargyllic}}{\text{H—}\underset{|}{\overset{|}{\text{C}}}\text{—C}\equiv\text{C}}$$

4. For alkyl groups in similar environments, shielding increases with the number of H's on the C. The order for δ values is

$$\underset{\text{Methine}}{\text{—}\underset{|}{\overset{|}{\text{C}}}\text{—H}} > \underset{\text{Methylene}}{\text{—}\underset{|}{\overset{\text{H}}{\text{C}}}\text{—H}} > \underset{\text{Methyl}}{\text{—}\underset{\text{H}}{\overset{\text{H}}{\text{C}}}\text{—H}}$$

5. Electropositive atoms, such as Si, shield H's; hence the reference property of TMS.
6. H's attached to a cyclopropane ring and those situated *in* the π cloud of an aromatic system are strongly shielded. Some may have negative δ values.
7. H's which participate in H-bonding—for instance, OH and NH—exhibit variable δ values over a wide range, depending mainly on sample concentrations. H-bonding diminishes shielding. Hence, in a non-H-bonding solvent, the OH-signal for ROH (or NH-signal for RNH_2) moves downfield as the sample

TABLE 12.3 Proton Chemical Shifts

δ, ppm	CHARACTER OF UNDERLINED PROTON	
0.2	Cyclopropane:	(cyclopropyl)—H
0.9	Primary:	R—CH_3
1.3	Secondary:	R_2CH_2
1.5	Tertiary:	R_3—CH
1.7	Allylic:	—C=C—CH_3
2.0–4.0	Iodide: α H	I—C—H
2.2–2.0	Ester (II): α H to C=O	H—C—C(=O)OR
2.6–2.0	Carboxylic acid: α H	H—C—C(=O)OH
2.7–2.0	Carbonyl: α H	—C(=O)—C—H
3–2	Acetylenic:	—C≡C—H
3–2.2	Benzylic:	(phenyl)—C—H
3.3–4.0	Ether: α H	R—O—C—H
4–2.5	Bromide:	Br—C—H

δ, ppm	CHARACTER OF UNDERLINED PROTON	
4–3	Chloride:	Cl—C—H
4–3.4	Alcohol:	HO—C—H
4–4.5	Fluoride:	F—C—H
4.1–3.7	Ester (I): α H to alkyl O	R—C(=O)—O—C—H
5.0–1.0	Amine:	R—NH_2
5.5–1.0	Hydroxyl:	RO—H
5.9–4.6	Olefinic:	—C=C—H
8.5–6.0	Aromatic:	(phenyl)—H (Ar—H)
10.0–9.0	Aldehyde:	—C(=O)—H
12.0–10.5	Carboxyl:	R—C(=O)—O—H
12.0–4.0	Phenolic:	(phenyl)—C—H
15.0–17.0	Enolic:	—C=C—O—H

is concentrated, because H-bonding is enhanced. H-bonding is accompanied by exchange of H's between the ROH molecules, resulting in broadening of the signals.

Problem 12.14 Use Table 12.3 to assign approximate δ values for the chemical shift of the one type of H in (*a*) $(CH_3)_2C{=}C(CH_3)_2$, (*b*) $(CH_3)_2C{=}O$, (*c*) benzene, (*d*) $O{=}CH{-}CH{=}O$.

(*a*) 1.7 ppm, (*b*) 2.3 ppm, (*c*) 7.2 ppm, (*d*) 9.5 ppm.

Problem 12.15 Give the numbers of kinds of H's present in (*a*) CH_3CH_3, (*b*) $CH_3CH_2CH_3$, (*c*) $(CH_3)_2CHCH_2CH_3$, (*d*) $H_2C{=}CH_2$, (*e*) $CH_3CH{=}CH_2$, (*f*) $C_6H_5NO_2$, (*g*) $C_6H_5CH_3$.

(*a*) *One* (all equivalent).
(*b*) *Two*: $CH_3^aCH_2^bCH_3^a$.
(*c*) *Four*: $(CH_3^a)_2CH^bCH_2^cCH_3^d$.
(*d*) *One* (all equivalent).
(*e*) *Four*:

$$\begin{array}{ccc} CH_3^a & & H^c \\ & \diagdown\ \ \diagup & \\ & C{=}C & \\ & \diagup\ \ \diagdown & \\ H^b & & H^d \end{array}$$

The $=CH_2$ H's are not equivalent, since one is *cis* to the CH_3 and the other is *trans*. Replacement of H^c by X gives the *cis*-diastereomer. Replacement of H^d gives the *trans*-diastereomer.
(*f*) *Three*: Two *ortho*, two *meta*, and one *para*.
(*g*) Theoretically there are three kinds of aromatic H's, as in (*f*). Actually the ring H's are little affected by alkyl groups and are equivalent. There are two kinds: $C_6H_5^aCH_3^b$.

Problem 12.16 How many kinds of equivalent H's are there in the following?

(*a*) $CH_3CHClCH_2CH_3$ (*b*) *p*-$CH_3CH_2{-}C_6H_4{-}CH_2CH_3$ (*c*) $Br_2CHCH_2CH_2CH_2Br$

(*a*) *Five* as shown:

$$\begin{array}{ccccc} & H^b & & H^c & \\ & | & & | & \\ CH_3^a{-} & C^* & {-} & C & {-}CH_3^e \\ & | & & | & \\ & Cl & & H^d & \end{array}$$

The two H's of CH_2 are *not* equivalent, because of the presence of a chiral C. Replacing H^c and H^d separately by X gives two diastereomers; H^c and H^d are diasteriomeric H's.
(*b*) *Three*: All four aromatic H's are equivalent, as are the six in the two CH_3's and the four in the CH_2's.
(*c*) *Four*: $Br_2CH^aCH_2^bCH_2^cCH_2^dBr$.

Problem 12.17 How many kinds of H's are there in the isomers of dimethylcyclopropane?

Dimethylcyclopropane has three isomers, shown with labeled H's to indicate differences and equivalencies.

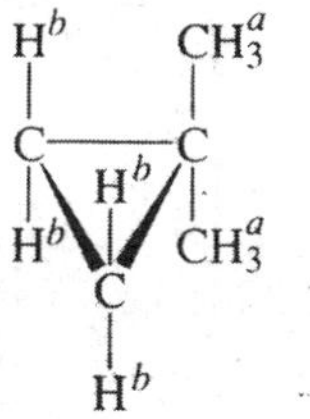

1,1-Dimethylcyclopropane (I): ring C bearing H^b and H^b; ring C bearing CH_3^a and CH_3^a; ring C bearing H^b and H^b.

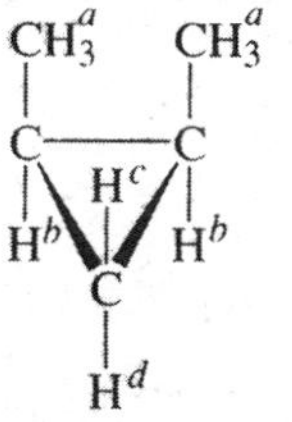

cis-1,2-Dimethylcyclopropane (II): ring C bearing CH_3^a and H^b; ring C bearing CH_3^a and H^b; ring CH_2 bearing H^c and H^d.

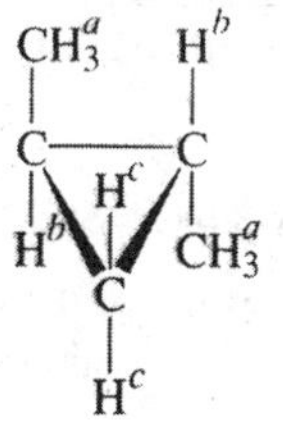

trans-1,2-Dimethylcyclopropane (III): ring C bearing CH_3^a and H^b; ring C bearing H^b and CH_3^a; ring CH_2 bearing H^c and H^c.

In II, H^c and H^d are different, since H^c is *cis* to the CH_3's and H^d is *trans*. In III, the CH_2 H's are equivalent; they are each *cis* to a CH_3 and *trans* to a CH_3.

Relative Peak Areas; H-Counting

The area under a signal graph is directly proportional to the number of equivalent H's giving the signal. For example, the compound $C_6H_5^aCH_2^bC(CH_3^c)_3$ has five aromatic protons (*a*), two benzylic protons (*b*), and nine equivalent CH_3 protons (*c*). Its nmr spectrum shows three peaks for the three different kinds of H, which appear at: (*a*) $\delta = 7.1$ ppm (aromatic H), (*b*) $\delta = 2.2$ ppm (benzylic H), (*c*) $\delta = 0.9$ ppm (1° H). The relative areas under the peaks are $a:b:c = 5:2:9$. The nmr instrument integrates the areas as follows: When no signal is present, it draws a horizontal line. When the signal is reached, the line ascends and levels off when the signal ends. The relative distance from plateau to plateau gives the relative area. See Fig. 12.10 for typical nmr spectra showing integration.

Problem 12.18 (*a*) Suggest a structure for a compound C_9H_{12} showing low-resolution nmr signals at δ values of 7.1, 2.2, 1.5, and 0.9 ppm. (*b*) Give the relative signal areas for the compound.

(*a*) The value 7.1 ppm indicates H's on a benzene ring. The formula shows three more C's, which might be attached to the ring, as shown below (assuming that, since this is an alkylbenzene, all aromatic H's are equivalent):

(1) 3 CH_3's in trimethylbenzene, $(CH_3^a)_3C_6H_3^b$ (2) a CH_3 and a CH_2CH_3 in $CH_3^aC_6H_4^bCH_2^cCH_3^d$
(3) a $CH_2CH_2CH_3$ in $C_6H_5^aCH_2^bCH_2^cCH_3^d$ (4) a $CH(CH_3)_2$ in $C_6H_5^aCH^b(CH_3^c)_2$

Compounds (1) and (4) can be eliminated because they would give two and three signals, respectively, rather than the four observed signals. Although (2) has four signals, H^a and H^c are different benzylic H's and the compound should have two signals in the region 3.0–2.2 ppm rather than the single observed signal. Hence, (2) can be eliminated. Only (3) can give the four observed signals with the proper chemical shifts.
(*b*) 5: 2: 2: 3.

Problem 12.19 What compound C_7H_8O has nmr signals at $\delta = 7.3$, 4.4, and 3.7 ppm, with relative areas 7: 2.9: 1.4, respectively?

The relative areas of the three different kinds of H become 5:2:1 on dividing by 1.4. That is, five H's contribute to the $\delta = 7.3$, 2 H's to $\delta = 4.4$ and 1 H to $\delta = 3.7$, for a total of eight H's, which is consistent with the formula. The five H's at $\delta = 7.2$ are aromatic, indicating a C_6H_5 compound. The remaining portion of the formula accounts for the CH_2OH group. The H at $\delta = 3.7$ is part of OH. The two H's at $\delta = 4.4$ are benzylic and alpha to OH. The compound is $C_6H_5CH_2OH$, benzyl alcohol.

Problem 12.20 The pmr (proton magnetic resonance) spectrum of $CH_3OCH_2CH_2OCH_3$ shows chemical shifts of 3.4 and 3.2 ppm, with corresponding peak areas in the ratio 2 : 3. Are these numbers consistent with the given structure?

Yes. Since the CH_3's and CH_2's are each equivalent, only two signals appear. Both are shifted downfield by the O, the CH_2-signal more than the CH_3-signal. Integration provides the correct ratio, $2:3 = 4:6$, the actual numbers of H's engendering the signals.

Peak-Splitting; Spin-Spin Coupling

Because of **spin-spin coupling**, most nmr spectra do not show simple single peaks but rather *groups* of *peaks* that tend to cluster about certain δ values. To see how this coupling arises, we examine the molecular fragment

$$-\overset{|}{C}H^{a}-\overset{|}{C}H_2^{b}$$

present in a very large number of like molecules. The signal for H^b is shifted slightly upfield or downfield depending on whether the spin of H^a is aligned against or with the applied field. Since in about half of the molecules the H^a are spinning $\uparrow$ and in half $\downarrow$, H^b gives rise to a doublet instead of a singlet. The effect is reciprocal: The two H^b's split the signal of H^a. There are four spin states of approximately equal probability for the two H^b's:

$$\uparrow\uparrow;\quad \underbrace{\uparrow\downarrow,\quad \downarrow\uparrow}; \quad \downarrow\downarrow$$

Because the middle two spin states have the same effect, the signal of H^a is split into a triplet with relative intensities 1:2:1.

In the molecular fragment

$$-\overset{|}{C}H^{a}-CH_3^{b}$$

the three H^b's produce a *doublet*, due to the effect of the single H^a. H^a, however, yields a *quartet*, due to the effects of the three H^b's, which may be spinning as follows:

$$\text{Intensities}\quad \underset{1}{\uparrow\uparrow\uparrow};\quad \underbrace{\uparrow\uparrow\downarrow,\ \uparrow\downarrow\uparrow,\ \downarrow\uparrow\uparrow}_{3};\quad \underbrace{\uparrow\downarrow\downarrow,\ \downarrow\downarrow\uparrow,\ \downarrow\uparrow\downarrow}_{3};\quad \underset{1}{\downarrow\downarrow\downarrow}$$

The entire quartet integrates for one H.

Spin-spin coupling usually, but not always, occurs between *nonequivalent* H's on adjacent atoms. In general, if n equivalent H's are affecting the peak of H's on an adjacent C, the peak is split into $n + 1$ peaks. A symmetrical multiplet is an ideal condition and not always observed in practice.

Problem 12.21 In which of the following molecules does spin-spin coupling occur? If splitting is observed, give the multiplicity of each kind of H.

(*a*) $ClCH_2CH_2Cl$ (*b*) $ClCH_2CH_2I$ (*c*) $CH_3-C(CH_3)_2-CH_2Br$

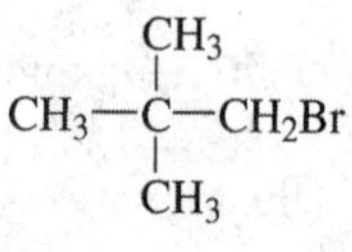

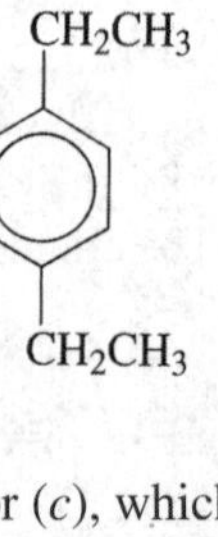

(*d*) H(Br)C=C(H)(Br) [H and Br on each C; H's cis] (*e*) H(Br)C=C(Cl)(H) (*f*) I(Cl)C=C(H)(H) (*g*) 1,4-$C_6H_4(CH_2CH_3)_2$

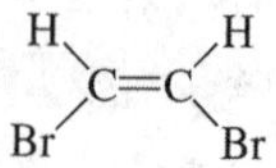

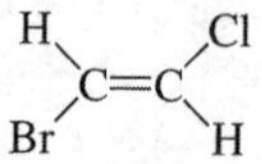

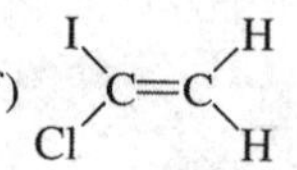

Splitting is not observed for (*a*) or (*d*), which each have only equivalent H's, or for (*c*), which has no nonequivalent H's on *adjacent* C's. The H's of CH_2 in (*b*) are non-equivalent, and each signal is split into a triplet

($n = 2; 2 + 1 = 3$). In (*e*), the two H's are not equivalent, and each generates a doublet. The vinyl H's in (*f*) are nonequivalent, since one is *cis* to Cl and the other is *cis* to I; each gives rise to a doublet. In this case, the interacting H's are on the same C. Compound (*g*) gives a singlet for the equivalent uncoupled aromatic H's, quartet for the H's of the two equivalent CH_2 group coupled with CH_3, and a triplet for the two equivalent CH_3 groups coupled with CH_2.

Problem 12.22 Why is splitting observed in 2-methylpropene but not in 1-chloro-2,2-dimethylpropane?

See Fig. 12.5. In (*d*), H^a is more downfield than H^b because Cl is more electron-withdrawing than Br. In $(CH_3^a)_3C{-}CH_2^bCl$, H^a and H^b are not on adjacent C's and are too far away from one another to couple. In

$$(CH_3^a)_2C{=}C(H^b)_2$$

although H^a and H^b are not on adjacent C's, they are close enough to couple because of the shorter C=C bond.

Problem 12.23 Sketch the nmr spectra of (*a*) 1,1-dichloroethane, (*b*) 1,1,2-trichloroethane, (*c*) 1,1,2,2-tetrachloroethane, and (*d*) 1-bromo-2-chloroethane. In each case, indicate the "staircase" curve of relative areas.

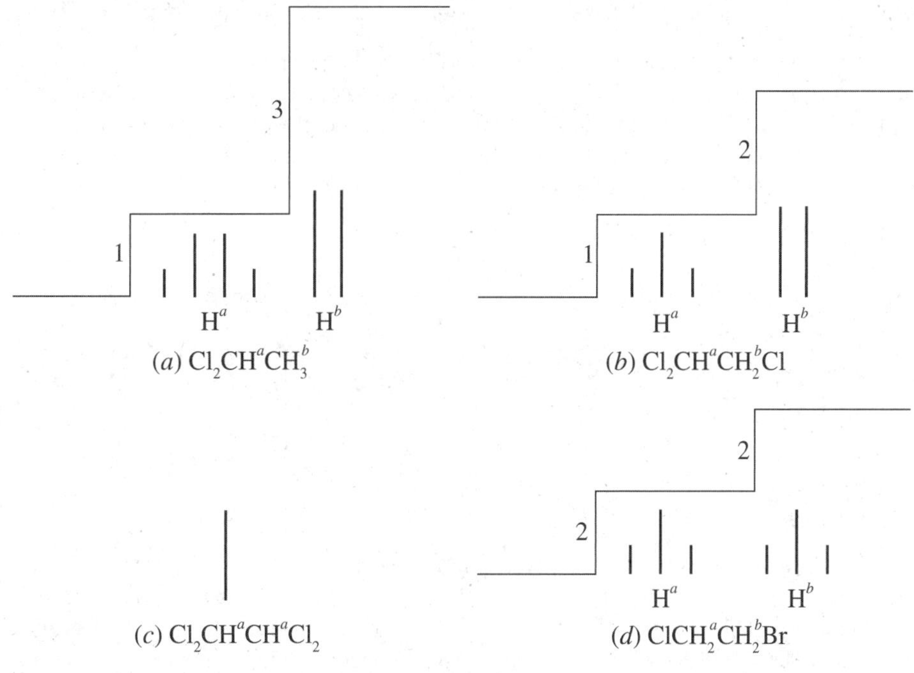

Figure 12.5

Problem 12.24 F's couple H's in the same way as do other H's. Predict the splitting in the nmr spectrum of 2,2-difluoropropane.

In $CH_3^aCF_2CH_3^a$, the two F's split the H^a-signal into a 1 : 2 : 1 triplet. The F-signal, when detected by a special probe, would be a septet.

Problem 12.25 Deuterium does not give a signal in the proton nmr spectrum, nor does it split signals of nearby protons. Thus D's might just as well not be there. What is the difference between nmr spectra of CH_3CH_2Cl and CH_3CHDCl?

The spectrum of CH_3CH_2Cl is given by: (2 H^a : 3 H^b; $X{-}CH_2^aCH_3^b$). $CH_3^aCH^bDCl$ has a *doublet* for H^a and, more downfield, a *quartet* for H^b.

Problem 12.26 The stable *anti* conformer of CH_3CH_2Cl shows a nonequivalency of the CH_3 H's:

H* is *anti* to the Cl, while the H_Δ's are *gauche*. Why does H* *not* give a signal different from the H_Δ's? (Instead, the three H's produce an equivalent triplet.)

Rotation around the C—C bond is rapid. Detection by the nmr spectrometer is slower. The spectrometer therefore detects the average condition, which is the same for each H; 1/3 *anti* and 2/3 *gauche*.

Problem 12.27 What information can you deduce from the fact that one signal in the nmr spectrum of 2,2,6,6-tetradeuterobromocyclohexane changes to two smaller signals when the spectrum is taken at low temperatures?

As the ring changes its conformation from one chair form to another, the Br—C—H proton changes its position from axial to equatorial (Fig. 12.6). An axial H and an equatorial H have different chemical shifts. But at room temperature, the ring "flips" too fast for the instrument to detect the difference; it senses the average condition. At low temperatures, this process becomes slow enough so that the instrument can pick up the two different H_{ax} and H_{eq} signals. D's are used to ensure that the H under study is a singlet.

Figure 12.6

Coupling Constants

Figure 12.7 summarizes the nmr spectrum of CH_3CH_2Cl by using a vertical line segment for each peak.

The spacing between lines within a multiplet is typically constant; furthermore, the *spacing in each coupled multiplet* is constant. This constant distance, which is independent of H_0, is called the **coupling constant**, J, and is expressed in Hz. The value of the coupling constant depends on the structural relationship of the coupled H's, and becomes a valuable tool for structure proof. Some typical values are given in Table 12.4.

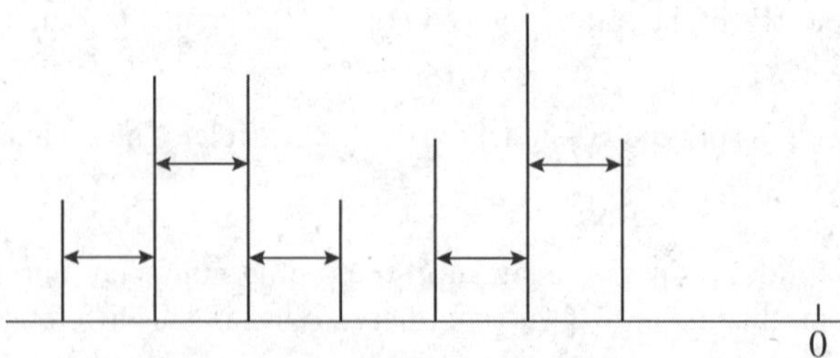

Figure 12.7 nmr Spectrum (CH_3Ch_2Cl)

TABLE 12.4

TYPE OF H's		J, Hz
H—C—C—H (free rotation of C—C)		~ 7
H(C=C)H (H on opposite carbons, opposite sides)	*trans*	13–18
H(C=C)H (H on opposite carbons, same side)	*cis*	7–12
=C(H)H		0–3
Phenyl H's *ortho* *meta* *para*		 6–9 1–3 0–1

Problem 12.28 A compound, C_2H_2BrCl, has two doublets, $J = 16$ Hz. Use Table 12.4 to suggest a structure.

The three possibilities showing J values for two doublets are

Br(Cl)C=CH(H): *gem*-vinyl H's, $J = 0$–3 Hz
H(Br)C=CH(Cl): *cis* H's, $J = 7$–12 Hz
Br(H)C=CH(Cl): *trans* H's, $J = 13$–18 Hz

The *trans* isomer fits the data.

12.5 ^{13}C NMR (CMR)

Although ^{12}C is not nmr-active, the ^{13}C isotope, which has a natural abundance of about 1%, is nmr-active because it has an odd number of neutrons. Modern techniques of **Fourier transform spectroscopy** are capable of detecting signals of this isotope in concentrated solution. Since H and ^{13}C absorb at different frequencies, proton signals do not appear in the ^{13}C spectra. However, spin-spin coupling between attached H's and a ^{13}C (**proton-coupling**) is observed. The same $(n + 1)$-rule that is used for H's on adjacent C's is used to analyze the ^{13}C-H coupling pattern. Thus, the ^{13}C of a $^{13}CH_3$ group gives rise to a quartet. To avoid this coupling, a **proton-decoupled** spectrum can be taken, in which each different ^{13}C nucleus appears as a sharp singlet. Both types of spectra are useful. The decoupled spectrum permits the counting of different ^{13}C's in the molecule; the coupled spectrum allows the determination of the number of H's attached to each ^{13}C. Figure 12.8 shows (*a*) the proton-coupled and (*b*) the proton-decoupled spectrum of dichloroacetic acid, $CHCl_2COOH$. Each ^{13}C has a characteristic chemical shift δ, as listed in Table 12.5. Note that the ^{13}C of $^{13}C{=}O$ is the most downfield, with aromatic ^{13}C's somewhat more upfield. Integration of the peak areas affords the relative number of ^{13}C's each peak represents, as shown in Fig. 12.9 for p-$BrC_6H_4COCH_3$.

Problem 12.29 Why is spin-spin coupling between adjacent ^{13}C's not observed?

Since the natural abundance of this isotope is so low, the chance of finding two ^{13}C's next to each other is practically nil. However, if a compound were synthesized with only ^{13}C's, then coupling would be observed.

Problem 12.30 How many peaks would be evidenced in the decoupled spectrum of (*a*) methylcyclohexane? (*b*) cyclohexene? (*c*) 1-methylcyclohexene?

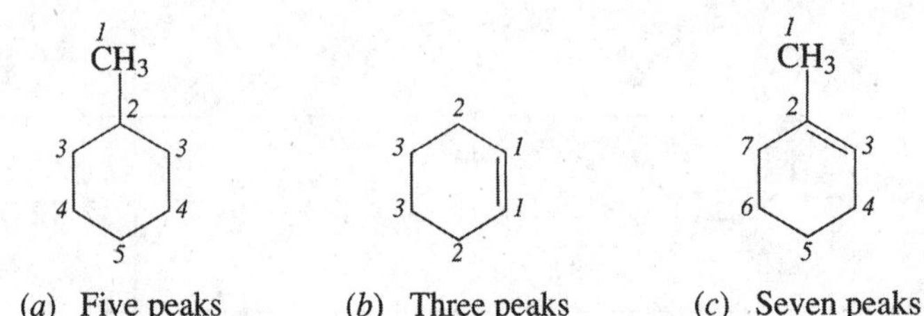

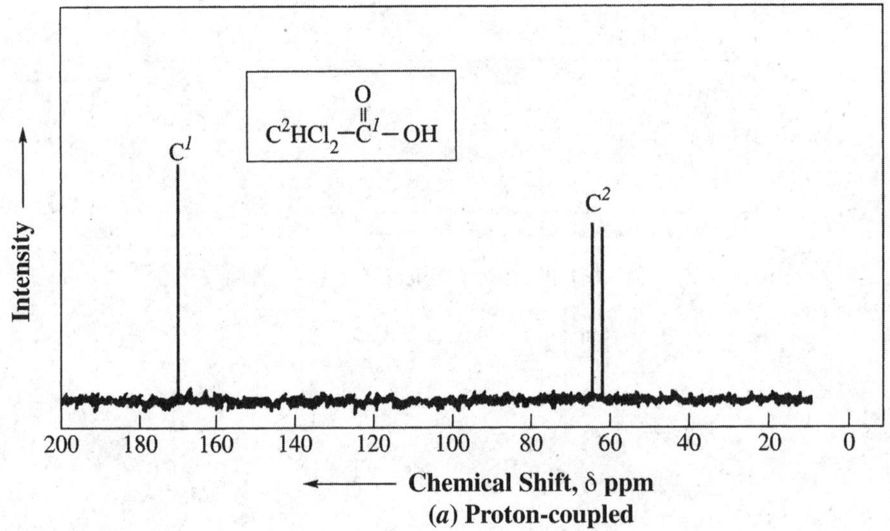

(*a*) Proton-coupled

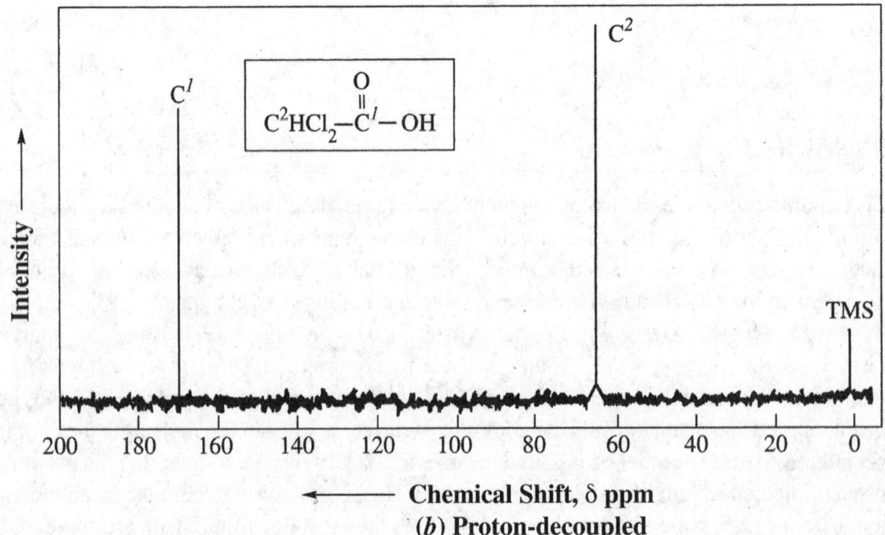

(*b*) Proton-decoupled

Figure 12.8

TABLE 12.5

TYPE OF ^{13}C	δ, ppm	TYPE OF ^{13}C	δ, ppm
C—I	0–40	$-$C	100–150
C—Br	25–65	C$-$O	40–80
C—Cl	35–80	C=O	170–210
CH_3	8–30		110–160
CH_2	15–55	C$-$N	30–65
CH	20–60		
≡C	65–85		

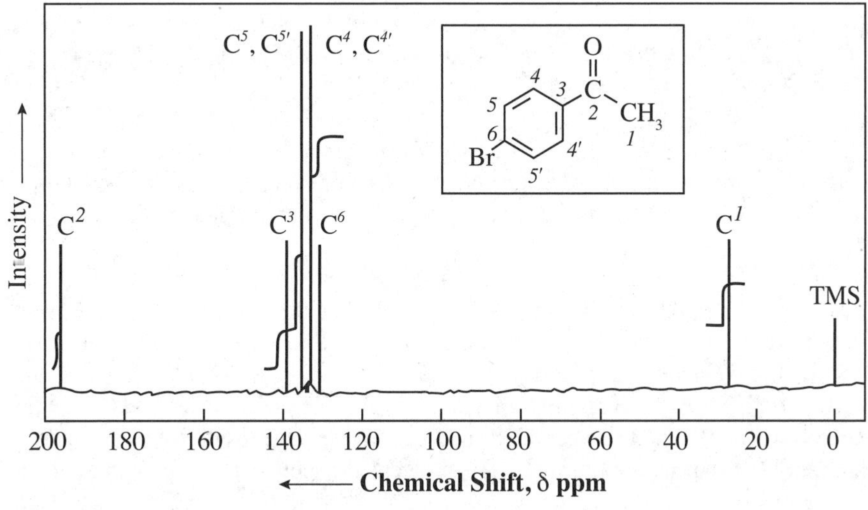

Figure 12.9

12.6 Mass Spectroscopy

When exposed to sufficient energy, a molecule may lose an electron to form a cation-radical, which then may undergo fragmentation of bonds. These processes make **mass spectroscopy** (**ms**) a useful tool for structure proof. Very small concentrations of the parent molecules (RS), in the vapor state, are ionized by a beam of energetic electrons (e^-),

$$\underset{\text{Parent molecule}}{R{:}S + e^-} \longrightarrow \underset{\text{Parent cation-radical}}{2e^- + R{\cdot}S^+} \longrightarrow \underset{\text{Fragmention ions}}{R\cdot + S^+ \quad \text{or} \quad R^+ + S\cdot}$$

and a number of parent cation-radicals (RS^+) may then fragment to give other cations and neutral species. Fragment ions can undergo further bond-cleavage to give even smaller cations and neutral species. In a mass spectrogram, sharp peaks appear at the valules of m/e (the mass-to-charge ratio) for the various cations. The relative heights (intensities) of the peaks represent relative abundances of the cations. Most cations have a charge of $+1$, and therefore most peaks record the masses of the cations. Fragmentation tends to give the more stable cations. The most abundant cations are the most stable ones.

The standard against which all peak intensities in a given mass spectrogram are measured is the most intense peak, called the **base peak**, which is arbitrarily assigned the value 100. If few parent molecules fragment—not a typical situation—the parent cation will furnish the base peak.

Unless all the parent ions fragment, which rarely happens, the largest observed *m/e* value is the molecular weight of the parent (RS) molecule. This generalization overlooks the presence of naturally occurring isotopes in the parent. Thus, the chances of finding a ^{13}C atom in an organic molecule are 1.11%; the chances of finding two are negligible. Therefore, the instrument detects a small peak at $m_{RS} + 1$, owing to the ^{13}C-containing parent. The chances of finding ^2H in a molecule are negligible.

The masses and possible structures of fragment cations, especially the more stable ones, are clues to the structure of the original molecule. However, rearrangement of cations complicates the interpretation.

Mass spectra, like other spectra, are unique properties used to identify known and unknown compounds.

Problem 12.31 (*a*) What molecular formulas containing only C and H can be assigned to a cation with *m/e* equal to (i) 43, (ii) 65, (iii) 91? (Assume that $e = +1$.) (*b*) What combination of C, H, and N can account for an *m/e* of (i) 43, (ii) 57? (Assume that $e = +1$.)

(*a*) Divide by 12 to get the number of C's; the remainder of the weight is due to H's. (i) $C_3H_7^+$, (ii) $C_5H_5^+$, (iii) $C_7H_7^+$.

(*b*) (i) If one N is present, subtracting 14 leaves a mass of 29, which means 2 C's (mass of 24) are present. Therefore, the formula is $C_2H_5N^+$. If 2 N's are present, it is $CH_3N_2^+$. (ii) $CH_3N_3^+$, $C_2H_5N_2^+$ or $C_3H_7N^+$.

Problem 12.32 (*a*) Do parent (molecular) ions, RS+, of hydrocarbons ever have odd *m/e* values? (*b*) If an RS^+ contains only C, H, and O, may its *m/e* value be either odd or even? (*c*) If an RS^+ contains only C, H, and N, may its *m/e* value be either odd or even? (*d*) Why can an ion, $m/e = 31$, not be $C_2H_7^+$? What might it be?

(*a*) No. Hydrocarbons, and their parent ions, must have an even number of H's: C_nH_{2n+2}, C_nH_{2n}, C_nH_{2n-2}, C_nH_{2n-6}, and so on. Since the atomic weight of C is even (12), the *m/e* values must be even.

(*b*) The presence of O in a formula does not change the ratio of C to H. Since the mass of O is even (16), the mass of RS^+ with C, H, and O *must be even.*

(*c*) The presence of each N ($m = 14$) requires an additional H ($C_nH_{2n+3}N$, $C_nH_{2n+1}N$, $C_nH_{2n-1}N$). Therefore, if the number of N's is odd, an odd number of H's and an odd *m/e* value result. An even number of N's requires an even number of H's and an even *m/e* value. These statements apply *only* to parent ions, *not* to fragment ions.

(*d*) The largest number of H's for two C's is six (C_2H_6). Some possibilities are CH_3O^+ and CH_5N^+.

Problem 12.33 Write equations involving the electron-dot formulas for each fragmentation used to explain the following: (*a*) Isobutane, a typical branched-chain alkane, has a lower-intensity RS^+ peak than does *n*-butane, a typical unbranched alkane. (*b*) All 1° alcohols, RCH_2CH_2OH, have a prominent fragment cation at $m/e = 31$. (*c*) All $C_6H_5CH_2R$-type hydrocarbons have a prominent fragment cation at $m/e = 91$. (*d*) Alkenes of the type $H_2C{=}CHCH_2R$ have a prominent fragment cation at $m/e = 41$. (*e*) Aldehydes,

$$\mathrm{R{-}\overset{\overset{\large H}{|}}{C}{=}O}$$

show intense peaks at $(m/e)_{RS} - 1$ and $m/e = 29$.

(*a*) Cleavage of C—C is more likely than cleavage of (the stronger) C—H. Fragmentation of RS^+ for isobutane,

$$\mathrm{(H_3C)_2\overset{\overset{\large H}{|}}{C}{\dot{+}}CH_3 \longrightarrow (H_3C)_2\overset{\overset{\large H}{|}}{C}^{+} + \cdot CH_3}$$

gives a 2° R^+, which is more stable than the 1° R^+ from n-butane:

$$H_3C\overset{H}{\underset{H}{C}}\!:\!\overset{H}{\underset{H}{C}}CH_3 \longrightarrow H_3C\overset{H}{\underset{H}{C}}^{+} + \cdot CH_2CH_3$$

Hence, RS^+ of isobutane undergoes fragmentation more readily than does RS^+ of n-butane, and fewer RS^+ fragments of isobutane survive. Consequently, isobutane, typical of branched-chain alkanes, has a low-intensity RS^+ peak compared with n-butane.

$$(b)\quad R-\overset{H}{\underset{H}{\overset{\beta}{C}}}\!:\!\overset{\alpha}{C}H_2\ddot{O}H^{+\cdot} \longrightarrow R-\overset{H}{\underset{H}{\overset{\beta}{C}}}\cdot + \overset{+}{\underset{\alpha}{C}}H_2\ddot{O}H \longleftrightarrow H_2C=\overset{+}{\ddot{O}}H \qquad m/e = 31$$

A—C^+— next to an O is stabilized by extended π bonding (resonance). The RS^+ species of alcohols generally undergoes cleavage of the bond

$$\overset{\beta}{C} \,\wr\, \overset{\alpha}{C}-OH$$

$$(c)\quad C_6H_5:\overset{H}{\underset{H}{C}}\overset{+}{\cdot}R \longrightarrow R\cdot + C_6H_5:CH_2^{+} \longrightarrow \text{(tropylium ion)}$$

stable benzyl R^+ — a more stable aromatic cycloheptatrienyl cation, $m/e = 91$

$$(d)\quad H_2\overset{\cdot+}{C}-CH-\overset{H}{\underset{H}{C}}\!:\!R \longrightarrow H_2C=CH-\overset{+}{C}H_2 + \cdot R$$

stable allylic cation, $m/e = 41$

$$(e)\quad R\!:\!\overset{H\ (1)}{C}=\overset{+\cdot}{\ddot{O}}: \;\xrightarrow{(1)\ -H\cdot}\; R-\overset{+}{C}=\ddot{O}: \longleftrightarrow R-C\equiv\overset{+}{O} \qquad (m/e = (m/e)_{RS} - 1)$$

$$\xrightarrow{(2)\ -R\cdot}\; \underset{H}{\overset{+}{C}}=\ddot{O}: \longleftrightarrow H-C\equiv\overset{+}{O} \qquad (m/e = 29)$$

stable acylonium ions

Problem 12.34 Why is less than 1 mg of parent compound used for mass spectral analysis?

A relatively small number of molecules are taken to prevent collision and reaction between fragments. Combination of fragments might lead to ions with larger masses than RS^+, making it impossible to determine the molecular weight. The fragmentation pattern would also become confusing.

Problem 12.35 Give the structure of a compound, $C_{10}H_{12}O$, whose mass spectrum shows m/e values of 15, 43, 57, 91, 105, and 148.

The value 15 suggests a $^+CH_3$. Because 43 − 15 = 28, the mass of a C=O group, the valule of 43 could mean an acetyl, CH_3CO, group in the compound. The highest value, 148, gives the molecular weight. Cleaving an acetyl group ($m/e = 43$) from 148 gives 105, which is an observed peak. Next, below 105 is 91, a difference of 14; this suggests a CH_2 attached to CH_3CO. So far we have CH_3COH_2 adding up to 57, leaving $148 - 57 = 91$ to be accounted for. This peak is likely to be $C_7H_7^+$, whose precursor is the stable benzyl cation, $C_6H_5\overset{+}{C}H_2$. The structure is $CH_3—\underset{\underset{O}{\|}}{C}—CH_2—CH_2—C_6H_5$.

Problem 12.36 How could mass spectroscopy distinguish among the three deuterated forms of ethyl methyl ketone?

(1) $DCH_2CH_2COCH_3$ (2) $CH_3CH_2COCH_2D$ (3) $CH_3CHDCOCH_3$

The expected peaks for each compound are shown in Table 12.6; each has a different combination of peaks.

TABLE 12.6

m/e	$DCH_2CH_2COCH_3$	$CH_3CH_2COCH_2D$	$CH_3CHDCOCH_3$
15	CH_3^+	CH_3^+	CH_3^+
16	DCH_2^+	DCH_2^+	—
29	—	$CH_3CH_2^+$	—
30	$DCH_2CH_2^+$	—	CH_3CHD^+
43	CH_3CO^+	—	CH_3CO^+
44	—	DCH_2CO^+	—

SUPPLEMENTARY PROBLEMS

Problem 12.37 Match the type of spectrometer with the kind of information which it can provide the chemist.

1. Mass	A. Functional groups
2. Infrared	B. Molecular weights
3. Ultraviolet	C. Proton environment
4. Nuclear magnetic resonance	D. Conjugation

1. B, 2. A, 3. D, 4. C.

Problem 12.38 Why do colored organic compounds, such as β-carotene (the orange pigment isolated from carrots), have extended conjugation?

The more effective electron delocalization in molecules with extended conjugation narrows the energy gap between the HOMO and the LUMO. Thus, visible radiation of lower frequency (longer wavelength) is absorbed in the HOMO → LUMO electron transition.

Problem 12.39 (*a*) Account for the fact that benzene absorbs at 254 nm in the uv, and phenol, C_6H_5OH, absorbs at 280 nm. (*b*) Where would one expect 1,3,5-hexatriene to absorb, relative to benzene?

(*a*) The *p* orbital on O, housing a pair of *n* electrons, overlaps with the cyclic π system of benzene, thereby extending the electron delocalization and decreasing the energy for the HOMO → LUMO transition. Consequently, phenol absorbs at longer wavelengths.

(*b*) The triene absorbs at a longer λ_{max} (275 nm). Since the cyclic π system of benzene has a lower energy than the linear π system of the triene, benzene absorbs radiation of shorter wavelength.

Problem 12.40 Which peaks in the ir spectra distinguish cyclohexane from cyclohexene?

One of the C—H stretches in cyclohexene, is above 3000 cm^{-1} (C_{sp^2}—H); in cyclohexane the C—H stretches are below 3000 cm^{-1} (C_{sp^3}—H). Cyclohexene has a C═C stretch at about 1650 cm^{-1}.

Problem 12.41 The mass spectrum of a compound containing C, H, O, and N gives a maximum *m/e* of 121. Its ir spectrum shows peaks at 700, 750, 1520, 1685, and 3100 cm^{-1}, and a twin peak at 3440 cm^{-1}. What is a reasonable structure for the compound?

The molecular weight is 121. Since the mass is odd, there must be an odd number of N's [Problem 12.32(*c*)]. The ir data indicate the following groups to be present:

1520 cm^{-1}:	Aromatic ring (1450–1600 cm^{-1} range)
1685 cm^{-1}:	C═O stretch of amide structure —CO—N(1630–1690 cm^{-1} range)
3100 cm^{-1}:	Aromatic C—H bond (3000–3100 cm^{-1} range)
3440 cm^{-1}:	—N—H in amine or amide (3300–3500 cm^{-1} range)
700, 750 cm^{-1}:	Monosubstituted phenyl

A twin peak due to symmetric and antisymmetric N—H stretches means an NH_2 group. By putting the pieces together, we find that the compound is benzamide:

$$C_6H_5—\overset{\overset{\large O}{\|}}{C}—NH_2$$

Problem 12.42 The ir spectrum of methyl salicylate, *o*-$HOC_6H_4COOCH_3$, has peaks at 3300, 1700, 3050, 1540, 1590, and 2990 cm^{-1}. Correlate these peaks with the following structures: (*a*) CH_3, (*b*) C═O, (*c*) OH group on the ring, and (*d*) aromatic ring.

(*a*) 2990 cm^{-1}; (*b*) 1700 cm^{-1}; (*c*) 3300 cm^{-1}; (*d*) 3050, 1540, 1590 cm^{-1}.

Problem 12.43 Calculate ε_{max} for a compound whose maximum absorbance is A_{max} = 1.2. The cell length (l) is 1.0 cm, and the concentration is 0.076 g/L. The mass spectrum of the compound has the largest *m/e* value at 100.

The molecular weight is 100 g/mol; therefore, $c = 7.6 \times 10^{-4}$ mol/L

and
$$\varepsilon_{max} = \frac{A_{max}}{Cl} = \frac{1.2}{(7.6 \times 10^{-4})(1.0)} = 1600$$

Problem 12.44 Methanol is a good solvent for uv but not for ir determinations. Why?

Methanol absorbs in the uv in 183 nm, which is below 190 nm, the cutoff for most spectrophotometers, and therefore it doesn't interfere. Its ir spectrum has bands in most regions, and therefore it cannot be used. Solvents such as CCl_4 and CS_2 have few interfering bands and are preferred for ir determinations.

Problem 12.45 A compound, C_3H_6O, contains a C═O group. How could nmr establish whether this compound is an aldehyde or a ketone?

If an aldehyde, the compound is CH_3CH_2CHO, with three multiplet peaks and a downfield signal for

$$\overset{\text{H}}{\underset{|}{}}\!\!-\text{C}{=}\text{O}$$

(δ = 9–10 ppm). If a ketone, it is $(CH_3)_2C{=}O$, with one singlet.

Problem 12.46 The nmr spectrum of a dichloropropane shows a quintuplet and, downfield, a triplet of about twice the intensity. Is the isomer 1,1-, 1,2-, 1,3- or 2,2-dichloropropane?

We would expect the following signals:

1,1-Dichloropropane, $Cl_2CH^aCH_2^bCH_3^c$: a triplet (H^c), a complex multiplet more downfield (H^b), and a triplet still more downfield (H^a).

1,2-Dichloropropane, $ClCH_2^aCH^b(Cl)CH_3^c$: a doublet (H^c), another doublet more downfield (H^a), a complex multiplet most downfield (H^b)

1,3-Dichloropropane, $ClCH_2^aCH_2^bCH_2^aCl$: a quintuplet (H^b), and, downfield, a triplet (H^a).

2,2-Dichloropropane, $CH_3^a{-}CCl_2{-}CH_3^a$: a singlet (H^a)

The compound is 1,3-dichloropropane.

Problem 12.47 Consider the coupled ^{13}C nmr spectra of (*a*) 1,3,5-trimethylbenzene, (*b*) 1,2,3-trimethylbenzene, (*c*) *n*-propylbenzene. (i) Write each structure and label the different kinds of C's by numerals 1,2. … (ii) Show the splitting for each peak by the letters s for singlet, d for doublet, t for triplet, and q for quartet.

(*i*)

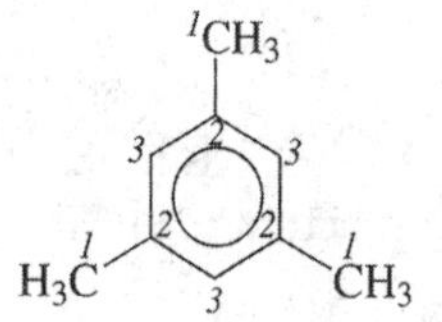
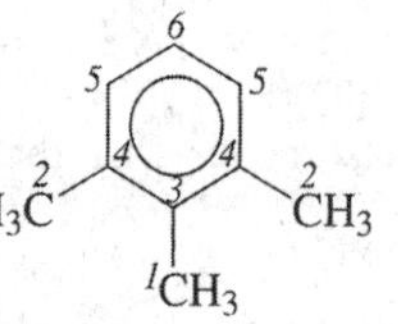
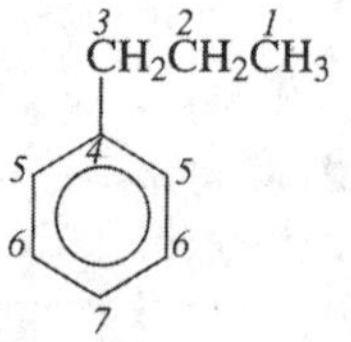

	(*a*)	(*b*)	(*c*)
(*ii*)	1(q), 2(s), 3(d)	1(q), 2(q), 3(s), 4(s), 5(d), 6(d)	1(q), 2(t), 3(t), 4(s), 5(d), 6(d), 7(d)

Problem 12.48 Indicate whether the following statements are True or False, and give a reason in each case. (*a*) The ir spectra are identical for the enantiomers

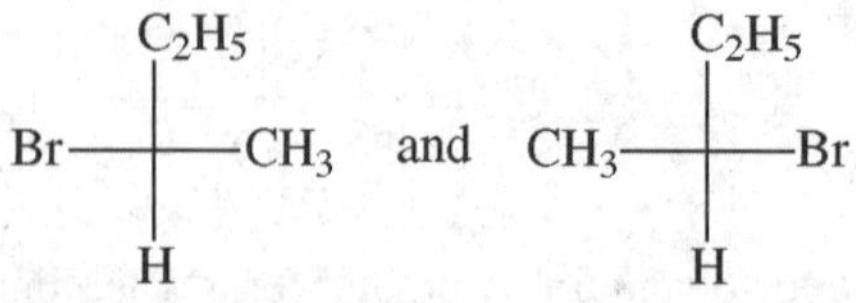

(*b*) The nmr spectra of the compounds in (*a*) are also identical. (*c*) The ir spectrum of 1-hexene has more peaks than the uv spectrum. (*d*) Compared to CH_3CH_2CHO, the n → π* for $H_2C{=}CHCHO$ has shifted to a shorter wavelength (**blue shift**).

(*a*) and (*b*) True. The compounds are enantiomers which have identical vibrational modes and proton resonances.

(*c*) True. The ir spectrum has peaks for stretching and bending of all bonds, while the uv spectrum has only one peak for excitation of a π electron.

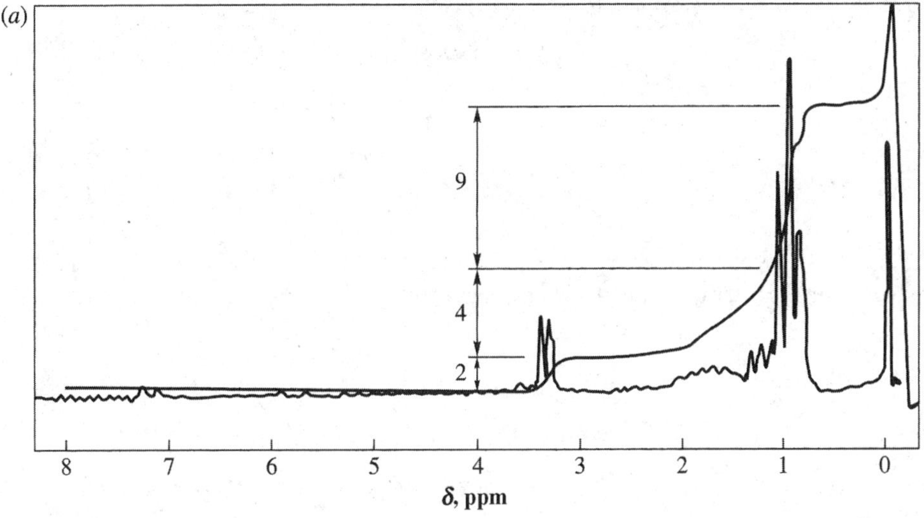
(a)
9
4
2
8 7 6 5 4 3 2 1 0
δ, ppm

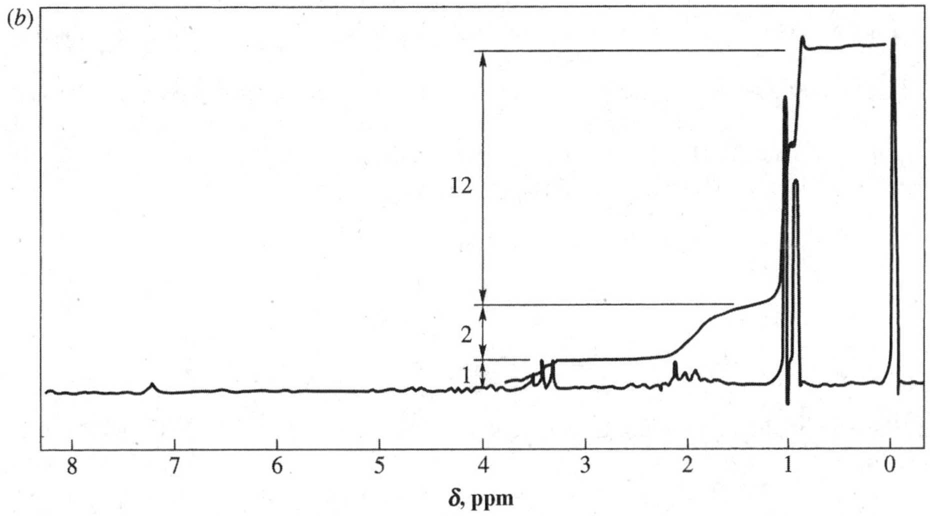
(b)
12
2
1
8 7 6 5 4 3 2 1 0
δ, ppm

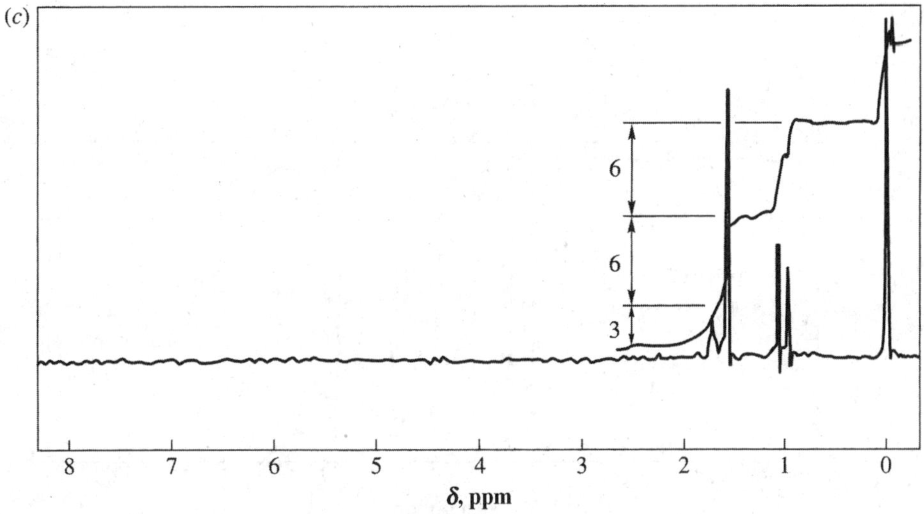
(c)
6
6
3
8 7 6 5 4 3 2 1 0
δ, ppm

Figure 12.10

(*d*) False. The shift is to longer wavelength (**red shift**), since

$$\mathrm{H_2C{=}CH\underset{\displaystyle H}{\underset{|}{C}}{=}O}$$

is a conjugated system.

Problem 12.49 Assign the nmr spectra shown in Fig. 12.10 to the appropriate monochlorination products of 2,4-dimethylpentane ($C_7H_{15}Cl$), and justify your assignment. Note the integration assignments drawn in the spectra.

The three possible structures are

$$\mathrm{ClCH_2{-}\overset{\displaystyle CH_3}{\overset{|}{C}H}{-}CH_2{-}\overset{\displaystyle CH_3}{\overset{|}{C}H}{-}CH_3}\qquad \mathrm{\overset{1}{C}H_3{-}\underset{\displaystyle Cl}{\underset{|}{\overset{\displaystyle CH_3}{\overset{|}{\overset{2}{C}}}}}{-}\overset{3}{C}H_2{-}\overset{\displaystyle CH_3}{\overset{|}{\overset{4}{C}H}}{-}\overset{5}{C}H_3}\qquad \mathrm{CH_3{-}\overset{\displaystyle CH_3}{\overset{|}{C}H}{-}\underset{\displaystyle Cl}{\underset{|}{C}H}{-}\overset{\displaystyle CH_3}{\overset{|}{C}H}{-}CH_3}$$

1-Chloro-2,4 dimethylpentane (I)   2-Chloro-2, 4-dimethylpentane (II)   3-Chloro-2, 4-dimethylpentane (III)

The best clue is the most downfield signal arising from the H's closest to Cl. In spectrum (*a*), the signal with the highest δ value is a *doublet*, integrating for two H's, that corresponds only to structure I ($ClCH_2$—). This is confirmed by the nine H's of the 3 CH_3's that are most upfield and the four 2° and 3° H's with signals between these.

In spectrum (*b*), the most downfield signal is a triplet, for one H, which arises from the

$$\mathrm{{-}\overset{\displaystyle H}{\overset{|}{C}}{-}\underset{\displaystyle \underline{H}}{\underset{|}{\overset{\displaystyle Cl}{\overset{|}{C}}}}{-}\overset{\displaystyle H}{\overset{|}{C}}{-}}$$

grouping in III. In addition, the most upfield signal is a doublet, integrating for 12 H's, which is produced by the H's of the 4 CH_3's split by the 3° H.

This leaves II for spectrum (*c*). The most downfield group of irregular signals, integrating for three H's, comes from the two 2° and one 3° H on C^3 and C^4, respectively. The most upfield doublet, integrating for six H's, arises from the two equivalent CH_3's on C^4 split by the C^4 3° H. The two CH_3's on C^2 give rise (six H's) to the singlet of median δ value.

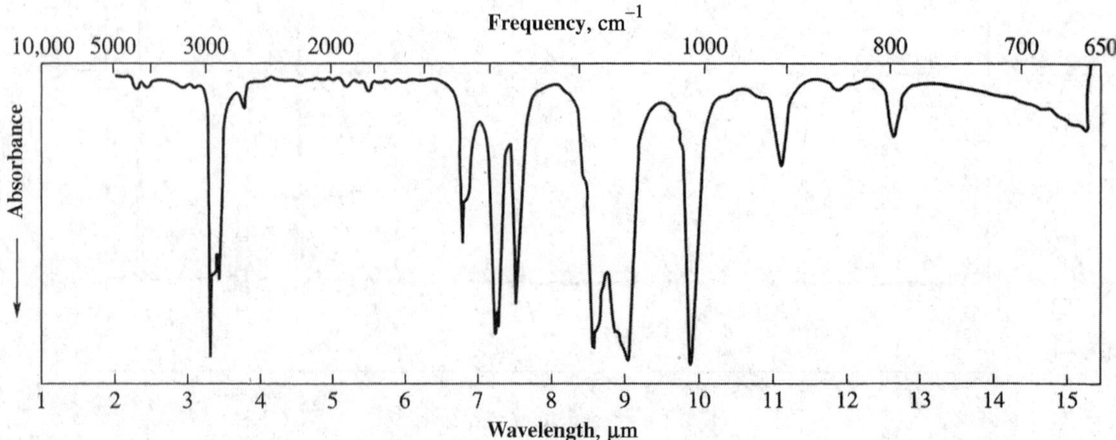

Figure 12.11 Infrared Spectrum

Problem 12.50 Deduce structures for the compound whose spectral data are presented in Fig. 12.11, Table 12.7, and Fig. 12.12. Assume an O is present in the molecule. There was no uv absorption above 180 nm.

TABLE 12.7 **Mass Spectrum**

m/e	26	27	29	31	39	41	42	43	44	45	59	87	102
Relative intensity, % of base peak	3	18	6	4	11	17	6	61	4	100	11	21	0.63

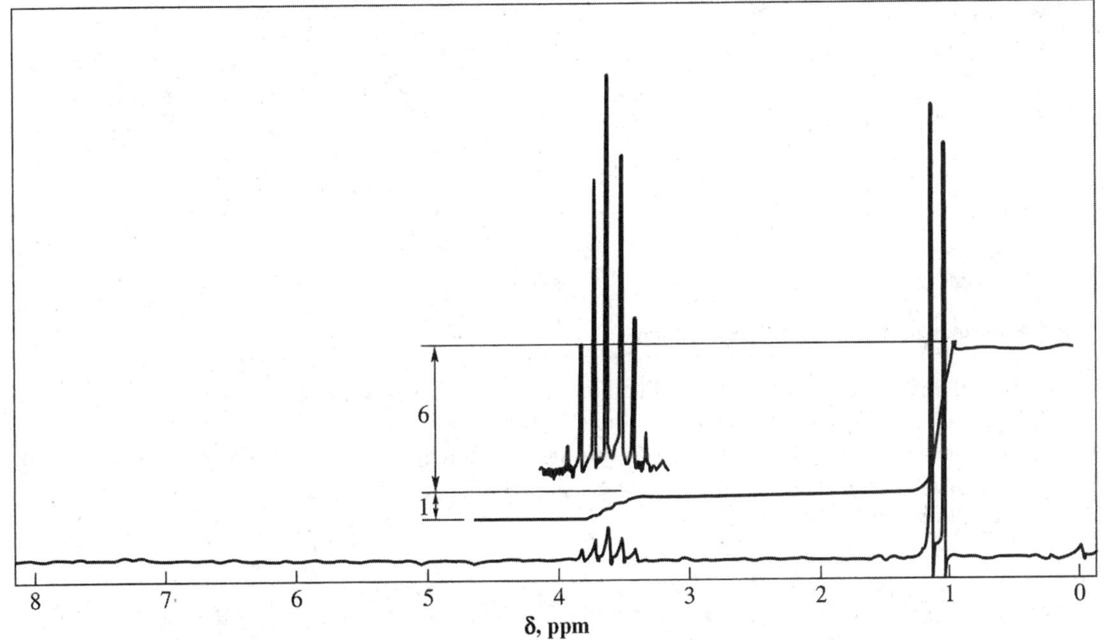

Figure 12.12 nmr Spectrum ($CDCL_3$)

From the mass spectrum, we know the RS^+ has molecular weight 102. The C, H portion of the compound has a mass of 102 − 16 (for O) = 86. Dividing by 12 gives 6 C's, leaving a remainder of 14 for 14 H's. The molecular formula is $C_6H_{14}O$. (Had we chosen 7 C's, we would have had C_7H_2O, which is an impossible formula.) The compound has no degrees of unsaturation. This fact is consistent with (but not proven by) the absence of uv absorption above 180 nm. Note also the absence of C—H stretch above 3000 cm^{-1} (H—C_{sp^2} or H—C_{sp}); the O must be present as C—O—H (an alcohol) or as C—O—C (an ether). The absence of a peak in the ir at 3300–3600 cm^{-1} precludes the presence of an O—H group. The strong band at about 1110 cm^{-1} represents the C—O stretch.

The structure of the alkyl groups R—O—R of the ether is best revealed by the nmr spectrum. The downfield septet (see blown-up signal) and the upfield doublet integrate 1:6. This arrangement is typical for a $CH(CH_3)_2$ grouping. Both R groups are isopropyl, since no other signals are present. The compound is $(CH_3)_2CHOCH(CH_3)_2$.

The significant peaks in the mass spectrum that are consistent with our assignment are *m/e* = 102 − 15 (CH_3) = 87, which is $(CH_3)_2CHO\overset{+}{C}HCH_3$; 43 is $(CH_3)_2\overset{+}{C}H$; 41 would be the allyl cation $H_2C{=}CHCH_2^+$ formed from fragmentation of $(CH_3)_2CH$. The most abundant peak, *m/e* = 45, probably comes from rearrangement of fragment ions. It could have the formula C_2H_5O. The presence of this peak convinces us that O was indeed present in the compound. The peak cannot be from a fragment with only C and H; $C_3H_9^+$ is impossible.

Alcohols and Thiols

A. ALCOHOLS

13.1 Nomenclature and H-Bonding

ROH is an **alcohol**, and ArOH is a **phenol** (Chapter 19). Some alcohols have common names, usually made up of the name of the alkyl group attached to the OH and the word "alcohol," for example, *ethyl alcohol*, C_2H_5OH. More generally, the IUPAC method is used, in which the suffix **-ol** replaces the **-e** of the alkane to indicate the OH. The longest chain with the OH group is used as the parent, and the C bonded to the OH is called the **carbinol** carbon.

Problem 13.1 Give a common name for each of the following alcohols and classify them as 1°, 2°, or 3°:

$$(a)\ CH_3CH_2CH_2OH \quad (b)\ CH_3CH_2\underset{\underset{OH}{|}}{C}HCH_3 \quad (c)\ CH_3\underset{\underset{CH_3}{|}}{C}HCH_2OH \quad (d)\ CH_3-\overset{\overset{CH_3}{|}}{\underset{\underset{CH_3}{|}}{C}}-OH$$

$$(e)\ CH_3-\overset{\overset{CH_3}{|}}{C}H-OH \quad (f)\ CH_3-\overset{\overset{CH_3}{|}}{\underset{\underset{CH_3}{|}}{C}}-CH_2OH \quad (g)\ C_6H_5CH_2OH$$

(*a*) *n*-propyl alcohol, 1°; (*b*) *sec*-butyl alcohol, 2°; (*c*) isobutyl alcohol, 1°; (*d*) *t*-butyl alcohol, 3°; (*e*) isopropyl alcohol, 2°; (*f*) neopentyl alcohol, 1°; (*g*) benzyl alcohol, 1°.

Problem 13.2 Name the following alcohols by the IUPAC method:

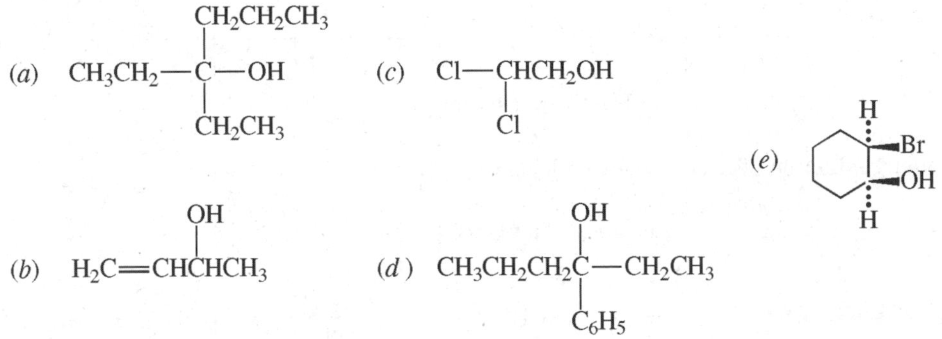

(*a*) $CH_3CH_2—C(CH_2CH_2CH_3)(CH_2CH_3)—OH$

(*b*) $H_2C=CHCH(OH)CH_3$

(*c*) $Cl—CH(Cl)CH_2OH$

(*d*) $CH_3CH_2CH_2C(OH)(C_6H_5)—CH_2CH_3$

(*e*) cyclohexane ring with H, Br and H, OH

(*a*) 3-Ethyl-3-hexanol
(*b*) 3-Buten-2-ol
(*c*) 2,2-Dichloroethanol
(*d*) 3-Phenyl-3-hexanol
(*e*) *cis*-2-Bromocyclohexanol

Note that, in IUPAC, the OH is given a lower number than C=C or Cl.

Problem 13.3 Explain why (*a*) propanol boils at a higher temperature than the corresponding hydrocarbon; (*b*) propanol, unlike propane or butane, is soluble in H_2O; (*c*) *n*-hexanol is not soluble in H_2O; (*d*) dimethyl ether (CH_3OCH_3) and ethyl alcohol (CH_3CH_2OH) have the same molecular weight, yet dimethyl ether has a lower boiling point (−24 °C) than ethyl alcohol (78 °C).

(*a*) Propanol can H-bond intermolecularly. $C_3H_7—O(H)\cdots H—O(C_3H_7)$ There is also a less important dipole–dipole interaction.

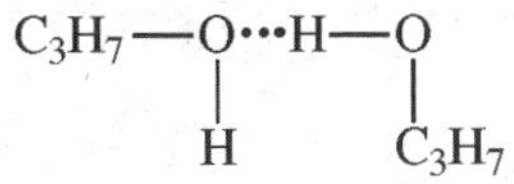

(*b*) Propanol can H-bond with H_2O. $C_3H_7—O(H)\cdots H—O(H)$

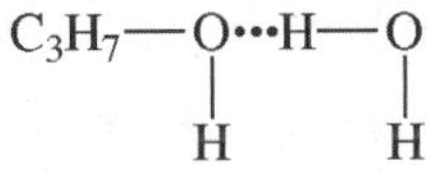

(*c*) As the R group becomes larger, ROH resembles the hydrocarbon more closely. There is little H-bonding between H_2O and *n*-hexanol. When the ratio of C to OH is more than 4, alcohols have little solubility in water.

(*d*) The ether CH_3OCH_3 has no H on O and cannot H-bond; only the weaker dipole-dipole interaction exists.

Problem 13.4 The ir spectra of *trans*- and *cis*-1,2-cyclopentanediol show a broad band in the region 3450–3570 cm^{-1}. On dilution with CCl_4, this band of the *cis* isomer remains unchanged, but the band of the *trans* isomer shifts to a higher frequency and becomes sharper. Account for this difference in behavior.

The OH's of the *cis* isomer participate in *intramolecular* H-bonding, Fig. 13.1(*a*), which is not affected by dilution. In the *trans* isomer, the H-bonding is *intermolecular*, Fig. 13.1(*b*), and dilution breaks these bonds, causing disappearance of the broad band and its replacement by a sharp OH band at higher frequency.

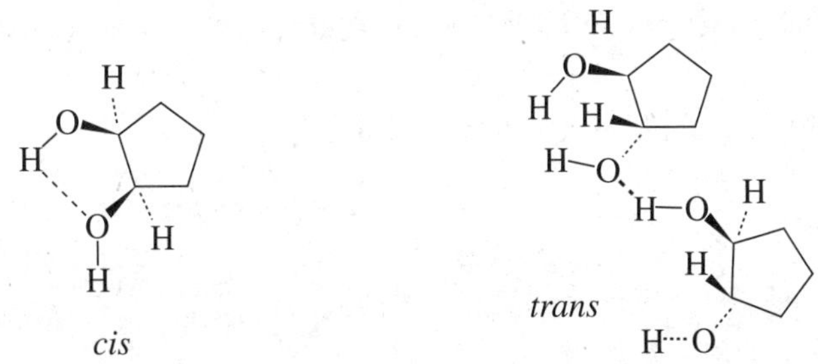

cis
Intramolecular H-bond
(*a*)

trans
Intermolecular H-bond
(*b*)

Figure 13.1

13.2 Preparation

1. $RX + OH^- \xrightarrow{H_2O} ROH + X^-$ (**S_N2 or S_N1 Displacement,** Table 7.1)
2. **Hydration of Alkenes** [see Problem 6.19(*d*)]
3. **Hydroboration-Oxidation of Alkenes** [see Problem 6.19(*f*)]

Treatment of the alkylboranes with H_2O_2 in OH^- replaces —B— with OH.

$$RCH{=}CH_2 + (BH_3)_2 \longrightarrow \underset{\text{trialkylborane}}{(R\underset{|}{\overset{}{C}}HCH_2)_3\text{—}B} \xrightarrow[OH^-]{H_2O_2} RCHCH_2$$

The net addition of H—OH to alkenes is *cis*, anti-Markovnikov, and free from rearrangement.

4. **Oxymercuration-Demercuration of Alkenes**

$$RCH{=}CH_2 \xrightarrow[H_2O,\ 25^\circ C]{Hg(OCOCH_3)_2} \underset{\textit{not isolated}}{\underset{OH\quad HgOCOCH_3}{RCH\text{—}CH_2}} \xrightarrow{NaBH_4} \underset{HO\quad H}{RCHCH_2}$$

The net addition of H—OH is Markovnikov and is free from rearrangement.

Problem 13.5 Give structures and IUPAC names of the alcohols formed from $(CH_3)_2CHCH{=}CH_2$ by reaction with (*a*) dilute H_2SO_4; (*b*) B_2H_6, then H_2O_2, OH^-; (*c*) $Hg(OCOCH_3)_2$, H_2O, then $NaBH_4$.

(*a*) The expected product is 3-methyl-2-butanol, $(CH_3)_2CHCHOHCH_3$, from a Markovnikov addition of H_2O. However, the major product is likely to be 2-methyl-2-butanol, $(CH_3)_2COHCH_2CH_3$, formed by rearrangement of the intermediate R^+.

$$(CH_3)_2CHCH{=}CH_2 \xrightarrow[(H_2O)]{+H^+} \underset{(2^\circ)}{(CH_3)_2CH\overset{+}{C}HCH_3} \xrightarrow{\sim H:} \underset{(3^\circ)}{(CH_3)_2\overset{+}{C}CH_2CH_3} \xrightarrow[-H^+]{+H_2O} (CH_3)_2COHCH_2CH_3$$

(*b*) Anti-Markovnikov HOH addition forms $(CH_3)_2CHCH_2CH_2OH$, 3-methyl-1-butanol.
(*c*) Markovnikov HOH addition with no rearrangment gives $(CH_3)_2CHCHOHCH_3$, 3-methyl-2-butanol.

Problem 13.6 Give the structure and IUPAC name of the product formed on hydroboration-oxidation of 1-methylcyclohexene.

H and OH add *cis* and therefore CH_3 and OH are *trans*.

$$\text{1-methylcyclohexene} \xrightarrow[2.\ OH^-,\ H_2O_2]{1.\ B_2H_6} \textit{trans}\text{-2-Methylcyclohexanol}$$

5. **Carbonyl Compounds and Grignard Reagents (Increase in Number of Carbons)**

Grignard reagents, RMgX and ArMgX, are reacted with aldehydes or ketones and the intermediate products hydrolyzed to alcohols.

$$\boxed{R}MgX + \underset{\text{formaldehyde}}{H{-}\overset{\displaystyle H}{\overset{|}{C}}{=}O} \longrightarrow \boxed{R}{-}CH_2{-}\bar{O}(MgX)^+ \xrightarrow{H_2O} \boxed{R}{-}CH_2{-}OH \quad (1^\circ \text{ alcohol})$$

$$\boxed{R}MgX + \underset{\text{aldehyde}}{R'{-}\overset{\displaystyle H}{\overset{|}{C}}{=}O} \longrightarrow R'{-}\underset{\boxed{R}}{\overset{H}{C}}{-}\bar{O}(MgX)^+ \xrightarrow{H_2O} R'{-}\underset{\boxed{R}}{\overset{H}{C}}{-}OH \quad (2^\circ \text{ alcohol})$$

$$\boxed{Ar}MgX + \underset{\text{ketone}}{R'{-}\overset{\displaystyle R''}{\overset{|}{C}}{=}O} \longrightarrow R'{-}\underset{\boxed{Ar}}{\overset{R''}{C}}{-}\bar{O}(MgX)^+ \xrightarrow{H_2O} R'{-}\underset{\boxed{Ar}}{\overset{R''}{C}}{-}OH \quad (3^\circ \text{ alcohol})$$

alkoxide salts

The boxed group in the alcohol comes from the Grignard; the remainder comes from the carbonyl compound.

$$\underset{\text{electrophilic site}}{>C{=}\ddot{O}{:}} + \underset{\text{nucleophile}}{R{:}^-(MgX)^+} \longrightarrow \underset{\substack{\text{an alkoxide:}\\ \text{a strong conjugate}\\ \text{base of an alcohol}}}{\left[>\overset{R}{C}{-}\ddot{O}{:}\right]^-(MgX)^+} \xrightarrow{H{:}OH} \underset{\text{alcohol}}{>\overset{R}{C}{-}OH} + Mg(X)(OH)$$

ArMgX is best made from ArBr in ether or from ArCl in tetrahydrofuran (ArCl is not reactive in ether).

$$ArBr + Mg \xrightarrow{\text{ether}} ArMgBr$$
$$ArCl + Mg \xrightarrow{\text{THF}} ArMgCl$$

Problem 13.7 Give 3 combinations of RMgX and a carbonyl compound that could be used to prepare:

$$C_6H_5CH_2\underset{OH}{\overset{CH_3}{C}}{-}CH_2CH_3$$

This 3° alcohol is made from RMgX and a ketone, R′COR″. The possibilities are:

(1) $\boxed{C_6H_5CH_2}{-}\underset{OH}{\overset{CH_3}{C}}{-}CH_2CH_3$ from $\boxed{C_6H_5CH_2}MgCl$ and $CH_3\overset{O}{\overset{\|}{C}}CH_2CH_3$

(2) $C_6H_5CH_2{-}\underset{OH}{\overset{CH_3}{C}}{-}\boxed{CH_2CH_3}$ from $\boxed{CH_3CH_2}MgBr$ and $C_6H_5CH_2\overset{O}{\overset{\|}{C}}CH_3$

(3) $C_6H_5CH_2{-}\underset{OH}{\overset{\boxed{CH_3}}{C}}{-}CH_2CH_3$ from $\boxed{CH_3}MgI$ and $C_6H_5CH_2\overset{O}{\overset{\|}{C}}CH_2CH_3$

The best combination is usually the one in which the two reactants share the C content as equally as possible. In this case, (1) is best.

Problem 13.8 Give four limitations of the Grignard reaction.

(1) The halide cannot possess a functional group with an acidic H, such as OH, COOH, NH, SH, or C≡C—H, because then the carbanion of the Grignard group would remove the acidic H and be reduced. For example:

$$HOCH_2CH_2Br + Mg \longrightarrow \underset{unstable}{[HOCH_2CH_2MgBr]} \longrightarrow (BrMg)^{+}\,{}^{-}OCH_2CH_2\mathbf{H}$$

(2) If the halide also has a C═O (or C═N—, C≡N, N═O, S═O, C≡CH) group, it reacts inter- or intramolecularly with itself.

(3) The reactant cannot be a *vic*-dihalide, because it would undergo dehalogenation:

$$BrCH_2CH_2Br + Mg \longrightarrow H_2C{=}CH_2 + MgBr_2$$

(4) A ketone with two bulky R groups, for example, —$C(CH_3)_3$, would be too sterically hindered to react with an organometallic compound with a bulky R′ group.

Problem 13.9 Prepare 1-butanol from (*a*) an alkene, (*b*) 1-chlorobutane, (*c*) 1-chloropropane and (*d*) ethyl bromide.

(*a*) $CH_3CH_2CH{=}CH_2 \xrightarrow[2.\ H_2O_2,\ OH^-]{1.\ (BH_3)_2} CH_3CH_2CH_2CH_2OH$

(*b*) $HO^- + CH_3CH_2CH_2CH_2Cl \xrightarrow{H_2O} CH_3CH_2CH_2CH_2OH + Cl^-$ (S_N2)

(*c*) 1-Chloropropane has one less C than the needed 1° alcohol. The Grignard reaction is used to lengthen the chain by adding $H_2C{=}O$ (formaldehyde).

$$CH_3CH_2CH_2Cl \xrightarrow[ether]{Mg} CH_3CH_2CH_2MgCl \xrightarrow[2.\ H_3O^+]{1.\ HCH{=}O} CH_3CH_2CH_2CH_2OH$$

(*d*) 1-Butanol is a 1° alcohol with two C's more than CH_3CH_2Br. Reaction of CH_3CH_2MgBr with ethylene oxide followed by hydrolysis gives 1-butanol.

$$CH_3CH_2Br \xrightarrow[ether]{Mg} \boxed{CH_3CH_2}MgBr \xrightarrow[(S_N2)]{CH_2-CH_2\ (\text{O bridged})} \boxed{CH_3CH_2}CH_2CH_2\bar{O}(\overset{+}{MgBr}) \xrightarrow{H_2O} CH_3CH_2CH_2CH_2OH$$

Problem 13.10 For the following pairs of halides and carbonyl compounds, give the structure of each alcohol formed by the Grignard reaction. (*a*) Bromobenzene and acetone. (*b*) *p*-Chlorophenol and formaldehyde. (*c*) Isopropyl chloride and benzaldehyde. (*d*) Chlorocyclohexane and methyl phenyl ketone.

(*a*) $C_6H_5Br \xrightarrow[ether]{Mg} \boxed{C_6H_5}MgBr \xrightarrow[2.\ H_2O]{1.\ CH_3-\overset{O}{\overset{\|}{C}}-CH_3} \boxed{C_6H_5}C(CH_3)_2-OH$

(*b*) The weakly acidic OH in *p*-chlorophenol prevents formation of the Grignard reagent.

(*c*) $C_6H_5\overset{H}{\overset{|}{C}}{=}O \xrightarrow[2.\ H_2O]{1.\ \boxed{(CH_3)_2CH}MgCl} C_6H_5\overset{H}{\overset{|}{\underset{OH}{\underset{|}{C}}}}-\boxed{CH(CH_3)_2}$

(*d*) $C_6H_{11}Cl$ (chlorocyclohexane) $\xrightarrow[ether]{Mg} C_6H_{11}MgCl \xrightarrow[2.\ H_2O]{1.\ C_6H_5\overset{O}{\overset{\|}{C}}CH_3} C_6H_5\overset{CH_3}{\overset{|}{\underset{OH}{\underset{|}{C}}}}-C_6H_{11}$ + Mg(OH)Cl

Methylcyclohexylphenylcarbinol

6. Addition of :H^- or H_2 to $-\overset{|}{C}=O$ (Fixed Number of Carbons)

The :H^- is best provided by sodium borohydride, $NaBH_4$, in protic solvents such as ROH or H_2O.

$$-\overset{|}{C}=O \xrightarrow[\text{in ethanol}]{NaBH_4} -\overset{|}{\underset{H}{C}}-OH$$

Lithium aluminum hydride, $LiAlH_4$, in anhydrous ether can also be used.

Problem 13.11 How do the alcohols from $LiAlH_4$ or catalytic reduction of ketones differ from those derived from aldehydes?

Ketones yield 2° alcohols while aldehydes give 1° alcohols.

$$\text{Ketone:}\quad R-\underset{O}{\underset{\|}{C}}-R' \xrightarrow[2.\ H_2O]{1.\ LiAlH_4} \underset{\text{a 2° alcohol}}{R-\overset{H}{\overset{|}{\underset{OH}{\underset{|}{C}}}}-R'} \xleftarrow{H_2/Pt} R\underset{O}{\underset{\|}{C}}R'$$

$$\text{Aldehyde:}\quad R-\underset{O}{\underset{\|}{C}}-H \xrightarrow[2.\ H_2O]{1.\ LiAlH_4} \underset{\text{a 1° alcohol}}{R-\overset{H}{\overset{|}{\underset{OH}{\underset{|}{C}}}}-H} \xleftarrow{H_2/Pt} R\underset{O}{\underset{\|}{C}}H$$

Problem 13.12 (*a*) What is the expected product from catalytic hydrogenation of acetophenone, $C_6H_5COCH_3$? (*b*) One of the products of the reaction in (*a*) is $C_6H_5CH_2CH_3$. Explain its formation.

(*a*) $C_6H_5CHOHCH_3$. (*b*) The initial product, typical of benzylic alcohols:

$$C_6H_5\overset{R}{\overset{|}{\underset{H}{\underset{|}{C}}}}-OH$$

can be further reduced with H_2. This reaction is a **hydrogenolysis** (bond-breaking by H_2).

$$C_6H_5CHOHCH_3 + H_2 \xrightarrow{Pd} C_6H_5CH_2CH_3 + H_2O$$

Problem 13.13 Reduction of $H_2C=CHCHO$ with $NaBH_4$ gives a product different from that of catalytic hydrogenation (H_2/Ni). What are the products?

$$\underset{\text{selective reduction of } >C=O}{H_2C=CHCH_2OH} \xleftarrow{NaBH_4} \underset{\text{Acrolein}}{H_2C=CHCHO} \xrightarrow{H_2/Ni} \underset{\text{nonselective reduction}}{H_3CCH_2CH_2OH}$$

7. Reduction of RCOOH or RCOOR with $LiAlH_4$

The initial product is a lithium alkoxide salt, which is then hydrolyzed to a 1° alcohol.

$$C_2H_5\overset{\overset{O}{\|}}{C}-OH \text{ (Propionic acid)} \xrightarrow{LiAlH_4} C_2H_5CH_2O^-Li^+ \text{ (a lithium alkoxide)} \xrightarrow{+H_2O} C_2H_5CH_2OH \text{ (Propanol)}$$

$$C_2H_5\overset{\overset{O}{\|}}{C}-OCH_3 \text{ (Methyl propionate)} \xrightarrow{LiAlH_4} C_2H_5CH_2O^-Li^+$$

13.3 Reactions

1. **The electron pairs on O make alcohols Lewis bases.**
2. H **of OH is very weakly acidic.** The order of decreasing acidity is

$$H_2O > ROH(1°) > ROH(2°) > ROH(3°) > RC{\equiv}CH >> RCH_3.$$

3. **1° and 2° alcohols have at least one H on the carbinol C and are oxidized to carbonyl compounds.** They also lose H_2 in the presence of Cu (300°C) to give carbonyl compounds.
4. **Formation of alkyl halides** (Problem 7.3(*a*)–(*d*)).
5. **Intramolecular dehydration to alkenes** (Problems 6.11 through 6.13).
6. **Intermolecular dehydration to ethers.**

$$2ROH \xrightarrow[130°C]{H_2SO_4} ROR + H_2O$$

7. **Ester formation.**

$$ROH + R'\underset{\underset{O}{\|}}{C}OH \xrightarrow{H_2SO_4} R'\underset{\underset{O}{\|}}{C}OR + H_2O$$
$$\text{ester}$$

8. **Inorganic ester formation.** With cold conc. H_2SO_4, sulfate esters are formed.

$$ROH + H_2SO_4 \xrightarrow{cold} H_2O + \underset{\text{an alkyl acid sulfate}}{ROSO_3H} \xrightarrow{ROH} \underset{\text{a dialkyl sulfate}}{ROSO_2OR}$$

Similarly, alkyl phosphates are formed with H_3PO_4, phosphoric acid.

Problem 13.14 Supply equations for the formation from phosphoric acid of (*a*) alkyl phosphate esters by reactions with an alcohol, in which each acidic H is replaced and H_2O is eliminated; (*b*) phosphoric anhydrides on heating to eliminate H_2O.

(*a*)
$$\underset{\text{Phosphoric acid}}{O{=}P(OH)_3} \xrightarrow[-H_2O]{+ROH} \underset{\text{an alkyl dihydrogen phosphate}}{O{=}P(OR)(OH)_2} \xrightarrow[-H_2O]{+ROH} \underset{\text{a dialkyl hydrogen phosphate}}{O{=}P(OR)_2(OH)} \xrightarrow[-H_2O]{+ROH} \underset{\text{a trialkyl phosphate}}{O{=}P(OR)_3}$$

(*b*) $HO{-}P(=O)(OH){-}O\,\boxed{H + HO}{-}P(=O)(OH){-}OH \xrightarrow{-H_2O} HO{-}P(=O)(OH){-}O{-}P(=O)(OH){-}OH \xrightarrow[-H_2O]{+H_3PO_4} HO{-}P(=O)(OH){-}O{-}P(=O)(OH){-}O{-}P(=O)(OH){-}OH$

Phosphoric acid — Diphosphoric acid — Triphosphoric acid

Problem 13.15 Alkyl esters of di- and triphosphoric acid are important in biochemistry because they are stable in the aqueous medium of living cells and are hydrolyzed by enzymes to supply the energy needed for muscle contraction and other processes. (*a*) Give structural formulas for these esters. (*b*) Write equations for the hydrolysis reactions that are energy-liberating.

(*a*) $RO{-}P(=O)(OH){-}O{-}P(=O)(OH){-}OH$ an alkyl trihydrogen diphosphate

$RO{-}P(=O)(OH){-}O{-}P(=O)(OH){-}O{-}P(=O)(OH){-}OH$ an alkyl tetrahydrogen triphosphate

(*b*) Hydrolysis occurs at the ester (reaction I) or the anhydride bonds (reactions II and III).

$$H_2O + R{-}O{-}P(=O)(OH){-}O{-}P(=O)(OH){-}O{-}P(=O)(OH){-}OH \xrightarrow{I} ROH + HO{-}P(=O)(OH){-}O{-}P(=O)(OH){-}O{-}P(=O)(OH){-}OH$$

$$\xrightarrow{II} RO{-}P(=O)(OH){-}OH + HO{-}P(=O)(OH){-}O{-}P(=O)(OH){-}OH$$

$$\xrightarrow{III} RO{-}P(=O)(OH){-}O{-}P(=O)(OH){-}OH + HO{-}P(=O)(OH){-}OH$$

9. **Alkyl sulfonates (esters of sulfonic acids),** $RSO_2R'(Ar)$

$$RS(=O)_2{-}Cl + HOR' \longrightarrow RS(=O)_2{-}OR' + HCl$$

a sulfonyl chloride — a sulfonate ester

Problem 13.16 How does sulfonate ester formation from sulfonyl chloride resemble nucleophilic displacements of alkyl halides?

The alcohol acts as a nucleophile and halide ion is displaced.

$$R'{-}\ddot{\underset{\cdot\cdot}{O}}{-}H + R{-}S(=O)_2{-}Cl \xrightarrow{\text{pyridine}} R{-}S(=O)_2{-}\ddot{\underset{\cdot\cdot}{O}}{-}R' + (\text{pyridine H})^+\ Cl^-$$

Sulfonyl chlorides are prepared from the sulfonic acid or salt with PCl_5.

$$\left.\begin{matrix} RSO_2OH \\ RSO_2ONa \end{matrix}\right\} + PCl_5 \longrightarrow RSO_2Cl + POCl_3 + \left\{\begin{matrix} HCl \\ NaCl \end{matrix}\right.$$

Aromatic sulfonyl chlorides are also formed by ring chlorosulfonation with chlorosulfonic acid, $HOSO_2Cl$.

$$ArH + HOSO_2Cl \longrightarrow \underset{\text{a sulfonyl chloride}}{ArSO_2Cl} + H_2O$$

10. **Oxidation.** Alcohols with at least one H on the carbinol carbon (1° and 2°) are oxidized to carbonyl compounds.

$$\underset{1^\circ}{RCH_2OH} \longrightarrow \underset{\text{an aldehyde}}{RCH{=}O} \quad \text{and} \quad \underset{2^\circ}{R_2CHOH} \longrightarrow \underset{\text{a ketone}}{R_2C{=}O}$$

Aldehydes are further oxidized to carboxylic acids, RCOOH. To get aldehydes, milder reagents, such as the **Jones** (diluted chromic acid in acetone) or **Collins** reagent (a complex of CrO_3 with 2 mol of pyridine), are used.

Problem 13.17 Write equations to show why alcohols cannot be used as solvents with Grignard reagents or with $LiAlH_4$.

Strongly basic $R\bar{:}$ and $H\bar{:}$ react with weakly acidic alcohols.

$$CH_3OH + CH_3\overset{-}{C}H_2\overset{+}{M}gCl \longrightarrow CH_3CH_3 + (CH_3O)^-(MgCl)^+$$

$$4CH_3OH + LiAlH_4 \longrightarrow 4H_2 + LiAl(OCH_3)_4$$

Problem 13.18 Give the main products of reaction of 1-propanol with (*a*) alkaline aq. $KMnO_4$ solution during distillation; (*b*) hot Cu shavings; (*c*) CH_3COOH, H^+.

(*a*) CH_3CH_2CHO. Since aldehydes are oxidized further under these conditions, CH_3CH_2COOH is also obtained. Most of the aldehyde is removed before it can be oxidized.
(*b*) CH_3CH_2CHO. The aldehyde can't be oxidized further.
(*c*) $CH_3\underset{\underset{O}{\|}}{C}{-}OCH_2CH_2CH_3$, an ester.

Problem 13.19 Explain the relative acidity of liquid 1°, 2°, and 3° alcohols.

The order of decreasing acidity of alcohols, $CH_3OH > 1° > 2° > 3°$, is attributed to electron-releasing R's. These intensify the charge on the conjugate base, RO^-, and destabilize this ion, making the acid weaker.

Problem 13.20 Give simple chemical tests to distinguish (*a*) 1-pentanol and *n*-hexane; (*b*) *n*-butanol and *t*-butanol; (*c*) 1-butanol and 2-buten-1-ol; (*d*) 1-hexanol and 1-bromohexane.

(*a*) Alcohols such as 1-pentanol dissolve in cold H_2SO_4. Alkanes such as *n*-hexane are insoluble. (*b*) Unlike *t*-butanol (a 3° alcohol), *n*-butanol (a 1° alcohol) can be oxidized under mild conditions. The analytical reagent is chromic anhydride in H_2SO_4. A positive test is signaled when this orange-red solution turns a deep green because of the presence of Cr^{3+}. (*c*) 2-Buten-1-ol decolorizes Br_2 in CCl_4 solution; 1-butanol does not. (*d*) 1-Hexanol reduces orange-red CrO_3 to green Cr^{3+}; alkyl halides such as 1-bromohexane do not. The halide on warming with $AgNO_3$(EtOH) gives AgBr.

Problem 13.21 Draw the structure of $C_4H_{10}O$ if the compound: (1) reacts with Na but fails to react with a strong oxidizing agent such as $K_2Cr_2O_7$; (2) gives a negative iodoform test; and (3) gives a positive Lucas test in 4 minutes.

(1) Because the compound reacts with Na, it must be an alcohol. Furthermore, because the compound does not react with a strong oxidizing agent, it must be a tertiary (3°) alcohol. Therefore, the structure of $C_4H_{10}O$ is *tert*-butyl alcohol:

$$CH_3-\underset{\underset{CH_3}{|}}{\overset{\overset{CH_3}{|}}{C}}-OH$$

(2) A negative iodoform test would occur for the primary four-carbon alcohol, *n*-butyl alcohol:

$$CH_3CH_2CH_2CH_2OH$$

(3) In the Lucas test, the Lucas reagent reacts with 1°, 2°, and 3° alcohols. The alcohols are distinguished by their reactivity with the Lucas reagent: 3° alcohols react immediately; 2° alcohols react within 5 minutes; and 1° alcohols react poorly at room temperature. Because the compound reacts with the Lucas reagent in 4 minutes, then the structure of $C_4H_{10}O$, a 2° alcohol, is *sec*-butyl alcohol:

$$CH_3CH_2\underset{\underset{OH}{|}}{C}HCH_3$$

Problem 13.22 Write balanced ionic equations for the following redox reaction:

$$CH_3CHOHCH_3 + K_2Cr_2O_7 + H_2SO_4 \xrightarrow{\text{heat}} CH_3—CO—CH_3 + Cr_2(SO_4)_3 + H_2O + K_2SO_4$$

Write partial equations for the oxidation and the reduction. Then: (1) **Balance charges** by adding H^+ in acid solutions or OH^- in basic solutions. (2) **Balance the number of** O's by adding H_2O's to one side. (3) **Balance the number of** H's by adding H's to one side. The number added is the **number of equivalents** of oxidant or reductant.

	OXIDATION	REDUCTION
	$(CH_3)_2CHOH \longrightarrow (CH_3)_2C{=}O$	$Cr_2O_7^{2-} \longrightarrow 2Cr^{3+}$
(1) In acid, balance charges with H^+:	*(no change)*	$Cr_2O_7^{2-} + 8H^+ \longrightarrow 2Cr^{3+}$
(2) Balance O with H_2O:	*(no change)*	$Cr_2O_7^{2-} + 8H^+ \longrightarrow 2Cr^{3+} + 7H_2O$
(3) Balance H:	$(CH_3)_2CHOH \longrightarrow (CH_3)_2C{=}O + 2H$	$Cr_2O_7^{2-} + 8H^+ + 6H \longrightarrow 2Cr^{3+} + 7H_2O$

(4) Balance equivalents:

$$3(CH_3)_2CHOH \longrightarrow 3(CH_3)_2C{=}O + 6H$$

$$Cr_2O_7^{2-} + 8H^+ + 6H \longrightarrow 2Cr^{3+} + 7H_2O$$

(5) Add:

$$3(CH_3)_2CHOH + Cr_2O_7^{2-} + 8H^+ \longrightarrow 3(CH_3)_2C{=}O + 2Cr^{3+} + 7H_2O$$

Problem 13.23 How can the difference in reactivity of 1°, 2°, and 3° alcohols with HCl be used to distinguish among these kinds of alcohols, assuming the alcohols have six or less C's?

The **Lucas test** uses conc. HCl and $ZnCl_2$ (to increase the acidity of the acid).

$$\underset{\textit{soluble, by assumption}}{3°ROH} + HCl \longrightarrow \underset{\textit{insoluble liquid}}{3°RCl} + H_2O$$

The above reaction is immediate; a 2° ROH reacts within 5 minutes; a 1° ROH does not react at all at room temperature.

Problem 13.24 In CCl_4 as solvent, the nmr spectrum of CH_3OH shows two singlets. In $(CH_3)_2SO$, there is a doublet and a quartet. Explain in terms of the "slowness" of nmr detection.

In CCl_4, CH_3OH H-bonds intermolecularly, leading to a rapid interchange of the H of O—H. The instrument senses an average situation, and therefore, there is no coupling between CH_3 and OH protons. In $(CH_3)_2SO$, H-bonding is with solvent, and the H stays on the O of OH. Now coupling occurs. This technique can be used to distinguish among RCH_2OH, R_2CHOH, and R_3COH, whose signals for H of OH are a triplet, a doublet, and a singlet, respectively.

13.4 Summary of Alcohol Chemistry

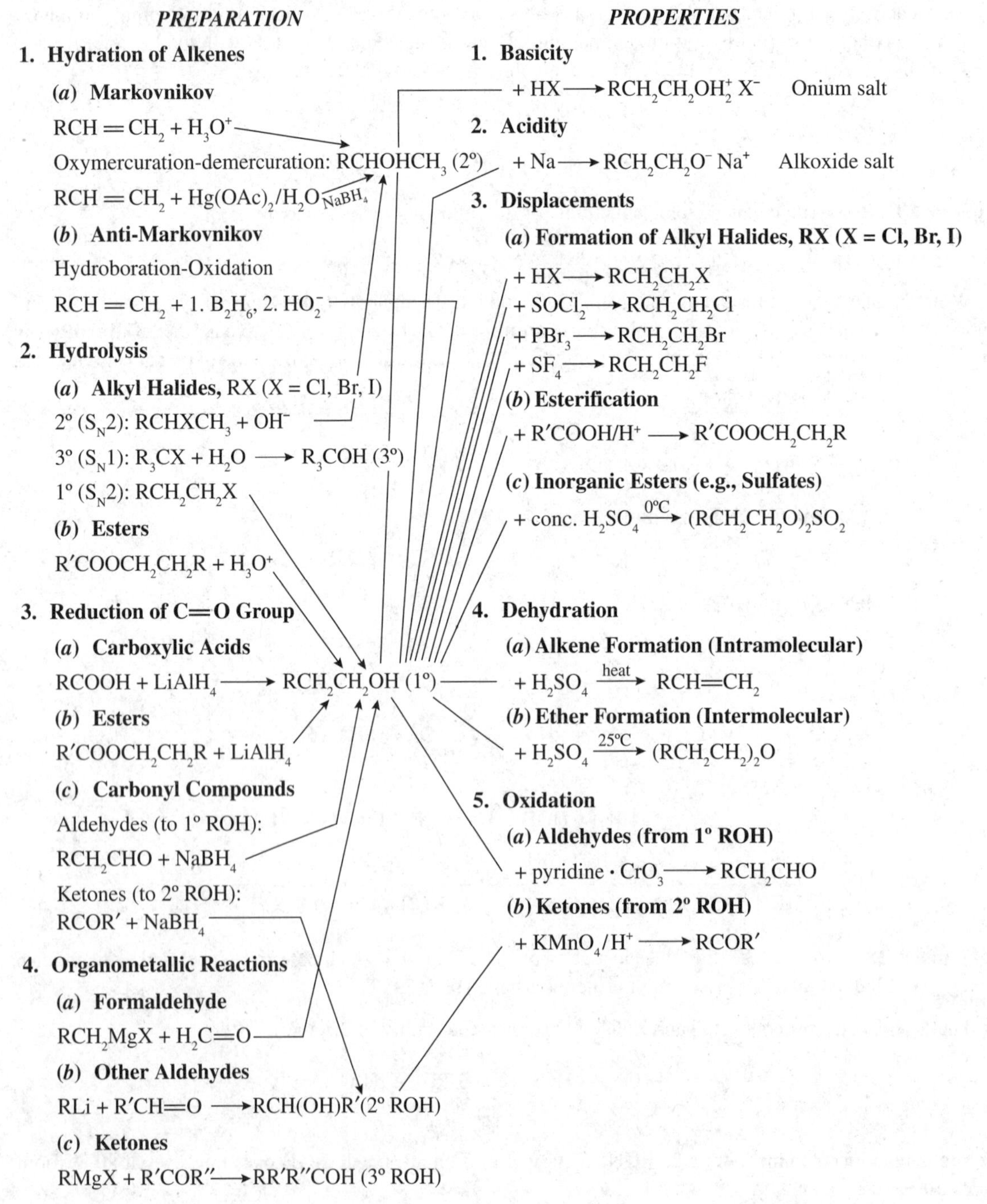

B. THIOLS

13.5 General

Also called **mercaptans**, these compounds, with the generic formula RSH, are sulfur analogs of alcohols, just as H_2S is of H_2O. The O atom of the OH has been replaced by an S atom, and the —SH group is denominated **sulfhydryl**.

A physical characteristic is their odor; 2-butene-1-thiol, $CH_3CH{=}CHCH_2SH$, and 3-methyl-1-butanethiol, $CH_3CH(CH_3)CH_2CH_2SH$ contribute to the skunk smell.

Thiols are biochemically important because they are oxidized with mild reagents to disulfides, which are found in insulin and proteins.

$$\underset{\text{a thiol}}{2\text{R—S—H}} \underset{\text{Zn, H}^+}{\overset{\text{Br}_2}{\rightleftharpoons}} \underset{\text{a disulfide}}{\text{R—S—S—R}}$$

Problem 13.25 How are thiols prepared in good yield?

The preparation of thiols by S_N2 attack of nucleophilic HS^- on an alkyl halide gives poor yields because the mercaptan loses a proton to form an anion, RS^-, which reacts with a second molecule of alkyl halide to form a thioether.

$$\text{H}\ddot{\text{S}}{:}^- + \text{R—}\ddot{\text{X}}{:} \longrightarrow {:}\ddot{\text{X}}{:}^- + \text{R}\ddot{\text{S}}{:}\text{H} \xrightarrow{-\text{H}^+} \text{R}\ddot{\text{S}}{:}^- \xrightarrow{+\text{R—X}} \underset{\text{thioether}}{\text{R—}\ddot{\text{S}}\text{—R}} + {:}\ddot{\text{X}}{:}^-$$

Dialkylation is minimized by using an excess of $:SH^-$ and is avoided by using thiourea to form an alkylisothiourea salt that is then hydrolyzed.

$$\underset{\text{Thiourea}}{(\text{H}_2\text{N})_2\text{C}{=}\ddot{\text{S}}{:}} \xrightarrow{+\text{R—X}} \left[(\text{H}_2\text{N})_2\text{C}{=}\overset{+}{\ddot{\text{S}}}\text{—R}\right]\text{X}^- \xrightarrow[\text{NaOH}]{\text{H}_2\text{O}} \underset{\text{Urea}}{(\text{H}_2\text{N})_2\text{C}{=}\text{O}} + \text{R—SH}$$

Problem 13.26 Show steps in the synthesis of ethyl ethanesulfonate, $CH_3CH_2SO_2OCH_2CH_3$, from CH_3CH_2Br and any inorganic reagents.

The alkyl sulfonic acid is made by oxidizing the thiol, which in turn comes from the halide.

$$CH_3CH_2Br \xrightarrow[\text{HS}^-]{\text{excess}} CH_3CH_2SH \xrightarrow{\text{KMnO}_4} CH_3CH_2SO_3H$$

$$CH_3CH_2Br \xrightarrow{\text{OH}^-} CH_3CH_2OH \qquad CH_3CH_2SO_3H \xrightarrow{\text{PCl}_5} CH_3CH_2SO_2Cl$$

$$CH_3CH_2OH + CH_3CH_2SO_2Cl \xrightarrow[\textit{a base}]{\text{pyridine}} CH_3CH_2SO_2OCH_2CH_3$$

Problem 13.27 Why are mercaptans (*a*) more acidic ($K_a \approx 10^{-11}$) than alcohols ($K_a \approx 10^{-17}$) and (*b*) more nucleophilic than alcohols?

(*a*) There are more and stronger H-bonds in alcohols, thus producing an acid-weakening effect. Also, in the conjugate bases RS^- and RO^-, the charge is more dispersed over the larger S, thereby making RS^- the weaker base and RSH the stronger acid (Section 3.11). (*b*) The larger S is more easily polarized than the smaller O and therefore is more nucleophilic. For example, RSH participates more rapidly in S_N2 reactions than ROH. Recall that among the halide anions, nucleophilicity also increases as size increases: $F^- < Cl^- < Br^- < I^-$.

Problem 13.28 Offer mechanisms for

$$CH_3CH{=}CH_2 \begin{cases} \xrightarrow{H_2S,\ H^+} CH_3CH(SH)CH_3 \\ \xrightarrow[h\nu]{H_2S} CH_3CH_2CH_2SH \end{cases}$$

Acid-catalyzed addition has an ionic mechanism (Markovnikov):

$$CH_3CH{=}CH_2 + H^+ \longrightarrow CH_3\overset{+}{C}HCH_3 \xrightarrow[-H^+]{H_2S} CH_3\underset{}{\overset{SH}{\overset{|}{C}}}HCH_3$$

The peroxide- or light-catalyzed reaction has a free-radical mechanism (anti-Markovnikov):

$$H{-}S{-}H \xrightarrow{h\nu} HS\cdot + \cdot H$$
$$CH_3CH{=}CH_2 + \cdot SH \longrightarrow CH_3\dot{C}HCH_2SH$$
$$CH_3\dot{C}HCH_2SH \xrightarrow{H_2S} CH_3CH_2CH_2SH + \cdot SH$$

13.6 Summary of Thiol Chemistry

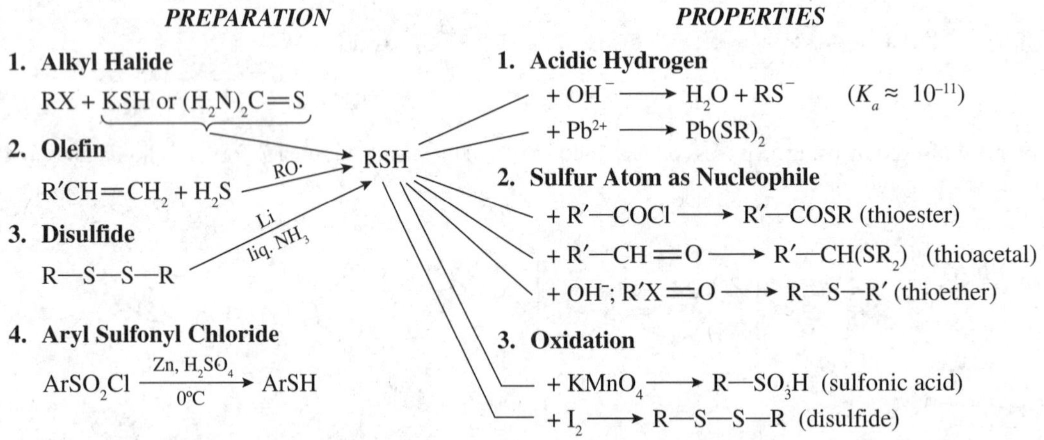

PREPARATION		*PROPERTIES*
1. Alkyl Halide $RX + KSH$ or $(H_2N)_2C{=}S$ → RSH		**1. Acidic Hydrogen** $+ OH^- \longrightarrow H_2O + RS^-$ ($K_a \approx 10^{-11}$); $+ Pb^{2+} \longrightarrow Pb(SR)_2$
2. Olefin $R'CH{=}CH_2 + H_2S \xrightarrow{RO\cdot}$ RSH	RSH	**2. Sulfur Atom as Nucleophile** $+ R'{-}COCl \longrightarrow R'{-}COSR$ (thioester); $+ R'{-}CH{=}O \longrightarrow R'{-}CH(SR_2)$ (thioacetal); $+ OH^-;\ R'X{=}O \longrightarrow R{-}S{-}R'$ (thioether)
3. Disulfide $R{-}S{-}S{-}R \xrightarrow[\text{liq. } NH_3]{Li}$ RSH		**3. Oxidation** $+ KMnO_4 \longrightarrow R{-}SO_3H$ (sulfonic acid); $+ I_2 \longrightarrow R{-}S{-}S{-}R$ (disulfide)
4. Aryl Sulfonyl Chloride $ArSO_2Cl \xrightarrow[0°C]{Zn,\ H_2SO_4} ArSH$		

SUPPLEMENTARY PROBLEMS

Problem 13.29 Give the IUPAC names for each of the following alcohols. Which are 1°, 2°, and 3°?

(*a*) $CH_3{-}CH_2{-}CH(CH_3){-}CH_2{-}OH$ (*b*) $HO{-}CH_2{-}CH(CH_3){-}CH_2C_6H_5$ (*c*) 1-methylcyclopentanol (cyclopentane ring bearing CH_3 and OH on the same carbon)

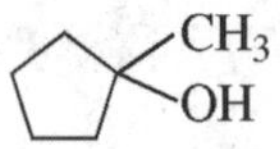

(*d*) $CH_3C(CH_2CH_3)(OH){-}CH_2CH_3$ (*e*) $Cl{-}CH(C(CH_3)_2C_6H_4\text{-}m\text{-}Cl)CH_2CH{=}CHCH_2OH$, i.e. Cl—CHCH₂CH=CHCH₂OH with CH₃—C—CH₃ bearing C_6H_4-*m*-Cl attached to the CH

(*a*) 2-Methyl-1-butanol, 1°. (*b*) 2-Methyl-3-phenyl-1-propanol, 1°. (*c*) 1-Methyl-1-cyclopentanol, 3°. (*d*) 3-Methyl-3-pentanol, 3°. (*e*) 5-Chloro-6-methyl-6-(3-chlorophenyl)-2-hepten-1-ol. (The longest chain with

OH has seven C's and the prefix is **hept-**. Numbering begins at the end of the chain with OH; therefore, -1-ol. The aromatic ring substituent has Cl at the 3 position counting from the point of attachment, and it is put in parentheses to show that the entire ring is attached to the chain at C^6. Cl on the chain is at C^5.) 1°.

Problem 13.30 Write condensed structural formulas and give IUPAC names for (*a*) vinylcarbinol, (*b*) diphenylcarbinol, (*c*) dimethylethylcarbinol, (*d*) benzylcarbinol.

(*a*) $H_2C{=}CHCH_2OH$ 2-Propen-1-ol (Allyl alcohol)

(*b*) $(C_6H_5)_2CH{-}OH$ Diphenylmethanol (Benzhydrol)

(*c*) $(CH_3)_2C(CH_2CH_3){-}OH$ 2-Methyl-2-butanol

(*d*) $C_6H_5{-}\overset{\beta}{C}H_2\overset{\alpha}{C}H_2OH$ 2-Phenylethanol (β-Phenylethanol)

Problem 13.31 The four isomeric C_4H_9OH alcohols are

(i) $(CH_3)_3COH$ (ii) $CH_3CH_2CH_2CH_2OH$ (iii) $(CH_3)_2CHCH_2OH$ (iv) $CH_3CH(OH)CH_2CH_3$

Synthesize each, using a different reaction from among a reduction, an S_N2 displacement, a hydration, and a Grignard reaction.

Synthesis of the 3° isomer, (i), has restrictions: the S_N2 displacement of a 3° halide cannot be used because elimination would occur; nor is there any starting material that can be reduced to a 3° alcohol. Either of the two remaining methods can be used; arbitrarily the Grignard is chosen.

$$CH_3MgI + (CH_3)_2C{=}O \xrightarrow[\text{hydrolysis}]{\text{after}} (CH_3)_3COH$$

The S_N2 displacement on the corresponding RX is best for 1° alcohols such as (ii) and (iii); let us choose (ii) for this synthesis.

$$CH_3CH_2CH_2CH_2Cl + OH^- \xrightarrow{-Cl^-} CH_3CH_2CH_2CH_2OH$$

The 1° alcohol, (iii), and the 2° alcohol, (iv), can be made by either of the two remaining syntheses. However, the one-step hydration with H_3O^+ to give (iv) is shorter than the two-step hydroboration-oxidation to give (iii).

$$CH_3CH_2CH{=}CH_2 + H_3O^+ \xrightarrow{-H^+} CH_3CH(OH)CH_2CH_3$$

Finally, (iii) is made by reducing the corresponding RCH$=$O or RCOOH.

$$(CH_3)_2CH_2COOH \xrightarrow[2.\ H^+]{1.\ LiAlH_4} (CH_3)_2CH_2CH_2OH$$

Problem 13.32 Prepare ethyl *p*-chlorophenylcarbinol by a Grignard reaction.

Prepare this 2° alcohol, *p*-$ClC_6H_4CHOHCH_2CH_3$, from RCHO and R′MgX. Since the groups on the carbinol C are different, there are two combinations possible:

(1) $p\text{-}ClC_6H_4C(H){=}O \xrightarrow[2.\ H_3O^+]{1.\ CH_3CH_2MgCl} p\text{-}ClC_6H_4CHOHCH_2CH_3$
(ring Cl has not interfered)

(2) $p\text{-}ClC_6H_4MgBr \xrightarrow[2.\ H_3O^+]{1.\ CH_3CH_2CHO} p\text{-}ClC_6H_4CHOHCH_2CH_3$

Br is more reactive than Cl when making a Grignard of *p*-ClC_6H_4Br.

Problem 13.33 Give the hydroboration-oxidation product from (*a*) cyclohexene, (*b*) *cis*-2-phenyl-2-butene, (*c*) *trans*-2-phenyl-2-butene.

Addition of H_2O is *cis* anti-Markovnikov. See Fig. 13.2. In Fig. 13.2(*c*), the second pair of conformations show the eclipsing of the H's with each other and of the Me's with each other. The more stable staggered conformations are not shown.

(*a*)

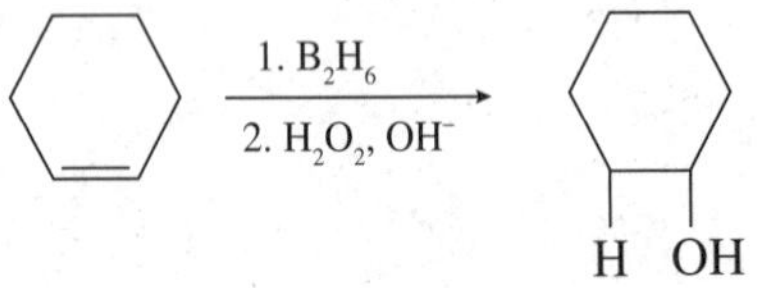

1. B_2H_6
2. H_2O_2, OH^-

H OH

(*b*)

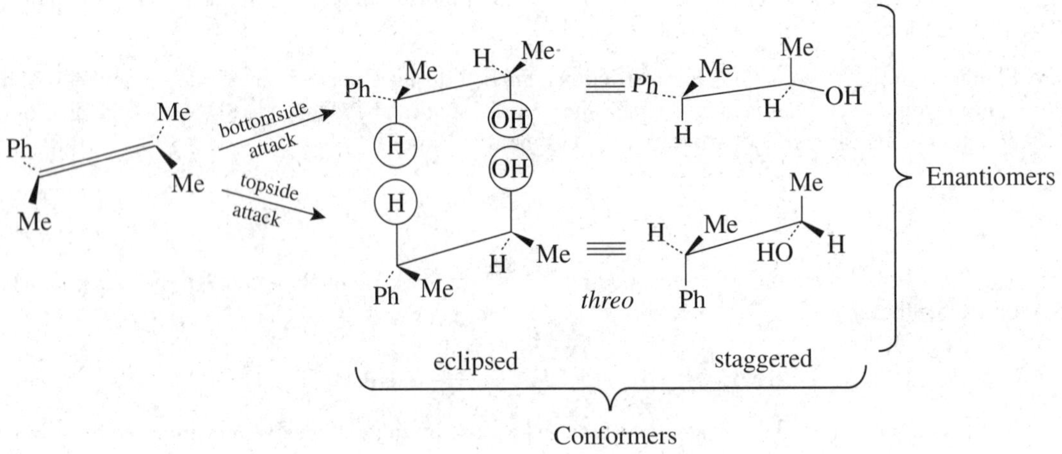

Ph Me Me Me

bottomside attack

topside attack

threo

eclipsed staggered

Conformers

Enantiomers

(*c*)

Ph Me Me H

topside attack

bottomside attack

erythro

Conformers

Enantiomers

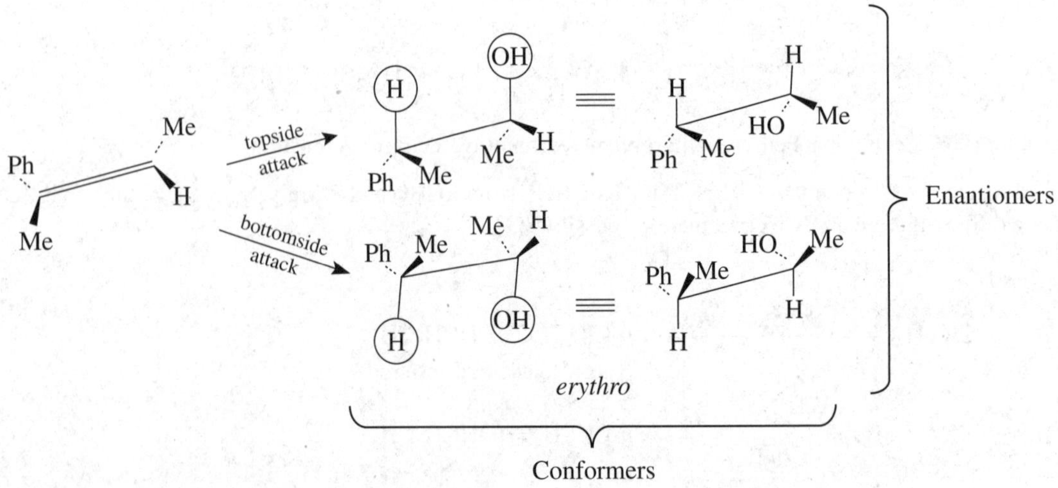

Figure 13.2

Problem 13.34 The following Grignard reagents and aldehydes or ketones are reacted and the products hydrolyzed. What alcohol is produced in each case? (*a*) Benzaldehyde ($C_6H_5CH{=}O$) and C_2H_5MgBr. (*b*) Acetaldehyde and phenyl magnesium bromide. (*c*) Acetone and benzyl magnesium bromide. (*d*) Formaldehyde and cyclohexyl magnesium bromide. (*e*) Acetophenone ($C_6H_5\underset{\|}{\overset{}{C}}CH_3$, with C=O) and ethyl magnesium bromide.

(*a*) $C_6H_5CH{=}O + C_2H_5MgBr \longrightarrow C_6H_5{-}C(H)(C_2H_5){-}OH$ 1-Phenyl-1-propanol

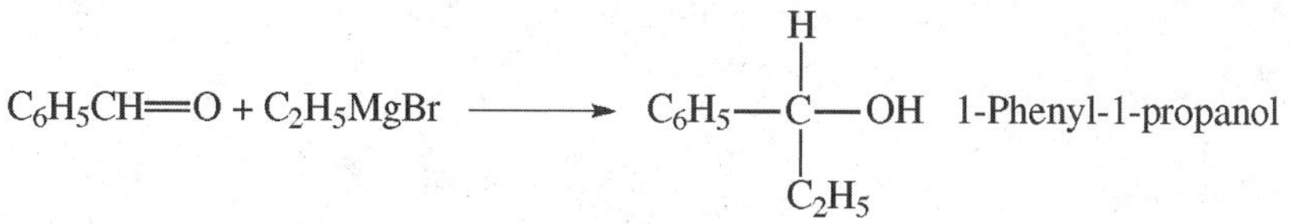

(*b*) $CH_3CH{=}O + C_6H_5MgBr \longrightarrow C_6H_5{-}C(H)(CH_3){-}OH$ 1-Phenyl-1-ethanol

(*c*) $CH_3{-}C({=}O){-}CH_3 + C_6H_5CH_2MgBr \longrightarrow C_6H_5CH_2{-}C(CH_3)(OH){-}CH_3$ 1-Phenyl-2-methyl-2-propanol

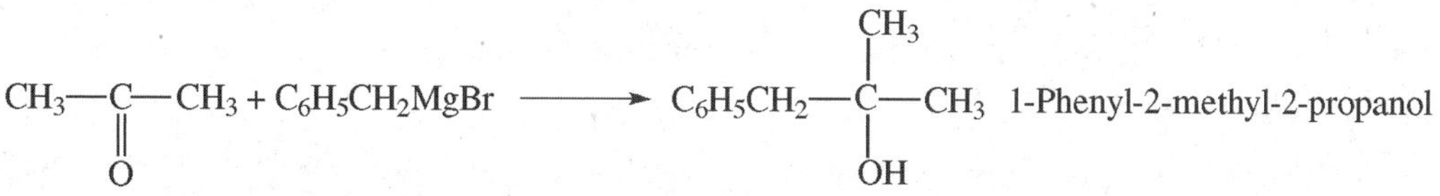

(*d*) $H_2C{=}O + C_6H_{11}MgBr \longrightarrow C_6H_{11}CH_2OH$ Cyclohexylcarbinol

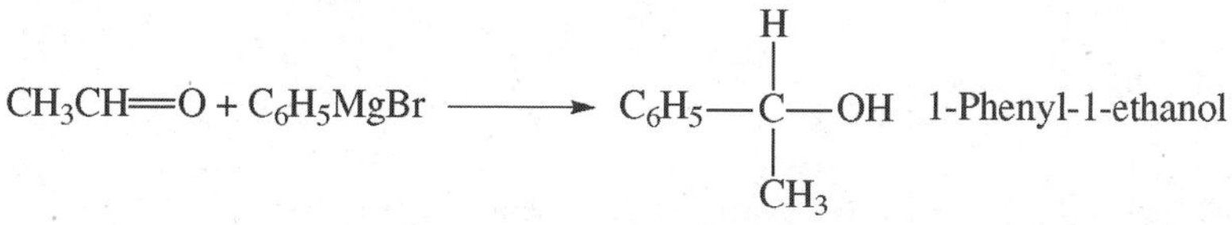

(*e*) $C_6H_5{-}C({=}O){-}CH_3 + C_2H_5MgBr \longrightarrow C_6H_5{-}C(CH_3)(OH){-}CH_2CH_3$ 2-Phenyl-2-butanol

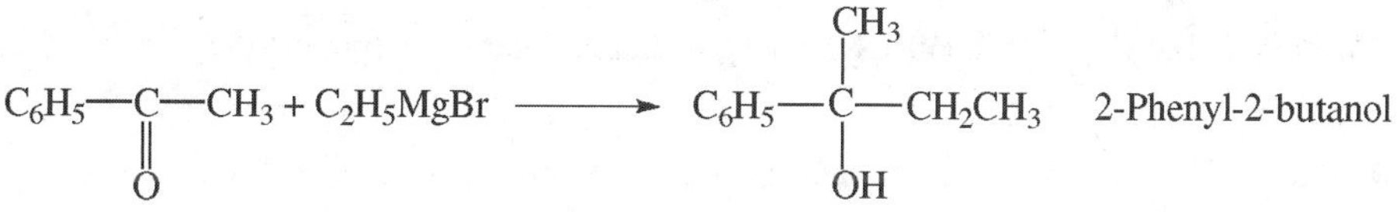

Problem 13.35 Give the mechanism in each case:

(*a*) $CH_3CH_2CH_2CH_2CH_2OH \xrightarrow{HCl} CH_3CH_2CH_2CH_2CH_2Cl$ (*b*) $CH_3CH(CH_3){-}CH(OH)CH_3 \xrightarrow{HCl} CH_3C(CH_3)(Cl)CH_2CH_3$

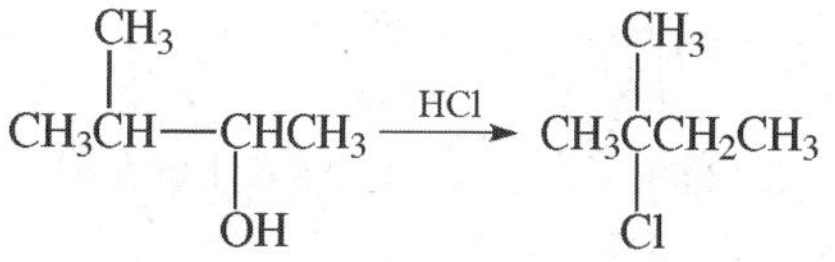

(*c*) $CH_3{-}C(CH_3)(OH){-}CH_2CH_3 \xrightarrow{HCl} CH_3{-}C(CH_3)(Cl){-}CH_2CH_3$

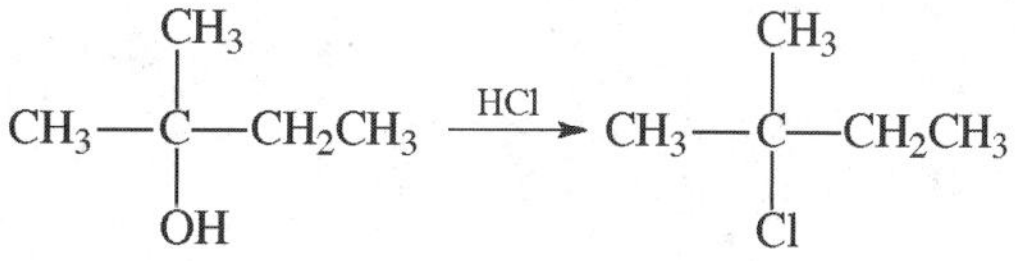

Why did rearrangement occur only in (*b*)?

(*a*) The mechanism is S_N2, since we are substituting Cl for H_2O from 1° ROH_2^+.
(*b*) The mechanism is S_N1.

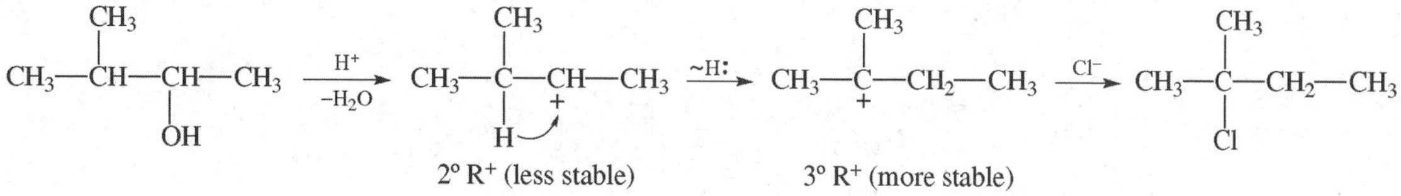

$$CH_3{-}CH(CH_3){-}CH(OH){-}CH_3 \xrightarrow[-H_2O]{H^+} CH_3{-}C(CH_3)(H){-}\overset{+}{C}H{-}CH_3 \xrightarrow{\sim H:} CH_3{-}\overset{+}{C}(CH_3){-}CH_2{-}CH_3 \xrightarrow{Cl^-} CH_3{-}C(CH_3)(Cl){-}CH_2{-}CH_3$$

2° R^+ (less stable) 3° R^+ (more stable)

(*c*) S_N1 mechanism. The stable 3° $(CH_3)_2\overset{+}{C}CH_2CH_3$ reacts with Cl^- with no rearrangement.

Problem 13.36 Why does dehydration of 1-phenyl-2-propanol in acid form 1-phenyl-1-propene rather than 1-phenyl-2-propene?

1-Phenyl-1-propene, $PhCH{=}CHCH_3$, is a more highly substituted alkene and therefore more stable than 1-phenyl-2-propene, $PhCH_2CH{=}CH_2$. Even more important, it is more stable because the double bond is conjugated with the ring.

Problem 13.37 Write the structural formulas for the alcohols formed by oxymercuration-demercuration from (*a*) 1-heptene, (*b*) 1-methylcyclohexene, (*c*) 3,3-dimethyl-1-butene.

The net addition of H_2O is Markovnikov.

(*a*) $CH_3(CH_2)_4CHOHCH_3$ 2-Heptanol

(*b*) (cyclohexane ring bearing OH and CH_3 on the same carbon) 1-Methylcyclohexanol

(*c*) $(CH_3)_3CCHOHCH_3$ 3,3-Dimethyl-2-butanol (no rearrangement occurs)

Problem 13.38 List the alcohols and acids that compose the inorganic esters (*a*) $(CH_3)_3COCl$ and (*b*) $CH_3CH_2ONO_2$. Name the esters.

Conceptually hydrolyze the O to the heteroatom bond while adding an **H** to the O and an **OH** to the heteroatom. (*a*) $(CH_3)_3$**COH** and **HOCl**, ***t*-butyl hypochlorite**. (*b*) CH_3CH_2**OH** and **HONO**$_2$, **ethyl nitrate**. Tert-butyl hypochlorite is used to chlorinate hydrocarbons by free-radical chain mechanisms.

Problem 13.39 Starting with isopropyl alcohol as the only available organic compound, prepare 2,3-dimethyl-2-butanol.

This 3° alcohol, $(CH_3)_2COHCH(CH_3)_2$, is prepared from a Grignard reagent and a ketone.

$$CH_3CH(OH)CH_3 \xrightarrow{H^+,\ MnO_4^-} CH_3C(=O)CH_3$$

$$CH_3CH(OH)CH_3 \xrightarrow{PBr_3} CH_3CHBrCH_3 \xrightarrow[\text{ether}]{Mg} CH_3CH(MgBr)CH_3$$

$$CH_3C(=O)CH_3 + CH_3CH(MgBr)CH_3 \longrightarrow CH_3\!-\!C(CH_3)(\overset{-}{O}\overset{+}{M}gBr)\!-\!CH(CH_3)\!-\!CH_3 \xrightarrow{H_3O^+} CH_3C(CH_3)(OH)\!-\!CH(CH_3)\!-\!CH_3 + Mg^{2+} + Br^-$$

Problem 13.40 Alcohols such as Ph_2CHCH_2OH rearrange on treatment with acid; they can be dehydrated by heating their methyl xanthates (**Tschugaev reaction**). The pyrolysis proceeds by a cyclic transition state. Outline the steps using Ph_2CHCH_2OH.

$$Ph_2CHCH_2OH \xrightarrow{Na} Ph_2CHCH_2O^-Na^+ \xrightarrow{S{=}C{=}S} Ph_2CHCH_2{-}O{-}C({=}S){-}S^-Na^+$$

a xanthate

$$Ph_2CHCH_2{-}O{-}C({=}S){-}S^-Na^+ \xrightarrow{CH_3I} Ph_2C(H)\!-\!CH_2\!-\!O\!-\!C(=S)\!-\!SCH_3 \longrightarrow Ph_2C{=}CH_2 + S{=}C{=}O + HSCH_3$$

an *s*-methylxanthate

The elimination is *cis*.

Problem 13.41 Methyl ketones give the **haloform test**:

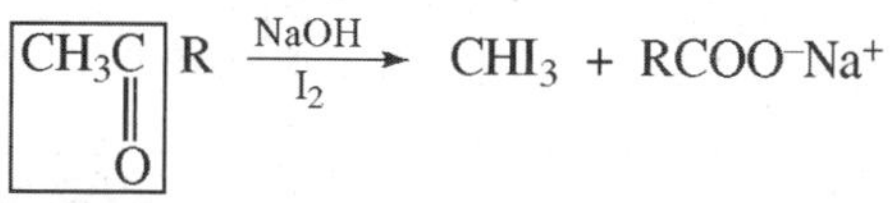

$$\boxed{\underset{\substack{\| \\ O}}{CH_3C}}R \xrightarrow[I_2]{NaOH} CHI_3 + RCOO^-Na^+$$

a methyl ketone — yellow precipitate

I_2 can also oxidize 1° and 2° alcohols to carbonyl compounds. Which butyl alcohols give a positive haloform test?

Alcohols with the

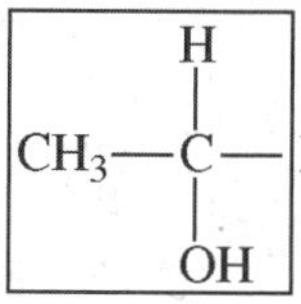

$$\boxed{CH_3-\overset{H}{\underset{OH}{C}}-}R$$

groups are oxidized to

$$CH_3\underset{\substack{\| \\ O}}{C}-R$$

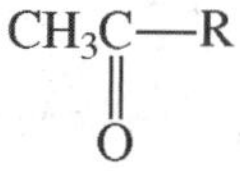

and give a positive test. The only butyl alcohol giving a positive test is

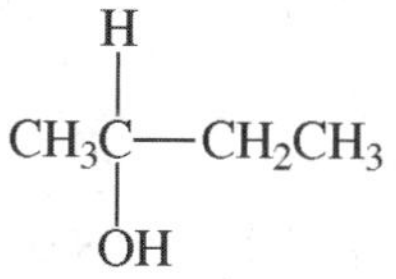

$$CH_3\overset{H}{\underset{OH}{C}}-CH_2CH_3$$

Problem 13.42 A compound, $C_9H_{12}O$, is oxidized under vigorous conditions to benzoic acid. It reacts with CrO_3 and gives a positive iodoform test (Problem 13.41). Is this compound chiral?

Since benzoic acid is the product of oxidation, the compound is a monosubstituted benzene, C_6H_5G. Subtracting C_6H_5 from $C_9H_{12}O$ gives C_3H_7O as the formula for a saturated side chain. A positive CrO_3 test means a 1° or 2° OH. Possible structures are:

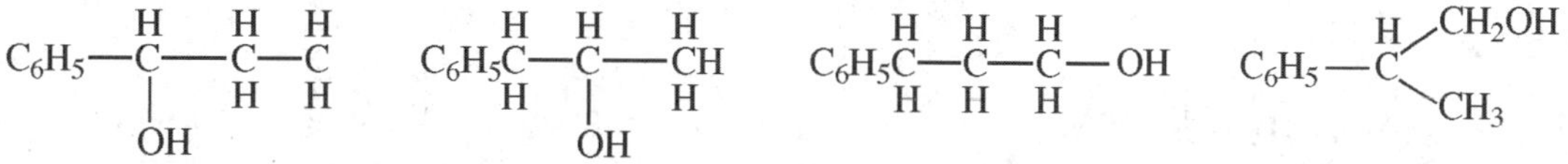

$$C_6H_5-\overset{H}{\underset{OH}{C}}-\overset{H}{\underset{H}{C}}-\overset{H}{\underset{H}{C}} \quad \text{I}$$

$$C_6H_5\overset{H}{\underset{H}{C}}-\overset{H}{\underset{OH}{C}}-\overset{H}{\underset{H}{CH}} \quad \text{II}$$

$$C_6H_5\overset{H}{\underset{H}{C}}-\overset{H}{\underset{H}{C}}-\overset{H}{\underset{H}{C}}-OH \quad \text{III}$$

$$C_6H_5-C(H)(CH_2OH)(CH_3) \quad \text{IV}$$

Only II has the $-CH(OH)CH_3$ needed for a positive iodoform test. II is chiral.

Problem 13.43 Suggest a possible industrial preparation for (*a*) *t*-butyl alcohol, (*b*) allyl alcohol, (*c*) glycerol ($HOCH_2CHOHCH_2OH$).

$$(a)\quad \underset{\text{Isobutane}}{CH_3-\overset{CH_3}{CH}-CH_3} \xrightarrow[\text{heat}]{\text{catalyst}} H_2 + \underset{\text{Isobutylene}}{CH_3-\overset{CH_3}{C}=CH_2} \xrightarrow{H_3O^+} CH_3-\overset{CH_3}{\underset{OH}{C}}-CH_3$$

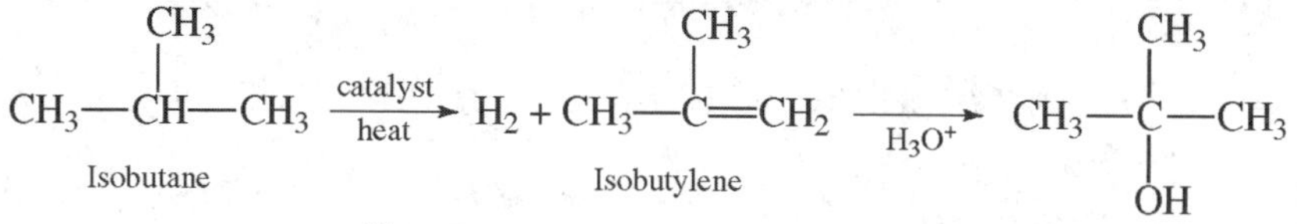

$$(b)\quad \underset{\text{Propene}}{CH_3-CH=CH_2} \xrightarrow[\text{heat}]{Cl_2} \underset{\text{Allyl chloride}}{ClCH_2-CH=CH_2} \xrightarrow{HOH} \underset{\text{Allyl alcohol}}{HOCH_2CH=CH_2}$$

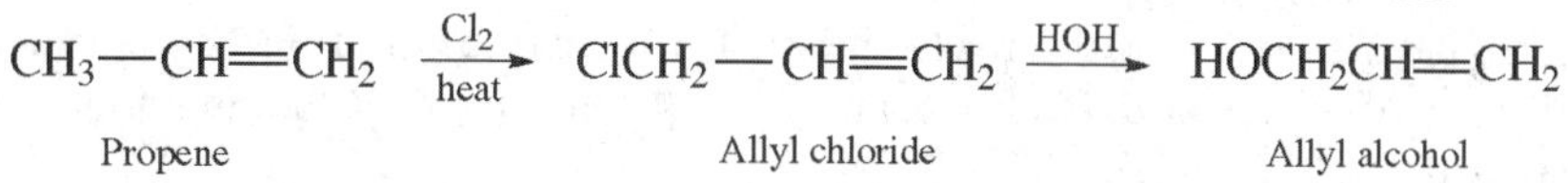

(c) $ClCH_2—CH{=}CH_2 + HOCl$ (Allyl chloride) → $CH_2(Cl)—CH(Cl)—CH_2(OH)$ and $CH_2(Cl)—CH(OH)—CH_2(Cl)$; each $\xrightarrow{OH^-}$ $CH_2(OH)—CH(OH)—CH_2(OH)$ (Glycerol)

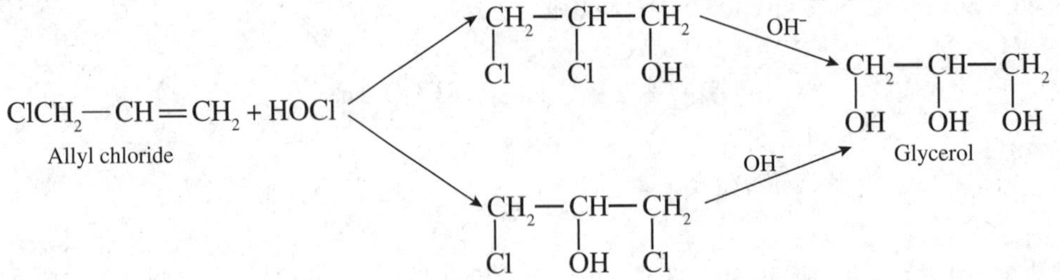

Problem 13.44 Assign numbers, from (1) for the LOWEST to (5) for the HIGHEST, to indicate *relative reactivity* with HBr in forming benzyl bromides from the following benzyl alcohols: (*a*) *p*-Cl—C_6H_4—CH_2OH, (*b*) $(C_6H_5)_2CHOH$, (*c*) *p*-O_2N—C_6H_4—CH_2OH, (*d*) $(C_6H_5)_3COH$, (*e*) $C_6H_5CH_2OH$.

Differences in reaction rates depend on the relative abilities of the protonated alcohols to lose H_2O to form R^+. The stability of R^+ affects the $\Delta H^‡$ for forming the incipient R^+ in the transition state and determines the overall rate.

Electron-attracting groups such as NO_2 and Cl in the *para* position destabilize R^+ by intensifying the positive charge. NO_2 of (*c*) is more effective since it destabilizes by both resonance and induction, while Cl of (*a*) destabilizes only by induction. The more C_6H_5's on the benzyl C, the more stable is R^+.

(*a*) 2 (*b*) 4 (*c*) 1 (*d*) 5 (*e*) 3

Problem 13.45 Supply structural formulas and stereochemical designations for the organic compounds (A) through (H).

(*a*) (A) + conc. H_2SO_4 $\xrightarrow{\text{cold}}$ $CH_3—CH(OSO_3H)CH_2CH_3$ $\xrightarrow[H^+]{H_2O}$ (B)

(*b*) (R)-$CH_3CHOHCH_2CH{=}CH_2$ $\xrightarrow[H_2SO_4]{H_2O}$ (C) + (D)

(*c*) (R)-$CH_3CH_2—CH(OH)—CH{=}CH_2$ + HBr ⟶ (E) + (F) + (G)

(*a*) (A) is *cis* or *trans*-$CH_3CH{=}CHCH_3$ or $H_2C{=}CHCH_2CH_3$; (B) is *rac*-$CH_3CHOHCH_2CH_3$. The intermediate $CH_3C^+HCH_2CH_3$ can be attacked from either side by HSO_4^- to give an optically inactive racemic hydrogen sulfate ester which is hydrolyzed to *rac*-2-butanol.

(*b*) (C) is (R,R)-CH_3—C(OH)(H)—CH_2—C(H)(OH)—CH_3; (D) is (R,S)-CH_3—C(OH)(H)—CH_2—C(OH)(OH)—CH_3

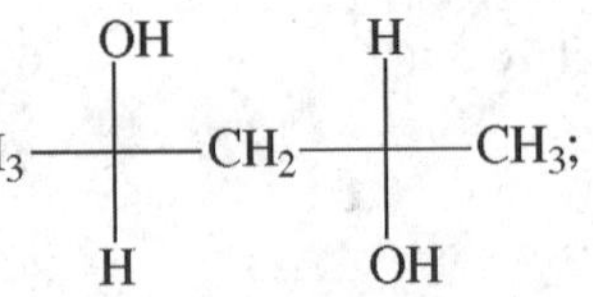

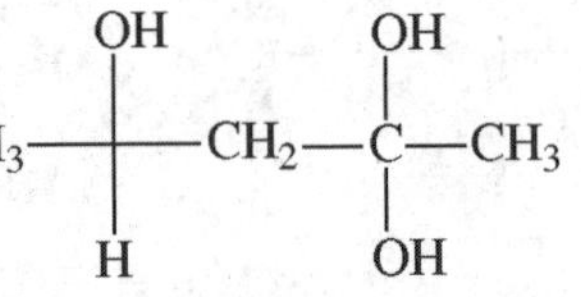

optically active *meso*

Hydration follows Markovnikov's rule, and a new similar chiral C is formed. The original chiral C remains R, but the new chiral C may be R or S. The two diastereomers are not formed in equal amounts [see Problem 5.20(*b*)].

(*c*) (E) is *rac*-$CH_3CH_2CHBrCH{=}CH_2$; (F) is *trans*- and (G) *cis*-$CH_3CH_2CH{=}CHCH_2Br$. The intermediate R^+ in this S_N1 reaction is a resonance-stabilized (charge-delocalized) allylic cation

$$[CH_3CH_2\overset{+}{C}HCH{=}CH_2 \longleftrightarrow CH_3CH_2CH{=}CH\overset{+}{C}H_2] \text{ or } CH_3CH_2\overset{\delta+}{C}H{\cdots}CH{\cdots}\overset{\delta+}{C}H_2$$

In the second step, Br^- attacks either of the positively charged C's to give one of the three products. Since R^+ is flat, the chiral C in (E) can be R or S; (E) is racemic. (F) is the major product because it is most stable (*trans* and disubstituted).

Problem 13.46 How does the Lewis theory of acids and bases explain the functions of (*a*) $ZnCl_2$ in the Lucas reagent? (*b*) ether as a solvent in the Grignard reagent?

(*a*)
$$\underset{\textit{Lewis acid}}{ZnCl_2} + 2HCl \longrightarrow \underset{\text{(a stronger acid than HCl)}}{H_2ZnCl_4} \xrightarrow{ROH} ROH_2^+ \xrightarrow{Cl^-} RCl + H_2O$$

(*b*) R′MgX acts as a Lewis acid because Mg can coordinate with one unshared electron pair of each O of two ether molecules to form an addition compound:

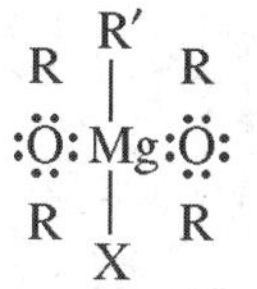

that is soluble in ether.

Problem 13.47 Draw Newman projections of the conformers of the following substituted ethanols and predict their relative populations: (*a*) FCH_2CH_2OH, (*b*) $H_2NCH_2CH_2OH$, (*c*) $BrCH_2CH_2OH$.

If the substituents F, H_2N, and Br are designated by Z, the conformers may be generalized as *anti* or *gauche*.

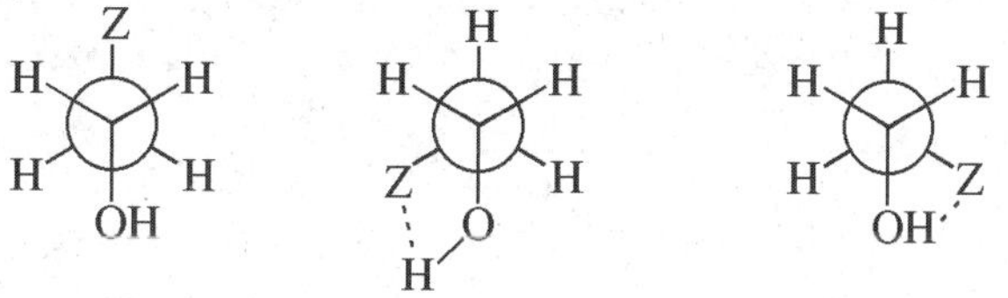

anti *gauche*

For (*a*) and (*b*), the *gauche* is the more stable conformer and has a greater population because of H-bonding with F and N. The *anti* conformer is more stable in (*c*) because there is no H-bonding with Br and dipole-dipole repulsion causes OH and Br to lie as far from each other as possible.

Problem 13.48 Deduce the structure of a compound, $C_4H_{10}O$, which gives the following nmr data: $\delta = 0.8$ (doublet, six H's), $\delta = 1.7$ (complex multiplet, one H), $\delta = 3.2$ (doublet, two H's) and $\delta = 4.2$ (singlet, one H; disappears after shaking sample with D_2O).

The singlet at $\delta = 4.2$ which disappears after shaking with D_2O is for O<u>H</u> (Problem 13.28). The compound must be one of the four butyl alcohols. Only isobutyl alcohol, $(CH_3)_2CHCH_2OH$, has six equivalent H's (two CH_3's), accounting for the six-H doublet at $\delta = 0.8$, the one-H multiplet at $\delta = 1.7$, and the two-H doublet which is further downfield at $\delta = 3.2$ because of the electron-attracting O.

Problem 13.49 The attempt to remove water from ethanol by fractional distillation gives 95% ethanol, an **azeotrope** that boils at a constant temperature of 78.15°C. It has a lower boiling point than either water (100°C) or ethanol (78.3°C). A liquid mixture is an azeotrope if it gives a vapor of the same composition. How does boiling 95% ethanol with Mg remove the remaining H_2O?

$$Mg + H_2O \longrightarrow H_2 + \underset{\text{insoluble}}{Mg(OH)_2}$$

The dry ethanol, called **absolute**, is now distilled from the insoluble $Mg(OH)_2$.

Problem 13.50 Explain why the most prominent (base) peak of 1-propanol is at $m/e = 31$, while that of allyl alcohol is at $m/e = 57$.

$CH_3CH_2—CH_2OH^+$ cleaves mainly into

$$CH_3CH_2\cdot + \overset{+}{C}H_2\ddot{O}H \longleftrightarrow H_2C{=}\overset{+}{\ddot{O}}H$$

($m/e = 31$) rather than $CH_3CH_2\overset{+}{C}HOH$ + H·, because C—C is weaker than C—H. In allyl alcohol:

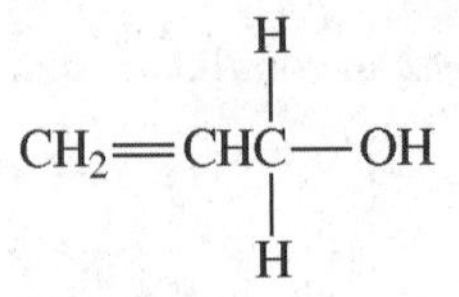

$$CH_2{=}CH\overset{H}{\underset{H}{C}}—OH$$

the C—H bond cleaves to give

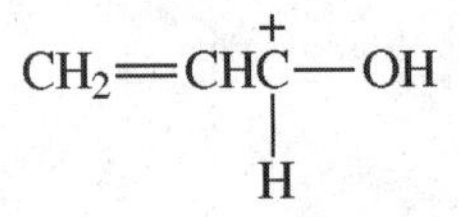

$$CH_2{=}CH\overset{+}{\underset{H}{C}}—OH$$

($m/e = 57$). This cation is stabilized by both the $CH_2{=}\underset{|}{C}H$ and the O.

Problem 13.51 Inorganic acids such as H_2SO_4, H_3PO_4, and HOCl (hypochlorous acid) form esters. Write structural formulas for (*a*) dimethyl sulfate, (*b*) tribenzyl phosphate, (*c*) diphenyl hydrogen phosphate, (*d*) *t*-butyl nitrite, (*e*) lauryl hydrogen sulfate (lauryl alcohol is *n*-$C_{11}H_{23}CH_2OH$), (*f*) sodium lauryl sulfate.

Replacing the H of the OH of an acid gives an ester.

(*a*) $CH_3O—\overset{O}{\underset{O}{S}}—OCH_3$ (*b*) $(PhCH_2O)_3P—O$ (*c*) $(PhO)_2\overset{O}{P}—OH$ (*d*) $(CH_3)_3CONO$

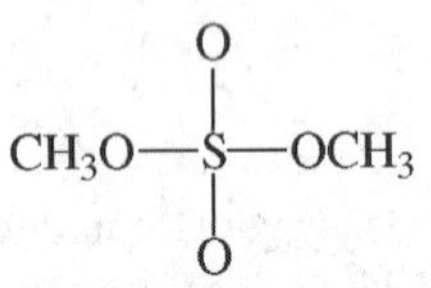

(*e*) *n*-$C_{11}H_{23}CH_2O\overset{O}{\underset{O}{S}}—OH$ (*f*) *n*-$C_{11}H_{23}CH_2OSO_2O^-Na^+$ (a detergent)

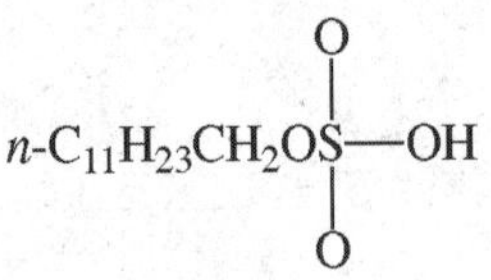

Problem 13.52 Explain why trialkyl phosphates are readily hydrolyzed with OH^- to dialkyl phosphate salts, whereas dialkyl hydrogen phosphates and alkyl dihydrogen phosphates resist alkaline hydrolysis.

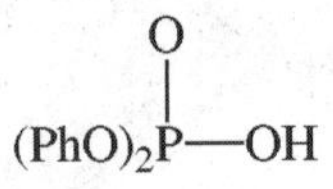

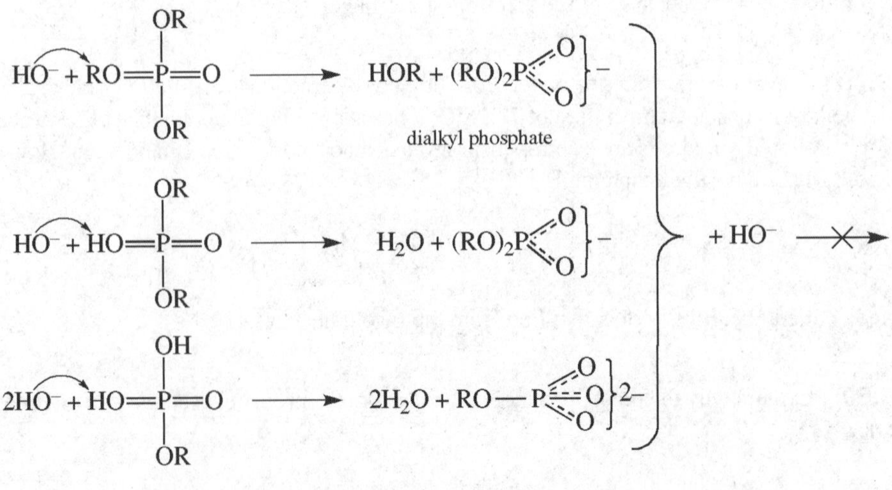

$$HO^- + RO{=}\overset{OR}{\underset{OR}{P}}{=}O \longrightarrow HOR + (RO)_2P\langle\!\begin{smallmatrix}O\\O\end{smallmatrix}\!\rangle^-$$

dialkyl phosphate

$$HO^- + HO{=}\overset{OR}{\underset{OR}{P}}{=}O \longrightarrow H_2O + (RO)_2P\langle\!\begin{smallmatrix}O\\O\end{smallmatrix}\!\rangle^-$$

$$2HO^- + HO{=}\overset{OH}{\underset{OR}{P}}{=}O \longrightarrow 2H_2O + RO—P\langle\!\begin{smallmatrix}O\\O\\O\end{smallmatrix}\!\rangle^{2-}$$

(all three products) $+ HO^- \not\longrightarrow$

The hydrogen phosphates are moderately strong acids and react with bases to form anions (conjugate bases). Repulsion between negatively charged species prevents further reaction between these anions and OH^-.

Problem 13.53 The ir spectrum of RSH shows a weak —S—H stretching band at about 2600 cm^{-1} that does not shift significantly with concentration or nature of solvent. Explain the difference in behavior of S—H and O—H bonds.

The S—H bond is weaker than the O—H bond and therefore absorbs at a lower frequency. There is little or no H-bonding in S—H and, unlike the O—H bond, there is little shifting of absorption frequency on dilution.

Ethers, Epoxides, Glycols, and Thioethers

A. ETHERS

14.1 Introduction and Nomenclature

Simple (symmetrical) ethers have the general formula R—O—R or Ar—O—Ar; **mixed (unsymmetrical) ethers** are R—O—R′ or Ar—O—Ar′ or Ar—O—R. The derived system names R and Ar as separate words and adds the word "ether." In the IUPAC system, ethers (ROR) are named as alkoxy- (RO-) substituted alkanes. Cyclic ethers have at least one O in a ring.

Problem 14.1 Give a derived and IUPAC name for the following ethers: (*a*) $CH_3OCH_2CH_2CH_2CH_3$, (*b*) $(CH_3)_2CHOCH(CH_3)CH_2CH_3$, (*c*) $C_6H_5OCH_2CH_3$, (*d*) *p* -$NO_2C_6H_4OCH_3$, (*e*) $CH_3OCH_2CH_2OCH_3$.

(*a*) Methyl *n*-butyl ether, 1-methoxybutane. (*b*) *sec*-Butyl isopropyl ether, 2-isopropoxybutane. (Select the longest chain of C's as the alkane root.) (*c*) Ethyl phenyl ether, ethoxybenzene (commonly called **phenetole**). (*d*) Methyl *p*-nitrophenyl ether, 4-nitromethoxybenzene (or *p*-**nitroanisole**). (*e*) 1,2- dimethoxyethane.

Problem 14.2 Account for the following: (*a*) Ethers have significant dipole moments (≈ 1.18 D). (*b*) Ethers have lower boiling points than isomeric alcohols. (*c*) The water solubilities of isomeric ethers and alcohols are comparable.

(*a*) The C—O—C bond angle is about 110° and the dipole moments of the two C—O bonds do not cancel. (*b*) The absence of OH in ethers precludes H-bonding and therefore there is no strong intermolecular force of attraction between ether molecules as there is between alcohol molecules. The weak polarity of ethers has no appreciable effect. (*c*) The O of ethers is able to undergo H-bonding with H of H_2O.

R₂O---H—O—H (R, R, O---H, O, H, H-bond)

14.2 Preparation

Simple Ethers

1. **Intermolecular Dehydration of Alcohols** (see Section 13.3)
2. **2° Alkyl Halides with Silver Oxide**

$$2(CH_3)_2CHCl + Ag_2O \longrightarrow (CH_3)_2CHOCH(CH_3)_2 + 2AgCl$$

Mixed Ethers

1. ***Williamson Synthesis***

$$\underset{\text{alkoxide}}{Na^+R'\ddot{\underset{\cdot\cdot}{O}}:^-} + \underset{1°}{R(:X)} \xrightarrow{(S_N2)} R'OR + Na^+ + X^- \qquad X = Cl, Br, I, OSO_2R, OSO_2Ar$$

Problem 14.3 Specify and account for your choice of an alkoxide and an alkyl halide to prepare the following ethers by the Williamson reaction: (*a*) $C_2H_5OC(CH_3)_3$, (*b*) $(CH_3)_2CHOCH_2CH{=}CH_2$.

(*a*) $C_2H_5X + Na^+\ ^-OC(CH_3)_3$ (*b*) $(CH_3)_2CHO^-Na^+ + XCH_2CH{=}CH_2$

The 2° and 3° alkyl halides readily undergo E2 eliminations with strongly basic alkoxides to form alkenes. Hence, to prepare mixed ethers such as (*a*) and (*b*), the 1° alkyl groups should come from RX and the 2° and 3° alkyl groups should come from the alkoxide.

Problem 14.4 Show how dimethyl sulfate, $MeOSO_2OMe$, is used in place of alkyl halides in the Williamson syntheses of methyl ethers.

Alkyl sulfates are conjugate bases of the very strongly acidic alkyl sulfuric acids and are very good leaving groups. Dimethyl sulfate is less expensive than CH_3I, the only liquid methyl halide at room temperature. Liquids are easier to use than gases in laboratory syntheses.

$$:R\ddot{\underset{\cdot\cdot}{O}}:^- + \underset{\text{Dimethyl sulfate (very toxic)}}{Me(:OSO_2OMe)} \longrightarrow ROMe + MeOSO_3^-$$

2. **Alkoxymercuration-demercuration** (Section 13.2 for ROH preparation)

Whereas mercuration-demercuration of alkenes in the presence of water gives alcohols, in alcohol solvents (free from H_2O), ethers result. These reactions in the presence of nucleophilic solvents such as water and alcohols are examples of **solvomercuration**. The mercuric salts usually used are the acetate, $Hg(OAc)_2$ (—OAc is an abbreviation for $-\underset{\substack{\|\\O}}{C}CH_3$ written —OCCH₃ with C=O) or the trifluoracetate, $Hg(OCOCF_3)_2$.

$$RCH{=}CH_2 + R'OH + Hg(O\overset{O}{\overset{\|}{C}}{-}CF_3)_2 \longrightarrow \underset{R'O\quad HgOCOCF_3}{RCHCH_2} \xrightarrow{NaBH_4} \underset{R'O\quad H}{RCHCH_2}$$

Problem 14.5 Suggest a mechanism consistent with the following observations for the solvomercuration of $RCH{=}CH_2$ with R′OH in the presence of $Hg(OAc)_2$, leading to the formation of $RCH(OR')CH_2Hg(OAc)$: (i) no rearrangement, (ii) Markovnikov addition, (iii) *anti* addition, and (iv) reaction with nucleophilic solvents.

Absence of rearrangement excludes a carbocation intermediate. Stereoselective *anti* addition is reminiscent of a bromonium-ion-type intermediate, in this case a three-membered ring **mercurinium** ion. The *anti* regioselective Markovnikov addition requires an S_N1-type backside attack by the nucleophilic solvent on the more substituted C of the ring—the C bearing more of the partial + charge (δ^+). Frontside attack is blocked by the large HgOAc group.

$$RCH{=}CH_2 + Hg(OAc)_2 \longrightarrow RCH\overset{+}{\text{---}}CH_2\ (\text{bridged by HgOAc}) \xrightarrow[-H^+]{HOR'} RCH(OR')CH_2HgOAc$$

Although the mercuration step is stereospecific, the reductive demercuration step is not, and therefore neither is the overall reaction.

Problem 14.6 Give the alkene and alcohol needed to prepare the following ethers by alkoxymercuration-demercuration: (*a*) diisopropyl ether, (*b*) 1-methyl-1-methoxycyclopentane, (*c*) 1-phenyl-1-ethoxypropane, (*d*) di-*t*-butyl ether.

(*a*) $CH_2{=}CHCH_3$ and $CH_3CH(OH)CH_3$ give $(CH_3)_2CHOCH(CH_3)_2$.

(*b*) CH_3 and CH_3OH give H_3C OCH_3.

(*c*) $PhCH{=}CHCH_3$ and CH_3CH_2OH give $PhCH(OCH_2CH_3)CH_2CH_3$. Although each double-bonded C is 2°, ROH preferentially bonds with the more positively charged benzylic C.

(*d*) Ethers with two 3° alkyl groups cannot be synthesized in decent yields because of severe steric hindrance.

Problem 14.7 Give (*a*) S_N2 and (*b*) S_N1 mechanisms for formation of ROR from ROH in conc. H_2SO_4.

(*a*) (1) $\underset{base_1}{ROH} + \underset{acid_2}{H_2SO_4} \longrightarrow \underset{acid_1}{R{:}OH_2^+} + \underset{base_2}{HSO_4^-}$

(2) $ROH + R{:}OH_2^+ \longrightarrow R\overset{+}{O}(H)R + H_2O$ (conjugate acid of an ether)

(3) $R\overset{+}{O}(H)R + HSO_4^-(\text{or } ROH) \longrightarrow ROR + H_2SO_4\ (\text{or } ROH_2^+)$

(*b*) (1) Same as (1) of (*a*) (2) $R{:}OH_2^+ \longrightarrow R^+ + H_2O$

(3) $ROH + R^+ \longrightarrow R\overset{+}{O}(H)R$ (4) Same as (3) of (*a*)

Problem 14.8 Compare the mechanisms for the formation of an ether by intermolecular dehydration of (*a*) 1°, (*b*) 3°, and (*c*) 2° alcohols.

(*a*) For 1° alcohols, the mechanism is S_N2, with alcohol as the attacking nucleophile and water as the leaving group. There would be no rearrangements.

(*b*) The mechanism for 3° alcohols is S_N1. However, a 3° carbocation such as Me_3C^+ cannot react with Me_3COH, the parent 3° alcohol, or any other 3° ROH because of severe steric hindrance. It can react with a 1° RCH_2OH, if such an alcohol is present.

$$Me_3C^+ + RCH_2OH \xrightarrow{-H^+} Me_3COCH_2R \quad \text{(a mixed ether)}$$

The 3° carbocation can also readily eliminate H^+ to give an alkene, $Me_2C{=}CH_2$.

(*c*) 2° alcohols react either way. Rearrangements may occur when they react by the S_N1 mechanism, because the intermediate is a carbocation.

Problem 14.9 List the ethers formed in the reaction between concentrated H_2SO_4, and equimolar quantities of ethanol and (*a*) methanol and (*b*) *tert*-butanol.

(*a*) These 1° alcohols react by S_N2 mechanisms to give a mixture of three ethers: $C_2H_5OC_2H_5$ from $2C_2H_5OH$, CH_3OCH_3 from $2CH_3OH$, and $C_2H_5OCH_3$ from C_2H_5OH and CH_3OH.

(*b*) This is an S_N1 reaction.

$$(CH_3)_3COH \xrightarrow{H^+} (CH_3)_3C\overset{+}{O}H_2 \xrightarrow{-H_2O} (CH_3)_3\overset{+}{C} \xrightarrow[-H^+]{CH_3CH_2OH} \underset{\text{Ethyl } tert\text{-butyl ether}}{(CH_3)_3C{-}O{-}CH_2CH_3}$$

Reaction between $(CH_3)_3C^+$ and $(CH_3)_3COH$ is sterically hindered and occurs much less readily.

Problem 14.10 Use any needed starting material to synthesize the following ethers, selecting from among intermolecular dehydration, Williamson synthesis, and alkoxymercuration-demercuration. Justify your choice of method.

(*a*) $CH_3(CH_2)_3OCH_2CH_3$ (*b*) $CH_3CH_2\underset{\displaystyle CH_3}{\underset{|}{C}}HOCH_2CH_2CH_3$ (*c*) dicyclohexyl ether

(*a*) Use the Williamson synthesis: $CH_3(CH_2)_3Cl + C_2H_5O^-Na^+$. Since alkoxymercuration is a Markovnikov addition, it cannot be used to prepare an ether in which both R's are 1°. Unless one of the R's can form a stable R^+, intermolecular dehydration cannot be used to synthesize a mixed ether.

(*b*)
$$\overset{4}{C}H_3\overset{3}{C}H_2\overset{2}{C}H{=}\overset{1}{C}H_2 + Hg(OCOCF_3)_2 + n\text{-}C_3H_7OH \longrightarrow \overset{4}{C}H_3\overset{3}{C}H_2\overset{2}{C}H(\overset{1}{C}H_2HgOCOCF_3){-}OC_3H_7\text{-}n \xrightarrow{NaBH_4} \overset{4}{C}H_3\overset{3}{C}H_2\overset{2}{C}H(\overset{1}{C}H_3)OC_3H_7\text{-}n$$

This is better than the Williamson synthesis because there is no competing elimination reaction.

(*c*) Dehydration; Cyclohexyl—OH $\xrightarrow{H_2SO_4}$ $(Cyclohexyl)_2O$. This is a simple ether.

Problem 14.11 (*R*)-2-Octanol and its ethyl ether are levorotatory. Predict the configuration and sign of rotation of the ethyl ether prepared from this alcohol by: (*a*) reacting with Na and then C_2H_5Br; (*b*) reacting in a solvent of low dielectric constant with concentrated HBr and then with $C_2H_5O^-Na^+$.

(*a*) No bond to the chiral C of the alcohol is broken in this reaction; hence, the *R* configuration is unchanged but rotation is indeterminant. (*b*) These conditions for the reaction of the alcohol with HBr favor an S_N2

mechanism of the chiral C is inverted. Attack by RO^- is also S_N2 and the net result of two inversions is retention of configuration. Again, rotation is unpredictable.

$$\text{H}-\overset{\displaystyle C_6H_{13}}{\underset{\displaystyle CH_3}{\underset{(R)}{C}}}-\text{OH} \xrightarrow{\text{HBr}} \text{Br}-\overset{\displaystyle C_6H_{13}}{\underset{\displaystyle CH_3}{\underset{(S)}{C}}}-\text{H} \xrightarrow{Na^+\bar{O}C_2H_5} \text{H}-\overset{\displaystyle C_6H_{13}}{\underset{\displaystyle CH_3}{\underset{(R)}{C}}}-O-C_2H_5$$

14.3 Chemical Properties

Brönsted and Lewis Basicity

Ethers are among the most unreactive functional groups toward reagents used in organic chemistry (for an exceptional subclass, see Section 14.6). This property, together with their ability to dissolve nonpolar compounds, makes them good solvents for organic compounds. Because of the unshared pairs of electrons, —Ö— is a basic site, and ethers are protonated by strong acids, such as conc. H_2SO_4, to form oxonium cations, R_2OH^+. Ethers also react with Lewis acids.

Problem 14.12 (*a*) Why do ethers dissolve in cold concentrated H_2SO_4 and separate out when water is added to the solution? (*b*) Why are ethers used as solvents for BF_3 and the Grignard reagent?

(*a*) Water is a stronger base than ether and removes the proton from the protonated ether.

Step 1 (solution)

$$\underset{\text{base}_1}{ROR} + \underset{\text{acid}_2}{H_2SO_4} \longrightarrow \underset{\text{acid}_1}{\overset{\displaystyle H}{\overset{|}{ROR^+}}} + \underset{\text{base}_2}{HSO_4^-}$$

Step 2 (dissolution)

$$\underset{\text{acid}_1}{R_2OH^+} + \underset{\text{base}_2}{H_2O} \longrightarrow \underset{\text{base}_1}{ROR} + \underset{\text{acid}_2}{H_3O^+}$$

(*b*) BF_3 and RMgX are Lewis acids which share a pair of electrons on the —Ö— of ethers.

$$R_2\ddot{O}: + BF_3 \longrightarrow R_2\overset{+}{\ddot{O}}:\overset{-}{B}F_3 \qquad 2R_2\ddot{O}: + R'MgX \longrightarrow R_2\overset{+}{\ddot{O}}:\overset{\displaystyle X}{\underset{\displaystyle R'}{Mg^{2-}}}:\overset{+}{\ddot{O}}R_2$$

Notice that two ether molecules coordinate with one Mg atom.

Cleavage

Ethers are cleaved by concentrated HI (ROR + HI → ROH + RI). With excess HI, 2 mol of RI are formed (ROR + 2HI → 2RI).

Problem 14.13 Identify the ethers that are cleaved with excess HI to yield (*a*) $(CH_3)_3CI$ and $CH_3CH_2CH_2I$, (*b*) cyclohexyl and methyl iodides, (*c*) $I(CH_2)_5I$.

(*a*) $(CH_3)_3COCH_2CH_2CH_3$

(*b*) (cyclohexyl)—O—CH_3

(*c*) The diiodide with I's on the terminal C's indicate a cyclic ether, (tetrahydropyran ring with O)

Problem 14.14 (*a*) Show how the cleavage of ethers with HI can proceed by an S_N2 or an S_N1 mechanism. (*b*) Why is HI a better reagent than HBr for this type of reaction? (*c*) Why do reactions with excess HI afford two moles of RI?

(*a*) **Step 1** $\underset{\text{base}_1}{R—O—R'} + \underset{\text{acid}_2}{HI} \longrightarrow \underset{\text{acid}_1}{R—\overset{+}{O}(H)—R'} + \underset{\text{base}_2}{I^-}$

Step 2 for S_N2 $I^- + R—\overset{+}{O}(H)—R' \xrightarrow{\text{slow}} RI + HOR'$ (R is 1°)

Step 2 for S_N1 $R\overset{+}{O}(H)R' \xrightarrow{\text{slow}} R^+ + R'OH$ (R is 3°)

Step 3 for S_N1 $R^+ + I^- \longrightarrow RI$

(*b*) HI is a stronger acid than HBr and gives a greater concentration of the oxonium ion

$$R—\overset{+}{O}(H)—R$$

I^- is also a better nucleophile in the S_N2 reactions than is Br^-.

(*c*) The first-formed ROH reacts in typical fashion with HI to give RI.

Problem 14.15 Account for the following observations:

$$(CH_3)_3COCH_3 \xrightarrow[\text{(1)}]{\text{anhyd. HI, ether}} CH_3I + (CH_3)_3COH$$

$$(CH_3)_3COCH_3 \xrightarrow[\text{(2)}]{\text{aq. HI}} CH_3OH + (CH_3)_3CI$$

The high polarity of the solvent (H_2O) in reaction (2) favors an S_N1 mechanism, giving the 3° R^+.

$$CH_3\overset{+}{O}(H)C(CH_3)_3 \longrightarrow CH_3OH + (CH_3)_3C^+ \xrightarrow{I^-} (CH_3)_3CI$$

The low polarity of solvent (ether) in reaction (1) favors the S_N2 mechanism and the nucleophile, I^-, attacks the 1° C of CH_3.

$$I^- + CH_3\overset{+}{O}(H)C(CH_3)_3 \longrightarrow CH_3I + HOC(CH_3)_3$$

Free-Radical Substitution

Ethers readily undergo free-radical substitution of an H on the α C, $-\overset{\displaystyle H}{\underset{|}{\overset{|}{C^{\alpha}}}}-O-$.

Problem 14.16 Can resonance account for the preferential radical substitution at the α-C of ethers?

The intermediate α radical $R\dot{C}HOR$ is not stabilized by delocalization of electron density by the adjacent O through extended π bonding. One resonance structure would have 9 electrons on O.

$$\left[R\dot{C}(H)-\ddot{\underset{..}{O}}-R \;\not\leftrightarrow\; RC(H)=\dot{\underset{..}{O}}-R \right]$$

Problem 14.17 (*a*) Give a mechanism for the formation of the explosive solid hydroperoxides, for example:

$$R\underset{}{\overset{OOH}{\overset{|}{C}}}HOCH_2R$$

from ethers and O_2. (*b*) Why should ethers be purified before distillation?

(*a*) **Initiation Step** $\quad RCH_2OCH_2R + \cdot\ddot{O}-\ddot{O}\cdot \longrightarrow R\dot{C}HOCH_2R + H-\ddot{O}-\ddot{O}\cdot$

Propagation Step 1 $\quad R\dot{C}HOCH_2R + \cdot\ddot{O}-\ddot{O}\cdot \longrightarrow RCH(O-O\cdot)OCH_2R$

Propagation Step 2 $\quad RCH(O-O\cdot)OCH_2R + RCH_2OCH_2R \longrightarrow RCH(O-OH)OCH_2R + R\dot{C}HOCH_2R$

(*b*) An ether may contain hydroperoxides which concentrate as the ether is distilled and which may then explode. Ethers are often purified by mixing with $FeSO_4$ solution, which reduces the hydroperoxides to the nonexplosive alcohols (ROOH → ROH).

Problem 14.18 Does peroxide formation occur more rapidly with $(RCH_2)_2O$ or $(R_2CH)_2O$?

With $(R_2CH)_2O$ because the 2° radical is more stable and forms faster.

Electrophilic Substitution of Aryl Ethers

The—OR group is a moderately activating *op*-directing group (Problem 11.8).

Problem 14.19 Give the main products of (*a*) mononitration of *p*-methylphenetole, (*b*) monobromination of *p*-methoxyphenol.

(*a*) Since OR is a stronger activating group than R groups, the NO_2 attacks *ortho* to $-OC_2H_5$; the product is 2-nitro-4-methylphenetole. (*b*) OH is a stronger activating group than OR groups; the product is 2-bromo-4-methoxyphenol.

14.4 Cyclic Ethers

The chemistry of cyclic ethers such as those shown below is similar to that of open chain ethers.

Tetrahydrofuran (Oxacyclopentane) | Tetrahydropyran (THP) (Oxacyclohexane) | Pyran | 1,4-Dioxane (1,4-Dioxacyclohexane)

Problem 14.20 Which alcohol would undergo dehydration to give (*a*) THP? (*b*) 1,4-dioxane?

(*a*) Since the product is a cyclic ether, the starting material must be a diol with OH groups on the terminal C's. The diol must have five CH_2 groups to match the number in THP. The alcohol used in $HO(CH_2)_5OH$, 1,5-pentanediol. This is an *intramolecular* dehydration to form an ether.

(*b*) 1,4-Dioxane has two ether groups, requiring dehydration between two pairs of OH groups. Again, the starting alcohol must be a diol, but now the dehydration is *intermolecular*. The alcohol used is ethylene glycol, $HOCH_2CH_2OH$.

Problem 14.21 (*a*) With the aid of the mechanism show why DHP, unlike typical alkenes, readily undergoes the following reaction:

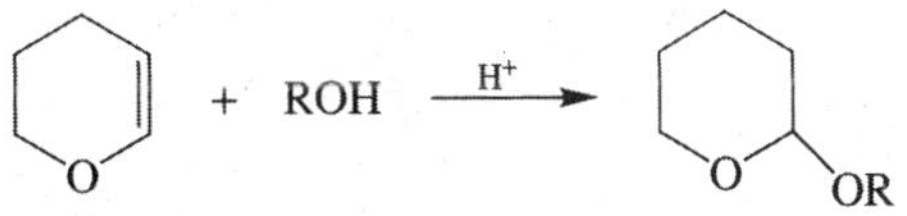

$$\text{DHP} + \text{ROH} \xrightarrow{H^+} \text{THP–OR}$$

2,3-Dihydro-4H-pyran (DHP) a tetrahydropyranyl (THP) ether

(*b*) Why do THP ethers, unlike ordinary ethers, cleave under *mildly* aqueous acidic conditions?

(*a*) The H^+ adds to C═C to generate a carbocation with the positive charge on the C that is α to the —Ö— of the ring. This is a fairly stable cation because the positive charge is stabilized by delocalization of electron density from the O atom.

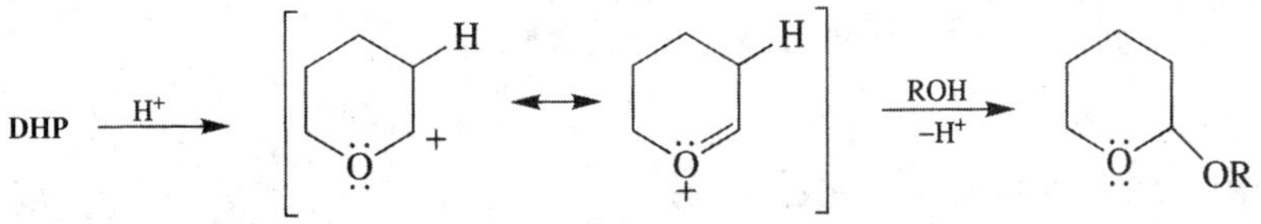

$$\text{DHP} \xrightarrow{H^+} [\text{carbocation} \longleftrightarrow \text{oxonium ion}] \xrightarrow[-H^+]{\text{ROH}} \text{THP–OR}$$

The nucleophilic site (—Ö—) of ROH then bonds to the C^+ of the carbocation, forming an onium ion of the ether, which loses a proton to the solvent alcohol (ROH) and becomes the ether product.

(*b*) Aqueous acid reverses the reaction of (*a*), reforming the same intermediate carbocation. This loses a proton to give the C═C, rather than reacting with water to give the very unstable alcohol-analog of the ether.

Problem 14.22 Since THP ethers are, like most ethers, stable in base, their formation can be used to protect the OH group from reacting under basic conditions. Using this fact, show how to convert $HOCH_2CH_2Cl$ to $HOCH_2CH_2D$ via the Grignard reagent.

$HOCH_2CH_2Cl$ cannot be converted directly to the Grignard because of the presence of the acidic OH group; the product obtained would be $HOCH_2CH_3$. The desired reaction is achieved by protecting the OH group, as shown schematically:

$$\text{DHP} + HOCH_2CH_2Cl \rightarrow ClCH_2CH_2O\text{—THP} \rightarrow DCH_2CH_2O\text{—THP} \rightarrow DCH_2CH_2OH + \text{DHP}$$

To generalize on Problem 14.22, a good **protecting group** (i) is easily attached, (ii) permits the desired chemistry to occur, and (iii) is easily, removed. Other methods for protecting OH groups involve **benzyl** and **silyl** ethers:

$$ROH + BrCH_2Ph \xrightarrow[-AgBr]{Ag^+} \underset{\text{a benzyl ether}}{ROCH_2Ph} \xrightarrow{H_2/Pt} ROH + CH_3Ph$$

$$ROH + \underset{\text{Chlorotrimethylsilane}}{Cl—Si(CH_3)_3} \xrightarrow[-HCl]{\text{pyridine}} \underset{\text{a trimethylsilyl ether}}{ROSi(CH_3)_3} \xrightarrow{H_3O^+} ROH + HOSi(CH_3)_3$$

Crown ethers are large-ring cyclic ethers with several O atoms. A typical example is 18-crown-6 ether [Fig. 14.1(*a*)]. The first number in the name is the total number of atoms in the ring; the second number is the number of O atoms. Crown ethers are excellent solvaters of cations of salts through formation of ion-dipole bonds. 18-Crown-6 ether strongly complexes and traps K^+, from, for instance, KF, as shown in Fig. 14.1(*b*).

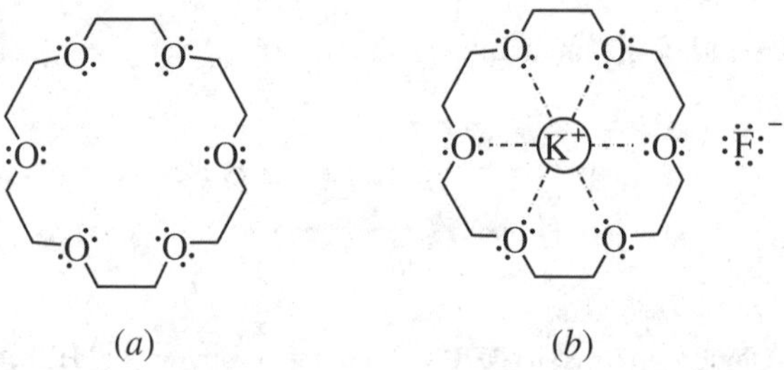

Figure 14.1

Problem 14.23 Suggest two important synthetic uses of crown ethers.

(1) They enable inorganic salts to be used in nonpolar solvents, a media with which salts are typically incompatible. (2) The cation of the salt is complexed in the center of the crown ether, leaving the anion "bare" and enhanced in reactivity. These effects of crown ethers are similar to those achieved with phase-transfer agents. The "bare" anion is also present when polar aprotic solvents are used.

14.5 Summary of Ether Chemistry

PREPARATION

1. Intermolecular Dehydration

$2ROH + H_2SO_4$ (1° or 2°) → $R\ddot{O}R\ (R')$

2. S_N2 Displacement (Williamson)

$RO^- + RX$ (1°) → $R\ddot{O}R\ (R')$

3. Alkoxymercuration-demercuration

$RCH{=}CH_2 + R'OH \longrightarrow RCH(OR')CH_3$

PROPERTIES

1. Basicity

$+ H_2SO_4 \longrightarrow R\overset{H}{\ddot{O}}R^+\ HSO_4^-$

$+ BF_3 \longrightarrow R_2OBF_3$

2. Cleavage

$+ 2HI \longrightarrow 2RI \text{ or } RI + R'I$

3. Substitution on α C

$+ O_2 \longrightarrow \overset{\alpha}{C}HOR$ (with OOH on the α C)

B. EPOXIDES

14.6 Introduction

Unique among cyclic ethers are those with three-membered rings, the **epoxides** or **oxiranes**. *Their large ring strains make them highly reactive.*

14.7 Synthesis

1. From Alkenes and a Peroxyacid, RC(=O)OOH (Problem 6.28) **(Mainly, *m*-Chloroperoxybenzoic Acid)**

Problem 14.24 (*a*) Give the structural formula of the epoxide formed when *m*-chloroperoxybenzoic acid reacts with (i) *cis*-2-butene and (ii) *trans*-2-butene. (The epoxides are stereoisomers.) (*b*) What can you say about the stereochemistry of the epoxidation? (*c*) Why is a carbocation *not* an intermediate?

(*a*)

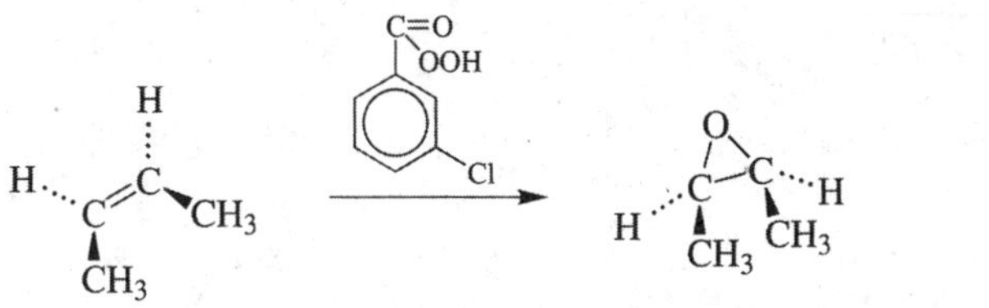

(i) *cis*-2-Butene → (m-ClC₆H₄C(=O)OOH) → cis-2,3-Dimethyloxirane

(ii) *trans*-2-Butene → (m-ClC₆H₄C(=O)OOH) → *trans*-2,3-Dimethyloxirane

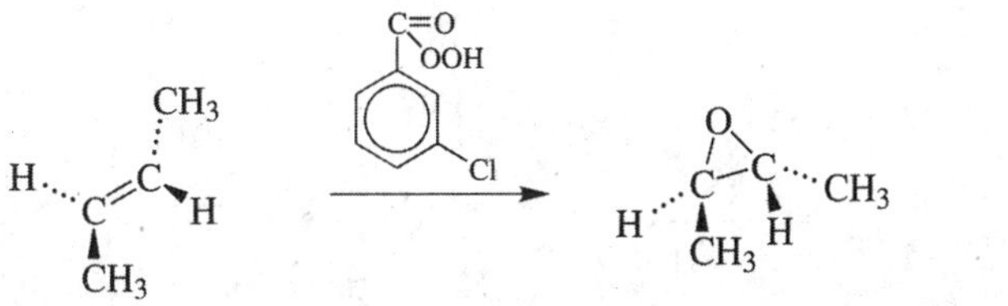

(*b*) The stereochemistry of the alkene is retained in the epoxide. The reaction is a stereospecific *cis* addition.

(*c*) The *cis*- and *trans*-alkenes would give the same carbocation, which would go on to give the same product(s). The mechanism probably involves a one-step transfer of the O to the double bond, without intermediates.

2. From Halohydrins by Intramolecular S_N2-Type Reaction

Halohydrins, formed by electrophilic addition of HO—Cl(Br) to alkenes (Problem 6.27), are treated with a base to give epoxides.

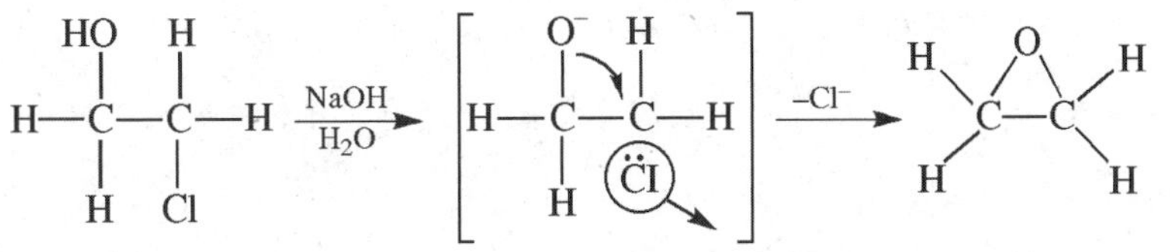

$$\mathrm{H_2C(OH)\text{—}CH_2Cl} \xrightarrow[H_2O]{NaOH} [\mathrm{H_2C(O^-)\text{—}CH_2Cl}] \xrightarrow{-Cl^-} \text{ethylene oxide}$$

intermediate alkoxide

Problem 14.25 Why does *trans*-2-chlorocyclohexanol give a very good yield of 1,2-epoxycyclohexane but the *cis* isomer gives no epoxide?

The nucleophilic O^- group displaces the Cl atom (as Cl^-) by an intramolecular S_N2-type process that requires a backside attack. In the *trans* isomer, the O^- and Cl are properly positioned for such a displacement, and the epoxide is formed. In the *cis* isomer, a backside attack cannot occur and the epoxide is not formed.

This role of the O^-, called **neighboring-group participation**, always leads to an inversion of configuration if the attacked C is a chiral center (stereocenter).

14.8 Chemistry

1. S_N2 Ring-Opening

Problem 13.9(*d*) illustrates such a reaction.

Problem 14.26 Outline the S_N2 mechanism for acid- and base-catalyzed addition to ethylene oxide, and give the structural formulas of the products of addition of the following: (*a*) H_2O, (*b*) CH_3OH, (*c*) CH_3NH_2, (*d*) CH_3CH_2SH.

In acid, O is first protonated.

$$\text{H:A} + \underset{\text{O}}{-\overset{|}{\text{C}}-\overset{|}{\text{C}}-} \longrightarrow \text{:A}^- + \underset{\overset{+}{\text{O}}\text{H}}{-\overset{|}{\text{C}}-\overset{|}{\text{C}}-} \longrightarrow -\underset{\text{A}}{\overset{|}{\text{C}}}-\underset{\text{OH}}{\overset{|}{\text{C}}}-$$

The protonated epoxide can also react with nucleophilic solvents such as CH_3OH.

$$\underset{\text{H}}{\text{CH}_3\ddot{\text{O}}\text{:}} + \underset{\overset{+}{\text{O}}\text{H}}{-\overset{|}{\text{C}}-\overset{|}{\text{C}}-} \longrightarrow -\underset{\text{CH}_3\overset{+}{\text{O}}\text{H}}{\overset{|}{\text{C}}}-\underset{\text{OH}}{\overset{|}{\text{C}}}- \xrightarrow{-\text{H}^+} -\underset{\text{CH}_3\text{O}}{\overset{|}{\text{C}}}-\underset{\text{OH}}{\overset{|}{\text{C}}}-$$

In the base, the ring is cleaved by attack of the nucleophile on the less substituted C to form an alkoxide anion, which is then protonated. Reactivity is attributed to the highly strained three-membered ring, which is readily cleaved.

$$\text{B:}^- + \underset{\text{O}}{-\overset{|}{\text{C}}-\overset{|}{\text{C}}-} \longrightarrow -\underset{|}{\overset{\text{B}}{\text{C}}}-\underset{\text{O}^-}{\overset{|}{\text{C}}}- \xrightarrow{\text{H}^+} -\underset{|}{\overset{\text{B}}{\text{C}}}-\underset{\text{OH}}{\overset{|}{\text{C}}}-$$

(*a*) $HOCH_2CH_2OH$ (*b*) $CH_3OCH_2CH_2OH$ (*c*) $CH_3NHCH_2CH_2OH$ (*d*) $CH_3CH_2SCH_2CH_2OH$

Base-induced ring-openings require a strong base because the strongly basic O^- is displaced as part of the alkoxide. Acid-induced ring-openings are achieved with weak bases, such as nucleophilic solvents, because now the very weakly basic OH, formed by protonation of the O atom, is displaced as part of the alcohol portion of the product.

Problem 14.27 (*a*) Give the product of the S_N2-type addition of C_2H_5MgBr to ethylene oxide. (*b*) What is the synthetic utility of the reaction of Grignard reagents and ethylene oxide?

(*a*) $\boxed{C_2H_5}MgBr + H_2C\text{—}CH_2$ (epoxide, O bridging) $\xrightarrow{(S_N2)}$ $\boxed{C_2H_5}\text{—}CH_2\text{—}CH_2\overset{-}{O}\overset{+}{M}gBr \xrightarrow{H_2O} C_2H_5CH_2CH_2OH$

1-butanol

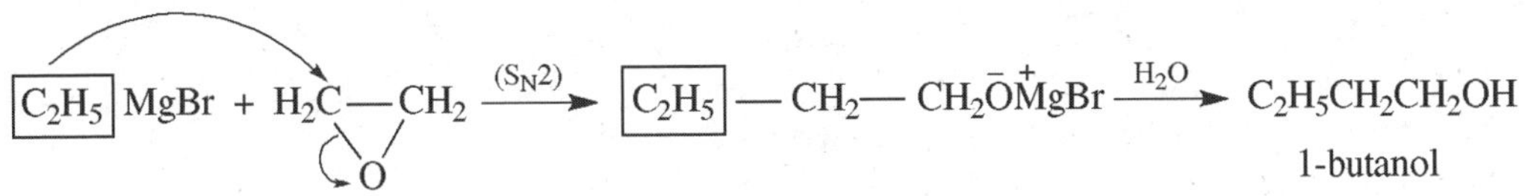

(*b*) It is a good method for extending the R group of the Grignard by $\text{—}CH_2CH_2OH$ in one step.

Problem 14.28 Account for the product from the following reaction of the ^{14}C-labeled chloroepoxide:

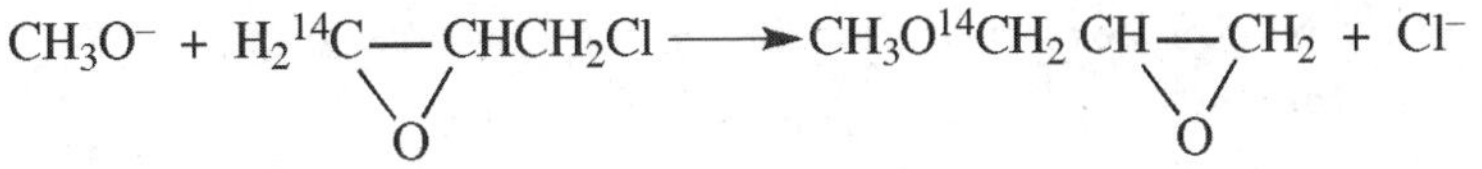

$$CH_3O^- + H_2{}^{14}C\text{—}CHCH_2Cl \text{ (epoxide, O bridging)} \longrightarrow CH_3O^{14}CH_2\,CH\text{—}CH_2 \text{ (epoxide, O bridging)} + Cl^-$$

S_N2 attack by CH_3O^- on the (less substituted) ^{14}C gives an intermediate alkoxide, $CH_3O^{14}CH_2C(O^-)HCH_2Cl$ which then displaces Cl^- by another S_N2 reaction, forming the new epoxide.

2. S_N1 Ring-Opening

In acid, the protonated epoxide may undergo ring-opening to give an intermediate carbocation.

Problem 14.29 Outline mechanisms to account for the different isomers formed from a reaction of

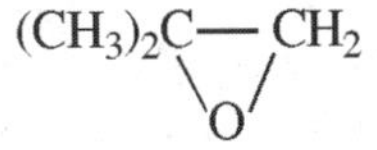

$$(CH_3)_2C\text{—}CH_2 \text{ (epoxide, O bridging)}$$

with CH_3OH in acidic (H^+) and in basic (CH_3O^-) media.

CH_3O^- reacts by an S_N2 mechanism attacking the less substituted C.

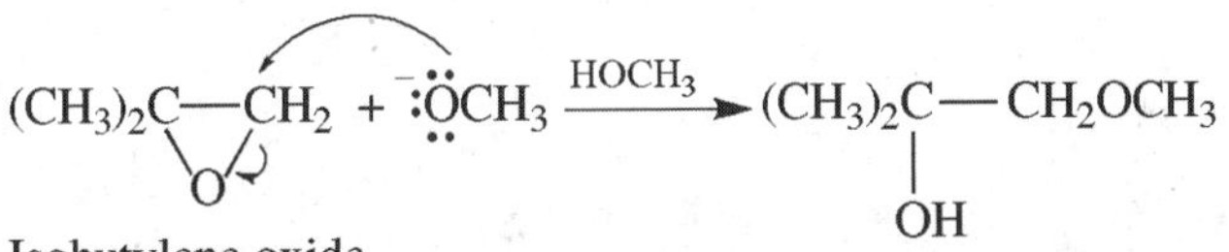

$$(CH_3)_2C\text{—}CH_2 \text{ (epoxide, O bridging)} + :\ddot{O}CH_3^- \xrightarrow{HOCH_3} (CH_3)_2C(OH)\text{—}CH_2OCH_3$$

Isobutylene oxide

In acid, the S_N1 mechanism produces the more stable 3° R^+, and the nucleophilic solvent forms a bond with the more substituted C.

$$(CH_3)_2\underset{\diagdown O \diagup}{C{-}CH_2} \xrightarrow{H^+} (CH_3)_2\underset{\diagdown {}^+OH \diagup}{C{-}CH_2} \qquad (CH_3)_2\overset{+}{C}CH_2OH \ \text{(a 3° R}^+\text{)}$$

$$\downarrow \!\!\!\times \qquad\qquad \xrightarrow[-H^+]{CH_3\ddot{O}H}$$

$$(CH_3)_2\underset{OH}{\underset{|}{C}}{-}\overset{+}{C}H_2 \ \text{(1° R}^+\text{)} \qquad (CH_3)_2\underset{OCH_3}{\underset{|}{C}}CH_2OH$$

Problem 14.30 Account for the fact that (R)-$CH_3CH_2CH\overset{O}{\diagdown\diagup}CH_2$ reacts with CH_3OH in acid to give the product with inversion and very little racemization.

When the protonated epoxide undergoes ring-opening, the CH_3OH molecule attacks from the backside of the C^+. The nearby, newly formed OH group hasn't moved out of the way and blocks the approach from the frontside. This leads to inversion at the chiral carbon. Since there was no change in group priorities, the configuration in the product is (S).

14.9 Summary of Epoxide Chemistry

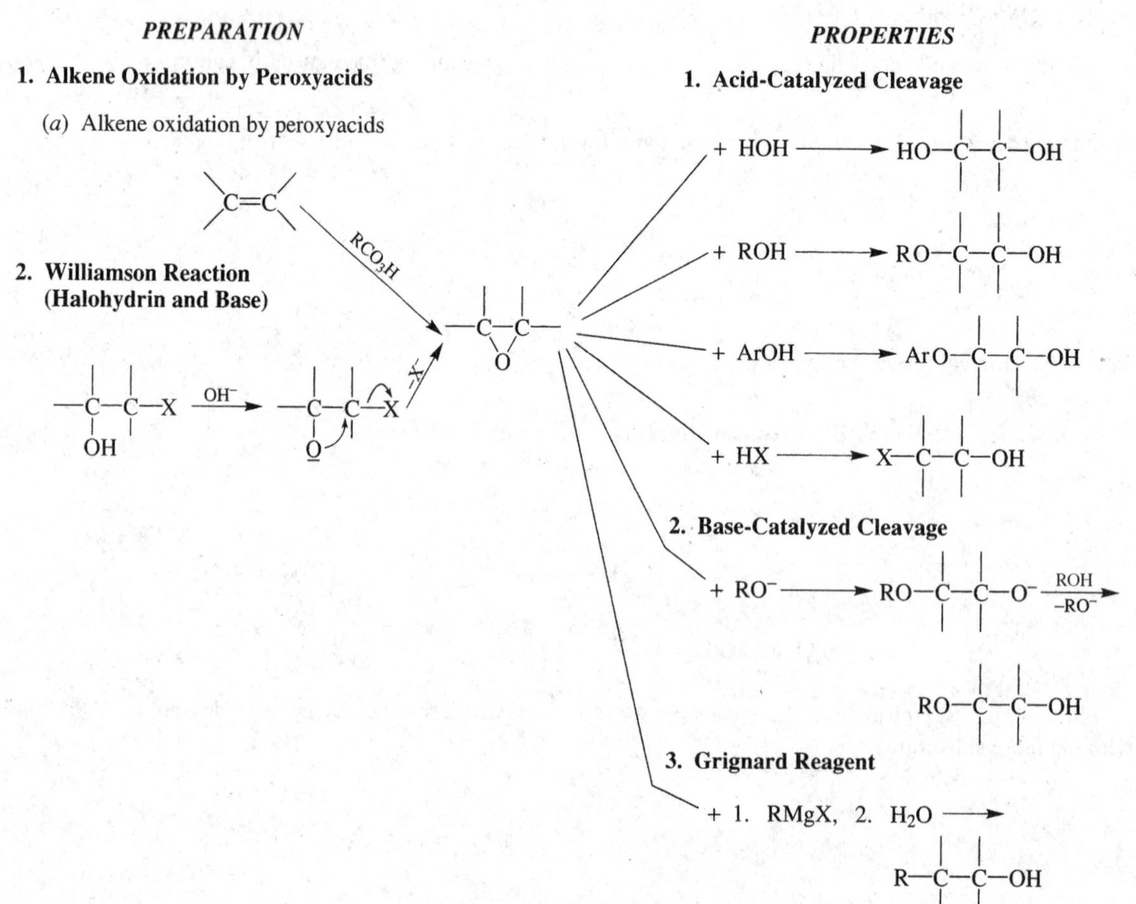

C. GLYCOLS

14.10 Preparation of 1,2-Glycols

1. **Oxidation of Alkenes (see Table 6.1 and Problem 6.28)**
2. **Hydrolysis of *vic*-Dihalides and Halohydrins**

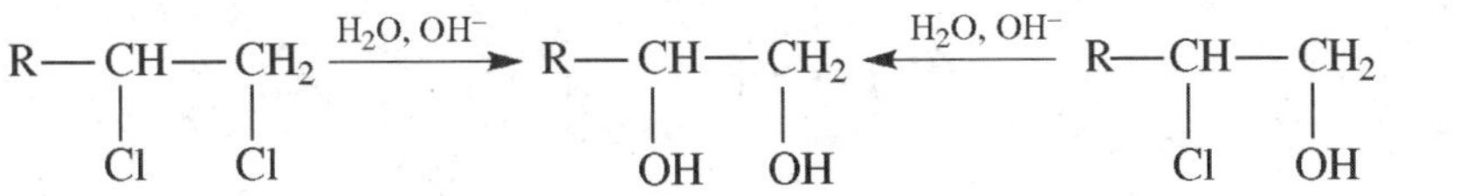

$$\underset{vic\text{-dihalide}}{\mathrm{R{-}CH(Cl){-}CH_2Cl}} \xrightarrow{H_2O,\ OH^-} \mathrm{R{-}CH(OH){-}CH_2OH} \xleftarrow{H_2O,\ OH^-} \underset{\text{1,2-haloalcohol (halohydrin)}}{\mathrm{R{-}CH(Cl){-}CH_2OH}}$$

3. **Hydrolysis of Epoxides**

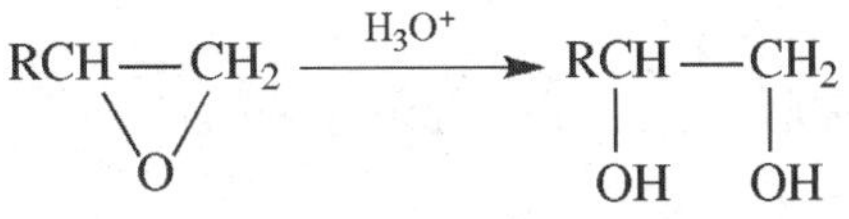

$$\mathrm{RCH{-}CH_2}\ (\text{epoxide, bridged by O}) \xrightarrow{H_3O^+} \mathrm{RCH(OH){-}CH_2OH}$$

4. **Reductive Dimerization of Carbonyl Compounds**

Symmetrical 1,2-glycols, known as **pinacols**, are prepared by bimolecular reduction of aldehydes or ketones.

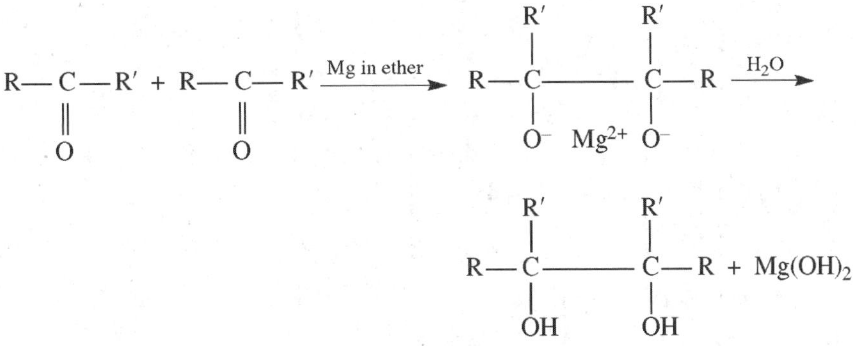

$$\mathrm{R{-}\overset{O}{\overset{\|}{C}}{-}R' + R{-}\overset{O}{\overset{\|}{C}}{-}R'} \xrightarrow{\text{Mg in ether}} \mathrm{R{-}C(R')(O^-){-}C(R')(O^-){-}R}\ \ \mathrm{Mg^{2+}} \xrightarrow{H_2O}$$

$$\underset{a\ pinacol}{\mathrm{R{-}C(R')(OH){-}C(R')(OH){-}R}} + \mathrm{Mg(OH)_2}$$

Problem 14.31 What compounds would you use to prepare 2,3-diphenyl-2,3-butanediol:

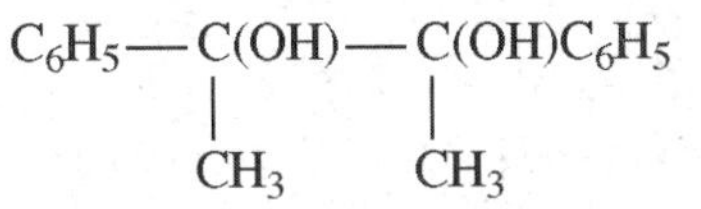

$$\mathrm{C_6H_5{-}C(OH)(CH_3){-}C(OH)(CH_3)C_6H_5}$$

by (*a*) halide hydrolysis and (*b*) reductive dimerization of a carbonyl compound?

(*a*) $C_6H_5C(CH_3)ClC(CH_3)ClC_6H_5$ or $C_6H_5C(CH_3)ClC(CH_3)OHC_6H_5$, (*b*) $C_6H_5COCH_3$.

14.11 Unique Reactions of Glycols

1. Periodic Acid (HIO_4) or Lead Tetracetate ($Pb(Oac)_4$), Oxidative Cleavage

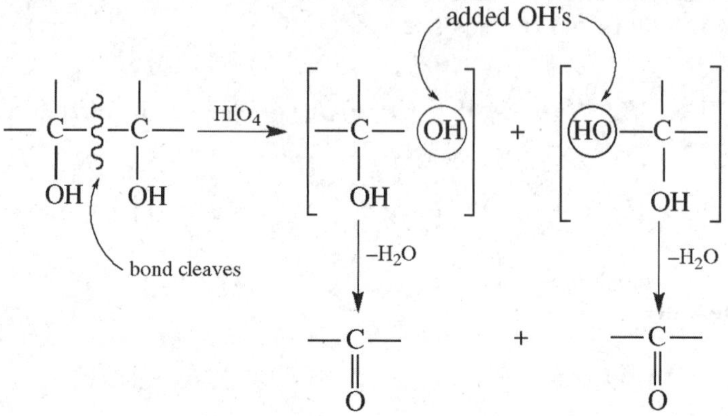

$$-\underset{OH}{\overset{|}{C}}\wr\underset{OH}{\overset{|}{C}}- \xrightarrow{HIO_4} \left[-\underset{OH}{\overset{|}{C}}-(OH)\right] + \left[(HO)-\underset{OH}{\overset{|}{C}}-\right] \xrightarrow{-H_2O} -\underset{O}{\underset{\|}{C}}- + -\underset{O}{\underset{\|}{C}}-$$

(bond cleaves; added OH's)

A 1° OH yields $H_2C{=}O$; a 2° OH an aldehyde, RCHO; a 3° OH a ketone, $R_2C{=}O$. In polyols, if two vicinal OH's are termed an "adjacency," the number of moles of HIO_4 consumed is the number of such adjacencies.

Problem 14.32 Give the products and the number of moles of HIO_4 consumed in the reaction with 2,4-dimethyl-2,3,4,5-hexanetetrol. Indicate the adjacencies with zigzag lines.

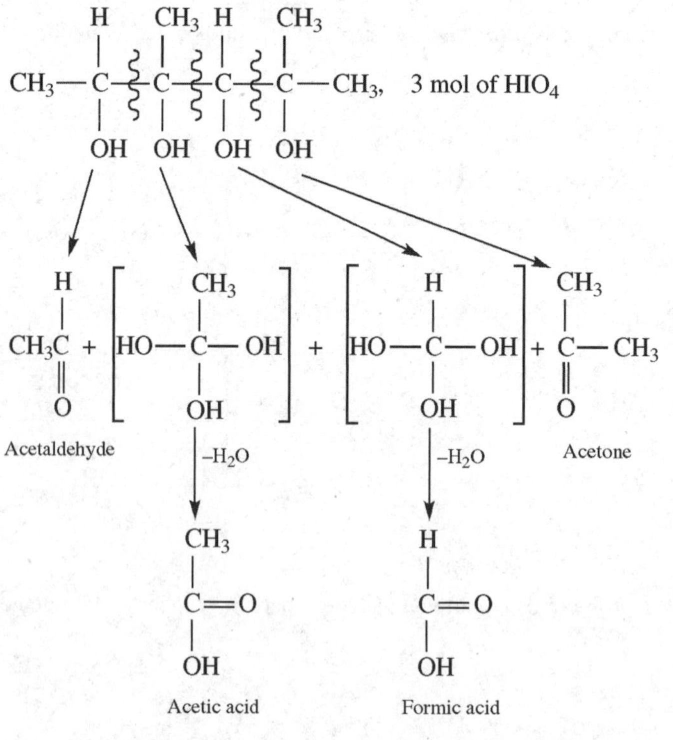

$$CH_3-\underset{OH}{\overset{H}{C}}\wr\underset{OH}{\overset{CH_3}{C}}\wr\underset{OH}{\overset{H}{C}}\wr\underset{OH}{\overset{CH_3}{C}}-CH_3, \quad 3 \text{ mol of } HIO_4$$

$$\underset{\text{Acetaldehyde}}{CH_3\overset{H}{C}{=}O} + \left[HO-\underset{OH}{\overset{CH_3}{C}}-OH\right] + \left[HO-\underset{OH}{\overset{H}{C}}-OH\right] + \underset{\text{Acetone}}{CH_3-\underset{O}{\underset{\|}{C}}-CH_3}$$

$$\left[HO-\underset{OH}{\overset{CH_3}{C}}-OH\right] \xrightarrow{-H_2O} \underset{\text{Acetic acid}}{CH_3-C({=}O)-OH} \qquad \left[HO-\underset{OH}{\overset{H}{C}}-OH\right] \xrightarrow{-H_2O} \underset{\text{Formic acid}}{H-C({=}O)-OH}$$

Note that the middle —C—OH's are oxidized to a —COOH because C—C bonds on both sides are cleaved.

2. Pinacol Rearrangement

Acidification of glycols produces an aldehyde or a ketone by rearrangement. There are four steps: (1) protonation of an OH; (2) loss of H_2O to form an R^+; (3) 1,2-shift of :H, :R, or :Ar to form a more stable cation; (4) loss of H^+ to give a product.

$$\underset{\text{pinacol}}{CH_3-\overset{CH_3}{\underset{OH}{C}}-\overset{CH_3}{\underset{OH}{C}}-CH_3} \xrightarrow[(1)]{H^+} CH_3-\overset{CH_3}{\underset{\overset{+}{O}H_2}{C}}-\overset{CH_3}{\underset{OH}{C}}-CH_3 \xrightarrow[(2)]{-H_2O} \underset{\text{a 3° }R^+}{CH_3-\overset{CH_3}{\underset{+}{C}}-\overset{CH_3}{\underset{OH}{C}}-CH_3} \xrightarrow[(3)]{\sim:CH_3}$$

$$CH_3-\overset{CH_3}{\underset{CH_3}{C}}-\overset{+}{\underset{:\ddot{O}H}{C}}-CH_3 \longleftrightarrow \underset{\substack{\text{protonated ketone}\\ \textit{more stable cation}}}{CH_3-\overset{CH_3}{\underset{CH_3}{C}}-\underset{\overset{+}{:O}H}{\overset{}{C}}-CH_3} \xrightarrow[(4)]{-H^+} \underset{\text{pinacolone}}{CH_3-\overset{CH_3}{\underset{CH_3}{C}}-\underset{O}{\overset{}{C}}-CH_3}$$

With unsymmetrical glycols, the product obtained is determined mainly by which OH is lost as H_2O to give the more stable R^+, and thereafter by which group migrates. The order of **migratory aptitude** is Ar > H or R.

Problem 14.33 Give the structural formula for the major product from the pinacol rearrangement of 1,1,2-triphenyl-1,2-propanediol. Indicate the protonated OH and the migrating group

$$C_6H_5-\overset{OH^a}{\underset{C_6H_5}{C}}-\overset{OH^b}{\underset{C_6H_5}{C}}-CH_3$$

Loss of OH^a yields the more stable $(C_6H_5)_2C^+-C(OH)(C_6H_5)CH_3$. C_6H_5 rather than CH_3 migrates to form

$$(C_6H_5)_3C-\underset{\underset{O}{\|}}{C}-CH_3$$

the major product. Migration of CH_3 would give

$$(C_6H_5)_2C(CH_3)\underset{\underset{O}{\|}}{C}-(C_6H_5)$$

which also arises from the loss of OH^b.

3. Reaction with Carbonyl Compounds (see Section 15.3)

Carbonyl compounds react with glycols in anhydrous acid to form 1,3-cyclic ethers. These are called **acetals** if formed from aldehydes, and **ketals** if formed from ketones.

$$\underset{H(R')}{\underset{|}{RC}}=O + HOCH_2CH_2OH \xrightarrow[\text{dry}]{H^+} \underset{\substack{\text{an acetal}\\ \text{(a ketal)}}}{\text{cyclic }O\text{–}CH_2CH_2\text{–}O\text{ with }C(R)(H(R'))} + H_2O$$

14.12 Summary of Glycol Chemistry

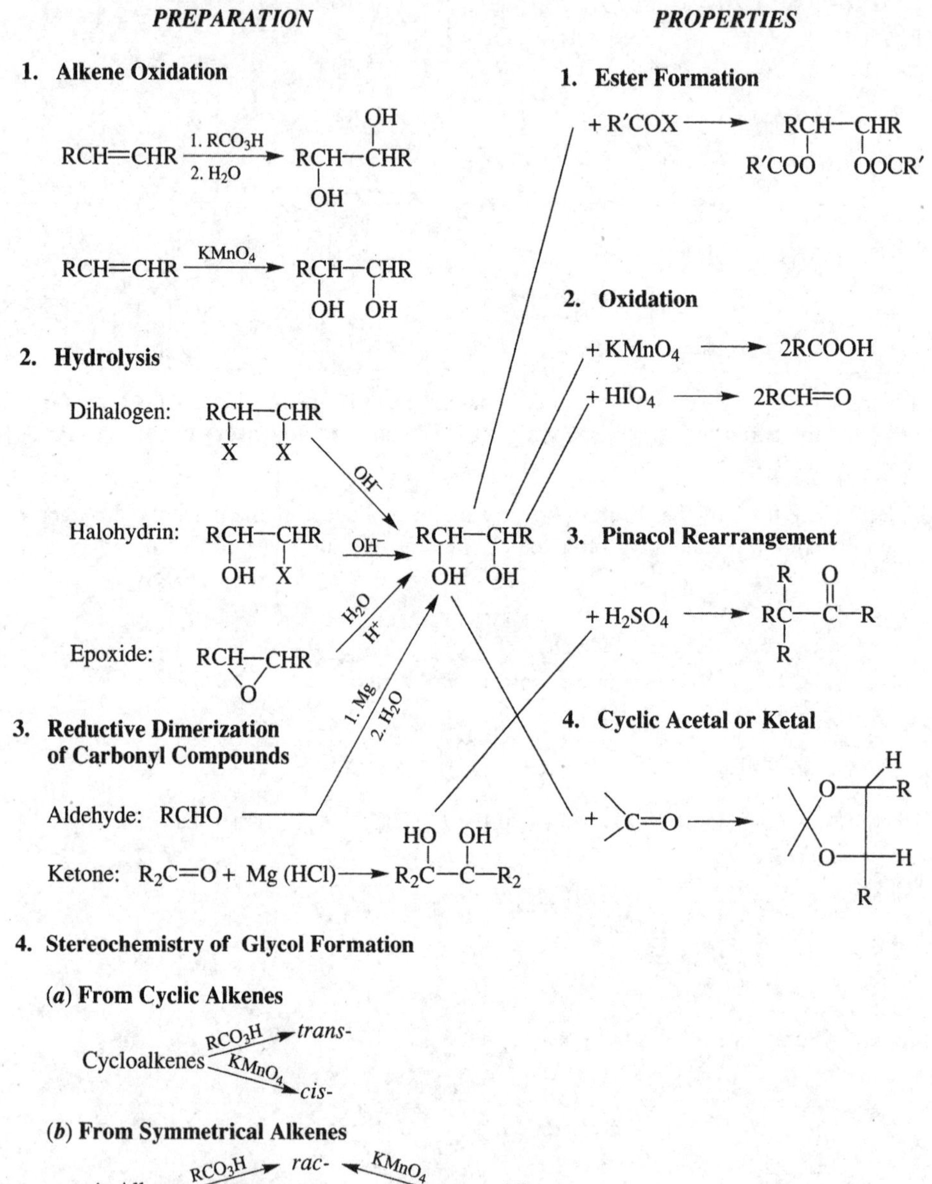

D. THIOETHERS

14.13 Introduction

Thioethers are the sulfur analogs of ethers, with the generic formula RSR. They are also known as **sulfides**.

14.14 Preparation

Thioethers are prepared mainly by Williamson-type S_N2 displacements of RS^- and R′X or $R'OSO_3Ar$ (aryl sulfonates). RS^- is formed from the acidic thiol RSH, with NaOH as the base.

$$CH_3S^-Na^+ + CH_3CH_2{-}Br \longrightarrow CH_3SCH_2CH_3 + NaBr$$

Ethyl methyl sulfide

Problem 14.34 Explain why the reaction of HS^- and RX is little used to prepare RSH.

Sulfur atoms in molecules and ions are very good nucleophilic sites. Hence, once formed in the base, RSH yields RS^-, which reacts with RX to give the thioether. For this reason, thiourea is used with RX to give thiols (Problem 13.25).

Problem 14.35 Give the expected principal organic products from the following reactions:

(*a*) $C_2H_5SH + (CH_3)_2CHCH_2CH_2CH_2Br \xrightarrow{OH^-}$

(*b*) $ICH_2CH_2I + HSCH_2CH_2SH \xrightarrow{OH^-}$

(*c*) Na_2S (1 mol) + $BrCH_2CH_2CH_2CH_2Br \longrightarrow$

(*a*) $C_2H_5{-}S{-}CH_2CH_2CH_2CH(CH_3)_2$

(*b*) 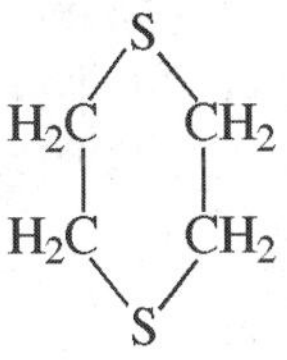

(*c*) 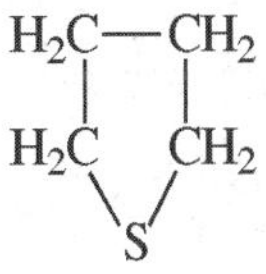

14.15 Chemistry

The chemistries of RSR and ROR are quite different, as indicated by the following reactions, which are not observed for ROR.

1. Reaction with R′X to Give Stable Sulfonium Salts, $R_2R'S{:}^+\ X^-$

$$Me_2S{:} + CH_3CH_2{:}Br \longrightarrow [Me_2SCH_2CH_3]^+\ Br^-$$

Dimethylethylsulfonium bromide

Ethers do not undergo this reaction, because —Ö— is a much weaker nucleophilic site than —S̈—.

2. Reduction (Hydrogenolysis)

The reaction requires the **Raney nickel catalyst** (H_2 is absorbed on Ni).

$$R{-}S{-}R' + H_2 \xrightarrow{\text{Raney Ni}} RH + R'H + H_2S$$

Problem 14.36 What structural features must be present in order for an individual thioether to give a single alkane on hydrogenolysis?

The thioether must have the same two R groups, or it must be cyclic.

3. Oxidation at the S

The products are sulfoxides with one O and sulfones with two O atoms.

$$CH_3\ddot{S}CH_3 \xrightarrow{H_2O_2,\ 25°C} CH_3\overset{O}{\overset{\|}{\underset{\cdot\cdot}{S}}}CH_3 \quad \textit{a sulfoxide}$$

$$CH_3\ddot{S}CH_3 \xrightarrow{H_2O,\ 100°C} CH_3\overset{O}{\overset{\|}{\underset{\underset{O}{\|}}{S}}}CH_3 \quad \textit{a sulfone}$$

(The sulfoxide is also shown going on to the sulfone.)

Problem 14.37 Explain why sulfonium salts and sulfoxides having different R or Ar groups are resolvable into enantiomers.

The S atom in each of these molecules has three σ bonds and an unshared pair of e^-'s. According to the HON rule (Section 2.3), these S atoms use sp^3 HO's. If all the attached groups are different, the S is a chiral center. Notice in Fig. 14.2 that chirality prevails even though one of the sp^3 HO's houses an unshared pair of e^-'s, the fact that these species are resolvable indicates that their molecules do not undergo inversion of configuration, notwithstanding the fact that a lone pair of e^-'s is present. Such rigidity of configuration is characteristic of third-period elements (S, P), but not of second-period elements (C, N).

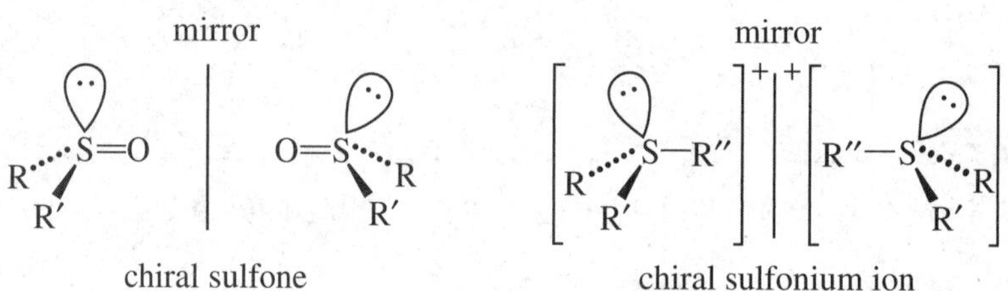

Figure 14.2

Problem 14.38 Since $(CH_3)_2\ddot{S}{=}\ddot{O}$: is an ambident nucleophile, a reaction with CH_3I could give $[(CH_3)_3S^+{=}O]I^-$ or $[(CH_3)_2S{=}O{-}CH_3]I^-$. (*a*) What type of spectroscopy can be used to distinguish between the two products? (*b*) Predict the major product.

(*a*) Use nmr spectroscopy. In $[(CH_3)_3\overset{+}{S}{=}O]I^-$, all H's are equivalent and a single peak is observed. Note that $^{32}_{16}S$ has even numbers of protons and neutrons and so shows *no* nuclear spin absorption. $[(CH^a_3)_2S{=}\overset{+}{O}{-}CH_3]I^-$ has two different kinds of H's, and therefore two peaks would be observed. (*b*) Since S is a far better nucleophile than O, $[(CH_3)_3\underset{+}{S}{=}O]I^-$ is the almost exclusive product.

SUPPLEMENTARY PROBLEMS

Problem 14.39 Give the structural formula and IUPAC name for (*a*) *n*-propyl propenyl ether, (*b*) isobutyl *tert*-butyl ether, and (*c*) 12-crown-4 ether.

(*a*) $CH_3CH_2CH_2—O—CH{=}CHCH_3$ 1-(*n*-Propoxy)-1-propene (*c*)

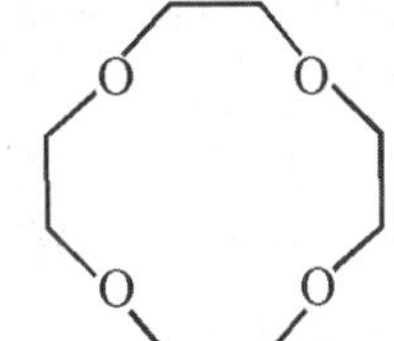

(*b*) $CH_3CH(CH_3)CH_2—O—C(CH_3)_2—CH_3$ 2-Isobutoxy-2-methylpropane

Problem 14.40 Give the structural formulas for (*a*) ethylene glycol, (*b*) propylene glycol, and (*c*) trimethylene glycol.

(*a*) $HOCH_2CH_2OH$ (*b*) $CH_3CHOHCH_2OH$ (*c*) $HOCH_2CH_2CH_2OH$.

Problem 14.41 Account for the fact that the C—O—C bond angle in dimethyl ether is greater than the H—O—H bond angle in water [112° versus 105°].

The repulsive van der Waals forces between the two CH_3 groups in dimethyl ether are greater than those between the two H's in water because the methyl groups are larger than the H's and have more electrons.

Problem 14.42 Distinguish between an ether and an alcohol by (*a*) chemical tests and (*b*) spectral methods.

(*a*) 1° and 2° alcohols are oxidizable and give positive tests with CrO_3 in acid (orange color turns green). All alcohols of moderate molecular weights evolve H_2 on addition of Na. Dry ethers are negative to both tests.
(*b*) The ir spectra of alcohols, but not ethers, show an O—H stretching band at about 3500 cm^{-1}. Comparing the ir spectra is the best method for distinguishing between these functional groups.

Problem 14.43 Why is di-*t*-butyl ether very easily cleaved by HI?

On treatment with HI, the ether is protonated. This oxonium ion cleaves readily to give *t*-butyl alcohol and the relatively stable *t*-butyl carbocation. Iodide ion adds to the carbocation, and the alcohol reacts with HI; both give *t*-butyl iodide.

$$(CH_3)_3C—\overset{+}{\ddot{O}}(H)—C(CH_3)_3 \longrightarrow (CH_3)_3C—OH + (CH_3)_3C^+$$

$$(CH_3)_3C—OH \xrightarrow{HI} (CH_3)_3C—I \xleftarrow{I^-} (CH_3)_3C^+$$

Problem 14.44 Give a chemical test to distinguish C_5H_{12} from $(C_2H_5)_2O$.

Unlike C_5H_{12}, $(C_2H_5)_2O$ is basic and dissolves in concentrated H_2SO_4.

$$(C_2H_5)_2O + H_2SO_4 \rightarrow (C_2H_5)_2OH^+ + HSO_4^-$$

Problem 14.45 Outline the mechanism for acid- and base-catalyzed additions to ethylene oxide, and give the structural formulas of the products of addition of the following: (*a*) H_2O, (*b*) CH_3OH, (*c*) CH_3NH_2, (*d*) CH_3CH_2SH.

In acid, O is first protonated.

$$H{:}A + -\underset{\backslash O/}{C-C}- \longrightarrow {:}A^- + -\underset{\backslash \overset{+}{O}H/}{C-C}- \xrightarrow[\text{or } S_N1]{S_N2} -\overset{A}{\underset{}{C}}-\underset{OH}{C}-$$

The protonated epoxide can also react with nucleophilic solvents such as CH_3OH.

$$CH_3\underset{H}{\ddot{O}}{:} + -\underset{\backslash \overset{+}{O}H/}{C-C}- \longrightarrow -\underset{CH_3\overset{+}{O}H}{C}-\underset{OH}{C}- \xrightarrow{-H^+} -\underset{CH_3O}{C}-\underset{OH}{C}-$$

In a base, the ring is cleaved by attack of the nucleophile on the less substituted C to form an alkoxide anion, which is then protonated. Reactivity is attributed to the highly strained three-membered ring, which is readily cleaved.

$$B{:}^- + -\underset{\backslash O/}{C-C}- \xrightarrow{S_N2} -\overset{B}{C}-\underset{O^-}{C}- \xrightarrow{H^+} -\overset{B}{C}-\underset{OH}{C}-$$

(*a*) $HOCH_2CH_2OH$ (*b*) $CH_3OCH_2CH_2OH$ (*c*) $CH_3NHCH_2CH_2OH$ (*d*) $CH_3CH_2SCH_2CH_2OH$

Problem 14.46 Supply structures for compounds (A) through (F).

$$H_2C{=}CH_2 + (A) \longrightarrow ClCH_2CH_2OH \xrightarrow[\text{heat}]{H_2SO_4} (B) \xrightarrow{\text{alc. KOH}} (C)$$

$$(CH_3)_3CBr + \text{alc. KOH} \longrightarrow (D) \xrightarrow{HOCl} (E) \xrightarrow{NaOH} (F)$$

(A) $HOCl$ (B) $ClCH_2CH_2{-}O{-}CH_2CH_2Cl$ (C) $H_2C{=}CH{-}O{-}CH{=}CH_2$

(D) $(CH_3)_2C{=}CH_2$ (E) $(CH_3)_2\underset{OH}{C}{-}CH_2Cl$ (F) $(CH_3)_2\underset{\backslash O/}{C{-}CH_2}$

Formation of (F), isobutylene oxide, is an internal S_N2 reaction.

$$(CH_3)_2\underset{OH}{C}{-}CH_2Cl \xrightarrow[NaOH]{\text{conc.}} \left[(CH_3)_2\underset{O^-}{C}{-}CH_2{-}Cl\right] \longrightarrow (CH_3)_2\underset{\backslash O/}{C{-}CH_2} + Cl^-$$

Problem 14.47 Are the *m/e* peaks 102, 87, and 59 (base peak) consistent for *n*-butyl ether (A) or methyl *n*-pentyl ether (B)? Give the structure of the fragments that justify your answer.

$$\overset{\beta'}{C}H_3\overset{\alpha'}{C}H_2O\overset{\alpha}{C}H_2\overset{\beta}{C}H_2CH_2CH_3 \text{ (A)} \qquad \overset{\alpha'}{C}H_3O\overset{\alpha}{C}H_2\overset{\beta}{C}H_2CH_2CH_2CH_3 \text{ (B)}$$

The parent P^+ is $m/e = 102$, the molecular weight of the ether. The other peaks arise as follows:

$$102 - 15(CH_3) = 87 \qquad 102 - 43(C_3H_7) = 59$$

Fragmentations of P^+ ions of ethers occur mainly at the $C^\alpha{-}C^\beta$ bonds. (A) fits these data for $H_2\overset{+}{C}{-}OCH_2CH_2CH_2CH_3$ ($m/e = 87$; $C^{\alpha'}{-}C^{\beta'}$ cleavage) and $CH_3CH_2O{-}\overset{+}{C}H_2$ ($m/e = 59$; $C^\alpha{-}C^\beta$ cleavage). Cleavage of the $C^\beta{-}C^\alpha$ bonds in (B) would give a cation, $CH_3O{=}\overset{+}{C}H_2$ ($m/e = 45$), but this peak *was not observed.*

Problem 14.48 Prepare the following ethers starting with benzene, toluene, phenol (C_6H_5OH), cyclohexanol, any aliphatic compound of three C's or less and any solvent or inorganic reagent: (*a*) dibenzyl ether, (*b*) di-*n*-butyl ether, (*c*) ethyl isopropyl ether, (*d*) cyclohexyl methyl ether, (*e*) *p*-nitrophenyl ethyl ether, (*f*) divinyl ether, (*g*) diphenyl ether.

(*a*) $C_6H_5CH_3 \xrightarrow[\text{light}]{Cl_2} C_6H_5CH_2Cl \xrightarrow[H_2O]{OH^-} C_6H_5CH_2OH \xrightarrow[-H_2O]{H_2SO_4} (C_6H_5CH_2)_2O$

(*b*) $\boxed{CH_3CH_2}\text{—}MgBr + \underset{\text{(ethylene oxide, O bridging)}}{CH_2CH_2} \xrightarrow[\text{2. } H_2O]{\text{1. reaction}} \boxed{CH_3CH_2}\text{—}CH_2CH_2OH \xrightarrow[-H_2O]{H_2SO_4} (CH_3CH_2CH_2CH_2)_2O$

(*c*) $\underset{(1°)}{CH_3CH_2Br} + Na^+\ \bar{O}CH(CH_3)_2 \longrightarrow CH_3CH_2OCH(CH_3)_2 + Na^+\ Br^-$

Use the 1° RX to minimize the competing E2 elimination reaction or use

$CH_3CH{=}CH_2 + CH_3CH_2OH \xrightarrow{Hg(OCOCF_3)} CH_3CH(OCH_2CH_3)CH_2HgOCOCF_3 \xrightarrow{NaBH_4} \text{product}$

(*d*) $C_6H_{11}OH + CH_2N_2 \xrightarrow{H^+} C_6H_{11}OCH_3 + N_2$ or $C_6H_{11}OH \xrightarrow{Na} C_6H_{11}O^-Na^+ \xrightarrow{CH_3I} C_6H_{11}OCH_3$

(*e*) $C_6H_5OH \xrightarrow{NaOH} C_6H_5O^-Na^+ \xrightarrow{C_2H_5Br} C_6H_5OC_2H_5 \xrightarrow[H_2SO_4]{HNO_3} p\text{-}NO_2C_6H_4OC_2H_5$

Williamson synthesis of an aryl alkyl ether requires the Ar to be part of the nucleophile ArO^- and *not the halide*, since ArX does not readily undergo S_N2 displacements. Note that since ArOH is much more acidic than ROH, it is converted to ArO^- by OH^- instead of by Na as required for ROH.

(*f*) See Problem 14.46, compounds (A), (B), and (C). Vinyl alcohol, $H_2C{=}CHOH$, cannot be used as a starting material because it is not stable and rearranges to CH_3CHO. The double bond must be introduced after the ether bond is formed.

(*g*) $C_6H_6 \xrightarrow[Fe]{Br_2} C_6H_5Br \xrightarrow[\text{Cu(>200°C), no solvent}]{C_6H_5O^-Na^+} (C_6H_5)_2O$

Phenols do not undergo intermolecular dehydration. Although aryl halides cannot be used as substrates in typical Williamson syntheses, they do undergo a modified Williamson-type synthesis at a higher temperature in the presence of Cu.

Problem 14.49 Prepare ethylene glycol from the following compounds: (*a*) ethylene, (*b*) ethylene oxide, (*c*) 1,2-dichloroethane.

(*a*) Oxidation: $H_2C{=}CH_2 \xrightarrow{\text{dil. aq. } KMnO_4} HOCH_2\text{—}CH_2OH$

(*b*) Acid hydrolysis $\underset{\text{(O bridging)}}{CH_2\text{—}CH_2} \xrightarrow{H_2O,\ H^+} HOCH_2\text{—}CH_2OH$

(*c*) Alkaline hydrolysis: $ClCH_2\text{—}CH_2Cl \xrightarrow{H_2O,\ OH^-} HOCH_2\text{—}CH_2OH$

Problem 14.50 Outline the steps and give the product of pinacol rearrangement of: (*a*) 3-phenyl-1,2-propanediol, (*b*) 2,3-diphenyl-2,3-butanediol.

(*a*) $C_6H_5CH_2\text{—}\underset{OH}{\overset{2°}{CH}}\text{—}\underset{OH}{\overset{1°}{CH_2}} \xrightarrow[(2)\ -H_2O]{(1)\ +H^+} C_6H_5CH_2\text{—}\overset{+}{CH}\text{—}\underset{OH}{\overset{H}{C}}\text{—}H$

$\big\downarrow$ ~:H

$C_6H_5CH_2CH_2\text{—}\underset{O}{\overset{\|}{C}}\text{—}H \xleftarrow{-H^+} C_6H_5CH_2\text{—}CH_2\text{—}\underset{OH}{\overset{+}{C}}\text{—}H$

The 2° OH is protonated and lost as H_2O in preference to the 1° OH.

(*b*) $C_6H_5C(CH_3)(OH)-C(CH_3)(OH)-C_6H_5 \xrightarrow{H^+} C_6H_5-C(CH_3)(OH)-C(CH_3)(\overset{+}{O}H_2)-C_6H_5 \xrightarrow{-H_2O} C_6H_5-C(CH_3)(OH)-\overset{+}{C}(CH_3)-C_6H_5$

$\xrightarrow{\sim :C_6H_5} \overset{+}{C}(CH_3)(OH)-C(CH_3)(C_6H_5)-C_6H_5 \xrightarrow{-H^+} CH_3-C(=O)-C(CH_3)(C_6H_5)-C_6H_5$

Since the symmetrical pinacol can give only a single R^+, the product is determined by the greater migratory aptitude of C_6H_5.

Problem 14.51 Show how ethylene oxide is used to manufacture the following water-soluble organic solvents:

(*a*) Carbitol ($C_2H_5OCH_2CH_2OCH_2CH_2OH$) (*b*) Diethylene glycol ($HOCH_2CH_2OCH_2CH_2OH$)

(*c*) Diethanolamine ($HOCH_2CH_2N(H)CH_2CH_2OH$) (*d*) 1,4-Dioxane (O◯O)

(*a*) $C_2H_5\ddot{O}H + H_2C\!-\!CH_2\ (\text{epoxide}) \xrightarrow{H^+} C_2H_5OCH_2CH_2\ddot{O}H \xrightarrow[H^+]{H_2C-CH_2\ (\text{epoxide})} C_2H_5OCH_2CH_2OCH_2CH_2OH$

(*b*) $H_2\ddot{O} + H_2C\!-\!CH_2\ (\text{epoxide}) \xrightarrow{H^+} HOCH_2CH_2\ddot{O}H \xrightarrow[H^+]{H_2C-CH_2\ (\text{epoxide})} HOCH_2CH_2OCH_2CH_2OH$

(*c*) $\ddot{N}H_3 + H_2C\!-\!CH_2\ (\text{epoxide}) \longrightarrow HOCH_2CH_2\ddot{N}H_2 \xrightarrow{H_2C-CH_2\ (\text{epoxide})} HOCH_2CH_2\!-\!NH\!-\!CH_2CH_2OH$

(*d*)

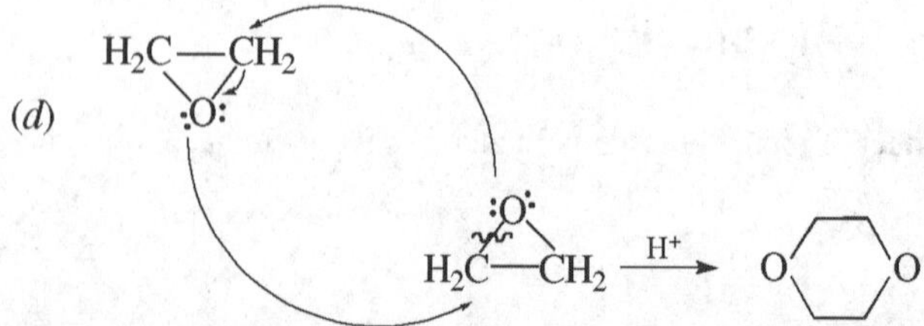

Problem 14.52 Ethers, especially those with more than one ether linkage, are also named by the **oxa method**. The ether O's are counted as C's in determining the longest hydrocarbon chain. The O is designated by the prefix **oxa-**, and a number indicates its position. Use this method to name the following:

(*a*) $(CH_3)_3COCH_2CH(CH_3)_2$ (*b*) $C_2H_5OCH_2CH_2OCH_2CH_2OH$ (*c*) Tetrahydrofuran (*d*) Dioxane

(*a*) $\overset{1}{C}H_3-\overset{2}{C}(CH_3)_2-\overset{3}{O}-\overset{4}{C}H_2-\overset{5}{C}H(CH_3)-\overset{6}{C}H_3$ 2,2,5-Trimethyl-3-oxahexane

(*b*) $CH_3CH_2OCH_2CH_2OCH_2CH_2OH$ 3,6-Dioxa-1-octanol

(*c*) Oxacyclopentane (*d*) 1,4-Dioxacyclohexane

Problem 14.53 Outline mechanisms to account for the different isomers formed from the reaction of

$$(CH_3)_2C\overset{O}{—}CH_2$$

with CH_3OH in acidic (H^+) and in basic (CH_3O^-) media.

CH_3O^- reacts by an S_N2 mechanism attacking the less substituted C.

$$(CH_3)_2C\overset{O}{—}CH_2 + {}^-\!:\ddot{O}CH_3 \xrightarrow{HOCH_3} (CH_3)_2—C(OH)—CH_2OCH_3$$

Isobutylene oxide

In acid, the S_N1 mechanism produces the more stable 3° R^+.

$$(CH_3)_2C\overset{O}{—}CH_2 \xrightarrow{H^+} (CH_3)_2C\overset{^+OH}{—}CH_2 \longrightarrow (CH_3)_2\overset{+}{C}CH_2OH \text{ (a 3° } R^+) \xrightarrow{CH_3\ddot{O}H,\ -H^+} (CH_3)_2C(OCH_3)CH_2OH$$

$$(CH_3)_2C\overset{^+OH}{—}CH_2 \not\longrightarrow (CH_3)_2C(OH)—\overset{+}{C}H_2 \text{ (1° } R^+)$$

Problem 14.54 Prepare mustard gas, $(ClCH_2CH_2)_2S$, from ethylene.

$$H_2C{=}CH_2 + S_2Cl_2 \rightarrow (ClCH_2CH_2)S + S$$

Problem 14.55 A compound, $C_3H_8O_2$, gives a negative test with HIO_4. List all possible structures, and show how ir and nmr spectroscopy can distinguish among them. (Note that *gem*-diols can be disregarded since they are usually not stable.)

There are no degrees of unsaturation and hence no rings or multiple bonds. The O's must be present as C—O—H and/or C—O—C. The compound can be a diol, a hydroxyether or a diether. A negative test with HIO_4 rules out a *vic*-diol. Possible structures are a diol, $^1HOCH^2_2CH^3_2CH^2_2OH^1$ (A); two hydroxyethers, $^1HOCH^2_2CH^3_2OCH^4_3$ (B) and $^1HOCH^2_2OCH^3_2CH^4_3$ (C); and a diether (an acetal), $CH_3OCH_2OCH_3$ (D). (D) is pinpointed by ir; it has no OH, there is no O—H stretch and peaks are not observed at greater than 2950 cm^{-1}. (A) can be differentiated from (B) and (C) by nmr. (A) has only three kinds of equivalent H's, as labeled, while (B) and (C) each have four. In dimethyl sulfoxide, the nmr spectrum of (C) shows all H peaks to be split: H^3, a quartet, couples H^4, a triplet; H^2, a doublet, couples H^1, a triplet. The nmr spectrum of (B) in DMSO shows a sharp singlet for H^4 integrating for three H's. Other differences may be observed, but those described are sufficient for identification. The DMSO used is deuterated, $(CD_3)_2SO$, to prevent interference with the spectrum.

Carbonyl Compounds: Aldehydes and Ketones

15.1 Introduction and Nomenclature

Carbonyl compounds have only H, R, or Ar groups attached to the **carbonyl group**.

$$\gt C{=}O$$

Aldehydes have at least one H bonded to the carbonyl group; **ketones** have only R's or Ar's.

Aldehydes

IUPAC names the longest continuous chain including the C of —CH=O and replaces **-e** of the alkane name by the suffix **-al**. The C of CHO is number 1. For compounds with two —CHO groups, the suffix **-dial** is added to the alkane name. When other functional groups have naming priority, —CHO is called **formyl**.

Common names replace the suffix **-ic** (**-oic** or **-oxylic**) and the word **acid** of the corresponding carboxylic acids by **-aldehyde**. Locations of substituents *on chains* are designated by Greek letters; for example:

$$-\overset{\varepsilon}{C}-\overset{\delta}{C}-\overset{\gamma}{C}-\overset{\beta}{C}-\overset{\alpha}{C}-\underset{\displaystyle H}{\underset{|}{C}}{=}O$$

The terminal C of a long chain is designated ω (omega).

The compound is named as an aldehyde (or **carbaldehyde**) whenever —CHO is attached to a ring [see Problem 15.3(*c*)].

Ketones

Common names use the names of R or Ar as separate words, along with the word **ketone**. The **IUPAC system** replaces the **-e** of the name of the longest chain by the suffix **-one**.

In molecules with functional groups, such as —COOH, that have a higher naming priority, the carbonyl group is indicated by the prefix **keto-**. Thus, CH_3—CO—CH_2—CH_2—COOH is 4-ketopentanoic acid. Groups like

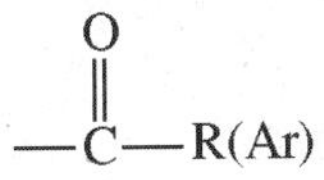

$$-\overset{\overset{\displaystyle O}{\|}}{C}-R(Ar)$$

are called **acyl** groups; for example, $-\overset{\overset{O}{\|}}{C}CH_3$ is the acetyl group. Phenyl ketones are often named as the acyl group followed by the suffix **-phenone** [see Problem 15.1(*e*)].

Problem 15.1 Give the common and IUPAC names for (*a*) CH_3CHO, (*b*) $(CH_3)_2CHCH_2CHO$, (*c*) $CH_3CH_2CH_2CHClCHO$, (*d*) $(CH_3)_2CHCOCH_3$, (*e*) $CH_3CH_2COC_6H_5$, (*f*) $H_2C{=}CHCOCH_3$.

(*a*) Acetaldehyde (from acetic acid), ethanal;

(*b*) $\underset{4}{\overset{\gamma}{CH_3}}-\underset{3}{\overset{\overset{\displaystyle CH_3}{\beta|}}{C}}H\underset{2}{\overset{\alpha}{C}}H_2\underset{1}{C}HO$ β-methylbutyraldehyde, 3-methylbutanal;

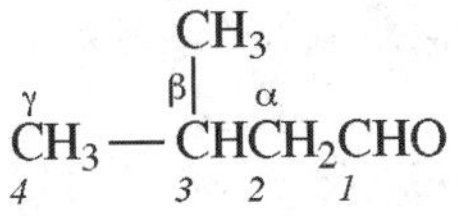

(*c*) α-chlorovaleraldehyde, 2-chloropentanal;
(*d*) methyl isopropyl ketone, 3-methyl-2-butanone;
(*e*) ethyl phenyl ketone, 1-phenyl-1-propanone (propiophenone);
(*f*) methyl vinyl ketone, 3-buten-2-one.

The C=O group has numbering priority over the C=C group.

Problem 15.2 Give structural formulas for (*a*) methyl isobutyl ketone, (*b*) phenylacetaldehyde, (*c*) 2-methyl-3-pentanone, (*d*) 3-hexenal, (*e*) β-chloropropionaldehyde.

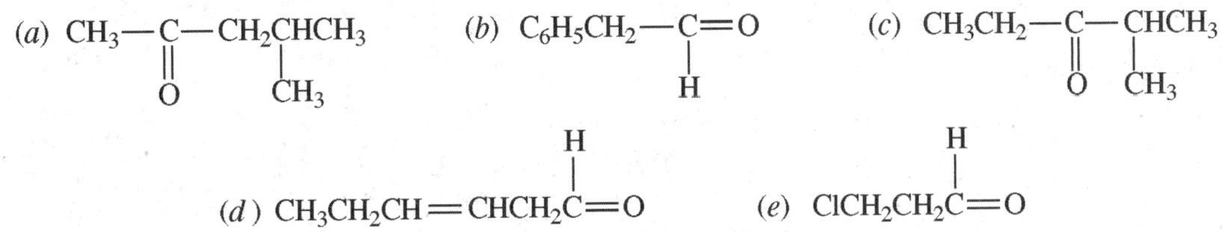

(*a*) $CH_3-\underset{\underset{\displaystyle O}{\|}}{C}-CH_2\underset{\underset{\displaystyle CH_3}{|}}{C}HCH_3$ (*b*) $C_6H_5CH_2-\underset{\underset{\displaystyle H}{|}}{C}{=}O$ (*c*) $CH_3CH_2-\underset{\underset{\displaystyle O}{\|}}{C}-\underset{\underset{\displaystyle CH_3}{|}}{C}HCH_3$

(*d*) $CH_3CH_2CH{=}CHCH_2\overset{\overset{\displaystyle H}{|}}{C}{=}O$ (*e*) $ClCH_2CH_2\overset{\overset{\displaystyle H}{|}}{C}{=}O$

Problem 15.3 Name the following compounds: (*a*) $OHCCH_2CH_2CH_2CH(CH_3)CHO$, (*b*) *p*-$OHCC_6H_4SO_3H$, (*c*) $H_3C\overset{\Delta}{-}$ CHO, (*d*) *o*-BrC_6H_4CHO.

(*a*) 2-Methyl-1,6-hexanedial. (*b*) —SO_3H takes priority over —CHO; thus, *p*-formylbenzenesulfonic acid. (*c*) The corresponding acid is a cyclopropanecarb*oxylic acid*, and *-oxylic acid* is replaced by *-aldehyde*: 2-methylcyclopropanecarbaldehyde. (*d*) The *-oic acid* in benz*oic acid* is replaced by *-aldehyde: o-bromo*-benzaldehyde (also called 2-bromobenzenecarbaldehyde).

Problem 15.4 (*a*) Draw (i) an atomic orbital representation of the carbonyl group and (ii) resonance structures. (*b*) What is the major difference between the C=O and C=C groups?

(*a*) (i) The C uses sp^2 HO's and its three σ bonds are coplanar, with bond angles near 120°. Each pair of unshared electrons is in a nonbonding (n) orbital. The π bond formed by lateral overlap of *p* AO's of C and O is in a plane perpendicular to the plane of the σ bonds.

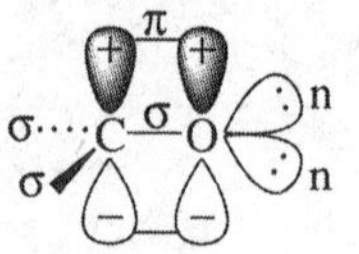

ii $\rangle C{=}\ddot{O}: \longleftrightarrow \rangle C^{+}{-}\ddot{O}:^{-}$

In this case, the polar resonance structure makes a considerable contribution to the hybrid and has a profound effect on the chemistry of the C$=$O group.

(*b*) The C$=$C group has no significant polar character, and its π bond acts as a nucleophilic site. The polarity of the π bond in C$=$O causes the C to be an electrophilic site and the O to be a nucleophilic site.

Problem 15.5 Account for the following: (*a*) *n*-butyl alcohol boils at 118°C and *n*-butyraldehyde boils at 76°C, yet their molecular weights are close, 74 and 72, respectively; (*b*) the C$=$O bond (0.122 nm) is shorter than the C—O (0.141 nm) bond; (*c*) the dipole moment of propanal (2.52 D) is greater than that of 1-butene (0.3 D); (*d*) carbonyl compounds are more soluble in water than the corresponding alkanes.

(*a*) H-bonding between alcohol molecules is responsible for the higher boiling point.
(*b*) The sharing of two pairs of electrons in C$=$O causes the double bond to be shorter and stronger.
(*c*) The polar contributing structure [Problem 15.4(*a*)(ii)] induces the large dipole moment of the aldehyde.
(*d*) H-bonding between carbonyl oxygen and water renders carbonyl compounds more water-soluble than hydrocarbons.

Problem 15.6 Compare aldehydes and ketones as to stability and reactivity.

As was the case for alkenes, alkyl substituents lower the enthalpy of the unsaturated molecule. Hence, ketones with two R's have lower enthalpies than aldehydes with one R. The electron-releasing R's diminish the electrophilicity of the carbonyl C, lessening the chemical reactivity of ketones. Furthermore, the R's, especially large bulky ones, make approach of reactants to the C more difficult.

Problem 15.7 Draw up a table of corresponding sequential oxidation levels of hydrocarbons and organic Cl, O, and N compounds.

See Table 15.1.

TABLE 15.1

OXIDATION →

Hydrocarbon	CH_3CH_3	$H_2C{=}CH_2$	$HC{\equiv}CH$		
Halogen compounds	—	CH_3CH_2Cl	CH_3CHCl_2	CH_3CCl_3	CCl_4
Oxygen compounds	—	CH_3CH_2OH	$CH_3CH{=}O$	CH_3COOH	CO_2
Nitrogen compounds	—	$CH_3CH_2NH_2$	$CH_3CH{=}NH$	$CH_3C{\equiv}N$	$H_2N{-}C{\equiv}N$

← REDUCTION

15.2 Preparation

As shown in Table 15.1, $-\overset{|}{\underset{|}{C}}=$ is at an oxidation level between COOH and $-\overset{|}{\underset{|}{C}}$OOH. Hence, aldehydes, RCHO, and ketones, $R_2C{=}O$, are made by *oxidizing* the corresponding 1° RCH_2OH and 2° R_2CHOH, respectively. RCHO, but not $R_2C{=}O$, can also be prepared by *reducing* the corresponding RCOOH or its derivative RCOX. Hydrolysis (overall reaction with water) of the other groups in the same oxidation level as $-\underset{|}{C}{=}O$. ($-C{\equiv}C-$, $-CCl_2-$, and $-\underset{|}{C}{=}NH$) will also give the $-\underset{|}{C}{=}O$ group.

By Oxidation

1. 1° $RCH_2OH \rightarrow$ RCHO **and** 2° $R_2CHOH \rightarrow R_2CO$. See Section 13.3.

Alcohols are the most important precursors in the synthesis of carbonyl compounds, being readily available. More complex aicohols are prepared by reaction of Grignard reagents with simpler carbonyl compounds. Ordinarily MnO_4^- and $Cr_2O_7^{2-}$ in acid are used to oxidize 2° R_2CHOH to R_2CO. However, oxidizing 1° RCH_2OH to RCHO without allowing the ready oxidation of RCHO to RCOOH requires special reagents. These include (*a*) pyridinium chlorochromate (PCC)

$$C_5H_5NH^+(CrO_3Cl)^- \quad \text{(the best method)}$$

(*b*) hot Cu (only with easily vaporized ROH); (*c*) MnO_2 (mild, only with allylic-type, $RCH{=}CHCH_2OH$, or benzylic-type, $ArCH_2OH$, alcohols); (*d*) $Na_2Cr_2O_7/H_2SO_4$/acetone (**Jones reagent**, which may permit RCOOH).

2. 1° RCH_2X (X=Cl, Br, I) **or** $-OSO_2R'(Ar) + Me_2S{=}O \rightarrow$ RCHO

To prevent overoxidation of aldehydes, the very mild oxidant dimethyl sulfoxide, or DMSO, is used to react with 1° halides or sulfonates to give aldehydes. These reactants are in the same oxidation level as alcohols:

$$CH_3(CH_2)_6I + CH_3\overset{\overset{O}{\|}}{S}CH_3 \xrightarrow{HCO_3^-} CH_3(CH_2)_5CHO + CH_3SCH_3 + HI$$

Problem 15.8 Suggest a mechanism for the reaction of RCH_2Cl with DMSO.

A C—O bond is formed, and the Cl^- is displaced in Step 1 by an S_N2 attack. The C=O bond results from an E2 β-elimination of H^+ and Me_2S, a good leaving group as indicated in Step 2.

Step 1 $$Me_2S^+{-}\ddot{\underset{..}{O}}{:}^- + RCH_2\text{Cl} \longrightarrow Me_2S^+{-}O{-}\underset{\underset{H}{|}}{C}HR + Cl^-$$

an alkoxysulfonium salt

Step 2 $$Me_2S^+{-}O{-}\underset{\underset{H}{|}}{C}HR \xrightarrow[-H^+]{\text{base}} O{=}CHR + Me_2S$$

3. Alkyl Arenes: $ArCH_3 \rightarrow ArCHO$, $ArCH_2R \rightarrow ArCOR$

Benzylic CH_3 and CH_2 groups can be oxidized to groups at the same oxidation level as C=O. These groups are then hydrolyzed to the C=O group.

$$ArCH_2CH_3 \xrightarrow{NBS} ArCBr_2CH_3 \xrightarrow{H_2O} ArCOCH_3$$

$$ArCH_3 \xrightarrow[(CH_3CO)_2O]{CrO_3} ArCH(O{-}CO{-}CH_3)_2 \xrightarrow[H^+]{H_2O} ArCHO + 2CH_3COOH$$

a *gem*-diacetate

4. **Alkylboranes**

(See Problem 6.19(*f*) for hydroboration.)

$$2\ \text{1-methylcyclohexene} \xrightarrow{B_2H_6} [\text{2-methylcyclohexyl}]_2BH \xrightarrow[H_3O^+]{CrO_3} 2\ \text{2-methylcyclohexanone} \quad (78\%\ \text{yield})$$

The vinyl C with more H's is converted into C=O. Alkenes can also be transformed into dialkyl carbonyls by a carbonylation-oxidation procedure.

$$3\ \text{cyclopentene} \xrightarrow{B_2H_6} [\text{cyclopentyl}]_3B \xrightarrow[2.\ H_2O_2,\ NaOH]{1.\ CO + H_2O} 2\ (C_5H_9)_2C{=}O\ (90\%\ \text{yield}) + C_5H_9OH$$

5. **Oxidative Cleavages**

Ozonolysis of alkenes (end of Section 6.4) and cleavage of glycols (Section 14.11) afford carbonyl compounds. These reactions, once used for structure determinations, have been superseded by spectral methods.

By Reductions of Acid Derivatives, RCOX, or Nitriles, RC≡N

Acid chlorides, R(Ar)COCl, are reduced to R(Ar)CHO by H_2/Pd(S), a moderate catalyst that does not reduce RCHO to RCH_2OH (**Rosenmund reduction**). Acid chlorides, esters (R(Ar)COOR), and nitriles (RC≡N) are reduced with lithium tri-*t*-butoxyaluminum hydride, $LiAlH[OC(CH_3)_3]_3$, at very low temperatures, followed by H_2O. The net reaction is a displacement of X^- by $:H^-$,

$$RCOCl + :H^- \longrightarrow RCHO + Cl^-$$

See Section 16.3 for preparation of acid derivatives.

By Hydrolysis and Hydration of Compounds at —C=O Oxidation Level

1. **$—CX_2—$, $—C(OCOR)_2$, and $—C(OR)_2$ (Acetal or Ketal)**

These groupings are hydrolizable to the —C=O group. (See Section 15.4 for acetal chemistry.)

2. **Alkynes**

See Section 8.2 for direct hydration and for net hydration through formation of vinylboranes by hydroboration.

Problem 15.9 Which is the only aldehyde that can be prepared by $HgSO_4$-catalyzed hydration of an alkyne?

Since the addition of H_2O to C=C is Markovnikov regiospecific, RC≡CH or RC≡CR must give ketones. Only HC≡CH is hydrated to give an aldehyde, CH_3CHO.

By Friedel-Craft Acylations or Formylations of Arenes

Friedel-Crafts acylations of arenes with RCOCl or anhydrides ($R\overset{O}{\overset{\|}{C}}—O—\overset{O}{\overset{\|}{C}}R$) in the presence of $AlCl_3$ give good yields of ketones.

Problem 15.10 Suggest a mechanism for acylation of ArH with RCOCl in $AlCl_3$.

The mechanism is similar to that of alkylation:

$$\text{(1)}\quad RCOCl + AlCl_3 \longrightarrow \underset{\text{acylonium ion}}{R\overset{+}{C}{=}\ddot{O}:} + AlCl_4^-$$

$$\text{(2)}\quad R\overset{+}{C}{=}\ddot{O}: + ArH \longrightarrow \left[Ar\begin{matrix} H \\ C(=O)—R \end{matrix} \right] \xrightarrow{-H^+} ArC(=O)—R$$

Problem 15.11 Can formylation of an arene, ArH, with an acid chloride be employed to prepare ArCHO?

No. The needed acid chloride is the hypothetical "formyl chloride," HCOCl. But this compound cannot be realized; attempts to prepare it from formic acid (HCOOH + $SOCl_2$) yield only mixtures of HCl and carbon monoxide, :C≡O:.

Arenes can be formylated by generating the active intermediate, $:\ddot{O}{\equiv}\overset{+}{C}—H$, from reagents other than HCOCl. The **Gatterman–Koch reaction** uses a high-pressure gaseous mixture of CO and HCl.

$$CO + HCl \xrightarrow[-Cl^-]{AlCl_3,\ CuCl} O{=}\overset{+}{C}—H \xrightarrow[-H^+]{ArH} ArCHO$$

By Acylation or Hydroformylation of Alkenes

1. Oxo Process

This is an industrial hydroformylation for synthesizing aliphatic aldehydes, RCHO.

$$RCH{=}CH_2 + CO + H_2 \xrightarrow{CO_2(CO)_8} RCH_2—CH_2CHO + R\underset{}{\overset{CHO}{\overset{|}{C}H}}—CH_3$$

2. Acylation

$$R\underset{\underset{O}{\|}}{C}—Cl + H_2C{=}CHR' \xrightarrow{BF_3} \left[R—\overset{O}{\overset{\|}{C}}—CH_2—\overset{R'}{\overset{|}{C}}HCl \right] \xrightarrow{-HCl} R—\overset{O}{\overset{\|}{C}}—CH{=}CHR'$$

This is a Markovnikov addition initiated by $RC^+{=}\ddot{O}:$, an acylonium cation.

By Coupling Reactions

1. Carboxylic Acids and Their Derivatives, with Organometallics

(*a*) $R'-\underset{\underset{O}{\|}}{C}-Cl + R_2CuLi \longrightarrow R'-\underset{\underset{O}{\|}}{C}-R$ (cf. Corey–House reaction, Section 4.3)

an acid chloride → a ketone

(*b*) $\boxed{C_6H_5}-MgBr + C_6H_{11}-C{\equiv}N \xrightarrow[2.\ H_3O^+]{1.\ mix} C_6H_{11}-\underset{\underset{O}{\|}}{C}-\boxed{C_6H_5} + NH_4^+$

a nitrile → cyclohexyl phenyl ketone

(*c*) $R'-\underset{\underset{O}{\|}}{C}-OH + 2RLi \longrightarrow R'-\underset{\underset{O}{\|}}{C}-R + RH + 2LiOH$

a carboxylic acid → a ketone

Problem 15.12 (*a*) Why doesn't reaction of RMgX with R′COCl give a ketone? (*b*) Account for the different behaviors of RMgX and R_2CuLi. (*c*) What is the relationship between the reactivity of an organometallic and the activity of the metal?

(*a*) The ketone RCOR′ is formed initially, but once formed, since it is more reactive than RCOCl, it reacts further with RMgX to give the 3° alcohol $R'R_2COH$. (*b*) The C-to-Mg bond has much more ionic character than has the C-to-Cu bond. Therefore, the R group in RMgX is more like $R:^-$ and is much more reactive. (*c*) The more active the metal, the more apt it is to carry a + charge and the more apt is the C to carry a − charge.

2. Alkylation of 1,3-Dithianes with 1° RX or $ROSO_2Ar$

1,3-Dithiane, prepared from $H_2C{=}O$ and 1,3-propanedithiol, $HSCH_2CH_2CH_2SH$ [Problem 15.13(*c*)], can be alkylated at the acidic $-S-CH_2-S-$ group and then hydrolyzed to give the aldehyde. The acidity of this group results from the delocalization of the negative charge of the carbanion to each S by *p*-*d* π bonding (Section 3.11).

1,3-Dithiane (ring S, S with C(H)(H)) $\xrightarrow{\text{n-butyl Li}}$ carbanion (C:⁻ with H) $\xrightarrow{RX}$ alkylated product (C(H)(R)) $\xrightarrow[H_2O]{HgCl_2,\ MeOH}$ RCHO

Ketones can be prepared by (*a*) dialkylating 1,3-dithiane before hydrolysis or (*b*) forming the dithiane of RCHO and then monoalkylating.

Problem 15.13 Synthesize: (*a*) *p*-methoxybenzaldehyde from benzene; (*b*) cyclohexylethanal by hydroboration and oxidation; (*c*) phenylacetaldehyde, using 1,3-dithiane; (*d*) phenyl *n*-propyl ketone from a dithiane; (*e*) cyclohexyl phenyl ketone from PhCOOH and RLi; (*f*) 2-heptanone, using a cuprate.

(*a*) $C_6H_6 \xrightarrow[AlCl_3]{Cl_2} C_6H_5Cl \xrightarrow[350°C]{NaOH} C_6H_5O^-Na^+ \xrightarrow{(CH_3)_2SO_4} C_6H_5OCH_3 \xrightarrow[AlCl_3]{CO,\ HCl} p\text{-}CH_3OC_6H_4CH{=}O$

(*b*) Cyclohexylacetylene C_6H_{11}—C≡CH $\xrightarrow[2.\ H_2O_2,\ NaOH]{1.\ R_2BH}$ [C_6H_{11}—CH=CH—OH] (*unstable enol*) ⟶ C_6H_{11}—CH_2—CH=O

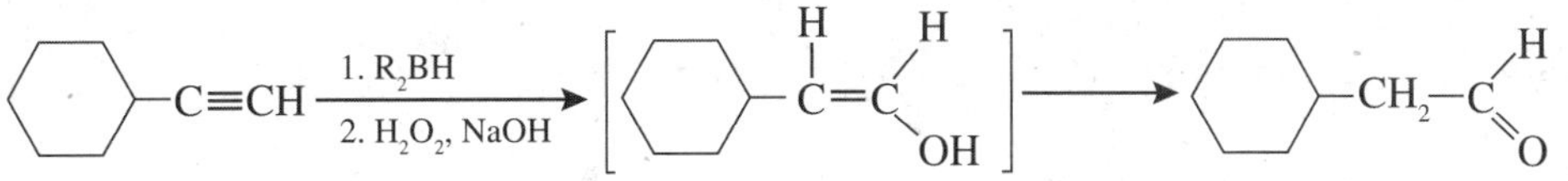

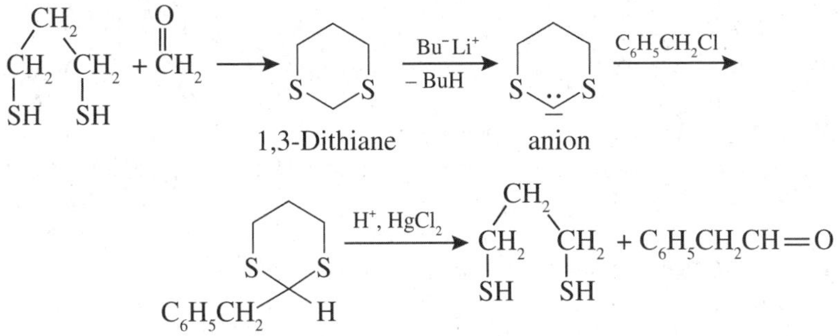

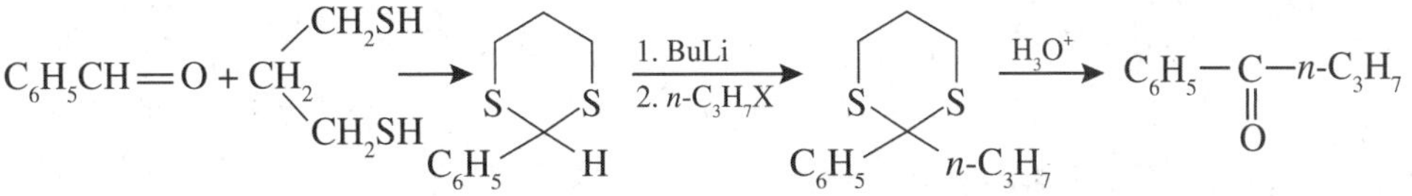

(*c*) $HSCH_2CH_2CH_2SH + CH_2{=}O$ ⟶ 1,3-Dithiane $\xrightarrow[-\ BuH]{Bu^-Li^+}$ anion $\xrightarrow{C_6H_5CH_2Cl}$

2-($C_6H_5CH_2$)-1,3-dithiane $\xrightarrow{H^+,\ HgCl_2}$ $HSCH_2CH_2CH_2SH + C_6H_5CH_2CH{=}O$

(*d*) $C_6H_5CH{=}O + CH_2(CH_2SH)_2$ ⟶ 2-C_6H_5-1,3-dithiane $\xrightarrow[2.\ n\text{-}C_3H_7X]{1.\ BuLi}$ 2-C_6H_5-2-(*n*-C_3H_7)-1,3-dithiane $\xrightarrow{H_3O^+}$ C_6H_5—C(=O)—*n*-C_3H_7

(*e*) $C_6H_5COOH \xrightarrow{LiOH} C_6H_6COO^-Li^+ \xrightarrow[2.\ H_3O^+]{1.\ \text{excess cyclohexyl lithium}} C_6H_5$—C(=O)—$C_6H_{11}$

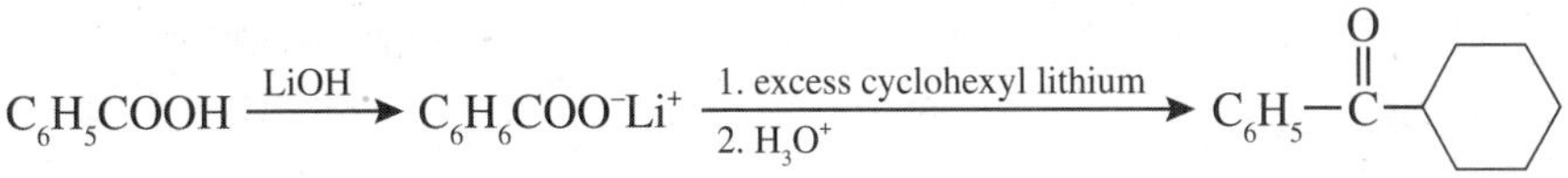

(*f*) $CH_3(CH_2)_4COOH \xrightarrow{PCl_3} CH_3(CH_2)_4COCl \xrightarrow[2.\ H_3O^+]{1.\ (CH_3)_2CuLi} CH_3(CH_2)_4\overset{O}{\overset{\|}{C}}CH_3$

By Pinacol-Pinacolone Rearrangement (Section 14.11)

Problem 15.14 What products are formed in the following reactions? (*a*) CH_3CH_2OH, $Cr_2O_7^{2-}$, H^+; (*b*) $CH_3CHOHCH_3$, $Cr_2O_7^{2-}$, H^+ (60 °C); (*c*) CH_3COCl, LiAl(O—*t*-$C_4H_9)_3H$; (*d*) CH_3COCl, C_6H_6, $AlCl_3$; (*e*) CH_3COCl, $C_6H_5NO_2$, $AlCl_3$

(*a*) CH_3CHO (some oxidation to CH_3COOH occurs).

(*b*) CH_3—C(=O)—CH_3 (*c*) CH_3—C(H)=O

(*d*) $C_6H_5COCH_3$.
(*e*) No reaction; acylation like alkylation does not occur because NO_2 deactivates the ring.

Problem 15.15 Show the substances needed to prepare the following compounds by the indicated reactions:

(*a*) $CH_3CH_2\underset{O}{\underset{\|}{C}}CH_2CH_2C_6H_5$ (Grignard) (*b*) $C_6H_5CH_2CH{=}CH\text{—}\underset{O}{\underset{\|}{C}}\text{—}CH_2C_6H_5$ (Acylation of an alkene)

(*c*) 2,4-$Cl_2C_6H_3COC_6H_5$ (Friedel-Crafts acylation)

(*a*) R′C≡N + RMgX. The carbonyl C in RCOR′ and one alkyl group (R′) come from R′—C≡N; the other R from RMgX. The two possible combinations are

$CH_3CH_2C{\equiv}N + ClMgCH_2CH_2C_6H_5$ or $CH_3CH_2MgBr + N{\equiv}CCH_2CH_2C_6H_5$

(*b*) The R attached to C$=$C is part of the alkene. O$=$CR′ comes from R′COCl.

$$C_6H_5CH_2CH{=}CH_2 + ClCOCH_2C_6H_5 \xrightarrow{BF_3} \text{product}$$

(*c*) $2,4\text{-}Cl_2C_6H_3COCl + C_6H_6 \xrightarrow{AlCl_3} \text{product}$

C_6H_5COCl and $1,3\text{-}C_6H_4Cl_2$ cannot react with $AlCl_3$ because the two aryl Cl's deactivate the ring.

Problem 15.16 Prepare the following compounds from benzene, toluene, and alcohols of four or fewer C's: (*a*) 2-methylpropanal (isobutyraldehyde), (*b*) *p*-chlorobenzaldehyde, (*c*) *p*-nitrobenzophenone ($p\text{-}NO_2C_6H_4COC_6H_5$), (*d*) benzyl methyl ketone, (*e*) *p*-methylbenzaldehyde.

(*a*) $(CH_3)_2CHCH_2OH \xrightarrow[250°C]{Cu} (CH_3)_2CHCHO$ (RCHO is not oxidized further.)

(*b*) $C_6H_5CH_3 \xrightarrow[Fe]{Cl_2} p\text{-}ClC_6H_4CH_3 \xrightarrow[2.\ H_3O^+]{1.\ CrO_3,\ \text{acetic anhydride}} p\text{-}ClC_6H_4CHO$

(*c*) $C_6H_5CH_3 \xrightarrow[H_2SO_4]{HNO_3} p\text{-}O_2NC_6H_4CH_3 \xrightarrow[H^+]{KMnO_4}$

$$p\text{-}O_2NC_6H_4COOH \xrightarrow{SOCl_2} p\text{-}O_2C_6H_4COCl \xrightarrow[AlCl_3]{C_6H_6} p\text{-}O_2NC_6H_4COC_6H_5$$

We cannot acylate $C_6H_5NO_2$ with C_6H_5COCl because NO_2 deactivates the ring.

(*d*) $C_6H_6 \xrightarrow[Fe]{Br_2} C_6H_5Br \xrightarrow{Mg} C_6H_5MgBr \xrightarrow{H_2C\text{—}CH_2\ (\text{epoxide, O})} C_6H_5CH_2CH_2OH \xrightarrow{KMnO_4}$

$$C_6H_5CH_2COOH \xrightarrow{SOCl_2} C_6H_5CH_2COCl \xrightarrow{(CH_3)_2CuI} C_6H_5CH_2COCH_3$$

$$CH_3CH_2OH \xrightarrow{H_2SO_4} H_2C{=}CH_2 \xrightarrow[Ag]{O_2} H_2C\text{—}CH_2\ (\text{epoxide, O})$$

$$CH_3OH \xrightarrow{HBr} CH_3Br \xrightarrow{Li} CH_3Li \xrightarrow{CuI} (CH_3)_2CuLi$$

(*e*) $C_6H_5CH_3 + CO, HCl \xrightarrow[CuCl]{AlCl_3} p\text{-}CH_3C_6H_4CHO$

15.3 Oxidation and Reduction

OXIDATION

1. To Carboxylic Acids

Aldehydes undergo the oxidation:

$$R\text{—}CH{=}O \xrightarrow{KMnO_4\ \text{or}\ K_2Cr_2O_7, H^+} R\text{—}COOH$$

A mild oxidant is **Tollens' reagent,** $Ag(NH_3)_2^+$ (from Ag^+ and NH_3)

$$R\text{—}\underset{\|\atop O}{C}\text{—}H + 2Ag(NH_3)_2^+ + 3OH^- \longrightarrow R\text{—}COO^- + 2H_2O + 4NH_3 + \underset{(mirror)}{2Ag}$$

Formation of the shiny Ag mirror is a positive test for aldehydes. The RCHO must be soluble in aqueous alcohol. This mild oxidant permits —CHO to be oxidized in a molecule having groups more difficult to oxidize, such as 1° or 2° OH's.

Ketones resist mild oxidation, but with strong oxidants at high temperatures, they undergo cleavage of C—C bonds on either side of the carbonyl group to give a mixture of carboxylic acids.

$$RCH_2 \overset{(a)}{—} \underset{\underset{O}{\|}}{C} \overset{(b)}{—} CH_2R' \xrightarrow{\text{oxid.}} \underbrace{RCOOH + R'CH_2COOH}_{\text{from cleavage of bond }(a)} + \underbrace{RCH_2COOH + R'COOH}_{\text{from cleavage of bond }(b)}$$

2. **Via the Haloform Reaction**

Methyl ketones:

$$CH_3\underset{\underset{O}{\|}}{C}—R$$

are readily oxidized by NaOI (NaOH + I_2) to iodoform, CHI_3, and $RCOO^-Na^+$. (See Problem 13.41.)

3. **With Peroxyacids**

In the Baeyer-Villiger reaction, a **ketone** is oxidized to an ester by persulfuric acid, H_2SO_5.

$$Ar—\underset{\underset{O}{\|}}{C}—R \xrightarrow{H_2SO_5} \underset{\text{ester}}{Ar—O—\underset{\underset{O}{\|}}{C}—R} \quad (\textit{very little } Ar—\underset{\underset{O}{\|}}{C}—O—R)$$

When an aryl alkyl ketone is oxidized, the R remains attached to the carbonyl carbon and Ar is bonded to O of the ester group.

Reduction

1. **To Alcohols by Metal Hydrides or H_2/Catalyst**

$$H:^- \text{ (form NaBH}_4) + \rangle C=O \longrightarrow H—\overset{|}{\underset{|}{C}}—O^- \xrightarrow{H_2O} H—\overset{|}{\underset{|}{C}}—OH$$

2. **To Methylene**

$$\rangle C=O \longrightarrow \rangle CH_2 \text{ (also see Problem 15.39)}$$

$$R—\overset{\overset{O}{\|}}{C}—R' \xrightarrow[\text{or } H_2NNH_2 + KOH \text{ (Wolff-Kishner)}]{\text{Zn-Hg + HCl (Clemmensen)}} RCH_2R'$$

The Clemmensen reaction is used mainly with aryl alkyl ketones, $Ar\underset{\underset{O}{\|}}{C}R \longrightarrow ArCH_2R$.

Problem 15.17 Give the products of reaction for (*a*) benzaldehyde + Tollens' reagent; (*b*) cyclohexanone + HNO_3, heat; (*c*) acetaldehyde + dilute $KMnO_4$; (*d*) phenylacetaldehyde + $LiAlH_4$; (*e*) methyl vinyl ketone + H_2/Ni; (*f*) methyl vinyl ketone + $NaBH_4$; (*g*) cyclohexanone + C_6H_5MgBr and then H_3O^+; (*h*) methyl ethyl ketone + strong oxidant; (*i*) methyl ethyl ketone + $Ag(NH_3)_2^+$.

(*a*) $C_6H_5COO^-NH_4^+$, Ag°
(*b*) $HOOC(CH_2)_4COOH$
(*c*) CH_3COOH

(*d*) $C_6H_5CH_2CH_2OH$
(*e*) CH_3—CH(OH)CH_2CH_3 (C=O and C=C are reduced)
(*f*) CH_3—CH(OH)CH=CH_2 (only C=O is reduced, not —C=C—)

(*g*) cyclohexanone (=O) + C_6H_5—MgBr $\xrightarrow[2.\ H_3O^+]{1.\ Et_2O}$ 1-phenylcyclohexanol (OH)

(*h*) $CH_3 \overset{(a)}{\wr} \underset{\|}{\overset{}{C}} \overset{(b)}{\wr} CH_2CH_3$ (C=O) $\xrightarrow{\text{hot acid } KMnO_4}$ $\underbrace{CO_2 + CH_3CH_2COOH}_{\text{from cleavage at } (a)}$ + $\underbrace{CH_3COOH}_{\text{from cleavage at } (b)}$

(*i*) no reaction.

3. Disproportionation. Cannizzaro Reaction

Aldehydes with no H on the α C undergo self-redox (disproportionation) in hot concentrated alkali.

$$2HCHO \xrightarrow[\text{heat}]{50\%\ NaOH} CH_3OH + HCOO^- Na^+$$

$$2C_6H_5CHO \xrightarrow[\text{heat}]{50\%\ NaOH} C_6H_5CH_2OH + C_6H_5COO^- Na^+$$

$$\underset{\textit{always used}}{C_6H_5CHO + HCHO} \xrightarrow[\text{heat}]{50\%\ NaOH} \underset{\textit{always formed}}{C_6H_5CH_2OH + HCOO^- Na^+} \ \textbf{(crossed-Cannizzaro)}$$

Problem 15.18 Devise a mechanism for the Cannizzaro reaction from the reactions

$$2ArCDO \xrightarrow[H_2O]{OH^-} ArCOO^- + ArCD_2OH \qquad 2ArCHO \xrightarrow[D_2O]{OD^-} ArCOO^- + ArCH_2OH$$

The D's from OD^- and D_2O (solvent) are not found in the products. The molecule of ArCDO that is oxidized must transfer its D to the molecule that is reduced. A role must also be assigned to OH^-.

$$Ar{-}\underset{D}{C}{=}O + {:}OH^- \longrightarrow Ar{-}\underset{(D)}{\overset{O-H}{C}}{-}O^- + Ar{-}\overset{D}{C}{=}O \longrightarrow$$

$$Ar{-}\overset{O-(H)}{C}{=}O + Ar{-}\underset{D}{\overset{D}{C}}{-}O^- \longrightarrow Ar{-}\overset{O^-}{C}{=}O + Ar{-}\underset{D}{\overset{D}{C}}{-}OH$$

Problem 15.19 For the Cannizzaro reaction, indicate (*a*) why the reaction cannot be used with aldehydes having an α H, —CHCHO; (*b*) the role of OH^- and OD^- (Problem 15.18); (*c*) the reaction product with ethanedial, O=CH—CH=O; (*d*) the reaction products of a crossed-Cannizzaro reaction between (i) formaldehyde and benzaldehyde, and (ii) benzaldehyde and *p*-chlorobenzaldehyde.

(*a*) An α H is acidic and is removed by OH^-, leaving a carbanion that undergoes other reactions.
(*b*) They are strong nucleophiles that attack the electrophilic C of C=O to give a tetrahedral intermediate. This intermediate reestablishes the resonance-stabilized C=O group by transferring an $:H^-$ to the C=O of another aldehyde molecule.

$$HO{:}^- + R{-}\overset{O}{\overset{\|}{C}}{-}H \longrightarrow \left[\begin{array}{c} O^- \\ R\cdots C \\ HO \quad H \end{array}\right] \xrightarrow[\text{(transfer)}]{-{:}H^-} R{-}\overset{O}{\overset{\|}{C}}{-}OH \xrightarrow[-H^+]{+OH^-} R{-}\overset{O}{\overset{\|}{C}}{-}O^-$$

(*c*) An internal Cannizzaro yields hydroxyacetic acid, $HOCH_2COOH$.

(*d*) (i) H_2CO is mainly attacked by OH^- because it is more electrophilic than PhCHO, whose Ph group delocalizes the electron deficiency of the C of C═O. (ii) There is little difference in the reactivities of the two aldehydes, and both sets of products are found, PhCOOH and $PhCH_2OH$ mixed with *p*-ClC_6H_4COOH and *p*-$ClC_6H_4CH_2OH$.

15.4 Addition Reactions of Nucleophiles to >C═O

The C of the carbonyl group is electrophilic:

$$>C{=}\ddot{O}: \longleftrightarrow >\overset{+}{C}{-}\ddot{\underset{\cdot\cdot}{O}}:^-$$

[Problem 2.24(*b*)] and initially forms a bond with strong nucleophiles:

$$\underset{\text{Electrophile}}{\overset{sp^2}{>C{=}\ddot{O}:}} + \underset{\substack{\text{Nucleophile}\\\text{(strong)}}}{:Nu^-} \longrightarrow \underset{\substack{\text{Transition}\\\text{state}}}{\left[\overset{\delta-}{Nu}\cdots C\cdots\overset{\delta-}{O}\right]^{\ddagger}} \longrightarrow \overset{sp^3}{>\underset{Nu}{C}{-}\ddot{\underset{\cdot\cdot}{O}}:^-} \xrightarrow{H^+} -\underset{Nu}{\overset{|}{C}}{-}OH$$

For example, $:Nu^-$ can be $:R'^-$ of R′MgX or $:H^-$ of $NaBH_4$. With $:NuH_2$, the adduct loses water to give —C═Nu.

$$-\underset{|}{C}{=}O + :NuH_2 \longrightarrow \underset{\textit{unstable adduct}}{\left[HO{-}\underset{|}{\overset{|}{C}}{-}NuH\right]} \xrightarrow{-H_2O} -\underset{|}{C}{=}Nu$$

$:NuH_2$ is most often a 1° amine, RNH_2, or one of its derivatives, such as $HONH_2$ (hydroxylamine).

Acid increases the rate of addition of weak nucleophiles by first protonating the O of C═O, thereby enhancing the electrophilicity of the C of C═O.

$$>C{=}O \xrightarrow{H^+} \left\{>\overset{+}{C}{-}\ddot{O}H \updownarrow >C{=}\overset{+}{\ddot{O}}H\right\} \xrightarrow{\text{weak HNu}} \underset{\text{Transition state}}{\left[>\overset{\delta+}{C}{-}OH \atop \overset{|}{HNu^{\delta+}}\right]^{\ddagger}} \longrightarrow >\underset{\overset{|}{\underset{+}{HNu}}}{C}{-}OH \xrightarrow{-H^+} >\underset{\overset{|}{Nu}}{C}{-}OH$$

The reactivity of the carbonyl group decreases with increasing size of R's and with electron donation by R. Electron-attracting R's increase the reactivity of C═O.

Problem 15.20 The order of reactivity in nucleophilic addition is

$$CH_2{=}O > RCH{=}O > R_2C{=}O > R\underset{\underset{O}{\|}}{C}{=}\ddot{Y}$$

Account for this order in terms of steric and electronic factors.

A change from a trigonal sp^2 to a tetrahedral sp^3 C in the transition state is accompanied by crowding of the four groups on C. Crowding and destabilization of the transition state is in the order

$$CH_2{=}O < RCH{=}O < R_2C{=}O$$

Also, the electron-releasing R's intensify the – charge developing on O, which destabilizes the transition state and decreases reactivity.

In $\underset{\underset{O}{\|}}{RC}Y$, extended π bonding between —Y and C=O

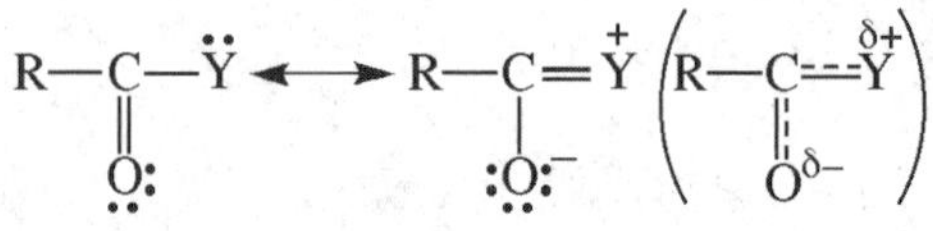

$$R{-}\underset{\underset{\ddot{O}:}{\|}}{C}{-}\ddot{Y} \longleftrightarrow R{-}\underset{\underset{:\ddot{O}:^-}{|}}{C}{=}\overset{+}{Y} \quad \left(R{-}\underset{\underset{O^{\delta-}}{\vdots}}{C}{\cdots}\overset{\delta+}{Y}\right)$$

lowers the enthalpy of the ground state, raises $\Delta H^\ddagger$ and decreases the reactivity of C=O toward nucleophilic attack. Hence, acid derivatives RCOY, in which

$$Y = -\ddot{\underset{..}{X}}:, \; -\ddot{N}H_2, \; -\ddot{\underset{..}{O}}R, \; -\ddot{\underset{..}{O}}{-}\underset{\underset{O}{\|}}{C}{-}R$$

are less reactive than RCHO or R_2CO.

Problem 15.21 Explain the order of reactivity $ArCH_2COR > R_2C{=}O > ArCOR > Ar_2CO$ in nucleophilic addition.

When attached to C=O, Ar's, like —Y: (Problem 15.20), are electron-releasing by extended π bonding (resonance) and deactivate C=O. Two Ar's are more deactivating than one Ar. In $ArCH_2COR$, only the electron-withdrawing inductive effect of Ar prevails; consequently, $ArCH_2$ increases the reactivity of C=O.

Problem 15.22 Why is cyanohydrin formation useful in synthesis?

The cyanohydrin not only adds an additional C at the site of the C=O but also introduces two new functional groups, OH and CN, which can be used to introduce other functional groups. The OH can be used to form an alkene (C=C), an ether (—RO), or a halogen compound (C—X); the C≡N can be reduced to an amine (CH_2NH_2), be hydrolyzed to a carboxyl (COOH) group, or react with Grignard reagents if the OH is protected.

Problem 15.23 $NaHSO_3$ reacts with RCHO in EtOH to give a solid adduct. (*a*) Write an equation for the reaction. (*b*) Explain why only RCHO, methyl ketones ($RCOCH_3$) and cyclic ketones react. (*c*) If the carbonyl compound can be regenerated on treating the adduct with acid or base, explain how this reaction with $NaHSO_3$ can be used to separate RCHO from noncarbonyl compounds such as RCH_2OH.

(*a*) HSO_3^- can protonate RCHO.

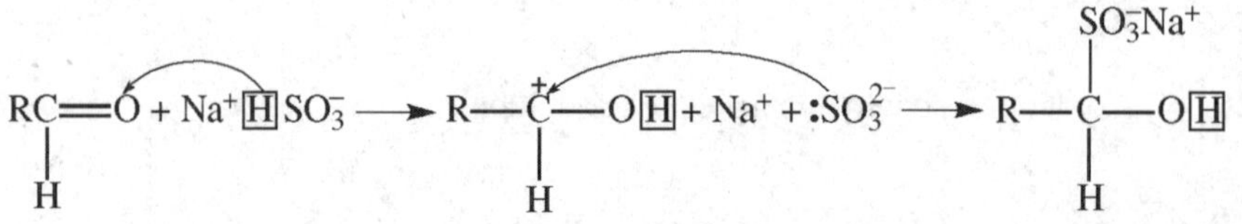

$$\underset{\underset{H}{|}}{RC}{=}\ddot{O} + Na^+\boxed{H}SO_3^- \longrightarrow R{-}\underset{\underset{H}{|}}{\overset{+}{C}}{-}O\boxed{H} + Na^+ + :SO_3^{2-} \longrightarrow R{-}\underset{\underset{H}{|}}{\overset{\overset{SO_3^-Na^+}{|}}{C}}{-}O\boxed{H}$$

Sodium bisulfite adduct (solid)

A C—S bond is formed because S is a more nucleophilic site than O.

(*b*) SO_3^{2-} is a large ion and reacts only if C=O is not sterically hindered, as is the case for RCHO, $RCOCH_3$, and cyclic ketones.

(*c*) The solid adduct is filtered from the ethanolic solution of unreacted RCH_2OH and then is decomposed by acid or base:

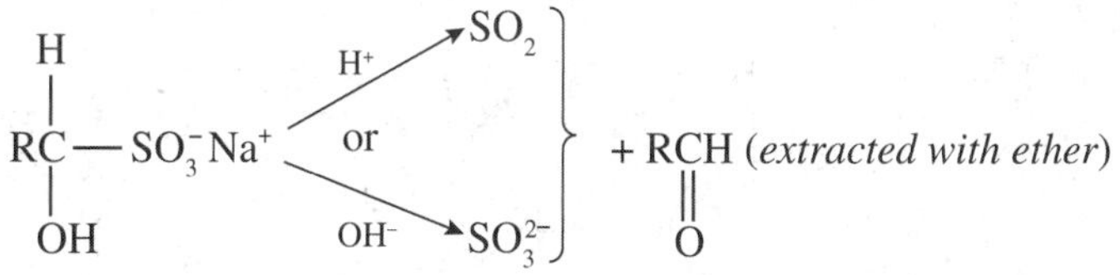

$$\text{RC(H)(OH)}-SO_3^-Na^+ \xrightarrow[\text{or } OH^-]{H^+} \left.\begin{matrix} SO_2 \\ SO_3^{2-} \end{matrix}\right\} + \text{RCH{=}O} \text{ (extracted with ether)}$$

Problem 15.24 Write the formula for the solid derivative formed when an aldehyde or ketone reacts with each of the following ammonia derivatives:

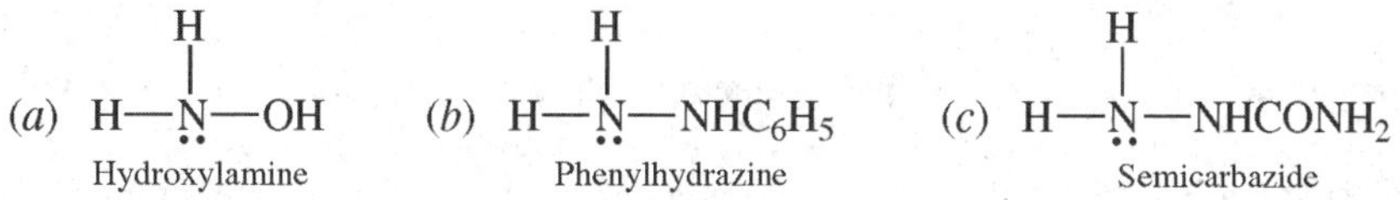

(*a*) $H-\ddot{N}(H)-OH$ Hydroxylamine (*b*) $H-\ddot{N}(H)-NHC_6H_5$ Phenylhydrazine (*c*) $H-\ddot{N}(H)-NHCONH_2$ Semicarbazide

Since these nucleophiles are of the $:NuH_2$ type, addition is followed by dehydration.

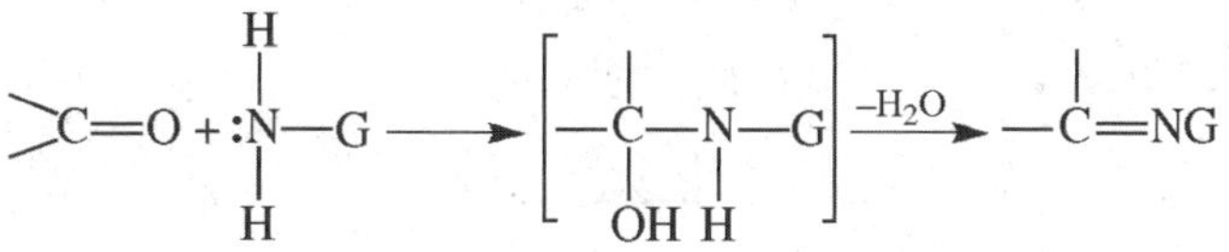

$$\rangle C{=}O + :N(H)_2-G \longrightarrow \left[-\overset{|}{\underset{OH}{C}}-\underset{H}{N}-G\right] \xrightarrow{-H_2O} -\overset{|}{C}{=}NG$$

(*a*) $G = -OH$; $\rangle C{=}N-OH$ (Oxime).

(*b*) $G = -NHC_6H_5$; $\rangle C{=}NNHC_6H_5$ (Phenylhydrazone).

(*c*) $G = -NHCONH_2$; $\rangle C{=}NNHCONH_2$ (Semicarbazone).

The melting points of these solid derivatives are used to identify carbonyl compounds.

Problem 15.25 Why do carbonyl compounds having an α H react with R_2NH (2°) to yield **enamines**, $-\overset{|}{C}{=}\overset{|}{C}-NR_2$, but give **imines**, $\rangle\overset{|}{C}-\overset{|}{C}{=}NR$, with RNH_2 (1°)?

After protonation of the O, the nucleophilic RNH_2 adds to the C and the adduct loses H^+, to give the **carbinolamine**. Dehydration proceeds by protonation of the O of OH, loss of H_2O, and then loss of H^+, to give the imine.

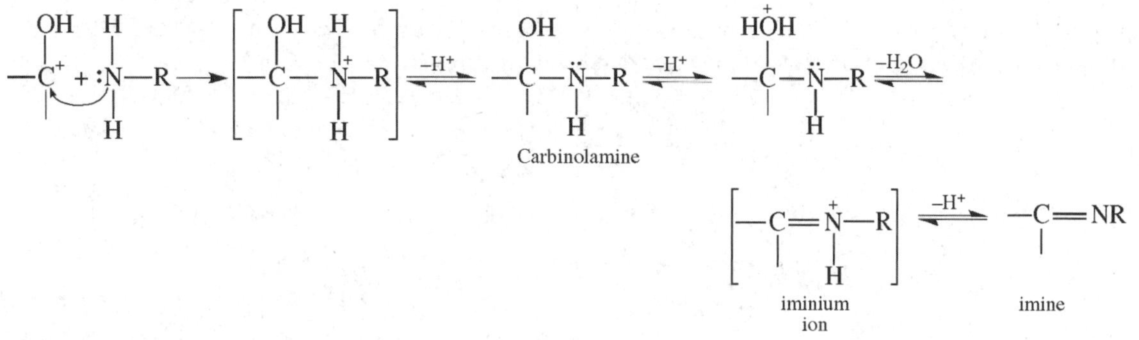

$$-\overset{OH}{\overset{|}{C^+}} + :\underset{H}{\overset{H}{N}}-R \longrightarrow \left[-\overset{OH}{\overset{|}{C}}-\overset{+}{\underset{H}{\overset{H}{N}}}-R\right] \underset{}{\overset{-H^+}{\rightleftharpoons}} \underset{\text{Carbinolamine}}{-\overset{OH}{\overset{|}{C}}-\underset{H}{\ddot{N}}-R} \overset{-H^+}{\rightleftharpoons} -\overset{\overset{+}{H}OH}{\overset{|}{C}}-\underset{H}{\ddot{N}}-R \overset{-H_2O}{\rightleftharpoons}$$

$$\underset{\text{iminium ion}}{\left[-\overset{|}{C}{=}\overset{+}{\underset{H}{N}}-R\right]} \overset{-H^+}{\rightleftharpoons} \underset{\text{imine}}{-\overset{|}{C}{=}NR}$$

The carbinolamine formed from R_2NH lacks an H on N, and its dehydration involves instead loss of the acidic α H to give the resonance-stabilized eneamine.

$$-\overset{H}{\underset{|}{\overset{|}{C}}}-\overset{O-H}{\underset{|}{\overset{\|}{C}}} + \ddot{N}R_2 \longrightarrow \left[-\overset{H}{\underset{|}{\overset{|}{C}}}-\overset{OH}{\underset{|}{\overset{|}{C}}}-\ddot{N}R_2\right] \xrightarrow{-H_2O} -\underset{|}{C}=\underset{|}{C}-\ddot{N}R_2$$

Carbinolamine eneamine

Problem 15.26 Reaction of 1 mole of semicarbazide with a mixture of 1 mol each of cyclohexanone and benzaldehyde precipitates cyclohexanone semicarbazone, but after a few hours, the precipitate is benzaldehyde semicarbazone. Explain.

The C=O of cyclohexanone is not deactivated by the electron-releasing C_6H_5 and does not suffer from steric hindrance. The semicarbazone of cyclohexanone is the kinetically controlled product. Conjugation makes PhCH=NNHCONH$_2$ more stable, and its formation is thermodynamically controlled. In such reversible reactions, the equilibrium shifts to the more stable product (Fig. 15.1).

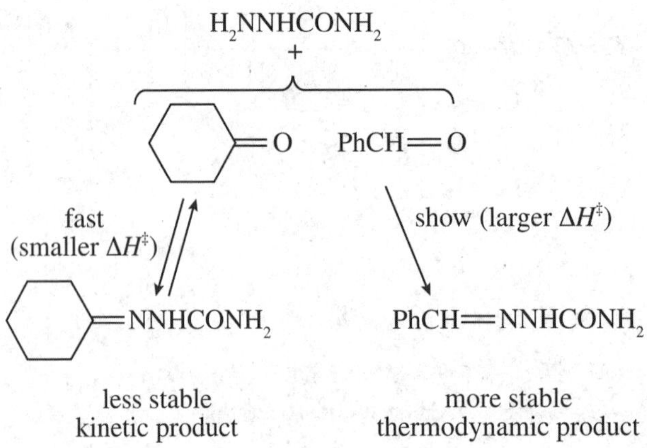

Figure 15.1

Problem 15.27 Symmetrical ketones, $R_2C{=}O$, form a single oxime, but aldehydes and unsymmetrical ketones may form two isomeric oximes. Explain.

The π bond in

$$\rangle C=N\langle$$

prevents free rotation, and therefore, geometric isomerism occurs if the groups on the carbonyl C are dissimilar. The old terms *syn* and *anti* are also used in place of *cis* and *trans*, respectively.

$$\begin{array}{cccc} R\quad H & R\quad H & R\quad R' & R\quad R' \\ \diagdown\ \diagup & \diagdown\ \diagup & \diagdown\ \diagup & \diagdown\ \diagup \\ C & C & C & C \\ \| & \| & \| & \| \\ :N & N: & :N & N: \\ \diagdown OH & HO \diagup & \diagdown OH & HO \diagup \end{array}$$

trans (anti) *cis (syn)*

15.5 Addition of Alcohols: Acetal and Ketal Formation

$$\text{R}'\text{—}\overset{\text{H}}{\overset{|}{\text{C}}}\text{=O} + 2\text{ROH} \underset{\text{H}_3\text{O}^+}{\overset{\text{dry HCl}}{\rightleftharpoons}} \text{R}'\text{—}\underset{\text{OR}}{\overset{\text{H}}{\text{C}}}\text{—OR} + \text{H}_2\text{O}$$

an acetal (*gem*-diether)

In H_3O^+, R′CHO is regenerated because acetals undergo acid-catalyzed cleavage much more easily than do ethers. Since acetals are stable in neutral or basic media, they are used to protect the —CH=O group. Unhindered ketones form ketals, $R_2C(OR')_2$. RSH forms thioacetals, $RCH(SR')_2$, and thioketals, $R_2C(SR')_2$.

Problem 15.28 Give mechanisms for (*a*) acid-catalyzed acetal formation:

$$(a)\ \text{R—}\overset{\text{H}}{\overset{|}{\text{C}}}\text{=O} + \text{R}'\text{OH} \overset{\text{dry HCl}}{\rightleftharpoons} \left[\text{R—}\underset{\text{O—R}'}{\overset{\text{H}}{\text{C}}}\text{—O—H}\right] \overset{\text{R}'\text{OH}}{\rightleftharpoons} \text{R—}\underset{\text{O—R}'}{\overset{\text{H}}{\text{C}}}\text{—O—R}' + \text{H}_2\text{O}$$

hemiacetal — acetal

(*b*) base induced hemiacetal formation with OR^- in ROH.

$$(a)\ \text{R—}\overset{\text{H}}{\overset{|}{\text{C}}}\text{=O} \underset{-\text{H}^+}{\overset{+\text{H}^+}{\rightleftharpoons}} \text{R—}\overset{\text{H}}{\underset{+}{\text{C}}}\text{—OH} \overset{\text{R}'\ddot{\text{O}}\text{H}}{\rightleftharpoons} \text{R—}\underset{\text{H}\overset{+}{\text{O}}\text{R}'}{\overset{\text{H}}{\text{C}}}\text{—OH} \underset{+\text{H}^+}{\overset{-\text{H}^+}{\rightleftharpoons}} \left[\text{R—}\underset{\text{OR}'}{\overset{\text{H}}{\text{C}}}\text{—OH}\right] \underset{-\text{H}^+}{\overset{+\text{H}^+}{\rightleftharpoons}} \text{R—}\underset{\text{OR}'}{\overset{\text{H}}{\text{C}}}\text{—}\overset{+}{\text{O}}\text{H}_2$$

hemiacetal

From the protonated hemiacetal, the mechanism is similar to that for the formation of ethers from alcohols [Problem 14.7(*b*)].

$$\text{R—}\underset{\text{OR}'}{\overset{\text{H}}{\text{C}}}\text{—}\overset{+}{\text{O}}\text{H}_2 \underset{+\text{H}_2\text{O}}{\overset{-\text{H}_2\text{O}}{\rightleftharpoons}} \text{R—}\underset{:\ddot{\text{O}}\text{R}'}{\overset{\text{H}}{\text{C}^+}} \longleftrightarrow \text{R}\overset{\text{H}}{\text{C}}\text{=}\overset{+}{\ddot{\text{O}}}\text{R}' \overset{\text{R}'\text{OH}}{\rightleftharpoons} \text{R—}\underset{\text{OR}'}{\overset{\text{H}}{\text{C}}}\text{—}\overset{+}{\text{O}}\text{(H)R}' \underset{+\text{H}^+}{\overset{-\text{H}^+}{\rightleftharpoons}} \text{R—}\underset{\text{OR}'}{\overset{\text{H}}{\text{C}}}\text{—OR}'$$

protonated hemiacetal — protonated acetal — acetal

$$(b)\ \rangle\text{C=}\ddot{\text{O}}\text{:} + {}^-\text{OR}' \longrightarrow \rangle\underset{\text{OR}'}{\text{C}}\text{—}\ddot{\text{O}}\text{:}^- \xrightarrow{\text{H:OR}'} \rangle\underset{\text{OR}'}{\text{C}}\text{—OH} + {}^-\text{:}\ddot{\text{O}}\text{:R}'$$

Problem 15.29 Show how a C=O group can be protected by acetal formation in the conversion of $OHCCH_2C{\equiv}CH$ to $OHCCH_2C{\equiv}CCH_3$.

The introduction of CH_3 requires that the terminal alkyne C first become a carbanion and then be methylated. Such a carbanion, acting like the R group of RMgX, would react with the C=O group of another molecule before it could be methylated. To prevent this, C=O is protected by acetal formation before the carbanion is formed. The acetal is stable under the basic conditions of the methylation reactions. The aldehyde is later unmasked by acid-catalyzed hydrolysis.

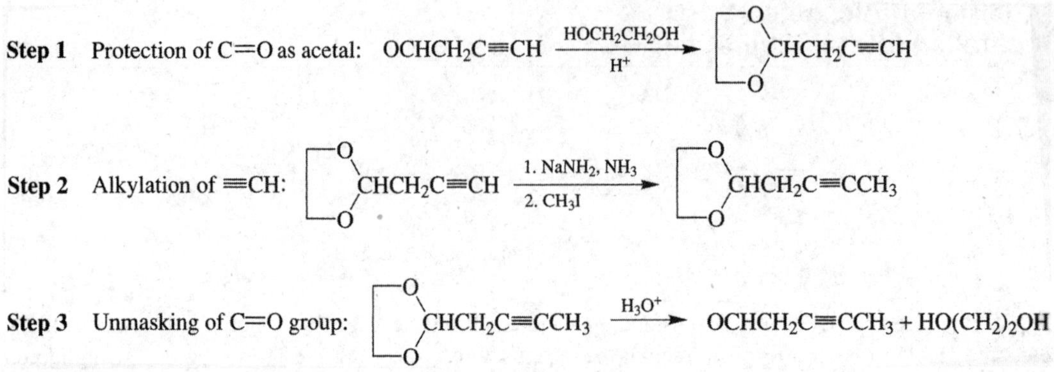

Step 1 Protection of C=O as acetal: $OCHCH_2C{\equiv}CH \xrightarrow[H^+]{HOCH_2CH_2OH}$ (cyclic acetal, 1,3-dioxolane) $CHCH_2C{\equiv}CH$

Step 2 Alkylation of ≡CH: (1,3-dioxolane) $CHCH_2C{\equiv}CH \xrightarrow[2.\ CH_3I]{1.\ NaNH_2,\ NH_3}$ (1,3-dioxolane) $CHCH_2C{\equiv}CCH_3$

Step 3 Unmasking of C=O group: (1,3-dioxolane) $CHCH_2C{\equiv}CCH_3 \xrightarrow{H_3O^+} OCHCH_2C{\equiv}CCH_3 + HO(CH_2)_2OH$

Problem 15.30 In acid, most aldehydes form nonisolable hydrates (*gem*-diols). Two exceptions are the stable chloral hydrate, $Cl_3CCH(OH)_2$, and ninhydrin:

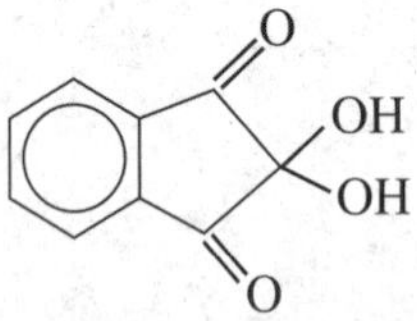

(*a*) Given the bond energies 749, 464, and 360 kJ/mol for C=O, O—H, and C—O, respectively, show why the equilibrium typically lies toward the carbonyl compound. (*b*) Account for the exceptions.

(*a*) Calculating ΔH for

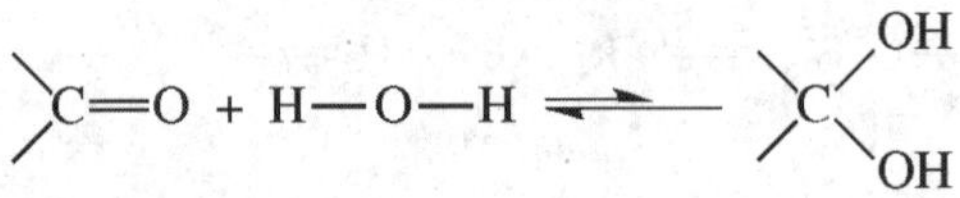

$$\rangle C{=}O + H{-}O{-}H \rightleftharpoons \rangle C(OH)_2$$

we obtain

$$\underset{\substack{\text{cleavages}\\ \textit{endothermic}}}{[\underset{(C=O)}{749} + \underset{(O-H)}{2(464)}]} + \underset{\substack{\text{formations}\\ \textit{exothermic}}}{[\underset{(C-O)}{2(-360)} + \underset{(O-H)}{2(-464)}]} = \Delta H$$

or $\Delta H = +29$ kJ/mol. Hydrate formation is endothermic and not favored. The carbonyl side is also favored by entropy because two molecules:

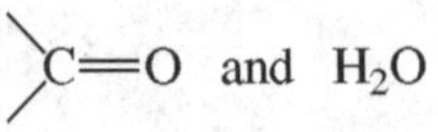

$$\rangle C{=}O \text{ and } H_2O$$

are more random than 1 *gem*-diol molecule.

(*b*) Strong electron-withdrawing groups on an α C destabilize an adjacent carbonyl group because of repulsion of adjacent + charges. Hydrate formation overcomes the forces of repulsion.

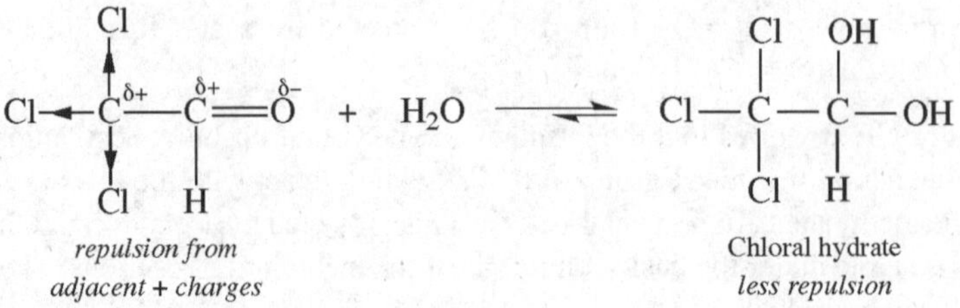

$$Cl{\leftarrow}\overset{\delta+}{C}Cl_2{-}\overset{\delta+}{C}H{=}\overset{\delta-}{O} + H_2O \rightleftharpoons Cl{-}CCl_2{-}CH(OH){-}OH$$

repulsion from adjacent + charges — Chloral hydrate *less repulsion*

Hydration of the middle carbonyl group of ninhydrin removes both pairs of repulsions.

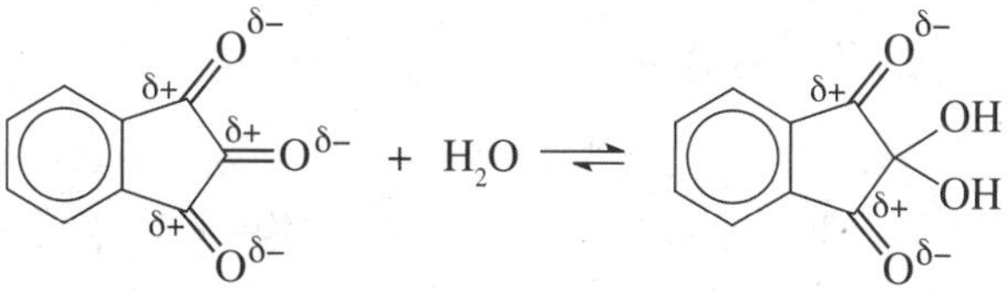

$$\text{ninhydrin (C=O}^{\delta-}\text{ groups with } \delta+ \text{ carbons)} + H_2O \rightleftharpoons \text{hydrate (C(OH)}_2\text{)}$$

Problem 15.31 Show steps in the synthesis of **cyclooctyne**, the smallest ring with a triple bond, from $C_2H_5OOC(CH_2)_6COOC_2H_5$.

The 1,8-diester is converted to an eight-membered ring **acyloin**, which is then changed to the alkyne.

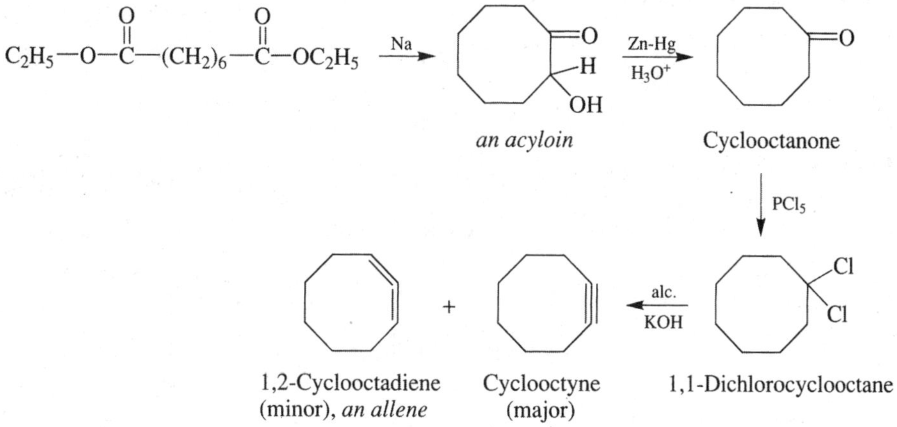

$$C_2H_5{-}O{-}\overset{O}{\overset{\|}{C}}{-}(CH_2)_6{-}\overset{O}{\overset{\|}{C}}{-}OC_2H_5 \xrightarrow{Na} \underset{\textit{an acyloin}}{\text{cyclooctane ring with C=O and CH(OH)}} \xrightarrow[H_3O^+]{Zn\text{-}Hg} \underset{\text{Cyclooctanone}}{\text{cyclooctane ring C=O}}$$

$$\xrightarrow{PCl_5} \underset{\text{1,1-Dichlorocyclooctane}}{\text{ring }CCl_2} \xrightarrow{\text{alc. KOH}} \underset{\text{(major)}}{\text{Cyclooctyne}} + \underset{\text{(minor), } \textit{an allene}}{\text{1,2-Cyclooctadiene}}$$

15.6 Attack by Ylides; Wittig Reaction

A carbanion C can form a *p-d* π bond (Section 3.11) with an adjacent P or S. The resulting charge delocalization is especially effective if P or S, furnishing the empty d orbital, also has a + charge. Carbanions with these characteristics are called **ylides**; for example:

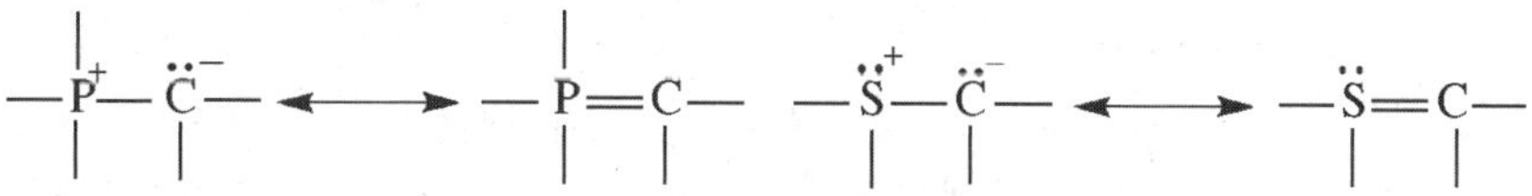

$$-\overset{|}{\underset{|}{P^+}}-\overset{\cdot\cdot\,-}{\underset{|}{C}}- \longleftrightarrow -\overset{|}{\underset{|}{P}}=\underset{|}{C}- \qquad -\overset{\cdot\cdot\,+}{\underset{|}{S}}-\overset{\cdot\cdot\,-}{\underset{|}{C}}- \longleftrightarrow -\overset{\cdot\cdot}{\underset{|}{S}}=\underset{|}{C}-$$

The **Wittig reaction** uses P ylides to change O of the carbonyl group to

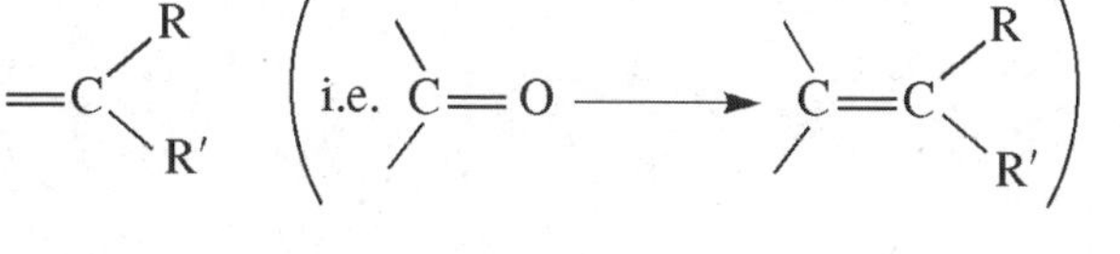

$$=C\begin{matrix}R\\R'\end{matrix} \quad \left(\text{i.e. } \gt C=O \longrightarrow \gt C=C\begin{matrix}R\\R'\end{matrix}\right)$$

The carbanion portion of the ylide replaces the O.

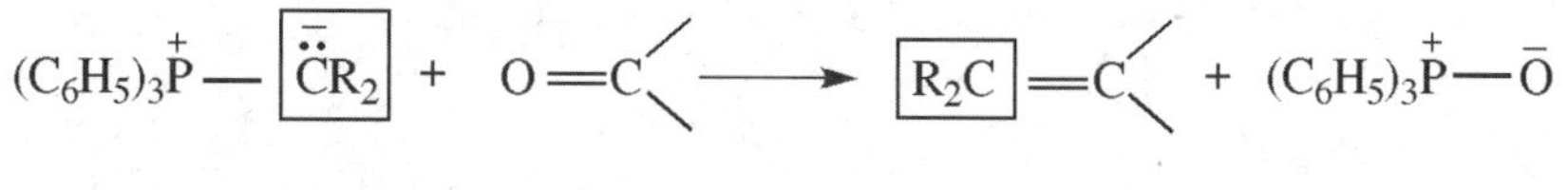

$$(C_6H_5)_3\overset{+}{P}-\boxed{\overset{\cdot\cdot\,-}{C}R_2} + O=C\lt \longrightarrow \boxed{R_2C}=C\lt + (C_6H_5)_3\overset{+}{P}-\overset{-}{O}$$

The ylide is prepared in two steps from RX.

$$Ph_3P{:} + RCH_2{-}X \xrightarrow{S_N2} [Ph_3\overset{+}{P}CH_2R]\,X^- \xrightarrow{C_4H_9^-Li^+} Ph_3\overset{+}{P}\ddot{C}^-HR + C_4H_{10} + \overset{+}{Li}X^-$$

a phosphine

Sulfur ylides react with aldehydes and ketones to form epoxides (oxiranes):

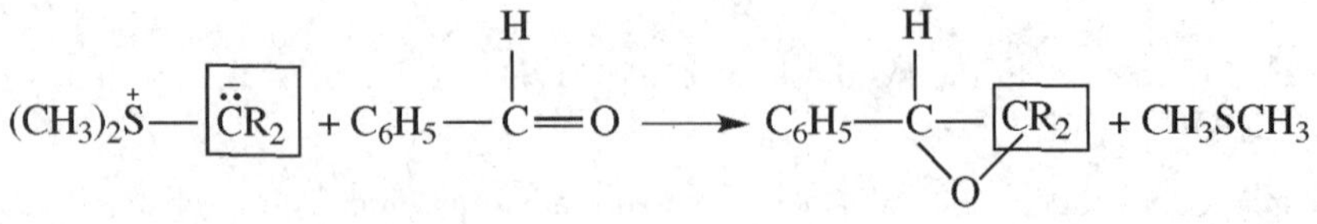

$$(CH_3)_2\overset{+}{S}{-}\boxed{\ddot{C}^-R_2} + C_6H_5{-}\underset{}{\overset{H}{C}}{=}O \longrightarrow C_6H_5{-}\overset{H}{C}{-}\boxed{CR_2}\ (\text{epoxide, bridged by O}) + CH_3SCH_3$$

The sulfur ylide is formed from the sulfonium salt:

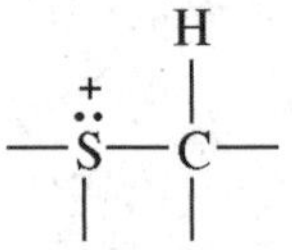

$$-\overset{+}{\ddot{S}}-\overset{H}{\underset{|}{C}}-$$

with a strong base, such as sodium dimethyloxosulfonium methylide:

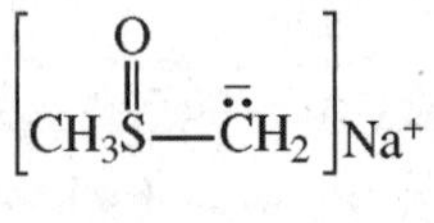

$$\left[CH_3\overset{O}{\overset{\|}{S}}{-}\ddot{C}^-H_2\right]Na^+$$

Problem 15.32 Which alkenes are formed from the following ylide-carbonyl compound pairs? (*a*) 2-butanone and $CH_3CH_2CH_2CH{=}P(C_6H_5)_3$, (*b*) acetophenone and $(C_6H_5)_3P{=}CH_2$, (*c*) benzaldehyde and $C_6H_5{-}CH{=}P(C_6H_5)_3$, (*d*) cyclohexanone and $(C_6H_5)_3P{=}C(CH_3)_2$. (Disregard stereochemistry.)

The boxed portions below come from the ylide.

(*a*) $CH_3CH_2\overset{CH_3}{C}{=}\boxed{CHCH_2CH_2CH_3}$ (*b*) $C_6H_5{-}\overset{CH_3}{C}{=}\boxed{CH_2}$

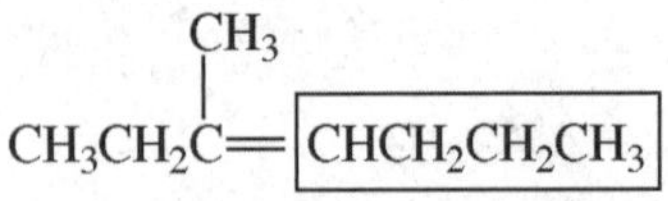

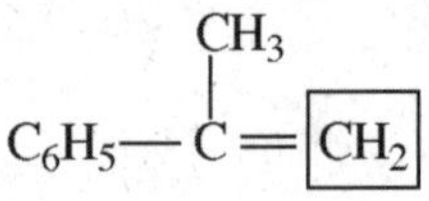

(*c*) $C_6H_5{-}CH{=}\boxed{CH{-}C_6H_5}$ (*d*) cyclohexylidene$=\boxed{C(CH_3)_2}$

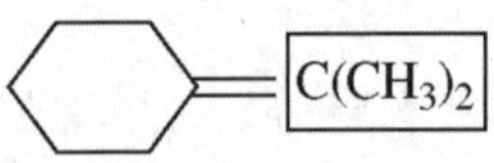

Problem 15.33 Give structures of the ylide and carbonyl compound needed to prepare

(*a*) $C_6H_5CH{=}CHCH_3$ (*b*) methylenecyclopentane, cyclopentylidene$=CH_2$ (*c*) $CH_3CH_2\underset{\underset{CHC_6H_5}{\|}}{C}{-}CH(CH_3)_2$

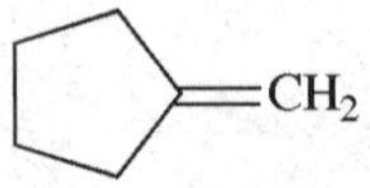

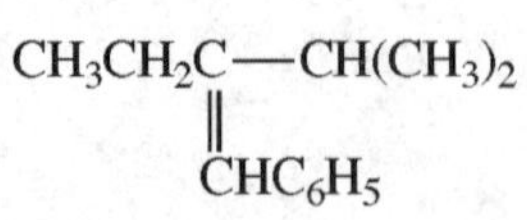

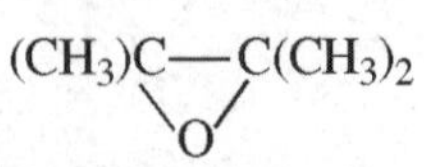

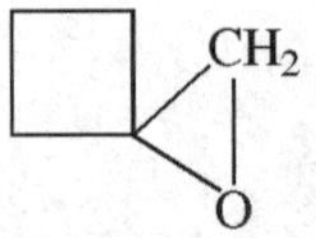

(*d*) $(CH_3)C{-}C(CH_3)_2$ (bridged by O, epoxide) (*e*) cyclobutane spiro-epoxide with CH_2 and O

(*a*) $Ph_3\overset{+}{P}\overset{\cdot\cdot}{\bar{C}}HCH_3 + C_6H_5\overset{H}{C}{=}O$ or $Ph_3\overset{+}{P}\overset{\cdot\cdot}{\bar{C}}HC_6H_5 + CH_3\overset{H}{C}{=}O$

The *cis-trans* geometry of the alkene is influenced by the nature of the substituents, solvent, and dissolved salts. Polar protic or aprotic solvents favor the *cis* isomer.

(*b*) cyclopentanone (cyclopentylidene $=O$) $+ Ph_3\overset{+}{P}\overset{\cdot\cdot}{\bar{C}}H_2$ or cyclopentyl carbanion $\overset{-}{C}{-}\overset{+}{P}Ph_3 + O{=}CH_2$

(*c*) $CH_3CH_2\underset{\underset{O}{\|}}{C}CH(CH_3)_2 + Ph_3\overset{+}{P}\overset{\cdot\cdot}{\bar{C}}HC_6H_5$ or $CH_3CH_2\underset{\underset{^+PPh_3}{|}}{\overset{\cdot\cdot}{\bar{C}}}CH(CH_3)_2 + C_6H_5CHO$

(*d*) $(CH_3)_2C{=}O + Ph_2\overset{+}{S}{-}\overset{\cdot\cdot}{\bar{C}}(CH_3)_2$

(*e*) cyclobutanone $+ (CH_3)_2\overset{+}{S}{-}\overset{\cdot\cdot}{\bar{C}}H_2$

15.7 Miscellaneous Reactions

1. Conversion to Dihalides

$$\text{cyclopentanone} + PCl_5 \longrightarrow \text{1,1-dichlorocyclopentane} + POCl_3$$

$$\text{cyclohexanone} + SF_4 \longrightarrow \text{1,1-difluorocyclohexane} + SOF_2$$

2. Reformatsky Reaction

Ketones or aldehydes can be reacted to form β-hydroxyesters.

$$\begin{matrix}R\\ \\R'\end{matrix}\!\!>C{=}O + Zn + Br\overset{|}{\underset{|}{C}}{-}COOEt \xrightarrow[2.\ H_2O]{1.\ \text{dry ether}} \begin{matrix}R\\ \\R'\end{matrix}\!\!>\underset{OH}{\underset{|}{C}}{-}\overset{|}{\underset{|}{C}}{-}COOEt \text{ via } \left[Br\overset{\delta+}{Zn}{-}\overset{|\,\delta-}{\underset{|}{C}}{-}COOEt\right]$$

a β-hydroxyester

R′ and R may also be H or Ar.

Problem 15.34 Use the Reformatsky reaction to prepare

(*a*) $(CH_3)_2C(OH)CH_2COOC_2H_5$ (*b*) $Ph\underset{CH_3}{\underset{|}{C}}(OH)\underset{CH_3}{\underset{|}{C}}HCOOC_2H_5$ (*c*) $Ph\underset{CH_3}{\underset{|}{C}}{=}\underset{CH_3}{\underset{|}{C}}COOH$

The formed bond is

$$\left[-\overset{\text{OH}}{\underset{|}{\text{C}}}-\right]\downarrow\left(-\overset{|}{\underset{|}{\text{C}}}-\text{COOR}\right) \text{ or } \left[-\overset{|}{\underset{|}{\text{C}}}-\right]\downarrow\left(-\overset{|}{\underset{|}{\text{C}}}-\text{COOR}\right)$$

The structure in the box comes from the carbonyl compound (acceptor); the structure in the oval comes from the α-bromoester (carbanion source).

(*a*) $\left[CH_3-\overset{CH_3}{\underset{OH}{C}}-\right]\downarrow\left(CH_2COOC_2H_5\right) \longleftarrow CH_3-\overset{CH_3}{\underset{O}{\overset{|}{\underset{\|}{C}}}} + Zn + BrCH_2COOC_2H_5$

(*b*) $\left[Ph-\overset{OH}{\underset{CH_3}{C}}-\right]\downarrow\left(\overset{H}{\underset{CH_3}{C}}-COOC_2H_5\right) \longleftarrow Ph-\overset{CH_3}{\underset{O}{\overset{|}{\underset{\|}{C}}}} + Zn + Br\underset{CH_3}{C}HCOOC_2H$

(*c*) Product from (*b*) $\xrightarrow[-H_2O]{H^+} Ph\underset{CH_3}{C}=\underset{CH_3}{C}-COOH$

3. Reactions of the Aldehydic H

The chemistry of the aldehydic H, except for oxidation to OH, is meager. The C—H bond can be homolytically cleaved by participation of a free radical.

Problem 15.35 Propanal reacts with 1-butene in the presence of uv or free-radical initiators (peroxides, sources of RO·) to give $CH_3CH_2COCH_2CH_2CH_2CH_3$. Give steps for a likely mechanism.

The net reaction is in addition of $CH_3CH_2\underset{H}{C}=O$ to $H_2C=CHCH_2CH_3$.

Step 1 $CH_3CH_2\overset{H}{C}=O \xrightarrow[-H\cdot]{RO\cdot} CH_3CH_2\dot{C}=O$

Step 2 $CH_3CH_2\underset{O}{\underset{\|}{C}}\cdot + H_2C=CHCH_2CH_3 \longrightarrow CH_3CH_2\underset{O}{\underset{\|}{C}}-CH_2\dot{C}HCH_2CH_3$

Step 3 $CH_3CH_2COCH_2\dot{C}HCH_2CH_3 + CH_3CH_2\ddot{C}=O \longrightarrow CH_3CH_2COCH_2CH_2CH_2CH_3 + CH_3CH_2\dot{C}=O$

Step 1 is the initiation step. Steps 2 and 3 propagate the chain.

15.8 Summary of Aldehyde Chemistry

PREPARATION | *PROPERTIES*

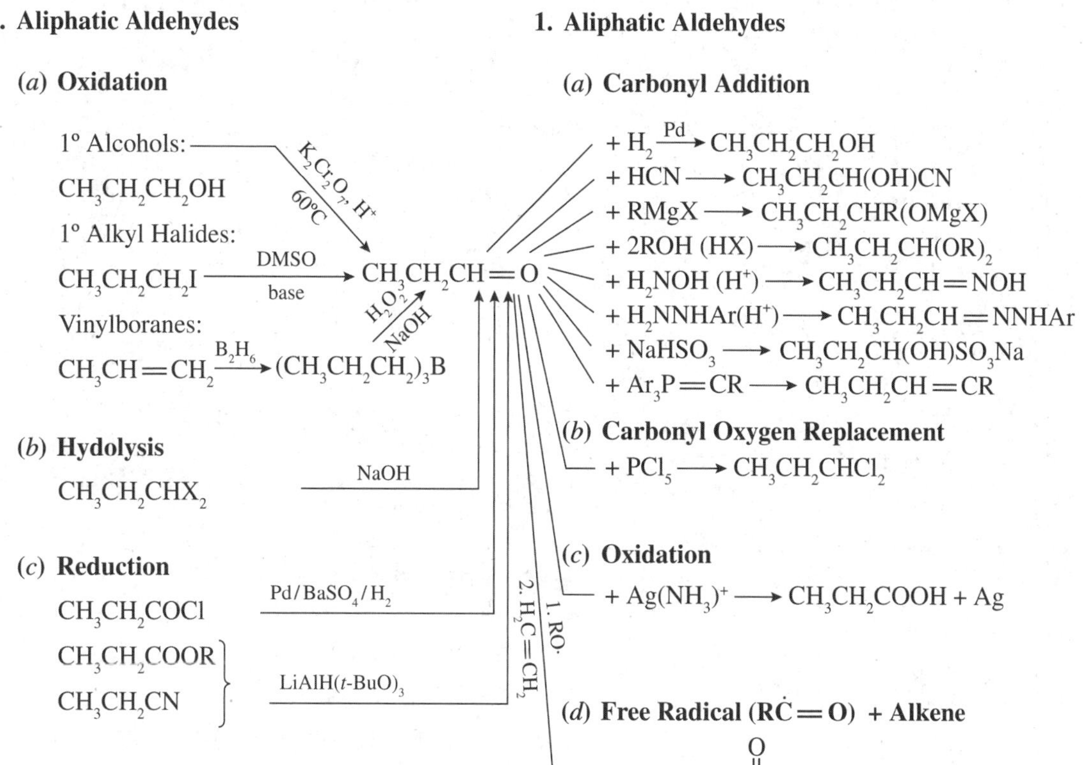
1. Aliphatic Aldehydes
(a) Oxidation
1° Alcohols:
CH3CH2CH2OH
K2Cr2O7, H+
60°C
1° Alkyl Halides:
CH3CH2CH2I
DMSO
base
CH3CH2CH=O
Vinylboranes:
CH3CH=CH2
B2H6
(CH3CH2CH2)3B
H2O2
NaOH
(b) Hydolysis
CH3CH2CHX2
NaOH
(c) Reduction
CH3CH2COCl
Pd/BaSO4/H2
CH3CH2COOR
CH3CH2CN
LiAlH(t-BuO)3
2. H2C=CH2
1. RO·
1. Aliphatic Aldehydes
(a) Carbonyl Addition
+ H2 —Pd→ CH3CH2CH2OH
+ HCN → CH3CH2CH(OH)CN
+ RMgX → CH3CH2CHR(OMgX)
+ 2ROH (HX) → CH3CH2CH(OR)2
+ H2NOH (H+) → CH3CH2CH=NOH
+ H2NNHAr(H+) → CH3CH2CH=NNHAr
+ NaHSO3 → CH3CH2CH(OH)SO3Na
+ Ar3P=CR → CH3CH2CH=CR
(b) Carbonyl Oxygen Replacement
+ PCl5 → CH3CH2CHCl2
(c) Oxidation
+ Ag(NH3)+ → CH3CH2COOH + Ag
(d) Free Radical (RĊ=O) + Alkene
CH3CH2—C(=O)—CH2CH3 (with H2C=CH2)

2. Aromatic Aldehydes | 2. Aromatic Aldehydes

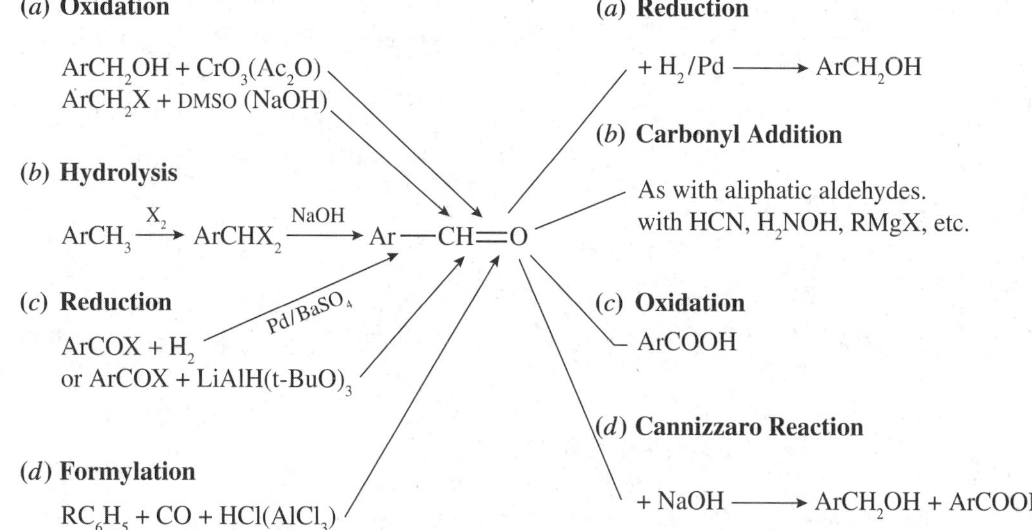
(a) Oxidation
ArCH2OH + CrO3(Ac2O)
ArCH2X + DMSO (NaOH)
(b) Hydrolysis
ArCH3 —X2→ ArCHX2 —NaOH→ Ar—CH=O
(c) Reduction
Pd/BaSO4
ArCOX + H2
or ArCOX + LiAlH(t-BuO)3
(d) Formylation
RC6H5 + CO + HCl(AlCl3)
(a) Reduction
+ H2/Pd → ArCH2OH
(b) Carbonyl Addition
As with aliphatic aldehydes.
with HCN, H2NOH, RMgX, etc.
(c) Oxidation
ArCOOH
(d) Cannizzaro Reaction
+ NaOH → ArCH2OH + ArCOOH

15.9 Summary of Ketone Chemistry

PREPARATION | *PROPERTIES*

1. Aliphatic and Aromatic Ketones | **1. Additions to Carbonyl Group**

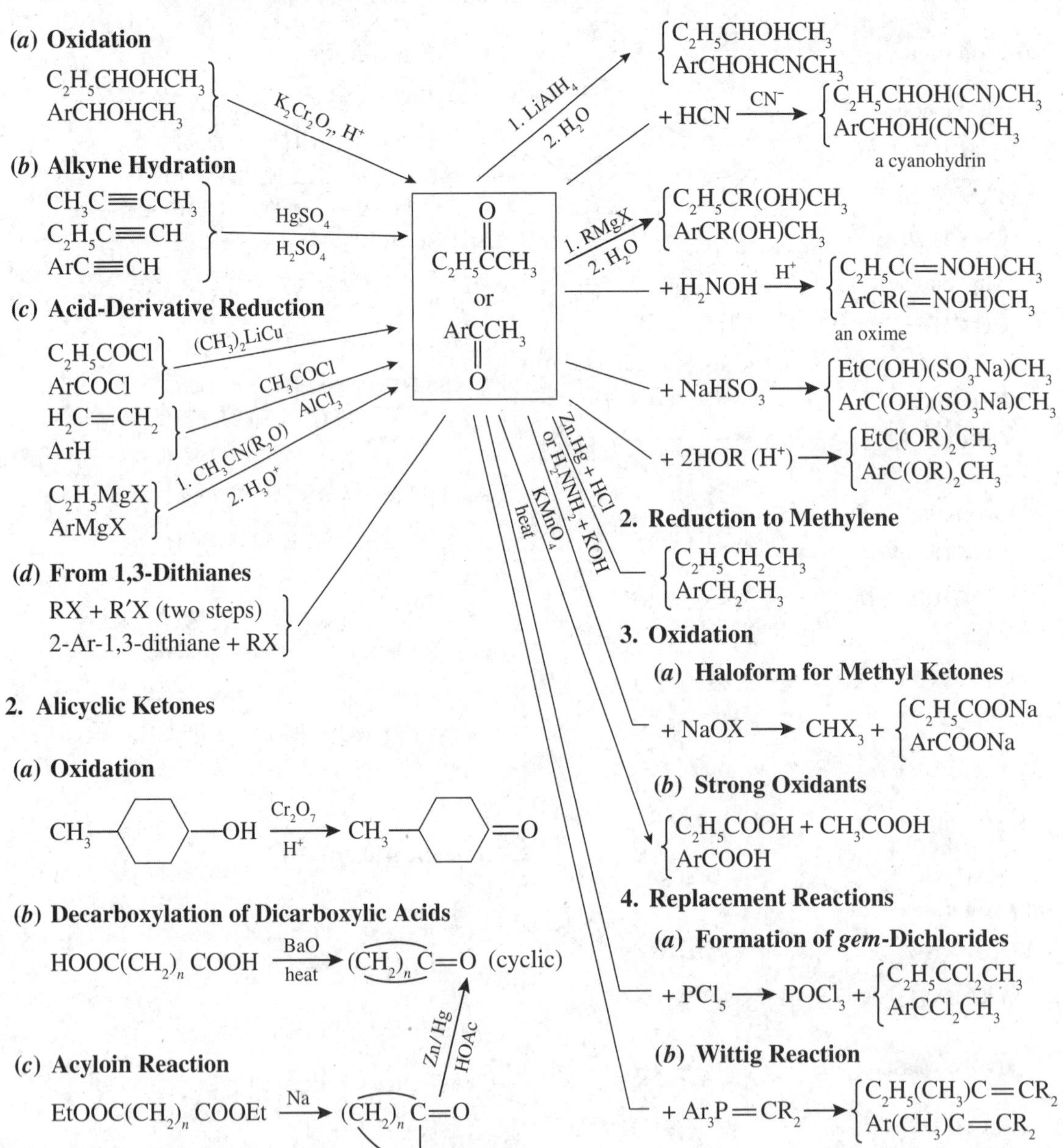

(d) From Cycloalkenes

Alkylborane Oxidation:

$$\text{R-cyclohexene} \xrightarrow{B_2H_6} \left[\text{R-cyclohexyl}\right]_2 BH \xrightarrow[H_2O]{CrO_3} \text{2-R-cyclohexanone}$$

SUPPLEMENTARY PROBLEMS

Problem 15.36 (*a*) What properties identify a carbonyl group of aldehydes and ketones? (*b*) How can aldehydes and ketones be distinguished?

(*a*) A carbonyl group (1) forms derivatives with substituted ammonia compounds such as H_2NOH, (2) forms sodium bisulfite adduct with $NaHSO_3$, (3) shows strong ir absorption at 1690–1760 cm^{-1} (C=O stretching frequency), (4) shows weak n-π^* absorption in uv at 289 nm. (*b*) The H—C bond in RCHO has a unique ir absorption at 2720 cm^{-1}. In nmr, the H of CHO has a very downfield peak at δ = 9–10 ppm. RCHO gives a positive Tollens' test.

Problem 15.37 What are the similarities and differences between C=O and C=C bonds?

Both undergo addition reactions. They differ in that the C of C=O is more electrophilic than a C of C=C, because O is more electronegative than C. Consequently, the C of C=O reacts with nucleophiles. The C=C is nucleophilic and adds mainly electrophiles.

Problem 15.38 Give another acceptable name for each of the following: (*a*) dimethyl ketone, (*b*) 1-phenyl-2-butanone, (*c*) ethyl isopropyl ketone, (*d*) dibenzyl ketone, (*e*) vinyl ethyl ketone.

(*a*) acetone or propanone, (*b*) benzyl ethyl ketone, (*c*) 2-methyl-3-pentanone, (*d*) 1,3-diphenyl-2-propanone, (*e*) 1-penten-3-one.

Problem 15.39 Identify the substances (I) through (V).

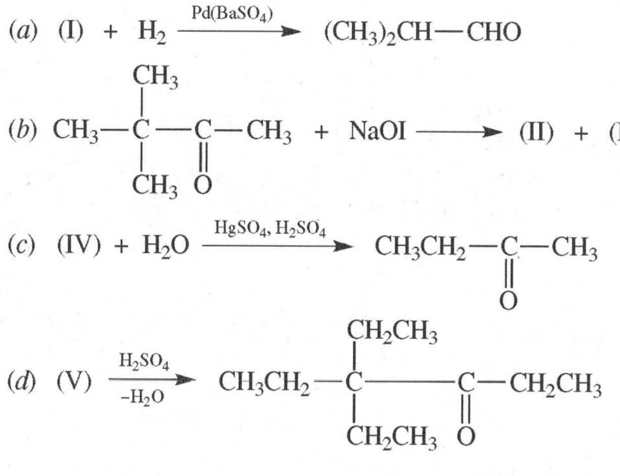

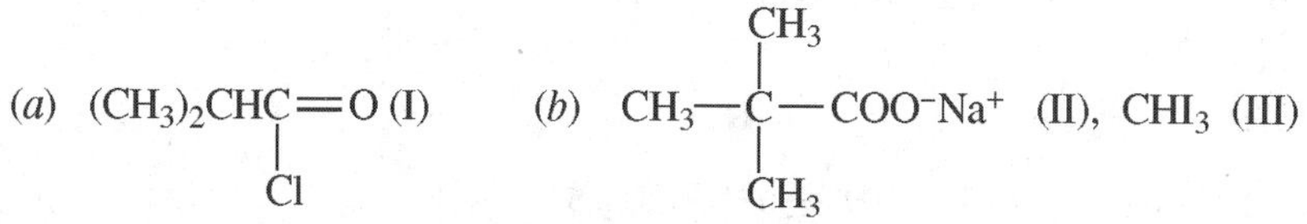

(*a*) $\text{(I)} + H_2 \xrightarrow{\text{Pd(BaSO}_4)} (CH_3)_2CH\text{—}CHO$

(*b*) $CH_3\text{—}C(CH_3)_2\text{—}C(=O)\text{—}CH_3 + NaOI \longrightarrow \text{(II)} + \text{(III)}$

(*c*) $\text{(IV)} + H_2O \xrightarrow{HgSO_4,\ H_2SO_4} CH_3CH_2\text{—}C(=O)\text{—}CH_3$

(*d*) $\text{(V)} \xrightarrow[-H_2O]{H_2SO_4} CH_3CH_2\text{—}C(CH_2CH_3)_2\text{—}C(=O)\text{—}CH_2CH_3$

(*a*) $(CH_3)_2CHC(Cl)=O$ (I) (*b*) $CH_3\text{—}C(CH_3)_2\text{—}COO^-Na^+$ (II), CHI_3 (III)

(*c*) $H\text{—}C\equiv C\text{—}CH_2CH_3$ or $CH_3\text{—}C\equiv C\text{—}CH_3$ (IV)

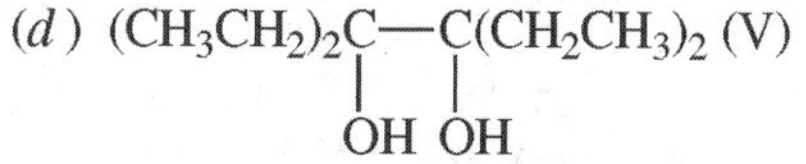

(*d*) $(CH_3CH_2)_2C(OH)\text{—}C(OH)(CH_2CH_3)_2$ (V)

Problem 15.40 By rapid test-tube reactions, distinguish between (*a*) pentanal and diethyl ketone, (*b*) diethyl ketone and methyl *n*-propyl ketone, (*c*) pentanal and 2,2-dimethylpropanal, (*d*) 2-pentanol and 2-pentanone.

(*a*) Pentanal, an aldehyde, gives a positive Tollens' test (Ag mirror). (*b*) Only the methyl ketone gives CHI_3 (yellow precipitate) on treatment with NaOI (iodoform test). (*c*) Unlike pentanal, 2,2-dimethylpropanal has no α H and so does not undergo an aldol condensation. Pentanal in base gives a colored solution. (*d*) Only the ketone 2-pentanone gives a solid oxime with H_2NOH. Additionally, 2-pentanol is oxidized by CrO_3 (color change is from orange-red to green). Both give a positive iodoform test.

Problem 15.41 Use benzene and any aliphatic and inorganic compounds to prepare (*a*) 1,1-diphenylethanol, (*b*) 4,4-diphenyl-3-hexanone.

(*a*) The desired 3° alcohol is made by reaction of a Grignard with a ketone by two possible combinations:

$$(C_6H_5)_2CO + CH_3MgBr \quad \text{or} \quad C_6H_5COCH_3 + C_6H_5MgBr$$

Since it is easier to make $C_6H_5COCH_3$ than $(C_6H_5)_2CO$ from C_6H_6, the latter pair is used. Benzene is used to prepare both intermediate products.

$$\left.\begin{matrix} C_6H_6 + CH_3COCl \xrightarrow{AlCl_3} C_6H_5COCH_3 \\ + \\ C_6H_6 \xrightarrow[Fe]{Br_2} C_6H_5Br \xrightarrow[Et_2O]{Mg} C_6H_5MgBr \end{matrix}\right\} \xrightarrow[2.\ H_2O]{1.\ \text{reaction}} (C_6H_5)_2C(OH)CH_3$$

(*b*) The 4° C of $CH_3CH_2CO\overset{4°\,C}{C}Ph_2CH_2CH_3$ is adjacent to C═O, and this suggests a pinacol rearrangement of

$$CH_3CH_2\underset{OH}{\underset{|}{C}}Ph\text{—}\underset{OH}{\underset{|}{C}}PhCH_2CH_3$$

which is made from CH_3CH_2COPh as follows:

$$C_6H_6 + CH_3CH_2COCl \xrightarrow{AlCl_3} PhCOCH_2CH_3 \xrightarrow[Et_2O]{Mg} CH_3CH_2\text{—}\underset{OH}{\underset{|}{\overset{Ph}{\overset{|}{C}}}}\text{—}\underset{OH}{\underset{|}{\overset{Ph}{\overset{|}{C}}}}\text{—}CH_2CH_3 \xrightarrow{H^+} \text{product}$$

Problem 15.42 Use butyl alcohols and any inorganic materials to prepare 2-methyl-4-heptanone.

The indicated bond

$$CH_3CH_2CH_2\underset{O}{\underset{\|}{C}}\overset{\downarrow}{\text{—}}CH_2CH(CH_3)_2$$

is formed from 2 four-carbon compounds by a Grignard reaction.

$$\underset{n\text{-Butyl alcohol}}{CH_3CH_2CH_2CH_2OH} \qquad CH_3\text{—}\overset{CH_3}{\overset{|}{C}}HCH_2Br \xleftarrow{PBr_3} \underset{\text{Isobutyl alcohol}}{CH_3\text{—}\overset{CH_3}{\overset{|}{C}}H\text{—}CH_2OH}$$

n-Butyl alcohol $\xrightarrow{\text{Cu, heat}}$ $CH_2CH_2CH_2CH{=}O$; $CH_3\text{—}CH(CH_3)CH_2Br \xrightarrow{Mg} CH_3\text{—}CH(CH_3)CH_2MgBr$

$$CH_2CH_2CH_2CH{=}O + CH_3\text{—}\overset{CH_3}{\overset{|}{C}}HCH_2MgBr \xrightarrow[2.\ H_2O,\ H^+]{1.\ \text{ether}} \underset{\text{2-Methyl-4-heptanol}}{CH_3CH_2CH_2\underset{OH}{\underset{|}{C}}H\overset{\downarrow}{\text{—}}CH_2\overset{CH_3}{\overset{|}{C}}HCH_3} \xrightarrow{KMnO_4} \text{product}$$

Problem 15.43 Compound (A), $C_3H_{10}O$, forms a phenylhydrazone, gives negative Tollens' and iodoform tests and is reduced to pentane. What is the compound?

Phenylhydrazone formation indicates a carbonyl compound. Since the negative Tollens' test rules out an aldehyde, (A) must be a ketone. A negative iodoform test rules out the $CH_3C{=}O$ group, and the reduction product, pentane, establishes the C's to be in a continuous chain. The compound is $CH_3CH_2COCH_2CH_3$.

Problem 15.44 A compound ($C_5H_8O_2$) is reduced to pentane. With H_2NOH, it forms a dioxime and also gives positive iodoform and Tollens' tests. Deduce its structure.

Reduction to pentane indicates 5 C's in a continuous chain. The dioxime shows two carbonyl groups. The positive CHI_3 test points to

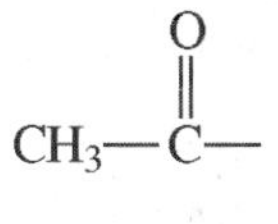

$$CH_3{-}\overset{\overset{\large O}{\|}}{C}{-}$$

while the positive Tollens' test establishes a $-CH{=}O$. The compound is

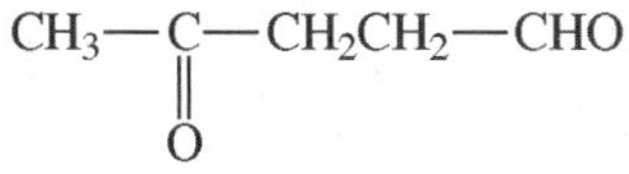

$$CH_3{-}\underset{\underset{\large O}{\|}}{C}{-}CH_2CH_2{-}CHO$$

Problem 15.45 The Grignard reagent of RBr(I) with CH_3CH_2CHO gives a 2° alcohol (II), which is converted to R′Br (III), whose Grignard reagent is hydrolyzed to an alkane (IV). (IV) is also produced by coupling (I). What are the compounds (I), (II), (III), and (IV)?

Since CH_3CH_2CHO reacts with the Grignard of (I) to give (II) after hydrolysis, (II) must be an alkyl ethyl carbinol

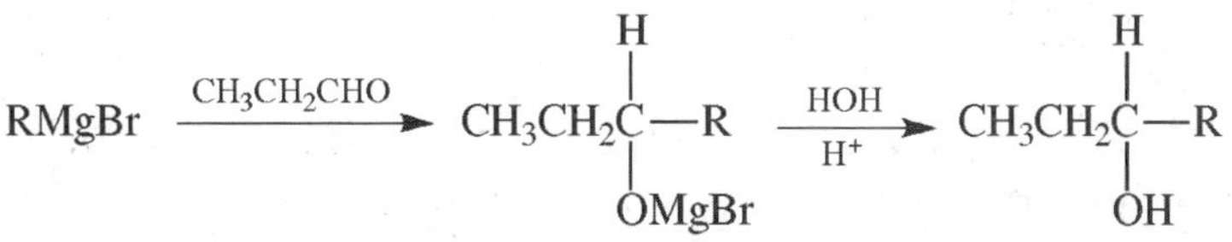

$$\underset{\text{Grignard of (I)}}{RMgBr} \xrightarrow{CH_3CH_2CHO} CH_3CH_2\underset{\underset{OMgBr}{|}}{\overset{\overset{H}{|}}{C}}{-}R \xrightarrow[H^+]{HOH} \underset{\text{(II)}}{CH_3CH_2\underset{\underset{OH}{|}}{\overset{\overset{H}{|}}{C}}{-}R}$$

The conversion of (II) to (IV) is

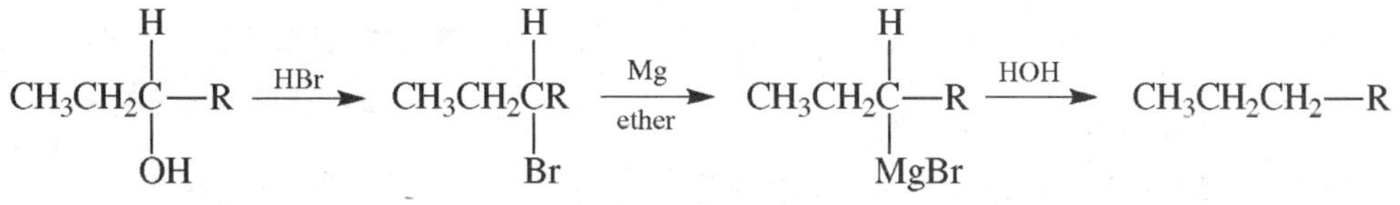

$$\underset{\text{(II)}}{CH_3CH_2\underset{\underset{OH}{|}}{\overset{\overset{H}{|}}{C}}{-}R} \xrightarrow{HBr} \underset{\text{(III)}}{CH_3CH_2\underset{\underset{Br}{|}}{\overset{\overset{H}{|}}{C}}R} \xrightarrow[\text{ether}]{Mg} CH_3CH_2\underset{\underset{MgBr}{|}}{\overset{\overset{H}{|}}{C}}{-}R \xrightarrow{HOH} \underset{\text{(IV)}}{CH_3CH_2CH_2{-}R}$$

(IV) must be symmetrical, since it is formed by coupling (I). R is therefore $-CH_2CH_2CH_3$. (I) is $CH_3CH_2CH_2Br$. (IV) is *n*-hexane. (II) is $CH_3CH_2CH(OH)CH_2CH_2CH_3$. (III) is $CH_3CH_2CHBrCH_2CH_2CH_3$.

Problem 15.46 Translate the following description into a chemical equation: Friedel-Crafts acylation of resorcinol (1,3-dihydroxybenzene) with $CH_3(CH_2)_4COCl$ produces a compound which on Clemmensen reduction yields the important antiseptic, hexylresorcinol.

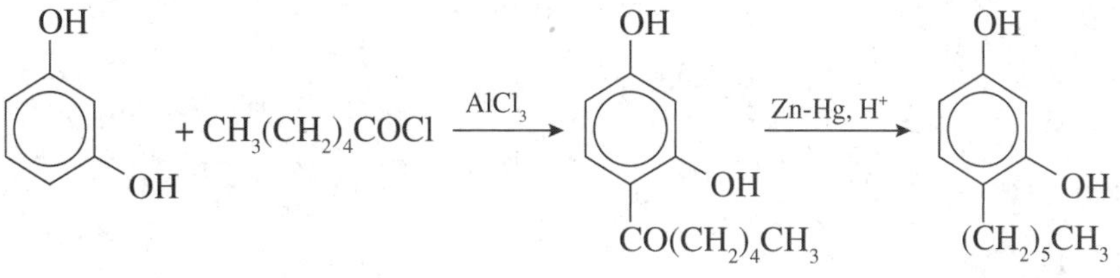

$$\text{resorcinol (C}_6\text{H}_4\text{(OH)}_2\text{)} + CH_3(CH_2)_4COCl \xrightarrow{AlCl_3} \text{C}_6\text{H}_3\text{(OH)}_2CO(CH_2)_4CH_3 \xrightarrow{Zn\text{-}Hg,\ H^+} \text{C}_6\text{H}_3\text{(OH)}_2(CH_2)_5CH_3$$

Problem 15.47 Treatment of benzaldehyde with HCN produces a mixture of two isomers that cannot be separated by very careful fractional distillation. Explain.

Formation of benzaldehyde cyanohydrin creates a chiral C and produces a racemic mixture, which cannot be separated by fractional distillation.

$$C_6H_5-\overset{\displaystyle H}{\overset{|}{C}}=O + HCN \longrightarrow C_6H_5-\overset{\displaystyle H}{\overset{|}{\underset{\displaystyle CN}{\underset{|}{C^*}}}}-OH$$

Problem 15.48 Prepare 1-phenyl-1-(*p*-bromophenyl)-1-propanol from benzoic acid, bromobenzene, and ethanol.

The compound is a 3° alcohol, conveniently made from a ketone and a Grignard reagent as shown.

$$C_6H_5COOH \xrightarrow{PCl_5} C_6H_5-\underset{\displaystyle O}{\underset{\|}{C}}-Cl \xrightarrow[AlCl_3]{C_6H_5Br} p\text{-}BrC_6H_4-\underset{\displaystyle O}{\underset{\|}{C}}-C_6H_5$$

$$+$$

$$C_2H_5OH \xrightarrow{HBr} C_2H_5Br \xrightarrow[ether]{Mg} C_2H_5MgBr$$

$$\downarrow \text{1. ether; 2. } NH_4^+ \text{ (mild acid, prevents } H_2O \text{ loss)}$$

$$p\text{-}BrC_6H_4-\overset{\displaystyle C_2H_5}{\overset{|}{\underset{\displaystyle OH}{\underset{|}{C}}}}-C_6H_5$$

Problem 15.49 Convert cinnamaldehyde, C_6H_5—CH═CH—CH═O, to 1-phenyl-1,2-dibromo-3-chloropropane, $C_6H_5CHBrCHBrCH_2Cl$.

We must add Br_2 to C═C and convert —CHO to —CH_2Cl. Since Br_2 oxidizes —CHO to —COOH, —CHO must be converted to CH_2Cl before adding Br_2.

$$C_6H_5-CH=CH-CH=O \xrightarrow[2.\ H_2O]{1.\ NaBH_4} C_6H_5-CH=CH-CH_2OH \xrightarrow{PCl_3}$$

$$C_6H_5-CH=CH-CH_2Cl \xrightarrow{Br_2} C_6H_5-\underset{\displaystyle Br}{\underset{|}{CH}}-\underset{\displaystyle Br}{\underset{|}{CH}}-CH_2Cl$$

Problem 15.50 Compounds "labeled" at various positions by isotopes such as ^{14}C (radioactive), D (deuterium), and ^{18}O are used in studying reaction mechanisms. Suggest a possible synthesis of each of the labeled compounds below, using $^{14}CH_3OH$ as the source of ^{14}C, D_2O as the source of D, and $H_2{}^{18}O$ as the source of ^{18}O. Once a ^{14}C-labeled compound is made, it can be used in ensuring syntheses. Use any other unlabeled compounds. (*a*) $CH_3{}^{14}CH_2OH$, (*b*) $^{14}CH_3CH_2OH$, (*c*) $^{14}CH_3CH_2CHO$, (*d*) $C_6H_5{}^{14}CHO$, (*e*) $^{14}CH_3CHDOH$, (*f*) $CH_3CH^{18}O$.

(*a*) The 1° alcohol with a labeled carbinol C suggests a Grignard reaction with $H_2{}^{14}C=O$.

$$^{14}CH_3OH \xrightarrow[300°C]{Cu} H_2{}^{14}C=O \xrightarrow[2.\ H_3O^+]{1.\ CH_3MgBr} CH_3{}^{14}CH_2OH$$

(*b*) Now the Grignard reagent is labeled instead of H_2CO.

$$^{14}CH_3OH \xrightarrow{HBr} {}^{14}CH_3Br \xrightarrow[Et_2O]{Mg} {}^{14}CH_3MgBr \xrightarrow[2.\ H_3O^+]{1.\ CH_2O} {}^{14}CH_3CH_2OH$$

(*c*) $^{14}CH_3MgBr$ [see (*b*)] + $H_2C\underset{O}{-}CH_2$ (ethylene oxide) $\xrightarrow[2.\ H_3O^+]{1.\ ether} {}^{14}CH_3CH_2CH_2OH \xrightarrow[heat]{Cu} {}^{14}CH_3CH_2CHO$

(*d*) $^{14}CH_3Br + C_6H_6 \xrightarrow{AlCl_3} C_6H_5{}^{14}CH_3 \xrightarrow[\text{acetic anhydride}\ \ 2.\ H_3O^+]{1.\ CrO_3} C_6H_5{}^{14}CHO$

(*e*) D on carbinol C is best introduced by reduction of a —CHO group with a D-labeled reductant. $^{14}CH_3CH_2OH$ from (*b*) $\xrightarrow[\text{heat}]{Cu}$ $^{14}CH_3CHO$ then D_2/Pt or $LiAlD_4$ $\longrightarrow$ $^{14}CH_3CHDOD \xrightarrow{H_2O} {}^{14}CH_3CHDOH$. D of OD is easily exchanged with excess H_2O.

(*f*) Add CH_3CHO to excess $H_2{}^{18}O$ with a trace of HCl.

$$H_2{}^{18}O + CH_3CHO \overset{H^+}{\rightleftharpoons} \left[\begin{array}{c} CH_3CH^{18}OH \\ | \\ OH \end{array}\right] \underset{H^+}{\rightleftharpoons} CH_3CH^{18}O + H_2O$$

hydrate

The unstable half-labeled hydrate can lose H_2O to give $CH_3CH^{18}O$.

Problem 15.51 Isopropyl chloride is treated with triphenylphosphine (Ph_3P) and then with NaOEt. CH_3CHO is added to the reaction product to give a compound, C_5H_{10}. When C_5H_{10} is treated with diborane and then CrO_3, a ketone is obtained. Give the structural formula for C_5H_{10} and the name of the ketone.

The series of reactions is

Formation of Ylide

$$(CH_3)_2CHCl + Ph_3P \longrightarrow [(CH_3)_2CH{-}\overset{+}{P}Ph_3]Cl^- \xrightarrow{NaOEt} (CH_3)_2C{=}PPh_3 + EtOH + Na^+Cl^-$$

Wittig Reaction

$$(CH_3)_2C{=}PPh_3 + CH_3{-}\overset{\overset{H}{|}}{C}{=}O \longrightarrow (CH_3)_2C{=}\overset{\overset{H}{|}}{C}{-}CH_3$$

Anti-Markovnikov Hydroboration-Oxidation

$$(CH_3)_2C{=}\underset{\underset{H}{|}}{C}{-}CH_3 \xrightarrow[2.\ CrO_3]{1.\ BH_3} (CH_3)_2\underset{\underset{H}{|}}{C}{-}\underset{\underset{O}{\|}}{C}{-}CH_3 \quad \text{3-Methyl-2-butanone}$$

Problem 15.52 Deduce the structure of a compound, C_4H_6O, with the following spectral data: (*a*) Electronic absorption at $\lambda_{max} = 213$ nm, $\varepsilon_{max} = 7100$ and $\lambda_{max} = 320$ nm, $\varepsilon_{max} = 27$. (*b*) Infrared bands, among others, at 3000, 2900, 1675 (most intense), and 1602 cm^{-1}. (*c*) Nmr singlet at $\delta = 2.1$ ppm (3 H's), three multiplets each integrating for 1 H at $\delta = 5.0$–6.0 ppm.

The formula C_4H_6O indicates two degrees of unsaturation and may represent an alkyne or some combination of two rings, C═C and C═O groups.

(*a*) λ_{max} at 213 nm comes from the $\pi \rightarrow \pi^*$ transition. It is more intense than the λ_{max} at 320 nm from the $n \rightarrow \pi^*$ transition. Both peaks are shifted to higher wavelengths than normal (190 and 280 nm, respectively), thus indicating an α,β-unsaturated carbonyl compound. The 2 degrees of unsaturation are a C═C and a C═O.

(*b*) The given peaks and their bonds are 3000 cm^{-1}, sp^2 C—H; 2900 cm^{-1}, sp^3 C—H; 1675 cm^{-1}, C═O (probably conjugated to C═C); 1602 cm^{-1}, C═C. All are stretching vibrations. Absence of a band at 2720 cm^{-1} means no aldehyde H. The compound is probably a ketone.

(*c*) The singlet at $\delta = 2.1$ ppm is from a

$$-\overset{\overset{O}{\|}}{C}{-}CH_3$$

There are also three nonequivalent vinylic H's ($\delta = 5.0$–6.0 ppm) that intercouple. The compound is

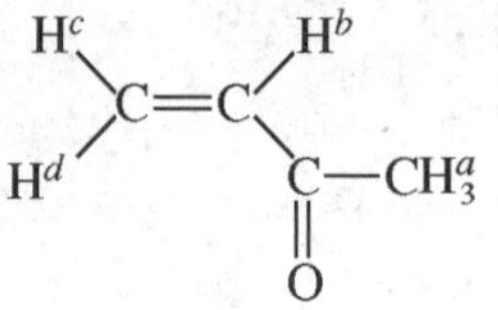

$$\underset{H^d}{\overset{H^c}{}}\!\!>\!C{=}C\!<\!\!\overset{H^b}{\underset{\underset{\underset{O}{\|}}{C}-CH_3^a}{}}$$

shown with nonequivalent H's.

Problem 15.53 A compound, $C_5H_{10}O$, has a strong ir band at about 1700 cm^{-1}. The nmr shows no peak at $\delta = 9$–10 ppm. The mass spectrum shows the base peak (most intense) at $m/e = 57$ and nothing at $m/e = 43$ or $m/e = 71$. What is the compound?

The strong ir band at 1700 cm^{-1} indicates a $C{=}O$, accounting for the one degree of unsaturation. The absence of a signal at $\delta = 9$–10 ppm means no

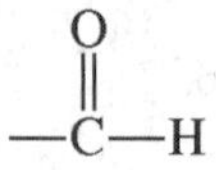

$$-\overset{\overset{O}{\|}}{C}-H$$

proton. The compound is a ketone, not an aldehyde. Nmr is the best way to differentiate between a ketone and an aldehyde.

Carbonyl compounds undergo fragmentation to give stable acylium ions:

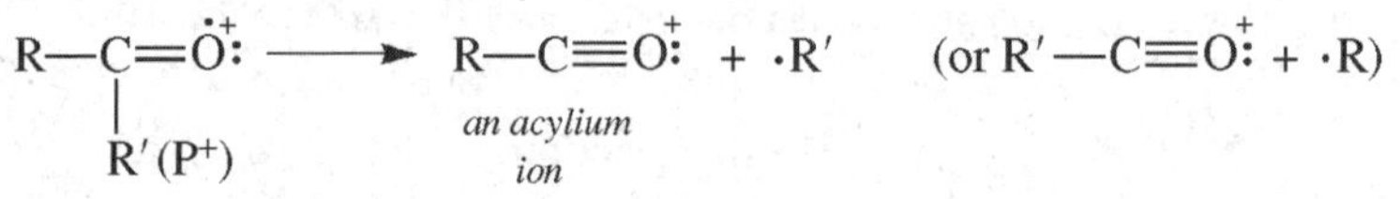

$$\underset{\underset{R'\,(P^+)}{|}}{R-C}{=}\overset{\bullet+}{\ddot{O}}: \longrightarrow \underset{\text{an acylium ion}}{R-C{\equiv}\overset{+}{O}:} + \cdot R' \qquad (\text{or } R'-C{\equiv}\overset{+}{O}: + \cdot R)$$

The possible ketones are

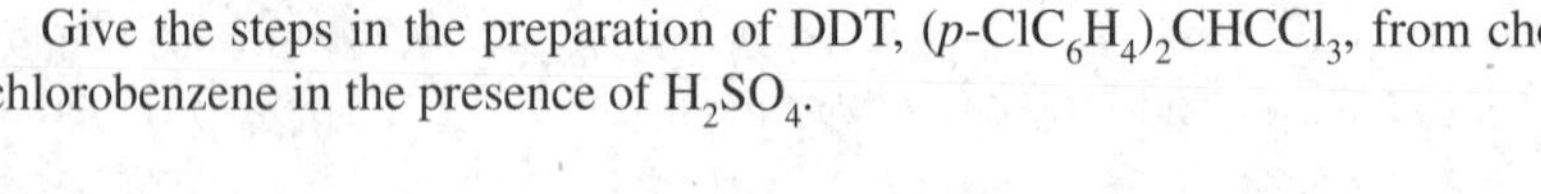

$$\underset{\text{(A)}}{CH_3CH_2\underset{\underset{O}{\|}}{C}CH_2CH_3} \qquad \underset{\text{(B)}}{CH_3\underset{\underset{O}{\|}}{C}CH_2CH_2CH_3} \qquad \underset{\text{(C)}}{CH_3\underset{\underset{O}{\|}}{C}CH(CH_3)_2}$$

Compounds (B) and (C) would both give some $CH_3C{\equiv}O{:}^+$ ($m/e = 43$) and $C_3H_7C{\equiv}O{:}^+$ ($m/e = 71$). These peaks were absent; therefore (A), which fragments to $CH_3CH_2C{\equiv}O{:}^+$ ($m/e = 57$), is the compound.

Problem 15.54 Give the steps in the preparation of DDT, $(p\text{-}ClC_6H_4)_2CHCCl_3$, from choral (trichloro-acetaldehyde) and chlorobenzene in the presence of H_2SO_4.

$$Cl_3CCHO \xrightarrow{H_2SO_4} Cl_3C\overset{+}{C}H(OH) \xrightarrow{PhCl} p\text{-}Cl-C_6H_4CHOHCCl_3 \xrightarrow{H_2SO_4} p\text{-}Cl-C_6H_4\overset{+}{C}HCCl_3 \xrightarrow{PhCl} DDT$$

Carboxylic Acids and Their Derivatives

16.1 Introduction and Nomenclature

Carboxylic acids (RCOOH or ArCOOH) have the **carboxyl** group, $-\overset{\displaystyle O}{\overset{\|}{C}}-OH$, which is an **acyl** group, $R-\overset{\displaystyle O}{\overset{\|}{C}}-$, bonded to OH.

Common names, such as **formic** (ant) and **butyric** (butter) acids, are based on the natural source of the acid. The positions of substituent groups are shown by Greek letters, α, β, γ, δ, etc. Some have names **derived** from acetic acid, for example, $(CH_3)_3CCOOH$ and $C_6H_5CH_2COOH$, are trimethylacetic acid and phenylacetic acid, respectively. Occasionally they are named as carboxylic acids. For instance:

[cyclohexane ring]—COOH

is cyclohexanecarboxylic acid.

For IUPAC, names replace the **-e** of the corresponding alkane with **-oic acid**: thus, CH_3CH_2COOH is propanoic acid. The C's are numbered; the C of COOH is numbered 1. C_6H_5COOH is benzoic acid. **Dicarboxylic** acids contain two COOH groups and are named by adding the suffix **-dioic** and the word **acid** to the name of the longest chain with the two COOH's.

Problem 16.1 Give a derived and IUPAC name for the following carboxylic acids. Note the common names. (*a*) $CH_3(CH_2)_4COOH$ (caproic acid); (*b*) $(CH_3)_3CCOOH$ (pivalic acid); (*c*) $(CH_3)_2CHCH_2CH_2COOH$ (γ-methylvaleric acid); (*d*) $C_6H_5CH_2CH_2COOH$ (β-phenylpropionic acid); (*e*) $(CH_3)_2C(OH)COOH$ (α-hydroxyisobutyric acid); (*f*) $HOOC(CH_2)_2COOH$ (succinic acid; no derived name for this one).

To get the IUPAC name, find the longest chain of C's, including the C from COOH, as shown below by a horizontal line. To get the derived name, find and name the groups attached to the α C:

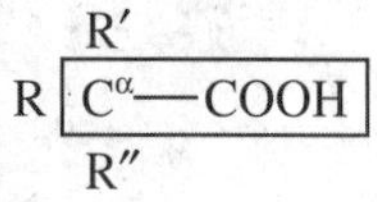

$$\mathrm{R}\,\boxed{\overset{\mathrm{R'}}{\underset{\mathrm{R''}}{\mathrm{C}^{\alpha}}}\text{—COOH}}$$

(*a*) $\mathrm{CH_3CH_2CH_2CH_2}\boxed{\mathrm{CH_2COOH}}$ *n*-Butylacetic acid, Hexanoic acid

(6 C's in longest chain)

(*b*) $\mathrm{CH_3}\boxed{\overset{\mathrm{CH_3}}{\underset{\mathrm{CH_3}}{\mathrm{C}}}\mathrm{COOH}}$ Trimethylacetic acid, 2,2-Dimethylpropanoic acid

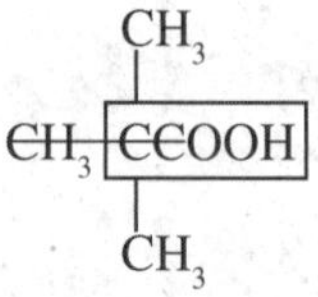

(3 C's in longest chain)

(*c*) $\mathrm{CH_3{-}\underset{\mathrm{CH_3}}{CH}CH_2}\boxed{\mathrm{CH_2COOH}}$ Isobutylacetic acid, 4-Methylpentanoic acid

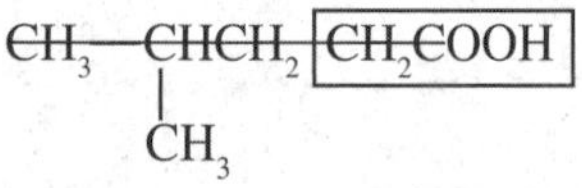

(*d*) $\mathrm{C_6H_5CH_2}\boxed{\mathrm{CH_2COOH}}$ Benzylacetic acid, 3-Phenylpropanoic acid

(*e*) $\mathrm{CH_3}\boxed{\overset{\mathrm{OH}}{\underset{\mathrm{CH_3}}{\mathrm{C}}}\mathrm{COOH}}$ Dimethylhydroxyacetic acid, 2-Hydroxy-2-methylpropanoic acid

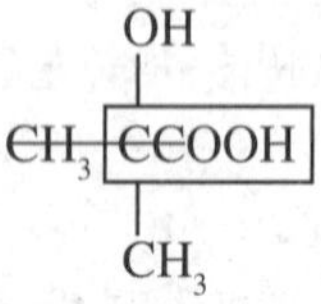

(*f*) $\mathrm{HOOCCH_2CH_2COOH}$ Butanedioic acid

Problem 16.2 Name the following aromatic carboxylic acids.

(*a*) benzene ring with COOH and NO_2 (para) (*b*) benzene ring with COOH, Br, Br (*c*) benzene ring with COOH and CHO (*d*) benzene ring with COOH and CH_3

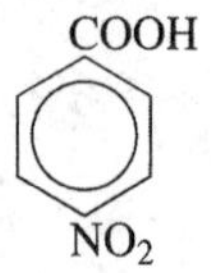

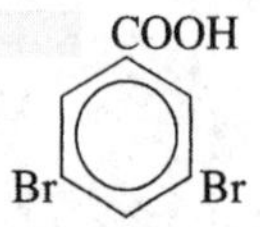

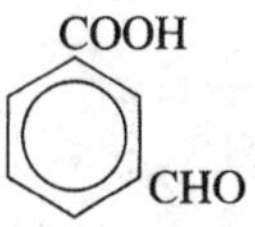

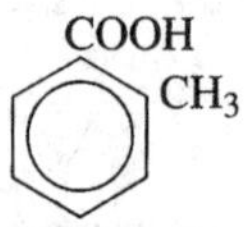

(*a*) *p*-nitrobenzoic acid; (*b*) 3,5-dibromobenzoic acid; (*c*) *m*-formylbenzoic acid (COOH takes priority over CHO, wherefore this compound is named as an acid, not as an aldehyde); (*d*) *o*-methylbenzoic acid, but more commonly called *o*-toluic acid (from toluene).

Problem 16.3 Account for the following physical properties of carboxylic acids. (*a*) Only RCOOH's with five or fewer C's are soluble in water, but many with six or more C's dissolve in alcohols. (*b*) Acetic acid in the vapor state has a molecular weight of 120, not 60. (*c*) Their boiling and melting points are higher than those of corresponding alcohols.

(*a*) RCOOH dissolves because the H of COOH can H-bond with H_2O. The R portion is nonpolar and hydrophobic; this effect predominates as R gets large (over five C's). Alcohols are less polar than water and are less antagonistic toward the less polar carboxylic acids of higher C content.

(*b*) CH_3COOH typically undergoes **dimeric intermolecular H-bonding**.

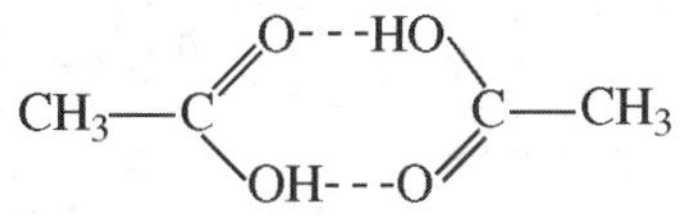

$$CH_3—C\begin{matrix}\nearrow O\text{- - }HO\searrow \\ \searrow OH\text{- - }O\nearrow\end{matrix}C—CH_3$$

(*c*) The intermolecular forces are greater for carboxylic acids.

Problem 16.4 Write resonance structures for the COOH group, and show how these and orbital hybridization account for: (*a*) polarity and dipole moments (1.7–1.9 D) of carboxylic acids; (*b*) their low reactivity toward nucleophilic additions, as compared to carbonyl compounds.

(*a*) The C of COOH uses sp^2 HO's to form the three coplanar σ bonds. A *p* AO of the O of the OH, accommodating a pair of electrons, overlaps with the π bond of C$=$O. In this extended π system, there is negative charge on the lone O and positive charge on the other O; the charge separation results in greater polarity and dipole moments.

$$—C(=\ddot{O}:)—\ddot{O}—H \longleftrightarrow —\overset{+}{C}(—\ddot{O}:^-)—\ddot{O}—H \longleftrightarrow —C(—\ddot{O}:^-)=\overset{+}{\ddot{O}}—H$$

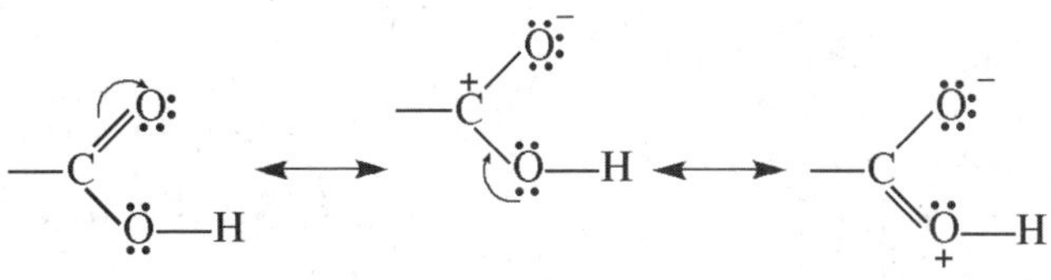

(*b*) The electron deficiency of the C of C$=$O, observed in carbonyl compounds, is greatly diminished in the C of —COOH by the attached OH group.

Carboxylic acid derivatives (or **acyl derivatives**), $R\overset{O}{\overset{\|}{C}}—G$, have the OH replaced by another electronegative functional group, G, such that it can be hydrolyzed back to the acid:

$$\underset{\text{a carboxylic acid derivative}}{R(Ar)—\underset{\underset{O}{\|}}{C}—G} + H_2O \longrightarrow \underset{\text{a carboxylic acid}}{R(Ar)—\underset{\underset{O}{\|}}{C}—OH} + HG$$

The common ones are given in Table 16.1, with conventions of nomenclature that involve changes of the name of the corresponding carboxylic acid; the G group is shown in bold type.

Problem 16.5 Name the following derivatives:

(*a*) C_6H_5COCl (*b*) $CH_3CH_2\underset{\underset{O}{\|}}{C}O\underset{\underset{O}{\|}}{C}CH_2CH_3$ (*c*) $CH_3CH_2CH_2\underset{\underset{O}{\|}}{C}—OCH_2CH_3$

(*d*) $C_6H_5CH_2\underset{\underset{O}{\|}}{C}NH_2$ (*e*) $C_6H_5\underset{\underset{O}{\|}}{C}—OC_6H_5$ (*f*) $C_6H_{11}—\underset{\underset{O}{\|}}{C}—NH_2$ (cyclohexyl)

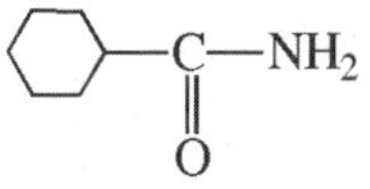

(*g*) $H\underset{\underset{O}{\|}}{C}N(CH_3)_2$ (*h*) $C_6H_5\underset{\underset{O}{\|}}{C}O\underset{\underset{O}{\|}}{C}CH_3$

(*a*) Benzoyl chloride. (*b*) Propionic (or propanoic) anhydride. (*c*) Ethyl butyrate (or butanoate). (*d*) Phenylacetamide. (*e*) Phenyl benzoate. (*f*) Cyclohexanecarboxamide. (*g*) N,N-Dimethylformamide. (*h*) Acetic benzoic anhydride.

TABLE 16.1

GENERAL FORMULA	TYPE	EXAMPLE	NAME	CHANGE
R–C(=O)–Cl*	Acid chloride	$CH_3C(=O)–Cl$	Acetyl chloride or Ethanoyl chloride	**-ic acid** to **-yl chloride**
R–C(=O)–OR′	Ester	$CH_3C(=O)–OCH_2CH_3$	Ethyl acetate or Ethyl ethanoate	Cite alkyl group attached to O; then change **-ic acid** to **-ate**
R–C(=O)–O–C(=O)–R′	Acid anhydride	$CH_3C(=O)–OC(=O)CH_3$	Acetic anhydride	**acid** to **anhydride**
R–C(=O)–N̈(H)–	Amide	$CH_3C(=O)–N–H$ (N–H) $CH_3CH_2C(=O)NHCH_3$	Acetamide or Ethanamide N-Methylpropanamide	**-ic** or **-oic acid** to **-amide** or **-carboxylic acid** to **carboxamide**
R–C(=O)–N̈(H)–C(=O)–R′	Imide	$CH_3CO–NH–OCCH_2CH_3$	Acetyl propionyl imide	**-ic** or **-oic acid** to **-imide**
$RC{\equiv}N$†	Nitrile	$CH_3C{\equiv}N$	Acetonitrile or Ethanenitrile	**-ic** or **-oic acid** to **nitrile** or add **-nitrile** to alkane name

* Some acid bromides are known.

† Although nitriles have no acyl group, they are grouped with acid derivatives because they are readily hydrolyzed to RCOOH.

Problem 16.6 Give structural formulas for the following acid derivatives: (*a*) propionitrile, (*b*) isopropyl-2-fluorobutanoate, (*c*) 3-phenylhexanoyl chloride, (*d*) 3-chloropropyl benzoate.

(*a*) $CH_3CH_2C{\equiv}N$ (*b*) $CH_3CH_2CHFCOOCH(CH_3)_2$

(*c*) $CH_3CH_2CH_2CH(C_6H_5)CH_2COCl$ (*d*) $C_6H_5COOCH_2CH_2CH_2Cl$

16.2 Preparation of Carboxylic Acids

1. Oxidation of 1° Alcohols, Aldehydes, and Arenes

$$R{-}CH_2OH \xrightarrow{KMnO_4} RCHO \xrightarrow{KMnO_4} RCOOH$$

$$ArCHR_2 \xrightarrow[H^+]{KMnO_4} ArCOOH$$

2. Oxidative Cleavage of Alkenes and Alkynes

$$\text{Cyclohexene} \xrightarrow{HNO_3} \underset{\substack{\text{Adipic acid} \\ \textit{a dicarboxylic acid}}}{HOOCCH_2CH_2CH_2CH_2COOH}$$

Problem 16.7 Account for the fact that on oxidative cleavage, all substituted alkynes give carboxylic acids, whereas some alkenes give ketones.

Imagine that the net effect of the oxidation is replacement of each bond and H on the multiple-bonded C by OH. Intermediates with several OH's on C are unstable and lose H_2O, leaving C═O.

$$[R_2C(OH)_2] \xrightarrow{-H_2O} R_2C{=}O \text{ and } [RC(OH)_3] \xrightarrow{-H_2O} RCOOH$$

Thus:

$$RC{\equiv}CR \xrightarrow{[O]} [RC(OH)_3] \xrightarrow{-H_2O} RCOOH$$

Alkenes with the structural unit $R_2C{=}\overset{|}{C}{-}$ would give $[R_2C(OH)_2] \xrightarrow{-H_2O} R_2C{=}O$. Multiple-bonded C's bonded only to H's would form $[C(OH)_4]$, which loses two molecules of H_2O to give CO_2.

3. Grignard Reagent and CO_2

$$\overset{-}{R}{-}\overset{+}{Mg}X + O{=}C{=}O \longrightarrow \underset{\substack{\textit{a carboxylate} \\ \textit{salt}}}{R{-}\underset{\substack{| \\ {}^{-}O(MgX)^{+}}}{C}{=}O} \xrightarrow[H_2O]{HX} R{-}\underset{\substack{| \\ OH}}{C}{=}O + Mg^{2+} + 2X^{-}$$

4. Hydrolysis of Acid Derivatives and Nitriles

$$R{-}\underset{\substack{\| \\ O}}{C}{-}G + H_3O^{+} \longrightarrow R{-}\underset{\substack{\| \\ O}}{C}{-}OH + HG \quad (\text{or } H_3O^{+} + Cl^{-}, \text{ for } G = Cl)$$

$$\underset{\text{nitrile}}{RC{\equiv}N} \xrightarrow{H_3O^{+}} RCOOH + NH_4^{+}$$

$$\underset{\text{nitrile}}{RC{\equiv}N} \xrightarrow[OH^{-}]{H_2O} RCOO^{-} + NH_3 \xrightarrow{H^{+}} RCOOH + NH_4^{+}$$

5. Haloform Reaction of Methyl Ketones

Although this reaction is used mainly for structure elucidation, it has synthetic utility when methyl ketones are readily prepared and halides are not.

Problem 16.8 (*a*) Why isn't 2-naphthoic acid made from 2-chloronaphthalene? (*b*) How is 2-naphthoic acid prepared in a haloform reaction?

(*a*) Since most electrophilic substitutions of naphthalene, including halogenation, occur at the 1-position, 2-chloronaphthalene is not a readily accessible starting material. (*b*) Acetylation with CH_3COCl in the presence of the solvent, nitrobenzene, occurs at the 2-position, which allows for the following synthesis:

$$\text{Naphthalene} + CH_3COCl \xrightarrow[\text{PhNO}_2]{\text{AlCl}_3} \text{2-naphthyl-}\overset{O}{\overset{\|}{C}}\text{—}CH_3 \xrightarrow{\text{NaOCl}} \text{2-naphthyl-COOH}$$

2-Acetonaphthalene (Methyl β-naphthyl ketone) — 2-Naphthoic acid (β-Naphthoic acid)

Problem 16.9 Prepare the following acids from alkyl halides or dihalides of fewer C's. (*a*) $C_6H_5CH_2COOH$, (*b*) $(CH_3)_3CCOOH$, (*c*) $HOCH_2CH_2CH_2COOH$, (*d*) $HOOCCH_2CH_2COOH$ (succinic acid).

Replace COOH by X to find the needed alkyl halide. The two methods for RX → RCOOH are

$$\text{RX}(1°, 2°, \text{or } 3°) \xrightarrow{\text{Mg}} \text{RMgX} \xrightarrow[2.\ H_3O^+]{1.\ CO_2} \boxed{\text{RCOOH}} \xleftarrow{H_3O^+} \text{RCN} \xleftarrow{CN^-} \text{RX}(1°)$$

(*a*) Either method can be used starting with $C_6H_5CH_2Br$ (a 1° RX).
(*b*) With the 3° $(CH_3)_3C$—Br, CN^- *cannot* be used because elimination rather than substitution would occur.
(*c*) $HOCH_2CH_2CH_2Br$ has an acidic H (O—H); hence, the Grignard reaction can't be used.
(*d*) $BrCH_2CH_2Br$ undergoes dehalogenation with Mg to form an alkene; therefore, use the nitrile method.

$$BrCH_2CH_2Br \xrightarrow{CN^-} N{\equiv}CCH_2CH_2C{\equiv}N \xrightarrow{H^+} \text{product}$$

16.3 Reactions of Carboxylic Acids

H of COOH Is Acidic

$$\underset{\text{acid}_1}{\text{RCOOH}} + \underset{\text{base}_2}{H_2O} \rightleftharpoons \underset{\text{base}_1}{\text{RCOO}^-} + \underset{\text{acid}_2}{H_3O^+} \qquad pK_a \approx 5 \text{ (Section 3.10)}$$

RCOOH forms **carboxylate salts** with bases; when R is large, these salts are called **soaps**.

$$\text{RCOOH} + \text{KOH} \longrightarrow \text{RCOO}^-\text{K}^+ + H_2O$$

$$2\text{RCOOH} + Na_2CO_3 \longrightarrow 2\text{RCOO}^-\text{Na}^+ + H_2O + CO_2$$

Problem 16.10 Use the concept of charge delocalization by extended π bonding (resonance) to explain why (*a*) RCOOH ($pK_a \approx 5$) is more acidic than ROH ($pK_a \approx 15$), and (*b*) peroxy acids, $R\overset{\|}{\underset{O}{C}}OOH$, are much weaker than RCOOH.

(*a*) It is usually best to account for relative strengths of acids in terms of relative stabilities of their conjugate bases. The weaker (more stable) base has the stronger acid. Since the electron density in $RCOO^-$ is dispersed to both O's:

$$\left[\text{R—C}\begin{smallmatrix}\nearrow O\\ \searrow O^-\end{smallmatrix} \longleftrightarrow \text{R—C}\begin{smallmatrix}\nearrow O^-\\ \searrow O\end{smallmatrix}\right] \text{ or } \left\{\text{R—C}\begin{smallmatrix}O\\ O\end{smallmatrix}\right\}^-$$

$RCOO^-$ is more stable and a weaker base than RO^-, whose charge is localized on only one O.

(*b*) There is no way to delocalize the negative charge of the anion, $RCOO^-$ (with C=O on the first O: $R\overset{\underset{\|}{O}}{C}OO^-$), to the C=O group, as can be done for $R\underset{\underset{O}{\|}}{C}O^-$.

Problem 16.11 Use the inductive effect (Section 3.11) to account for the following differences in acidity:

(*a*) $ClCH_2COOH > CH_3COOH$, (*b*) $FCH_2COOH > ClCH_2COOH$, (*c*) $ClCH_2COOH > ClCH_2CH_2COOH$, (*d*) $Me_3CCH_2COOH > Me_3SiCH_2COOH$, (*e*) $Cl_2CHCOOH > ClCH_2COOH$.

The influence of the inductive effect on acidity is best understood in terms of the conjugate base, $RCOO^-$, and can be summarized as follows:

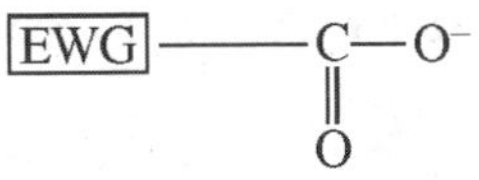

$$\boxed{EWG}\text{——}\underset{\underset{O}{\|}}{C}\text{—}O^-$$

Electron-withdrawing groups (EWG) stabilize $RCOO^-$ *and strengthen the acid.*

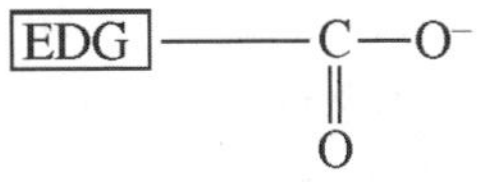

$$\boxed{EDG}\text{——}\underset{\underset{O}{\|}}{C}\text{—}O^-$$

Electron-donating groups (EDG) destabilize $RCOO^-$ *and weaken the acid.*

(*a*) Like all halogens, Cl is electronegative, electron-withdrawing, and acid-strengthening.
(*b*) Since F is more electronegative than Cl, it is a better EWG and a better acid-strengthener.
(*c*) Inductive effects diminish as the number of C's between Cl and the O's increases. $ClCH_2COO^-$ is a weaker base than $ClCH_2CH_2COO^-$, and so $ClCH_2COOH$ is the stronger acid.
(*d*) Si is electropositive and is an acid-weakening EDG.
(*e*) Two Cl's are more electron-withdrawing than one Cl. Cl_2CHCOO^- is the weaker base and $Cl_2CHCOOH$ is the stronger acid.

Resonance and induction, which affect ΔH, can be used with certainty to explain differences in acidities only when the K_a values are different by a factor of at least 10. Smaller differences may be accounted for by solvation, which affects ΔS.

Acid-strengthening ring substituents also retard electrophilic, and enhance nucleophilic, aromatic substitution. Conversely, the acid-weakening groups accelerate electrophilic, and retard nucleophilic, aromatic substitution.

Problem 16.12 Explain why highly branched carboxylic acids such as

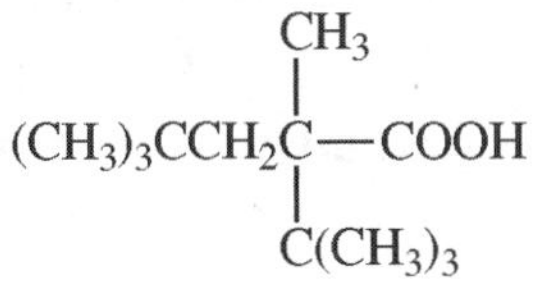

$$(CH_3)_3CCH_2\underset{\underset{C(CH_3)_3}{|}}{\overset{\overset{CH_3}{|}}{C}}\text{—}COOH$$

are less acidic than unbranched acids.

The $\text{—}CO_2^-$ group of the branched acid is shielded from solvent molecules and cannot be stabilized by solvation as effectively as can acetate anion.

Problem 16.13 Although *p*-hydroxybenzoic acid is less acidic than benzoic acid, salicyclic (*o*-hydroxybenzoic) acid ($K_a = 105 \times 10^{-5}$) is 15 times more acidic than benzoic acid. Explain.

The enhanced acidity is partly due to very effective H-bonding in the conjugate base, which decreases its basic strength.

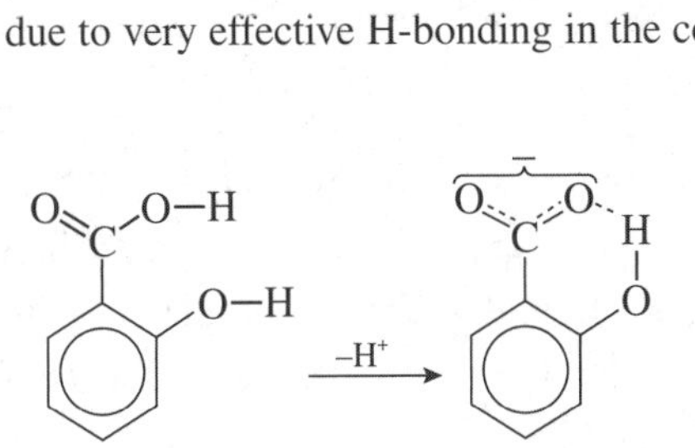

$\xrightarrow{-H^+}$

Problem 16.14 The K_2 for fumaric acid (*trans*-butaenedioic acid) is greater than that for maleic acid, the *cis* isomer. Explain by H-bonding.

Both dicarboxylic acids have two ionizable H's. The concern is with the second ionization step.

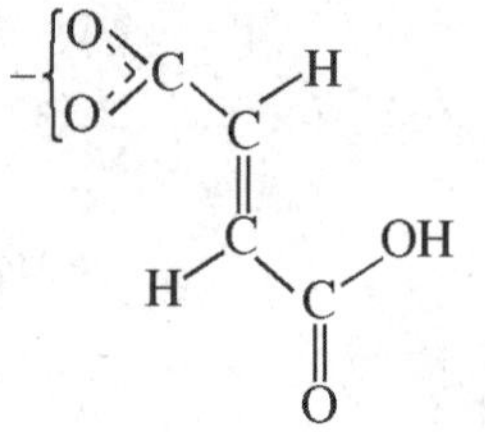

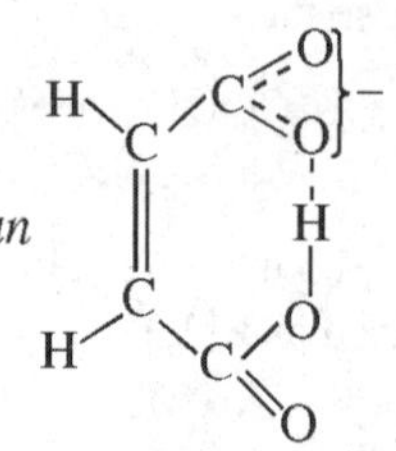

Fumarate monoanion (no H-bond) *is more acidic than* Maleate monoanion (H-bond)

Since the second ionizable H of maleate participates in H-bonding, more energy is needed to remove this H because the H-bond must be broken. The maleate monoanion is therefore the weaker acid.

In general, *H-bonding involving the acidic H has an acid-weakening effect; H-bonding in the conjugate base has an acid-strengthening effect.*

Nucleophilicity of Carboxylates

$RCOO^-$ acts as a nucleophile in S_N2 reactions with RX to give esters.

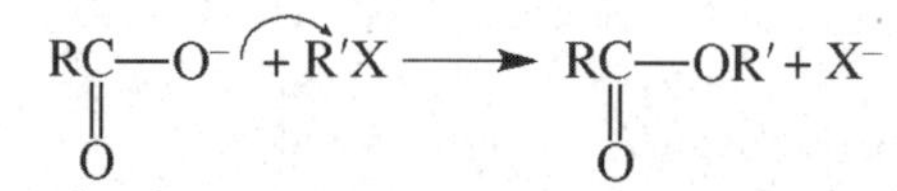

$$\underset{\text{a carboxylate}}{RC(=O)—O^-} + R'X \longrightarrow \underset{\text{an ester}}{RC(=O)—OR'} + X^-$$

Formation of Acid Derivatives (OH → G)

1. **Acyl Chloride** (RCOCl) **Formation**, OH → Cl

$$3R—COOH + PCl_3 \longrightarrow 3R—COCl + H_3PO_3(l)$$
$$R—COOH + PCl_5 \longrightarrow R—COCl + HCl(g) + POCl_3(l)$$
$$R—COOH + \underset{\text{Thionyl chloride}}{SOCl_2} \longrightarrow R—COCl + HCl(g) + SO_2(g)$$

Reaction with $SOCl_2$ is particularly useful because the two gaseous products SO_2 and HCl are readily separated from RCOCl.

2. **Ester** (RCOOR′) **Formation**, OH → OR′

$$R—COOH + R'OH \xrightarrow[\text{reflux}]{H_2SO_4} R—COOR' + H_2O$$

(In dilute acid, the reaction reverses.)

Problem 16.15 Give the mechanism for the acid-catalyzed esterification of RCOOH with R′OH.

In typical fashion, the O of C=O is protonated, which increases the electrophilicity of the C of C=O and renders it more easily attacked in the slow step by the weakly nucleophilic R′OH.

$$\mathrm{RCOOH + H^+} \xrightarrow[fast]{} \left[\underset{\underset{\mathrm{HO^+}}{\|}}{\mathrm{RC}}\mathrm{—OH} \longleftrightarrow \underset{\underset{\mathrm{HO}}{|}}{\overset{+}{\mathrm{RC}}}\mathrm{—OH}\right] \xrightarrow[slow]{\mathrm{R'OH}} \underset{\substack{\mathrm{HO}\\ \textit{tetrahedral}\\ \textit{intermediate}}}{\overset{\overset{+}{\mathrm{HO}}\mathrm{—R'}}{\mathrm{RC—OH}}} \longrightarrow \underset{\underset{\mathrm{O}}{\|}}{\mathrm{RC}}\mathrm{—OR' + H_3O^+}$$

The tetrahedral intermediate undergoes a sequence of fast deprotonations and protonations, the end result being the loss of H^+ and H_2O and the formation of the ester.

3. **Amide** ($RCONH_2$) **Formation**, OH → NH_2

$$\underset{\text{acid}}{\mathrm{R—COOH + NH_3}} \longrightarrow \underset{\text{ammonium salt}}{\mathrm{R—COO^-NH_4^+}} \xrightarrow{\text{heat}} \underset{\text{1° amide}}{\mathrm{R—CONH_2 + H_2O}}$$

$$\mathrm{RCOOH} + \underset{\text{1° amine}}{\mathrm{R'NH_2}} \longrightarrow \mathrm{RCOO^-\,R'NH_3^+} \xrightarrow{\text{heat}} \underset{\text{2° amide}}{\mathrm{RCONHR' + H_2O}}$$

$$\mathrm{RCOOH} + \underset{\text{2° amine}}{\mathrm{R'R''NH}} \longrightarrow \mathrm{RCOO^-R'R''NH_2^+} \xrightarrow{\text{heat}} \underset{\text{3° amide}}{\mathrm{RCONHR'R'' + H_2O}}$$

Problem 16.16 Use ethanol to prepare $CH_3COOC_2H_5$, an important commercial solvent.

$CH_3COOC_2H_5$, ethyl acetate, is the ester of CH_3CH_2OH and CH_3COOH. CH_3CH_2OH is oxidized to CH_3COOH.

$$\mathrm{CH_3CH_2OH} \xrightarrow{\mathrm{MnO_4^-/H^+}} \mathrm{CH_3COOH}$$

Ethanol and acetic acid are then refluxed with concentrated H_2SO_4.

$$\mathrm{C_2H_5OH + CH_3COOH} \underset{\text{heat}}{\overset{\mathrm{H_2SO_4}}{\rightleftharpoons}} \mathrm{CH_3COOC_2H_5 + H_2O}$$

With added benzene and a trace of acid, the reversible reaction is driven to completion by distilling off H_2O as an azeotrope (Problem 13.49).

Problem 16.17 The pain reliever acetaminophen is produced by reacting 4-aminophenol with acetic anhydride. Outline a synthesis of acetaminophen from 4-aminophenol including any needed inorganic reagents.

$$\underset{\text{4-aminophenol}}{\mathrm{HO{-}C_6H_4{-}NH_2}} + \underset{\text{acetic anhydride}}{\mathrm{CH_3{-}\overset{\overset{O}{\|}}{C}{-}O{-}\overset{\overset{O}{\|}}{C}{-}CH_3}} \xrightarrow[\mathrm{H_2O}]{\mathrm{NaOH}} \underset{\text{acetaminophen}}{\mathrm{C_6H_5{-}HN{-}\overset{\overset{O}{\|}}{C}{-}CH_3}} + \underset{\text{sodium acetate}}{\mathrm{CH_3{-}\overset{\overset{O}{\|}}{C}{-}O^-Na^+}}$$

Reduction of C=O of COOH (RCOOH → RCH_2OH)

Acids are best reduced to alcohols by $LiAlH_4$.

$$\mathrm{RCOOH} \xrightarrow[\text{2. } \mathrm{H_2O}]{\text{1. } \mathrm{LiAlH_4} \text{ (ether)}} \mathrm{RCH_2OH}$$

Halogenation of α H's. Hell-Volhard-Zelinksky (HVZ) Reaction

One or more α H's are replaced by Cl or Br by treating the acid with Cl_2 or Br_2, using phosphorus as catalyst.

$$RCH_2COOH \xrightarrow{X_2/P} \underset{\displaystyle X}{\underset{|}{RCHCOOH}} \xrightarrow{X_2/P} RCX_2COOH \qquad (X = Cl, Br)$$

α-Halogenated acids react like active alkyl halides and are convenient starting materials for preparing other α-substituted acids by nucleophilic displacement of halide anion.

$$X^- + R{-}\underset{\displaystyle OH}{\underset{|}{CH}}{-}COOH \xleftarrow[2.\ H^+]{1.\ NaOH} \boxed{R{-}\underset{\displaystyle X}{\underset{|}{CH}}{-}COOH} \xrightarrow[-H^+,\ X^-]{NH_3} R{-}\underset{\displaystyle NH_3^+}{\underset{|}{CH}}{-}COO$$

an α-hydroxy acid — an α-amino acid

Reaction of COOH. Decarboxylation

1. Arylcarboxylic Acids

$$ArCOOH \xrightarrow{\text{soda lime}} ArH \ (\textit{in poor yields}) + CO_2$$

2. β-Keto Acids and β-Dicarboxylic Acids

$$-\underset{\displaystyle O}{\underset{\|}{C}}-\underset{|}{\overset{|}{C}}-COOH \xrightarrow[\Delta]{-CO_2} -\underset{\displaystyle O}{\underset{\|}{C}}-\underset{|}{\overset{|}{C}}-H$$

Decarboxylation proceeds readily and in good yields when the C that is β to COOH is a C$=$O.

$$\underset{\displaystyle O}{\underset{\|}{RC}}-CH_2COOH \xrightarrow[\Delta]{-CO_2} \underset{\displaystyle O}{\underset{\|}{RC}}-\underset{\displaystyle H}{\underset{|}{CH_2}}$$

a β-ketocarboxylic acid — a ketone

$$\underset{\displaystyle O}{\underset{\|}{HOC}}-\underset{|}{\overset{|}{C}}-COOH \xrightarrow[\Delta]{-CO_2} H-\underset{|}{\overset{|}{C}}-COOH$$

a β-dicarboxylic (malonic) acid — a carboxylic acid

Problem 16.18 (*a*) Suggest a mechanism for a ready decarboxylation of malonic acid, $HOOCCH_2COOH$, that proceeds through an activated, intramolecular H-bonded, intermediate complex. (*b*) Give the decarboxylation products of (i) oxalic acid, HOOC—COOH, and (ii) pyruvic acid, $CH_3COCOOH$.

(*a*) $HO-C(=O)-CH_2-C(=O)-O-H$ ⟶ [$HO-C(=O)-CH_2-C(=O)-O\cdots H$] *complex* $\xrightarrow{-CO_2}$ [$HO-C(O-H)=CH_2$] *enol* ⟶ $O=C(OH)-CH_3$ Acetic acid

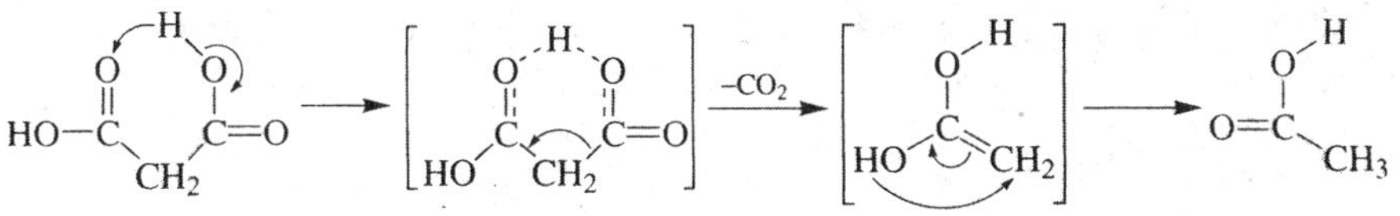

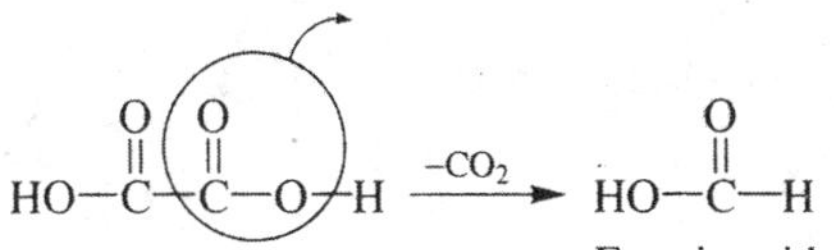

(*b*) (*i*) $HO-C(=O)-C(=O)-O-H$ $\xrightarrow{-CO_2}$ $HO-C(=O)-H$
Formic acid

(*ii*) $CH_3-C(=O)-C(=O)-O-H$ $\xrightarrow{-CO_2}$ $CH_3-C(=O)-H$
Acetaldehyde

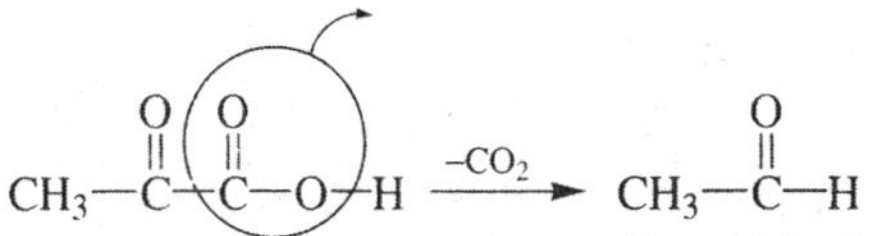

Problem 16.19 Prepare *n*-hexyl chloride from *n*-butylmalonic ester.

n-Butylmalonic ester is hydrolyzed with base and decarboxylated to hexanoic acid.

$$n\text{-}C_4H_9CH(COOC_2H_5)_2 \xrightarrow[1.\ H^+,\ \Delta]{1.\ OH^-} n\text{-}C_4H_9CH_2COOH$$

This acid and *n*-hexyl chloride have the same number of C's.

$$n\text{-}C_4H_9CH_2COOH \xrightarrow{LiAlH_4} n\text{-}C_5H_{11}CH_2OH \xrightarrow{SOCl_2} n\text{-}C_6H_{13}Cl$$

3. Conversion of RCOOH to RBr

The **Hunsdiecker reaction** treats heavy-metal (e.g., Ag^+) carboxylate salts with Br_2.

$$RCOO^-Ag^+ + Br_2 \longrightarrow RBr + CO_2 + AgBr$$

Problem 16.20 Suggest a typical free-radical mechanism for the Hunsdiecker reaction which requires the initial formation of an acyl hypobromite, $RC(=O)—OBr$ from the Ag^+ salt and Br_2.

Initiation $RC(=O)—OBr \longrightarrow RC(=O)—O\cdot + Br\cdot$

Propagation (1) $RC(=O)—O\cdot \longrightarrow R\cdot + CO_2$

(2) $R\cdot + RC(=O)—OBr \longrightarrow R—Br + RC(=O)—O\cdot$

Electrophilic Substitutions

During aromatic substitution with ArCOOH or ArCOG, the electron-attracting —COOH or —COG group is *meta*-directing and deactivating.

Problem 16.21 (*a*) Account for the fact that ArCOOH, with a strongly activating substituent *ortho* or *para* to COOH, loses CO_2 during attempted electrophilic substitution. (*b*) Write the equation for the reaction of *p*-aminobenzoic acid and Br_2.

(*a*) When the electrophile attacks the ring C bonded to COOH, the intermediate phenonium ion first loses an H^+ from COOH and then loses CO_2, which is a very good leaving group:

$$\left[\text{>N}\cdots\text{C}_6\text{H}_4(\text{E})(\text{COOH})\right]^+ \xrightarrow[-CO_2]{-H^+} H_2N\text{—C}_6\text{H}_4\text{—E}$$

(*b*) NH_2–C_6H_4–COOH $\xrightarrow[-H^+]{Br_2}$ 2,4,6-$Br_3C_6H_2NH_2$ $+ CO_2$

p-Aminobenzoic acid 1,3,5-Tribromoaniline

16.4 Summary of Carboxylic Acid Chemistry

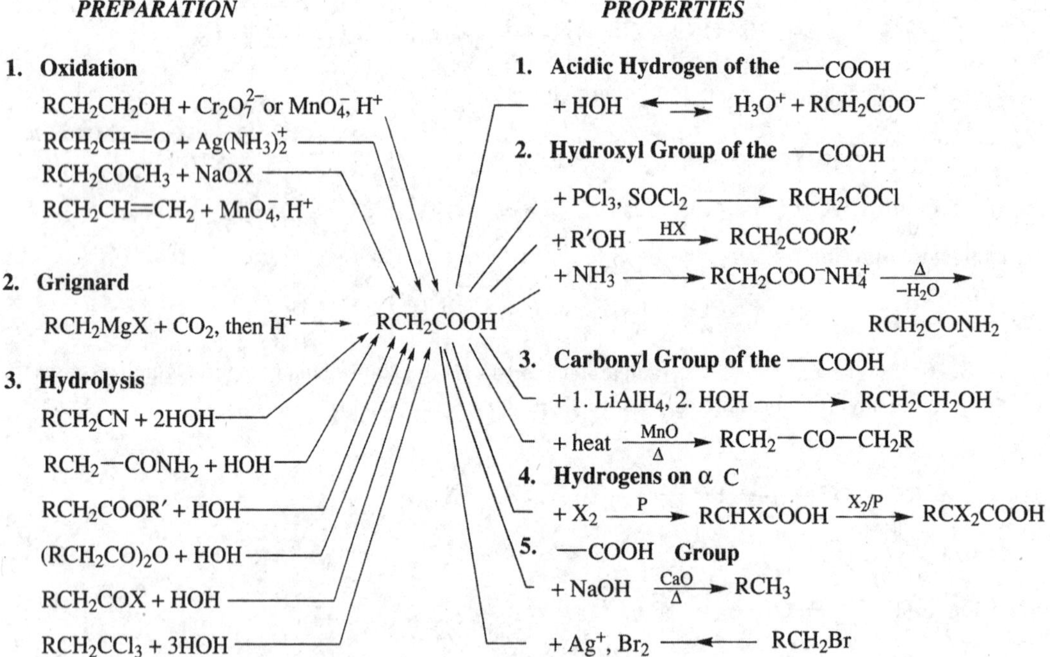

16.5 Polyfunctional Carboxylic Acids

Dicarboxylic Acids, $HOOC(CH_2)_n COOH$; Cyclic Anhydrides

The chemistry of dicarboxylic acids depends on the value of n. See Problem 16.18 for decarboxylations of oxalic acid ($n = 0$) and malonic acid ($n = 1$). When $n = 2$ or 3, the diacid forms cyclic anhydrides when heated. When n exceeds 3, acyclic anhydrides, often polymers, are formed.

Problem 16.22 Compare the products formed on heating the following dicarboxylic acids: (*a*) succinic acid, (*b*) glutaric acid (1,5-pentanedioic acid), (*c*) longer-chain $HOOC(CH_2)_nCOOH$.

(*a*) Intramolecular dehydration and ring formation:

$$\begin{array}{l} CH_2—COOH \\ | \\ CH_2—COOH \\ \quad n = 2 \end{array} \xrightarrow[-H_2O]{\text{heat}} \begin{array}{l} CH_2—C{=}O \\ | \qquad \;\; \searrow O \\ CH_2—C{=}O \end{array} \quad \text{Succinic anhydride}$$

(*b*) Intramolecular dehydration and ring formation:

$$\begin{array}{l} CH_2—COOH \\ CH_2 \\ CH_2—COOH \\ \quad n = 3 \end{array} \xrightarrow[-H_2O]{\text{heat}} \begin{array}{l} CH_2—C{=}O \\ CH_2 \qquad\;\; O \\ CH_2—C{=}O \end{array} \quad \text{Glutaric anhydride}$$

(*c*) Longer-chain α,ω-dicarboxylic acids usually undergo *inter*molecular dehydration on heating to form long-chain polymeric anhydrides (see Section 16.8).

$$\underset{n>3}{HOOC—(CH_2)_n—COOH} \xrightarrow{\text{heat}} HOOC—(CH_2)_n—\underset{\underset{O}{\|}}{C}—O—\underset{\underset{O}{\|}}{C}—(CH_2)_n—\underset{\underset{O}{\|}}{C}—\cdots$$

Problem 16.23 Show steps in the following syntheses, using any needed inorganic reagents:

(*a*) *o*-Xylene ⟶ Phthalic anhydride

(*b*) Tetrahydrofuran ⟶ $ClC(CH_2)_4CCl$ (with C=O at each end) Adipoyl dichloride

(*a*) *o*-Xylene ($C_6H_4(CH_3)_2$) $\xrightarrow[\text{catalytic oxidation}]{KMnO_4 \text{ or}}$ $C_6H_4(COOH)_2$ (Phthalic acid) $\xrightarrow{\text{heat}}$ phthalic anhydride + H_2O

(*b*) The chain is increased from 4 to 6 C's by forming a dinitrile.

Tetrahydrofuran $\xrightarrow{HCl}$ $HO(CH_2)_4Cl$ $\xrightarrow{HCl}$ $Cl(CH_2)_4Cl$ $\xrightarrow{CN^-}$ $NC(CH_2)_4CN$ $\xrightarrow{H_3O^+}$ $HOOC(CH_2)_4COOH$ $\xrightarrow{SOCl_2}$ $\underset{\underset{O}{\|}}{ClC}(CH_2)_4\underset{\underset{O}{\|}}{CCl}$

Hydroxyacids; Lactones

Reactions of hydroxycarboxylic acids, $HO(CH_2)_nCOOH$, also depend on the value of n. In acid solutions, γ-hydroxycarboxylic acid ($n = 3$) and δ-hydroxycarboxylic acid ($n = 4$) form cyclic esters (**lactones**) with, respectively, five-membered and six-membered rings.

Problem 16.24 Write the structure for the lactone formed on heating in the presence of acid (*a*) γ-hydroxybutyric acid, (*b*) δ-hydroxyvaleric acid.

Since an OH and a COOH are present in each compound, intramolecular dehydration gives lactones with 5- and 6-membered rings, respectively.

(*a*)
$$HOCH_2CH_2CH_2C(=O)OH \xrightarrow[-H_2O]{H^+} \text{ring: } H_2C{-}CH_2{-}CH_2{-}C(=O){-}O{-}$$

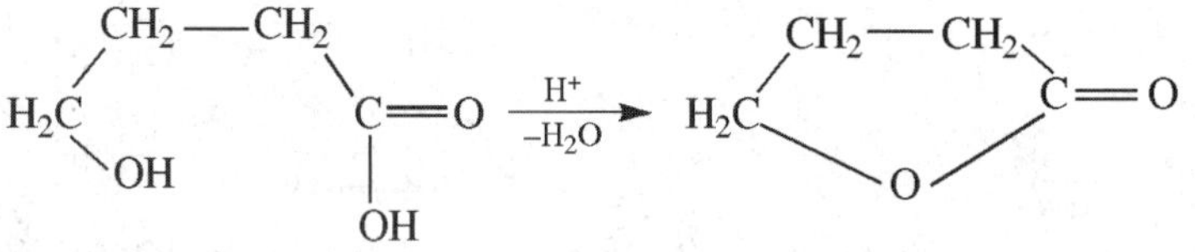

γ-Hydroxybutyric acid → γ-Butyrolactone (4-Butanolide)

(*b*)
$$HOCH_2CH_2CH_2CH_2C(=O)OH \xrightarrow[-H_2O]{H^+} \text{ring: } H_2C{-}CH_2{-}CH_2{-}C(=O){-}O{-}CH_2{-}$$

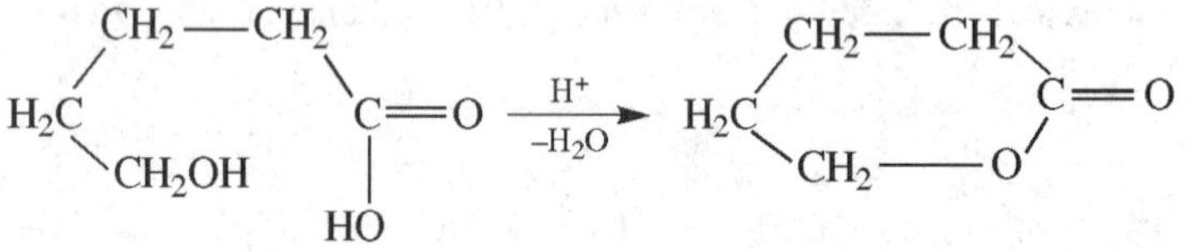

δ-Hydroxyvaleric acid → δ-Valerolactone (5-Pentanolide)

Problem 16.25 (*a*) When heated, 2 mol of an α-hydroxyacid ($n = 0$) lose 2 mol of H_2O to give a cyclic diester (a lactide). Give the structural formulas for two diastereomers obtained from lactic acid, $CH_3CHOHCOOH$, and select the diastereomer which is not resolvable. (*b*) Synthesize lactic acid from CH_3CHO.

(*a*) Two molecules of $CH_3CH(OH)COOH$ (OH and H circled for elimination) $\longrightarrow$ *cis,rac* lactide + *trans,meso* lactide

trans,meso: unresolvable — center of symmetry

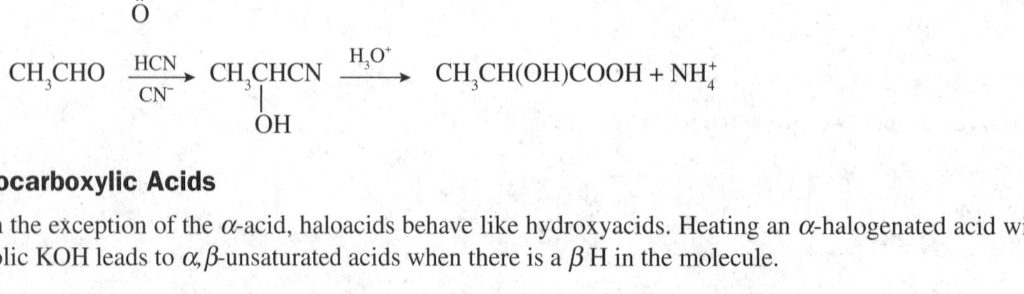

(*b*)
$$CH_3CHO \xrightarrow[CN^-]{HCN} CH_3CH(OH)CN \xrightarrow{H_3O^+} CH_3CH(OH)COOH + NH_4^+$$

Halocarboxylic Acids

With the exception of the α-acid, haloacids behave like hydroxyacids. Heating an α-halogenated acid with alcoholic KOH leads to α,β-unsaturated acids when there is a β H in the molecule.

$$R{-}CH_2CH(X){-}COOH \xrightarrow{\text{alc. KOH}} R{-}CH{=}CH{-}COO^-K^+ \xrightarrow{H^+} R{-}CH{=}CH{-}COOH$$

α, β-unsaturated acid salt

Problem 16.26 Prepare malonic acid (propanedioic acid, HOOC—CH_2—COOH) from CH_3COOH.

CH_3COOH is first converted to $ClCH_2COOH$. The acid is changed to its salt to prevent formation of the very poisonous HCN when replacing the Cl by CN. The C≡N group is then carefully hydrolyzed with acid instead of base, to prevent decarboxylation.

$$\mathrm{CH_3COOH} \xrightarrow{\mathrm{Cl_2/P}} \underset{\text{Chloroacetic acid}}{\mathrm{Cl{-}CH_2COOH}} \xrightarrow{\mathrm{NaOH}} \underset{\text{Sodium chloroacetate}}{\mathrm{Cl{-}CH_2CO\overset{-}{O}\overset{+}{N}a}} \xrightarrow{\mathrm{CN^-}}$$

$$\underset{\text{Sodium cyanoacetate}}{\mathrm{N{\equiv}C{-}CH_2CO\overset{-}{O}\overset{+}{N}a}} \xrightarrow{\mathrm{H_3O^+}} \mathrm{HOOC{-}CH_2{-}COOH} + \mathrm{NH_4^+}$$

Problem 16.27 What is the product when each of the following compounds is reacted with aqueous base (NaOH)? (*a*) 2-bromobutanoic acid, (*b*) 3-bromobutanoic acid, (*c*) 4-bromobutanoic acid, (*d*) 5-bromopentanoic acid. [In (*a*) and (*b*), the initially formed salt is acidified.]

(*a*) α-Hydroxyacid by S_N2 substitution.

$$\underset{\text{2-Bromobutanoic acid}}{\mathrm{CH_3CH_2\underset{\underset{Br}{|}}{C}H{-}COOH}} \xrightarrow[\text{2. H}^+]{\text{1. OH}^-} \underset{\text{2-Hydroxybutanoic acid}}{\mathrm{CH_3CH_2\underset{\underset{OH}{|}}{C}H{-}COOH}}$$

(*b*) Dehydrohalogenation to an α,β-unsaturated acid. The driving force for this easy reaction is the formation of a conjugated system.

$$\mathrm{CH_3\underset{\underset{Br}{|}}{C}HCH_2{-}\underset{\underset{OH}{|}}{C}{=}O} \xrightarrow[\text{2. H}^+]{\text{1. OH}^-} \underset{\text{2-Butenoic acid}}{\mathrm{CH_3CH{=}CH{-}\overset{\overset{OH}{|}}{C}{=}O}}$$

Elimination is a typical reaction of β-substituted carboxylic acid.

$$\mathrm{R\underset{\underset{Y}{|}}{C}HCH_2COOH} \xrightarrow{-\mathrm{HY}} \mathrm{RCH{=}CHCOOH} \quad (\mathrm{Y = Cl, Br, I, OH, NH_2})$$

(*c*) γ-Haloacids undergo intramolecular S_N2-type displacement of X^- initiated by the nucleophilic carboxylate anion, to yield γ-lactones.

$$\mathrm{BrCH_2CH_2CH_2COOH} \xrightarrow{\mathrm{NaOH}} \left[\mathrm{Br{-}CH_2CH_2CH_2{-}\underset{\underset{O}{\|}}{C}{-}O^-Na^+}\right] \longrightarrow \underset{\gamma\text{-Butyrolactone}}{\text{(ring: } \mathrm{H_2C{-}O,\ H_2C,\ C{=}O,\ CH_2)}} + \mathrm{NaBr}$$

(*d*) Similar to part (*c*) to give a δ-lactone.

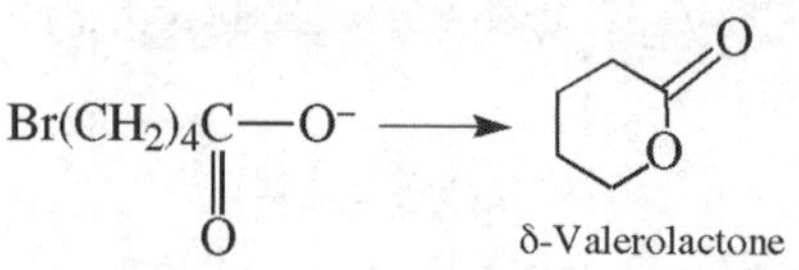

$$Br(CH_2)_4\overset{\overset{\displaystyle O}{\|}}{C}{-}O^- \longrightarrow \text{δ-Valerolactone}$$

δ-Valerolactone

Intramolecular nucleophilic displacements, such as those in lactone formation, have faster reaction rates than intermolecular S_N2 reactions because the latter require two species to collide. The neighboring participant is said to furnish **anchimeric assistance**.

16.6 Transacylation; Interconversion of Acid Derivatives

Transacylation is the transfer of the acyl group from one G group to another, resulting in the formation of various acid derivatives. Figure 16.1 summarizes the transacylation reactions. Notice that the more reactive derivatives are convertible to the less reactive ones. Because acetic anhydride reacts less violently, it is used instead of the more reactive acetyl chloride to make derivatives of acetic acid. In aqueous acid, the four kinds of carboxylic acid derivatives in the figure are hydrolyzed to RCOOH; in base, to $RCOO^-$.

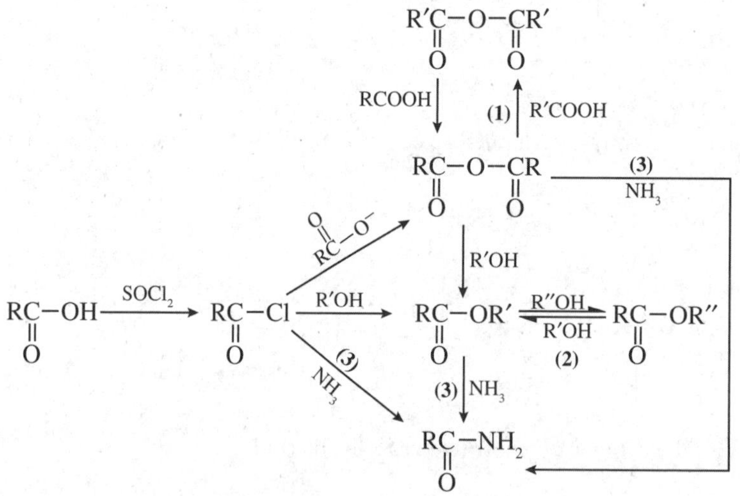

R′C(=O)−O−C(=O)R′

RCOOH ↓ (1) ↑ R′COOH

RC(=O)−O−C(=O)R —(3) NH_3→ RC(=O)−NH_2

RC(=O)−OH —$SOCl_2$→ RC(=O)−Cl —RC(=O)−O^-→ RC(=O)−O−C(=O)R

RC(=O)−Cl —R′OH→ RC(=O)−OR′ ⇌ (R″OH / R′OH) (2) RC(=O)−OR″

RC(=O)−O−C(=O)R —R′OH→ RC(=O)−OR′

RC(=O)−Cl —(3) NH_3→ RC(=O)−NH_2

RC(=O)−OR′ —(3) NH_3→ RC(=O)−NH_2

1. **Transanhydride formation** — an exchange of anhydrides.
2. **Transesterification** — an exchange of esters.
3. Also for $H_2NR' \longrightarrow$ RCONHR′ and HNR′R″ ⟶ RCONR′R″

Figure 16.1

Problem 16.28 (*a*) Give the mechanism for the reaction of RCOCl with $Nu{:}^-$. (*b*) Compare, and explain the difference in the reactivities of RCl and RCOCl with $Nu{:}^-$. (*c*) What is the essential difference between nucleophilic attack on C═O of a ketone or aldehyde and of an acid derivative?

(*a*) Nucleophilic substitutions of RCOG, such as RCOCl, occur in two steps. The first step (addition) resembles nucleophilic addition to ketones and aldehydes (Section 15.4), and the second step (elimination) is loss of G, in this case Cl as Cl^-.

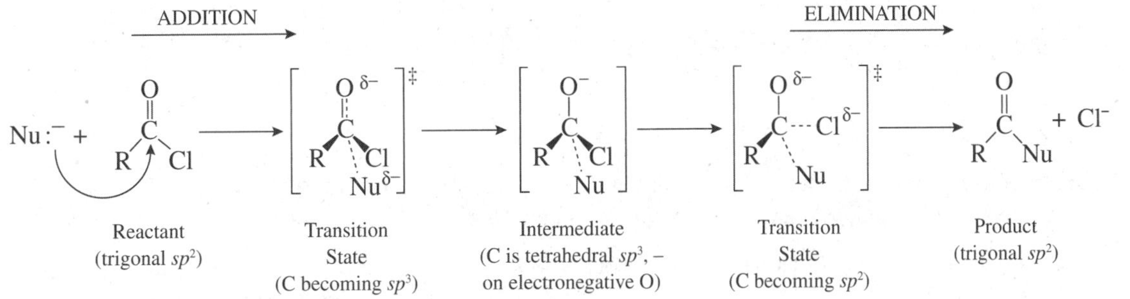

ADDITION → ELIMINATION →

$$Nu:^- + R{-}\overset{O}{\overset{\|}{C}}{-}Cl \longrightarrow \left[R{-}\overset{O^{\delta-}}{\overset{\vdots}{C}}(Cl)(Nu^{\delta-})\right]^{\ddagger} \longrightarrow \left[R{-}\overset{O^-}{\overset{|}{C}}(Cl)(Nu)\right] \longrightarrow \left[R{-}\overset{O^{\delta-}}{\overset{|}{C}}\cdots Cl^{\delta-}(Nu)\right]^{\ddagger} \longrightarrow R{-}\overset{O}{\overset{\|}{C}}{-}Nu + Cl^-$$

Reactant (trigonal sp^2) | Transition State (C becoming sp^3) | Intermediate (C is tetrahedral sp^3, – on electronegative O) | Transition State (C becoming sp^2) | Product (trigonal sp^2)

(*b*) Alkyl halides are much less reactive than acyl halides in nucleophilic substitution because nucleophilic attack on the tetrahedral C of RX involves a badly crowded transition state. Also, a σ bond must be partly broken to permit the attachment of the nucleophile. Nucleophilic attack on C $=$ O of RCOCl involves a relatively unhindered transition state leading to a tetrahedral intermediate. Most transacylations proceed by this mechanism, which is an example of addition-elimination reminiscent of aromatic nucleophilic substitution (Section 11.3).

(*c*) The addition step leading to the tetrahedral intermediate is the same in each case. However, the intermediates from the carbonyl compounds would have to eliminate an $R:^-$ (from a ketone) or an $H:^-$ (from an aldehyde) to restore the C $=$ O. These are very strong bases and are not eliminated. Instead, the intermediate accepts an H^+ to give the adduct. The intermediate from RCOG can eliminate $G:^-$ as the leaving group.

Problem 16.29 Account for the relative reactivities of RCOG with nucleophiles: RCOX $>$ $(RCO_2)_2O$ $>$ RCOOR′ $>$ $RCONH_2$.

Often this order of reactivity is related to the order of "leavability" of G^-, which is the reverse of its order of basicity; Cl^- is the best leaving group and the weakest base.

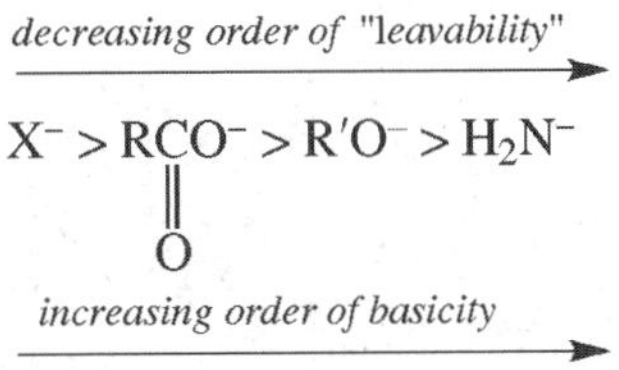

decreasing order of "leavability" →

$$X^- > R\overset{}{\underset{\underset{O}{\|}}{C}}O^- > R'O^- > H_2N^-$$

increasing order of basicity →

Although this rationale gives the correct answer, it overlooks an important feature of the mechanism—the step eliminating G^- cannot influence the reaction rate because it is faster than the addition step. The first step, the slow nucleophilic addition, determines the reaction rate, and here resonance stabilization of the $-\underset{\underset{O}{\|}}{C}-G$ group is important:

$$R{-}\underset{\underset{:O:}{\|}}{C}{-}G: \longleftrightarrow R{-}\underset{\underset{:\ddot{O}:^-}{|}}{C}{=}\overset{+}{G}$$

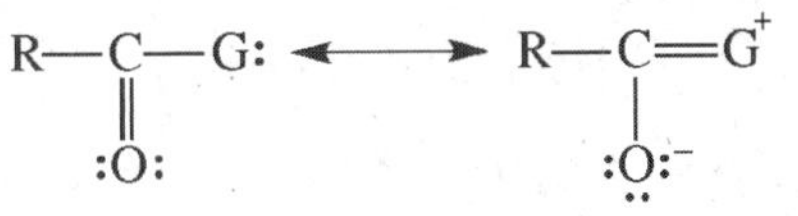

The greater the degree of resonance stabilization, the less reactive is RCOG. Thus, NH_2 has the greatest degree and $RCONH_2$ is the least reactive, while X has the smallest degree and RCOX is the most reactive. It so happens that the order of resonance stabilization correlates directly with the order of basicity; for instance, $-\underset{|}{\ddot{N}}:$ of amide is the best resonance stabilizer and $-\underset{|}{N}:^-$ is the strongest base. This relationship permits the use of relative basicities of the leaving groups to give correct answers, albeit for the wrong reasons.

Problem 16.30 Place **acyl azides**, $R{-}\overset{O}{\overset{\|}{C}}{-}N_3$, in the order of reactivities, if for hydrazoic acid, HN_3, $K_a = 2.6 \times 10^{-5}$. Recall that for CH_3COOH, $K_a = 1.8 \times 10^{-5}$.

Since HN_3 is only slightly more acidic than CH_3COOH, N_3^- is slightly less basic and is a slightly better leaving group than CH_3COO^-. $RCON_3$ is less reactive than RCOCl, but a little more reactive than the anhydride $R{-}\overset{\overset{\displaystyle O}{\|}}{C}{-}O{-}\overset{\overset{\displaystyle O}{\|}}{C}{-}R$.

Problem 16.31 Use the Hammond principle (Problem 11.8) to explain why strong bases such as OR^- (from esters) and NH_2^- (from amides) can be leaving groups in nucleophilic transacylations.

The elimination step for breaking the C—G bond is exothermic because it reestablishes the resonance-stabilized C=O. Its transition state resembles the reactant, in this case the intermediate. Consequently, there is little breaking of the C—G bond in the transition state of the elimination step and the basicity of the leaving group, G^-, has little or no influence.

Problem 16.32 Outline a mechansim for hydrolysis of acid derivaties with (*a*) H_3O^+, (*b*) NaOH.

(*a*) Protonation of carbonyl O makes C more electrophilic and hence more reactive toward weakly nucleophilic H_2O.

$$R{-}\overset{O}{\overset{\|}{C}}{-}G + H_3O^+ \longrightarrow \underset{\text{weaker nucleophile}}{H_2O} + R{-}\overset{\overset{+}{O}H}{\overset{\|}{C}}{-}G \longrightarrow R{-}\underset{{}^+OH_2}{\overset{OH}{C}}{-}G \xrightarrow{H_2O} R{-}\overset{O}{\overset{\|}{C}}{-}OH + 2H_3O^+ + G^- \text{ (or HG + } H_2O)$$

more electrophilic C

[If G^- is basic, for example, OR^-, we get HG(HOR).]

(*b*) Strongly basic OH^- readily attacks the carbonyl C. Unlike acid hydrolysis, this reaction is irreversible, because OH^- removes H^+ from —COOH to form resonance-stabilized $RCOO^-$.

$$R{-}\overset{O}{\overset{\|}{C}}{-}G + {:}OH^- \longrightarrow R{-}\underset{OH}{\overset{O^-}{C}}{-}G \xrightarrow{-G^-} R{-}\overset{O}{\overset{\|}{C}}{-}O{:}H \xrightarrow{:\ddot{O}H^-} H_2O + \left[R{-}C{\overset{O}{\underset{O}{\lessgtr}}}\right]^-$$

Problem 16.33 Does each of the following reactions take place easily? Explain.

(*a*) $CH_3COCl + H_2O \longrightarrow CH_3COOH + HCl$

(*b*) $CH_3COOH + NH_3 \longrightarrow CH_3CONH_2 + H_2O$

(*c*) $(CH_3CO)_2O + NaOH \longrightarrow CH_3COOH + CH_3COO^-Na^+$

(*d*) $CH_3COBr + C_2H_5OH \longrightarrow CH_3COOC_2H_5 + HBr$

(*e*) $CH_3CONH_2 + NaOH \longrightarrow CH_3COO^-Na^+ + NH_3$

(*f*) $CH_3COOCH_3 + Br^- \longrightarrow CH_3COBr + {}^-OCH_3$

Nucleophilic substitution of acyl compounds takes place readily if the incoming group (Nu:$^-$ or Nu:) is a stronger base than the leaving group (G:$^-$) or if the final product is a resonance-stabilized $RCOO^-$. (*a*) Yes. H_2O is a stronger base than Cl^- and reacts vigorously. (*b*) No. NH_3 reacts with RCOOH to form $RCOO^-NH_4^+$, which *does not react further*. Amides are prepared from RCOOH by strongly heating dry $RCOO^-NH_4^+$, because reaction is aided by acid catalysis by NH_4^+. (*c*) Yes. The leaving group $RCOO^-$ is a weaker base than OH^-. (*d*) Yes. Br^- is a much weaker base than C_2H_5OH. (*e*) Yes. Even though NH_2^- is a stronger base than OH^-, in basic solution the resonance-stabilized $RCOO^-$ is formed, and this shifts the reaction to completion. (*f*) No. Br^- is a weaker base than OCH_3^-.

16.7 More Chemistry of Acid Derivatives

ACYL CHLORIDES (see Section 16.3 for preparation of RCOCl)

The use of acyl chlorides in Friedel-Craft acylations of benzene rings, as well as their reactions with organometallics and reductions to aldehydes, has been discussed in Section 15.2.

Problem 16.34 Give the structure and name of the principal product formed when propionyl chloride reacts with: (*a*) H_2O, (*b*) C_2H_5OH, (*c*) NH_3, (*d*) $C_6H_6(AlCl_3)$, (*e*) $(n\text{-}C_3H_7)_2CuLi$, (*f*) aq. NaOH, (*g*) $LiAl(O\text{-}t\text{-}C_4H_9)_3$, (*h*) H_2NOH, (*i*) CH_3NH_2, (*j*) Na_2O_2 (sodium peroxide).

(*a*) CH_3CH_2COOH, propionic acid; (*b*) $CH_3CH_2COOC_2H_5$, ethyl propionate; (*c*) $CH_3CH_2CONH_2$, propanamide; (*d*) $C_6H_5COCH_2CH_3$, propiophenone or ethyl phenyl ketone; (*e*) $CH_3CH_2COC_3H_7$, 3-hexanone; (*f*) $CH_3CH_2COO^-Na^+$, sodium propionate; (*g*) CH_3CH_2CHO, propanal;

(*h*) $CH_3CH_2C(=O)—NHOH$ Propanehydroxamic acid (*i*) $CH_3CH_2CONHCH_3$ *N*-Methylpropanamide (a substituted amide) (*j*) $CH_3CH_2C(=O)—O—O—C(=O)—CH_2CH_3$ Propionyl peroxide

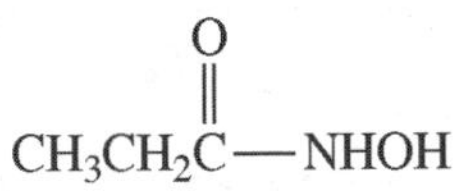

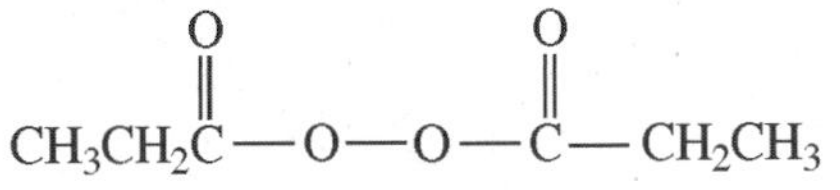

Acid Anhydrides

All carboxylic acids have anhydrides:

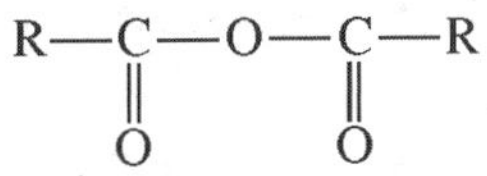

$$R—C(=O)—O—C(=O)—R$$

but the one most often used is acetic anhydride, prepared as follows:

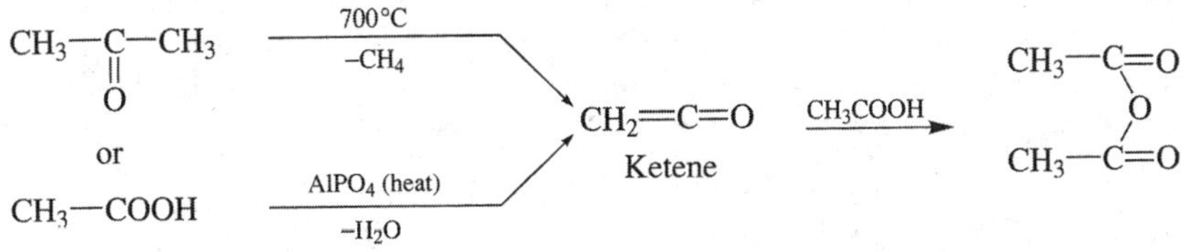

$$CH_3—C(=O)—CH_3 \xrightarrow[-CH_4]{700°C} CH_2{=}C{=}O \text{ (Ketene)}$$

or

$$CH_3—COOH \xrightarrow[-H_2O]{AlPO_4\text{ (heat)}} CH_2{=}C{=}O \text{ (Ketene)} \xrightarrow{CH_3COOH} (CH_3—C{=}O)_2O \text{ Acetic anhydride}$$

Heating dicarboxylic acids, $HOOC(CH_2)_nCOOH$ (n = 2 or 3), forms cyclic anhydrides by intramolecular dehydration [Problem 16.22(*a*), (*b*)]. Anhydrides resemble acid halides in their reactions. Because acetic anhydride reacts less violently, it is often used in place of acetyl chloride. Acid anhydrides can also be used to acylate aromatic rings in electrophilic substitutions.

Problem 16.35 Give the products formed when acetic anhydride reacts with (*a*) H_2O, (*b*) NH_3, (*c*) C_2H_5OH, (*d*) C_6H_6 with $AlCl_3$.

(*a*) $2CH_3COOH$, (*b*) $CH_3CONH_2 + CH_3CO_2^-NH_4^+$. (*c*) $CH_3COOC_2H_5 + CH_3COOH$. This is a good way to form acetates. (*d*) $C_6H_5COCH_3$. Friedel-Crafts acetylation.

Problem 16.36 Give the structural formula and name for the product formed when 1 mol of succinic anhydride, the cyclic anhydride of succinic acid (Problem 16.22), reacts with (*a*) 1 mol of CH_3OH; (*b*) 2 mol of NH_3 and (*c*) 1 mol of C_6H_6 with $AlCl_3$.

Products are formed in which half of the anhydride forms the appropriate derivative and the other half becomes a COOH.

$$\text{Succinic anhydride} \xrightarrow{CH_3OH} \begin{matrix} CH_2COOCH_3 \\ | \\ CH_2{-}COOH \end{matrix} \quad \text{Methyl hydrogen succinate} \qquad (a)$$

$$\text{Succinic anhydride} \xrightarrow{2NH_3} \begin{matrix} CH_2CONH_2 \\ | \\ CH_2CO\bar{O}\overset{+}{N}H_4 \end{matrix} \quad \text{Ammonium succinamate} \qquad (b)$$

$$\text{Succinic anhydride} \xrightarrow{1.\ C_6H_6,\ AlCl_3\quad 2.\ H^+} \begin{matrix} CH_2COC_6H_5 \\ | \\ CH_2COOH \end{matrix} \quad \beta\text{-Benzoylpropionic acid} \qquad (c)$$

The monoamides of dicarboxylic acids, $HOOC(CH_2)_nCONH_2$ ($n = 2{,}3$), ($n = 2, 3$), form cyclic imides on heating.

$$\text{Glutaric anhydride} \xrightarrow[-H_2O]{NH_3} \text{Glutaramic acid} \xrightarrow[-H_2O]{\Delta} \text{Glutarimide}$$

$$\text{Phthalic anhydride} \xrightarrow[-H_2O]{NH_3} \text{Phthalamic acid} \xrightarrow[-NH_3]{\Delta} \text{Phthalimide}$$

Esters

The mechanism for esterification given in Problem 16.16 is reversible, the reverse being the mechanism for acid-catalyzed hydrolysis of esters. As an example of the principle of microscopic reversibility, the forward and reverse mechanisms proceed through the same intermediates and transition states.

$$RCOOH + \left\{\begin{matrix} HOR' \\ \text{or} \\ HOAr \end{matrix}\right. \underset{\text{hydrol.}}{\overset{\text{ester.}}{\rightleftharpoons}} \left.\begin{matrix} RCOOR' \\ \text{or} \\ RCOOAr \end{matrix}\right\} + H_2O$$

For the role of steric hindrance, see Problem 16.37.

Esters react with the Grignard reagent:

$$\underset{\displaystyle \text{OR}'}{\text{R—C=O}} \xrightarrow{\boxed{\text{R}''}\text{MgX}} \text{Mg(OR}')\text{X} + \underset{\textit{not isolated}}{\left[\text{R—}\overset{}{\underset{\underset{\text{O}}{\|}}{\text{C}}}\text{—}\boxed{\text{R}''}\right]} \xrightarrow{\text{R}''\ \text{MgX}} \text{R—}\overset{\boxed{\text{R}''}}{\underset{\text{OMgX}}{\text{C}}}\text{—}\boxed{\text{R}''} \xrightarrow{\text{H}_2\text{O}} \text{R—}\overset{\boxed{\text{R}''}}{\underset{\text{OH}}{\text{C}}}\text{—}\boxed{\text{R}''}$$

a 3° alcohol with at least 2 like R″'s

Reduction of esters gives alcohols:

$$\text{RCOOR}' \xrightarrow[\text{2. H}_2\text{O in ether}]{\text{1. LiAlH}_4} \text{RCH}_2\text{OH} + \text{R}'\text{OH}$$

On pyrolysis, esters give alkenes:

$$\text{RCH}_2\text{CH}_2\text{O}\underset{\underset{\text{O}}{\|}}{\text{C}}\text{—R}' \xrightarrow{400^\circ\text{C}} \text{RCH=CH}_2 + \text{R}'\text{COOH}$$

Problem 16.37 Use the mechanism of esterification to explain the lower rates of both esterification and hydrolysis of esters when the alcohol, the acid, or both have branched substituent groups.

The carbonyl C of RCOOH and RCOOR′ is trigonal sp^2-hybridized, but that of the intermediate is tetrahedral sp^3-hybridized. If R′ in R′OH or R in RCOOH is extensively branched, formation of the unavoidably crowded transition state has to occur with greater difficulty and more slowly.

Problem 16.38 Write mechanisms for the reactions of RCOOR′ with (*a*) aqueous OH^- to form $RCOO^-$ and R′OH (**saponification**), (*b*) NH_3 to form $RCONH_2$, (*c*) R″OH in acid, HA, to form a new ester RCOOR″ (trans-esterification; Section 16.6).

(*a*)
$$\text{R—}\overset{\overset{\ddot{\text{O}}:}{\|}}{\text{C}}\text{—}\ddot{\text{O}}\text{R}' + \text{OH}^- \rightleftharpoons \text{R—}\overset{:\ddot{\text{O}}:^-}{\underset{\text{OH}}{\text{C}}}\text{—}\ddot{\text{O}}\text{R}' \longrightarrow \text{R—}\overset{\overset{\text{O}}{\|}}{\text{C}}\text{—OH} + :\ddot{\text{O}}\text{R}'^- \longrightarrow \left[\text{R—C}\begin{smallmatrix}\text{O}\\ \text{O}\end{smallmatrix}\right]^- + \text{HOR}'$$

The last step is irreversible and drives the reaction to completion.

(*b*)
$$\text{R—}\overset{\overset{\text{O}}{\|}}{\text{C}}\text{—OR}' + :\text{NH}_3 \rightleftharpoons \text{R—}\overset{\text{O}^-}{\underset{\underset{\text{H}}{|}}{\underset{\overset{+}{\text{H}}\text{NH}}{\text{C}}}}\text{—OR}' \xrightarrow{+\text{H}^+} \text{R—}\overset{\text{O}}{\underset{\text{NH}_2}{\text{C}}}\text{—}\underset{\text{H}}{\overset{+}{\text{O}}}\text{R}' \longrightarrow \text{R—}\overset{\overset{\text{O}}{\|}}{\text{C}}\text{—NH}_2 + \text{HOR}'$$

(*c*)
$$\text{R—}\overset{\overset{\text{O}}{\|}}{\text{C}}\text{—OR}' \underset{-\text{A}^-}{\overset{\text{H:A}}{\rightleftharpoons}} \text{R—}\overset{\overset{\overset{+}{\text{O}}\text{H}}{\|}}{\text{C}}\text{—OR}' \underset{\text{HOR}''}{\rightleftharpoons} \text{R—}\overset{\text{O—H}}{\underset{\underset{+}{\text{HOR}''}}{\text{C}}}\text{—OR}' \overset{\text{A}^-}{\rightleftharpoons} \text{R—}\overset{\overset{\text{O}}{\|}}{\text{C}}\text{—OR}'' + \text{R}'\text{OH} + \text{HA}$$

To drive the reaction to completion, a large excess of R″OH is used, and when R′OH is lower-boiling than R″OH, R′OH is removed by distillation.

Problem 16.39 Assign numbers from (1) for LEAST to (4) for MOST to show relative rates of alkaline hydrolysis of compounds I through IV and point out the factors determining the rates.

	I	II	III	IV
(*a*)	$CH_3COOCH(CH_3)_2$	CH_3COOCH_3	$CH_3COOC(CH_3)_3$	$CH_3COOC_2H_5$
(*b*)	$HCOOCH_3$	$(CH_3)_2CHCOOCH_3$	CH_3COOCH_3	$(CH_3)_3CCOOCH_3$
(*c*)	$O_2N-C_6H_4-COOCH_3$	$CH_3O-C_6H_4-COOCH_3$	$C_6H_5-COOCH_3$	$Cl-C_6H_4-COOCH_3$

See Table 16.2.

TABLE 16.2

	RANKS				
	I	II	III	IV	RATE-DETERMINING FACTORS
(*a*)	2	4	1	3	Steric effects (branching an alcohol portion)
(*b*)	4	2	3	1	Steric factor (branching on acid portion)
(*c*)	4	1	2	3	Electron-attracting groups disperse, developing negative charge in transition state and increasing reactivity

Problem 16.40 Acid-catalyzed hydrolysis with $H_2{}^{18}O$ of an ester of an optically active 3° alcohol, RCOOC*R′R″R‴, yields the partially racemic alcohol containing ^{18}O, R′R″R‴$C^{18}OH$. Similar hydrolyses of esters of 2° chiral alcohols, RCOOC*HR′R″, produce no change in the optical activity of the alcohol, and ^{18}O is found in $RC^{18}O_2H$. Explain these observations.

Hydrolyses of esters of most 2° and 1° alcohols occur by cleavage of the O—acyl bond:

$$R-\overset{O}{\overset{\|}{C}} \wr\, O-\overset{*}{C}HR'R''$$

Since no bond to C* is broken, no racemization occurs. However, with 3° alcohols, there is an S_N1 O—alkyl cleavage:

$$R-\overset{O}{\overset{\|}{C}}-O \wr\, CR'R''R'''$$

producing RCOOH and a 3° carbocation, $^{+}CR'R''R'''$, which reacts with the solvent ($H_2{}^{18}O$) to form $R'R''R'''C^{18}OH$. This alcohol is partially racemized because $^{+}CR'R''R'''$ is partially racemized.

$$R-\overset{\overset{\displaystyle O}{\|}}{C}-O-CR'R''R''' + H{:}A \xrightleftharpoons{-A^-} R-\overset{\overset{\displaystyle O}{\|}}{C}-\underset{+}{\overset{\overset{\displaystyle H}{|}}{O}}-CR'R''R''' \xrightleftharpoons{S_N1}$$

$$RCOOH + R'R''R'''C^+ \xrightleftharpoons{H_2{}^{18}O} R'R''R'''C^{18}OH + H^+$$

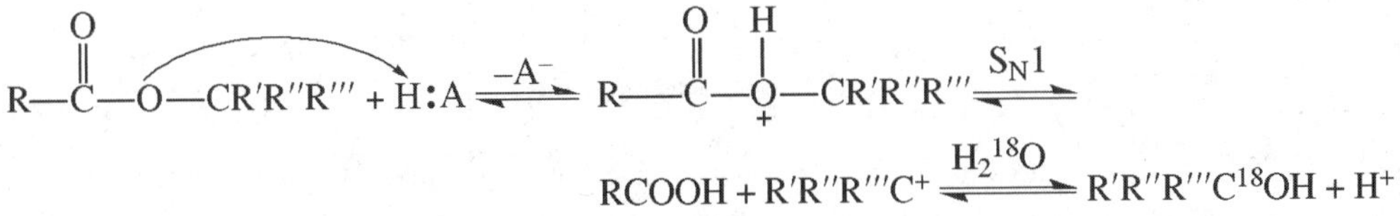

Problem 16.41 Write structures of the organic products for the following reactions:

(*a*) $(R)\text{—}CH_3COOCH(CH_3)CH_2CH_3 + H_2O \xrightarrow{NaOH}$
(*b*) $(R)\text{—}CH_3COOCH(CH_3)CH_2CH_3 + H_3O^+ \longrightarrow$
(*c*) $C_6H_5COOC_2H_5 + NH_3 \longrightarrow$
(*d*) $C_6H_5COOC_2H_5 + n\text{-}C_4H_9OH \xrightarrow{H^+}$
(*e*) $C_6H_5COOC_2H_5 + LiAlH_4 \longrightarrow$
(*f*) $C_6H_5\overset{\overset{\displaystyle O}{\|}}{C}-O-O-\overset{\overset{\displaystyle O}{\|}}{C}-C_6H_5 + CH_3O^-Na^+ \longrightarrow$

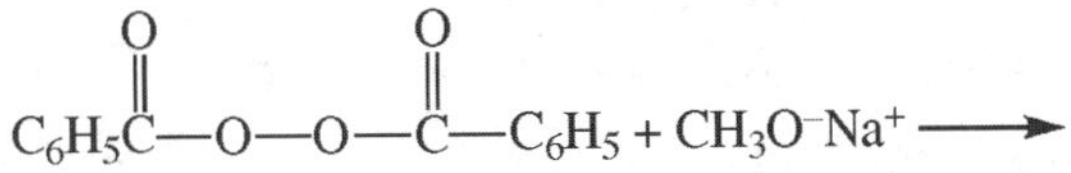

(*a*) $CH_3COO^-Na^+$ + (R)-$HOCH(CH_3)CH_2CH_3$. The alcohol is 2°; we find O—acyl cleavage and no change in configuration of alcohol. $RCOO^-Na^+$ forms in basic solution.
(*b*) CH_3COOH + (R)-$HOCH(CH_3)CH_2CH_3$. Again O—acyl cleavage occurs.
(*c*) $C_6H_5CONH_2 + C_2H_5OH$.
(*d*) $C_6H_5COO\text{-}n\text{-}C_4H_9 + C_2H_5OH$. Acid-catalyzed transesterification.
(*e*) $C_6H_5CH_2OH + C_2H_5OH$.
(*f*) $C_6H_5\overset{\overset{\displaystyle O}{\|}}{C}-O-O^-Na^+ + C_6H_5COOCH_3$
Sodium peroxybenzoate

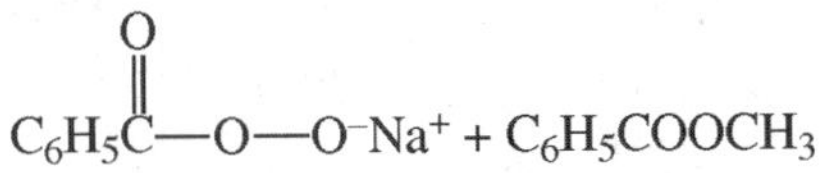

Problem 16.42 Supply a structural formula for the alcohol formed from

(*a*) $CH_3CH_2COOCH_3 + 2C_3H_7MgBr$ (*b*) $C_6H_5CH_2COOCH_3 + 2C_6H_5MgBr$

Two R or Ar groups bonded to carbinol C come from the Grignard reagent, while the carbonyl C becomes the carbinol C.

(*a*) $(CH_3CH_2)COOCH_3 + 2\,\boxed{C_3H_7}MgBr \longrightarrow (CH_3CH_2)-\overset{\overset{\displaystyle \boxed{C_3H_7}}{|}}{\underset{\underset{\displaystyle OH}{|}}{C}}-\boxed{C_3H_7}$

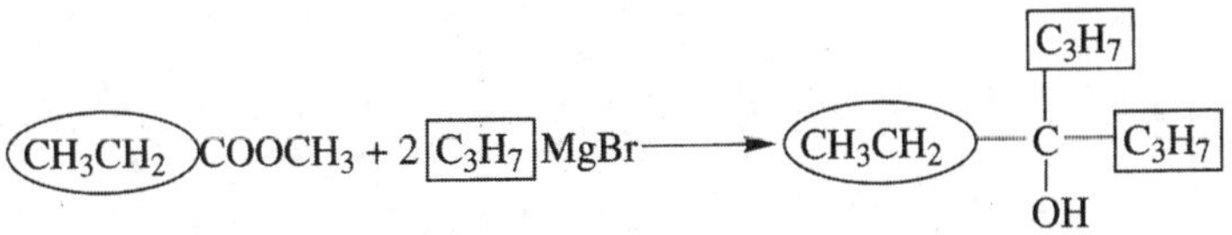

(*b*) $(C_6H_5CH_2)COOCH_3 + 2\,\boxed{C_6H_5}MgBr \longrightarrow (C_6H_5CH_2)-\overset{\overset{\displaystyle \boxed{C_6H_5}}{|}}{\underset{\underset{\displaystyle OH}{|}}{C}}-\boxed{C_6H_5}$

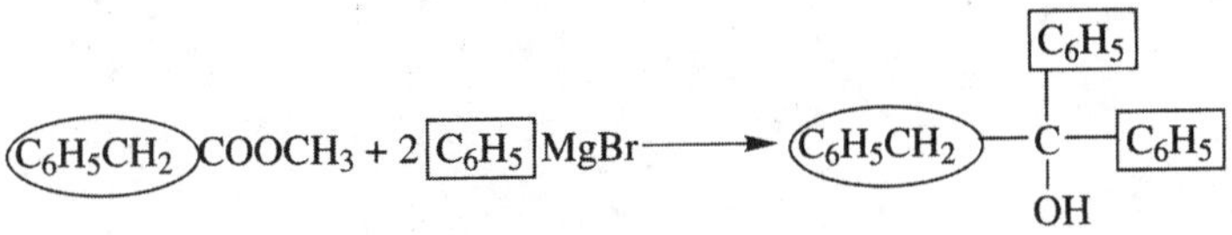

Problem 16.43 Methyl esters can be prepared on a small scale from RCOOH and CH_2N_2. Suggest a mechanism involving S_N2 displacement of N_2.

Diazomethane is a resonance hybrid:

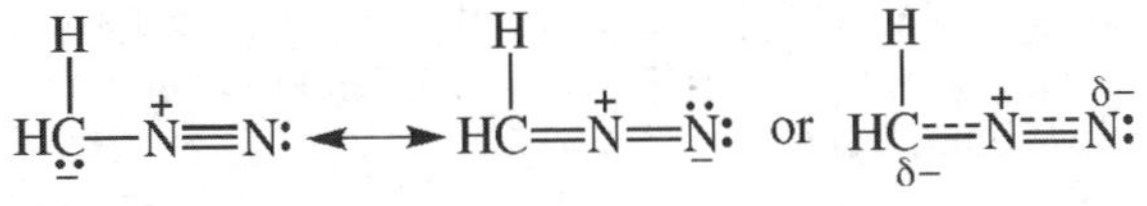

$$H\overset{\overset{\displaystyle H}{|}}{\underset{\bullet\bullet}{C}}-\overset{+}{N}\equiv N{:} \longleftrightarrow H\overset{\overset{\displaystyle H}{|}}{C}=\overset{+}{N}=\ddot{N}{:} \quad \text{or} \quad H\overset{\overset{\displaystyle H}{|}}{\underset{\delta-}{C}}\text{=-=}\overset{+}{N}\text{=-=}\overset{\delta-}{N}{:}$$

and we have

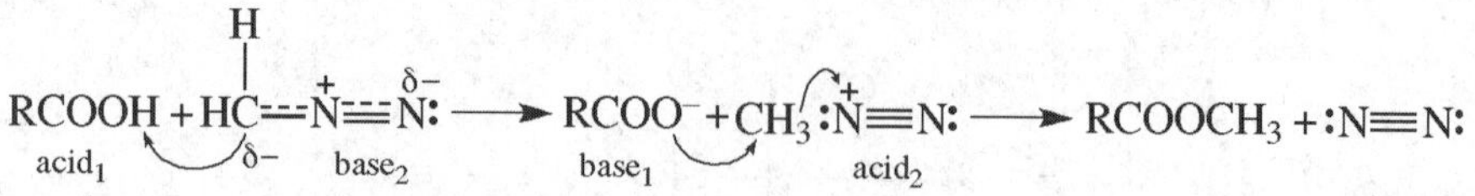

$$\underset{\text{acid}_1}{\text{RCOOH}} + \text{H}\overset{\text{H}}{\text{C}}\overset{\delta-}{=}\overset{+}{\text{N}}\equiv\overset{\delta-}{\text{N}}\text{:}\ (\text{base}_2) \longrightarrow \underset{\text{base}_1}{\text{RCOO}^-} + \text{CH}_3\text{:}\overset{+}{\text{N}}\equiv\text{N:}\ (\text{acid}_2) \longrightarrow \text{RCOOCH}_3 + \text{:N}\equiv\text{N:}$$

Problem 16.44 In living cells, alcohols are converted to acetate esters (acetylated) by the thiol ester CH_3COS—(CoA), acetyl coenzyme A. CoA is an abbreviation for a very complex piece. Illustrate this reaction using glycerol-1-phosphate.

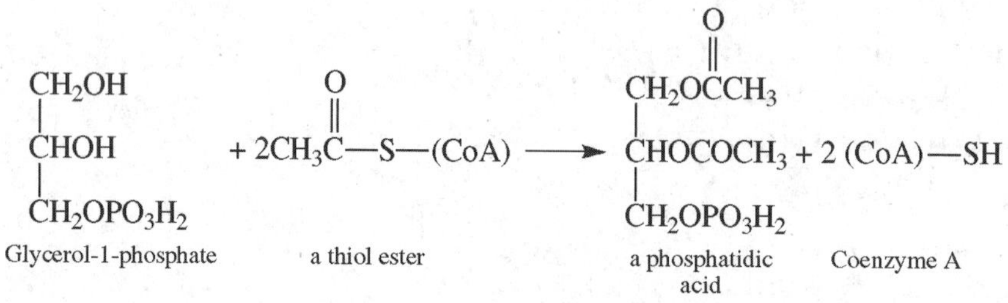

$$\underset{\text{Glycerol-1-phosphate}}{\begin{matrix}\text{CH}_2\text{OH}\\ |\\ \text{CHOH}\\ |\\ \text{CH}_2\text{OPO}_3\text{H}_2\end{matrix}} + \underset{\text{a thiol ester}}{2\text{CH}_3\overset{\text{O}}{\overset{\|}{\text{C}}}\text{—S—(CoA)}} \longrightarrow \underset{\text{a phosphatidic acid}}{\begin{matrix}\text{CH}_2\text{O}\overset{\text{O}}{\overset{\|}{\text{C}}}\text{CH}_3\\ |\\ \text{CHOCOCH}_3\\ |\\ \text{CH}_2\text{OPO}_3\text{H}_2\end{matrix}} + \underset{\text{Coenzyme A}}{2\ \text{(CoA)—SH}}$$

The reactivity of thiol esters, $\text{R}\underset{\text{O}}{\underset{\|}{\text{C}}}\text{—SR}'$ lies between that of anhydrides and esters.

Fats and Oils

Fats and oils are mixtures of esters of glycerol, $HOCH_2CHOHCH_2OH$, with acyl groups from carboxylic acids, usually with long carbon chains. These **triacylglycerols**, also called **triglycerides**, are types of **lipids** because they are naturally occurring and soluble only in nonpolar solvents. The acyl groups may be identical, or they may be different. Fats are *solid* esters of *saturated* carboxylic acids; oils, with the exception of palm and coconut, are esters of *unsaturated* acids with *cis* C=C bonds, which prevent close-packing of the molecules.

Problem 16.45 (*a*) Write a formula for a fat (found in butter) of butanoic acid. (*b*) Alkaline hydrolysis of a fat of a high-molecular-weight acid gives a carboxylate salt (a **soap**). Write the equation for the reactions of the fat of palmitic acid, $n\text{-}C_{15}H_{31}COOH$, with aqueous NaOH.

(*a*)
$$\begin{matrix} n\text{-C}_3\text{H}_7\text{—}\overset{\text{O}}{\overset{/\!/}{\text{C}}}\text{—O—CH}_2\\ \quad\quad\quad\quad\quad\quad |\\ n\text{-C}_3\text{H}_7\text{—}\overset{\text{O}}{\overset{/\!/}{\text{C}}}\text{—O—CH}\\ \quad\quad\quad\quad\quad\quad |\\ n\text{-C}_3\text{H}_7\text{—}\overset{\text{O}}{\overset{/\!/}{\text{C}}}\text{—O—CH}_2\end{matrix}$$

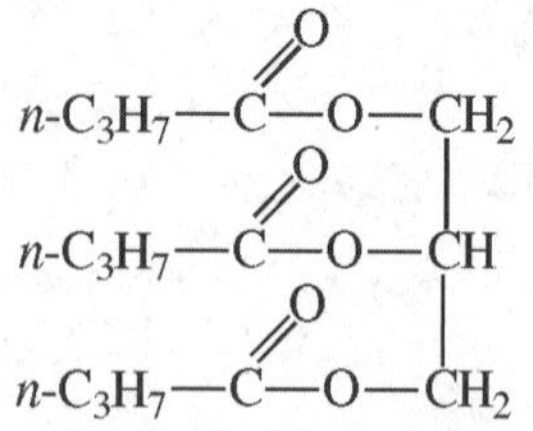

(*b*)
$$\begin{matrix} n\text{-C}_{15}\text{H}_{31}\text{COO—CH}_2\\ |\\ n\text{-C}_{15}\text{H}_{31}\text{COO—CH}\\ |\\ n\text{-C}_{15}\text{H}_{31}\text{COO—CH}_2\end{matrix} + 3\text{NaOH} \longrightarrow \underset{a\ soap}{3\ n\text{-C}_{15}\text{H}_{31}\text{COO}^-\text{Na}^+} + \text{HOCH}_2\text{CHOHCH}_2\text{OH}$$

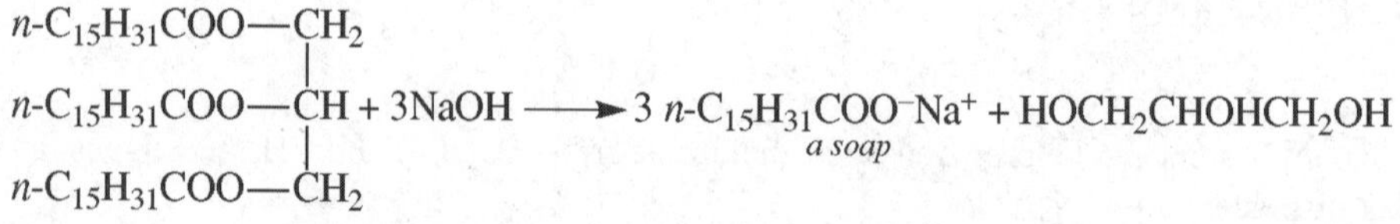

Problem 16.46 Two isomeric triglycerides are hydrolyzed to 1 mol of $C_{17}H_{33}COOH$ (A) and 2 mol of $C_{17}H_{35}COOH$ (B). Reduction of B yields *n*-stearyl alcohol, $CH_3(CH_2)_{16}CH_2OH$. Compound A adds 1 mol of H_2 to give B and is cleaved with O_3 to give nonanal, $CH_3(CH_2)_7CH{=}O$, and 9-oxononoic acid, $O{=}CH(CH_2)_7COOH$. What are the structures of the isomeric glycerides?

Compound B, which reduces to stearyl alcohol, is stearic acid, $CH_3(CH_2)_{16}COOH$. Compound A is a straight-chain unsaturated acid with one double bond, and its cleavage products suggest that the double bond is at C^9, making it oleic acid, $CH_3(CH_2)_7CH{=}CH(CH_2)_7COOH$. With 1 mol of oleic and 2 mol of stearic acid, there are two possible structures:

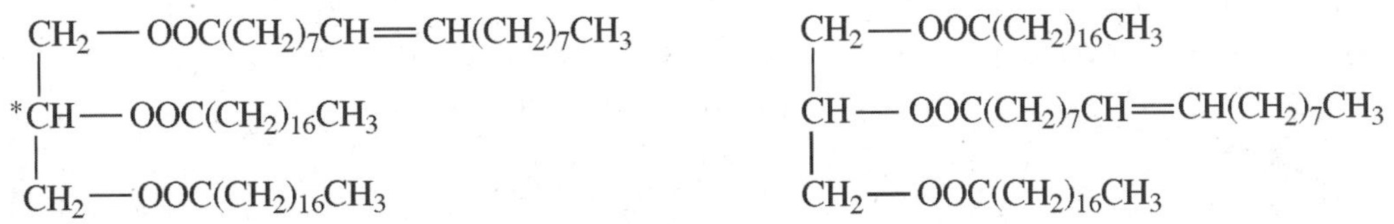

$$\begin{array}{l} CH_2{-}OOC(CH_2)_7CH{=}CH(CH_2)_7CH_3 \\ | \\ {}^{*}CH{-}OOC(CH_2)_{16}CH_3 \\ | \\ CH_2{-}OOC(CH_2)_{16}CH_3 \end{array} \qquad \begin{array}{l} CH_2{-}OOC(CH_2)_{16}CH_3 \\ | \\ CH{-}OOC(CH_2)_7CH{=}CH(CH_2)_7CH_3 \\ | \\ CH_2{-}OOC(CH_2)_{16}CH_3 \end{array}$$

1-Oleoyl-2,3-distearoylglycerol 2-Oleoyl-1,3-distearoylglycerol

Amides

In addition to transacylation reactions and the heating of ammonium carboxylates (Section 16.3), unsubstituted amides may be prepared by careful partial hydrolysis of nitriles:

$$H_2O + RC{\equiv}N \xrightarrow[\text{2. cold water}]{\text{1. cold } H_2SO_4} R{-}\underset{\underset{O}{\|}}{C}{-}NH_2$$

Amides are slowly hydrolyzed under either acidic or basic conditions. The mechanisms are those shown in Problem 16.32. Unsubstituted amides are converted to RCOOH with HNO_2.

$$R\underset{\underset{O}{\|}}{C}NH_2 \xrightarrow[(NaNO_2 + \text{aq. } HCl)]{HONO} R\underset{\underset{O}{\|}}{C}{-}OH + N_2$$

and are dehydrated to RCN with P_4O_{10}:

$$R\underset{\underset{O}{\|}}{C}{-}NH_2 \xrightarrow[-H_2O]{P_4O_{10}} \underset{\textit{a nitrile}}{RCN}$$

In the Hofmann degradation, $RCONH_2$ goes to RNH_2 (Section 18.2):

$$RCONH_2 + Br_2 + 4KOH \longrightarrow RNH_2 + K_2CO_3 + 2KBr + 2H_2O$$

Problem 16.47 Use the concepts of charge delocalization and resonance to account for the acidity of imides. (They dissolve in NaOH.)

The H on N of the imides is acidic because the negative charge on N of the conjugate base is delocalized to each O of the two C=O groups, thereby stabilizing the anion.

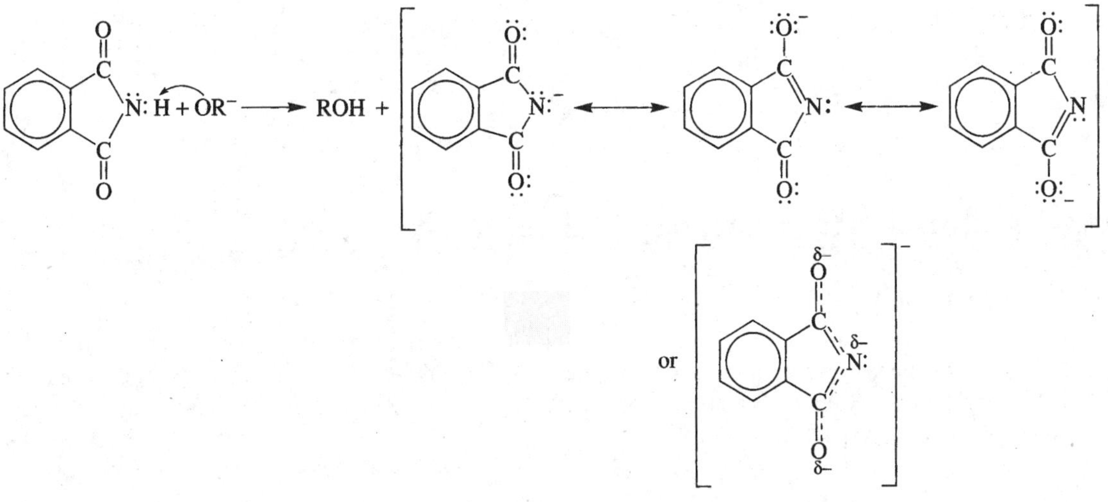

$$\text{(phthalimide) } \ddot{N}{:}H + OR^- \longrightarrow ROH + \left[\ddot{N}{:}^- \longleftrightarrow {:}\ddot{O}{:}^- \ldots \ddot{N}{:} \longleftrightarrow \ldots {:}\ddot{O}{:}^- \right]$$

or

$$\left[O^{\delta-} \ldots N^{\delta-} \ldots O^{\delta-} \right]^-$$

16.8 Summary of Carboxylic Acid Derivative Chemistry

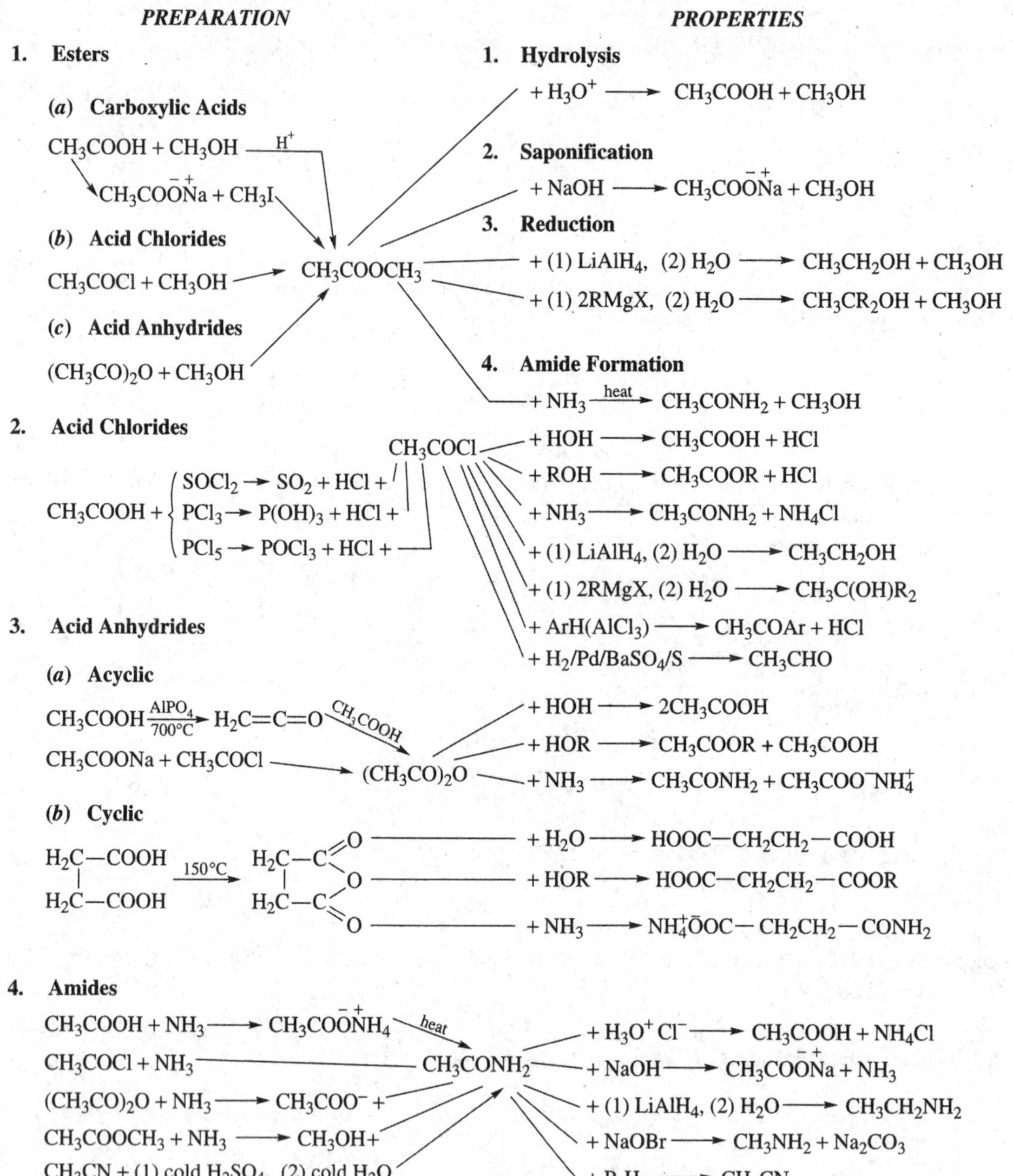

16.9 Analytical Detection of Acids and Derivatives

Neutralization Equivalent

Carboxylic acids dissolve in Na_2CO_3, thereby evolving CO_2. The **neutralization equivalent** or **equivalent weight** of a carboxylic acid is determined by titration with standard base; it is the number of grams of acid neutralized by one equivalent of base. If 40.00 mL of a 0.100 N base is needed to neutralize 0.500 g of an unknown acid, the number of equivalents of base is

$$\frac{40.00\,\text{mL}}{1000\,\text{mL/L}} \times 0.100\frac{\text{eq}}{\text{L}} = 0.00400\ \text{eq}$$

The neutralization equivalent is found by dividing the weight of acid by the number of equivalents of base:

$$\frac{0.500\ \text{g}}{0.00400\ \text{eq}} = 125\ \text{g/eq}$$

Problem 16.48 A carboxylic acid has GMW = 118, and 169.6 mL of 1.000 N KOH neutralizes 10.0 g of the acid. When heated, 1 mol of this acid loses 1 mol of H_2O without loss of CO_2. What is the acid?

The volume of base (0.1696 L) multiplied by the normality of the base (1.000 eq/L) gives 0.1696 as the number of equivalents of acid titrated. Since the weight of acid is 10.0 g, one equivalent of acid weighs

$$\frac{10.0\ \text{g}}{0.1696\ \text{eq}} = 59.0\ \text{g / eq}$$

Because the molecular weight of the acid (118) is twice this equivalent weight, there must be two equivalents per mole. The number of equivalents gives the number of ionizable hydrogens. This carboxylic acid has two COOH groups.

The two COOH's weigh 90 g, leaving 118 g − 90 g = 28 g (two C's and four H's) as the weight of the rest of the molecule. Since no CO_2 is lost on heating, the COOH groups must be on separate C's. The compound is succinic acid, $HOOC—CH_2—CH_2—COOH$.

Spectroscopic Methods

1. Infrared

In the H-bonded dimeric state [Problem 16.3(*b*)], RCOOH has a strong O—H stretching band at 2500–3000 cm^{-1}. The strong C=O absorptions are at 1700–1725 cm^{-1} for aliphatic and 1670–1700 cm^{-1} for aromatic acids. See Table 12.1 for key absorptions of acid derivatives (G = X, OH, OR′, OCOR, $\overset{|}{N}$—).

2. Nmr

The H of COOH is weakly shielded and absorbs downfield at δ = 10.5–12.0 ppm. See Tables 12.4 and 12.6 for proton and ^{13}C nmr chemical shifts, respectively.

3. Mass Spectra

Carboxylic acids and their derivatives are cleaved into stable acylium ions and free radicals.

$$R—\overset{\ddot{O}:}{\overset{\|}{C}}—G \xrightarrow{-e^-} R—\overset{\cdot\ddot{O}^+}{\overset{\|}{C}}\wr G \longrightarrow R—\overset{\ddot{O}^+}{\overset{|||}{C}} + \cdot G \quad (G = Cl, OH, OR, OCOR, \overset{|}{N}—)$$

Like other carbonyl compounds, carboxylic acids undergo β cleavage and γH transfer.

$$\underset{\text{H—RCH—CH}_2\text{—CH}_2}{}—\overset{:\ddot{O}}{\overset{\|}{C}}—OH \xrightarrow{-e^-} \overset{\gamma}{\text{H—RCH}}—\underset{\beta}{CH_2}\wr\underset{\alpha}{CH_2}—\overset{+\ddot{O}}{\overset{\|}{C}}—OH \longrightarrow RCH{=}CH_2 + \left[\cdot CH_2—\overset{HO^+}{\overset{\|}{C}}—OH\right]$$

16.10 Carbonic Acid Derivatives

Problem 16.49 The following carboxylic acids are unstable, and their decomposition products are shown in parentheses: carbonic acid, $(HO)_2C{=}O$ ($CO_2 + H_2O$); carbamic acid, H_2NCOOH ($CO_2 + NH_3$); and chlorocarbonic acid, ClCOOH (CO_2 + HCl). Indicate how the *stable* compounds below are derived from one or more of these unstable acids. Name those for which a common name is not given.

(*a*) $Cl_2C{=}O$ Phosgene
(*b*) $(H_2N)_2C{=}O$ Urea
(*c*) $ClC(=O)OCH_3$
(*d*) $(CH_3O)_2C{=}O$
(*e*) $H_2NC(=O){-}OCH_3$
(*f*) $HN{=}C{=}O$ Isocyanic acid

(*a*) The acid chloride of chlorocarbonic acid; (*b*) amide of carbamic acid; (*c*) ester of chlorocarbonic acid, methyl chlorocarbonate; (*d*) diester of carbonic acid, methyl carbonate; (*e*) ester of carbamic acid, methyl carbamate (called a **urethane**); (*f*) dehydrated carbamic acid.

Problem 16.50 What products are formed when 1 mol of urea reacts with (*a*) 1 mol, (*b*) a second mol of methyl acetate (or acetyl chloride)?

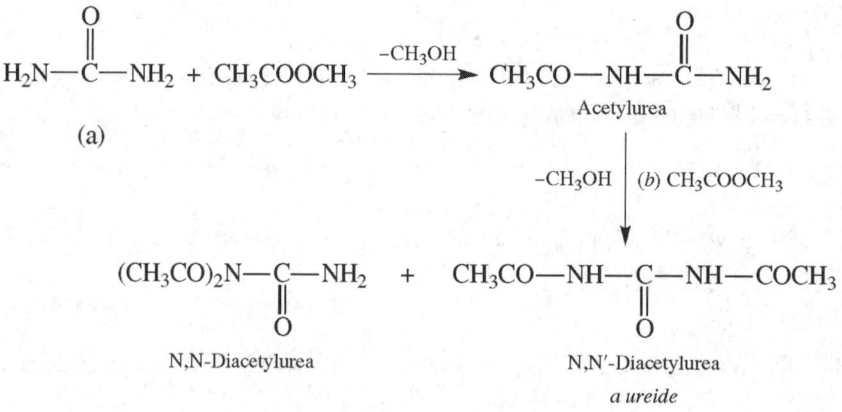

$$H_2N{-}C(=O){-}NH_2 + CH_3COOCH_3 \xrightarrow{-CH_3OH} CH_3CO{-}NH{-}C(=O){-}NH_2 \text{ (Acetylurea)}$$

(a)

$$\text{Acetylurea} \xrightarrow[-CH_3OH]{(b)\ CH_3COOCH_3} (CH_3CO)_2N{-}C(=O){-}NH_2 + CH_3CO{-}NH{-}C(=O){-}NH{-}COCH_3$$

N,N-Diacetylurea; N,N′-Diacetylurea, *a ureide*

Problem 16.51 Barbiturates are sedative-hypnotic varieties of 5,5-dialkyl substituted barbituric acids. Write the reaction for the formation of Veronal (5,5-diethylbarbituric acid) from the condensation of urea with diethylmalonic ester. [See Problem 17.11(*a*).]

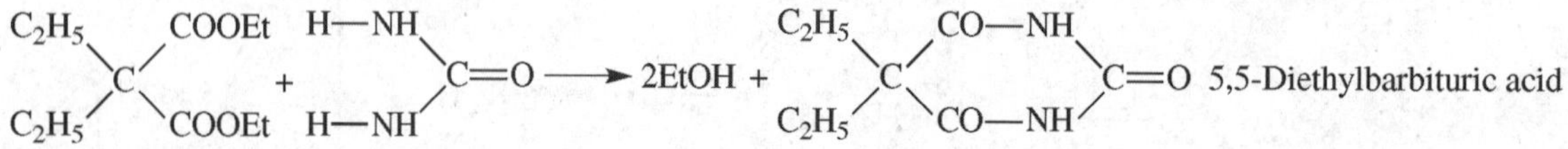

$$(C_2H_5)_2C(COOEt)_2 + (H{-}NH)_2C{=}O \longrightarrow 2EtOH + (C_2H_5)_2C(CO{-}NH)_2C{=}O$$

5,5-Diethylbarbituric acid

Pentobarbital is 5-ethyl-5-(1-methylbutyl)-barbituric acid.

16.11 Summary of Carbonic Acid Derivative Chemistry

PREPARATION | ***PROPERTIES***

1. Esters

(a) Phosgene

Cl—CO—Cl + 2HOR → $O{=}C(OR)_2$

(b) Alkyl Halide

$2RX + Ag_2CO_3$ → $O{=}C(OR)_2$

Properties of $O{=}C(OR)_2$:

1. Ammonolysis

+ NH_3 → $2ROH + O{=}C(NH_2)_2$

2. Saponification

+ $2NaOH$ → $2ROH + Na_2CO_3$

2. Acid Halides

(a) Chlorocarbonic Ester

Cl—CO—Cl + ROH → RO—CO—Cl

(b) Carbonyl Chloride

$CO + Cl_2$ → Cl—CO—Cl

3. Transacylate

RO—CO—Cl:

+ HOH → $ROH + CO_2 + HCl$

+ ROH → HCl + RO—CO—OR

+ NH_3 → NH_4Cl + RO—CO—NH_2

Cl—CO—Cl:

+ HOH → $2HCl + CO_2$

+ ROH → 2HCl + RO—CO—OR

+ NH_3 → $NH_4Cl + H_2N$—CO—NH_2

3. Carbamic Acid Derivatives

(a) Ester (Urethane)

RO—CO—OR + NH_3 → RO—CO—NH_2 (Alkyl Carbamate)

RO—CO—Cl + NH_3 → RO—CO—NH_2 (Alkyl Carbamate)

RO—CO—NH_2 + HOH → $ROH + CO_2 + NH_3$

(b) Chloride

Cl—CO—Cl + $NH_4^+Cl^-$ → H_2N—CO—Cl (Carbamyl Chloride) $\xrightarrow{60°C}$ H—N=C=O + HCl

4. Carbamide (Urea)

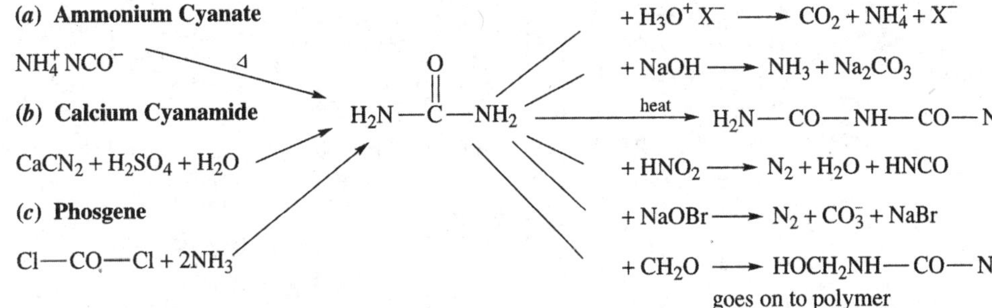

(a) Ammonium Cyanate

$NH_4^+NCO^-$ $\xrightarrow{\Delta}$ H_2N—C(=O)—NH_2

(b) Calcium Cyanamide

$CaCN_2 + H_2SO_4 + H_2O$ → H_2N—C(=O)—NH_2

(c) Phosgene

Cl—CO—Cl + $2NH_3$ → H_2N—C(=O)—NH_2

4. Miscellaneous

$+ H_3O^+X^-$ → $CO_2 + NH_4^+ + X^-$

+ NaOH → $NH_3 + Na_2CO_3$

$\xrightarrow{heat}$ H_2N—CO—NH—CO—NH_2

+ HNO_2 → $N_2 + H_2O$ + HNCO

+ NaOBr → $N_2 + CO_3^= $ + NaBr

+ CH_2O → $HOCH_2NH$—CO—NH_2 goes on to polymer

5. Ureides (Acylureas)

R—CO—X + H_2N—CO—NH_2 → HX + R—CO—NH—CO—NH_2

6. Alkylureas

$RNH_3^+X^- + K^+NCO^-$ → RNH—CO—NH_2

7. Guanidines (Iminoureas)

(a) Alkyl Orthocarbonates

$C(OR)_4 + NH_3$ → H_2N—C(=NH)—NH_2

Amine and Cyanamide

$RNH_2 + H_2N$—CN → RNH—C(=NH)—NH_2

16.12 Synthetic Condensation Polymers

Condensation polymers are prepared by reactions in which the monomeric units are joined by *intermolecular* elimination of small molecules such as water and alcohol. Among the most important kinds are **polyesters** and **polyamides. Polyurethanes** are addition polymers of acid derivatives.

Problem 16.52 Indicate the reactions involved and show the structures of the following condensation polymers obtained from the indicated reactants: (*a*) Nylon 66 from adipic acid and hexamethylene diamine: (*b*) Nylon 6 from ε-caprolactam; (*c*) Dacron from methyl terephthalate and ethylene glycol; (*d*) Glyptal from glycerol and terephthalic acid; (*e*) polyurethane from diisocyanates and ethylene glycol.

(*a*) Nylon 66 is a polyamide produced by reaction of both COOH groups of adipic acid with both NH_2 groups of hexamethylene diamine; —CONH— bonds are formed by H_2O elimination. The initial reaction gives a nylon salt, which is then heated.

$$HOOC(CH_2)_4COOH + H_2N(CH_2)_6NH_2 \xrightarrow[mer]{} {}^-OOC(CH_2)_4COO^- \ {}^+H_3N(CH_2)_6NH_3^+ \xrightarrow{heat}$$

$$-\underset{\underset{O}{\|}}{C}-(CH_2)_4-\underset{\underset{O}{\|}}{C}-NH-(CH_2)_6-NH-\underset{\underset{O}{\|}}{C}-(CH_4)_4-\underset{\underset{O}{\|}}{C}-NH-(CH_2)_6-NH-$$

Poly(hexamethylene adipamide)

(*b*) Nylon 6 is also a polyamide but is made from the monomer ε-caprolactam, which is a cyclic amide of ε-aminocaproic acid. Heat opens the lactam ring to give the amino acid salt, which forms amide bonds with other molecules by eliminating water.

$$\text{ε-Caprolactam} \xrightarrow[H_2O]{heat} H_3\overset{+}{N}-(CH_2)_5-\underset{\underset{O}{\|}}{C}-O^- \xrightarrow[-H_2O]{heat}$$

$$\overbrace{-NH-(CH_2)_5-\underset{\underset{O}{\|}}{C}}^{mer}\overbrace{-NH-(CH_2)_5-\underset{\underset{O}{\|}}{C}}^{mer}\overbrace{-NH-(CH_2)_5-\underset{\underset{O}{\|}}{C}-}^{mer}$$

Poly(6-aminohexanoic acid amide)

(*c*) Dacron is a condensation polyester formed by a transesterification reaction between dimethyl terephthalate and ethylene glycol.

$$HO-CH_2CH_2-OH + H_3COOC-C_6H_4-COOCH_3 + HO-CH_2CH_2-HO$$

Dimethyl terephthalate

$$+ H_3COOC-C_6H_4-COOCH_3 \xrightarrow[-CH_3OH]{heat}$$

$$-O-CH_2CH_2-O-\overbrace{\underset{\underset{O}{\|}}{C}-C_6H_4-\underset{\underset{O}{\|}}{C}-OCH_2CH_2-O}^{mer}-\underset{\underset{O}{\|}}{C}-C_6H_4-\underset{\underset{O}{\|}}{C}-$$

Poly(ethylene terephthalate)

(*d*) Glyptal is also a polyester condensation product, but glycerol ($HOCH_2CHOHCH_2OH$) produces a **cross-linked** thermosetting resin. In the first stage, a linear polymer is formed with the more reactive *primary* OH groups.

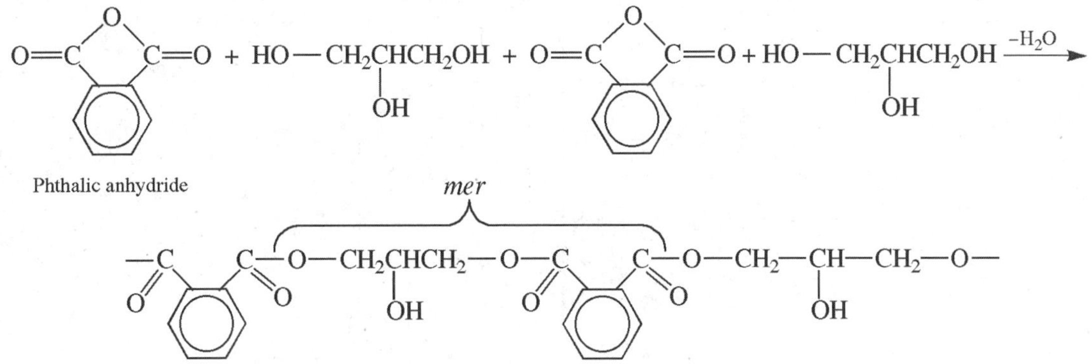

Phthalic anhydride + $HO—CH_2CHCH_2OH$ (OH) + Phthalic anhydride + $HO—CH_2CHCH_2OH$ (OH) $\xrightarrow{-H_2O}$

Phthalic anhydride

mer

$—C(=O)—C_6H_4—C(=O)—O—CH_2CH(OH)CH_2—O—C(=O)—C_6H_4—C(=O)—O—CH_2—CH(OH)—CH_2—O—$

The free 2° OH's are then cross-linked with more molecules of phthalic anhydride.

(*e*) Urethanes [Problem 16.49(*e*)] are made by the rapid exothermic reaction of an isocyanate with an alcohol or phenol.

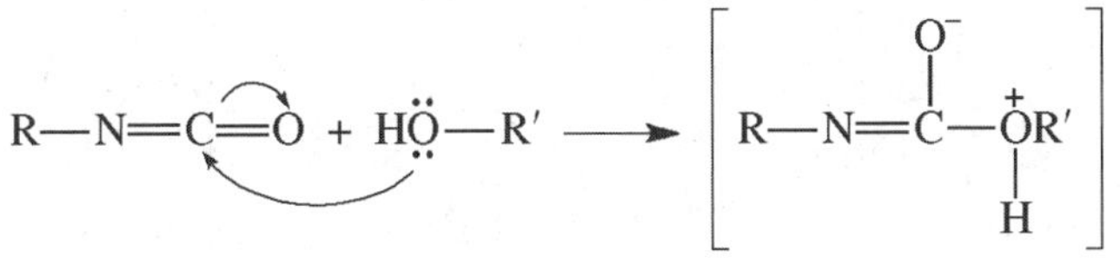

$$R—N{=}C{=}\ddot{O} + H\ddot{O}—R' \longrightarrow \left[R—N{=}\underset{\substack{|\\ }}{\overset{O^-}{\overset{|}{C}}}—\overset{+}{\underset{\substack{|\\H}}{O}}R' \right] \longrightarrow R—\overset{H}{\overset{|}{N}}—\overset{O}{\overset{\|}{C}}—OR'$$

Polyurethanes are formed from a diol (e.g., $HOCH_2CH_2OH$) and a diisocyanate, a compound with two $—N{=}C{=}O$ groups (example, toluene diisocyanate).

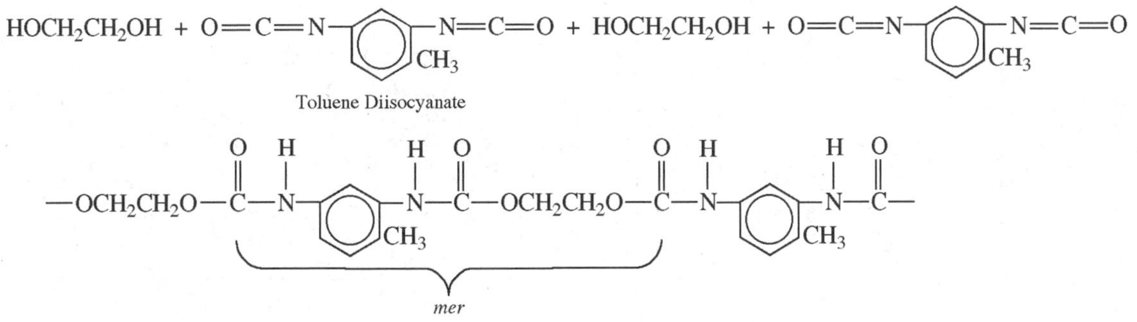

$$HOCH_2CH_2OH + O{=}C{=}N—C_6H_3(CH_3)—N{=}C{=}O + HOCH_2CH_2OH + O{=}C{=}N—C_6H_3(CH_3)—N{=}C{=}O$$

Toluene Diisocyanate

$$—OCH_2CH_2O—\overset{O}{\overset{\|}{C}}—\overset{H}{\overset{|}{N}}—C_6H_3(CH_3)—\overset{H}{\overset{|}{N}}—\overset{O}{\overset{\|}{C}}—OCH_2CH_2O—\overset{O}{\overset{\|}{C}}—\overset{H}{\overset{|}{N}}—C_6H_3(CH_3)—\overset{H}{\overset{|}{N}}—\overset{O}{\overset{\|}{C}}—$$

mer

16.13 Derivatives of Sulfonic Acids

Sulfonic acids, $R(Ar)SO_3H$, form derivatives similar to those of carboxylic acids (see Table 16.3). These are sulfonyl chlorides, sulfonates (esters), and sulfonamides. The transsulfonylation reactions are similar to the transacylation reactions, except that the ester and amide cannot be made directly from the acid. See Problem 13.16 for preparation of sulfonyl chlorides and esters.

Sulfonate esters can be prepared from optically active alcohols without inversion of configuration of the chiral carbinol C. The reason is that reaction involves cleavage of the H—O bond of the alcohol.

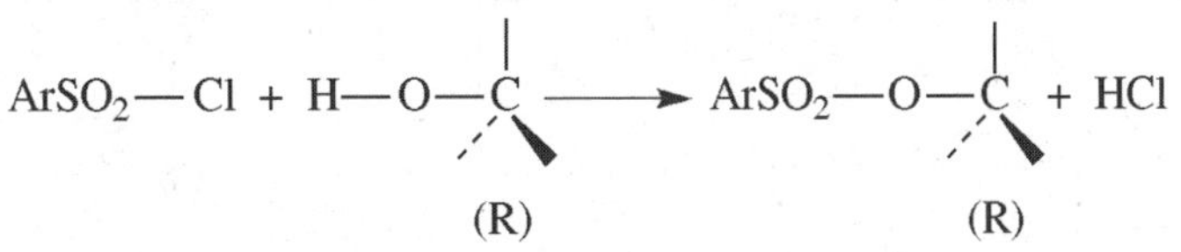

$$ArSO_2—Cl + H—O—C \longrightarrow ArSO_2—O—C + HCl$$

(R) (R)

TABLE 16.3 **Comparison of Sulfonic and Carboxylic Acid Chemistry**

	SULFONIC	CARBOXYLIC
Acids	Ar(R)SO_3H	RCO_2H
1. Acid strength	Strong	Weak
2. Formation of derivatives	Indirect ($ArSO_2Cl$)	Direct
3. Nucleophilic displacement on anion (occurs only with $ArSO_3^-$)	By OH^-, CN^-	None
4. Solubility in H_2O	Soluble	Insoluble, except acids or low MW
Esters	$ArSO_2OR$	R′COOR
1. Preparation	From $ArSO_2Cl$	From R′COOH, R′COCl or R′—C(=O)OC(=O)—R′
2. Hydrolysis with $H_2{}^{18}O$	$ArSO_3H + R^{18}OH$ Cleavage of alkyl—oxygen bond	$R'CO^{18}OH + ROH$ Cleavage of acyl—oxygen bond
3. Reaction with nucleophiles	At alkyl C with inversion (like RX)	At acyl C with retention or occasionally racemization of R. Intermediate is sp^3 and has an octet
Acid Chlorides	$ArSO_2Cl$	RCOCl
1. Formation of acids, esters, and amides	Slow; requires base	Rapid (ArCOCl requires base)
2. Reduction	To sulfinic acid, $ArSO_2H$ (Zn, HCl), and thiophenols, ArSH	To RCHO
Amides	$ArSO_2NH_2$	$RCONH_2$
1. Hydrolysis	Only by acids; slow	By acids or bases; rapid
2. Formation from acyl halides	Slow	Rapid
3. Acidity of H on N	Forms salts with OH^-	No salt formation

Reduction of sulfonyl chlorides with Zn and acid yields first sulfinic acids and then thiophenols.

$$C_6H_5SO_2Cl \xrightarrow{\text{Zn, HCl}} \underset{\text{Benzenesulfinic acid}}{C_6H_5SO_2H} \xrightarrow{\text{Zn, HCl}} C_6H_5SH$$

Problem 16.53 Name (*a*) $C_6H_5SO_2OCH_3$, (*b*) $C_6H_5SO_2NH_2$, (*c*) *p*-$BrC_6H_4CO_2Cl$.

(*a*) methyl benzenesulfonate, (*b*) benzenesulfonamide, (*c*) *p*-bromobenzenesulfonyl chloride (brosyl chloride)

Problem 16.54 Give the product formed when $PhSO_2Cl$ is treated with (*a*) phenol, (*b*) aniline, (*c*) water, (*d*) excess Zn and HCl.

(*a*) phenyl benzenesulfonate, $PhSO_2OPh$; (*b*) N-phenylbenzenesulfonamide, $PhSO_2NHPh$; (*c*) benzenesulfonic acid, $PhSO_2OH$, (*d*) thiophenol, C_6H_5SH.

Problem 16.55 Prepare (*a*) tosylamide (Ts = tosyl = *p*-$CH_3C_6H_4SO_2$—) from toluene, (*b*) PhCOOH from $PhSO_2Cl$, (*c*) *o*-methylthiophenol from $PhCH_3$.

(*a*) $C_6H_5CH_3 \xrightarrow{H_2SO_4} p\text{-}CH_3C_6H_4SO_3H \xrightarrow{PCl_5} p\text{-}CH_3C_6H_4SO_2Cl \xrightarrow{NH_3} p\text{-}CH_3C_6H_4SO_2NH_2$ (or $ClSO_3H$ directly from $C_6H_5CH_3$ to $p\text{-}CH_3C_6H_4SO_2Cl$)

(*b*) $PhSO_2Cl \xrightarrow{NaOH} PhSO_3Na \xrightarrow[\text{fuse}]{NaCN} PhCN \xrightarrow{H_3O^+} PhCOOH$

(*c*) $PhCH_3 \xrightarrow[0°C]{H_2S_2O_7} \underset{\text{rate-controlled product}}{o\text{-}CH_3C_6H_4SO_3H} \xrightarrow{PCl_5} o\text{-}CH_3C_6H_4SO_2Cl \xrightarrow{\text{Zn, HCl}} o\text{-}CH_3C_6H_4SH$

Problem 16.56 Write structures for compounds (A) through (C) in the synthesis of saccharin.

Toluene ($C_6H_5CH_3$) $\xrightarrow{(A)}$ *o*-$CH_3C_6H_4SO_2Cl$ $\xrightarrow{NH_3}$ (B) $\xrightarrow{(C)}$ Saccharin (benzene ring fused with C(=O)–NH–SO_2)

Saccharin

(A) $ClSO_3H$ (B) *o*-$CH_3C_6H_4SO_2NH_2$ (C) $KMnO_4$

Problem 16.57 Prepare methyl 2-methylpropanesulfonate from 1-chloro-2-methylpropane.

Form the C—S bond by using the hydrogen sulfite anion as the nucleophile in an S_N2 reaction. S: is a more nucleophilic site than is O^-, even though O has a negative charge.

$$Na^+\,^-O\text{—}\ddot{S}(=O)OH + (CH_3)_2CHCH_2Cl \longrightarrow (CH_3)_2CHCH_2SO_3^-Na^+$$

$$\downarrow PCl_5$$

$$(CH_3)_2CHCH_2S(=O)_2\text{—}OCH_3 \xleftarrow{CH_3O^-} (CH_3)_2CHCH_2S(=O)_2\text{—}Cl$$

Notice that the strong base, CH_3O^-, rather than the much weaker base, CH_3OH, must be used to form the ester. This is an important difference between the very reactive acyl chloride (RCOCl) and the much less reactive sulfonyl chloride (RSO_2Cl).

SUPPLEMENTARY PROBLEMS

Problem 16.58 Write formulas for each of the following: (*a*) phenylacetic acid, (*b*) phenylethanoic acid, (*c*) 2-methylpropenoic acid, (*d*) (*E*)-butenedioic acid, (*e*) ethanedioic acid, (*f*) 3 -methylbenzenecarboxylic acid.

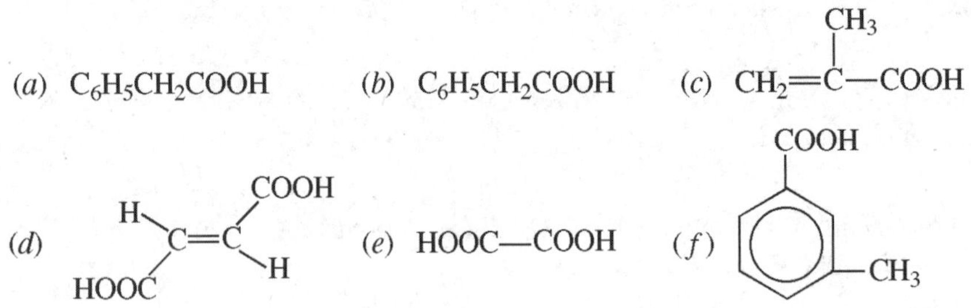

(*a*) $C_6H_5CH_2COOH$ (*b*) $C_6H_5CH_2COOH$ (*c*) $CH_2{=}C(CH_3){-}COOH$

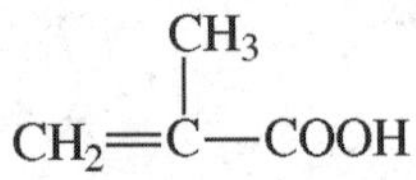

(*d*) H(HOOC)C=C(COOH)H (*e*) HOOC—COOH (*f*) benzene ring with COOH and CH_3 (meta)

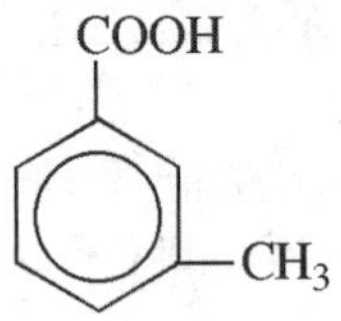

Problem 16.59 Name the following compounds:

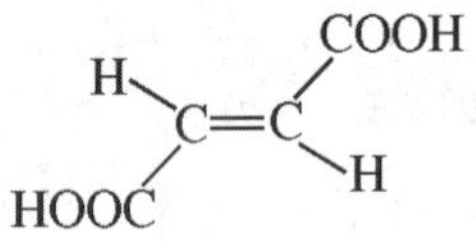

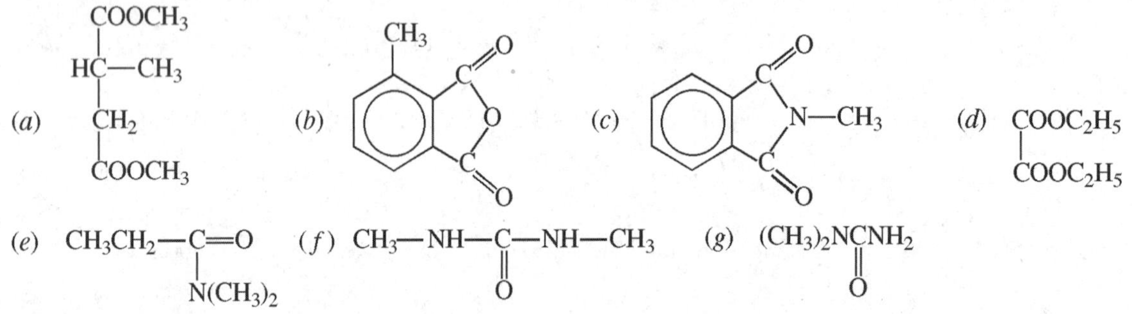

(*a*) $CH_3OOC{-}CH(CH_3){-}CH_2{-}COOCH_3$ (*b*) 3-methylphthalic anhydride structure (*c*) N-methylphthalimide structure (*d*) $COOC_2H_5{-}COOC_2H_5$

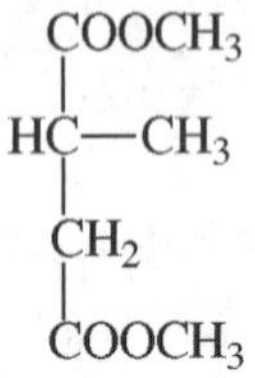

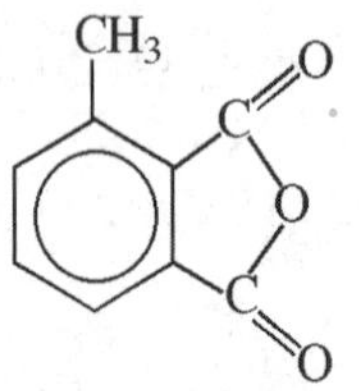

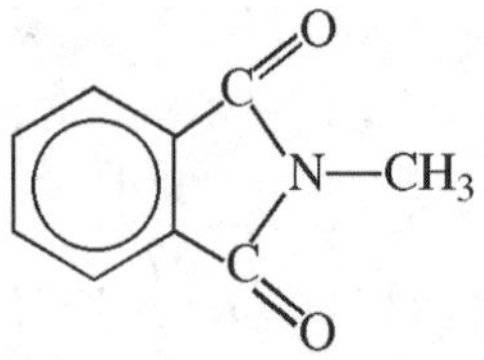

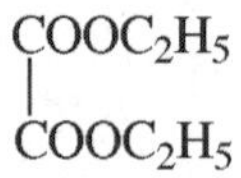

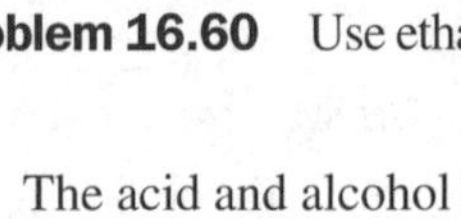

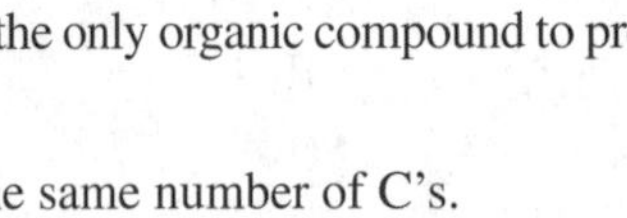

(*e*) $CH_3CH_2{-}C({=}O){-}N(CH_3)_2$ (*f*) $CH_3{-}NH{-}C({=}O){-}NH{-}CH_3$ (*g*) $(CH_3)_2NC({=}O)NH_2$

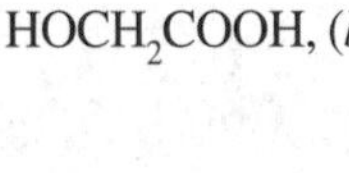

(*a*) dimethyl α-methylsuccinate (dimethyl 2-methylbutanedioate), (*b*) 3-methylphthalic anhydride, (*c*) N-methylphthalimide, (*d*) diethyl oxalate (diethyl ethanedioate), (*e*) *N,N*-dimethylpropanamide, (*f*) N, N′-dimethylurea, (*g*) *N,N*-dimethylurea.

Problem 16.60 Use ethanol as the only organic compound to prepare (*a*) $HOCH_2COOH$, (*b*) $CH_3CHOHCOOH$.

(*a*) The acid and alcohol have the same number of C's.

$$CH_3CH_2OH \xrightarrow[H^+]{KMnO_4} CH_3COOH \xrightarrow{Cl_2/P} CH_2ClCOOH \xrightarrow[2.\ H^+]{1.\ OH^-} HOCH_2COOH$$

(*b*) Now the acid has one more C. A one-carbon "step-up" is needed before introducing the OH.

$$CH_2CH_5OH \xrightarrow{PCl_3} C_2H_5Cl \xrightarrow{KCN} C_2H_5CN \xrightarrow{H_3O^+} CH_3CH_2COOH \xrightarrow[\text{part }(a)]{\text{as in}} CH_3CHOHCOOH$$

Problem 16.61 Prepare α-methylbutyric acid from ethanol.

Introduce COOH of $CH_3CH_2CH(CH_3)COOH$ through a Cl and build up the needed 4-carbon skeleton

$$\mathrm{C{-}C{-}\underset{\displaystyle Cl}{\underset{|}{C}}{-}C}$$

by a Grignard reaction.

(1) $C_2H_5OH \xrightarrow{PBr_3} C_2H_5Br \xrightarrow[\text{ether}]{Mg} C_2H_5MgBr$ (use in step 3)

(2) $C_2H_5OH \xrightarrow[\Delta,\ Cu]{\text{oxid.}} CH_3CH{=}O$ (use in step 3)

(3) $CH_3CH{=}O + C_2H_5MgBr \longrightarrow \underset{\displaystyle ^-O(MgBr)^+}{CH_3\underset{|}{C}HCH_2CH_3} \xrightarrow{H_3O^+} \underset{\displaystyle OH}{CH_3\underset{|}{C}HCH_2CH_3} \xrightarrow{PCl_3} \underset{\displaystyle Cl}{CH_3\underset{|}{C}HCH_2CH_3} \xrightarrow{Mg} \underset{\displaystyle MgCl}{CH_3\underset{|}{C}HCH_2CH_3} \xrightarrow[2.\ H_3O^+]{1.\ CO_2} \underset{\displaystyle COOH}{CH_3\underset{|}{C}HCH_2CH_3}$

Problem 16.62 Convert 2-chlorobutanoic acid into 3-chlorobutanoic acid.

2-Chlorobutanoic acid is dehydrohalogenated to 2-butenoic acid and HCl is added. H^+ adds to give a β-carbocation which bonds to Cl^- to form the β-chloroacid.

$$\underset{\displaystyle Cl}{CH_3CH_2\underset{|}{C}H}{-}COOH \xrightarrow{\text{alc. KOH}} CH_3CH{=}CH{-}COOH \xrightarrow{\boxed{H^+}} CH_3\overset{+}{C}H\overset{\boxed{H}}{\overset{|}{C}}HCOOH \ (\text{a } \beta\text{-}R^+) \xrightarrow{Cl^-} \underset{\displaystyle Cl}{CH_3\underset{|}{C}H}{-}CH_2{-}COOH$$

The α-carbocation is not formed because its + charge would be next to the positive C of COOH.

$$CH_3\overset{\boxed{H}}{\overset{|}{C}}H\overset{+}{C}H\overset{O^{\delta-}}{\overset{\|}{C}}{}^{\delta+}{-}OH$$

Problem 16.63 Write structures for the compounds (A) through (D).

$$\underset{\displaystyle Br}{\underset{|}{CH_2}}{-}COOC_2H_5 \xrightarrow{Zn} (A) \xrightarrow[2.\ H_2O]{1.\ \text{acetone}} (B) \xrightarrow[2.\ \Delta]{1.\ H_3O^+} (C) \xrightarrow{H_2/Pt} (D)$$

This is a Reformatsky reaction (see Problem 15.43).

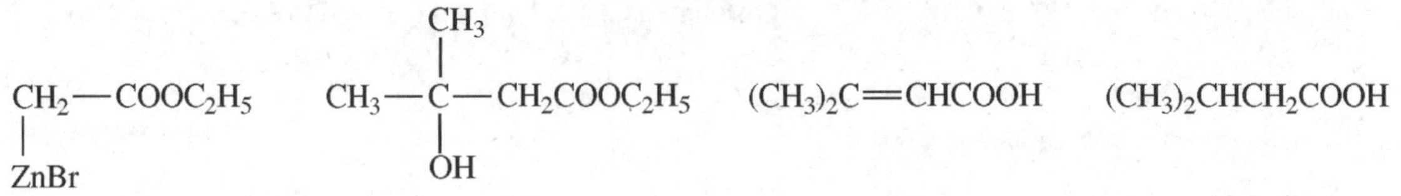

$$\underset{\text{(A)}}{\underset{\text{ZnBr}}{\overset{|}{CH_2}}\text{—}COOC_2H_5} \qquad \underset{\text{(B)}}{CH_3\text{—}\overset{\overset{CH_3}{|}}{\underset{\underset{OH}{|}}{C}}\text{—}CH_2COOC_2H_5} \qquad \underset{\text{(C)}}{(CH_3)_2C{=}CHCOOH} \qquad \underset{\text{(D)}}{(CH_3)_2CHCH_2COOH}$$

Problem 16.64 Use simple, rapid, test tube reactions to distinguish among hexane, hexanol, and hexanoic acid.

Only hexanoic acid liberates CO_2 from aqueous Na_2CO_3. Na reacts with hexanol to liberate H_2. Hexane is inert.

Problem 16.65 Describe the electronic effect of C_6H_5 on acidity, if the acid strengths of C_6H_5COOH and HCOOH are 6.3×10^{-5} and 1.7×10^{-4}, respectively.

The weaker acidity of C_6H_5COOH shows that the electron-releasing resonance effect of C_6H_5 outweighs its electron-attracting inductive effect.

Problem 16.66 What compounds are formed on heating (*a*) 1,2,2-cyclohexanetricarboxylic acid, and (*b*) 1,1,2-cyclobutanetricarboxylic acid?

(*a*) 1,2,2-Cyclohexanetricarboxylic acid (COOH, COOH, COOH) $\xrightarrow[-CO_2]{\Delta}$ *cis*- and *trans*-Cyclohexane-1,2-dicarboxylic acid (COOH, COOH) $\xrightarrow[-H_2O]{\Delta}$ *cis*- and *trans*-anhydride

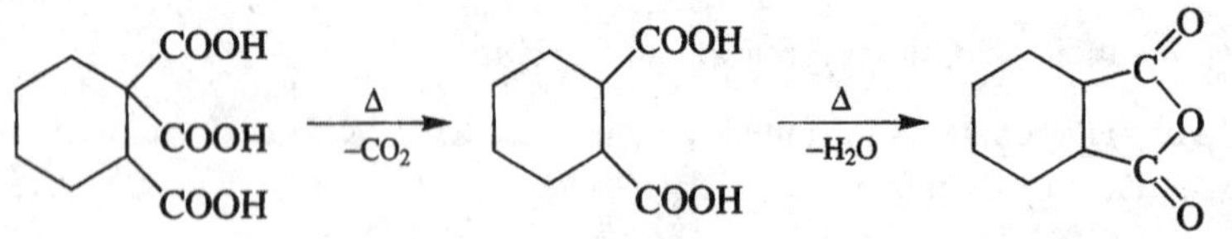

(*b*) 1,1,2-Cyclobutanetricarboxylic acid (COOH, COOH, COOH) $\xrightarrow[-CO_2,\ -H_2O]{\Delta}$ *trans*-1,2-Cyclobutanedicarboxylic acid (COOH, COOH) + *cis*-anhydride

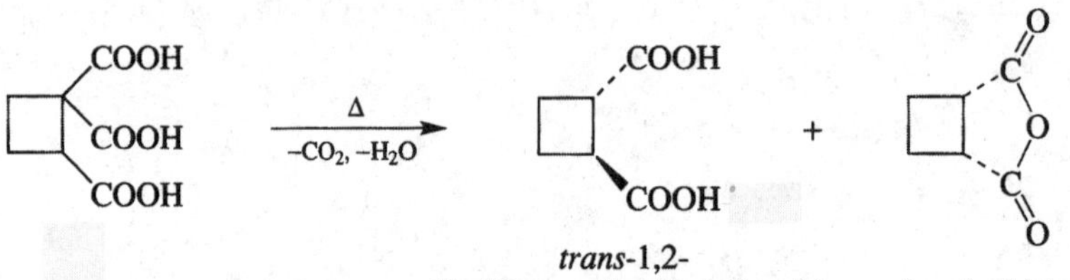

The *trans*-dicarboxylic acid cannot form the anhydride because a five- and a four-membered ring cannot be fused *trans*.

Problem 16.67 Write structural formulas for the products formed from reaction of δ-valerolactone with: (*a*) $LiAlH_4$, then H_2O; (*b*) NH_3; (*c*) CH_3OH and H_2SO_4 catalyst.

(*a*) $HOCH_2CH_2CH_2CH_2CH_2OH$, (*b*) $HOCH_2CH_2CH_2CH_2CONH_2$, (*c*) $HOCH_2CH_2CH_2CH_2COOCH_3$.

Problem 16.68 Prepare the **mixed anhydride** of acetic and propionic acids.

Mixed anhydrides:

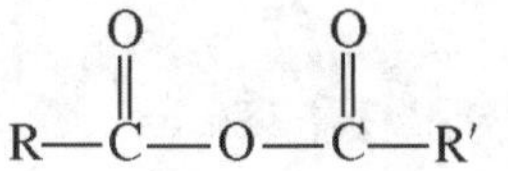

$$R\text{—}\overset{\overset{O}{\|}}{C}\text{—}O\text{—}\overset{\overset{O}{\|}}{C}\text{—}R'$$

are made by reacting the acid chloride of one of the acid portions with the carboxylate salt of the other. Use CH_3COCl and CH_3CH_2COONa, or CH_3COONa and CH_3CH_2COCl. For example:

$$CH_3{-}\overset{O}{\overset{\|}{C}}{-}Cl + Na^{+}{}^{-}O{-}\overset{O}{\overset{\|}{C}}{-}CH_2CH_3 \longrightarrow CH_3{-}\overset{O}{\overset{\|}{C}}{-}O{-}\overset{O}{\overset{\|}{C}}{-}CH_2CH_3 + Na^{+}Cl^{-}$$

Problem 16.69 Use $^{14}CH_3CH_2OH$ to synthesize $^{14}CH_3CONH_2$.

$$^{14}CH_3CH_2OH \xrightarrow[KMnO_4]{oxid.} {}^{14}CH_3COOH \xrightarrow{PCl_5} {}^{14}CH_3COCl \xrightarrow{NH_3} {}^{14}CH_3CONH_2$$

Problem 16.70 Distinguish by chemical tests (*a*) CH_3COCl from $(CH_3CO)_2O$, (*b*) nitrobenzene from benzamide.

(*a*) With H_2O, CH_3COCl liberates HCl, which is detected by giving a white precipitate of AgCl on adding $AgNO_3$. (*b*) Refluxing the amide with aqueous NaOH releases NH_3, detected by odor and with moist litmus or pH paper.

Problem 16.71 Name the main organic product(s) formed in the following reactions:

(*a*) $C_6H_5COOCH_3$ + excess $CH_3CH_2CH_2CH_2MgBr$, then H_3O^+
(*b*) $(CH_3)_2CHCH_2CH_2OH + C_6H_5COCl$
(*c*) $HCOOCH_2(CH_2)_4CH_3 + NH_3$
(*d*) $CH_3CH_2CH_2COCl + CH_3CH_2COONa$
(*e*) $C_6H_5COBr + 2C_6H_5MgBr$, then H_3O^+

(*a*) $C_6H_5C(OH)(CH_2CH_2CH_3)_2$
5-Phenyl-5-nonanol

(*b*) $C_6H_5COOCH_2CH_2CH(CH_3)_2$
Isoamyl benzoate

(*c*) $HCONH_2 + HO(CH_2)_5CH_3$
Formamide 1-Hexanol

(*d*) $CH_3CH_2CH_2{-}\overset{O}{\overset{\|}{C}}{-}O{-}\overset{O}{\overset{\|}{C}}{-}CH_2CH_3$
Butyric propionic anhydride

(*e*) $(C_6H_5)_3COH$
Triphenylcarbinol

Problem 16.72 Name and write the structures of the main organic products: (*a*) $H_2C{=}CHCH_2I$ heated with NaCN; (*b*) $CH_3CH_2CONH_2$ heated with P_4O_{10}; (*c*) *p*-iodobenzyl bromide reacted with CH_3COOAg; (*d*) nitration of benzamide.

(*a*) allyl cyanide or 3-butenenitrile, $CH_2{=}CH{-}CH_2{-}CN$; (*b*) propionitrile or ethyl cyanide, CH_3CH_2CN (an intramolecular dehydration); (*c*) *p*-iodobenzyl acetate; *p*-$IC_6H_4CH_2OCOCH_3$; (*d*) *m*-nitrobenzamide, *m*-$NO_2C_6H_4CONH_2$ (carboxylic acid derivatives orient *meta* during electrophilic substitution).

Problem 16.73 Give steps for the following preparations: (*a*) 1-phenylpropane from β-phenylpropionic acid, (*b*) β-benzoylpropionic acid from benzene and succinic acid.

(*a*) The net change is COOH $\longrightarrow$ CH_3, a reduction.

$$\underset{\beta\text{-Phenylpropionic acid}}{PhCH_2CH_2COOH} \xrightarrow{LiAlH_4} PhCH_2CH_2CH_2OH \xrightarrow[-H_2O]{H_2SO_4} PhCH_2CH{=}CH_2 \xrightarrow{H_2/Pt} \underset{\text{1-Phenylpropane}}{PhCH_2CH_2CH_3}$$

(*b*)

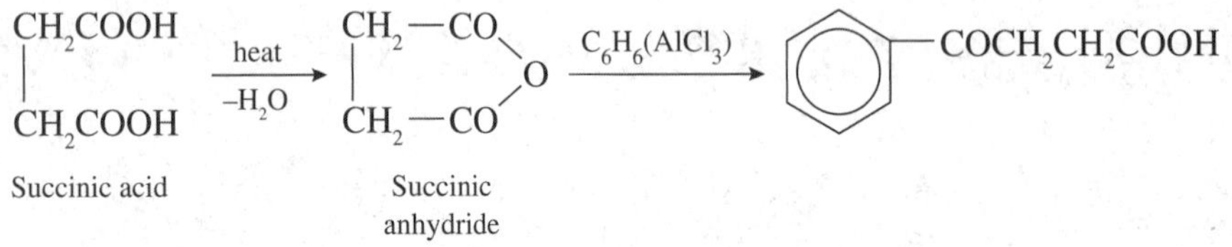

$$\underset{\text{Succinic acid}}{\begin{matrix} CH_2COOH \\ | \\ CH_2COOH \end{matrix}} \xrightarrow[-H_2O]{heat} \underset{\text{Succinic anhydride}}{\begin{matrix} CH_2{-}CO \\ | \quad\quad O \\ CH_2{-}CO \end{matrix}} \xrightarrow{C_6H_6(AlCl_3)} C_6H_5{-}COCH_2CH_2COOH$$

Problem 16.74 Identify the substances (A) through (E) in the sequence

$$C_6H_5COOH \xrightarrow{PCl_5} (A) \xrightarrow{(B)} C_6H_5CONH_2 \xrightarrow{P_4O_{10}} (C) \xrightarrow{H_2/Ni} (D) \xrightarrow{(E)} C_6H_5CH_2NH{-}\overset{\overset{\displaystyle O}{\|}}{C}{-}NHCH_2C_6H_5$$

(A) C_6H_5COCl, (B) NH_3, (C) C_6H_5CN, (D) $C_6H_5CH_2NH_2$, (E) $COCl_2$.

Problem 16.75 Give the products from reaction of benzamide, $PhCONH_2$, with (*a*) $LiAlH_4$, then H_3O^+; (*b*) P_4H_{10}; (*c*) hot aqueous NaOH; (*d*) hot aqueous HCl.

(*a*) $PhCH_2NH_2$; (*b*) PhCN, benzonitrile (phenyl cyanide); (*c*) PhCOONa + NH_3; (*d*) PhCOOH + NH_4Cl.

Problem 16.76 Suggest a synthesis of $(CH_3)_3CCH{=}CH_2$ from $(CH_3)_3CCHOHCH_3$.

An attempt to dehydrate the alcohol directly would lead to rearrangement of the R^+ intermediate. The major product would be $(CH_3)_2C{=}C(CH_3)_2$. To avoid this, pyrolyze the acetate ester of this alcohol.

$$CH_3COCl + (CH_3)_3CCHOHCH_3 \longrightarrow \underset{\displaystyle OCOCH_3}{(CH_3)_3C\underset{|}{C}HCH_3} \xrightarrow{\Delta} \text{product}$$

Problem 16.77 Assign numbers from (1) for LEAST to (3) for MOST to show the relative ease of acid-catalyzed esterification of:

	I	II	III
(*a*) $CH_3CH_2CH_2OH$ by:	C_6H_5COOH	2,4,6-$(CH_3)_3C_6H_2COOH$	2,4-$(CH_3)_2C_6H_3COOH$
(*b*) CH_3CH_2OH by:	$CH_3{-}CH(CH_3){-}COOH$	CH_3CH_2COOH	$(CH_3)_3CCOOH$
(*c*) C_6H_5COOH by:	$CH_3{-}C(CH_3)(OH){-}CH_2CH_3$	$CH_3CHOHCH_2CH_3$	$CH_3CH_2CH_2OH$

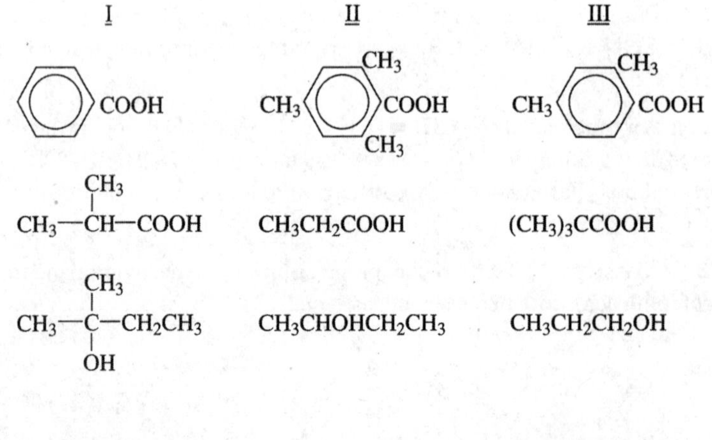

Steric factors are chiefly responsible for relative reactivities.

	I	II	III
(*a*)	3	1	2
(*b*)	2	3	1
(*c*)	1	2	3

Problem 16.78 The ^{18}O of $R^{18}OH$ appears in the ester, not in the water, when the alcohol reacts with a carboxylic acid. Offer a mechanism consistent with this finding.

The O from ROH must bond to the C of COOH and the OH of COOH ends up in H_2O. The acid catalyst protonates the O of C═O to enhance nucleophilic attack by ROH.

$$R-\overset{O}{\overset{\|}{C}}-OH \xrightarrow{H^+} R-\overset{^+OH}{\overset{\|}{C}}-OH \xrightarrow{R'^{18}OH} R-\underset{\underset{R'}{|}}{\underset{^{18}\overset{+}{O}H}{|}}{\overset{OH}{\overset{|}{C}}}-OH \longrightarrow R-\underset{\underset{R'}{|}}{\underset{^{18}O}{|}}{\overset{OH}{\overset{|}{C}}}-\overset{+}{O}H_2 \longrightarrow R-\overset{O}{\overset{\|}{C}}-{}^{18}O-R' + H_3O^+$$

Problem 16.79 Show how phosgene:

$$Cl-\underset{\underset{O}{\|}}{C}-Cl$$

is used to prepare (*a*) urea, (*b*) methyl carbonate, (*c*) ethyl chlorocarbonate, (*d*) ethyl N-ethylcarbamate (a urethane), (*e*) ethyl isocyanate ($C_2H_5-N=C=O$).

(*a*) $COCl_2 \xrightarrow{NH_3} NH_2CONH_2$

(*b*) $COCl_2 \xrightarrow{2CH_3OH} CH_3-O-\underset{\underset{O}{\|}}{C}-O-CH_3$

(*c*) $COCl_2 \xrightarrow{C_2H_5OH \text{ (1 mole)}} C_2H_5O-\underset{\underset{O}{\|}}{C}-Cl$

(*d*) $COCl_2 \xrightarrow{C_2H_5NH_2} C_2H_5\overset{H}{\overset{|}{N}}-\underset{\underset{O}{\|}}{C}-Cl \xrightarrow{C_2H_5OH} C_2H_5\overset{H}{\overset{|}{N}}-\underset{\underset{O}{\|}}{C}-OC_2H_5$

(*e*) $COCl_2 \xrightarrow{C_2H_5NH_2} C_2H_5\overset{H}{\overset{|}{N}}-\underset{\underset{O}{\|}}{C}-Cl \xrightarrow{\text{heat}} C_2H_5-N=C=O$

Problem 16.80 What carboxylic acid (A) has a neutralization equivalent (NE) of 52 and decomposes on heating to yield CO_2 and a carboxylic acid (B) with an NE of 60?

The NE of a carboxylic acid is its equivalent weight (molecular weight divided by the number of COOH groups). Since CO_2 is lost on heating, there are at least 2 COOH's in acid (A), and loss of CO_2 can produce a monocarboxylic acid, (B). Since one mole of COOH weighs 45 g, the rest of (B) weighs 60 g − 45 g = 15 g, which is a CH_3 group.

(B) is CH_3COOH, and (A) is malonic acid ($HOOC—CH_2—COOH$), whose NE is

$$\frac{104 \text{ g}}{2 \text{ eq}} = 52 \text{ g/eq}$$

Problem 16.81 An acyclic compound, $C_6H_{12}O_2$, has strong ir bands at 1740, 1250, and 1060 cm^{-1}, and no bands at frequencies greater than 2950 cm^{-1}. The nmr spectrum has two singlets at $\delta = 3.4$ (1 H) and $\delta = 1.0$ (3 H). What is the compound?

The one degree of unsaturation is due to a carbonyl group, indicated by the ir band at 1740 cm^{-1}. The lack of bands above 2950 cm^{-1} shows the absence of an OH group. Hence, the compound is not an alcohol or a carboxylic acid. That the compound is probably an ester is revealed by the ir bands at 1250 and 1060 cm^{-1} (C—O stretch). Two singlets in the nmr means two kinds of H's. The integration of 1 : 3 means the 12 H's are in the ratio of 3 : 9. The signal at $\delta = 3.4$ indicates a CH_3—O group. The nine equivalent protons at $\delta = 1.0$ are present in three CH_3's not attached to an electron-withdrawing group. A *t*-butyl group, $(CH_3)_3C$—, fits these requirements. The compound is $(CH_3)_3CCOOCH_3$, methyl trimethylacetate (methyl pivalate).

Problem 16.82 Predict the base (most prominent) peak in the mass spectrum of the compound in Problem 16.81.

Parent ions of esters resemble those of other acid derivatives and carboxylic acids in that they cleave into an acylium ion.

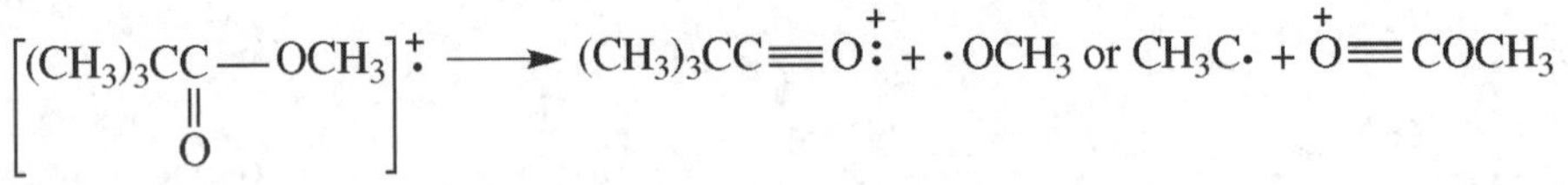

$$\left[(CH_3)_3C\overset{\displaystyle O}{\overset{\|}{C}}—OCH_3\right]^{\dot{+}} \longrightarrow (CH_3)_3CC{\equiv}\overset{+}{O}\!: + \cdot OCH_3 \text{ or } CH_3C\cdot + \overset{+}{O}{\equiv}COCH_3$$

The base peak should be *m/e* = 85 or 59.

Problem 16.83 Identify the compound $C_9H_{11}O_2SCl$ (A), and also compounds (B), (C), and (D), in the following reactions:

$$AgCl(s) \xleftarrow[\text{rapid}]{AgNO_3} \boxed{C_9H_{11}O_2SCl\ (A)} \xrightarrow[2.\ H_3O^+]{1.\ NaOH} \text{water solutble (B)} \xrightarrow[\Delta]{H_3O^+}$$

$$C_9H_{12}\ (C) \xrightarrow[Fe]{Br_2} \text{one monobromo derivative (D)}$$

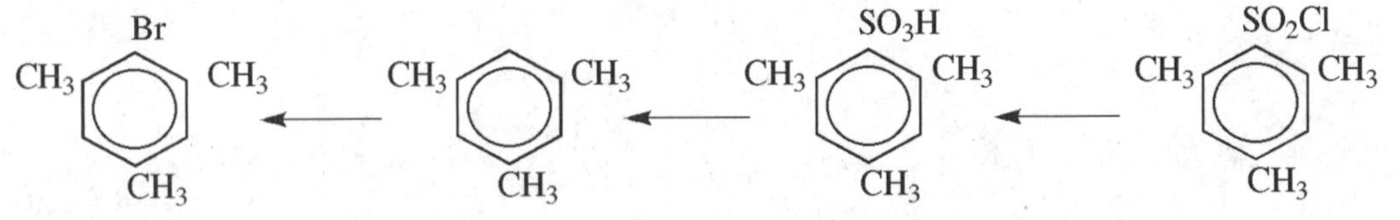

(D) 1-Bromo-2,4,6-trimethylbenzene ← (C) Mesitylene ← (B) 2,4,6-Trimethylbenzenesulfonic acid ← (A) 2,4,6-Trimethylbenzenesulfonyl chloride

Problem 16.84 (*a*) (S)-*sec*-C_4H_9OH, which has an optical rotation of + 13.8°, is reacted with tosyl chloride (see Problem 16.55), and the product is saponified. (*b*) Another sample of this alcohol is treated with benzoyl chloride, and the product is also hydrolyzed with base. What is the rotation of the *sec*-C_4H_9OH from each reaction?

(*a*) TsCl + HO—C(CH_3)(H)—C_2H_5 ⟶ TsO—C(CH_3)(H)—C_2H_5 $\xrightarrow[\text{inversion}]{OH^-}$ C_2H_5—C(CH_3)(H)—OH

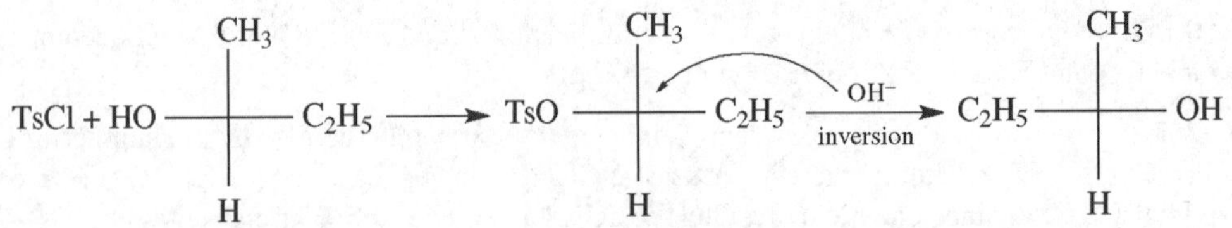

(S) (+ 13.8°) (S) (Tosyl ester) (R) (−13.8°)

(*b*) + 13.8°. Reaction with benzoyl chloride causes no change in configuration about the chiral C of the alcohol. Hydrolysis of PhCOOR occurs by attack at the carbonyl group, with retention of alcohol configuration.

Problem 16.85 Write structural formulas for the organic compounds (I) through (XI), and show the stereochemistry of (X) and (XI). (Write Ts for the tosyl group.)

(*a*) $ArSO_2Cl + CH_3OH \longrightarrow (I) \xrightarrow{H_3{}^{18}O^+} (II) + (III)$

(*b*) $Cl{-}\overset{\overset{O}{\|}}{C}{-}C_6H_4{-}SO_2NH_2 + CH_3NH_2 \longrightarrow (IV) \xrightarrow{H_3O^+X^-} (V) + (VI)$

(*c*) $C_6H_5SO_2Cl + n\text{-}C_4H_9OH \longrightarrow (VII) \xrightarrow{PhCH_2MgBr} (VIII) + (IX)$

(*d*) $TsCl + (S)\text{-}HO{-}\underset{\underset{CH_3}{|}}{\overset{\overset{C_6H_{13}}{|}}{C}}{-}H \longrightarrow (X) \xrightarrow{CH_3COO^-} (XI)$

(*a*) (I) $ArSO_2OCH_3$ (II) $ArSO_3H$ (III) $CH_3{}^{18}OH$

(*b*) (VI) $CH_3NH{-}\overset{\overset{O}{\|}}{C}{-}C_6H_4{-}SO_2NH_2$ (V) $CH_3\overset{+}{N}H_3X^-$ (VI) $HO{-}\overset{\overset{O}{\|}}{C}{-}C_6H_4{-}SO_2NH_2$

(*c*) (VI) $C_6H_5SO_2OC_4H_9\text{-}n$ (VII) $PhCH_2{-}C_4H_9\text{-}n$ (IX) $C_6H_5SO_3^-\ (MgX)^+$

(*d*) (X) $(S)\text{-}TsO{-}\underset{\underset{CH_3}{|}}{\overset{\overset{C_6H_{13}}{|}}{C}}{-}H$ (XI) $(R)\text{-}H{-}\underset{\underset{CH_3}{|}}{\overset{\overset{C_6H_{13}}{|}}{C}}{-}O\underset{\underset{O}{\|}}{C}CH_3$

Problem 16.86 Why are sulfonic acid derivatives less reactive toward nucleophilic substitutions than are the corresponding acyl derivatives?

Attack by Nu:$^-$ on the trigonal acyl C leads to a stable, uncrowded, tetrahedral intermediate (and transition state), with an octet of e^-'s on C. The sulfonyl S is already tetrahedral, and attack on Nu:$^-$ gives a less stable, more crowded, intermediate (and transition state), with a pentavalent S having 10 e^-'s, as shown:

$$Ar{-}\underset{\underset{O}{|}}{\overset{\overset{O}{|}}{S}}{-}G + Nu{:}^- \longrightarrow \left[Nu{-}\underset{\underset{Ar}{|}}{\overset{O\ \ O}{S}}{-}G\right]$$

pentavalent intermediate

Problem 16.87 (*a*) Show the mers of the following condensation polymers, formed from the two indicated monomers: (i) Polycarbonate (Lexan), phosgene, $Cl_2C{=}O + (HOC_6H_4)_2C(CH_3)_2$ (tough and optically clear); (ii) Kodel,

$$\text{dimethyl terephthalate} + HOCH_2{-}C_6H_{10}{-}CH_2OH$$

(strong fibers); (iii) Aramid, (terephthalic acid) + *p*-$H_2NC_6H_4NH_2$ (tire cords). (*b*) What small molecules are split off during growth of the polymer chain? (*c*) What type of functional group links the mer units?

(i) (*a*) $—OCH_2—C_6H_4—C(CH_3)_2—C_6H_4—CH_2O—C(=O)—$ (*b*) HCl (*c*) ester

(ii) (*a*) $—C(=O)—C_6H_4—C(=O)—OCH_2—C_6H_{10}—CH_2O—$ (*b*) CH_3OH (*c*) ester

(iii) (*a*) $—C(=O)—C_6H_4—C(=O)—N(H)—C_6H_4—N(H)—$ (*b*) H_2O (*c*) amide

Problem 16.88 Which would have the higher λ_{max} value in the uv, benzoic acid or cinnamic acid, $C_6H_5CH{=}CHCOOH$? Why?

Cinnamic acid, because it has a more extended π system.

Problem 16.89 Tell how propanal and propanoic acid can be distinguished by their (*a*) ir, (*b*) nmr, (*c*) cmr spectra.

(*a*) Propanal has a medium-sharp stretching band at 2720 cm^{-1} due to the aldehydic H. Propanoic acid has a very broad O—H band at 2500–3300 cm^{-1}. The C=O bands for both are in the same general region, about 1700 cm^{-1}.

(*b*) Propanal gives a triplet near $\delta = 9.7$ ppm (aldehydic H), a triplet near 0.9 (CH_3), and a multiplet near 2.5 (CH_2). Propanoic acid has a singlet in the range 10–13 (acidic H), a triplet near 0.9 (CH_3), and a quartet near 2.5 (CH_2).

(*c*) Carbonyl C's are sp^2-hybridized and attached to electronegative O's and, hence, suffer little shielding. They absorb farther downfield than any other kind of C: 190–220 for carbonyl compounds (propanal) and somewhat less downfield, 150–185, for carboxylic acids (propanoic acid).

Carbanion-Enolates and Enols

17.1 Acidity of H's α to C=O; Tautomerism

General

Three base-catalyzed reactions of such α H's are shown in Fig. 17.1. Since the reactions have the same rate expression, they have the same rate-determining step: the removal of an α H to form a stabilized **carbanion-enolate** anion. The name of this anion indicates that the resonance hybrid has negative charge on C (**carbanion**) and on O (**enolate**).

$$\underset{\text{acid}_1}{CH_3CH_2\text{—}\overset{\overset{CH_3}{|}}{\underset{\underset{H}{\cdot\cdot}}{C}}\text{—}\underset{\underset{H}{|}}{C}\text{=}\ddot{O}:} + \underset{\text{base}_2}{:B^-} \xrightleftharpoons{\text{slow}} \underset{\text{base}_1}{\left\{ \begin{matrix} CH_3CH_2\overset{\overset{CH_3}{|}}{C}\text{=}\underset{\underset{H}{|}}{C}\text{—}\ddot{\underset{\cdot\cdot}{O}}:^- \\ \updownarrow \\ CH_3CH_2\overset{\overset{CH_3}{|}}{\underset{\cdot\cdot}{C}}^-\text{—}\underset{\underset{H}{|}}{C}\text{=}\ddot{O}: \end{matrix} \quad \text{or} \quad CH_3CH_2\text{—}\overset{\overset{CH_3}{|}}{C}\text{---}\underset{\underset{H}{|}}{C}\text{---}O^{-} \right\}} + \underset{\text{acid}_2}{H\text{:}B}$$

stable carbanion-enolate ion

Stabilization of the anion by charge delocalization causes the α H of carbonyl compounds to be more acidic than H's of alkanes.

Problem 17.1 Show how the stable carbanion-enolate anion reacts to give the three products shown in Fig. 17.1.

H-D exchange. The anion accepts a D from D_2O, regenerating the catalyst OD^-.

$$[\text{anion}]^- + D_2O \rightarrow (\text{deuterated product}) + OD^-$$

Racemization. The negatively charged C of the anion is no longer chiral, as it was in the reactant aldehyde, because the extended π bond requires a flattening of the conjugated portion of the anion of which this C is a part.

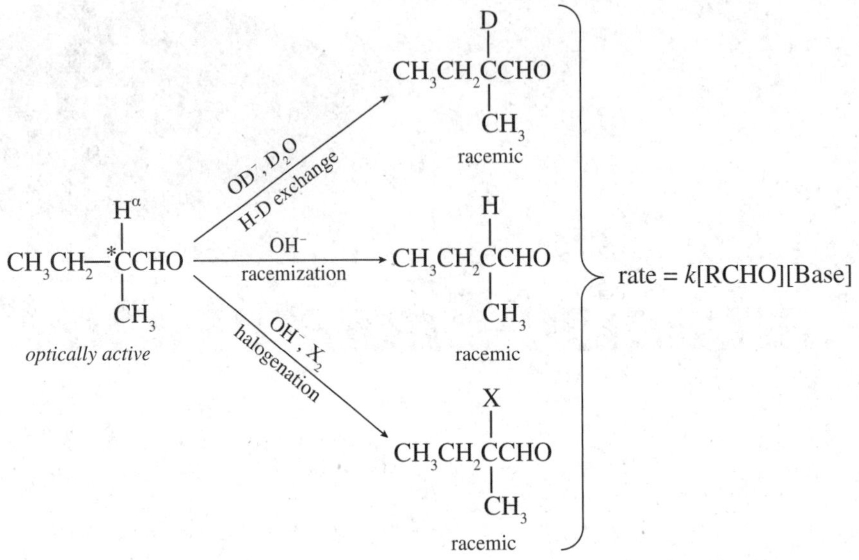

$$\underset{\textit{optically active}}{CH_3CH_2-\overset{H^\alpha}{\overset{|}{\underset{CH_3}{\underset{|}{^*C}}}}CHO} \xrightarrow{OD^-,\ D_2O\ \text{(H-D exchange)}} \underset{\text{racemic}}{CH_3CH_2\overset{D}{\overset{|}{\underset{CH_3}{\underset{|}{C}}}}CHO}$$

$$\xrightarrow[\text{racemization}]{OH^-} \underset{\text{racemic}}{CH_3CH_2\overset{H}{\overset{|}{\underset{CH_3}{\underset{|}{C}}}}CHO}$$

$$\xrightarrow{OH^-,\ X_2\ \text{(halogenation)}} \underset{\text{racemic}}{CH_3CH_2\overset{X}{\overset{|}{\underset{CH_3}{\underset{|}{C}}}}CHO}$$

rate = k[RCHO][Base]

Figure 17.1

Return of an H (or D) can occur equally well from the top face, to give one enantiomer, or from the bottom face, to give an equal amount of the other enantiomer; the product is racemic.

Halogenation.

$$[\text{anion}]^- + X_2 \rightarrow (\text{halogenated product}) + X^-$$

All three reactions are fast and, therefore, are not involved in the rate expression.

Groups other than C=O enhance the acidity of an H. Recall the 1,3-dithianes [see text preceding Problem 15.13] and see Table 17.1, which also indicates why the negative charge on the anionic C is stable.

Tautomerism

H^+ may return to C^-, to give the more stable carbonyl compound (the keto structure), or to O^-, to give the less stable enol.

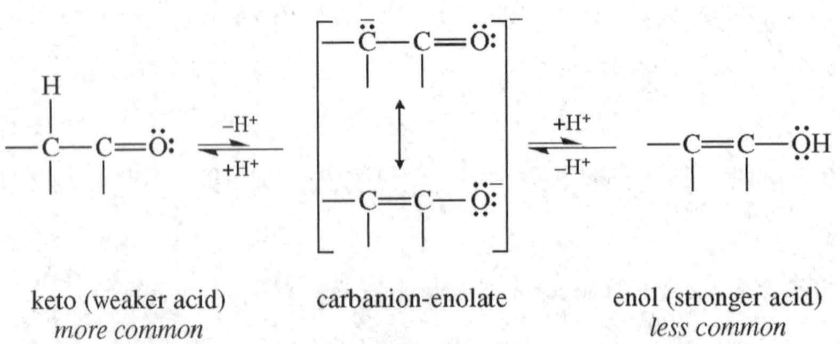

$$\underset{\substack{\text{keto (weaker acid)}\\ \textit{more common}}}{-\overset{H}{\overset{|}{\underset{|}{C}}}-\underset{|}{C}=\ddot{O}:} \underset{+H^+}{\overset{-H^+}{\rightleftharpoons}} \underset{\text{carbanion-enolate}}{\left[-\underset{|}{\ddot{C}}^{-}-\underset{|}{C}=\ddot{O}: \longleftrightarrow -\underset{|}{C}=\underset{|}{C}-\ddot{O}:^-\right]^-} \underset{-H^+}{\overset{+H^+}{\rightleftharpoons}} \underset{\substack{\text{enol (stronger acid)}\\ \textit{less common}}}{-\underset{|}{C}=\underset{|}{C}-\ddot{O}H}$$

With few exceptions, the keto structure rather than the enol is isolated from reactions (see Section 8.2, item 4).

Structural isomers existing in rapid equilibrium are **tautomers**, and the equilibrium reaction is **tautomerism**. The above is a keto-enol tautomerism.

Problem 17.2 Why is the keto form so much more stable (by about 46–59 kJ/mol) than the enol form?

TABLE 17.1

	OTHER STABLE CARBANIONS	REASON FOR STABILITY
(*a*)	$^{-}:CH_2-\overset{+}{N}(=\ddot{O}:)(\ddot{O}:^-) \longleftrightarrow CH_2=\overset{+}{N}(\ddot{O}:^-)_2$	*p-p* π bond
(*b*)	$^{-}:CH_2-C\equiv N: \longleftrightarrow CH_2=C=\ddot{N}:^-$	*p-p* π bond
(*c*)	$^{-}:CCl_3$ or $[C(Cl)_3]^-$ (C---Cl partial bonds)	*p-d* π bond
(*d*)	$CH_3C\equiv C:^-$	*sp* hybrid
(*e*)	Cyclopentadienyl anion, $C_5H_5^-$ (H on each carbon), or ring with circled (−)	aromaticity
(*f*)	$Ph-\ddot{C}H^--Ph \longleftrightarrow$ (−)ring$=CH-Ph \longleftrightarrow Ph-CH=$ring(−)	*p-p* π bond
(*g*)	$H_2\ddot{C}^-$–$C_6H_3(NO_2)$–$NO_2 \longleftrightarrow H_2C=$ring$=\overset{+}{N}(O^-)_2$ (NO₂) $\longleftrightarrow H_2C=$ring(=$\overset{+}{N}(O^-)_2$)–$NO_2$	*p-p* π bond
(*h*)	$RS-\ddot{C}H-SR \longleftrightarrow (RS\text{---}C\text{---}SR]^-$	*p-d* π bond
(*i*)	$Me_2\overset{+}{S}-\ddot{C}H_2^- \longleftrightarrow Me_2S=CH_2$	Electrostatic attraction and *p-d* π bond
(*j*)	$Ph_3\overset{+}{P}-\ddot{C}H_2^- \longleftrightarrow Ph_3P=CH_2$	Electrostatic attraction and *p-d* π bond

The resonance energy of the carbonyl group ($-\overset{|}{C}=\ddot{O}: \longleftrightarrow \overset{|}{C}^+-\ddot{O}:^-$) is greater than that of the enol ($-\overset{|}{C}=\overset{|}{C}-\ddot{O}-H \longleftrightarrow -\overset{|}{\ddot{C}}{}^--\overset{|}{C}=\ddot{O}^+-H$). Furthermore, the enol is a much stronger acid than the keto form. When equilibrium is established between the two forms, the weaker acid, the keto form, predominates.

Problem 17.3 Compare the mechanisms for (*a*) base-catalyzed and (*b*) acid-catalyzed keto-enol tautomerism.

(*a*) *base catalysis*

$$:B^- + -\overset{|}{\underset{|}{C}}(H)-\underset{|}{C}=\ddot{O}: \xrightarrow{\text{slow}} -\underset{|}{C}=\underset{|}{C}-\ddot{O}:^- + H:B \xrightarrow{\text{fast}} :B^- + -\underset{|}{C}=\underset{|}{C}-OH$$

(*b*) *acid catalysis*

$$-\overset{H}{\underset{|}{\ddot{C}}}-\overset{|}{C}=\ddot{O}: + H:A \xrightarrow{\text{fast}} :A^- + -\overset{H}{\underset{|}{\ddot{C}}}-\overset{|}{C}=\overset{+}{\ddot{O}}H \xrightarrow{\text{slow}} H:A + -\overset{|}{C}=\overset{|}{C}-\ddot{O}H$$

Problem 17.4 Racemization, D-exchange, and bromination of carbonyl compounds are also acid-catalyzed. (*a*) Suggest reasonable mechanisms in which enol is an intermediate. (*b*) In terms of your mechanisms, are the rate expressions of these reactions the same? (*c*) Why do enols not add X_2 as do alkenes?

Again, we have:

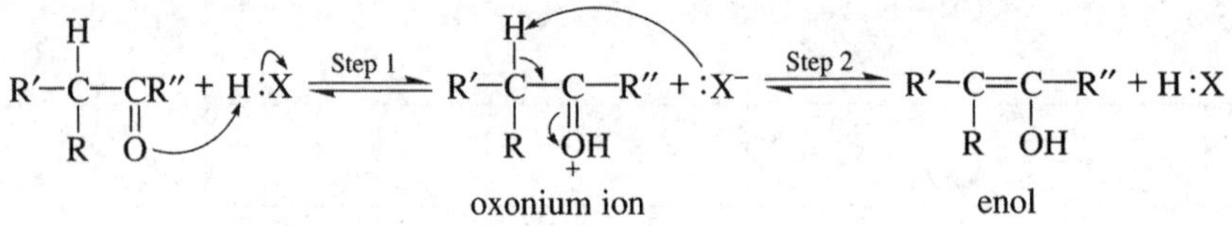

R′—CH(R)—C(=O)R″ + H:X ⇌ (Step 1) R′—CH(R)—C(=$\overset{+}{O}$H)—R″ (oxonium ion) + :X⁻ ⇌ (Step 2) R′—C(R)=C(OH)—R″ (enol) + H:X

(*a*)

Racemization

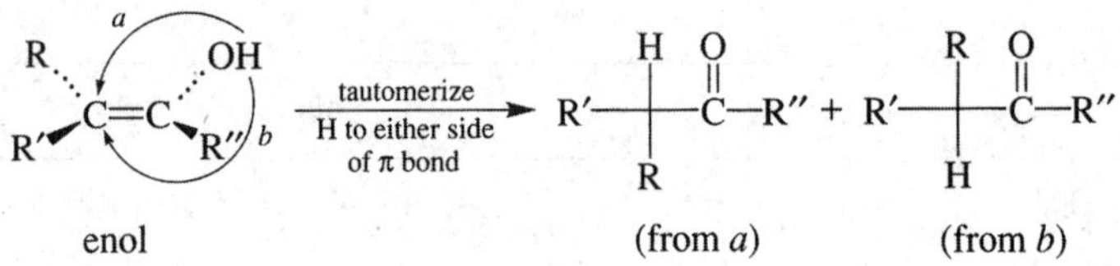

enol (R, R′C=C(OH)R″, attack at side *a* or *b*) —tautomerize, H to either side of π bond→ R′—C(H)(R)—C(=O)—R″ (from *a*) + R′—C(R)(H)—C(=O)—R″ (from *b*)

Deuterium Exchange

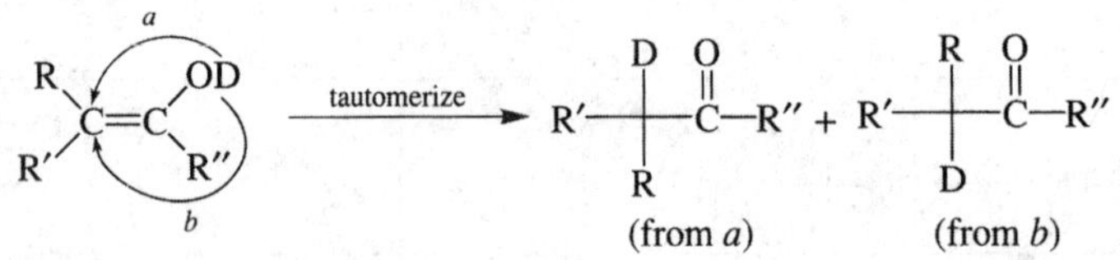

R, R′C=C(OD)R″ (sides *a*, *b*) —tautomerize→ R′—C(D)(R)—C(=O)—R″ (from *a*) + R′—C(R)(D)—C(=O)—R″ (from *b*)

The D that forms a bond to C more likely comes from the solvent, D_2O, than from the OD group.

Bromination

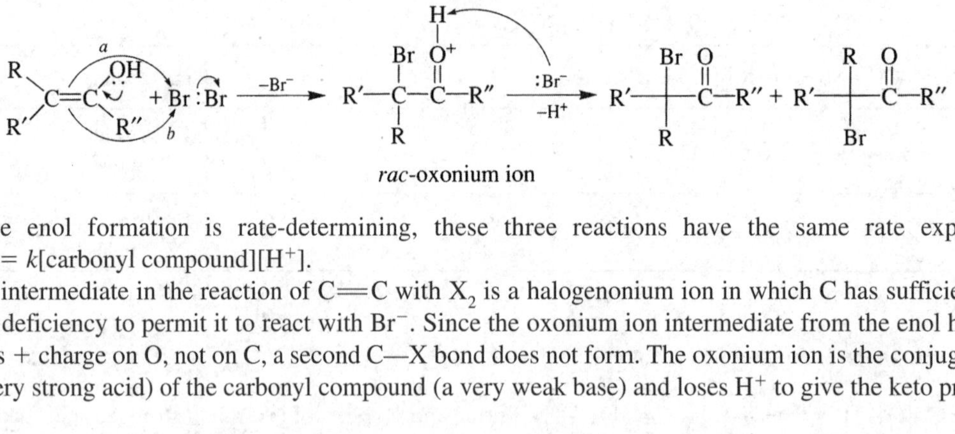

R, R′C=C(OH)R″ (sides *a*, *b*) + Br:Br —$-Br^-$→ R′—C(Br)(R)—C(=O^+H)—R″ (*rac*-oxonium ion) —:Br⁻, $-H^+$→ R′—C(Br)(R)—C(=O)—R″ + R′—C(R)(Br)—C(=O)—R″

(*b*) Since enol formation is rate-determining, these three reactions have the same rate expression, rate = *k*[carbonyl compound][H^+].

(*c*) The intermediate in the reaction of C=C with X_2 is a halogenonium ion in which C has sufficient electron deficiency to permit it to react with Br^-. Since the oxonium ion intermediate from the enol has most of its + charge on O, not on C, a second C—X bond does not form. The oxonium ion is the conjugate acid (a very strong acid) of the carbonyl compound (a very weak base) and loses H^+ to give the keto product.

Problem 17.5 (*a*) Show the tautomers of each of the following compounds, which are written as the more stable form: (1) CH_3CHO, (2) $C_6H_5COCH_3$, (3) CH_3NO_2, (4) Me_2C=NOH, and (5) CH_3CH=NCH_3. (*b*) Which two enols are in equilibrium with (i) 2-butanone, and (ii) 1-phenyl-2-butanone? Which is more stable?

(*a*) The grouping needed for tautomerism, X=Y—Z—H (a triple bond could also exist between X and Y), is encircled in each case.

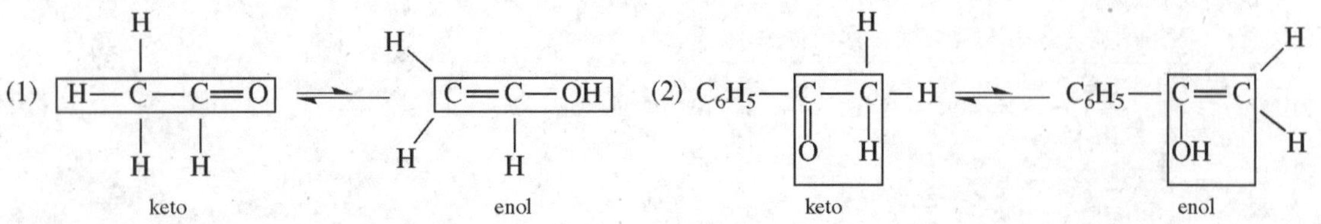

(1) [H—C—C=O] (keto) ⇌ H₂[C=C—OH]H (enol)

(2) C_6H_5—[C(=O)—C(H)]—H_2 (keto) ⇌ C_6H_5—[C(OH)=C]H_2 (enol)

(3) $\text{H—}\overset{\text{H}}{\underset{\text{H}}{\text{C}}}\text{—}\underset{\text{O}^-}{\text{N}^+}\text{=O} \rightleftharpoons \text{H}_2\text{C=}\underset{\text{O}^-}{\text{N}^+}\text{—O—H}$ (4) $\text{Me}_2\text{C=N—O—H} \rightleftharpoons \text{Me}_2\underset{\text{H}}{\text{C}}\text{—N=O}$

nitro form — *aci* form — oxime — nitroso

(5) $\text{H—}\overset{\text{H}}{\underset{\text{H}}{\text{C}}}\text{—}\underset{\text{H}}{\text{C}}\text{=}\underset{\text{CH}_3}{\text{N}} \rightleftharpoons \text{H}_2\text{C=}\underset{\text{H}}{\text{C}}\text{—}\underset{\text{CH}_3}{\text{N}}\text{—H}$

imine — enamine

(*b*) (i) $\text{H}_2\text{C=}\underset{\text{OH}}{\text{C}}\text{—CH}_2\text{CH}_3$ and $\text{CH}_3\text{—}\underset{\text{OH}}{\text{C}}\text{=CHCH}_3$

The latter is more stable because it has a more substituted double bond.

(ii) $\text{C}_6\text{H}_5\text{CH=}\underset{\text{OH}}{\text{C}}\text{CH}_2\text{CH}_3$ and $\text{C}_6\text{H}_5\text{CH}_2\underset{\text{OH}}{\text{C}}\text{=CHCH}_3$

The former is more stable because the C=C is conjugated with the benzene ring.

Problem 17.6 (*a*) Write the structural formulas for the stable keto and enol tautomers of ethyl acetoacetate. (*b*) Why is this enol much more stable than that of a simple ketone? (*c*) How can the enol be chemically detected?

(*a*) $\text{CH}_3\underset{\text{O}}{\underset{\|}{\text{C}}}\text{CH}_2\underset{\text{O}}{\underset{\|}{\text{C}}}\text{—OC}_2\text{H}_5 \rightleftharpoons \text{CH}_3\underset{\text{O—H}}{\underset{|}{\text{C}}}\text{=CH}\underset{\cdots\text{O}}{\underset{\|}{\text{C}}}\text{—OC}_2\text{H}_5$ (H-bond)

keto form — enol form

(*b*) There is a stable conjugated C=C—C=O linkage; moreover, intramolecular H-bonding (chelation) adds stability to the enol.

(*c*) The enol decolorizes a solution of Br_2 in CCl_4.

Problem 17.7 Reaction of 1 mol each of Br_2 and $PhCOCH_2CH_3$ in basic solution yields 0.5 mol of $PhCOCBr_2CH_3$ and 0.5 mol of unreacted $PhCOCH_2CH_3$. Explain.

Substitution by one Br gives $PhCOCHBrCH_3$. The electron-withdrawing Br increases the acidity of the remaining α H, which reacts more rapidly than, and is substituted before, the H's on the unbrominated ketone.

$$\underset{(0.5\text{ mol})}{\text{PhCOCH}_2\text{CH}_3} \xrightarrow[\text{OH}^-]{0.5\text{ mol Br}_2} \underset{\substack{(0.5\text{ mol})\\ \textit{not isolated}}}{[\text{PhCOCHBrCH}_3]} \xrightarrow[\text{OH}^-]{0.5\text{ mol Br}_2} \underset{(0.5\text{ mol})}{\text{PhCOCBr}_2\text{CH}_3}$$

(less acidic: H of CH_2 in $PhCOCH_2CH_3$; more acidic: H of CHBr in $PhCOCHBrCH_3$)

17.2 Alkylation of Simple Carbanion-Enolates

General

Carbanion-enolates are nucleophiles that react with alkyl halides (or sulfonates) by typical S_N2 reactions. Carbanion-enolates are best formed using lithium diisopropylamide (LDA), $(i\text{-Pr})_2N^-Li^+$, in tetrahydrofuran.

This base is very strong and converts all the substrate to the anion. Furthermore, it is too sterically hindered to react with RX.

$$-\underset{|}{\text{CH}}-\underset{|}{\text{C}}=\text{O} + (i\text{-Pr})_2\text{N}^- \text{Li}^+ \longrightarrow \left[-\underset{|}{\ddot{\text{C}}^-}-\underset{|}{\text{C}}=\ddot{\text{O}}: \longleftrightarrow -\underset{|}{\text{C}}=\underset{|}{\text{C}}-\ddot{\text{O}}:^-\right] + (i\text{-Pr})_2\text{NH}$$

carbanion-enolate   Diisopropylamine

Ketones, esters and nitriles but not aldehydes, which take a different route also can be reacted with LDA. See Table 17.1(*b*) for stabilization of negative charge on a C α to a C≡N group.

Since the carbanion-enolates are ambident ions with two different nucleophilic sites, they can be alkylated at C or at O.

$$\left[-\overset{\overset{\ddot{\text{O}}:}{\|}}{\text{C}}-\underset{|}{\ddot{\text{C}}^-}- \longleftrightarrow -\overset{\overset{:\ddot{\text{O}}:^-}{|}}{\text{C}}=\underset{|}{\text{C}}-\right] \xrightarrow{\text{RX}} \begin{cases} -\overset{\overset{\text{O}}{\|}}{\text{C}}-\underset{|}{\overset{\overset{\text{R}}{|}}{\text{C}}} & \text{an alkyl ketone (C-alkylation)} \\ -\overset{\overset{\text{OR}}{|}}{\text{C}}=\text{C}\langle & \text{a vinyl ether (O-alkylation)} \end{cases}$$

O-alkylation reduces the yield of the more usually desired C-alkylation product. Other drawbacks to the synthetic utility of this reaction are: (1) di- and tri-alkylation produces mixtures if more than a single H is present on the α C; (2) ketones with H's on more than one α C will give a mixture of alkylation products.

Problem 17.8 (*a*) Use carbanion-enolate alkylations to synthesize: (i) 2-ethylbutanenitrile, (ii) 3-phenyl-2-pentanone, (iii) 2-benzylcyclopentanone. (*b*) Why can the unsymmetrical ketone in part (ii) be alkylated in good yield?

(*a*) (i) $$CH_3CH_2CH_2-C\equiv N \xrightarrow[2.\ C_2H_5I]{1.\ LDA} CH_3CH_2\overset{\overset{C_2H_5}{|}}{C}H-C\equiv N$$

(ii) $$C_6H_5-CH_2-CO-CH_3 \xrightarrow[2.\ C_2H_5I]{1.\ LDA} C_6H_5\overset{\overset{C_2H_5}{|}}{C}H-CO-CH_3$$

(iii) $$\text{cyclopentanone} \xrightarrow[2.\ C_6H_5CH_2Br]{1.\ LDA} \text{2-(}CH_2-C_6H_5\text{)cyclopentanone}$$

(*b*) The benzylic C forms the carbanion $Ph\overset{-}{\ddot{C}}HCOCH_3$ because it is more stabilized through charge delocalization to the benzene ring.

Enamine Alkylations

This reaction, designed by Gilbert Stork, fosters monoalkylation. **Enamines** $-\overset{|}{C}=\overset{|}{C}-\underset{\cdot\cdot}{\overset{|}{N}}-$ [see Problem 17.5(*a*)(5)], of ketones are monoalkylated with reactive halides, such as benzyl and allyl, in good yield at the α C. The enamines are made from the ketone and preferably a 2° amine, R_2NH.

Cyclohexanone + Pyrrolidine (2° amine) $\xrightarrow{-H_2O}$ [an enamine] $\xrightarrow{PhCH_2Cl}$ an immonium compound $\xrightarrow{+H_2O}$ pyrrolidinium chloride + 2-benzylcyclohexanone

Enamines also can be acylated on the α C with acid chlorides.

Problem 17.9 Use enamines in the following conversions: (*a*) cyclohexanone to (i) 2-allylcyclohexanone and (ii) 2-acetylcyclohexanone, and (*b*) 3-pentanone to 2-methyl-1-phenyl-3-pentanone.

(*a*) Pyrrolidine is used to convert cyclohexanone into an enamine, which in (i) is alkylated with $H_2C{=}CHCH_2Cl$ and in (ii) is acylated with CH_3COCl.

(i) $Cl{-}CH_2CH{=}CH_2$; then H_3O^+ → pyrrolidinium Cl^- + 2-allylcyclohexanone ($CH_2CH{=}CH_2$)

(ii) $ClCOCH_3$; then H_3O^+ → pyrrolidinium Cl^- + 2-acetylcyclohexanone ($O{=}C{-}CH_3$), a β-diketone

(*b*) Pyrrolidine + $O{=}C(CH_2CH_3)_2$ → enamine $N{-}C(=CHCH_3)CH_2CH_3$ $\xrightarrow{X-CH_2C_6H_5}$ $^+N{=}C(CH(CH_3)CH_2C_6H_5)(CH_2CH_3)$ X^- → $\overset{+}{N}H_2$ X^- + $O{=}C(CHCH_3CH_2C_6H_5)(CH_2CH_3)$

17.3 Alkylation of Stable Carbanion-Enolates

The acidity of an H is greatly enhanced when the C to which it is bonded is α to two C═O groups: O═C—CH—C═O (with single bonds on each C). The negative charge on the α C of such a β-dicarbonyl compound is now delocalized over two C═O groups: $[\text{O}{\overset{...}{=}}\text{C}{\overset{...}{=}}\text{C}{\overset{...}{=}}\text{C}{\overset{...}{=}}\text{O}]^-$ (with single bonds on each C).

Problem 17.10 Compare the relative acid strengths of 2,4-pentanedione, $CH_3COCH_2COCH_3$ (I); ethyl acetoacetate, $CH_3COCH_2COOC_2H_5$ (II); and diethyl malonate, $C_2H_5OOCCH_2COOC_2H_5$ (III). Explain your ranking.

I > II > III. All three compounds afford resonance-stabilized carbanions. However, COOEt has an electron-releasing O bonded to carbonyl C, which decreases resonance stabilization. There are two COOEt groups in III and one in II, while I has only ketonic carbonyl groups.

The compounds III and II are useful substrates for the synthesis of carboxylic acids and ketones, respectively.

Malonic Ester Synthesis of R-Substituted Acetic Acids, RCH_2COOH or R(R′) CHCOOH (R′ could≡R)

Step 1 A carbanion is formed with strong base (often NaOEt in EtOH).

$$\underset{\substack{\text{Malonic ester}\\ (\textit{Diethyl malonate})}}{\text{EtOOC—CH}_2\text{—COOEt}} + \overset{+}{\text{Na}}\overset{-}{\text{O}}\text{Et} \longrightarrow [\text{EtOOC—}\ddot{\overset{-}{\text{C}}}\text{H—COOEt}]\,\text{Na}^+ + \text{EtOH}$$

or

$$\text{C}_2\text{H}_5\text{O—}\overset{\overset{\text{O}}{\|}}{\text{C}}{\overset{...}{=}}\underset{\underset{\text{H}}{|}}{\text{C}}{\overset{...}{=}}\overset{\overset{\text{O}}{\|}}{\text{C}}\text{—OC}_2\text{H}_5 \quad (\text{charge } - \text{ delocalized over O, C, O})$$

stabilized carbanion

Step 2 The carbanion is alkylated by S_N2 reactions with unhindered RX or ROTs.

$$[\text{EtOOC—}\ddot{\overset{-}{\text{C}}}\text{H—COOEt}]\,\text{Na}^+ + \text{R:X} \longrightarrow \text{EtOOC—}\overset{\overset{\text{R}}{|}}{\text{C}}\text{H—COOEt} + \text{Na}^+\text{X}^-$$

For dialkylacetic acids, the second H of the α C is similarly replaced with another R of a different R′ group.

$$\text{EtOOC—}\underset{\underset{\text{R}}{|}}{\text{C}}\text{H—COOEt} \xrightarrow{\overset{+}{\text{Na}}\overset{-}{\text{O}}\text{Et}} [\text{EtOOC—}\ddot{\overset{-}{\text{C}}}\text{R—COOEt}]\text{Na}^+ \xrightarrow{\text{R}'\text{X}} \text{EtOOC—}\overset{\overset{\text{R}'}{|}}{\underset{\underset{\text{R}}{|}}{\text{C}}}\text{—COOEt} + :\text{X}^-$$

Step 3 Hydrolysis of the substituted malonic ester gives the malonic acid, which undergoes **decarboxylation** (loss of CO_2) to form a substituted acetic acid.

$$\text{EtOOC—}\underset{\underset{\text{R}}{|}}{\text{C}}\text{H—COOEt} \xrightarrow[\substack{1.\ \text{OH}^-\\ 2.\ \text{H}_3\text{O}^+}]{\text{hydrolysis}} \left[\text{HOOC}\underset{\underset{\text{R}}{|}}{\text{C}}\text{HCOOH}\right] \xrightarrow[\text{heat}]{-\text{CO}_2} \text{RCH}_2\text{COOH}$$

$$\text{EtOOC—}\overset{\overset{\text{R}'}{|}}{\underset{\underset{\text{R}}{|}}{\text{C}}}\text{—COOEt} \xrightarrow{\text{hydrolysis}} \left[\text{HOOC—}\overset{\overset{\text{R}'}{|}}{\underset{\underset{\text{R}}{|}}{\text{C}}}\text{—COOH}\right] \xrightarrow[\text{heat}]{-\text{CO}_2} \text{R}'\text{—}\underset{\underset{\text{R}}{|}}{\text{C}}\text{H—COOH}$$

For a general carboxylic acid, the parts are assembled as follows:

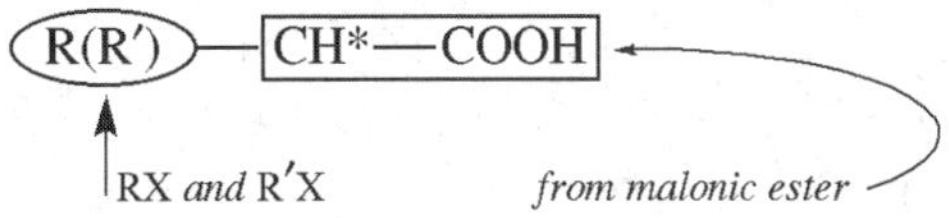

$$\boxed{R(R')}\text{—}\boxed{CH^*\text{—}COOH} \leftarrow \textit{from malonic ester}$$

from alkyl halides ↑ RX *and* R′X

where H* replaced COOEt.

Problem 17.11 Use malonic ester to prepare (*a*) 2-ethylbutanoic acid, (*b*) 3-methylbutanoic acid, (*c*) 2-methylbutanoic acid, (*d*) trimethylacetic acid.

The alkyl groups attached to the α C are introduced by the alkyl halides.

(*a*) In 2-ethylbutanoic acid

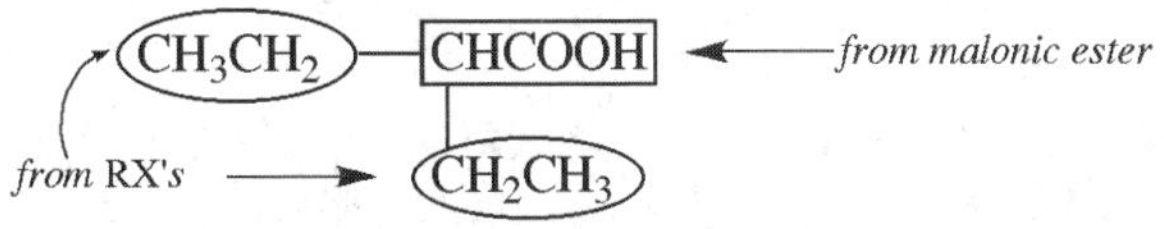

$$\boxed{CH_3CH_2}\text{—}\boxed{CHCOOH} \leftarrow \textit{from malonic ester}$$
$$\textit{from RX's} \rightarrow \boxed{CH_2CH_3} \text{ (attached to CH)}$$

the circled R's are both CH_2CH_3. Therefore, each α H is replaced sequentially by an ethyl group (R = R′ = Et), using CH_3CH_2Br.

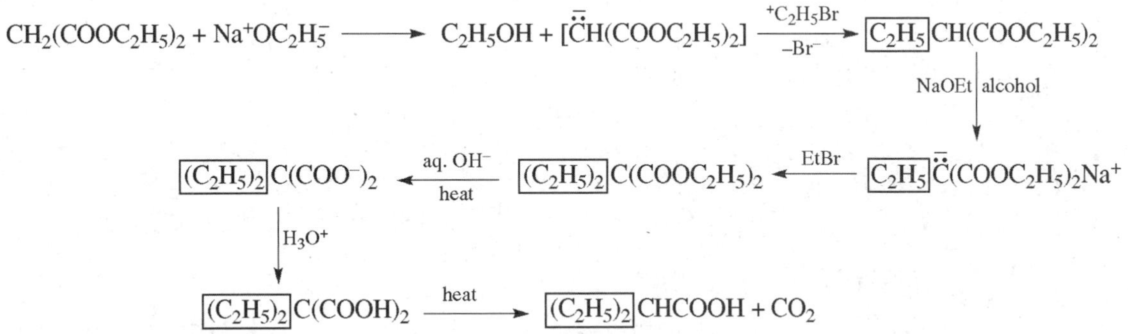

$$CH_2(COOC_2H_5)_2 + Na^+OC_2H_5^- \longrightarrow C_2H_5OH + [\ddot{\bar{C}}H(COOC_2H_5)_2] \xrightarrow[-Br^-]{+C_2H_5Br} \boxed{C_2H_5}CH(COOC_2H_5)_2$$

$$\boxed{C_2H_5}CH(COOC_2H_5)_2 \xrightarrow{\text{NaOEt | alcohol}} \boxed{C_2H_5}\ddot{\bar{C}}(COOC_2H_5)_2Na^+ \xrightarrow{EtBr} \boxed{(C_2H_5)_2}C(COOC_2H_5)_2 \xrightarrow[\text{heat}]{\text{aq. OH}^-} \boxed{(C_2H_5)_2}C(COO^-)_2$$

$$\boxed{(C_2H_5)_2}C(COO^-)_2 \xrightarrow{H_3O^+} \boxed{(C_2H_5)_2}C(COOH)_2 \xrightarrow{\text{heat}} \boxed{(C_2H_5)_2}CHCOOH + CO_2$$

(*b*) Only $(CH_3)_2CH$—Br is needed for a single alkylation (see product below).

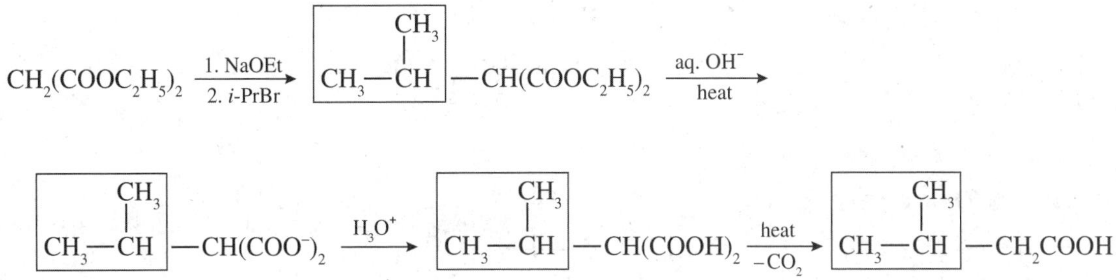

$$CH_2(COOC_2H_5)_2 \xrightarrow[\text{2. } i\text{-PrBr}]{\text{1. NaOEt}} \boxed{CH_3\text{—}\underset{}{CH}(CH_3)}\text{—}CH(COOC_2H_5)_2 \xrightarrow[\text{heat}]{\text{aq. OH}^-}$$

$$\boxed{CH_3\text{—}CH(CH_3)}\text{—}CH(COO^-)_2 \xrightarrow{H_3O^+} \boxed{CH_3\text{—}CH(CH_3)}\text{—}CH(COOH)_2 \xrightarrow[-CO_2]{\text{heat}} \boxed{CH_3\text{—}CH(CH_3)}\text{—}CH_2COOH$$

(*c*) To obtain

$$\boxed{CH_3CH_2}\,CHCOOH \text{ with } \boxed{CH_3} \text{ attached to CH}$$

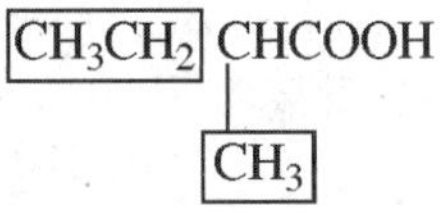

Alkylate first with CH_3CH_2Br and then with CH_3I. The larger R is introduced first to minimize steric hindrance in the second alkylation step.

$$CH_2(COOEt)_2 \xrightarrow[2.\ EtBr]{1.\ OEt^-} \boxed{CH_3CH_2}CH(COOEt)_2 \xrightarrow[2.\ CH_3I]{1.\ OEt^-} \boxed{CH_3CH_2}\underset{\boxed{CH_3}}{C}—(COOEt)_2$$

$$\xrightarrow{1.\ OH^-\quad 2.\ H_3O^+,\ heat} \boxed{CH_3CH_2}\underset{\boxed{CH_3}}{CH}COOH$$

(*d*) Trialkylacetic acids cannot be prepared from malonic ester. The product prepared from malonic ester must have at least one α H, which replaces the lost COOH.

Problem 17.12 Use $CH_3CH_2CH_2OH$ and $H_2C(COOC_2H_5)_2$ as the only organic reagents to synthesize valeramide, $CH_3CH_2CH_2CH_2CONH_2$.

Since valeric acid is *n*-propylacetic acid:

$$\boxed{CH_3CH_2CH_2}—CH_2COOH$$

$CH_3CH_2CH_2Br$ is used to alkylate malonic ester.

$$CH_3CH_2CH_2OH \xrightarrow{PBr_3} CH_3CH_2CH_2—Br$$

$$H_2C(COOC_2H_5)_2 \xrightarrow[2.\ EtOH]{1.\ NaOEt} [:\bar{C}H(COOC_2H_5)_2]$$

$$\longrightarrow CH_3CH_2CH_2—CH(COOC_2H_5)_2 \ (\text{formed bond}) \xrightarrow{1.\ OH^-\quad 2.\ H_3O^+,\ \Delta} CH_3CH_2CH_2CH_2COOH$$

$$CH_3CH_2CH_2CH_2COOH \xrightarrow{SOCl_2} CH_3(CH_2)_3COCl \xrightarrow{NH_3} CH_3(CH_2)_3CONH_2$$

Acetoacetic Ester (AAE) Synthesis

1. Carbanion Formation

Acetoacetic ester is acidic ($pK_a = 10.2$) and forms a resonance-stabilized carbanion whose negative charge is delocalized over one C and two O's

$$CH_3—\overset{O}{\overset{\|}{C}}—\overset{H}{\underset{H}{C}}—\overset{O}{\overset{\|}{C}}—OC_2H_5 + \overset{+}{Na}\overset{-}{O}C_2H_5 \rightleftharpoons C_2H_5OH + \overset{+}{Na}\left[CH_3—\overset{O}{\overset{\|}{C}}\text{---}\underset{H}{C}\text{---}\overset{O}{\overset{\|}{C}}—OC_2H_5\right]^-$$

2. Alkylation

As with malonic ester (Problem 17.11), either one or two R's can be introduced in acetoacetic ester.

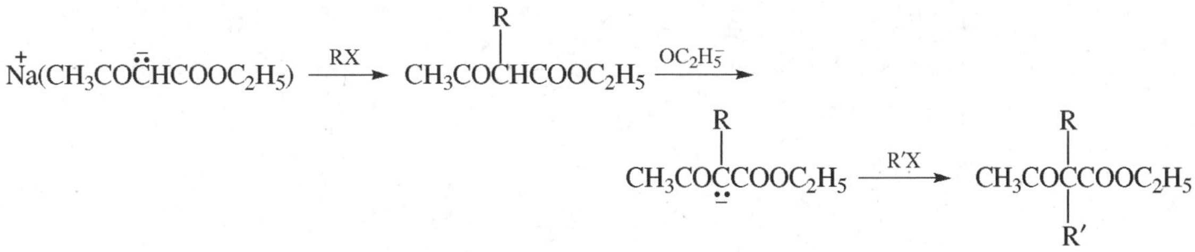

$$\overset{+}{Na}(CH_3CO\ddot{C}HCOOC_2H_5) \xrightarrow{RX} CH_3CO\overset{R}{\overset{|}{C}}HCOOC_2H_5 \xrightarrow{OC_2H_5^-} CH_3CO\overset{R}{\overset{|}{\underset{\bullet\bullet}{\overset{-}{C}}}}COOC_2H_5 \xrightarrow{R'X} CH_3CO\overset{R}{\overset{|}{\underset{R'}{\underset{|}{C}}}}COOC_2H_5$$

3. Hydrolysis and Decarboxylation

Dilute acid or *base* hydrolyzes the $COOC_2H_5$ group and forms acetoacetic acids, which decarboxylate to methyl ketones.

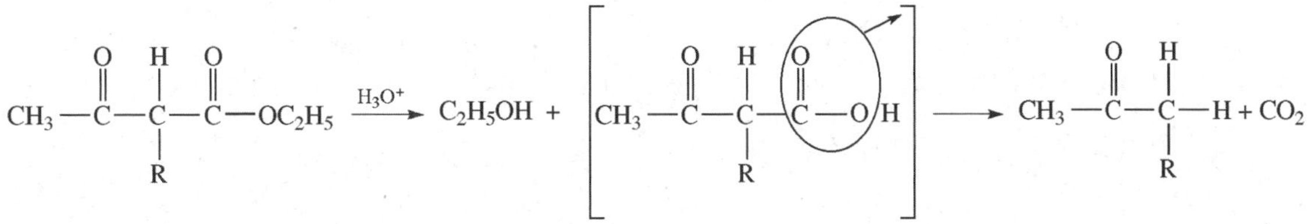

$$CH_3-\overset{O}{\overset{\|}{C}}-\overset{H}{\overset{|}{\underset{R}{\underset{|}{C}}}}-\overset{O}{\overset{\|}{C}}-OC_2H_5 \xrightarrow{H_3O^+} C_2H_5OH + \left[CH_3-\overset{O}{\overset{\|}{C}}-\overset{H}{\overset{|}{\underset{R}{\underset{|}{C}}}}-\overset{O}{\overset{\|}{C}}-O-H\right] \longrightarrow CH_3-\overset{O}{\overset{\|}{C}}-\overset{H}{\overset{|}{\underset{R}{\underset{|}{C}}}}-H + CO_2$$

This sequence of steps can be used to synthesize methyl ketones. For a general methyl ketone, the parts are assembled as follows:

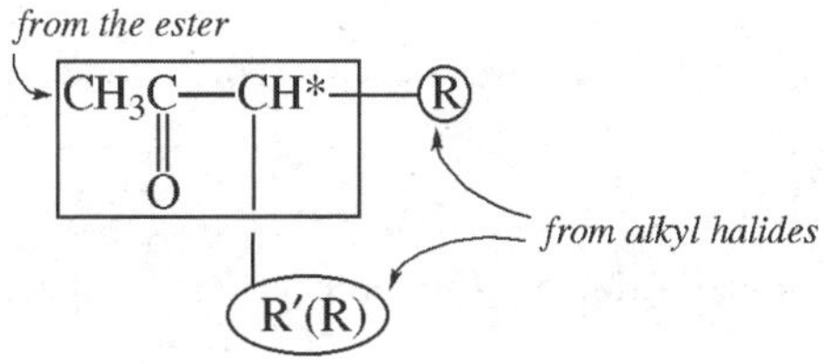

$$\boxed{CH_3\overset{}{\underset{\underset{O}{\|}}{C}}-CH^*}-\text{Ⓡ}, \quad CH^* - R'(R)$$

from the ester (boxed $CH_3C(=O)-CH^*$); *from alkyl halides* (Ⓡ and R′(R) attached to CH*)

The H* replaced COOEt.

Problem 17.13 Prepare 3-methyl-2-pentanone from acetoacetic ester.

In the product

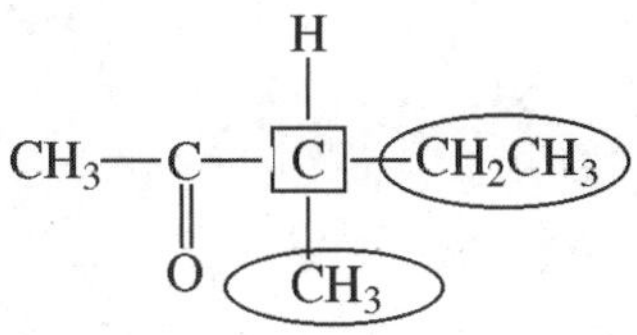

$$CH_3-\underset{\underset{O}{\|}}{C}-\overset{H}{\overset{|}{\underset{CH_3}{\underset{|}{\boxed{C}}}}}-CH_2CH_3$$

the $-CH_3$ and $-CH_2CH_3$ attached to $\boxed{C}$ are introduced by alkylation of the carbanion of acetoacetic ester with appropriate alkyl halides; in this case, use $BrCH_2CH_3$ and then CH_3I.

$$CH_3CO{-}\underset{H}{\overset{H}{C}}{-}COOC_2H_5 \xrightarrow{\overset{+}{Na}\bar{O}Et} CH_3CO\overset{..-}{C}HCOOC_2H_5 \xrightarrow{CH_3CH_2Br} CH_3CO\overset{CH_2CH_3}{C}HCOOC_2H_5$$

$$\xrightarrow{1.\ \overset{+}{Na}\bar{O}Et \quad 2.\ CH_3I} CH_3CO\underset{CH_2CH_3}{\overset{CH_3}{C}}COOC_2H_5 \text{ (Methylethylacetoacetic ester)} \xrightarrow[2.\ -CO_2]{1.\ \text{hydrolysis}} CH_3CO\underset{CH_2CH_3}{\overset{CH_3}{C}H} \text{ (3-Methyl-2-pentanone)}$$

Problem 17.14 Use AAE (acetoacetic ester) synthesis to prepare the β-diketone 2,4-hexanedione, $CH_3COCH_2COCH_2CH_3$.

The group attached to the α C of AAE is $COCH_2CH_3$. This acyl group is introduced with CH_3CH_2COCl. Because acyl halides react with ethanol, aprotic solvents are used. The carbanion is prepared with $:H^-$ from NaH.

$$CH_3{-}\overset{O}{\overset{\|}{C}}{-}CH_2{-}COOEt \xrightarrow[-H_2]{NaH} CH_3{-}\overset{O}{\overset{\|}{C}}{-}\overset{..-}{C}H{-}COOEt\ Na^+ \xrightarrow[-NaCl]{Cl{-}\overset{O}{\overset{\|}{C}}{-}CH_2CH_3} CH_3{-}\overset{O}{\overset{\|}{C}}{-}CH(COOEt){-}\overset{O}{\overset{\|}{C}}{-}CH_2CH_3$$

$$\xrightarrow{1.\ \overset{+}{Na}\bar{O}Et \quad 2.\ H_3O^+} \left[CH_3{-}\overset{O}{\overset{\|}{C}}{-}CH(COOH){-}\overset{O}{\overset{\|}{C}}{-}CH_2CH_3\right] \xrightarrow{-CO_2} CH_3{-}\overset{O}{\overset{\|}{C}}{-}CH_2{-}\overset{O}{\overset{\|}{C}}{-}CH_2CH_3$$

Acetoacetic ester is converted to a **dianion** by 2 moles of a very strong base.

$$CH_3COCH_2COOC_2H_5 \xrightarrow{2\ LDA} \underset{\text{a dianion}}{:\bar{C}H_2CO\underset{..}{\bar{C}}HCOOC_2H_5}$$

When treated with 1 mol of 1° RX, the more basic terminal carbanion is alkylated, not the less basic interior carbanion. The remaining carbanion-enolate can be protonated.

Problem 17.15 Devise a synthesis of $C_6H_5CH_2CH_2COCH_2COOH$ from acetoacetic ester.

Since the terminal methyl group of acetoacetic ester is alkylated, its dianion is reacted with $C_6H_5CH_2Cl$.

$$CH_3{-}\overset{O}{\overset{\|}{C}}{-}CH_2COOEt \xrightarrow{2\ LDA} :\bar{C}H_2{-}\overset{O}{\overset{\|}{C}}{-}\overset{..}{\bar{C}}H{-}COOEt \xrightarrow[-Cl^-]{+C_6H_5CH_2Cl} C_6H_5CH_2CH_2{-}CO{-}\overset{..}{\bar{C}}HCOOEt \xrightarrow{H_3O^+} C_6H_5CH_2CH_2{-}CO{-}CH_2{-}COOH$$

17.4 Nucleophilic Addition to Conjugated Carbonyl Compounds: Michael 3,4-Addition

α, β-Unsaturated carbonyl compounds add nucleophiles at the β C, leaving a – charge at the α C. This intermediate is a stable carbanion-enolate. These **3,4-Michael additions** compete with addition to the carbonyl group (1,2-addition; Section 15.4).

$$C_6H_5\overset{\beta}{C}H{=}\overset{\alpha}{C}H\underset{\underset{O}{\|}}{C}C_6H_5 \xrightarrow{CN^-} \left[C_6H_5\underset{\underset{CN}{|}}{C}H{-}CH{=}\underset{\underset{O^-}{|}}{C}C_6H_5 \longleftrightarrow C_6H_5\underset{\underset{CN}{|}}{C}H{-}\ddot{C}H{-}\underset{\underset{O}{\|}}{C}C_6H_5 \right] \xrightarrow{HCN} C_6H_5\underset{\underset{CN}{|}}{C}H{-}CH_2{-}\underset{\underset{O}{\|}}{C}{-}C_6H_5 + CN^-$$

Michael addition product

$$C_6H_5CH{=}CH{-}\underset{\underset{O}{\|}}{C}{-}C_6H_5 + \begin{cases} \underset{\text{more covalent}}{PhMgBr} \xrightarrow[2.\ H_2O]{1.\ ether} \underset{\text{major}}{C_6H_5\overset{\overset{Ph}{|}}{C}HCH_2\underset{\underset{O}{\|}}{C}C_6H_5} + \underset{\text{trace}}{C_6H_5{-}CH{=}CH{-}\overset{\overset{Ph}{|}}{\underset{\underset{OH}{|}}{C}}C_6H_5} \\ \underset{\text{more ionic}}{PhLi} \xrightarrow[2.\ H_2O]{1.\ ether} \underset{\substack{\text{minor}\\ \text{3,4-product}}}{C_6H_5\overset{\overset{Ph}{|}}{C}HCH_2\underset{\underset{O}{\|}}{C}C_6H_5} + \underset{\substack{\text{major}\\ \text{1,2-product}}}{C_6H_5{-}CH{=}CH{-}\overset{\overset{Ph}{|}}{\underset{\underset{OH}{|}}{C}}C_6H_5} \end{cases}$$

Stable carbanion-enolates can also serve as the nucleophile.

Problem 17.16 (*a*) Show the nucleophilic addition of the malonate carbanion to methyl vinyl ketone and the formation of the final keto product. (*b*) Show how this product can be converted to a δ-ketoacid.

(*a*)
$$CH_3{-}\overset{\overset{O}{\|}}{C}{-}CH{=}CH_2 + :\bar{C}H(COOR)_2 \longrightarrow CH_3{-}\overset{\overset{^-O}{|}}{C}{=}CHCH_2{-}CH(COOR)_2 \xrightarrow{HOEt} \left[CH_3{-}\overset{\overset{OH}{|}}{C}{=}CH{-}CH_2{-}CH(COOR)_2 \right]_{\text{enol}} \longrightarrow \underset{\text{keto}}{CH_3{-}\overset{\overset{O}{\|}}{C}{-}CH_2CH_2{-}CH(COOR)_2}$$

(*b*)
$$CH_3{-}\overset{\overset{O}{\|}}{C}{-}CH_2CH_2{-}CH(COOR)_2 \xrightarrow[2.\ H_3O^+]{1.\ NaOH} \left[CH_3{-}\overset{\overset{O}{\|}}{C}{-}CH_2CH_2{-}CH(COOH)_2 \right] \xrightarrow{-CO_2} CH_3\overset{\delta}{-}\overset{\overset{O}{\|}}{C}CH_2CH_2CH_2COOH$$

5-Oxohexanoic acid

Problem 17.17 **Cyanoethylation** is the replacement of an acidic α H of a carbonyl compound by a —CH_2CH_2CN group, using acrylonitrile (CH_2═CHCN) and base. Illustrate with cyclohexanone.

$$\text{cyclohexanone} \xrightarrow{OH^-} \text{carbanion} \xrightarrow{CH_2{=}CHCN} \text{2-}(CH_2\ddot{C}HCN)\text{cyclohexanone} \xrightarrow{+HOH} \text{2-}(CH_2CH_2CN)\text{cyclohexanone} + OH^-$$

17.5 Condensations

A condensation reaction leads to a product with a new C—C bond. Most often, the new bond results from a nucleophilic addition of a reasonably stable carbanion-enolate to the C═O group (acceptor) of an aldehyde; less frequently, the C═O group belongs to a ketone or acid derivative. Another acceptor is the C≡N group of a nitrile.

$$-\overset{O^-}{\overset{|}{C}}{=}\overset{|}{\underset{|}{C}} \longleftrightarrow -\overset{O}{\overset{\|}{C}}-\overset{|}{\underset{|}{C}}{:}^- + -\overset{|}{\underset{|}{C}}{=}O \longrightarrow -\overset{O}{\overset{\|}{C}}-\overset{|}{\underset{|}{C}}-\overset{|}{\underset{|}{C}}-O^-$$

carbanion-enolate acceptor new bond

Aldol Condensation

The addition of the nucleophilic carbanion-enolate, usually of an aldehyde, to the C═O group of *its parent compound* is called an **aldol condensation**. The product is a β-hydroxycarbonyl compound. In a **mixed aldol condensation,** the carbanion-enolate of an aldehyde or ketone adds to the C═O group of a molecule other than its parent. The more general condensation diagrammed above is termed an **aldol-type condensation**. Since the C, not the O, is the more reactive site in the hybrid, the enolate contributing structure is usually omitted when writing equations for these reactions. This is done even though the enolate is the more stable and makes the major contribution.

Net Reactions

$$CH_3CH_2CHO + (H)-CH(CH_3)CHO \underset{}{\overset{\text{base}}{\rightleftharpoons}} CH_3CH_2CH(OH)-CH(CH_3)CHO$$

Propanal → 2-Methyl-3-hydroxypentanal

$$(CH_3)_2C{=}O + (H)-CH_2COCH_3 \overset{\text{base}}{\rightleftharpoons} (CH_3)_2C(OH)-CH_2COCH_3$$

Acetone → Diacetone alcohol (4-Hydroxy-4-methyl-2-pentanone)

Aldol condensations are reversible, and with ketones the equilibrium is unfavorable for the condensation product. To effect condensations of ketones, the product is continuously removed from the basic catalyst. β-Hydroxycarbonyl compounds are readily dehydrated to give α,β-unsaturated carbonyl compounds. With Ar on the β carbon, only the dehydrated product is isolated.

Problem 17.18 Suggest a mechanism for the OH^--catalyzed aldol condensation of acetaldehyde.

Step 1 $H-\ddot{O}:^- + H:CH_2-CH=O \rightleftharpoons H_2O + :\bar{C}H_2-CH=\ddot{O}: \longleftrightarrow H_2C=CH-\ddot{O}:^-$

Step 2 $CH_3CH=O + :\bar{C}H_2-CH=O \rightleftharpoons CH_3-CH(\text{-}\ddot{O}:^-)-CH_2-CH=O$

alkoxide ion of the β-hydroxyaldehyde

Step 3 $CH_3-CH(\ddot{O}:^-)-CH_2-CH=O + H:OH \rightleftharpoons CH_3-CH(HO)-CH_2-CH=O + OH^-$

Problem 17.19 Aldehydes and ketones also undergo acid-catalyzed aldol condensations. Devise a mechanism for this reaction in which an enol is an intermediate.

$H-\overset{|}{\underset{|}{C}}-\overset{OH}{\underset{+}{C}}- \; + \; -C{=}\overset{OH}{C}- \rightleftharpoons \left[H-C-C(OH)-C-C^+(:\ddot{O}H) \longleftrightarrow H-C-C(OH)-C-C{=}\overset{+}{\ddot{O}}H \right] \xrightleftharpoons{-H^+} H-C-C(OH)-C-C{=}O$

protonated carbonyl compound (electrophile) — enol (nucleophile) — aldol

Problem 17.20 Write structural formulas for the β-hydroxycarbonyl compounds and their dehydration products formed by aldol condensations of: (*a*) butanal, (*b*) phenylacetaldehyde, (*c*) diethyl ketone, (*d*) cyclohexanone, (*e*) benzaldehyde.

Acceptor **Carbanion Source**

(*a*) $CH_3CH_2CH_2CH{=}O + (H)-CH(C_2H_5)-CH{=}O \longrightarrow CH_3CH_2CH_2CH(OH)-CH(C_2H_5)-CH{=}O \xrightarrow{-H_2O} CH_3CH_2CH_2CH{=}C(C_2H_5)-CH{=}O$

mixture of geometric isomers

Acceptor **Carbanion Source**

(*b*) $PhCH_2CH{=}O + (H)-CH(Ph)-CH{=}O \longrightarrow PhCH_2CH(OH)-CH(Ph)-CH{=}O \xrightarrow{-H_2O} (PhCH_2)(H)C{=}C(CHO)(Ph)$

(c) $CH_3CH_2C(CH_2CH_3){=}O + H{-}C(H)(CH_3){-}C({=}O){-}CH_2CH_3 \longrightarrow CH_3CH_2C(CH_2CH_3)(OH){-}C(H)(CH_3){-}C({=}O){-}CH_2CH_3$

$\xrightarrow{-H_2O} CH_3CH_2C(CH_2CH_3){=}C(CH_3){-}CCH_2CH_3({=}O)$

(d) O, (H) O, + H ⟶ HO, O, H ⟶ O

(*e*) No aldol condensation, because C_6H_5CHO has no α H. Aldol condensations require dilute NaOH at room temperatures. With concentrated NaOH at higher temperatures, C_6H_5CHO undergoes the Cannizzaro reaction (Problem 15.18).

Problem 17.21 Mixed aldol condensations are useful if (*a*) one of the two aldehydes has no α H, (*b*) a symmetrical ketone reacts with RCHO. Explain and illustrate.

(*a*) The aldehyde with no α H, for example, H_2CO and C_6H_5CHO, is only a carbanion acceptor, so that only two products are possible.

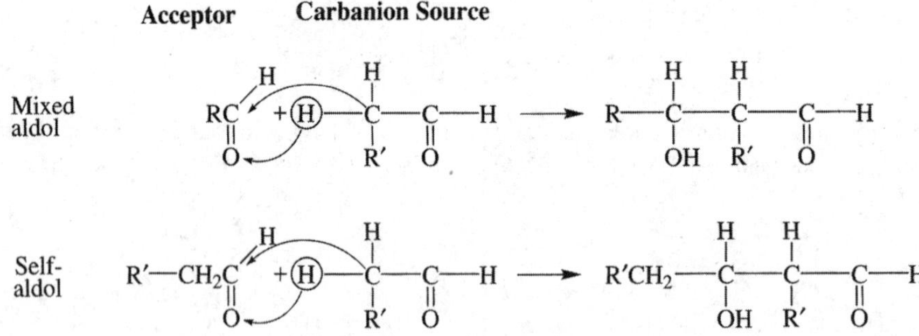

For reasonably good yields of mixed aldol product, the aldehyde with the α H should be added slowly to a large amount of the one with no α H.

(*b*) Ketones are poor carbanion acceptors but are carbanion sources. With symmetrical ketones and an RCHO having an α H, two products can be formed: (1) the self-aldol of RCHO and (2) the mixed aldol. If RCHO has no α H, only the mixed aldol results. As in part (*a*), the correct sequence of addition can give a good yield of the mixed aldol products.

Problem 17.22 Show how the following compounds are made from CH_3CH_2CHO. Do not repeat the synthesis of any compound needed in ensuing syntheses.

(*a*) $CH_3CH_2CH{=}C(CH_3)CHO$ (*b*) $CH_3CH_2CH_2CH(CH_3)CHO$ (*c*) $CH_3CH_2CH{=}C(CH_3)CH_2OH$
(*d*) $CH_3CH_2CH_2CH(CH_3)CH_2OH$ (*e*) $CH_3CH_2CH_2CH(CH_3)_2$ (*f*) $CH_3CH_2CH(OH)CH(CH_3)COOH$

Each product has six C's, which is twice the number of C's in CH_3CH_2CHO. This suggests an aldol condensation as the first step.

(*a*) $CH_3CH_2CHO \xrightarrow{OH^-} CH_3CH_2CH(OH)CH(CH_3)CHO \xrightarrow[-H_2O]{\Delta,\ H^+} CH_3CH_2CH{=}C(CH_3)CHO$ (A)

(*b*) —CHO can be protected by acetal formation to prevent its reduction when reducing C=C of (A).

$$(A) \xrightarrow[HCl]{CH_3OH} CH_3CH_2CH{=}C(CH_3)CH(OCH_3)_2 \xrightarrow{H_2/Pt} CH_3CH_2CH_2CH(CH_3)CH(OCH_3)_2 \xrightarrow[H_2O]{HCl} CH_3CH_2CH_2CH(CH_3)CHO$$

(There are specific catalysts that permit reduction only of C=C.)

(*c*) CHO is selectively reduced by $NaBH_4$.

$$(A) \xrightarrow{NaBH_4} CH_3CH_2CH{=}C(CH_3)CH_2OH$$

(*d*) $(A) \xrightarrow{H_2/Pt} CH_3CH_2CH_2CH(CH_3)CH_2OH$

(*e*) $(A) \xrightarrow[\Delta]{H_2NNH_2,\ OH^-} CH_3CH_2CH{=}C(CH_3)_2 \xrightarrow{H_2/Pt} CH_3CH_2CH_2CH(CH_3)_2$

(*f*) Tollens' reagent, $Ag(NH_3)_2^+$, is a specific oxidant for CHO ⟶ COOH.

$$CH_3CH_2CH(OH)CH(CH_3)CHO \xrightarrow[2.\ H^+]{1.\ Ag(NH_3)_2^+} CH_3CH_2CH(OH)CH(CH_3)COOH$$

Problem 17.23 Crotonaldehyde ($\overset{\gamma}{C}H_3\overset{\beta}{C}H{=}\overset{\alpha}{C}HCH{=}O$) undergoes an aldol condensation with acetaldehyde to form sorbic aldehyde ($CH_3CH{=}CH{-}CH{=}CH{-}CH{=}O$). Explain the reactivity and acidity of the γH.

Crotonaldehyde has C=C conjugated with C=O. On removal of the γH by base, the – charge on C is delocalized to O.

$$B\!: + H\!:\!\overset{|}{\underset{|}{C}}-\overset{|}{C}=\overset{|}{C}-\overset{|}{C}=O \longrightarrow B\!:\!H^+ + {}^-\!:\!\overset{|}{C}-\overset{|}{C}=\overset{|}{C}-\overset{|}{C}=\ddot{O}\!: \longleftrightarrow \overset{|}{C}=\overset{|}{C}-\overset{|}{C}=\overset{|}{C}-\ddot{O}\!:^- \text{ or } \overset{|}{C}{\cdots}\overset{|}{C}{\cdots}\overset{|}{C}{\cdots}\overset{|}{C}{\cdots}O$$

$$\text{or } \overset{|\,\delta-}{C}{\cdots}\overset{|}{C}{\cdots}\overset{|}{C}{\cdots}\overset{|}{C}{\cdots}\overset{\delta-}{O}$$

The nucleophilic carbanion adds to the carbonyl group of acetaldehyde.

$$CH_3\overset{H}{\underset{O}{C}} + {}^-\!:\!\overset{H}{\underset{H}{C}}-\overset{H}{C}=\overset{H}{C}-\overset{H}{C}=O \xrightarrow{H_2O} CH_3-\overset{H}{\underset{O-H}{C}}-\overset{H}{\underset{H}{C}}-\overset{H}{C}=\overset{H}{C}-\overset{H}{C}=O \xrightarrow{-H_2O} \text{product}$$

Problem 17.24 Use aldol condensations to synthesize the following useful compounds from cheap and readily available compounds: (*a*) the food preservative sorbic acid, $CH_3CH{=}CH{-}CH{=}CH{-}COOH$; (*b*) 2-ethyl-1-hexanol; (*c*) 2-ethyl-1,3-hexanediol, an insect repellant; (*d*) the humectant pentaerythritol, $C(CH_2OH)_4$.

(*a*) $2CH_3CH{=}O \xrightarrow[-H_2O]{OH^-} CH_3CH{=}CHCH{=}O \xrightarrow[OH^-]{CH_3CHO} CH_3CH{=}CHCH{=}CHCH{=}O \xrightarrow{\text{mild oxid.}} CH_3(CH{=}CH)_2COOH$

(*b*) $CH_3CH_2CH_2CH_2OH \xrightarrow[\text{heat}]{\text{Cu}} CH_3CH_2CH_2CH{=}O \xrightarrow{OH^-} CH_3CH_2CH_2CH(OH){-}CH(C_2H_5)CH{=}O \xrightarrow{-H_2O} CH_3CH_2CH_2CH{=}C(C_2H_5){-}CH{=}O \xrightarrow{H_2/Pt} CH_3CH_2CH_2CH_2{-}CH(C_2H_5){-}CH_2OH$

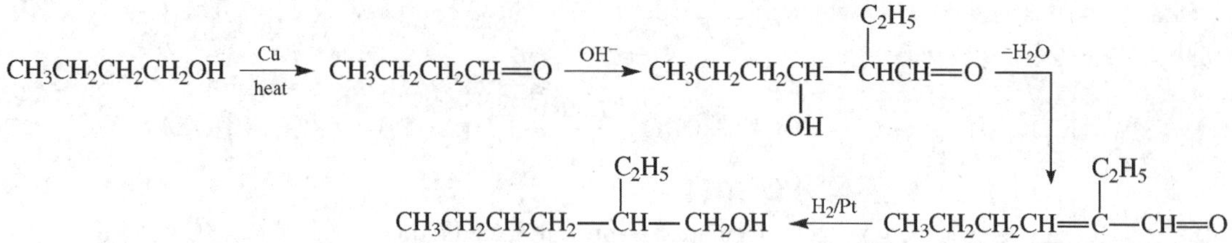

(*c*) As in (*b*) to

$$CH_3CH_2CH_2CH(OH){-}CH(C_2H_5){-}CH{=}O \xrightarrow{H_2/Pt} CH_3CH_2CH_2CH(OH){-}CH(C_2H_5){-}CH_2OH$$

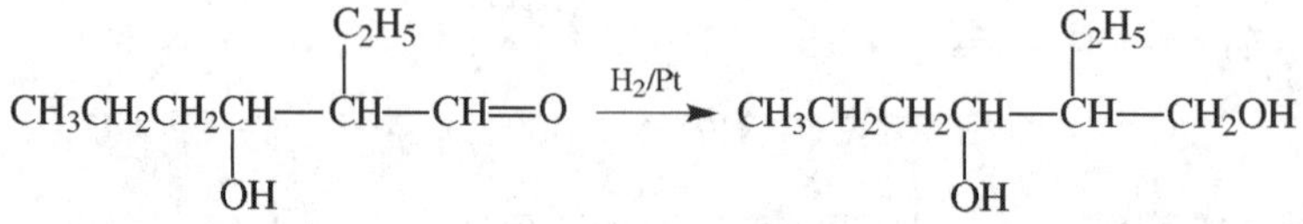

(*d*) One mole of CH_3CHO undergoes aldol condensation with 3 mol of H_2CO. A fourth mole of H_2CO then reacts with the product by a crossed-Cannizzaro reaction.

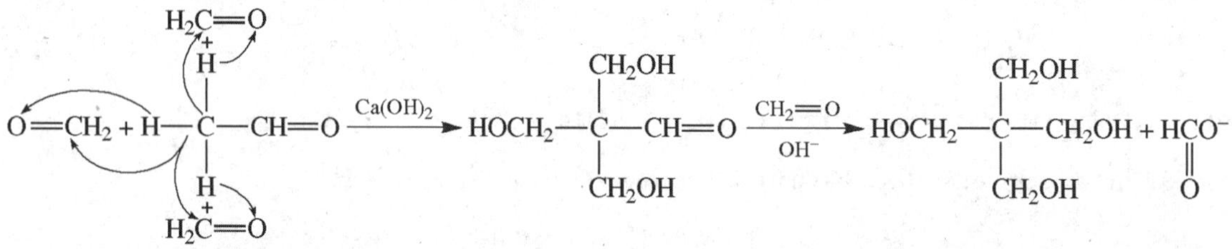

$$O{=}CH_2 + H{-}C(H)_2{-}CH{=}O \;(+\,2\,H_2C{=}O) \xrightarrow{Ca(OH)_2} HOCH_2{-}C(CH_2OH)_2{-}CH{=}O \xrightarrow[OH^-]{CH_2{=}O} HOCH_2{-}C(CH_2OH)_2{-}CH_2OH + HCO^-\,(\text{C}{=}O)$$

Problem 17.25 Which of the following alkanes can be synthesized from a self-aldol condensation product of an aldehyde [see Problem 17.24(*a*)] or a symmetrical ketone? (*a*) $CH_3CH_2CH_2CH_2CH(CH_3)CH_2CH_2CH_3$, (*b*) $CH_3CH_2CH_2CH_2CH(CH_3)CH_2CH_3$, (*c*) $(CH_3)_2CHCH_2CH_2CH_3$, (*d*) $(CH_3)_2CHCH_2C(CH_3)_3$, (*e*) $(CH_3)_2CHCH_2CH_2CH_2CH_2CH_3$, (*f*) $(CH_3CH_2)_2CHCH(CH_3)CH_2CH_2CH_3$.

The general formula for the aldol product from RR′CHCHO is

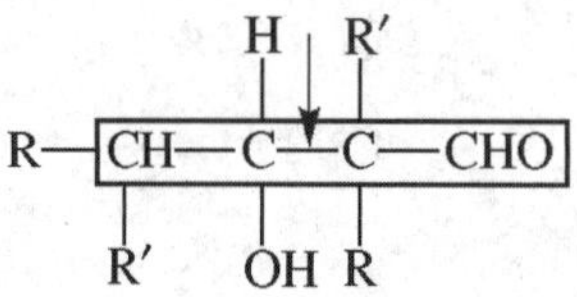

$$R{-}\boxed{CH(R'){-}CH(OH)\downarrow{-}C(R')(R){-}CHO}$$

The arrow points to the formed bond, and the α and C=O C's are in the rectangle. The alkane is

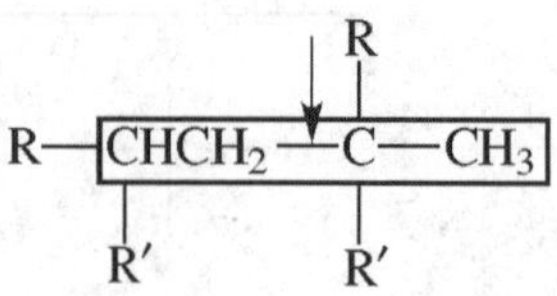

$$R{-}\boxed{CH(R')CH_2\downarrow{-}C(R)(R'){-}CH_3}$$

There is always a terminal CH_3 in this four-carbon sequence. From RR′CHCOCHRR′, the products are

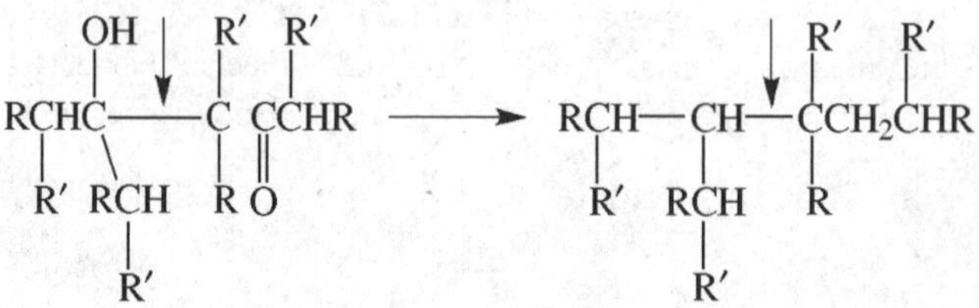

$$RCH(R'){-}C(OH)(CHRR')\downarrow{-}C(R')(R){-}C(=O)CHR(R') \longrightarrow RCH(R'){-}CH(CHRR')\downarrow{-}C(R')(R){-}CH_2CHR(R')$$

Each half must have the same skeleton of C's. Note that R and/or R′ can also be Ar or H. The alkane must always have an even number of C's (twice the number of C's of the carbonyl compound).

(*a*) No. There is an odd number of C's in the alkane.

(*b*) $CH_3CH_2\boxed{CH_2CH_2 \downarrow CHCH_3}$ with CH_2CH_3 on the CH

Yes. The four-carbon sequence has a terminal CH_3, and each half has the same sequence of C's. Use RR′CHCHO, where R = H and R′ = CH_2CH_3.

$$CH_3CH_2CH_2CHO \xrightarrow{OH^-} CH_3CH_2CH_2\underset{HO}{\underset{|}{C}}H\underset{CH_2CH_3}{\underset{|}{C}}HCHO \longrightarrow \text{alkane}$$

(*c*)

$$CH_3—\overset{CH_3}{\overset{|}{C}}H \downarrow CH_2CH_2CH_3$$

Yes. Each half has the same skeleton of C's. A ketone is needed; R = R′ = H.

$$(CH_3)_2C{=}O \xrightarrow{OH^-} (CH_3)_2\underset{OH}{\underset{|}{C}}—CH_2\underset{O}{\underset{||}{C}}—CH_3 \longrightarrow \text{alkane}$$

(*d*)

$$CH_3—\boxed{\overset{CH_3}{\overset{|}{\underset{H}{\underset{|}{C}}}}—CH_2—\overset{CH_3}{\overset{|}{\underset{CH_3}{\underset{|}{C}}}}—CH_3}$$

Yes. Use RR′CHCHO, R = R′ = CH_3.

$$(CH_3)_2CHCHO \xrightarrow{OH^-} CH_3—\overset{CH_3}{\overset{|}{\underset{H}{\underset{|}{C}}}}—\overset{H}{\overset{|}{\underset{HO}{\underset{|}{C}}}}—\overset{CH_3}{\overset{|}{\underset{CH_3}{\underset{|}{C}}}}—CHO \longrightarrow \text{alkane}$$

(*e*)

$$CH_3—\overset{CH_3}{\overset{|}{C}}H—CH_2 \downarrow CH_2CH_2CH_2CH_3$$

No. The formed bond is not part of a four-carbon sequence with a terminal CH_3, and the two halves do not have the same skeleton of C's; one half is branched, and the other half is not.

$$CH_3CH_2\overset{H}{\overset{|}{\underset{CH_2CH_3}{\underset{|}{C}}}}—\overset{H}{\overset{|}{\underset{CH_3}{\underset{|}{C}}}}—CH_2CH_2CH_3$$

Yes. Each half has the same skeleton of C's (5 C's in a row). Therefore, use a symmetrical ketone with 5 C's in a row (R = H, R′ = CH_3).

$$CH_3CH_2\overset{O}{\overset{||}{C}}CH_2CH_3 \xrightarrow{OH^-} CH_3CH_2\overset{H}{\overset{|}{\underset{CH_2CH_3}{\underset{|}{C}}}}—\overset{H}{\overset{|}{\underset{CH_3}{\underset{|}{C}}}}—\overset{O}{\overset{||}{C}}—CH_2CH_3 \longrightarrow \text{alkane}$$

Problem 17.26 Give the structural formulas for the products of the aldol-type condensations indicated in Table 17.2. See Table 17.1(*a*)–(*g*) for the carbanions of these condensations.

TABLE 17.2

	(*a*)	(*b*)	(*c*)	(*d*)	(*e*)	(*f*)	(*g*)
ACCEPTOR	PhCHO	PhCHO	Me_2CO	Me_2CO	Me_2CO	Ph_2CO	PhCHO
Base	OH^-	OH^-	OH^-	NH_2^-	OH^-	NH_2^-	NHR_2
Carbonion source	$CH_3\overset{+}{N}(O^-)=O$	$CH_3C{\equiv}N$	$CHCl_3$	$CH_3C{\equiv}CH$	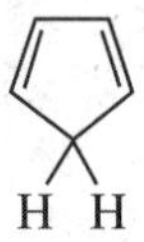 cyclopentadiene (CH_2, H H)	Ph_2CH_2	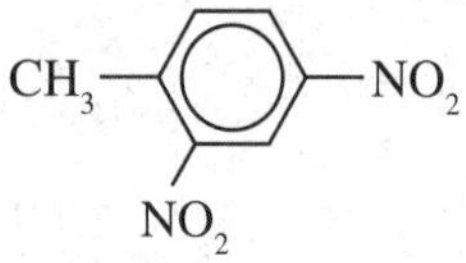 CH_3–C_6H_3–NO_2 (2,4-dinitrotoluene; NO_2, NO_2)

(*a*) *$PhCH{=}CHNO_2$ (*b*) *$PhCH{=}CHCN$ (*c*) $Me_2C(OH){-}CCl_3$ (*d*) $Me_2C(OH){-}C{\equiv}CCH_3$

(*e*) cyclopentadiene$={CMe_2}$ (*f*) $Ph_2C{=}CPh_2$ (*g*) *$PhCH{=}CH$–$C_6H_3(NO_2)$–NO_2 (2,4-dinitrophenyl)

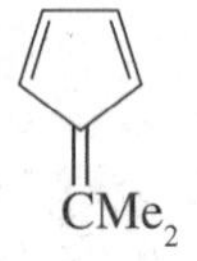

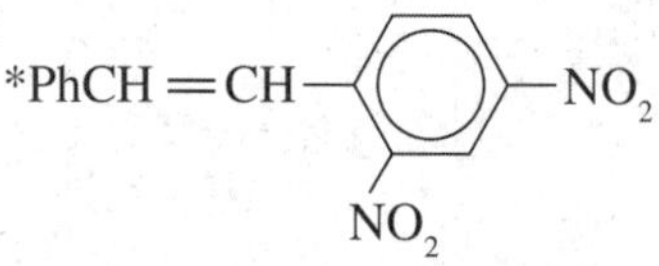

Problem 17.27 Give structures of the products from the following condensations:

(*a*) $p\text{-}CH_3C_6H_4CHO + (CH_3CH_2\overset{O}{\overset{\|}{C}})_2O \xrightarrow{CH_3CH_2COO^-Na^+}$

(*b*) Cyclohexanone + $CH_3CH_2NO_2 \xrightarrow{OH^-}$

(*c*) $C_6H_5CHO + C_6H_5CH_2C{\equiv}N \xrightarrow{OH^-}$

(*d*) Benzophenone + Cyclopentadiene $\xrightarrow{OH^-}$

(*e*) $CH_3COCH_3 + 2C_6H_5CHO \xrightarrow{OH^-}$

(*f*) cyclohexanone (C=O) $+ N{\equiv}CCH_2COOCH_3 \xrightarrow{CH_3COO^-NH_4^+}$

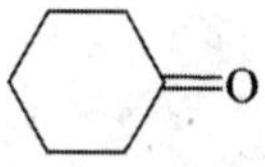

(*a*) This is a **Perkin condensation**.

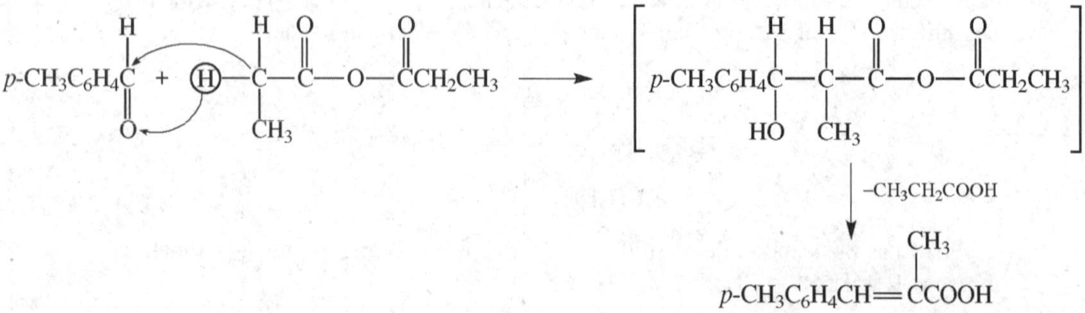

$$p\text{-}CH_3C_6H_4\overset{H}{\overset{|}{C}}{=}O + \text{(H)}{-}\underset{CH_3}{\overset{H}{C}}{-}\overset{O}{\overset{\|}{C}}{-}O{-}\overset{O}{\overset{\|}{C}}CH_2CH_3 \longrightarrow \left[p\text{-}CH_3C_6H_4\underset{HO}{\overset{H}{C}}{-}\underset{CH_3}{\overset{H}{C}}{-}\overset{O}{\overset{\|}{C}}{-}O{-}\overset{O}{\overset{\|}{C}}CH_2CH_3\right]$$

$$\xrightarrow{-CH_3CH_2COOH} p\text{-}CH_3C_6H_4CH{=}\overset{CH_3}{\overset{|}{C}}COOH$$

* The more stable *trans* product.

(*b*) cyclohexanone $=O$ + (H)$-\underset{CH_3}{\overset{H}{C}}-NO_2$ ⟶ [cyclohexane with OH and $CHNO_2$ (CH_3)] $\xrightarrow{-H_2O}$ cyclohexylidene$=\underset{CH_3}{C}NO_2$

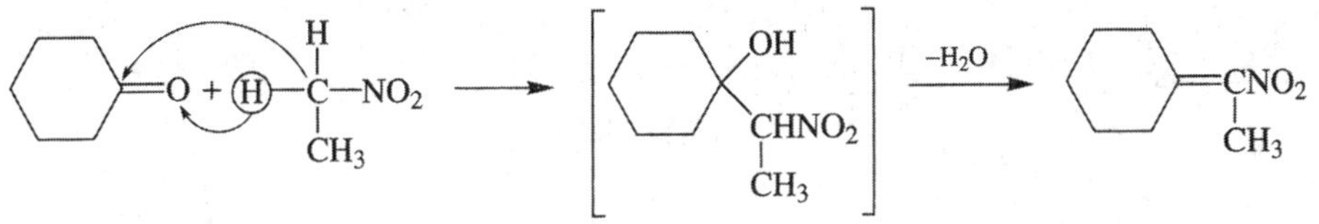

(*c*) $C_6H_5\overset{H}{C}=O$ + (H)$-\underset{C_6H_5}{\overset{H}{C}}-CN$ ⟶ $\left[C_6H_5\underset{OH}{\overset{H}{C}}-\underset{C_6H_5}{\overset{H}{C}}-CN\right]$ $\xrightarrow{-H_2O}$ $(C_6H_5)(H)C=C(CN)(C_6H_5)$

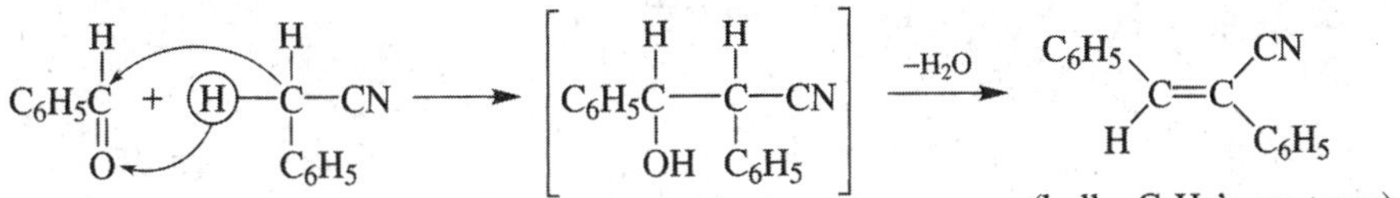

(bulky C_6H_5's are *trans*)

(*d*) $Ph_2C=O$ + (H)-cyclopentadiene ⟶ [cyclopentadiene with H and Ph_2C(OH)] $\xrightarrow{-H_2O}$ $Ph_2C=$cyclopentadienylidene

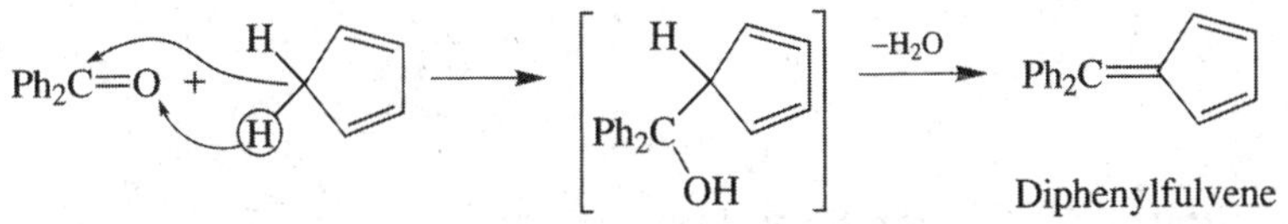

Diphenylfulvene

(*e*) Each CH_3 of $(CH_3)_2CO$ reacts with one PhCHO.

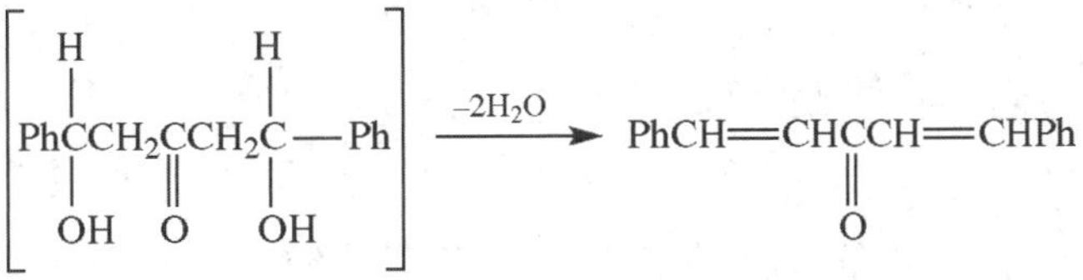

$$\left[Ph\underset{OH}{\overset{H}{C}}CH_2\underset{O}{\overset{\|}{C}}CH_2\underset{OH}{\overset{H}{C}}-Ph\right] \xrightarrow{-2H_2O} PhCH=CH\underset{O}{\overset{\|}{C}}CH=CHPh$$

(*f*) This is the **Cope reaction.**

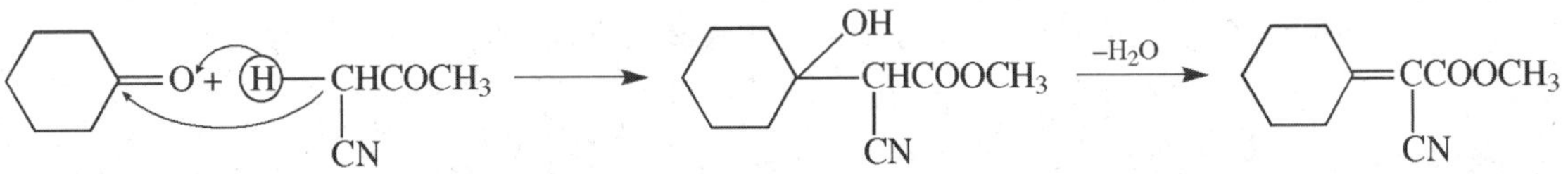

cyclohexanone $=O$ + (H)$-\underset{CN}{CH}COCH_3$ ⟶ cyclohexane(OH)$-\underset{CN}{CH}COOCH_3$ $\xrightarrow{-H_2O}$ cyclohexylidene$=\underset{CN}{C}COOCH_3$

Problem 17.28 In the **Knoevenagel reaction**, aldehydes or ketones condense with compounds having a reactive CH_2 between two C=O groups. The cocatalysts are *both* a weak base ($RCOO^-$) *and* a weak acid ($R_2NH_2^+$). Outline the reaction between $C_6H_5CH=O$ and $H_2C(COOEt)_2$.

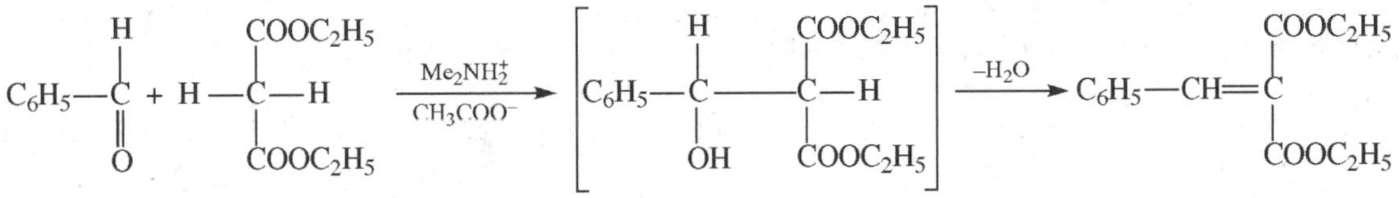

$$C_6H_5-\underset{O}{\overset{H}{C}}= + H-\underset{COOC_2H_5}{\overset{COOC_2H_5}{C}}-H \xrightarrow[CH_3COO^-]{Me_2NH_2^+} \left[C_6H_5-\underset{OH}{\overset{H}{C}}-\underset{COOC_2H_5}{\overset{COOC_2H_5}{C}}-H\right] \xrightarrow{-H_2O} C_6H_5-CH=\underset{COOC_2H_5}{\overset{COOC_2H_5}{C}}$$

Ethyl malonate

Problem 17.29 Prepare *trans*-cinnamic acid, $C_6H_5CH=CHCOOH$, by a Perkin condensation [Problem 17.27(*a*)]:

$$\underset{\text{Benzaldehyde}}{PhCHO} + \underset{\text{Acetic anhydride}}{(CH_3CO)_2O} \xrightarrow[\text{heat}]{NaOAc} PhCH=CHCOOH$$

Dehydration of the β-hydroxyester occurs on workup because the resulting C=C is conjugated with Ph.

Problem 17.30 The C of the $-C\equiv N$ group is an electrophilic site capable of being attacked by a carbanion. Show how nitriles like $CH_3CH_2C\equiv N$ undergo an aldol-type condensation (**Thorpe reaction**) with hindered bases.

$$CH_3-\underset{H}{\overset{H}{C}}-C\equiv N: + \xrightarrow[-H^+]{R_2N:^-Li^+} \left[CH_3-\overset{H}{\ddot{C}}-C\equiv N: \longleftrightarrow CH_3\overset{H}{C}=C=\ddot{N}:^-\right]Li^+ + R_2N:H$$

$$CH_3CH_2C\equiv N: + \left[\overset{CN}{:\overset{-}{C}HCH_3}\right]Li^+ \longrightarrow \left[CH_3CH_2\underset{:N^-}{\overset{}{C}}-\overset{CN}{C}HCH_3\right]Li^+ \xrightarrow{H:OH}$$

$$CH_3CH_2\underset{O}{\overset{}{C}}-\overset{CH_3}{C}HCN + NH_3 \xleftarrow[RT]{H_3O^+} LiOH + CH_3CH_2\underset{NH}{C}-\overset{CH_3}{C}H-CN$$

an iminonitrile

Claisen Condensation: Acylation of Carbanion-Enolates

In a Claisen condensation, the carbanion-enolate of an ester adds to the C=O group of its parent ester. The addition is followed by loss of the OR group of the ester to give a β-ketoester. In a **mixed Claisen condensation,** the carbanion-enolate adds to the C=O group of a molecule other than its parent.

Step 1 Formation of a stabilized α-carbanion.

$$RCH_2\underset{\ddot{O}:}{C}-OR' + ^-OR' \xrightarrow{-R'OH} \left[R\ddot{C}H-\underset{\ddot{O}:}{C}-OR' \longleftrightarrow RCH=\underset{:\ddot{O}:^-}{C}-OR'\right] \text{ or } RCH\text{---}\underset{O}{\overset{-}{C}}-OR$$

Step 2 Nucleophilic attack by α-carbanion on C=O of ester and displacement of $^-OR'$.

$$R\bar{\ddot{C}}HCOOR' + RCH_2\underset{:O}{C}-OR' \rightleftharpoons \left[RCH_2-\underset{:\ddot{O}:}{\overset{OR'}{C}}-\underset{R}{C}HCOOR'\right] \xrightarrow{-R'O^-} RCH_2-\underset{O}{C}-\underset{R}{C}HCOOR$$

This step is reminiscent of a transacylation (Section 16.6).

Step 3 The only irreversible step completes the reaction by forming a stable carbanion where negative charge is delocalized to both O's.

$$\underset{\text{a }\beta\text{-ketoester}}{RCH_2\underset{O}{C}-\underset{R}{C}H-\underset{O}{C}OR'} + ^-\ddot{O}R' \xrightarrow{-R'OH} RCH_2\underset{O}{C}-\underset{R}{\bar{\ddot{C}}}-\underset{O}{C}-OR' \text{ or } RCH_2\underset{O}{C}\text{---}\underset{R}{C}\text{---}\underset{O}{C}OR'$$

Acid is then added to neutralize the carbanion salt.

$$\underset{\substack{Na^+\ \textit{salt if}\\ NaOR'\ \textit{is used as base}}}{Na^+[RCH_2\underset{O}{C}-\bar{\ddot{C}}RCOOR']} \xrightarrow{HCl} \underset{\textit{a }\beta\textit{-ketoester}}{RCH_2\underset{O}{C}CHRCOOR'} + Na^+Cl^-$$

Problem 17.31 Write structural formulas for the products from the reaction of $C_2H_5O^-Na^+$ with the following esters:

(*a*) $CH_3CH_2COOC_2H_5$ (*b*) $C_6H_5COOC_2H_5 + CH_3COOC_2H_5$
(*c*) $C_6H_5CH_2COOC_2H_5 + O{=}C(OC_2H_5)_2$

In these Claisen condensations, $^-OC_2H_5$ is displaced from the $COOC_2H_5$ group by the α-carbanion formed from another ester molecule. Mixed Claisen condensations are feasible only if one of the esters has no α H.

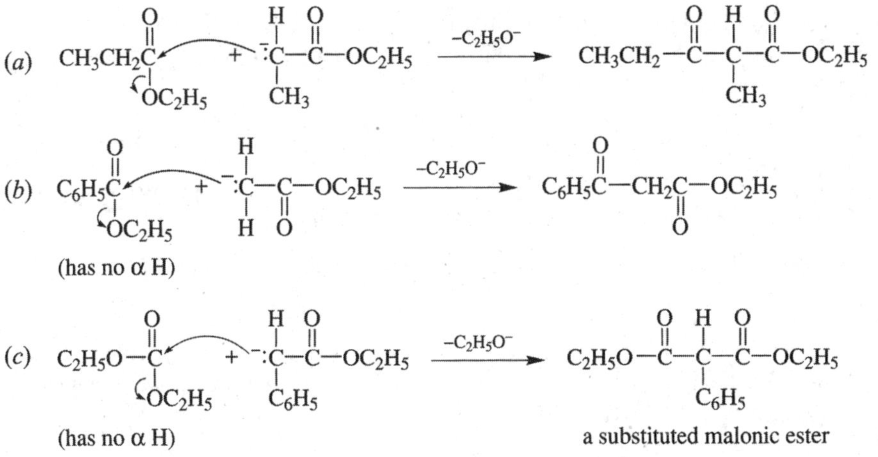

(*a*) $CH_3CH_2C(=O)(OC_2H_5) + {}^-:CH(CH_3)-C(=O)-OC_2H_5 \xrightarrow{-C_2H_5O^-} CH_3CH_2-C(=O)-CH(CH_3)-C(=O)-OC_2H_5$

(*b*) $C_6H_5C(=O)(OC_2H_5) + {}^-:CH_2-C(=O)-OC_2H_5 \xrightarrow{-C_2H_5O^-} C_6H_5C(=O)-CH_2C(=O)-OC_2H_5$

(has no α H)

(*c*) $C_2H_5O-C(=O)(OC_2H_5) + {}^-:CH(C_6H_5)-C(=O)-OC_2H_5 \xrightarrow{-C_2H_5O^-} C_2H_5O-C(=O)-CH(C_6H_5)-C(=O)-OC_2H_5$

(has no α H) a substituted malonic ester

Problem 17.32 Ethyl pimelate, $C_2H_5OOC(CH_2)_5COOC_2H_5$, react $C_2H_5O^-Na^+$ (**Dieckmann condensation**) to form a cyclic ketoester, $C_9H_{14}O_3$. Supply a mechanism for its formation and compare the yields in ethanol and ether as solvents.

We have

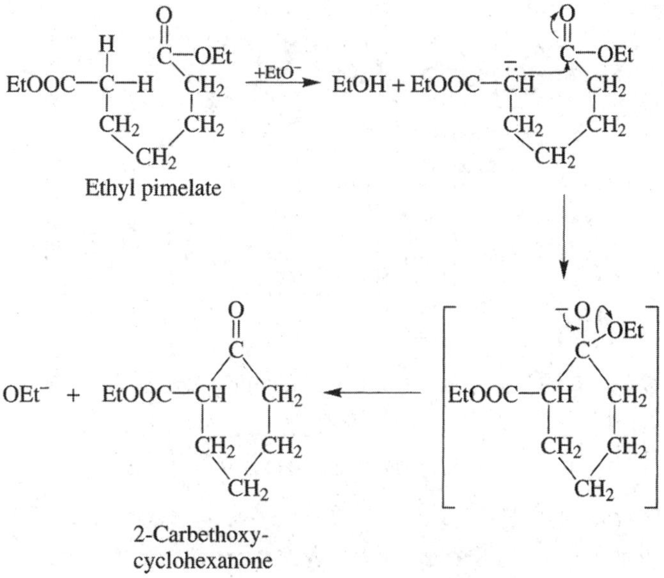

$EtOOC-CH_2-CH_2CH_2CH_2CH_2-C(=O)-OEt$ (Ethyl pimelate) $\xrightarrow{+EtO^-}$ $EtOH + EtOOC-\ddot{C}H^--CH_2CH_2CH_2CH_2-C(=O)-OEt$

$\downarrow$

$\left[\text{cyclic tetrahedral intermediate: } EtOOC-CH-C(O^-)(OEt)\text{ in six-membered ring with } (CH_2)_4\right] \longrightarrow OEt^- + \text{EtOOC-CH-C(=O) ring (2-Carbethoxycyclohexanone)}$

2-Carbethoxy-
cyclohexanone

The net reaction is $C_2H_5OOC(CH_2)_5COOC_2H_5 \rightarrow$ product + HOC_2H_5.

Since the reaction is reversible, yields are greater in ether than in alcohol because alcohol is a product (Le Chatelier principle).

Intramolecular Claisen cyclizations occur with ethyl adipate and pimelate because five- and six-membered rings are formed.

SUPPLEMENTARY PROBLEMS

Problem 17.33 Account for the fact that tricyanomethane, $(CN)_3CH$, is a strong acid ($K_a = 1$).

The conjugate base, $(CN)_3C\!:^-$, is extremely weak because its negative charge is delocalized, by extended π bonding, to the N of each CN group. Hence, the acid is strong.

Problem 17.34 Acetone reacts with LDA in THF and then with trimethylsilyl chloride, $(CH_3)_3SiCl$, at −78 °C, to give an **enolsilane**. (*a*) Give equations for the reactions. (*b*) Why does O- rather than C-silylation occur?

(*a*)
$$CH_3\underset{\underset{O}{\|}}{C}CH_3 \xrightarrow{\text{LDA}} \left[CH_3\underset{\underset{O^-}{|}}{C}{=}CH_2 \longleftrightarrow CH_3\underset{\underset{O}{\|}}{C}{-}\ddot{C}H_2^-\right] \xrightarrow{(CH_3)_3SiCl} CH_3\underset{\underset{OSi(CH_3)_3}{|}}{C}{=}CH_2$$

(*b*) The O—Si bond is much stronger than the C—Si bond because of *p-d* π bonding between the O (*p*) and Si (*d*) atoms.

Problem 17.35 Prepare 4-methyl-1-hepten-5-one from $(CH_3CH_2)_2C{=}O$ and other needed compounds.

$$CH_3CH_2\overset{\overset{O}{\|}}{C}{-}\underset{\underset{CH_3}{|}}{CH}(CH_2CH{=}CH_2)$$

has an allyl group substituted on the α C of diethyl ketone. This substitution is best achieved through the enamine reaction.

$$CH_3CH_2{-}\underset{\underset{C_2H_5}{|}}{C}{=}O + HN\langle\text{pyrrolidine}\rangle \longrightarrow CH_3\underset{\underset{H}{|}}{\ddot{C}^-}{-}\underset{\underset{C_2H_5}{|}}{C}{=}N^+\langle\text{pyrrolidine}\rangle \xrightarrow{ClCH_2CH=CH_2}$$

an enamine

$$CH_3\overset{\overset{CH_2CH=CH_2}{|}}{\underset{\underset{H}{|}}{C}}{-}\underset{\underset{C_2H_5}{|}}{C}{=}\overset{+}{N}\langle\text{pyrrolidine}\rangle\ Cl^- \xrightarrow[HCl]{H_2O} CH_3{-}\overset{\overset{CH-CH=CH_2}{|}}{\underset{\underset{H}{|}}{C}}{-}\underset{\underset{C_2H_5}{|}}{C}{=}O + \langle\text{pyrrolidine}\rangle\overset{+}{N}H_2\,Cl^-$$

Problem 17.36 Prepare 3-phenylpropenoic acid from malonic ester and $C_6H_5CH_2Br$.

$$CH_2(COOC_2H_5)_2 \xrightarrow[2.\ C_6H_5CH_2Br]{1.\ NaOEt} \boxed{C_6H_5CH_2}CH(COOC_2H_5)_2 \xrightarrow[2.\ H_3O^+,\ \Delta]{1.\ OH^-}$$

$$C_6H_5CH_2CH_2COOH \xrightarrow{Br_2/P} C_6H_5CH_2\underset{\underset{Br}{|}}{CH}{-}COOH \xrightarrow{\text{alc. KOH}} \underset{\text{Cinnamic acid}}{C_6H_5CH{=}CH{-}COOH}$$

Problem 17.37 Can the following ketones be prepared by the acetoacetic ester synthesis? Explain. (*a*) $CH_3COCH_2C_6H_5$, (*b*) $CH_3COCH_2C(CH_3)_3$.

(*a*) No. C_6H_5Br is aromatic and does not react in an S_N2 displacement. (*b*) No. $BrC(CH_3)_3$ is a 3° bromide which undergoes elimination rather than substitution.

Problem 17.38 Use acetoacetic ester (AAE) and any needed alkyl halide or dihalide to prepare: (*a*) $CH_3COCH_2CH_2COCH_3$, (*b*) cyclobutyl methyl ketone, (*c*) $CH_3COCH_2CH_2CH_2COCH_3$, and (*d*) 1,3-diacetylcyclopentane.

The portion of all these compounds that comes from AAE is

$$CH_3-\underset{\underset{O}{\|}}{C}-\overset{\overset{H}{|}}{\underset{|}{C}}-$$

Encircle this portion and obtain the rest of the molecule from the alkyl halide(s).

(*a*) $$\boxed{CH_3\underset{\underset{O}{\|}}{C}CH_2}-\boxed{CH_2\underset{\underset{O}{\|}}{C}CH_3}$$

Bond two molecules of AAE at the acidic CH_2 group of each with NaOEt and I_2.

$$2CH_3COCH_2COOEt \xrightarrow[2.\ I_2]{1.\ 2NaOEt} \begin{matrix} CH_3\overset{\overset{O}{\|}}{C}CHCOOEt \\ | \\ CH_3\underset{\underset{O}{\|}}{C}CHCOOEt \end{matrix} \xrightarrow[2.\ H_3O^+]{1.\ dil.\ OH^-} CH_3\underset{\underset{O}{\|}}{C}CH_2CH_2\underset{\underset{O}{\|}}{C}CH_3 + CO_2$$

(*b*) $$\square-\boxed{\overset{\overset{H}{|}}{C}-\underset{\underset{O}{\|}}{C}-CH_3}$$

We need three C's in the halogen compound, and the two terminal C's are bonded to the acidic CH_2 group of AAE. The halide is $BrCH_2CH_2CH_2Br$.

$$CH_3\underset{\underset{O}{\|}}{C}-\overset{COOEt}{\overset{|}{C}}H_2 + \begin{matrix} BrCH_2 \\ BrCH_2 \end{matrix}\!\!>CH_2 \xrightarrow{2NaOEt} CH_3-\underset{\underset{O}{\|}}{C}-\overset{COOEt}{\overset{|}{C}}\!\!<\begin{matrix} CH_2 \\ CH_2 \end{matrix}\!\!>CH_2 \xrightarrow[2.\ H_3O^+]{1.\ dil.\ NaOH} CH_3-\underset{\underset{O}{\|}}{C}-\overset{H}{C}\!\!<\!\square + CO_2$$

(*c*) $$\boxed{CH_3COCH_2}-CH_2-\boxed{CH_2COCH_3}$$

Join two molecules of AAE through the acidic CH_2 of each with two molecules of NaOEt and one molecule of $BrCH_2Br$.

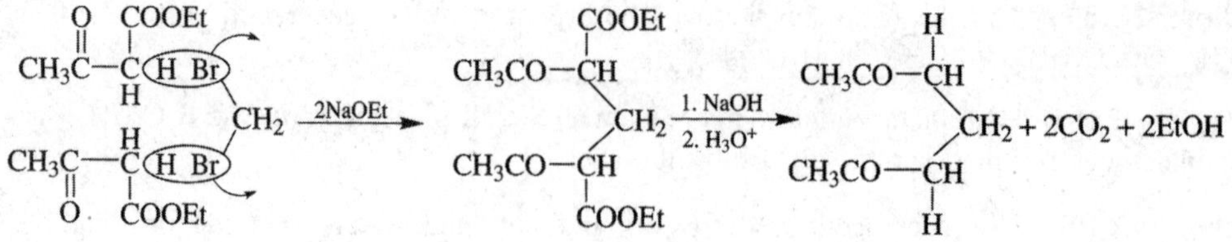

(*d*)

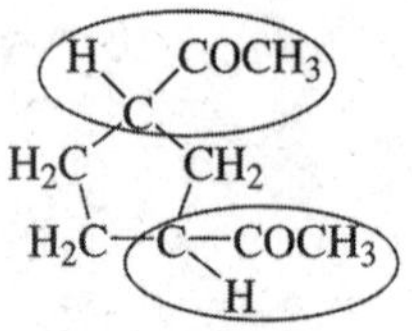

Two molecules of AAE are first bonded together with one molecule of $BrCH_2CH_2Br$, and then the ring is closed with $BrCH_2Br$.

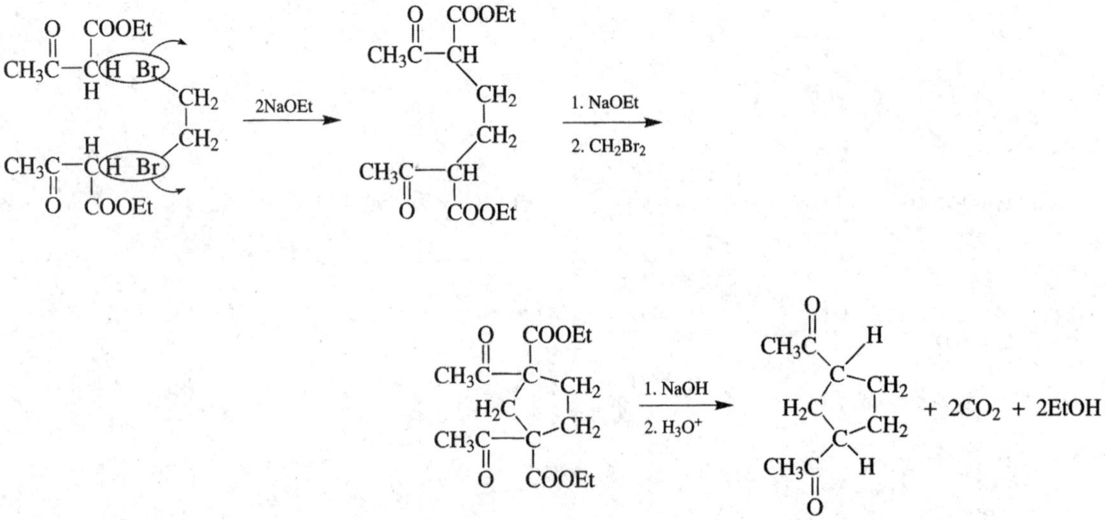

Problem 17.39 The **Robinson "annelation" reaction** for synthesizing fused rings uses Michael addition, followed by intramolecular aldol condensation. Illustrate with cyclohexanone and methyl vinyl ketone, $CH_2{=}CHCOCH_3$.

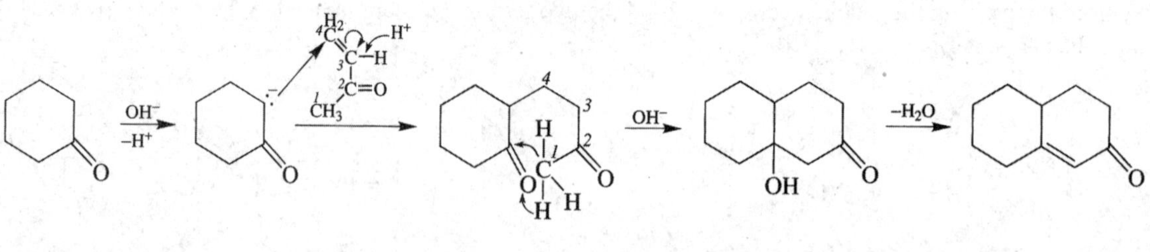

Problem 17.40 Give the ester or combination of esters needed to prepare the following by a Claisen condensation.

(*a*) $C_6H_5CH_2CH_2C(=O)—CH(CH_2C_6H_5)COOEt$ (*b*) $EtOC(=O)—C(=O)—CH(CH_3)COOEt$ (*c*) $H—C(=O)—CH(C_6H_5)COOEt$

In the Claisen condensation, the bond formed is between the carbonyl C and the C that is α to COOR. Work backwards by breaking this C—C bond and adding OR to the carbonyl C and adding H to the other C. Mixed Claisens are practical if one ester has no α H.

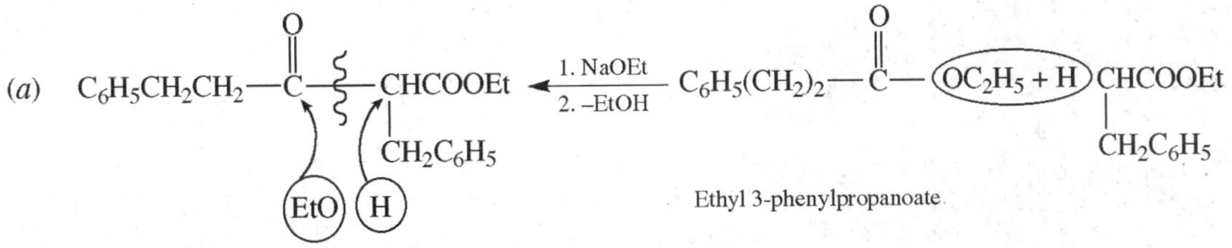

$$(a)\quad C_6H_5CH_2CH_2—\overset{O}{\overset{\|}{C}}—CHCOOEt \xleftarrow[2.\ -EtOH]{1.\ NaOEt} C_6H_5(CH_2)_2—\overset{O}{\overset{\|}{C}}—OC_2H_5 + H—CHCOOEt$$

with $CH_2C_6H_5$ on the CH of each; (EtO)(H) removed.

Ethyl 3-phenylpropanoate

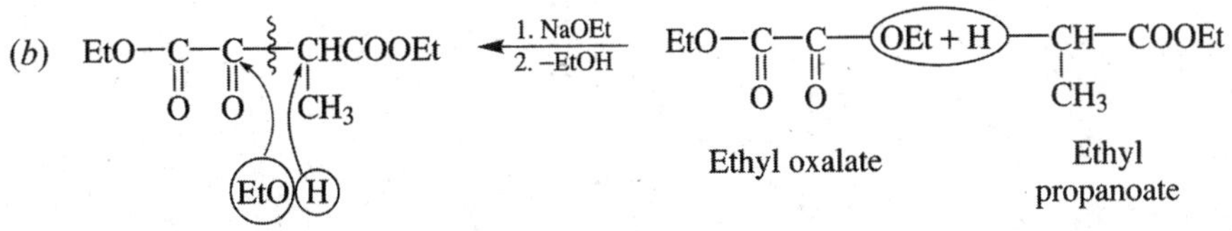

$$(b)\quad EtO—\underset{O}{\underset{\|}{C}}—\underset{O}{\underset{\|}{C}}—\underset{CH_3}{\underset{|}{C}}HCOOEt \xleftarrow[2.\ -EtOH]{1.\ NaOEt} EtO—\underset{O}{\underset{\|}{C}}—\underset{O}{\underset{\|}{C}}—OEt + H—\underset{CH_3}{\underset{|}{C}}H—COOEt$$

(EtO)(H)

Ethyl oxalate — Ethyl propanoate

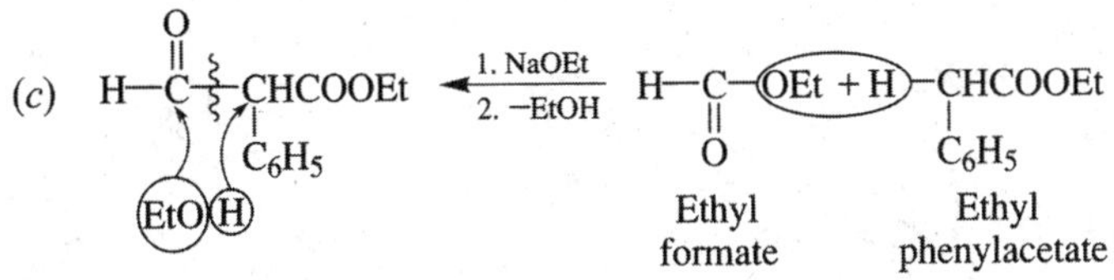

$$(c)\quad H—\overset{O}{\overset{\|}{C}}—\underset{C_6H_5}{\underset{|}{C}}HCOOEt \xleftarrow[2.\ -EtOH]{1.\ NaOEt} H—\underset{O}{\underset{\|}{C}}—OEt + H—\underset{C_6H_5}{\underset{|}{C}}HCOOEt$$

(EtO)(H)

Ethyl formate — Ethyl phenylacetate

Problem 17.41 In the biochemical conversion of the sugar glucose to ethanol (**alcoholic fermentation**), a key step is

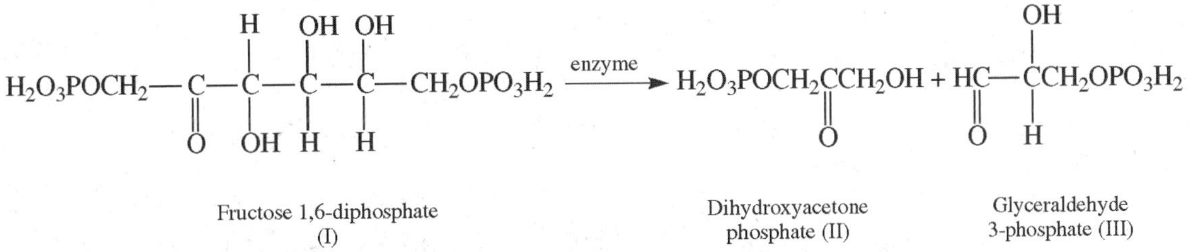

$$H_2O_3POCH_2—\underset{O}{\underset{\|}{C}}—\overset{H}{\underset{OH}{C}}—\overset{OH}{\underset{H}{C}}—\overset{OH}{\underset{H}{C}}—CH_2OPO_3H_2 \xrightarrow{\text{enzyme}} H_2O_3POCH_2\underset{O}{\underset{\|}{C}}CH_2OH + H\underset{O}{\underset{\|}{C}}—\overset{OH}{\underset{H}{C}}CH_2OPO_3H_2$$

Fructose 1,6-diphosphate (I) — Dihydroxyacetone phosphate (II) — Glyceraldehyde 3-phosphate (III)

Formulate this reaction as a reversal of an aldol condensation (**retroaldol condensation**).

(I) is a β-hydroxyketone. Loss of a proton from the C^{β}—OH affords an alkoxide (IV) that undergoes a retroaldol condensation by cleavage of the C^{α}—C^{β} bond.

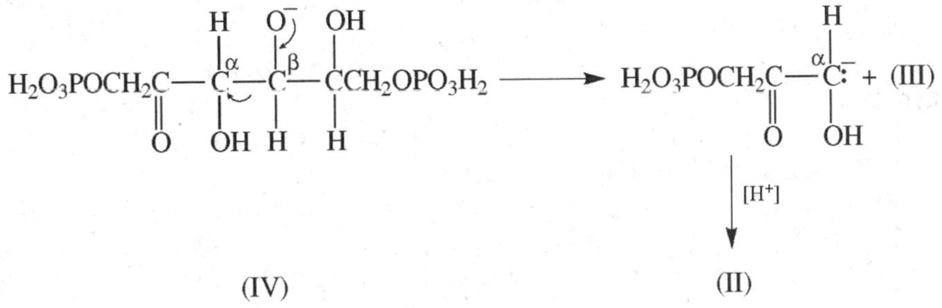

$$\underset{(IV)}{H_2O_3POCH_2\underset{O}{\underset{\|}{C}}—\overset{H}{\underset{OH}{C^{\alpha}}}—\overset{O^-}{\underset{H}{C^{\beta}}}—\overset{OH}{\underset{H}{C}}CH_2OPO_3H_2} \longrightarrow H_2O_3POCH_2\underset{O}{\underset{\|}{C}}—\overset{H}{\underset{OH}{\overset{\alpha}{C}:^-}} + (III)$$

$$\downarrow [H^+]$$

(II)

Amines

18.1 Nomenclature and Physical Properties

Amines are alkyl derivatives of NH_3. Replacing one, two, or three H's of NH_3 gives **primary** (1°), **secondary** (2°), and **tertiary** (3°) amines, respectively.

	1° amine	2° amine	3° amine
NH_3 Ammonia	RNH_2 (e.g. CH_3NH_2)	R'–NH–R, i.e. RNH with R' on N $\left(C_2H_5NH\text{–}CH_3\right)$	R_3N $\left(CH_3N(CH_3)CH_3\right)$
	$ArNH_2$ ($C_6H_5NH_2$)	$ArNH$–R $\left(C_6H_5NH\text{–}CH_3\right)$	$ArNR'$–R $\left(C_6H_5N(C_2H_5)CH_3\right)$

Amines are named by adding the suffix **-amine** to the name of (*a*) the alkyl group attached to N or (*b*) the longest alkane chain. The terminal *e* in the name of the parent alkane is dropped when "amine" follows but not when, for example, "diamine" follows [see Problem 18.1(*d*)]. Thus, $CH_3CH(NH_2)CH_2CH_3$ is named *sec*-butylamine or 2-butanamine. Amines, especially with other functional groups, are named by considering **amino**, *N*-**alkylamino** and *N,N*-**dialkylamino** as substituents on the parent molecule; *N* indicates substitution on nitrogen.

Aromatic and cyclic amines often have common names such as **aniline** (benzenamine), $C_6H_5NH_2$; *p*-**toluidine**, *p*-$CH_3C_6H_4NH_2$; and **piperidine** [Problem 18.1(*g*)].

Like the **oxa** method for naming ethers (Problem 14.61), the **aza** method is used for amines. Di-*n*-propylamine, $CH_3CH_2CH_2NHCH_2CH_2CH_3$, is 4-**azaheptane** and piperidine is **azacyclohexane**.

The four H's of NH_4^+ can be replaced to give a **quaternary** (4°) tetraalkyl (tetraaryl) ammonium ion.

$$C_6H_5CH_2\overset{+}{N}(CH_3)_3\ OH^- \qquad \text{(Triton-B)}$$

is **benzyltrimethylammonium hydroxide**.

Problem 18.1 Name and classify the following amines:

(*a*) $(CH_3)_3CNH_2$ (*b*) $(CH_3)_2NCH(CH_3)_2$ (*c*) $C_6H_5N(CH_3)_2$

(*d*) $H_2NCH_2CH_2CH_2NH_2$ (*e*) $CH_3NHCH(CH_3)CH_2CH_3$ (*f*) $CH_3NHCH_2NHCH_3$

(*g*) 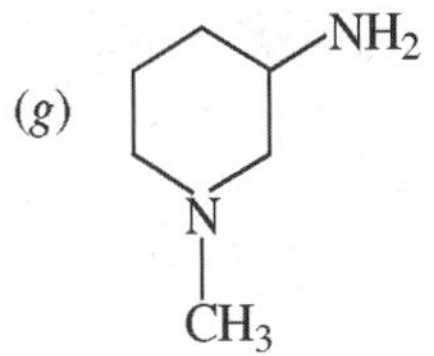

(*h*) $CH_3CH_2CH_2\overset{+}{N}(CH_3)_3Cl^-$

(*a*) *t*-butylamine or 2-methyl-2-propanamine, 1°; (*b*) dimethylisopropylamine, 3°; (*c*) *N,N*-dimethylaniline, 3°, (*d*) 1,3-propanediamine (or trimethylenediamine), both 1°; (*e*) 2-(*N*-methylamino)butane, 2°; (*f*) 2,5-diazahexane, both 2°; (*g*) 3-amino-*N*-methylpiperidine or 1-methyl-3-amino-1-azacyclohexane, 1° (N of NH_2) and 3°; (*h*) *n*-propyl-trimethylammonium chloride, 4°.

Problem 18.2 Give names for

(*a*) CH_3NHCH_3 (*b*) $CH_3NHCH(CH_3)_2$ (*c*) $CH_3CH_2CH(NH_2)COOH$

(*d*) $NHCH_3$, CH_3 (*e*) $^+N(CH_3)_3Br^-$ (*f*) $NHCH_3$, $NHCH_3$

(*a*) dimethylamine, (*b*) methylisopropylamine or 2-(*N*-methylamino)propane, (*c*) 2-aminobutanoic acid, (*d*) *N*-methyl-*m*-toluidine or 3-(*N*-methylamino)toluene, (*e*) trimethylanilinium bromide, (*f*) 3,4′-*N,N*′-methylaminobiphenyl (note the use of *N* and *N*′ to designate the different *N*'s on the separate rings).

Problem 18.3 Predict the orders of (*a*) boiling points, and (*b*) solubilities in water, for 1°, 2°, and 3° amines of identical molecular weights.

Both physical properties depend on the ability of the amino group to form H-bonds.

(*a*) Intermolecular H-bonding, as shown in Fig. 18.1 with four molecules of RNH_2, influences the boiling point. The more H's on N, the greater is the extent of H-bonding, the greater is the intermolecular attraction and the higher is the boiling point. A 1° amine, with two H's, can crosslink as in Fig. 18.1; a single amine molecule can H-bond with three other molecules. A 2° amine molecule can H-bond only with two other molecules. A 3° amine, has no H's on N and cannot form intermolecular H-bonds. The decreasing order of boiling points is RNH_2 (1°) > R_2NH (2°) > R_3N (3°).

```
        R          R          R
       /          /          /
----N—H-----N—H-----N—H----
    |          |          |
    H          H          H
    ¦          ¦          ¦
             R—N—H
               /
              H
```

Figure 18.1

(*b*) Water solubility depends on H-bonding between the amine and H_2O. Either the H of H_2O bonds with the N of the amine or the H on N bonds with the O of H_2O:

$$-\overset{|}{\underset{|}{N}}\text{---}H-O-H \quad \text{or} \quad -\overset{|}{N}-H\text{---}OH_2$$

All three kinds of amines exhibit the first type of H-bonding, which is thus a constant factor. The more H's on N, the more extensive is the second kind of H-bond and the more soluble is the amine. Thus, the order of water solubility is RNH_2 (1°) > R_2NH (2°) > R_3N (3°).

Problem 18.4 Does *n*-propylamine or 1-propanol have the higher boiling point?

Since N is less electronegative than O (3.1 < 3.5), amines form weaker H-bonds than do alcohols of similar molecular weights. The less effective intermolecular attraction causes the amine to have a lower boiling point than the alcohol (49°C < 97°C).

18.2 Preparation

By Nucleophilic Displacements

1. **Alkylation of NH_3, RNH_2 and R_2NH with RX**

Step 1 $$RX + NH_3 \longrightarrow \underset{\text{an ammonium salt}}{RNH_3^+X^-} \quad (\text{an } S_N2 \text{ reaction})$$

Step 2 $$RNH_3^+X^- + NH_3 \longrightarrow RNH_2 + NH_4^+X^-$$

Di-, tri-, and tetraalkylation:

$$\underset{1°}{RNH_2} \xrightarrow[-HX]{RX} \underset{2°}{R_2NH} \xrightarrow[-HX]{RX} \underset{3°}{R_3N} \xrightarrow[-HX]{RX} \underset{\text{4° ammonium salt}}{R_4N^+X^-}$$

When RX = MeI, the sequence is called **exhaustive methylation**.

Problem 18.5 (*a*) What complication arises in the synthesis of a 1° amine from the reaction of RX and NH_3? (*b*) How is this complication avoided? (*c*) Which halides cannot be used in this synthesis?

(*a*) Overalkylation: di- and tri-alkylation can occur, since both RNH_2 and R_2NH, once formed, can react with more RX. (*b*) Use a large excess of NH_3 to increase the probability of RX colliding with NH_3 to give RNH_2, rather than with RNH_2 to give R_2NH. (*c*) 3° RX undergoes elimination rather than S_N2 displacement. Aryl halides do not undergo S_N2 reactions. Some ArX compounds with properly substituted rings undergo nucleophilic aromatic substitution (Section 11.3) with amines.

Problem 18.6 Describe an efficient industrial preparation of allylamine, $H_2C{=}CHCH_2NH_2$, from propene, chlorine, and ammonia.

Propene is readily monochlorinated to allyl chloride, and the reactive Cl undergoes nucleophilic substitution with NH_3.

$$H_2C{=}CH{-}CH_3 \xrightarrow[400°C]{Cl_2} H_2C{=}CH{-}CH_2Cl \xrightarrow[-HCl]{NH_3} H_2C{=}CH{-}CH_2NH_2$$

2. **Alkylation of Imides; Gabriel synthesis of 1° Amines**

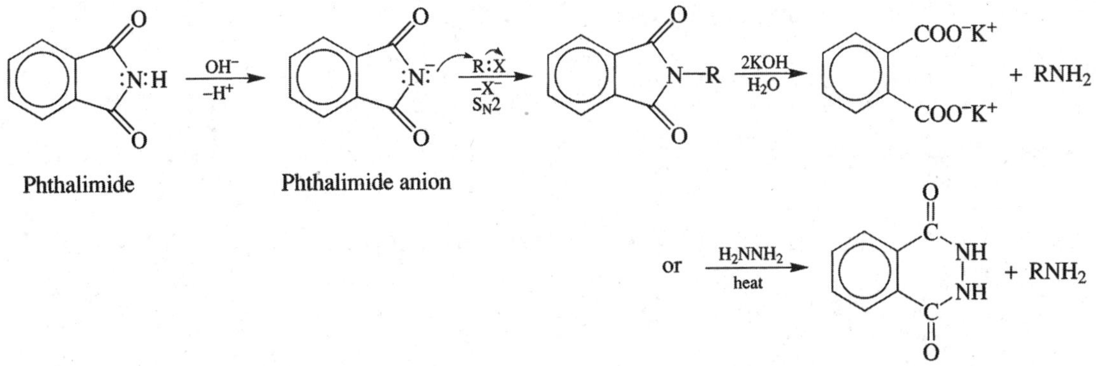

Phthalimide $\xrightarrow[-H^+]{OH^-}$ Phthalimide anion $\xrightarrow[-X^-,\ S_N2]{R:X}$ N-alkylphthalimide (N—R) $\xrightarrow[H_2O]{2KOH}$ $C_6H_4(COO^-K^+)_2$ + RNH_2

or $\xrightarrow[heat]{H_2NNH_2}$ Phthalhydrazide + RNH_2

See Problem 16.47 for the acidity of imides.

Reduction of N-Containing Compounds

1. **Nitro Compounds**

$$\underset{\text{Nitrobenzene}}{C_6H_5NO_2} \xrightarrow[\text{or } H_2/Pt]{1.\ Zn,\ HCl;\ 2.\ OH^-} \underset{\text{Aniline}}{C_6H_5NH_2}$$

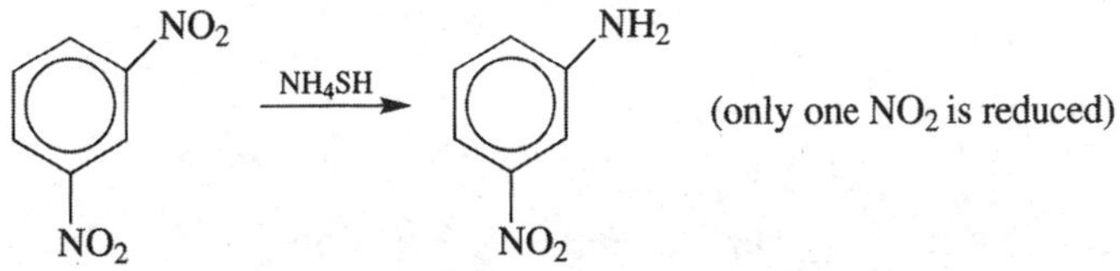

$$\underset{m\text{-Dinitrobenzene}}{C_6H_4(NO_2)_2} \xrightarrow{NH_4SH} \underset{m\text{-Nitroaniline}}{m\text{-}O_2NC_6H_4NH_2}$$ (only one NO_2 is reduced)

2. **Nitriles**

$$RCN \xrightarrow{LiAlH_4} RCH_2NH_2$$

3. **Amides**

$$RC(=O)—NR_2 \xrightarrow{LiAlH_4} RCH_2NR_2 \quad (R = H,\ \text{alkyl},\ \text{aryl})$$

4. **Oximes**

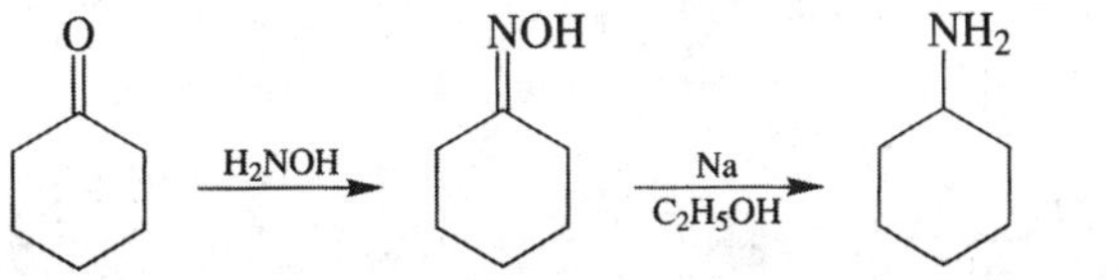

Cyclohexanone $\xrightarrow{H_2NOH}$ Oxime (C=NOH) $\xrightarrow[C_2H_5OH]{Na}$ Cyclohexylamine (NH_2)

5. **Azides**

$$\underset{\text{azide ion}}{RX + N_3^-} \xrightarrow{-X^-} \underset{\text{alkyl azide}}{RN_3} \xrightarrow[\text{or Na, EtOH}]{LiAlH_4 \text{ or } H_2/Pt} \underset{1^\circ \text{ amine}}{RNH_2}$$

This method is superior to the reaction of NH_3 and RX for the preparation of RNH_2 because no polyalkylation occurs. However, alkyl azides are explosive and must be carefully handled. Rather than being isolated, they should be kept in solution and used as soon as made.

6. Reductive Amination of Carbonyl Compounds

$$CH_3CH{=}O \xrightarrow[-H_2O]{NH_3,\ H_2/Ni} \underset{\text{an imine}}{[CH_3CH{=}NH]} \longrightarrow CH_3CH_2NH_2 \quad (NH_3 \longrightarrow 1^\circ \text{ amine})$$

$$CH_3CH_2CHO + CH_3CH_2NH_2 \xrightarrow{H_2/Ni} CH_3CH_2CH_2NHCH_2CH_3 \quad (1^\circ \text{ amine} \longrightarrow 2^\circ \text{ amine})$$

$$R_2C{=}O + R'_2NH \xrightarrow[H_2Ni]{-OH^-} [R_2C{=}\overset{+}{N}R'_2] \longrightarrow R_2CH{-}NR'_2 \quad (2^\circ \text{ amine} \longrightarrow 3^\circ \text{ amine})$$

$$RNH_2 + 2H_2C{=}O + 2HCOOH \longrightarrow RN(CH_3)_2 + 2H_2O + 2CO_2 \quad (\text{Dimethylation of } 1^\circ \text{ amine})$$

Hofmann Degradation of Amides

$RCONH_2 + Br_2 + 4KOH \rightarrow RNH_2 + K_2CO_3 + 2KBr + 2H_2O$ (The amine has one less C than the amide.)

Mechanism:

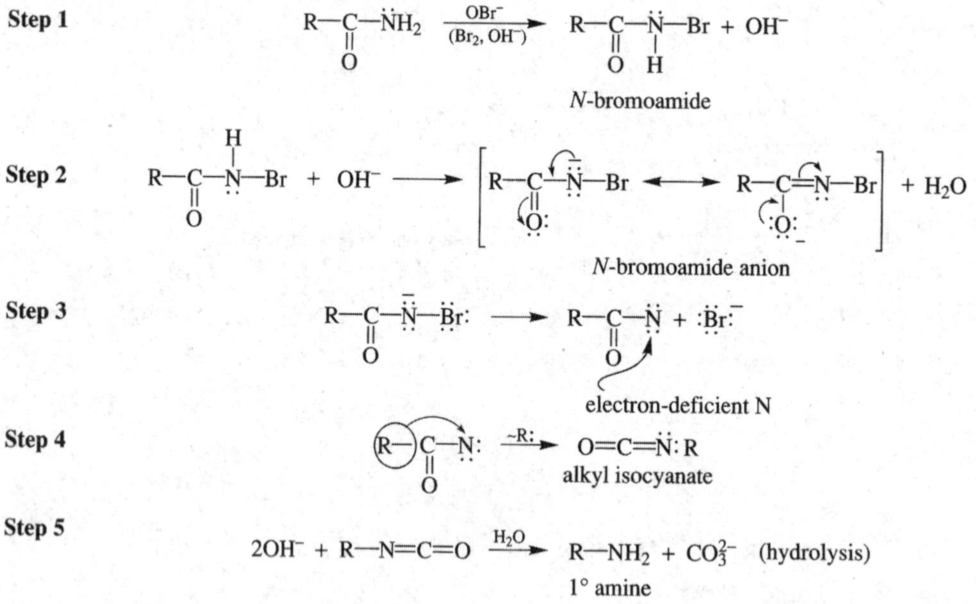

Step 1

$$R{-}\underset{\underset{O}{\|}}{C}{-}\ddot{N}H_2 \xrightarrow[(Br_2,\ OH^-)]{OBr^-} R{-}\underset{\underset{O}{\|}}{C}{-}\underset{\underset{H}{|}}{\ddot{N}}{-}Br + OH^-$$

N-bromoamide

Step 2

$$R{-}\underset{\underset{O}{\|}}{C}{-}\overset{\overset{H}{|}}{\underset{\cdot\cdot}{N}}{-}Br + OH^- \longrightarrow \left[R{-}\underset{\underset{\ddot{O}:}{\|}}{C}{-}\overset{-}{\underset{\cdot\cdot}{N}}{-}Br \longleftrightarrow R{-}\underset{\underset{:\ddot{O}:^-}{|}}{C}{=}\underset{\cdot\cdot}{N}{-}Br \right] + H_2O$$

N-bromoamide anion

Step 3

$$R{-}\underset{\underset{O}{\|}}{C}{-}\overset{-}{\underset{\cdot\cdot}{N}}{-}\ddot{B}r: \longrightarrow R{-}\underset{\underset{O}{\|}}{C}{-}\ddot{N} + :\ddot{B}r:^-$$

electron-deficient N

Step 4

$$R{-}\underset{\underset{O}{\|}}{C}{-}\ddot{N}: \xrightarrow{\sim R:} \underset{\text{alkyl isocyanate}}{O{=}C{=}\ddot{N}{:}R}$$

Step 5

$$2OH^- + R{-}N{=}C{=}O \xrightarrow{H_2O} \underset{1^\circ \text{ amine}}{R{-}NH_2} + CO_3^{2-} \quad (\text{hydrolysis})$$

Problem 18.7 Synthesize 2-phenylethanamine, $PhCH_2CH_2NH_2$, from styrene, $PhCH{=}CH_2$, by the azide reduction method.

$$PhCH{=}CH_2 \xrightarrow[\text{peroxide}]{HBr} PhCH_2CH_2Br \xrightarrow{NaN_3} PhCH_2CH_2N_3 \xrightarrow[EtOH]{Na} PhCH_2CH_2NH_2$$

Problem 18.8 Prepare ethylamine by (*a*) Gabriel synthesis, (*b*) alkyl halide amination, (*c*) nitrile reduction, (*d*) reductive amination, (*e*) Hofmann degradation.

(*a*) [potassium phthalimide] $\xrightarrow[-Br^-]{C_2H_5Br}$ [N-ethylphthalimide] $\xrightarrow[OH^-]{H_2O}$ [phthalic acid, o-$C_6H_4(COOH)_2$] + $C_2H_5NH_2$

(*b*) $C_2H_5Br \xrightarrow[NH_3]{excess} C_2H_5NH_2$

(*c*) $CH_3CN \xrightarrow[2.\ H_2O]{1.\ LiAlH_4} CH_3CH_2NH_2$

(*d*) $CH_3CH{=}O \xrightarrow[H_2/Ni]{NH_3} CH_3CH_2NH_2$

(*e*) $CH_3CH_2CONH_2 \xrightarrow[NH_3]{Br_2,\ KOH} CH_3CH_2NH_2$

Problem 18.9 Prepare *p*-toluidine from toluene.

[toluene, $C_6H_5CH_3$] $\xrightarrow[H_2SO_4]{HNO_3}$ [p-nitrotoluene, $p\text{-}CH_3C_6H_4NO_2$] $\xrightarrow[2.\ OH^-]{1.\ Sn,\ HCl}$ [p-toluidine, $p\text{-}CH_3C_6H_4NH_2$]

This is the best way of substituting an NH_2 on a phenyl ring.

Problem 18.10 Synthesize the following compounds from $n\text{-}C_{12}H_{25}COOH$ and inorganic reagents:

(*a*) $C_{14}H_{29}NH_2$ (*b*) $C_{13}H_{27}NH_2$ (*c*) $C_{12}H_{25}NH_2$

(*d*) $C_{12}H_{25}{-}CH(NH_2){-}C_{13}H_{27}$ (*e*) $C_{12}H_{25}NHC_{13}H_{27}$ (*f*) $(C_{13}H_{27})_2NH$

Do not repeat preparation of any needed compound.

First, note the change, if any, in carbon content.

(*a*) Chain length is increased by one C by reducing RCH_2CN ($R = n\text{-}C_{12}H_{25}$), prepared from RCH_2Br and CN^-.

$$n\text{-}C_{12}H_{25}COOH \xrightarrow{LiAlH_4} n\text{-}C_{12}N_{25}CH_2OH \xrightarrow{PBr_3} n\text{-}C_{12}H_{25}CH_2Br \xrightarrow{KCN} n\text{-}C_{13}H_{27}CN \xrightarrow{LiAlH_4} n\text{-}C_{14}H_{29}NH_2$$

(*b*) The chain length is unchanged.

$$n\text{-}C_{12}H_{25}COOH \xrightarrow{SOCl_2} n\text{-}C_{12}H_{25}COCl \xrightarrow{NH_3} n\text{-}C_{12}H_{25}CONH_2 \xrightarrow{LiAlH_4} n\text{-}C_{13}H_{27}NH_2$$

(*c*) Chain length is decreased by one C; use the Hofmann degradation.

$$n\text{-}C_{12}H_{25}CONH_2 \xrightarrow{NaOBr} n\text{-}C_{12}H_{25}NH_2$$

(*d*) The C content is doubled. The 1° amine is made from the corresponding ketone.

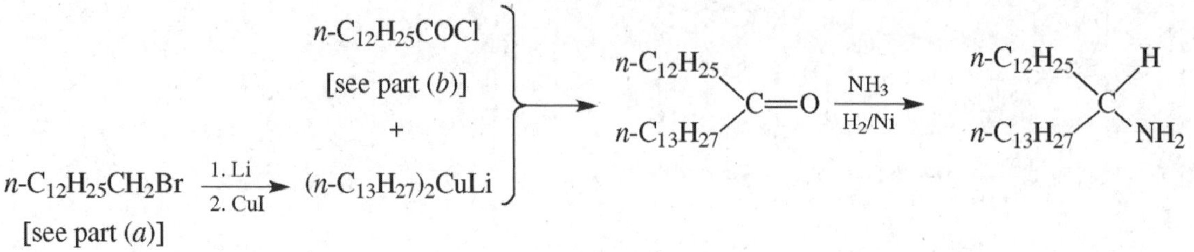

$$\left.\begin{array}{c} n\text{-}C_{12}H_{25}COCl \\ \text{[see part (b)]} \\ + \\ n\text{-}C_{12}H_{25}CH_2Br \xrightarrow[2.\ CuI]{1.\ Li} (n\text{-}C_{13}H_{27})_2CuLi \end{array}\right\} \longrightarrow (n\text{-}C_{12}H_{25})(n\text{-}C_{13}H_{27})C{=}O \xrightarrow[H_2/Ni]{NH_3} (n\text{-}C_{12}H_{25})(n\text{-}C_{13}H_{27})CH\text{-}NH_2$$

[see part (*a*)]

(*e*) Secondary amines may be prepared by reductive amination of an aldehyde (from RCOCl reduction), using a 1° amine.

$$n\text{-}C_{12}H_{25}COCl \xrightarrow[S,\ \Delta]{H_2,\ Pd/BaSO_4} n\text{-}C_{12}H_{25}CHO \xrightarrow[H_2/Ni]{n\text{-}C_{12}H_{25}NH_2} n\text{-}C_{13}H_{27}NHC_{12}H_{25}$$

(*f*) The acyl halide $C_{12}H_{25}COCl$ from (*e*) is reacted with the amine $C_{13}H_{27}NH_2$ from (*b*), to form an amide that is then reduced.

$$C_{12}H_{25}COCl + H_2NC_{13}H_{27} \longrightarrow C_{12}H_{25}\text{—}\overset{\overset{\displaystyle O}{\|}}{C}\text{—}NHC_{13}H_{27} \xrightarrow{LiAlH_4} (C_{13}H_{27})_2NH$$

Problem 18.11 There is no change in the configuration of the chiral C in *sec*-butylamine formed from the Hofmann degradation of (*S*)-2-methylbutanamide. Explain.

:R migrates with its electron pair to the electron-deficient :N̤ and configuration is retained because C—C is being broken at the same time that C—N is being formed in the transition state.

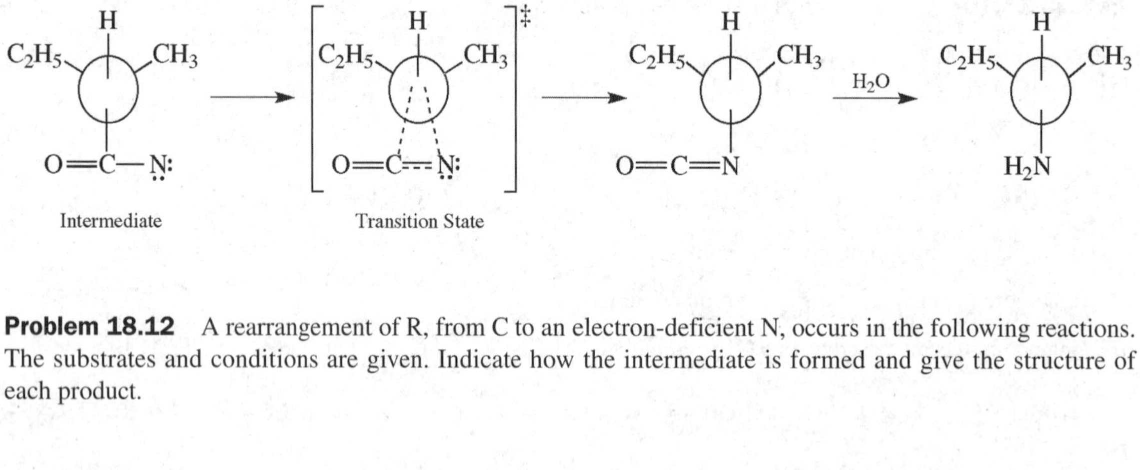

[Newman projections: Intermediate (H, C_2H_5, CH_3 on front carbon; O=C—N̤: behind) → Transition State [H, C_2H_5, CH_3; O=C---N̤:]‡ → (H, C_2H_5, CH_3; O=C=N) $\xrightarrow{H_2O}$ (H, C_2H_5, CH_3; H_2N)]

Intermediate

Transition State

Problem 18.12 A rearrangement of R, from C to an electron-deficient N, occurs in the following reactions. The substrates and conditions are given. Indicate how the intermediate is formed and give the structure of each product.

(*a*) **Curtius,** $R\overset{O}{\overset{\|}{C}}\text{—}N_3$ (an acylazide) with heat or **Schmidt,** RCOOH + HN_3 with H_2SO_4

(*b*) **Lossen,** $R\overset{O}{\overset{\|}{C}}\text{—}NHOH$ (a hydroxamic acid) with base

(*c*) **Beckmann,** $R\text{—}\underset{\underset{\displaystyle NOH}{\|}}{C}\text{—}R'$ with strong acid

(*a*) $R-\underset{\underset{O}{\|}}{C}-\bar{\ddot{N}}{:}\ \overset{+}{N}\equiv N{:} \xrightarrow{\Delta} {:}N\equiv N{:} + \left[R-\underset{\underset{O}{\|}}{C}-\ddot{N}{:}\right] \longrightarrow O=C=N-R \xrightarrow[H_2O]{OH^-} RNH_2 + CO_3^{2-}$

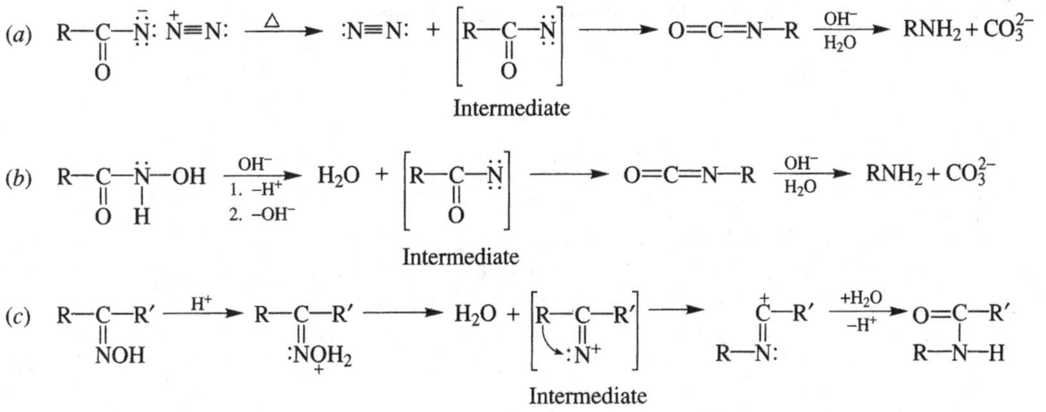

Intermediate

(*b*) $R-\underset{\underset{O}{\|}}{C}-\underset{\underset{H}{|}}{\ddot{N}}-OH \xrightarrow[1.\ -H^+\ \ 2.\ -OH^-]{OH^-} H_2O + \left[R-\underset{\underset{O}{\|}}{C}-\ddot{N}{:}\right] \longrightarrow O=C=N-R \xrightarrow[H_2O]{OH^-} RNH_2 + CO_3^{2-}$

Intermediate

(*c*) $R-\underset{\underset{NOH}{\|}}{C}-R' \xrightarrow{H^+} R-\underset{\underset{:\overset{}{N}OH_2^+}{\|}}{C}-R' \longrightarrow H_2O + \left[R-\underset{\underset{:N^+}{\|}}{C}-R'\right] \longrightarrow \underset{\underset{R-N:}{\|}}{\overset{+}{C}}-R' \xrightarrow[-H^+]{+H_2O} O=\underset{\underset{R-N-H}{|}}{C}-R'$

Intermediate

The group *trans* to the OH (R) migrates as H_2O leaves.

18.3 Chemical Properties

Stereochemistry

Amines with three different substituents and an unshared pair of electrons have enantiomers. However, in most cases an :NRR′R″-type amine *cannot be resolved*. The amine undergoes a very rapid **nitrogen inversion** similar to that for a C undergoing an S_N2 reaction (Fig. 18.2).

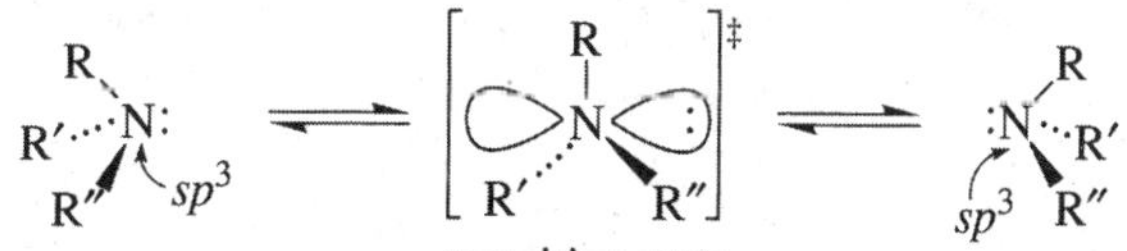

$$R,R',R''\text{-}N{:}\ (sp^3) \rightleftharpoons [\text{transition state}]^{\ddagger} \rightleftharpoons {:}N\text{-}R,R',R''\ (sp^3)$$

transition state

Figure 18.2

Simple carbanions, $\rangle C{:}^-$, behave similarly.

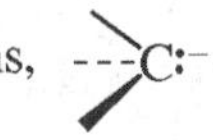

Problem 18.13 Which of the following compounds are (i) chiral, (ii) resolvable? Give reasons in each case.

(*a*) $[C_6H_5\overset{+}{N}(C_2H_5)_2(CH_3)]Br^-$ (*b*) $C_6H_5N(CH_3)(C_2H_5)$

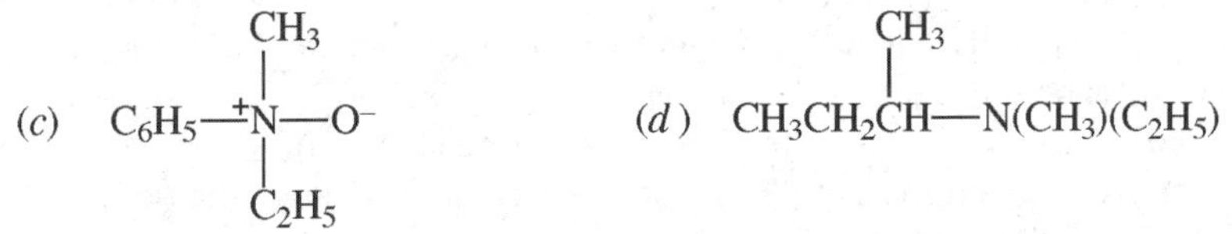

(*c*) $C_6H_5-\overset{+}{N}(CH_3)(C_2H_5)-O^-$ (*d*) $CH_3CH_2\overset{CH_3}{\overset{|}{C}}H-N(CH_3)(C_2H_5)$

(*a*) Not chiral and not resolvable, because N has two identical groups (C_2H_5).
(*b*) Chiral, but the low energy barrier (25 kJ/mol) to inversion of configuration prevents resolution of its enantiomers.
(*c*) Chiral and resolvable. N has four different substituents. The absence of an unshared pair of electrons on N prevents inversion as in part (*b*).
(*d*) Chiral and resolvable. An asymmetric C is present.

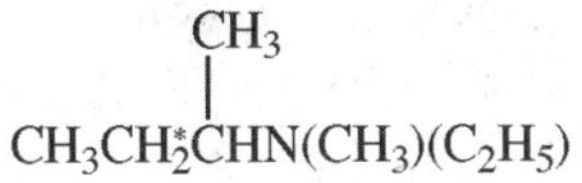

$$CH_3CH_2\overset{CH_3}{\overset{|}{C^*}}HN(CH_3)(C_2H_5)$$

Problem 18.14 Explain why 1,2,2-trimethylaziridine has isolable enantiomers.

$$\text{(structure: aziridine ring with } C(CH_3)_2 \text{ and } :N{-}CH_3\text{)}$$

This 3° amine is chiral because N has three different ligands and an unshared pair of e^-'s. Unlike typical amines, this molecule cannot undergo nitrogen inversion. The 3-membered ring requires ring angles of approximately 60° and restrains the N atom from attaining bond angles of 120°, as needed for the inversion transition state.

Basicity and Salt Formation [see Problem 3.25(*b*)]

The lone pair of electrons on the N atom of amines accounts for their base strength and nucleophilicity. They abstract protons from water, react with Lewis acids, and attack electrophilic sites such as carbonyl carbon.

Problem 18.15 (*a*) Why does aqueous CH_3NH_2 turn litmus blue? (*b*) Why does $C_6H_5NH_2$ dissolve in aqueous HCl?

(*a*) Methylamine ($pK_b = 3.36$) is a weak base, but a stronger base than water.

$$\underset{\text{base}_1}{CH_3NH_2} + \underset{\text{acid}_2}{H_2O} \rightleftharpoons \underset{\text{acid}_1}{CH_3NH_3^+} + \underset{\text{base}_2}{OH^-}$$

(*b*) A water-soluble salt forms.

$$C_6H_5NH_2 + H_3O^+ + Cl^- \longrightarrow \underset{\text{Anilinium chloride}}{C_6H_5NH_3^+Cl^-} + H_2O$$

Problem 18.16 Account for the following observations. (*a*) In the gas phase, the order of increasing basicity is $NH_3 < CH_3NH_2 < (CH_3)_2NH < (CH_3)_3N$. (*b*) In water, the order is $NH_3 < CH_3NH_2 \approx (CH_3)_3N < (CH_3)_2NH$.

(*a*) Me, a typical alkyl group, has an electron-donating inductive effect and is base-strengthening. Basicity increases with the increase in the number of Me groups.

(*b*) Water draws the acid-base equilibrium to the right by H-bonding more with the ammonium cation than with the free base. The greater the number of H's in the ammonium cation, the greater the shift to the right and the increase in basicity. This effect should make NH_3 the strongest base and the Me_3N the weakest base. However, opposing the shift is the enhanced inductive effect of the increasing number of Me groups. The inductive effect predominates for $MeNH_2$ and Me_2NH. In Me_3N, the loss of an H more than cancels out the additional inductive effect of the third Me. Consequently, the 3° amine is less basic in water than the 2° amine.

It is not unusual that relative basicities in the gas phase and water differ because of the solvation effects in H_2O.

Problem 18.17 (*a*) Assign numbers from 1 for LEAST to 4 for MOST to indicate relative base strengths of (i) $C_6H_5NH_2$, (ii) $(C_6H_5)_2NH$, (iii) $(C_6H_5)_3N$, (iv) NH_3. (*b*) Explain.

(*a*) (i) 3, (ii) 2, (iii) 1, (iv) 4. (*b*) $C_6H_5NH_2$, an *aromatic* amine, is much *less* basic than NH_3 because the electron density on N is delocalized to the ring, mainly to the *ortho* and *para* positions [Fig. 18.3(*a*)]. With more phenyls bonded to N, there is more delocalization and weaker basicity.

(a) (b)

Figure 18.3

Problem 18.18 Assign numbers from (1) for LEAST to (4) for MOST to indicate relative base strengths of the following:

	I	II	III	IV
(*a*)	$CH_3\ddot{N}H^-Na^+$	$C_2H_2NH_2$	$(i\text{-}C_3H_7)_3N$	CH_3CONH_2
(*b*)	$C_6H_5NH_2$	$p\text{-}NO_2C_6H_4NH_2$	$m\text{-}NO_2C_6H_4NH_2$	$p\text{-}H_3COC_6H_4NH_2$

(*a*)	I	II	III	IV
	4	3	2	1

CH_3NH_2 is a stronger base than NH_3 because the electron-donating effect of the Me group increases the electron density on N, the basic site. The three bulky *i*-propyl groups on N cause steric strain, but with an unshared electron pair on N, this strain is partially relieved by increasing the normal C—N—C bond angle (109°) to about 112°. If the unshared electron pair forms a bond to H, as in R_3NH^+, relief of strain by angle expansion is prevented. With bulky R groups, 3° amines therefore resist forming a fourth bond and suffer a decrease in basicity. Acyl *R*—C=O groups are strongly electron-withdrawing and base-weakening because electron density from N can be delocalized to O of the carbonyl group by extended π bonding [Fig. 18.3(*b*)].

(*b*)	I	II	III	IV
	3	1	2	4

The strongly electron-attracting NO_2 group *decreases* electron density on N. It thus also decreases *base strength* by an inductive effect in the *meta* position, and to a greater extent, by both extended π bonding and inductive effects, in *ortho* and *para* positions.

Since OCH_3 is electron-donating through extended π bonding, it *increases* electron density on N and the base strength of the amine because the ring accepts less electron density from N.

Problem 18.19 Account for the following order of decreasing basicity:

$$R\ddot{N}H_2 > R\ddot{N}{=}CHR' > RC{\equiv}N:$$

The hybrid atomic orbitals used by N to accommodate the lone pair of electrons in the above compounds are $RNH_2(sp^3)$, $RN{=}CHR'(sp^2)$, $RC{\equiv}N:(sp)$. The nitrile (RCN) N has the most *s* character and is the least basic. The 1° amine has the least *s* character and is the most basic [Problems 8.3 and 8.5(*b*)].

Reaction with Nitrous Acid, HONO

1. Primary Amines (Diazonium Ion Formation)

(*a*) Aromatic ($ArNH_2$)

$$C_6H_5NH_2 \xrightarrow[\text{HCl, 5°C}]{\text{HONO}} [C_6H_5{-}\overset{+}{N}{\equiv}N:]Cl^- \quad \text{Benzenediazonium chloride}$$

(*b*) Aliphatic (RNH_2).

$$CH_3CH_2CH_2NH_2 \xrightarrow[HCl]{HONO} [CH_3CH_2CH_2N_2]^+ Cl^-$$

(unstable)

$$N_2 + Cl^- + CH_3CH_2CH_2^+ \xrightarrow{\sim H:^-} CH_3\overset{+}{C}HCH_3$$

From $CH_3CH_2CH_2^+$: $\xrightarrow{H_2O}$ $CH_3CH_2CH_2OH$ (1-Propanol); $\xrightarrow{Cl^-}$ $CH_3CH_2CH_2Cl$ (1-Chloropropane); $\xrightarrow{-H^+}$ $CH_3CH{=}CH_2$ (Propene)

From $CH_3\overset{+}{C}HCH_3$: $\xrightarrow{-H^+}$ $CH_3CH{=}CH_2$ (Propene); $\xrightarrow{H_2O}$ $CH_3CH(OH)CH_3$ (2-Propanol); $\xrightarrow{Cl^-}$ $CH_3CH(Cl)CH_3$ (2-Chloropropane)

This reaction of RNH_2 has no synthetic utility, but the appearance of N_2 gas signals the presence of NH_2.

Problem 18.20 (*a*) Outline the steps in the formation of a diazonium cation from nitrous acid and a 1° amine. (*b*) Why are diazonium ions of aromatic amines more stable than those of aliphatic amines?

(*a*) Initially, a nitrosonium cation is formed from HNO_2 in acid solution.

$$H{-}\ddot{O}{-}\ddot{N}{=}O \xrightarrow{+H^+} H{-}\overset{H}{\overset{|}{O^+}}{-}\ddot{N}{=}O \longrightarrow H{-}\overset{H}{\overset{|}{\ddot{O}:}} + \overset{+}{\ddot{N}}{=}\ddot{O} \longleftrightarrow \ddot{N}{\equiv}\overset{+}{O}:$$

Nitrosonium cation

Nucleophilic attack by the electron pair of the amine on the nitrosonium ion produces an *N*-nitrosamine. It undergoes a series of proton transfers that produce an OH group on the second N—N bond.

$$R{-}\underset{H}{\overset{H}{N}}: + \overset{+}{N}{=}\ddot{O} \longrightarrow R{-}\underset{H}{\overset{H}{N^+}}{-}\ddot{N}{=}\ddot{O} \xrightarrow[-H^+]{+H_2O} R{-}\overset{H}{\ddot{N}}{-}\ddot{N}{=}\ddot{O} \xrightarrow{+H_3O^+} R{-}\overset{H}{\ddot{N}}{-}\ddot{N}{=}\overset{+}{\ddot{O}}{-}H$$

N-Nitrosamine

$$\xrightarrow[-H^+]{+H_2O} R{-}\ddot{N}{=}\ddot{N}{-}\ddot{O}H \xrightarrow{+H_3O^+} R{-}\ddot{N}{=}\ddot{N}{-}\overset{H}{\ddot{O}^+}{-}H \xrightarrow{-H_2O} R{-}\overset{+}{N}{\equiv}N:$$

a diazoic acid; Diazonium cation

(*b*) Aryldiazonium cations are stabilized by resonance involving the release of electrons from the *ortho* and *para* positions of the ring.

$$C_6H_5{-}\overset{+}{N}{\equiv}N: \longleftrightarrow \overset{+}{C_6H_5}{=}N{=}\ddot{N}:$$

2. Secondary Amines (Nitrosation)

$$Ar(R)NH \text{ (or } R_2NH) + HONO \longrightarrow \underset{\substack{\text{an } N\text{-nitrosamine} \\ \text{(insoluble in acid)}}}{Ar(R)N{-}NO \text{ (or } R_2N{-}NO)} + H_2O$$

N-Nitrosamines are cancer-causing agents (**carcinogens**).

3. Tertiary Amines

No useful reaction except for *N,N*-dialkyl arylamines.

$$(CH_3)_2N{-}C_6H_5 \xrightarrow[-H_2O]{HONO} (CH_3)_2N{-}C_6H_4{-}NO \quad \text{(attack by } NO^+)$$

Problem 18.21 Prepare p-$H_2NC_6H_4N(CH_3)_2$ from $C_6H_5N(CH_3)_2$.

$$C_6H_5N(CH_3)_2 \xrightarrow[-H_2O]{HONO} p\text{-}ON{-}C_6H_4{-}N(CH_3)_2 \xrightarrow{Zn,\ H^+} \text{product}$$

With HNO_3, polynitration would occur since NMe_2 is very activating. NO^+ is less electrophilic than NO_2^+.

Reactions with Carboxylic Acid Derivatives; Transacylations

$$RNH_2 \xrightarrow{R'COCl} R'CONHR + HCl$$

$$RNH_2 \xrightarrow{(R'CO)_2O} R'CONHR + R'COOH$$

$$RNH_2 \xrightarrow{R'COOR''} R'CONHR + R''OH$$

Reactions with Other Electrophilic Reagents

$$R'CH{=}O + RNH_2 \xrightarrow{-H_2O} R'CH{=}NR \quad \text{(\textbf{Schiff base}, imine or azomethine)}$$

$$Cl{-}\overset{O}{\overset{\|}{C}}{-}Cl + RNH_2 \longrightarrow 2HCl + RNH{-}\overset{O}{\overset{\|}{C}}{-}NHR \quad \text{(symmetrical disubstituted urea)}$$

$$\underset{\text{an isocyanate}}{R'{-}N{=}C{=}O} + H_2\ddot{N}R \longrightarrow R'{-}N{=}\overset{O^-}{\overset{|}{C}}{-}\overset{H}{\underset{H}{N^+}}{-}R \longrightarrow R'NH{-}\overset{O}{\overset{\|}{C}}{-}NHR \quad \text{(unsymmetrical disubstituted urea)}$$

$$\underset{\text{an isothiocyanate}}{R'{-}N{=}C{=}S} + H_2NR \longrightarrow R'NH{-}\overset{S}{\overset{\|}{C}}{-}NHR \quad \text{(a thiourea)}$$

Nucleophilic Displacements

1. Carbylamine Reactions of 1° Amines

$$RNH_2 + CHCl_3 + 3KOH \longrightarrow R{-}\overset{+}{N}{\equiv}\overset{-}{C}{:} + 3KCl + 3H_2O$$

an isocyanide
(foul smelling)

Nucleophilic RNH_2 attacks electrophilic intermediate [$:CCl_2$]

$$R{-}\ddot{N}H_2 + :CCl_2 \longrightarrow \left[R{-}N^{+}H_2{-}CCl_2^{:-}\right] \xrightarrow[-2HCl]{+2KOH} R{-}\overset{+}{N}{\equiv}\overset{-}{C}{:}$$

2. Hinsberg Reaction

$$C_6H_5SO_2Cl \xrightarrow{RNH_2} C_6H_5SO_2NHR \xrightarrow{Na^+OH^-} H_2O + Na^+\left[C_6H_5SO_2\overset{..}{N}R\right]^-$$

(acidic) (soluble in water)

$$C_6H_5SO_2Cl \xrightarrow{R_2NH} C_6H_5SO_2NR_2 \xrightarrow{Na^+OH^-} \text{(No reaction)}$$

(neutral)

$$C_6H_5SO_2Cl \xrightarrow{R_3N} \text{(No noticeable reaction)}$$

Problem 18.22 How can the Hinsberg test be used to distinguish among liquid RNH_2, R_2NH, and R_3N?

R_3N does not react; RNH_2 reacts to give a water solution of $[C_6H_5SO_2\bar{N}R]Na^+$; R_2NH reacts to give a solid precipitate, $C_6H_5SO_2NR_2$.

Problem 18.23 Outline two laboratory preparations of *sym*-diphenylurea.

$$C_6H_5NH_2 + Cl{-}\overset{O}{\overset{\|}{C}}{-}Cl + H_2NC_6H_5 \longrightarrow C_6H_5NH{-}\overset{O}{\overset{\|}{C}}{-}NHC_6H_5 + 2HCl$$

$$C_6H_5NH_2 + O{=}C{=}NC_6H_5 \longrightarrow C_6H_5NH\overset{O}{\overset{\|}{C}}NHC_6H_5$$

Problem 18.24 Condensation of $C_6H_5NH_2$ with $C_6H_5CH{=}O$ yields compound (A), which is hydrogenated to compound (B). What are compounds (A) and (B)?

$$C_6H_5NH_2 + O{=}CHC_6H_5 \xrightarrow{-H_2O} \underset{\text{Benzalaniline (A)}}{C_6H_5N{=}CHC_6H_5} \xrightarrow{H_2/Ni} \underset{\text{N-Benzalaniline (B)}}{C_6H_5NHCH_2C_6H_5}$$

Oxidation

Oxidations of 1° and 2° amines give mixtures of product and, therefore, are not synthetically useful. 3° amines are cleanly oxidized to 3° amine oxides.

$$R_3N \xrightarrow{H_2O_2} R_3\overset{+}{N}—\ddot{\underset{..}{O}}:^-$$

a 3° amine oxide

18.4 Reactions of Quaternary Ammonium Salts

Formation of 4° Ammonium Hydroxides

$$2R_4N^+X^- + Ag_2O + H_2O \longrightarrow 2R_4N^+OH^- + 2AgX$$

very strong bases; like NaOH

Hofmann Elimination of Quaternary Hydroxides

$$[(CH_3)_3NCH(CH_3)CH_2CH_3]^+OH^- \xrightarrow{\Delta} (CH_3)_3N + H_2C{=}CHCH_2CH_3 + H_2O$$

s–Butyltrimethylammonium hydroxide; 1 Butene

This E2 elimination (Table 7.3) gives the less substituted alkene (Hofmann product) rather than the more substituted alkene (Saytzeff product; Section 6.3).

Problem 18.25 Compare and account for the products obtained from thermal decomposition of (*a*) $[(CH_3)_3N^+(C_2H_5)]OH^-$, (*b*) $(CH_3)_4N^+OH^-$.

(*a*) $H_2C{=}CH_2$, $(CH_3)_3N$, H_2O. Alkenes are formed from C_2H_5 and larger R groups having an H on the β C.

(*b*) $H\,:\ddot{\underset{..}{O}}:^- + CH_3\overset{+}{N}(CH_3)_3 \xrightarrow{\Delta} HOCH_3 + :N(CH_3)_3$ (S_N2 reaction)

Alkene formation is impossible with four CH_3's on N.

Problem 18.26 Deduce the structures of the following amines from the products obtained from exhaustive methylation and Hofmann elimination. (*a*) $C_5H_{13}N$ (A) reacts with 1 mol of CH_3I and eventually yields propene; (*b*) $C_5H_{13}N$ (B) reacts with 2 mol of CH_3I and give ethene and a 3° amine. The latter reacts with 1 mol of CH_3I and eventually gives propene.

(*a*) (A) is a 3° amine because it reacts with only 1 mol of CH_3I. Since propene is eliminated, C_3H_7 can be *n*- or *iso*-; hence, (A) is $(C_3H_7)N(CH_3)_2$.

(*b*) (B) reacts with 2 mol of CH_3I; it is a 2° amine. Separate formation of C_2H_4 and C_3H_6 shows that the alkyl groups are C_3H_7 and C_2H_5. (B) is $C_3H_7NHC_2H_5$, where C_3H_7 is *n*-propyl or isopropyl.

Problem 18.27 Outline the reactions and reagents used to establish the structure of 4-methylpyridine by exhaustive methylation and Hofmann elimination.

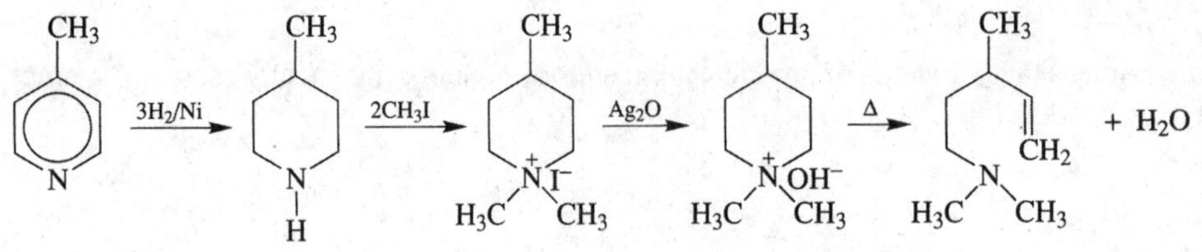

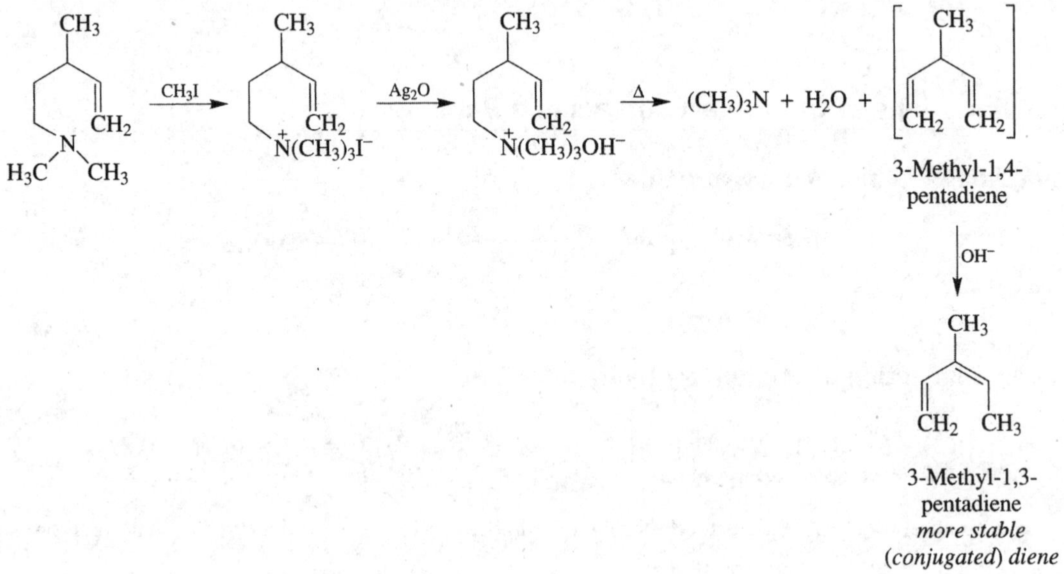

4-Methylpyridine $\xrightarrow{3H_2/Ni}$ 4-methylpiperidine (N—H) $\xrightarrow{2CH_3I}$ N,N-dimethyl-4-methylpiperidinium iodide ($\overset{+}{N}$ I^-, H_3C, CH_3) $\xrightarrow{Ag_2O}$ N,N-dimethyl-4-methylpiperidinium hydroxide ($\overset{+}{N}$ OH^-, H_3C, CH_3) $\xrightarrow{\Delta}$ ring-opened amine with $=CH_2$ and $N(CH_3)_2$ (H_3C—N—CH_3) + H_2O

Amine (CH_3, $=CH_2$, H_3C—N—CH_3) $\xrightarrow{CH_3I}$ $\overset{+}{N}(CH_3)_3I^-$ (with CH_3, $=CH_2$) $\xrightarrow{Ag_2O}$ $\overset{+}{N}(CH_3)_3OH^-$ (with CH_3, $=CH_2$) $\xrightarrow{\Delta}$ $(CH_3)_3N + H_2O +$ [CH_3; CH_2 CH_2]

3-Methyl-1,4-pentadiene

$\downarrow OH^-$

(CH_3; CH_2 CH_3)

3-Methyl-1,3-pentadiene
more stable
(conjugated) diene

Cope Elimination of 3° Amine Oxides

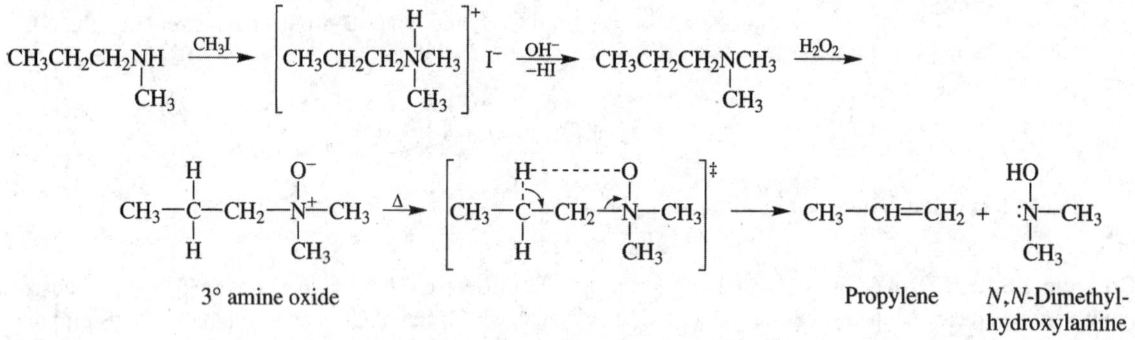

$$CH_3CH_2CH_2\underset{CH_3}{\underset{|}{N}}H \xrightarrow{CH_3I} \left[CH_3CH_2CH_2\underset{CH_3}{\underset{|}{\overset{H}{\overset{|}{N}}}}CH_3\right]^+ I^- \xrightarrow[-HI]{OH^-} CH_3CH_2CH_2\underset{CH_3}{\underset{|}{N}}CH_3 \xrightarrow{H_2O_2}$$

$$CH_3-\underset{H}{\underset{|}{\overset{H}{\overset{|}{C}}}}-CH_2-\underset{CH_3}{\underset{|}{\overset{O^-}{\overset{|}{N^+}}}}-CH_3 \xrightarrow{\Delta} \left[CH_3-\underset{H}{\underset{|}{\overset{H\cdots\cdots O}{C}}}-CH_2-\underset{CH_3}{\underset{|}{N}}-CH_3\right]^{\ddagger} \longrightarrow CH_3-CH{=}CH_2 + \text{:}\underset{CH_3}{\underset{|}{\overset{HO}{\overset{|}{N}}}}-CH_3$$

3° amine oxide — Propylene — *N*,*N*-Dimethylhydroxylamine

Elimination is *cis*—because of the cyclic transition state—and requires lower temperatures than pyrolysis of $[R_4N]^+OH^-$.

Phase-Transfer Catalysis (Problem 7.26)

18.5 Ring Reactions of Aromatic Amines

$-NH_2$, $-NHR$, and $-NR_2$ strongly activate the benzene ring toward *ortho-para* electrophilic substitution.

1. Halogenation

$PhNH_2$ with Br_2 (no catalyst) gives tribromination; 2,4,6-tribromoaniline is isolated. For monohalogenation, —NH_2 is first acetylated, because

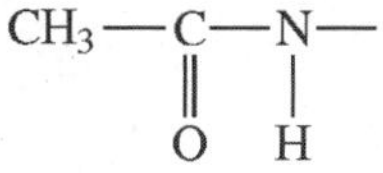

$$CH_3-\overset{\overset{\displaystyle O}{\|}}{C}-\underset{H}{\underset{|}{N}}-$$

is only moderately activating.

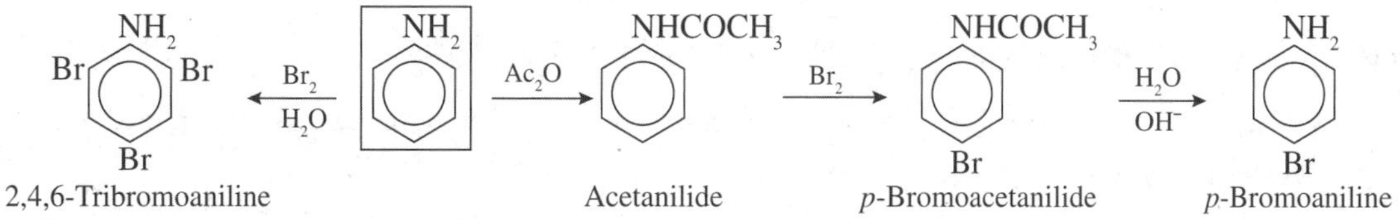

2,4,6-Tribromoaniline ← (Br_2, H_2O) — NH_2 (aniline) — Ac_2O → Acetanilide ($NHCOCH_3$) — Br_2 → *p*-Bromoacetanilide — H_2O, OH^- → *p*-Bromoaniline

2. Sulfonation

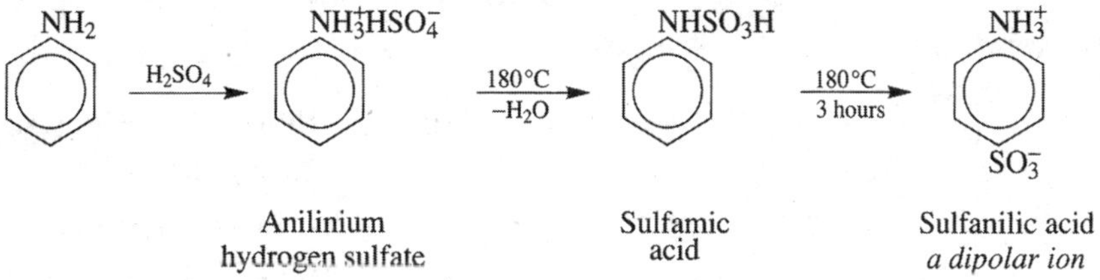

NH_2 — H_2SO_4 → $NH_3^+HSO_4^-$ (Anilinium hydrogen sulfate) — 180°C, $-H_2O$ → $NHSO_3H$ (Sulfamic acid) — 180°C, 3 hours → NH_3^+ / SO_3^- (Sulfanilic acid *a dipolar ion*)

Problem 18.28 How does the dipolar ion structure of sulfanilic acid account for its (*a*) high melting point, (*b*) insolubility in H_2O and organic solvents, (*c*) solubility in aqueous NaOH, (*d*) insolubility in aqueous HCl?

(*a*) Sulfanilic acid is ionic. (*b*) Because it is ionic, it is insoluble in organic solvents. Its insolubility in H_2O is typical of dipolar salts. Not all salts dissolve in H_2O. (*c*) The weakly acidic NH_3^+ transfers H^+ to OH^- to form a soluble salt, *p*-$H_2NC_6H_4SO_3^-Na^+$. (*d*) —SO_3^- is too weakly basic to accept H^+ from strong acids.

Problem 18.29 $H_3\overset{+}{N}CH_2COO^-$ exists as a dipolar ion whereas *p*-$H_2NC_6H_4COOH$ does not. Explain.

—COOH is too weakly acidic to transfer an H^+ to the weakly basic —NH_2 attached to the electron-withdrawing benzene ring. When attached to an aliphatic C, the NH_2 is sufficiently basic to accept H^+ from COOH.

3. Nitration

To prevent oxidation by HNO_3 and *meta* substitution of $C_6H_5NH_3^+$, amines are first acetylated.

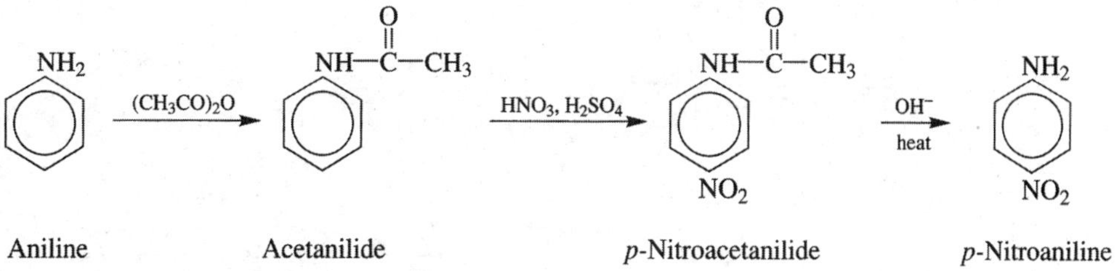

Aniline (NH_2) — $(CH_3CO)_2O$ → Acetanilide ($NH-\overset{O}{\overset{\|}{C}}-CH_3$) — HNO_3, H_2SO_4 → *p*-Nitroacetanilide (NO_2) — OH^-, heat → *p*-Nitroaniline (NH_2, NO_2)

18.6 Spectral Properties

The N—H stretching and NH_2 bending frequencies occur in the ir spectrum at 3050–3550 cm^{-1} and 1600–1640 cm^{-1}, respectively. In the N—H stretching region, 1° amines and unsubstituted amides show a pair of peaks for a symmetric and an antisymmetric vibration. In nmr, N—H proton signals of amines fall in a wide range (δ = 1–5 ppm) and are often very broad. The signals of N—H protons of amides are even broader, appearing at δ = 5–8 ppm. Mass spectra of amines show α,β-cleavage, like alcohols.

$$-\overset{|\beta}{\underset{|}{C}}-\overset{|\alpha}{\underset{|}{C}}-\underset{|}{\overset{\cdot+}{N}}- \longrightarrow -\overset{|\beta}{\underset{|}{C}}\cdot + \;>C=\underset{|}{\overset{+}{N}}-$$

Problem 18.30 Distinguish among 1°, 2°, and 3° amines by ir spectroscopy.

1° amine, two N—H stretching bands; 2° amine, one N—H stretching band; 3° amine, no N—H stretching band.

18.7 Reactions of Aryl Diazonium Salts

Displacement Reactions

$ArN_2^+X^-$
- $+ HPH_2O_2$ or $NaBH_4 \longrightarrow ArH + N_2$ (HPH_2O_2 is hypophosphorous acid)
- $+ KI \longrightarrow ArI + N_2$
- $+ CuCl\ (CuBr) \longrightarrow ArCl\ (ArBr) + N_2$ **(Sandmeyer reaction)**
- $+ HBF_4$ (from $H_3BO_3 + HF$) $\longrightarrow ArN_2^+BF_4^- \xrightarrow{\Delta} ArF + N_2 + BF_3$
- $+ HOH \xrightarrow{\Delta} ArOH + N_2$
- $+ HOC_2H_5 \xrightarrow{\Delta} ArOC_2H_5 + ArH + CH_3CHO + N_2$
- $+ CuCN \longrightarrow ArCN + N_2$
- $+ NaNO_2 + NaHCO_3 \xrightarrow[\text{or } Cu_2O]{Cu^{2+}} ArNO_2 + N_2$
- $+ NaHAsO_3 \longrightarrow ArAsO_3H_2$

Problem 18.31 Using C_6H_6, $C_6H_5CH_3$, via diazonium salts and other needed reagents, prepare (*a*) *o*-chlorotoluene, (*b*) *m*-chlorotoluene, (*c*) 1,3,5-tribromobenzene, (*d*) *m*-bromochlorobenzene, (*e*) *p*-iodotoluene, (*f*) *p*-dinitrobenzene, and (*g*) *p*-cyanobenzoic acid. Do not repeat the synthesis of intermediate products.

(*a*) $C_6H_5CH_3 \xrightarrow[H_2SO_4]{HNO_3}$ *p*-$CH_3C_6H_4NO_2$ $\xrightarrow[Fe]{Cl_2}$ 2-chloro-4-nitrotoluene (CH₃, Cl ortho, NO₂ para) $\xrightarrow[2.\ OH^-]{1.\ Sn,\ HCl}$ 4-amino-2-chlorotoluene (CH₃, Cl, NH₂) $\xrightarrow[HCl]{NaNO_2}$ diazonium salt (CH₃, Cl, $N_2^+Cl^-$) $\xrightarrow{HPH_2O_2}$ *o*-chlorotoluene (CH₃, Cl)

The —NO_2 is used to block the *para* position and it also directs *meta*, so that chlorination will occur only *ortho* to CH_3.

(*b*) [Reaction scheme: p-nitrotoluene → (1. Sn, HCl; 2. OH^-) → p-toluidine (CH_3, NH_2) → (Ac_2O) → CH_3/$NHCOCH_3$ → (Cl_2) → CH_3/Cl/$NHCOCH_3$ → (OH^-) → CH_3/Cl/NH_2 → (1. $NaNO_2$, HCl; 2. HPH_2O_2) → CH_3/Cl]

The acetylated —NH_2 is used to direct Cl into its *ortho* position, which is *meta* to CH_3; it is then removed.

(*c*) Aniline is rapidly and directly tribrominated and the NH_2 removed.

[Reaction scheme: C_6H_6 → (HNO_3/H_2SO_4) → NO_2 benzene → (H_2/Pt) → NH_2 benzene (Aniline) → (Br_2/H_2O) → Tribromoaniline (NH_2, Br, Br, Br) → ($NaNO_2$/HCl, 5°C) → $N_2^+Cl^-$ (Br, Br, Br) → (HPH_2O_2) → 1,3,5-Tribromobenzene]

(*d*) [Reaction scheme: NO_2 benzene → (Br_2/Fe) → NO_2/Br → (H_2/Pt) → NH_2/Br → ($NaNO_2$/HCl, 5°C) → $N_2^+Cl^-$/Br → (CuCl) → Cl/Br]

(*e*) [Reaction scheme: CH_3/NO_2 → (1. H_2/Pt; 2. $NaNO_2$, HCl, 5°C) → CH_3/$N_2^+Cl^-$ → (KI) → CH_3/I] (*f*) $C_6H_5NH_2 \xrightarrow{Ac_2O} C_6H_5NHCOCH_3 \xrightarrow{HNO_3} p\text{-}NO_2C_6H_4NHCOCH_3 \xrightarrow{OH^-} p\text{-}NO_2C_6H_4NH_2 \xrightarrow[HCl,\ 5°C]{NaNO_2} p\text{-}NO_2C_6H_4N_2^+Cl^- \xrightarrow[NaHCO_3,\ Cu_2O]{NaNO_2} p\text{-}C_6H_4(NO_2)_2$

(*g*) $p\text{-}NO_2C_6H_4CH_3 \xrightarrow[H^+]{KMnO_4} p\text{-}NO_2C_6H_4COOH \xrightarrow[2.\ NaNO_2,\ HCl,\ 5°C]{1.\ Zn,\ HCl} p\text{-}HOOCC_6H_4N_2^+Cl^- \xrightarrow{CuCN} p\text{-}HOOCC_6H_4CN$

Coupling; Formation of Diarylazo Compounds, Ar—N═N—Ar′

The aryl diazonium cation, ArN_2^+, is a weak electrophile that is nevertheless capable of attacking aromatic rings with strongly activating groups. Diarylazo compounds are thereby produced with no loss of N_2.

$$\underset{\text{a weak electrophile}}{ArN_2^+} + \underset{\text{a strongly activated ring}}{C_6H_5G} \longrightarrow \underset{\text{an azo compound mainly } para}{p\text{-}G{-}C_6H_4{-}N{=}N{-}Ar} \qquad \underset{\text{(electron-releasing group)}}{(G = OH,\ NR_2,\ NHR,\ NH_2)}$$

Problem 18.32 Explain the following conditions used in coupling reactions: (*a*) excess of mineral acid during diazotization of arylamines, (*b*) weakly acidic medium for coupling with $ArNH_2$, (*c*) weakly basic solution for coupling with ArOH.

(*a*) Acid prevents the coupling reaction

$$Ar\overset{+}{N}{\equiv}N{:} + H_2\ddot{N}Ar' \longrightarrow ArN{=}N{-}NHAr'$$

by converting $Ar'NH_2$ to its salt, $Ar'NH_3^+X^-$.

(*b*) In strong base, rather than coupling, $Ar\overset{+}{N}{\equiv}N$ reacts with OH^- to form $ArN{\equiv}N{-}OH$ (a diazoic acid), which reacts further to give a diazotate, $ArN{=}N{-}O^-$; neither of these couple. Strong acid converts $ArNH_2$ to $ArNH_3^+$, whose ring is deactivated towards coupling. It turns out that amines couple fastest in mildly acidic solutions.

(*c*) High acidity represses the ionization of ArOH and therefore decreases the concentration of the more reactive ArO^-. In a weak base, ArO^- is formed and $ArN{=}N{-}OH$ is not.

Azo compounds readily undergo reductive cleavage to form two aromatic 1° amines.

Problem 18.33 Deduce the structures of the azo compounds that yield the indicated aromatic amines on reduction with $SnCl_2$; (*a*) *p*-toluidine and *p*-NH_2-*N,N*-diMeaniline, (*b*) 1 mol of 4,4′-diaminobiphenyl and 2 mol of 2-hydroxy-5-aminobenzoic acid.

NH_2's originate from N's of the cleaved azo bond.

(*a*) $p\text{-}H_3CC_6H_4NH_2 + p\text{-}H_2NC_6H_4N(CH_3)_2 \longleftarrow p\text{-}H_3CC_6H_4{-}N{=}N{-}C_6H_4N(CH_3)_2\text{-}p$

(*b*) $HO(HOOC)C_6H_3NH_2 + H_2NC_6H_4{-}C_6H_4NH_2 + H_2NC_6H_3(COOH)OH$

$\longleftarrow HO(HOOC)C_6H_3{-}N{=}N{-}C_6H_4{-}C_6H_4{-}N{=}N{-}C_6H_3(COOH)OH$

18.8 Summary of Amine Chemistry

PREPARATION

1. Aliphatic Amines

(a) Primary

Alkylation by S_N2:

ammonia $RCH_2X + NH_3$ → RCH_2NH_2

imide o-$C_6H_4(CO)_2N^-K^+ + RCH_2X$ → RCH_2NH_2

Reduction:

$RC{\equiv}N$, $RCONH_2$, RCH_2NO_2, RCH_2N_3 $\xrightarrow{:H^-}$ RCH_2NH_2

Rearrangements:

$RCH_2CONH_2 + Br_2 + KOH$ → RCH_2NH_2

RCH_2CON_3 $\xrightarrow{\Delta}$ RCH_2NH_2

$RCH_2CONHOH + H^+$ → RCH_2NH_2

(b) Secondary

Alkylation:

$RNH_2 + R'X \longrightarrow RR'NH$

Reduction:

$R{-}N{\equiv}C{:} + H_2 \longrightarrow RNHCH_3$

$RCH{=}NR' + H_2 \longrightarrow RCH_2NHR'$

(c) Tertiary

Alkylation:

$3RX + NH_3 \longrightarrow R_3N$

(d) Quaternary

$R_3N + C_2H_5X \longrightarrow (R_3N{-}C_2H_5)^+X^-$

PROPERTIES

Primary (RCH_2NH_2):

1. Basic

$+ HOH \longrightarrow RCH_2NH_3^+OH^-$

$+ HX \longrightarrow RCH_2NH_3^+X^-$

$+ BF_3 \longrightarrow RCH_2NH_2BF_3$

2. Metal Cations

$+ Ag^+ \longrightarrow Ag(RCH_2NH_2)_2^+$

$+ Cu^{2+} \longrightarrow Cu(RCH_2NH_2)_4^{2+}$

3. Acidic

$+ K \longrightarrow RCH_2NH^-K^+ + \frac{1}{2}H_2$

$+ R'MgX \longrightarrow RCH_2NHMgX + R'H$

4. Electrophilic sites

$+ R'CH{=}O \longrightarrow R'CH{=}NCH_2R$

$+ R'COX$ or $(R'CO)_2O \longrightarrow R'CONHCH_2R$

$+ CHCl_3 + KOH \longrightarrow R{-}N{\equiv}C{:}$

$+ HNO_2 \longrightarrow N_2 + ROH + HOH$

$+ ArSO_2Cl \longrightarrow ArSO_2NHCH_2R$

Secondary ($RR'NH$):

1. Electrophilic sites

$+ R'COX$ or $(R'CO)_2O \longrightarrow R'CONHRR'$

$+ HNO_2 \longrightarrow RR'N{-}NO$

$+ ArSO_2Cl \longrightarrow ArSO_2NRR'$

Tertiary (R_3N) and quaternary:

$+ RX \longrightarrow R_4N^+X^-$

$+ O_2^{2-} \longrightarrow R_3N{-}O$

$(R_3N{-}C_2H_5)^+X^- + Ag_2O \longrightarrow (R_3NC_2H_5)^+OH^- \xrightarrow{\Delta} C_2H_4 + R_3N$

2. Aromatic Amines

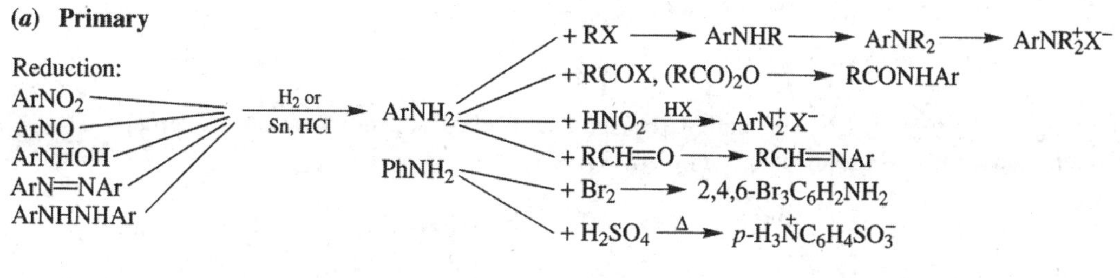

(b) Secondary

$ArNH_2 + RX \longrightarrow ArNHR \xrightarrow{HNO_2} ArNR(NO)$

SUPPLEMENTARY PROBLEMS

Problem 18.34 Write a structural formula for (*a*) *N,N′*-di-*p*-tolylthiourea, (*b*) 2,4-xylidine, (*c*) *N*-methyl-*p*-nitrosoaniline, (*d*) 4-ethyl-3′-methylazobenzene.

(*a*) CH_3–C_6H_4–NH–C(=S)–NH–C_6H_4–CH_3 (*b*) C_6H_3 ring with NH_2 (1), CH_3 (2), CH_3 (4)

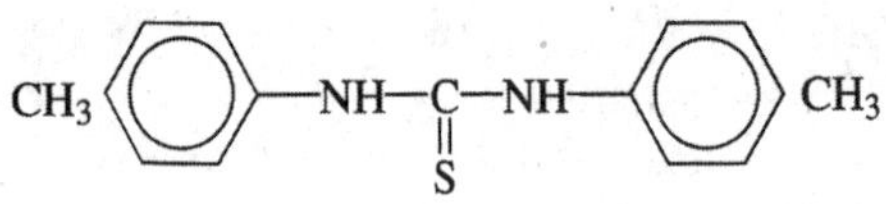

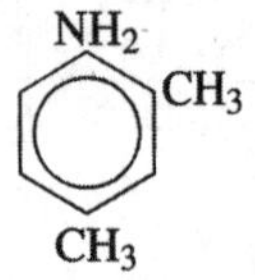

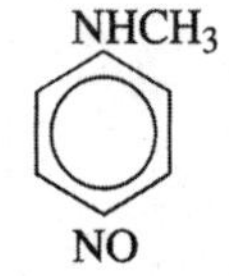

(*c*) C_6H_4 ring with $NHCH_3$ (top) and NO (para) (*d*) C_2H_5–C_6H_4–N=N–C_6H_4–CH_3 (meta)

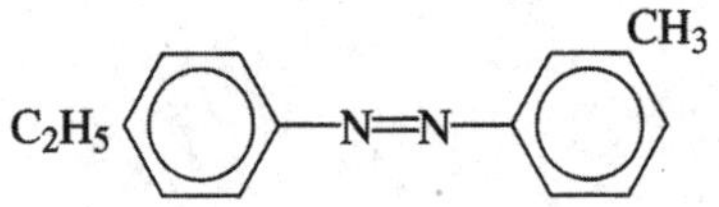

Problem 18.35 Name: (*a*) $C_6H_5CH_2CH_2NH_2$, (*b*) $CH_3NHCH_2CH_3$, (*c*) $NH_2CH_2CH_2CH(NH_2)CH_3$, (*d*) $NH_2CH_2CH_2OH$, (*e*) $(CH_3)_3CNHC(CH_3)_3$, (*f*) $NH_2CH_2CH_2NH_2$, (*g*) *p*-$C_6H_4(NH_2)_2$.

(*a*) 2-phenyl-1-aminoethane or β-phenylethylamine, (*b*) methylethylamine, (*c*) 1,3-diaminobutane, (*d*) 2-aminoethanol, (*e*) di-*tert*-butylamine, (*f*) 1,2-ethanediamine or ethylenediamine, (*g*) *p*-phenylenediamine or 1,4-benzenediamine.

Problem 18.36 Give the structure of (*a*) 3-(*N*-methylamino)-1-propanol, (*b*) ethyl 3-(*N*-methylamino)-2-butenoate, (*c*) 2-*N,N*-dimethylaminobutane, (*d*) allylamine.

(*a*) $CH_2(NHCH_3)CH_2CH_2OH$ (*b*) $CH_3C(NHCH_3){=}CHCOOC_2H_5$ (*c*) $CH_3CH(N(CH_3)_2)CH_2CH_3$ (*d*) $CH_2{=}CHCH_2NH_2$

Problem 18.37 Give the structures and names of five aromatic amines with the molecular formula C_7H_9N.

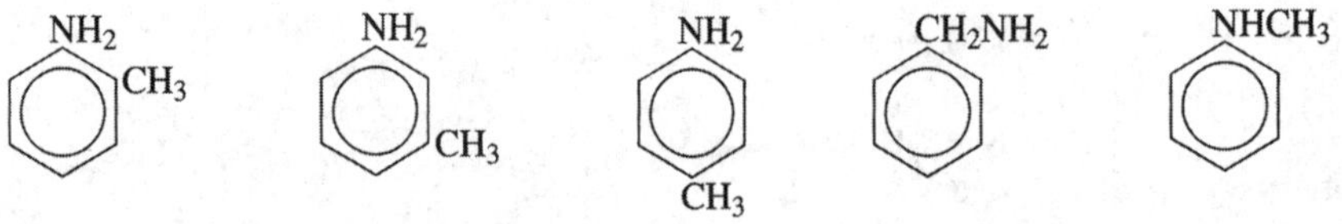

NH_2/CH_3 (ortho), NH_2/CH_3 (meta), NH_2/CH_3 (para): *o*-, *m*- and *p*-Toluidine; $C_6H_5CH_2NH_2$: Benzylamine; $C_6H_5NHCH_3$: *N*-Methylaniline

Problem 18.38 A compound C_3H_9N is a 2° amine. Deduce its structure.

Since the compound is a secondary amine, there must be two R groups and an H attached to an N:

$$R-N(H)-R$$

Removing N—H from C_3H_9N, we obtain C_3H_8, which must be divided between two R groups. These can only be a methyl and an ethyl group. The compound is

$$CH_3-N(H)-CH_2CH_3 \quad \text{Methylethylamine}$$

Problem 18.39 Give the product of reaction in each case:

(*a*) C_2H_5Br + excess NH_3 (*b*) $CH_2{=}CHCN + H_2/Pt$ (Acrylonitrile) (*c*) *n*-Butyramide + Br_2 + KOH

(*d*) Dimethylamine + HONO (*e*) Ethylamine + $CHCl_3$ + KOH

(*a*) $CH_3CH_2NH_2$ (*b*) $CH_3CH_2CH_2NH_2$ (*c*) $CH_3CH_2CH_2NH_2$ (*d*) $Me_2NN{=}O$

(*e*) $CH_3CH_2\overset{+}{N}{\equiv}\overset{-}{C}{:}$

Problem 18.40 What is the organic product when *n*-propylamine is treated with (*a*) $PhSO_2Cl$? (*b*) excess $CH_3CH_2CH_2Cl$? (*c*) chlorobenzene? (*d*) excess CH_3I, then Ag_2O and heat?

(*a*) *N*-(*n*-propyl)benzenesulfonamide, $PhSO_2NHCH_2CH_2CH_3$; (*b*) tetra-*n*-propylammonium chloride; (*c*) no reaction; (*d*) propene and trimethylamine.

Problem 18.41 What is the product of catalytic hydrogenation of (*a*) acetone oxime, (*b*) propane-1,3-dinitrile, (*c*) propanal and methylamine?

(*a*) isopropylamine (*b*) 1,5-diaminopentane (*c*) $CH_3CH_2CH_2NHCH_3$.

Problem 18.42 Show the steps in the following syntheses:

(*a*) Ethylamine → Methylethylamine
(*b*) Ethylamine → Dimethylethylaminc
(*c*) *n*-Propyl chloride → Isopropylamine
(*d*) Aniline → *p*-Aminobenzenesulfonamide (**sulfanilamide**)

(*a*) $CH_3CH_2NH_2 \xrightarrow[KOH]{CHCl_3} CH_3CH_2\overset{+}{N}\overset{-}{C} \xrightarrow{H_2/Pt} CH_3CH_2N(H)CH_3$

(*b*) $CH_3CH_2NH_2 \xrightarrow[HCOOH]{H_2C{=}O} CH_3CH_2N(CH_3)_2$

(*c*) $CH_3CH_2CH_2Cl \xrightarrow[KOH]{alc.} CH_3CH{=}CH_2 \xrightarrow{HBr} CH_3CHBrCH_3 \xrightarrow{NH_3} CH_3CH(NH_2)CH_3$

(*d*) $C_6H_5NH_2 \xrightarrow{(CH_3CO)_2O} C_6H_5NHCOCH_3 \xrightarrow[\text{(chlorosulfonic acid)}]{ClSO_2OH} p\text{-}CH_3CONHC_6H_4SO_2Cl \xrightarrow{NH_3} p\text{-}CH_3CONHC_6H_4SO_2NH_2 \xrightarrow[2.\ OH^-]{1.\ H_3O^+} p\text{-}H_2NC_6H_4SO_2NH_2$

Problem 18.43 Outline the steps in the syntheses of the following compounds from C_6H_6, $C_6H_5CH_3$ and any readily available aliphatic compound: (*a*) *p*-aminobenzoic acid, (*b*) *m*-nitroacetanilide, (*c*) 1-amino-1-phenylpropane, (*d*) 4-amino-2-chlorotoluene.

(*a*) $C_6H_5CH_3 \xrightarrow[H_2SO_4]{HNO_3} p\text{-}O_2NC_6H_4CH_3 \xrightarrow{KMnO_4} p\text{-}O_2NC_6H_4COOH \xrightarrow{H_2/Pt} p\text{-}H_2NC_6H_4COOH$

Side chain is oxidized when a deactivating group (NO_2) rather than an activating group (NH_2) is attached to ring.

(*b*) $C_6H_6 \xrightarrow[H_2SO_4]{HNO_3} C_6H_5NO_2 \xrightarrow[H_2SO_4,\ 100°C]{\text{fum. } HNO_3} m\text{-}C_6H_4(NO_2)_2 \xrightarrow{NaSH} m\text{-}O_2NC_6H_4NH_2 \xrightarrow{Ac_2O} m\text{-}O_2NC_6H_4NHCOCH_3$

m-Dinitrobenzene; *m*-Nitroaniline

(*c*) $C_6H_6 + CH_3CH_2COCl \xrightarrow{AlCl_3} C_6H_5 + COCH_2CH_3 \xrightarrow[H_2/Pt]{NH_3} C_6H_5CH(NH_2)CH_2CH_3$

(*d*) $C_6H_5CH_3 \xrightarrow[H_2SO_4]{HNO_3} p\text{-}O_2NC_6H_4CH_3 \xrightarrow[Fe]{Cl_2} \text{2-chloro-4-nitrotoluene} \xrightarrow[2.\ OH^-]{1.\ Sn,\ HCl} \text{4-amino-2-chlorotoluene}$

Problem 18.44 Use the **benzidine rearrangement** to synthesize 2,2′-dichlorobiphenyl from benzene and inorganic reagents.

$C_6H_6 \xrightarrow[H_2SO_4]{HNO_3} C_6H_5NO_2 \xrightarrow[FeCl_3]{Cl_2} o\text{-}ClC_6H_4NO_2 \xrightarrow[NaOH]{Sn} o\text{-}ClC_6H_4\text{—NHNH—}C_6H_4Cl\text{-}o \xrightarrow{1.\ HCl,\ \Delta\ 2.\ OH^-}$ 2,2′-Dichloro-4,4′-diaminobiphenyl $\xrightarrow[HCl,\ 5°C]{NaNO_2}$ *bis*-Diazonium salt ($Cl^-N_2^+$–C₆H₃Cl–C₆H₃Cl–$N_2^+Cl^-$) $\xrightarrow{HPH_2O_2}$ 2,2′-Dichlorobiphenyl

Problem 18.45 Use *o*- or *p*-nitroethylbenzene and any inorganic reagents to synthesize the six isomeric dichloroethylbenzenes. Do not repeat preparations of intermediate products.

The —NO_2 is used as a blocking group by the sequence —NO_2 → —NH_2 → —N_2^+ → —H or as a source of Cl by —N_2^+ → — Cl or is converted to —NHAc, whose directive effect supersedes that of C_2H_5.

(i) 2,3-Dichloroethylbenzene

Sn, HCl; Ac_2O; HNO_3/H_2SO_4

C_2H_5, NO_2 → C_2H_5, NH_2 → C_2H_5, NHAc →

C_2H_5, NHAc, O_2N —OH^-→ C_2H_5, NH_2, O_2N —Cl_2→ C_2H_5, NH_2, O_2N, Cl —HNO_2, 0°C→

C_2H_5, $N_2^+X^-$, O_2N, Cl —CuCl→ C_2H_5, Cl, O_2N, Cl —1. Sn, HCl; OH^- 2. HNO_2, 5°C 3. HPH_2O_2→ C_2H_5, Cl, Cl

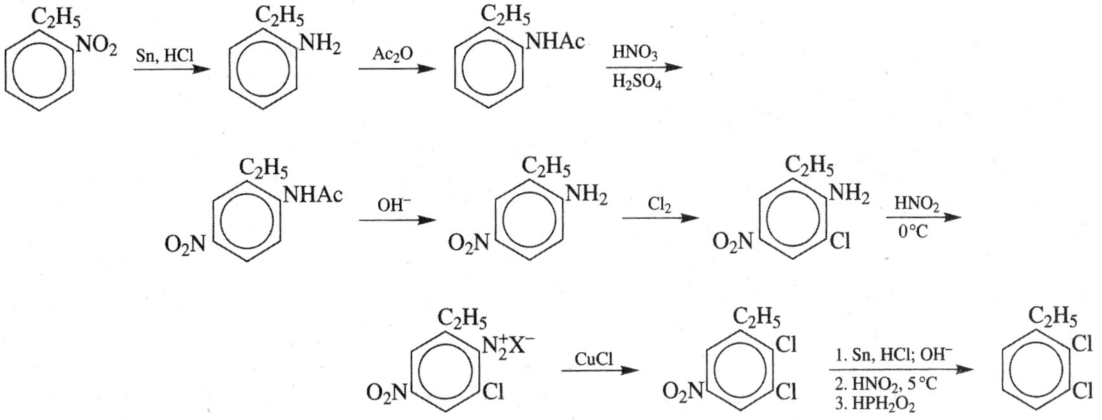

(ii) 2,4-Dichloroethylbenzene

C_2H_5, NO_2 → C_2H_5, Cl, NO_2 —1. Sn, HCl; OH^- 2. HNO_2, 5°C 3. CuCl→ C_2H_5, Cl, Cl

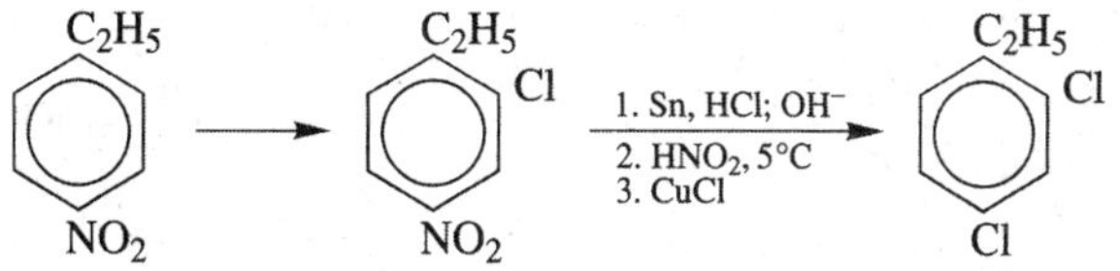

(iii) 2,5-Dichloroethylbenzene

C_2H_5, NHAc —Cl_2→ C_2H_5, NHAc, Cl —OH^-→ C_2H_5, NH_2, Cl —HNO_2, 5°C→ C_2H_5, $N_2^+X^-$, Cl —CuCl→ C_2H_5, Cl, Cl

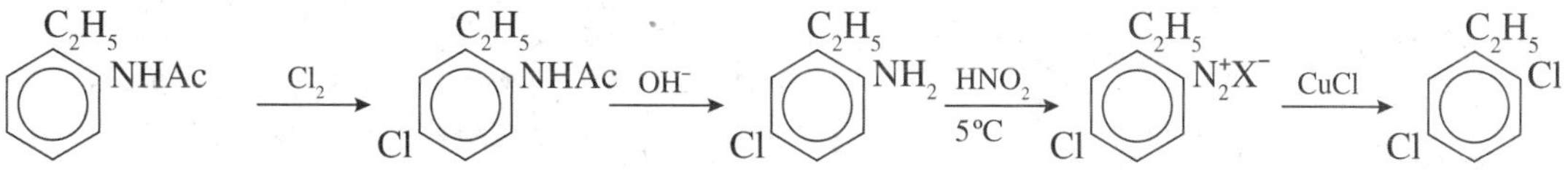

(iv) 2,6-Dichloroethylbenzene

C_2H_5, NO_2 —Cl_2, Fe→ C_2H_5, Cl, Cl, NO_2 —1. Sn, HCl; OH^- 2. HNO_2, 5°C 3. HPH_2O_2→ C_2H_5, Cl, Cl

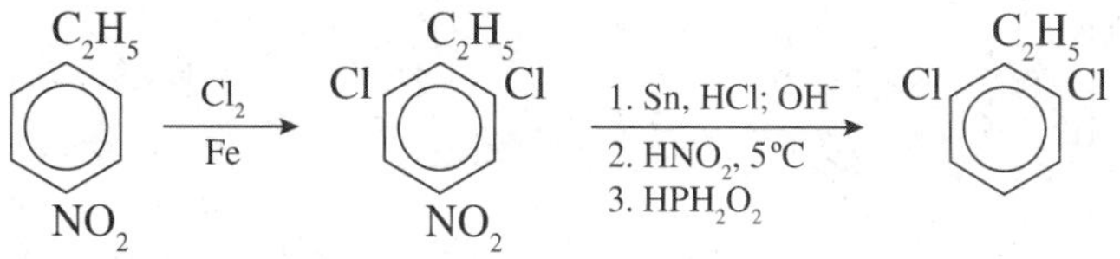

(v) 3,4-Dichloroethylbenzene

(*d*) C_2H_5, NO_2 —Sn, HCl→ C_2H_5, NH_2 —Ac_2O→ C_2H_5, NHAc —Cl_2→ C_2H_5, Cl, NHAc —1. OH^- 2. HNO_2, 5°C 3. CuCl→ C_2H_5, Cl, Cl

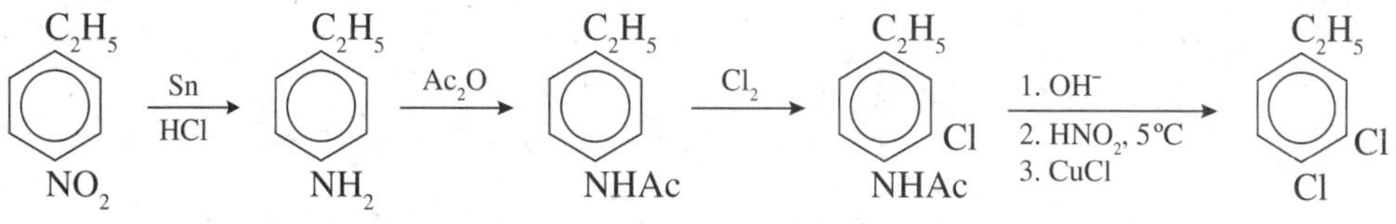

(vi) 3,5-Dichloroethylbenzene

C_2H_5, NH_2 —Cl_2→ C_2H_5, Cl, Cl, NH_2 —1. HNO_2, 5°C 2. HPH_2O_2→ C_2H_5, Cl, Cl

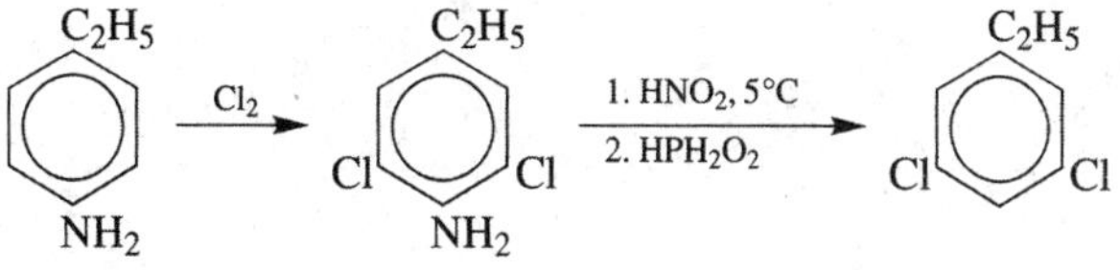

Problem 18.46 Deduce a possible structure for each of the following. (*a*) Compound (A), $C_6H_4N_2O_4$, is insoluble in both dilute acid and base and its dipole moment is zero. (*b*) Compound (B), C_8H_9NO, is insoluble in dilute acid and base. (B) is transformed by $KMnO_4$ in H_2SO_4 to compound (C), which is free of N, soluble in aqueous $NaHCO_3$ and gives only one mononitro substitution product. (*c*) Compound (D), $C_7H_7NO_2$, undergoes vigorous oxidation to form compound (E), $C_7H_5NO_4$, which is soluble in dilute aqueous $NaHCO_3$ and forms two isomeric monochloro substitution products.

(*a*) p-$O_2NC_6H_4NO_2$ (A)

(*b*) p-$CH_3C_6H_4CONH_2$ (B) ⟶ p-$HOOCC_6H_4COOH$ (C)

(*c*) p-$CH_3C_6H_4NO_2$ (D) ⟶ p-$HOOCC_6H_4NO_2$ (E)

Problem 18.47 Using any aliphatic and inorganic reagents, outline the syntheses of (*a*) m-$HOC_6H_4CH_3$ from toluene, (*b*) 4-bromo-4′-aminoazobenzene from aniline.

(*a*) Toluene $\xrightarrow[\text{(block } para)]{HNO_3,\ H_2SO_4}$ p-$CH_3C_6H_4NO_2$ $\xrightarrow[HCl]{Sn}$ p-$CH_3C_6H_4NH_2$ $\xrightarrow{Ac_2O}$ p-$CH_3C_6H_4NHAc$ $\xrightarrow[\text{(orient } meta \text{ to } CH_3)]{HNO_3}$ 4-CH_3-2-$NO_2C_6H_3NHAc$ $\xrightarrow{NaOH,\ \Delta}$ 4-CH_3-2-$NO_2C_6H_3NH_2$ $\xrightarrow[\text{2. } HPH_2O_2 \text{ (remove blocking group)}]{\text{1. } HNO_2,\ 5°C}$ m-$CH_3C_6H_4NO_2$ $\xrightarrow[\text{2. NaOH}]{\text{1. Sn, HCl}}$ m-$CH_3C_6H_4NH_2$ $\xrightarrow[\text{2. } H_2O,\ \Delta]{\text{1. } HNO_2,\ 5°C}$ m-$CH_3C_6H_4OH$

(*b*) $C_6H_5NH_2$ $\xrightarrow{Ac_2O}$ C_6H_5NHAc $\xrightarrow{Br_2}$ p-BrC_6H_4NHAc $\xrightarrow{\text{NaOH, heat}}$ p-$BrC_6H_4NH_2$ $\xrightarrow{HNO_2,\ 5°C}$ p-$BrC_6H_4N_2^+X^-$ $\xrightarrow{C_6H_5NH_2}$ p-$BrC_6H_4N{=}NC_6H_4NH_2$-p

Problem 18.48 Synthesize novocaine, p-$H_2NC_6H_4COOCH_2CH_2NEt_2$, from toluene and any aliphatic compound of four or fewer C's.

(1) $Et_2NH + H_2C{-}CH_2$ (ethylene oxide, bridged by O) ⟶ $HOCH_2CH_2NEt_2$

(2) $C_6H_5CH_3 \xrightarrow[H_2SO_4]{HNO_3} p\text{-}O_2NC_6H_4CH_3 \xrightarrow[H_3O^+,\ \Delta]{KMnO_4} p\text{-}O_2NC_6H_4COOH \xrightarrow{SOCl_2}$

$p\text{-}O_2NC_6H_4COCl \xrightarrow[\text{[from (1)]}]{HOCH_2CH_2NEt_2} p\text{-}O_2NC_6H_4COOCH_2CH_2N(C_2H_5)_2 \xrightarrow{H_2/Ni} p\text{-}H_2NC_6H_4COOCH_2CH_2N(C_2H_5)_2$

Problem 18.49 An optically active amine is subjected to exhaustive methylation and Hofmann elimination. The alkene obtained is ozonized and hydrolyzed to give an equimolar mixture of formaldehyde and butanal. What is the amine?

The alkene is 1-pentene, $CH_2{=}CHCH_2CH_2CH_3$ ($H_2C{=}O$ $O{=}CHCH_2CH_2CH_3$). The amine is chiral; $CH_3CH(NH_2)CH_2CH_2CH_3$. The other possibility, $H_2NCH_2CH_2CH_2CH_3$, is not chiral.

Problem 18.50 Draw a flow sheet to show the separation and recovery in almost quantitative yield of a mixture of the water-insoluble compounds benzaldehyde (A), *N,N*-dimethylaniline (B), chlorobenzene (C), *p*-cresol (D), benzoic acid (E).

See Fig. 18.4.

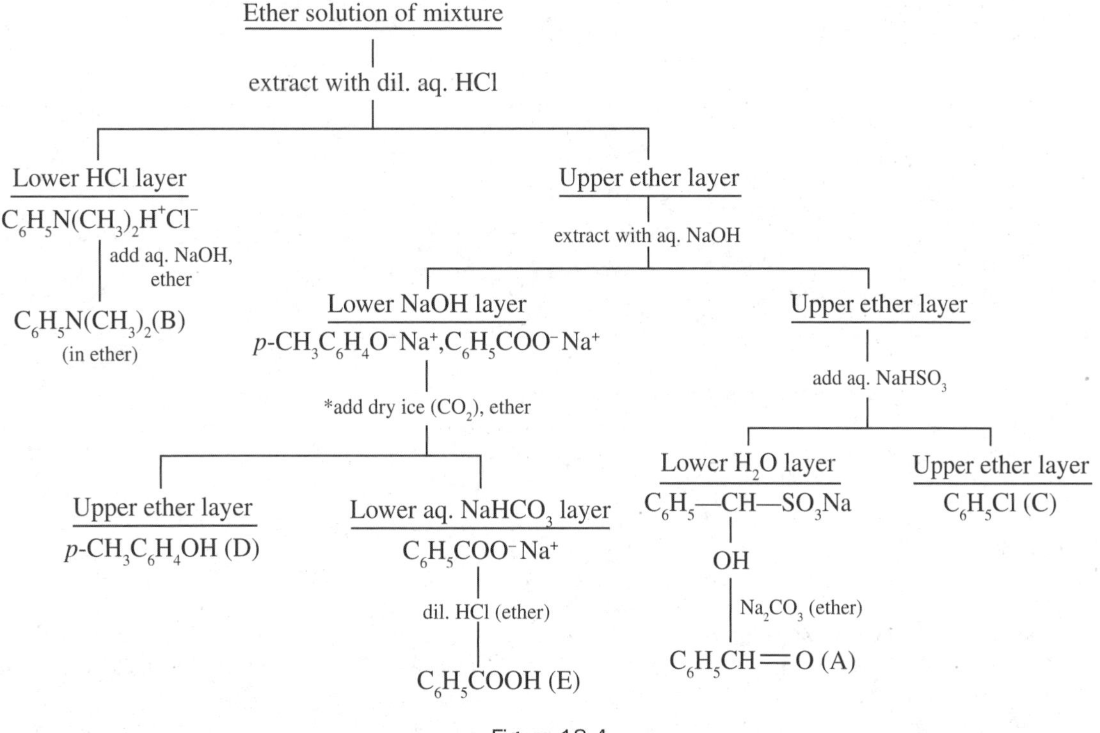

Figure 18.4

* NaOH + CO_2 gives $NaHCO_3$, in which carboxylic acids dissolve but phenols do not.

Problem 18.51 Synthesize the following compounds from alcohols of four or fewer C's, cyclohexanol and any needed solvents and inorganic reagents. (*a*) *n*-hexylamine, (*b*) triethylamine N-oxide, (*c*) 4-(N-methylamino)heptane, (*d*) cyclohexyldimethylamine, (*e*) cyclopentylamine, (*f*) 6-aminohexanoic acid.

(*a*) (1) $C_2H_5OH \xrightarrow{H_2SO_4} H_2C{=}CH_2 \xrightarrow[2.\ OH^-]{1.\ Br_2,\ H_2O} H_2C{-}CH_2$ (epoxide, bridged by O)

(2) $n\text{-}C_4H_9OH \xrightarrow{SOCl_2} n\text{-}C_4H_9Cl \xrightarrow{Mg} n\text{-}C_4H_9MgCl \xrightarrow[2.\ H_3O^+]{1.\ H_2C-CH_2 \text{ (epoxide) from (1)}}$

$n\text{-}C_6H_{13}OH \xrightarrow[heat]{Cu} n\text{-}C_5H_{11}CHO \xrightarrow[H_2/Pt]{Mg} n\text{-}C_6H_{13}NH_2$

(*b*) $C_2H_5OH \xrightarrow{HBr} C_2H_5Br \xrightarrow{NH_3} (C_2H_5)_3N \xrightarrow{H_2O_2} (C_2H_5)_3NO$

(*c*) (1) $CH_3OH \xrightarrow{PBr_3} CH_3Br \xrightarrow[NH_3]{excess} CH_3NH_2$

(2) $n\text{-BuOH} \xrightarrow[heat]{Cu} n\text{-PrCHO}$

$n\text{-PrOH} \xrightarrow{SO_2Cl} n\text{-PrCl} \xrightarrow{Mg} n\text{-PrMgCl}$

(both combine) $\longrightarrow (n\text{-Pr})_2\overset{H}{\overset{|}{C}}O^-(MgCl)^+ \xrightarrow{H_3O^+}$

$(n\text{-Pr})_2CHOH \xrightarrow[H_2SO_4]{Na_2Cr_2O_7} (n\text{-Pr})_2C{=}O \xrightarrow[CH_3NH_2 \text{ from (1)}]{H_2/Pt} (n\text{-Pr})_2CHNHCH_3$

(*d*) (1) $CH_3OH \xrightarrow[heat]{Cu} H_2CO \xrightarrow[2.\ H^+]{1.\ Ag(NH_3)_2^+} HCOOH$

Cyclohexanol $\xrightarrow[H_2SO_4]{Na_2Cr_2O_7}$ Cyclohexanone $\xrightarrow[H_2/Pt]{NH_3}$ Cyclohexylamine $\xrightarrow[HCOOH \text{ from (1)}]{H_2CO \text{ from (1)}}$ Cyclohexyldimethylamine

(*e*) Cyclohexanol $\xrightarrow{H_2SO_4}$ Cyclohexene $\xrightarrow[H^+,\ heat]{KMnO_4}$

$HOOC(CH_2)_4COOH \xrightarrow[heat]{BaO}$ Cyclopentanone $\xrightarrow[H_2/Pt]{NH_3}$ Cyclopentylamine

(*f*) Cyclohexanole $\xrightarrow[H_2SO_4]{Na_2Cr_2O_7}$ Cyclohexanone $\xrightarrow{H_2NOH}$

Cyclohexanone oxime $\xrightarrow{H_2SO_4}$ Caprolactam (seven-membered ring, C=O, NH) $\xrightarrow{H_3O^+}$ 6-Aminohexanoic acid

[See Problem 18.12(*c*).]

Problem 18.52 Use simple, rapid, test-tube reactions to distinguish between (*a*) $CH_3CONHC_6H_5$ and $C_6H_5CONH_2$, (*b*) $C_6H_5NH_3^+Cl^-$ and $p\text{-}ClC_6H_4NH_2$, (*c*) $(CH_3)_4N^+OH^-$ and $(CH_3)_2NCH_2OH$, (*d*) $p\text{-}CH_3COC_6H_4NH_2$ and $CH_3CONHC_6H_5$.

(*a*) With hot aqueous NaOH, only $C_6H_5CONH_2$ liberates NH_3. (*b*) Aqueous $AgNO_3$ precipitates AgCl from $C_6H_5NH_3^+Cl^-$. (*c*) CrO_3 is reduced to green Cr^{3+} by $(CH_3)_2NCH_2OH$. $(CH_3)_4N^+OH^-$ is strongly basic to litmus. (*d*) Cold dilute HCl dissolves $p\text{-}CH_3COC_6H_4NH_2$, which also gives a positive iodoform test with NaOI.

Problem 18.53 Synthesize from benzene, toluene, naphthalene (NpH), and any aliphatic or inorganic compounds (*a*) α-(*p*-nitrophenyl)ethylamine, (*b*) β-(*p*-bromophenyl)ethylamine, (*c*) 1-(α-aminomethyl)naphthalene, (*d*) 2-naphthylamine, (*e*) β-$NpCH_2NH_2$.

(*a*) $C_6H_6 \xrightarrow[AlCl_3]{C_2H_5Cl} C_6H_5CH_2CH_3 \xrightarrow[H_2SO_4]{HNO_3} p\text{-}O_2NC_6H_4CH_2CH_3 \xrightarrow[uv]{Cl_2} p\text{-}O_2NC_6H_4CHClCH_3 \xrightarrow{NH_3} p\text{-}O_2NC_6H_4CH(NH_2)CH_3$

(*b*) $C_6H_5CH_3 \xrightarrow[Fe]{Br_2} p\text{-}BrC_6H_4CH_3 \xrightarrow[uv]{Cl_2} p\text{-}BrC_6H_4CH_2Cl \xrightarrow{KCN} p\text{-}BrC_6H_4CH_2CN \xrightarrow{H_2/Ni} p\text{-}BrC_6H_4CH_2CH_2NH_2$

(*c*) $NpH \xrightarrow[H_2SO_4]{HNO_3} 1\text{-}NpNO_2 \xrightarrow[2.\ OH^-]{1.\ Sn,\ HCl} 1\text{-}NpNH_2 \xrightarrow[HCl,\ 5°C]{NaNO_2} 1\text{-}NpN_2^+Cl^- \xrightarrow{CuCN} 1\text{-}NpCN \xrightarrow[2.\ H_2O]{1.\ LiAlH_4} 1\text{-}NpCH_2NH_2$

(*d*) Naphthalene cannot be nitrated directly at the β-position.

$$NpH \xrightarrow[PhNO_2,\ AlCl_3]{CH_3COCl} 2\text{-}NpCOCH_3 \xrightarrow[2.\ H^+]{1.\ NaOH,\ I_2} 2\text{-}NpCOOH \xrightarrow[2.\ NH_3]{1.\ SOCl_2} 2\text{-}NpCONH_2 \xrightarrow{KOH,Br_2} 2\text{-}NpNH_2$$

(*e*) $NpH \xrightarrow[2.\ OH^-]{1.\ H_2SO_4} \beta\text{-}NpSO_3^-Na^+ \xrightarrow[fuse]{CN^-} \beta\text{-}NpCN \xrightarrow{LiAlH_4} \beta\text{-}NpCH_2NH_2$

Problem 18.54 Synthesize from naphthalene and any other reagents: (*a*) naphthionic acid (4-amino-1-naphthalenesulfonic acid), (*b*) 4-amino-1-naphthol, (*c*) 1,3-dinitronaphthalene, (*d*) 1,4-diaminonaphthalene, (*e*) 1,2-dinitronaphthalene. Do not repeat the synthesis of any compound.

(*a*) naphthalene $\xrightarrow[2.\ Sn,\ HCl;\ 3.\ OH^-]{1.\ HNO_3,\ H_2SO_4}$ 1-naphthylamine (NH_2) $\xrightarrow[2.\ heat]{1.\ H_2SO_4}$ 4-amino-1-naphthalenesulfonic acid (NH_3^+, SO_3^-)

(*b*) 1-naphthylamine (NH_2) $\xrightarrow[2.\ H_2O,\ heat]{1.\ HONO}$ 1-naphthol (OH) $\xrightarrow[5°C]{PhN_2^+Cl^-}$ 4-(N=NPh)-1-naphthol $\xrightarrow[2.\ OH^-]{1.\ Sn,\ H^+}$ 4-amino-1-naphthol (OH, NH_2) + $PhNH_2$

(*c*) 1-naphthylamine (NH_2) $\xrightarrow[H_2SO_4]{HNO_3}$ 1-amino-2,4-dinitronaphthalene (NH_2, NO_2, NO_2) $\xrightarrow[2.\ HPH_2O_2]{1.\ HONO,\ 5°C}$ 1,3-dinitronaphthalene (NO_2, NO_2)

(*d*) 1-naphthylamine (NH_2) $\xrightarrow{Ac_2O}$ $NHCOCH_3$ derivative $\xrightarrow{HNO_3}$ 4-nitro ($NHCOCH_3$, NO_2) $\xrightarrow[2.\ NaOH]{1.\ Sn,\ HCl}$ 1,4-diaminonaphthalene (NH_2, NH_2)

(*e*) NH_3^+ / SO_3^- naphthalene $\xrightarrow{HNO_3}$ NH_3^+, NO_2 / SO_3^- $\xrightarrow[\Delta]{50\%\ H_2SO_4}$ $NH_3^+HSO_4^-$, NO_2 $\xrightarrow{HONO,\ 0°C}$ $N_2^+HSO_4^-$, NO_2 $\xrightarrow{NaNO_2}$ NO_2, NO_2

—SO_3^- blocks C^4 position

Problem 18.55 In the presence of $C_6H_5NO_2$, 2 mol of $C_6H_5NH_2$ reacts with 1 mol of *p*-toluidine to give a triarylmethane that is converted to the dye **pararosaniline** (Basic Red 9) by reaction with PbO_2 followed by acid. Show the steps, indicating the function of (*a*) nitrobenzene, (*b*) PbO_2, (*c*) HCl.

(*a*) Nitrobenzene oxidizes the CH_3 of *p*-toluidine to CHO, whose O is eliminated with the *para* H's from 2 molecules of $C_6H_5NH_2$ to form *p*-triaminotriphenylmethane (a leuco base).

$$H_2N-C_6H_4-CH_3 \xrightarrow{PhNO_2} H_2N-C_6H_4-CH{=}O \xrightarrow{2PhNH_2} H_2N-C_6H_4-CH(-C_6H_4-NH_2)_2$$

(*b*) PbO_2 oxidizes the triphenylmethane to a triphenylmethanol.

$$\underset{\text{leuco base}}{(p\text{-}H_2N-C_6H_4)_3C-H} \xrightarrow{PbO_2} \underset{\text{color base}}{(p\text{-}H_2N-C_6H_4)_3C-OH}$$

(*c*) HCl protonates the OH, thus making possible the loss of H_2O to form Ar_3C^+, whose + charge is delocalized to the three N's. Delocalization of electrons is responsible for absorption of light in the visible spectrum, thereby producing color.

$$(p\text{-}H_2N-C_6H_4)_3C-OH \xrightarrow[-H_2O]{HCl} \underset{\text{dye}}{(p\text{-}H_2N-C_6H_4)_3C^+Cl^-} \longleftrightarrow \underset{\text{one of several contributing structures}}{(p\text{-}H_2N-C_6H_4)_2C{=}C_6H_4{=}\overset{+}{N}H_2 + Cl^-}$$

Problem 18.56 How can *N*-methylaniline and *o*-toluidine be distinguished by ir spectroscopy?

o-Toluidine is a 1° amine and has a pair of peaks (symmetric and antisymmetric stretches) in the N—H stretch region. *N*-Methylaniline is a 2° amine and has only one peak.

Problem 18.57 Amines A, B, D, and E each have their parent-ion peaks at $m/e = 59$. The most prominent peaks for each are at m/e values of 44 for A and B, 30 for D, and 58 for E. Give the structure for each amine and for the ion giving rise to the most prominent peak for each.

Since the parent peak is $m/e = 59$, the formula is C_3H_9N. The major fragmentation of amines is a bond to the α carbons,

$$-\overset{|}{N}-\overset{|}{\underset{|}{C}}{}^{\alpha}\wr$$

A C—C bond is weaker and breaks more easily and more often than a C—H bond. Amines A and B both lose a CH_3 ($m = 15$); $59 - 15 = 44$.

The two isomers are:

$$\underset{(A)}{CH_3\dot{N}HCH_2CH_3} \xrightarrow{-e^-} [CH_3\dot{N}HCH_2 \vdots \dot{C}H_3]^+ \longrightarrow CH_3\overset{H}{\overset{|}{N^+}}{=}CH_2 + \cdot CH_3$$

$$H_2N{-}\underset{CH_3}{\underset{|}{C}}H{-}CH_3 \; (B) \xrightarrow{-e^-} \left[H_2\dot{N}\underset{CH_3}{\underset{|}{C}}H \vdots \dot{C}H_3\right]^+ \longrightarrow H_2\overset{+}{N}{=}CHCH_3 + \cdot CH_3$$

Amine D loses CH_2CH_3 ($59 - 29 = 30$).

$$H_2NCH_2CH_2CH_3 \xrightarrow{-e^-} [H_2\dot{N}CH_2 \vdots \dot{C}H_2CH_3]^+ \longrightarrow H_2\overset{+}{N}{=}CH_2 + \cdot CH_2CH_3$$

Amine E loses H ($59 - 1 = 58$).

$$(CH_3)_3N \xrightarrow{-e^-} \left[(CH_3)_2\dot{N}{-}\overset{H}{\underset{H}{C}} \vdots H\right]^+ \longrightarrow (CH_3)_2\overset{+}{N}{=}CH_2 + H\cdot$$

Problem 18.58 What compound, C_3H_7NO, has the following nmr spectrum: $\delta = 6.5$, broad singlet (two H's); $\delta = 2.2$, quartet (two H's): and $\delta = 1.2$, triplet (three H's)?

The integration ratio 2 : 2 : 3 accounts for the seven H's. The peaks at $\delta = 2.2$ and $\delta = 1.2$ are from a $—CH_2CH_3$, as indicated by the splitting, and the group is attached to a C$=$O group, as shown by the $\delta = 2.2$ value for the H's on the CH_2. The broad singlet at $\delta = 6.5$ are H's of an amide. The compound is $CH_3CH_2CONH_2$, propanamide.

Problem 18.59 Determine the structure of a compound ($C_9H_{11}NO$) which is soluble in dilute HCl and gives a positive Tollens' test. Its ir spectrum shows a strong band at 1695 cm^{-1} but no bands in the 3300–3500 cm^{-1} region. The proton-decoupled ^{13}C spectrum shows six signals which would display the following splitting pattern in the coupled ^{13}C spectrum: one quartet, two singlets, and three doublets, one doublet being very downfield.

The positive Tollens' test, the band at 1695 cm^{-1}, and the downfield doublet are consistent with a CHO group. Basicity and absence of an N—H stretching band indicate a 3° amino group, $—N(CH_3)_2$ (the CH_3's give the quartet). The 5 degrees of unsaturation indicate a benzene ring and a C$=$O group, and the two singlets show that the ring is disubstituted. Since the other two doublets must arise from the other four ring carbons, there are two pairs of equivalent ring C's, and the two substituents must be *para*. The compound is *p*-dimethylaminobenzaldehyde, $(CH_3)_2NC_6H_4CHO$.

Problem 18.60 What compound results from treating the diazonium salt of *p*-toluidine with copper bronze powder?

4,4′-dimethylbiphenyl (**Gatterman reaction**).

Phenolic Compounds

19.1 Introduction

Nomenclature

Phenols (ArOH) and alcohols (ROH) are similar in properties, but they differ sufficiently, so phenols may be considered as a separate homologous series.

Problem 19.1 Name the following phenols by the IUPAC system:

(*a*) OH Phenol (*b*) OH CH_3 *m*-Cresol (*c*) OH OH Resorcinol

(*d*) OH OH Catechol (*e*) OH OH Hydroquinone (*f*) OH COOH Salicylic acid

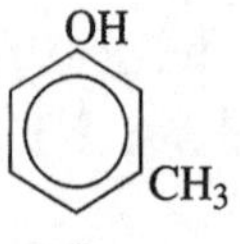

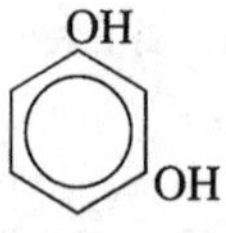

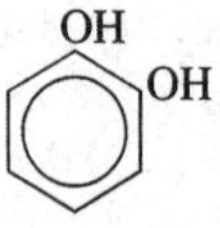

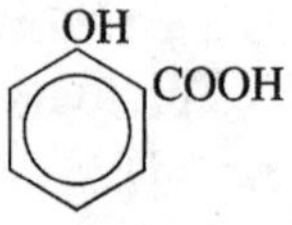

(*a*) hydroxybenzene, (*b*) *m*-hydroxytoluene, (*c*) 1,3-dihydroxybenzene, (*d*) 1,2-dihydroxybenzene, (*e*) 1,4-dihydroxybenzene, (*f*) *o*-hydroxybenzoic acid.

Problem 19.2 Name the following compounds:

(*a*) CH_2CH_3 OCH_3 (*b*) $NHCOCH_3$ OH (*c*) OH $CH_2CH{=}CH_2$ (*d*) COO^-Na^+ $OCCH_3$ (C=O) (*e*) OCH_2CH_3

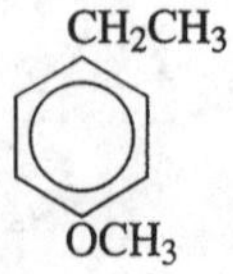

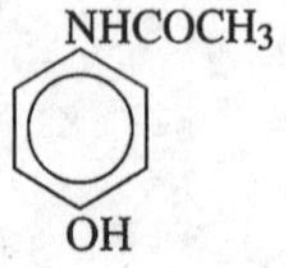

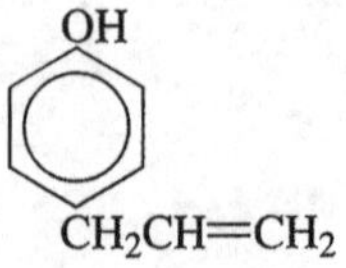

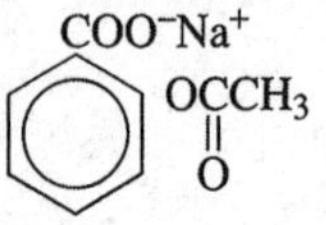

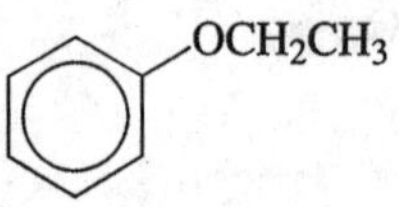

(*a*) *p*-methoxyethylbenzene, (*b*) *p*-hydroxyacetanilide, (*c*) *p*-allylphenol, (*d*) sodium acetylsalicylate (sodium salt of aspirin), (*e*) ethoxybenzene or phenetole.

Physical Properties

Problem 19.3 Compared to toluene, phenol (*a*) has a higher boiling point and (*b*) is more soluble in H_2O. Explain.

$$\text{Ph—O—H}\cdots\text{O(H)—Ph} \qquad\qquad \text{Ph—O—H}\cdots\text{OH}_2$$

(*a*) Intermolecular H-bonding (*b*) H-bonding with H_2O

Problem 19.4 Account for the lower boiling point and decreased H_2O solubility of *o*-nitrophenol and *o*-hydroxybenzaldehyde as compared with their *m* and *p* isomers.

In some *ortho*-substituted phenols, intramolecular H-bonding (chelation) forms a six-membered ring. This inhibits H-bonding with water and reduces solubility in H_2O. Since chelation diminishes the *inter*molecular H-bonding attraction present in the *para* and *meta* isomers, the boiling point is decreased.

o-Nitrophenol *o*-Hydroxybenzaldehyde

chelation

The greater H_2O solubility of the *meta* and *para* isomers is due to coassociation with water molecules through H-bonding.

19.2 Preparation

Industrial Methods

1. **Dow Process by Benzene Mechanism** (Section 11.3)

$$\underset{\text{Chlorobenzene}}{C_6H_5Cl} + 2\,NaOH \xrightarrow[320\text{ atm}]{360^\circ C} H_2O + NaCl + \underset{\text{Sodium phenoxide}}{C_6H_5O^-Na^+} \xrightarrow{H^+} \underset{\text{Phenol}}{C_6H_5OH}$$

2. **From Cumene Hydroperoxide**

$$C_6H_6 + \underset{\text{Propene}}{CH_2{=}CHCH_3} \xrightarrow{H_2SO_4} \underset{\text{Cumene}}{C_6H_5\text{—}CH(CH_3)_2} \xrightarrow{O_2} \underset{\text{Cumene hydroperoxide}}{C_6H_5\text{—}C(CH_3)_2\text{—O—O—H}} \xrightarrow{H_3O^+} \underset{\text{Phenol}}{C_6H_5OH} + \underset{\text{Acetone}}{(CH_3)_2C{=}O}$$

Problem 19.5 Give a mechanism for the acid-catalyzed rearrangement of cumene hydroperoxide involving an intermediate with an electron-deficient O (like R^+).

$(CH_3)_2C(Ph)-\ddot{O}-\ddot{O}-H \xrightarrow{H^+} (CH_3)_2C(Ph)-O-\overset{+}{O}H_2$ (an oxonium ion) $\xrightarrow{-H_2O:} [(CH_3)_2C(Ph)-\overset{+}{O}:]$ (electron-deficient intermediate) $\xrightarrow{\sim Ph:} [(CH_3)_2C=\overset{+}{O}-Ph \longleftrightarrow (CH_3)_2\overset{+}{C}-O-Ph]$ (carbocation) $\xrightarrow{H_2O} [(CH_3)_2C(\overset{+}{O}H_2)-O-Ph] \xrightarrow{-H^+} [(CH_3)_2C(OH)(OPh)]$ (Hemiacetal) $\longrightarrow (CH_3)_2C=O + PhOH$ (Acetone, Phenol)

The rearrangment of Ph may be synchronous with loss of H_2O.

3. **Alkali Fusion of Arylsulfonate Salts**

Laboratory Methods

1. **Hydrolysis of Diazonium Salts** (Section 18.7)
2. **Aromatic Nucleophilic Substitution of Nitro Aryl Halides** (See Section 11.3)
3. **Ring Oxidation** (with trifluoroperoxyacetic acid, $F_3C\overset{O}{\overset{\|}{C}}-OOH$)

Mesitylene (an activated ring) $\xrightarrow[BF_3]{F_3CCOOH}$ Mesitol (electrophilic substitution by OH^+)

Problem 19.6 Outline reactions and reagents for industrial syntheses of the following from benzene, and naphthalene (NpH) and inorganic reagents: (*a*) catechol, (*b*) resorcinol, (*c*) picric acid (2,4,6-trinitrophenol), (*d*) β-naphthol (β-NpOH).

(*a*) Benzene $\xrightarrow[2.\ 2NaOH]{1.\ 2H_2S_2O_7} C_6H_5SO_3^-Na^+ \xrightarrow[2.\ H_3O^+]{1.\ NaOH\ fusion} C_6H_5OH \xrightarrow{1.\ H_2SO_4,\ 20°C \quad 2.\ NaOH}$ o-$(O^-Na^+)C_6H_4(SO_3^-Na^+)$ $\xrightarrow[2.\ H_3O^+]{1.\ NaOH\ fusion}$ o-$C_6H_4(OH)_2$ (catechol)

(b) Benzene $\xrightarrow[\text{2. NaOH}]{\text{1. } H_2S_2O_7}$ m-benzenedisulfonate ($SO_3^-Na^+$, $SO_3^-Na^+$) $\xrightarrow[\text{2. } H_3O^+]{\text{1. NaOH fusion}}$ m-dihydroxybenzene (OH, OH)

(c) Benzene $\xrightarrow[\text{Fe}]{Cl_2}$ chlorobenzene $\xrightarrow[H_2SO_4]{HNO_3}$ 2,4-Dinitrochlorobenzene $\xrightarrow[\text{2. } H_3O^+]{\text{1. NaOH}}$ 2,4-Dinitrophenol $\xrightarrow[H_2SO_4]{HNO_3}$ Picric acid

Direct nitration of phenol leads to excessive oxidation and destruction of material because HNO_3 is a strong oxidizing agent and the OH activates the ring.

(d) $\text{NpH} \xrightarrow[\text{2. } OH^-]{\text{1. } H_2SO_4\ (140°C)} \beta\text{-NpSO}_3^- \xrightarrow[\text{fuse}]{OH^-} \beta\text{-NpOH}$

Problem 19.7 Devise practical laboratory syntheses of the following phenols from benzene or toluene and any inorganic or aliphatic compounds: (*a*) *m*-iodophenol, (*b*) 3-chloro-4-methylphenol, (*c*) 2-bromo-4-methylphenol.

(a) Benzene $\xrightarrow[H_2SO_4]{HNO_3}$ nitrobenzene (NO_2) $\xrightarrow[H_2SO_4]{HNO_3}$ m-dinitrobenzene (NO_2, NO_2) $\xrightarrow{(NH_4)_2S}$ m-nitroaniline (NO_2, NH_2) $\xrightarrow{\text{1. } HNO_2,\ 5°C \mid \text{2. KI}}$ m-iodonitrobenzene (NO_2, I) $\xrightarrow[\text{2. } HNO_2,\ 5°C]{\text{1. Sn, HX}}$ m-iodobenzenediazonium salt ($N_2^+X^-$, I) $\xrightarrow[\text{heat}]{H_2O}$ *m*–Iodophenol (OH, I)

(b) Toluene (CH_3) $\xrightarrow[H_2SO_4]{HNO_3}$ p-nitrotoluene (CH_3, NO_2) $\xrightarrow[\text{Fe}]{Cl_2}$ (CH_3, Cl, NO_2) $\xrightarrow[\text{2. } HNO_2,\ 5°C]{\text{1. Sn, HX}}$ (CH_3, Cl, $N_2^+X^-$) $\xrightarrow{H_2O}$ 3-Chloro-4-methylphenol (CH_3, Cl, OH)

(c) Toluene (CH_3) $\xrightarrow[H_2SO_4]{HNO_3}$ (NO_2, CH_3) $\xrightarrow[\text{2. } Ac_2O]{\text{1. Sn, HCl}}$ (NHAc, CH_3) $\xrightarrow{Br_2}$ (NHAc, Br, CH_3) $\xrightarrow[\text{2. } HNO_2,\ 5°C;\ \text{3. } H_2O]{\text{1. } H_3O^+}$ 2-Bromo-4-methylphenol (OH, Br, CH_3)

19.3 Chemical Properties

Reactions of H of The OH Group

1. Acidity

Phenols are weak acids (pK_a = 10):

$$\text{ArOH} + H_2O \rightleftharpoons \text{ArO}^- + H_3O^+$$

They form salts with aqueous NaOH but not with aqueous $NaHCO_3$.

Problem 19.8 Why does aqueous $NaHCO_3$ solution dissolve RCOOH but not PhOH?

In both cases, the product would be carbonic acid ($pK_a = 6$), which is a stronger acid than phenols ($pK_a = 10$) but weaker than carboxylic acids ($pK_a = 4.5$). Acid-base equilibria lie toward the weaker acid and weaker base.

$$\underset{\text{stronger acid}_1}{RCOOH} + \underset{\text{stronger base}_2}{HCO_3^-} \rightleftharpoons \underset{\text{weaker base}_1}{RCOO^-} + \underset{\text{weaker acid}_2}{H_2CO_3}$$

$$\underset{\text{weaker acid}_2}{ArOH} + \underset{\text{weaker base}_1}{HCO_3^-} \rightleftharpoons \underset{\text{stronger base}_2}{ArO^-} + \underset{\text{stronger acid}_1}{H_2CO_3}$$

Problem 19.9 What are the effects of (*a*) electron-attracting and (*b*) electron-releasing substituents on the acid strength of phenols?

(*a*) Electron-attracting substituents disperse negative charges and therefore stabilize ArO^- and increase acidity of ArOH. (*b*) Electron-releasing substituents concentrate the negative charge on O, destabilize ArO^-, and decrease acidity of ArOH.

Electron-Attracting
($-NO_2, -CN, -CHO, -COOH, -NR_3^+, -X$)

OH-C6H4-A (More acidic) + :B ⟶ B:H⁺ + O=C6H4(−)-A (Charge dispersed; ion stabilized)

Electron-Releasing
($-CH_3, -C_2H_5, -OR, -NR_2$)

OH-C6H4-R (Less acidic) + B: ⟶ B:H⁺ + O⁻-C6H4-R (Charge concentrated; ion destabilized)

Problem 19.10 In terms of resonance and inductive effects, account for the following relative acidities.

(*a*) $p\text{-}O_2NC_6H_4OH > m\text{-}O_2NC_6H_4OH > C_6H_5OH$
(*b*) $m\text{-}ClC_6H_4OH > p\text{-}ClC_6H_4OH > C_6H_5OH$

(*a*) The $-NO_2$ is electron-withdrawing and acid-strengthening. Its resonance effect, which occurs only from *para* and *ortho* positions, predominates over its inductive effect, which occurs also from the *meta* position. Other substituents in this category are

$$\rangle C{=}O \qquad -CN \qquad -COOR \qquad -SO_2R$$

(*b*) Cl is electron-withdrawing by induction. This effect diminishes with increasing distance between Cl and OH. The *meta* is closer than the *para* position, and *m*-Cl is more acid-strengthening than *p*-Cl. Other substituents in this category are F, Br, I, $\overset{+}{N}R_3$.

Problem 19.11 Assign numbers from (1) for LEAST to (4) for MOST to indicate the relative acid strengths in the following groups: (*a*) phenol, *m*-cholorophenol, *m*-nitrophenol, *m*-cresol; (*b*) phenol, benzoic acid, *p*-nitrophenol, carbonic acid; (*c*) phenol, *p*-chlorophenol, *p*-nitrophenol, *p*-cresol; (*d*) phenol, *o*-nitrophenol, *m*-nitrophenol, *p*-nitrophenol; (*e*) phenol, *p*-chlorophenol, 2,4,6-trichlorophenol, 2,4-dichlorophenol; (*f*) phenol, benzyl alcohol, benzenesulfonic acid, benzoic acid.

(*a*) 2, 3, 4, 1. Because

$$-\overset{+}{N}{\overset{\large O}{\underset{\large O^-}{\lessgtr}}}$$

has + on N, it has a greater electron-withdrawing inductive effect than has Cl.

(*b*) 1, 4, 2, 3.

(*c*) 2, 3, 4, 1. The resonance effect of *p*-NO_2 exceeds the inductive effect of *p*-Cl. *p*-CH_3 is electron-releasing.

(*d*) 1, 3, 2, 4. Intramolecular H-bonding makes the *o*-isomer weaker than the *p*-isomer.

(*e*) 1, 2, 4, 3. Increasing the number of electron-attracting groups increases the acidity.

(*f*) 2, 1, 4, 3.

Formation of Ethers

1. Williamson Synthesis

$$ArOH \xrightarrow{OH^-} ArO^- \begin{cases} \xrightarrow{R'X\ (1^\circ \text{ or } 2^\circ)} Ar{-}O{-}R' + X^- \\ \xrightarrow{(CH_3)_2SO_4} Ar{-}O{-}CH_3 + CH_3OSO_3^- \end{cases}$$

2. Aromatic Nucleophilic Substitution

$$\text{1-Cl-2,4-}(NO_2)_2C_6H_3 + Na^+OC_2H_5^- \longrightarrow \text{1-}OC_2H_5\text{-2,4-}(NO_2)_2C_6H_3 + Cl^-$$

2,4-Dinitrophenetole
(2,4-Dinitrophenyl ethyl ether)

Problem 19.12 Why does cleavage of aryl ethers, ArOR, with HI yield only ArOH and RI?

An S_N2 displacement on the onium ion of the ether occurs only on the C of the R group, not on the C of the Ar group.

$$ArOR + HI \longrightarrow Ar\overset{+}{O}(H)R + I^- \longrightarrow ArOH + RI$$

Formation of Esters

Phenyl esters (RCOOAr) are not formed directly from RCOOH. Instead, acid chlorides or anhydrides are reacted with ArOH in the presence of strong base.

$$(CH_3CO)_2O + C_6H_5OH + NaOH \longrightarrow \underset{\text{Phenyl acetate}}{CH_3COOC_6H_5} + CH_3COO^-Na^+ + H_2O$$

$$C_6H_5COCl + C_6H_5OH + NaOH \longrightarrow \underset{\text{Phenyl benzoate}}{C_6H_5COOC_6H_5} + Na^+Cl^- + H_2O$$

OH^- converts ArOH to the more nucleophilic ArO^- and also neutralizes the acids formed.

Problem 19.13 Phenyl acetate undergoes the **Fries rearrangement** with $AlCl_3$ to form *ortho-* and *para,*-hydroxyacetophenone. The *ortho* isomer, is separated from the mixture by its volatility with steam.

$$C_6H_5O{-}\overset{O}{\overset{\|}{C}}{-}CH_3 \xrightarrow{AlCl_3(CS_2)} o\text{-}HOC_6H_4\overset{O}{\overset{\|}{C}}{-}CH_3 + p\text{-}HOC_6H_4\overset{O}{\overset{\|}{C}}{-}CH_3$$

o-Hydroxyaceto-phenone   *p*-Hydroxyaceto-phenone

(*a*) Account for the volatility in steam of the *ortho* but not the *para* isomer. (*b*) Why does the *para* isomer predominate at low, and the *ortho* at higher, temperatures? (*c*) Apply this reaction to the synthesis of the antiseptic 4-*n*-hexylresorcinol, using resorcinol, aliphatic compounds, and any needed inorganic reagents.

(*a*) The *ortho* isomer has a higher vapor pressure because of chelation, O—H---O═C (see Problem 19.4). In the *para* isomer there is intermolecular H-bonding with H_2O.

(*b*) The *para* isomer (rate-controlled product) is the exclusive product at 25°C because it has a lower $\Delta H^\ddagger$ and is formed more rapidly. Its formation is reversible, unlike that of the *ortho* isomer, which is stabilized by chelation. Although it has a higher $\Delta H^\ddagger$, the *ortho* isomer (equilibrium-controlled product) is the chief product at 165°C because it is more stable.

(*c*) Two activating OH groups in *meta* positions reinforce each other in electrophilic substitution and permit Friedel-Crafts reactions of resorcinol directly with RCOOH and $ZnCl_2$.

$$\text{Resorcinol} \xrightarrow[ZnCl_2]{CH_3(CH_2)_4\overset{O}{\overset{\|}{C}}OH} 2,4\text{-}(HO)_2C_6H_3\overset{O}{\overset{\|}{C}}(CH_2)_4CH_3 \xrightarrow[HCl]{Zn(Hg)} 2,4\text{-}(HO)_2C_6H_3(CH_2)_5CH_3$$

Displacement of OH Group

Phenols resemble aryl halides in that the functional group resists displacement. Unlike ROH, phenols do not react with HX, $SOCl_2$, or phosphorus halides. Phenols are reduced to hydrocarbons, but the reaction is used for structure proof and not for synthesis.

$$ArOH + Zn \xrightarrow{\Delta} ArH + ZnO \quad \textit{(poor yields)}$$

Bucherer Reaction for Interconverting β-NH_2 and β-OH Naphthalenes

$$\beta\text{-Naphthol} \underset{NaHSO_3,\ H_2O,\ \Delta}{\overset{NH_3,\ (NH_4)_2SO_3,\ \Delta}{\rightleftharpoons}} \beta\text{-Naphthylamine}$$

The ready availability of β-naphthol [see Problem 19.6(*d*)] makes this a good method for synthesizing β-naphthylamine.

Reactions of the Benzene Ring

1. Hydrogenation

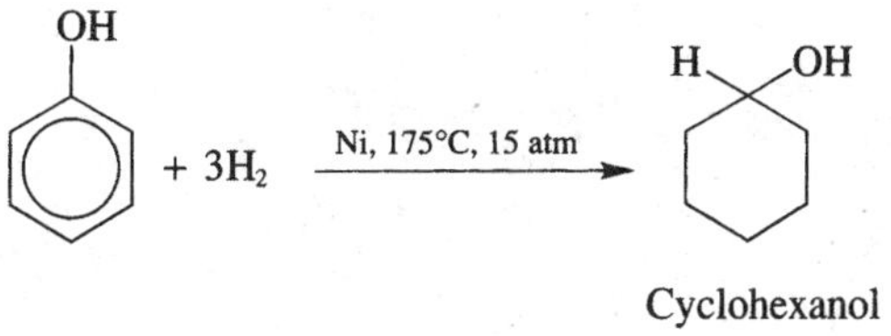

Phenol + $3H_2$ $\xrightarrow{\text{Ni, 175°C, 15 atm}}$ Cyclohexanol

Cyclohexanol

2. Oxidation to Quinones

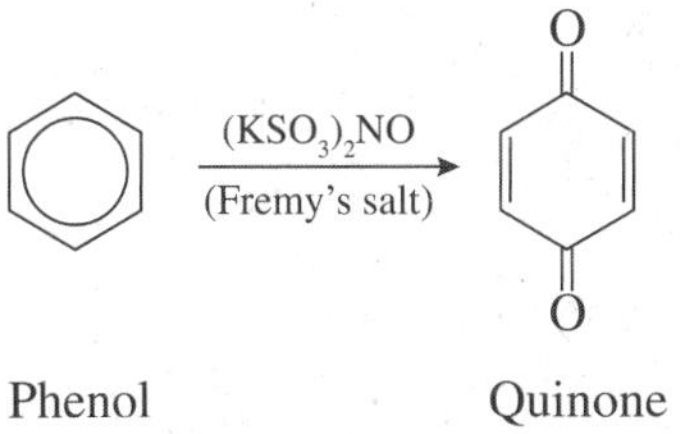

Phenol $\xrightarrow[\text{(Fremy's salt)}]{(KSO_3)_2NO}$ Quinone

Phenol Quinone

3. Electrophilic Substitution

The —OH and even more so the —O^- (**phenoxide**) are strongly activating and *op*-directing.

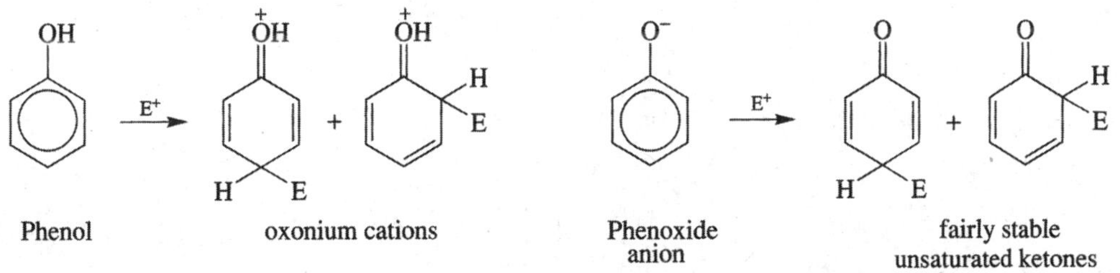

Phenol $\xrightarrow{E^+}$ oxonium cations; Phenoxide anion $\xrightarrow{E^+}$ fairly stable unsaturated ketones

Phenol oxonium cations Phenoxide anion fairly stable unsaturated ketones

Special mild conditions are needed to achieve electrophilic monosubstitution in phenols because their high reactivity favors both *polysubstitution* and *oxidation*.

(*a*) Halogenation.

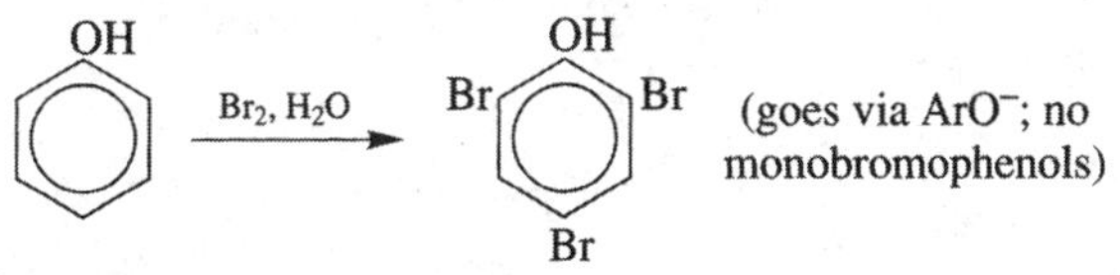

Phenol $\xrightarrow{Br_2, H_2O}$ 2,4,6-tribromophenol (goes via ArO^-; no monobromophenols)

Monobromination is achieved with nonpolar solvents such as CS_2 to decrease the electrophilicity of Br_2 and also to minimize phenol ionization.

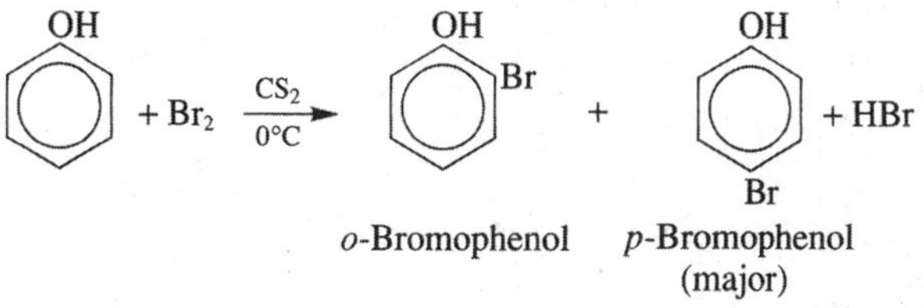

Phenol + Br_2 $\xrightarrow[0°C]{CS_2}$ *o*-Bromophenol + *p*-Bromophenol (major) + HBr

o-Bromophenol *p*-Bromophenol (major)

(*b*) Nitrosation.

$$C_6H_5OH \xrightarrow[7.5°C]{HONO} p\text{-}ON\text{-}C_6H_4\text{-}OH \xrightleftharpoons{\text{tautomerize}} O{=}C_6H_4{=}NOH$$

p-Nitroso-phenol Quinone monoxime

(*c*) Nitration. Low yields of *p*-nitrophenol are obtained from direct nitration of PhOH because of ring oxidation.

(*d*) Sulfonation.

C_6H_5OH Phenol $\xrightarrow[15\text{–}20°C]{H_2SO_4}$ *o*-Phenolsulfonic acid (rate-controlled) $\xrightarrow[100°C]{H_2SO_4}$ *p*-Phenolsulfonic acid (equilibrium-controlled)

C_6H_5OH Phenol $\xrightarrow[100°C]{H_2SO_4}$ *p*-Phenolsulfonic acid (equilibrium-controlled)

(*e*) Diazonium salt coupling to form azophenols (Section 18.7).

(*f*) Ring alkylation.

$$C_6H_5OH + \begin{cases} CH_3CH{=}CH_2 \\ (CH_3)_2CHOH \end{cases} \xrightarrow[\text{or HF}]{H_2SO_4} o\text{- and } p\text{–}HOC_6H_4CH(CH_3)_2 + H_2O$$

RX and $AlCl_3$ give poor yields because $AlCl_3$ coordinates with O.

(*g*) Ring acylation. Phenolic ketones are best prepared by the Fries rearrangement (Problem 19.13).

(*h*) **Kolbe synthesis** of phenolic carboxylic acids.

$$C_6H_5O^-Na^+ + O{=}C{=}O \xrightarrow[6\text{ atm}]{125°C} \underset{\text{Sodium salicylate}}{o\text{-}HOC_6H_4COO^-\,Na^+} \xrightarrow{H_3O^+} \underset{\text{Salicylic acid}}{o\text{-}HOC_6H_4COOH}$$

(*i*) **Reimer-Tiemann synthesis** of phenolic aldehydes.

$$C_6H_5OH + HCCl_3 + OH^- \longrightarrow o\text{-}HOC_6H_4CHO$$

(j) Condensations with carbonyl compounds; phenol-formaldehyde resins. Acid or base catalyzes electrophilic substitution of carbonyl compounds in *ortho* and *para* positions of phenols to form phenol alcohols (**Lederer-Manasse reaction**).

base catalysis: $\text{R-CHO} + \text{PhO}^- \xrightarrow[\text{2. H}^+]{\text{1. OH}^- \text{ (on PhOH)}} p\text{-HOC}_6\text{H}_4\text{-CH(OH)R}$

acid catalysis: $\text{PhOH} + [\text{RCH=}\overset{+}{\text{O}}\text{H} \leftrightarrow \text{R}\overset{+}{\text{C}}\text{H-OH}]$ (from $\text{RCH=O} + \text{H}^+$) $\longrightarrow p\text{-HOC}_6\text{H}_4\text{-RCHOH}$

(k) Rearrangements from O to ring.

(1) Fries rearrangement of phenolic esters to phenolic ketones (Problem 19.13).

(2) Claisen rearrangement. The reaction is *intramolecular* and has a cyclic mechanism.

$$\text{C}_6\text{H}_5\text{-O-CH}_2\text{CH=CH}_2 \xrightarrow[100\%]{200^\circ\text{C}} [\text{cyclohexadienone with H and CH}_2\text{CH=CH}_2] \longrightarrow o\text{-HOC}_6\text{H}_4\text{-CH}_2\text{CH=CH}_2$$

o-Allylphenol

(3) Alkyl phenyl ethers.

$$\text{C}_6\text{H}_5\text{-O-C}_2\text{H}_5 \xrightarrow[150^\circ\text{C}]{\text{AlCl}_3} o\text{-C}_2\text{H}_5\text{C}_6\text{H}_4\text{O-H} + p\text{-C}_2\text{H}_5\text{C}_6\text{H}_4\text{O-H}$$

Phenetole   *o*-Ethylphenol   *p*-Ethylphenol

Problem 19.14 Outline a mechanism for the (a) Kolbe reaction, (b) Reimer-Tiemann reaction.

(a) Phenoxide carbanion adds at the electrophilic carbon of CO_2.

$$\text{C}_6\text{H}_5\text{O}^- \longleftrightarrow [\text{cyclohexadienone carbanion}] + \text{O=C=O} \longrightarrow [\text{cyclohexadienone with H and -C(=O)-O}^-] \longrightarrow o\text{-HOC}_6\text{H}_4\text{-C(=O)-O}^- \xrightarrow{\text{H}^+} o\text{-HOC}_6\text{H}_4\text{-C(=O)-OH}$$

The conjugated ketonic diene tautomerizes to reform the more stable benzenoid ring.

(*b*) The electrophile is the carbene $:CCl_2$

$$:\ddot{O}H^- + H:CCl_3 \longrightarrow H:OH + [^-:CCl_3] \longrightarrow :Cl^- + :CCl_2$$

a benzal chloride

Problem 19.15 Use phenol and any inorganic or aliphatic reagents to synthesize (*a*) aspirin (acetylsalicyclic acid), (*b*) oil of wintergreen (methyl salicylate). Do not repeat the synthesis of any compound.

(*a*) $C_6H_5OH \xrightarrow[2.\ H_3O^+]{1.\ CO_2,\ KOH} o\text{-}HOC_6H_4COOH \xrightarrow{Ac_2O} o\text{-}CH_3COOC_6H_4COOH$

(*b*) $o\text{-}HOC_6H_4COOH \xrightarrow{CH_3OH/H_2SO_4} o\text{-}HOC_6H_4COOCH_3$

Problem 19.16 Predict the product of the Claisen rearrangement of (*a*) allyl-3-^{14}C phenyl ether, (*b*) 2,6-dimethylphenyl allyl-3-^{14}C ether.

(*a*) In this concerted intramolecular rearrangement, the ends of the allyl system interchange so that the γ C is bonded to the *ortho* C.

$$C_6H_5O-\overset{\alpha}{C}H_2-\overset{\beta}{C}H=\overset{14}{\underset{\gamma}{C}}H_2 \longrightarrow [\text{dienone, } H, \overset{14}{C}H_2CH=CH_2] \longrightarrow o\text{-}HOC_6H_4\overset{14}{C}H_2CH=CH_2$$

(*b*) When the *ortho* position is blocked, the allyl group migrates to the *para* position by two consecutive rearrangements and the ^{14}C is in the γ position of the product.

$$2,6\text{-}(CH_3)_2C_6H_3O-\overset{\alpha}{C}H_2-\overset{\beta}{C}H=\overset{14\gamma}{C}H_2 \longrightarrow [\text{dienone, } CH_3, \overset{14}{C}H_2-CH=CH_2] \longrightarrow 2,6\text{-}(CH_3)_2\text{-}4\text{-}(CH_2-CH=\overset{14}{C}H_2)C_6H_2OH$$

19.4 Analytical Detection of Phenols

Phenols are soluble in NaOH but not in $NaHCO_3$. With Fe^{3+}, they produce complexes whose characteristic colors are green, red, blue, and purple.

Infrared stretching bands of phenols are 3200–3600 cm^{-1} for the O—H (like alcohols), but 1230 cm^{-1} for the C—O (alcohols: 1050–1150 cm^{-1}). Nmr absorption of OH depends on H-bonding and the range is $\delta = 4$–12 ppm.

19.5 Summary of Phenolic Chemistry

PREPARATION

1. Industrial

(*a*) Dow Process ArCl + NaOH → ArONa

(*b*) Raschig

Ar—H + HCl + O_2 ⟶ H_2O + Ar—Cl (→ ArCl + NaOH)

(*c*) NaOH Fusion

$ArSO_3Na$ + NaOH ⟶ Na_2SO_3 + ArONa (→ ArOH, H^+)

(*d*) Cumene Peroxide

$$\text{Ar—CH(CH}_3)_2 \xrightarrow{O_2} \text{Ar—C(CH}_3)_2\text{—O—O—H} \xrightarrow{H^+} (\text{CH}_3)_2\text{C=O} + \text{ArOH}$$

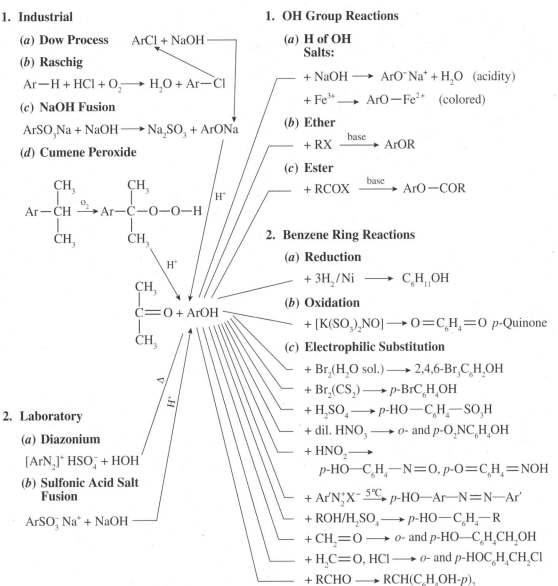

2. Laboratory

(*a*) Diazonium

$[ArN_2]^+ HSO_4^-$ + HOH ⟶ (Δ) ArOH

(*b*) Sulfonic Acid Salt Fusion

$ArSO_3^- Na^+$ + NaOH ⟶ (H^+) ArOH

PROPERTIES (of ArOH)

1. OH Group Reactions

(*a*) H of OH
Salts:

+ NaOH ⟶ ArO^-Na^+ + H_2O (acidity)

+ Fe^{3+} ⟶ ArO—Fe^{2+} (colored)

(*b*) Ether

+ RX $\xrightarrow{\text{base}}$ ArOR

(*c*) Ester

+ RCOX $\xrightarrow{\text{base}}$ ArO—COR

2. Benzene Ring Reactions

(*a*) Reduction

+ $3H_2$/Ni ⟶ $C_6H_{11}OH$

(*b*) Oxidation

+ $[K(SO_3)_2NO]$ ⟶ O=C_6H_4=O *p*-Quinone

(*c*) Electrophilic Substitution

+ Br_2(H_2O sol.) ⟶ 2,4,6-$Br_3C_6H_2OH$

+ $Br_2(CS_2)$ ⟶ *p*-BrC_6H_4OH

+ H_2SO_4 ⟶ *p*-HO—C_6H_4—SO_3H

+ dil. HNO_3 ⟶ *o*- and *p*-$O_2NC_6H_4OH$

+ HNO_2 ⟶
p-HO—C_6H_4—N=O, *p*-O=C_6H_4=NOH

+ $Ar'N_2^+X^-$ $\xrightarrow{5°C}$ *p*-HO—Ar—N=N—Ar′

+ ROH/H_2SO_4 ⟶ *p*-HO—C_6H_4—R

+ CH_2=O ⟶ *o*- and *p*-HO—$C_6H_4CH_2OH$

+ H_2C=O, HCl ⟶ *o*- and *p*-$HOC_6H_4CH_2Cl$

+ RCHO ⟶ $RCH(C_6H_4OH\text{-}p)_2$

19.6 Summary of Phenolic Ethers and Esters

$$\text{ArO}^-\text{Na}^+ + \text{X—CH}_2\text{CH=CH}_2 \longrightarrow \text{ArO—CH}_2\text{CH=CH}_2 \xrightarrow{\text{heat}} o\text{-HOC}_6\text{H}_4\text{CH}_2\text{CH=CH}_2 \quad (\text{Ar=Ph})$$

$$\text{ArO}^-\text{Na}^+ + \text{X—CH}_2\text{CH(CH}_3\text{)—CH}_3 \longrightarrow \text{ArO—CH}_2\text{CH(CH}_3)_2 \xrightarrow{\text{AlCl}_3} p\text{-HO—C}_6\text{H}_4\text{C(CH}_3)_3 \quad (\text{Ar=Ph})$$

$$\text{ArO}^-\text{Na}^+ + \text{X—C(=O)—CH}_3 \longrightarrow \text{ArO—C(=O)—R} + \text{AlCl}_3 \longrightarrow p\text{- or } o\text{-HO—C}_6\text{H}_4\text{C(=O)—R} \quad (\text{Ar=Ph})$$

SUPPLEMENTARY PROBLEMS

Problem 19.17 Name the following compounds:

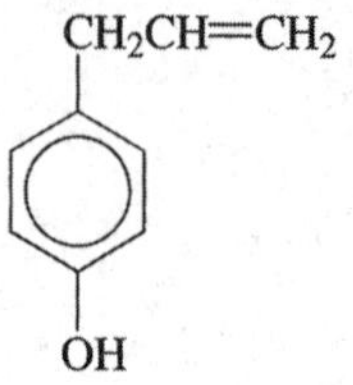

(*a*) *m*-ethoxytoluene, (*b*) methylhydroquinone, (*c*) *p*-allylphenol.

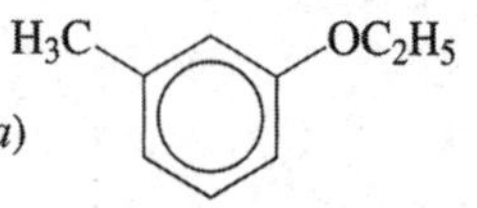

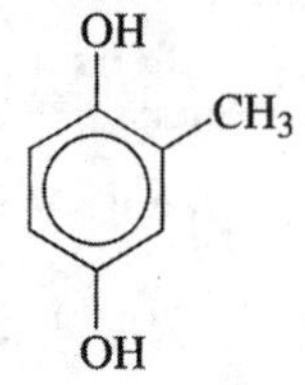

Problem 19.18 Write the structure for (*a*) phenoxyacetic acid, (*b*) phenyl acetate, (*c*) 2-hydroxy-3-phenylbenzoic acid, (*d*) *p*-phenoxyanisole.

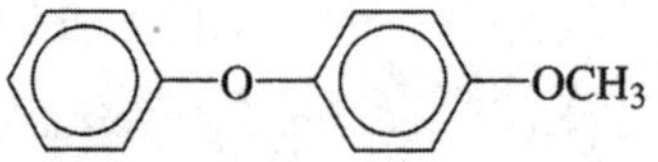

Problem 19.19 Draw a flow sheet for the separation of a mixture of PhOH, $PhCH_2OH$, and PhCOOH.

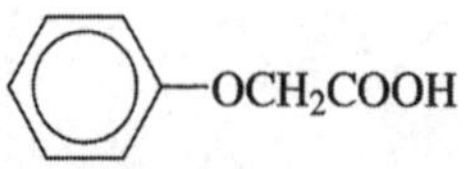

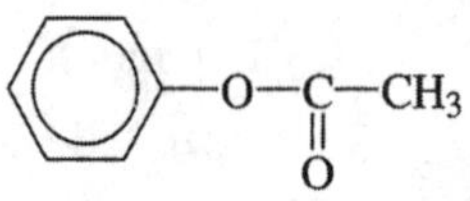

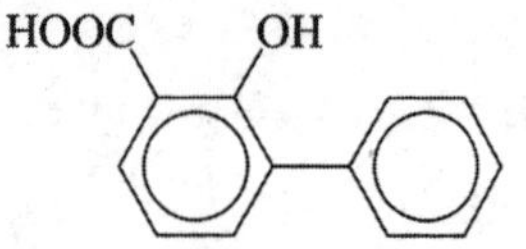

See Fig. 19.1.

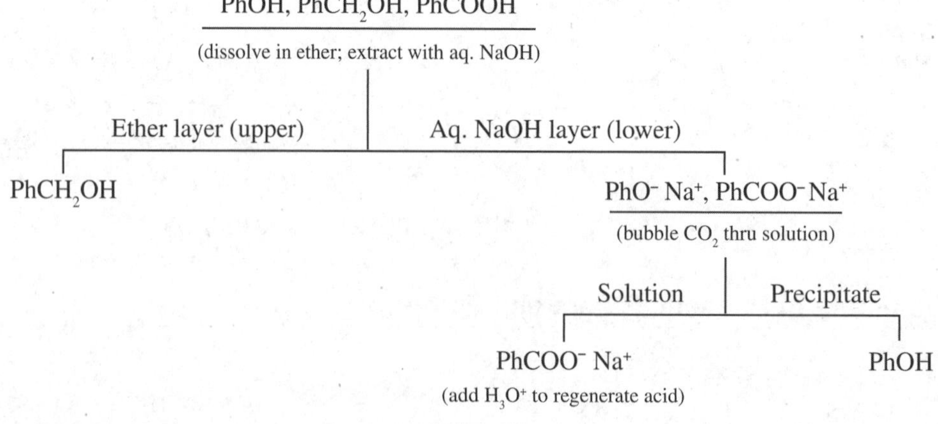

Figure 19.1

Problem 19.20 What product is formed when *p*-cresol is reacted with (*a*)$(CH_3CO)_2O$, (*b*) $PhCH_2Br$ and base, (*c*) aqueous NaOH, (*d*) aqueous $NAHCO_3$, (*e*) bromine water?

(*a*) *p*-$CH_3C_6H_4OCOCH_3$, *p*-cresyl acetate; (*b*) *p*-$CH_3C_6H_4OCH_2Ph$, *p*-tolyl benzyl ether; (*c*) *p*-$CH_3C_6H_4O^-Na^+$, sodium *p*-cresoxide; (*d*) no reaction; (*e*) 2,6-dibromo-4-methylphenol.

Problem 19.21 Use simple test tube reactions to distinguish (*a*) *p*-cresol from *p*-xylene, (*b*) salicylic acid from aspirin (acetylsalicylic acid).

(*a*) Aqueous NaOH dissolves the cresol. (*b*) Salicylic acid is a phenol which gives a color (purple in this case) with $FeCl_3$.

Problem 19.22 Identify compounds (A) through (D).

p-nitrophenol (NO₂ para to OH) $\xrightarrow[OH^-]{(C_2H_5)_2SO_4}$ (A) $\xrightarrow{Zn,\ HCl}$ (B) $\xrightarrow[5°C]{NaNO_2,\ HCl}$ (C) $\xrightarrow{C_6H_5OH}$ (D)

(A) p-$C_2H_5OC_6H_4NO_2$ (B) p-$C_2H_5OC_6H_4NH_2$ (*p*-phenetidine) (C) p-$C_2H_5OC_6H_4\overset{+}{N}_2Cl^-$

(D) C_2H_5O—C_6H_4—N=N—C_6H_4—OH

Problem 19.23 Prepare (*a*) 2-bromo-4-hydroxytoluene from toluene, (*b*) 2-hydroxy-5-methylbenzaldehyde from *p*-toluidine, (*c*) *m*-methoxyaniline from benzenesulfonic acid.

(*a*) toluene $\xrightarrow[H_2SO_4]{HNO_3}$ p-nitrotoluene (CH_3, NO_2) $\xrightarrow[Fe]{Br_2}$ 2-bromo-4-nitrotoluene (CH_3, Br, NO_2) $\xrightarrow[HCl]{Zn}$ 2-bromo-4-aminotoluene (CH_3, Br, NH_2) $\xrightarrow[HCl]{NaNO_2}$ (CH_3, Br, $\overset{+}{N}_2Cl^-$) $\xrightarrow[\Delta]{H_2O}$ (CH_3, Br, OH)

(*b*) p-toluidine (CH_3, NH_2) $\xrightarrow[HCl]{NaNO_2}$ (CH_3, $\overset{+}{N}_2Cl^-$) $\xrightarrow[\Delta]{H_2O}$ p-cresol (CH_3, OH) $\xrightarrow[2.\ H_3O^+]{1.\ CHCl_3,\ NaOH}$ (CH_3, CHO, OH)

(*c*) benzenesulfonic acid (SO_3H) $\xrightarrow[H_2SO_4]{HNO_3}$ m-nitrobenzenesulfonic acid (SO_3H, NO_2) $\xrightarrow{1.\ NaOH;\ 2.\ fuse\ with\ NaOH;\ 3.\ H^+}$ m-nitrophenol (OH, NO_2) $\xrightarrow[OH^-]{(CH_3)_2SO_4}$ (OCH_3, NO_2) $\xrightarrow{Zn,\ HCl}$ (OCH_3, NH_2)

Problem 19.24 Use the Bucherer reaction to prepare 2-(N-methyl)- and 2-(N-phenyl)naphthylamines.

CH_3NH_2 and $C_6H_5NH_2$ replace NH_3.

$$\text{2-}C_{10}H_7NHCH_3 \xleftarrow[\text{2. OH}^-]{\text{1. }CH_3NH_2,\ NaHSO_3} \boxed{\text{2-}C_{10}H_7OH} \xrightarrow[\text{2. OH}^-]{\text{1. }C_6H_5NH_2,\ NaHSO_3} \text{2-}C_{10}H_7NHC_6H_5$$

Problem 19.25 (*a*) Give products, where formed, for reaction of $PhN_2^+Cl^-$ with (i) α-naphthol, (ii) β-naphthol, (iii) 4-methyl-1-naphthol, (iv) 1-methyl-2-naphthol. (*b*) How can these products be used to make the corresponding aminonaphthols?

(*a*) Structural formulas are given below. Coupling occurs (i) *para* (α) to OH; (ii) *ortho* (α, not β) to OH; (iii) *ortho* (β) to OH, since *para* (α) position is blocked. (iv) No reaction. An activating β-substituent cannot activate other β positions.

(i) 4-Phenylazo-1-naphthol (OH at C-1, N=NPh at C-4)

(ii) 1-Phenylazo-2-naphthol (N=NPh at C-1, OH at C-2)

(iii) 4-Methyl-2-Phenylazo-1-naphthol (OH at C-1, N=NPh at C-2, CH_3 at C-4)

(iv) *no reaction* (CH_3 at C-1, OH at C-2)

(*b*) Reduction of the azo compounds with $LiAlH_4$ or $Na_2S_2O_4$ or Sn, HCl yields the amines.

4-Amino-1-naphthol *from (i)* (OH at C-1, NH_2 at C-4)

1-Amino-2-naphthol *from (ii)* (NH_2 at C-1, OH at C-2)

2-Amino-4-methyl-1-naphthol *from (iii)* (OH at C-1, NH_2 at C-2, CH_3 at C-4)

Problem 19.26 From readily available phenolic compounds, synthesize the antioxidant food preservatives (*a*) BHA (*tert*-**b**utylated**h**ydroxy**a**nisole, a mixture of 2- and 3-*tert*-butyl-4-methoxyphenol; (*b*) BHT (*tert*-**b**utylated **h**ydroxy**t**oluene).

(*a*) *p*-$CH_3OC_6H_4OH + (CH_3)_2C{=}CH_2$ gives the mixture. This Friedel-Craft monoalkylation occurs both *ortho* to the OCH_3 and to the OH group. (*b*) *p*-cresol, *p*-$HOC_6H_4CH_3$, is *tert*-butylated with $(CH_3)_2CH{=}CH_2$.

Problem 19.27 The acid-base indicator phenolphthalein is made by using anhydrous $ZnCl_2$ to condense 2 mol of phenol and 1 mol of phthalic anhydride by eliminating 1 mol of H_2O. What is its formula?

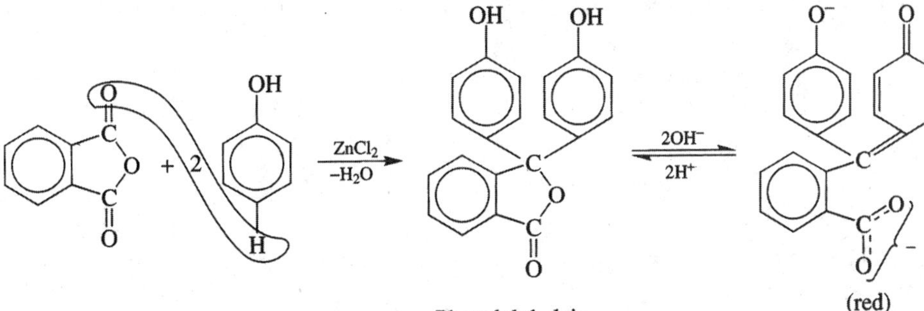

Phthalic anhydride + 2 phenol $\xrightarrow[-H_2O]{ZnCl_2}$ Phenolphthalein (colorless) $\underset{2H^+}{\overset{2OH^-}{\rightleftharpoons}}$ (red)

(red) $\underset{OH^-}{\overset{H^+}{\rightleftharpoons}}$ (colorless)

Problem 19.28 From phenol, prepare (*a*) *p*-benzoquinone, (*b*) *p*-benzoquinone dioxime, (*c*) quinhydrone (a 1 : 1 complex of *p*-benzoquinone and hydroquinone).

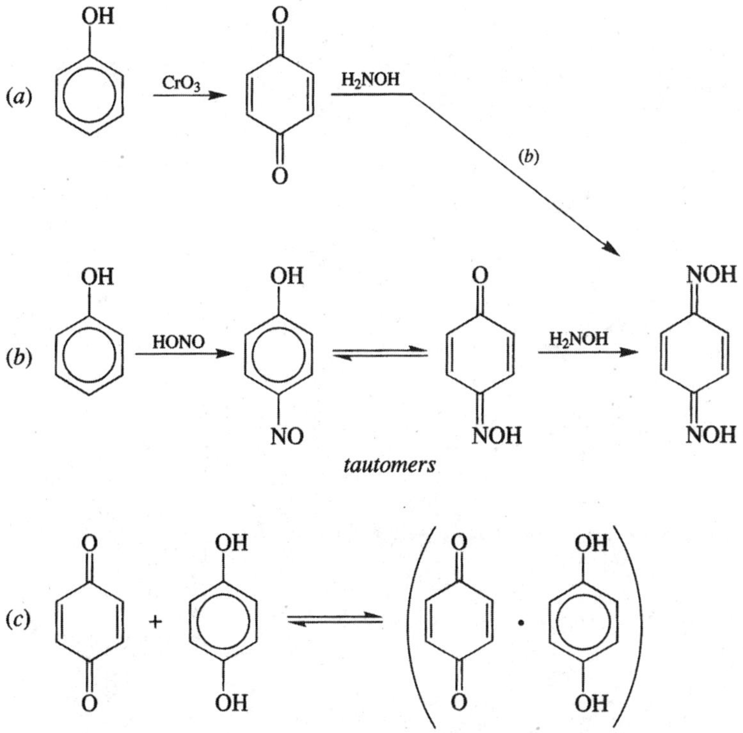

(*a*) Phenol $\xrightarrow{CrO_3}$ *p*-benzoquinone $\xrightarrow{H_2NOH}$ (*b*)

(*b*) Phenol $\xrightarrow{HONO}$ *p*-nitrosophenol $\rightleftharpoons$ quinone monoxime (*tautomers*) $\xrightarrow{H_2NOH}$ *p*-benzoquinone dioxime

(*c*) *p*-benzoquinone + hydroquinone $\rightleftharpoons$ (quinone · hydroquinone)

a charge-transfer complex

Problem 19.29 Write a structural formula for the product of the Diels-Alder reaction of *p*-benzoquinone with:

(*a*) Butadiene (*b*) 1,3-Cyclohexadiene (*c*) 1,1′-Bicyclohexadienyl

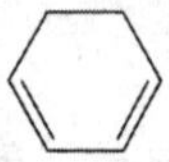

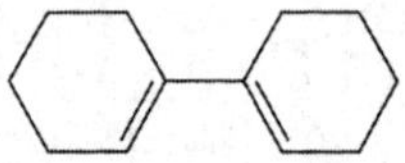

(*a*) [tautomers] (*b*) 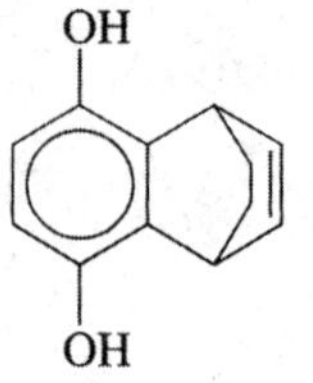 (*c*)

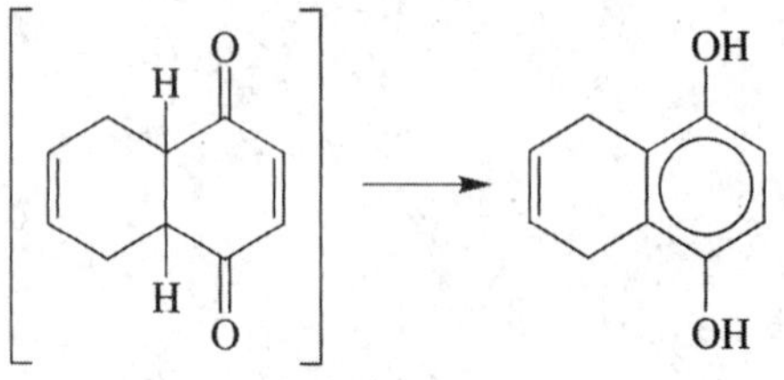

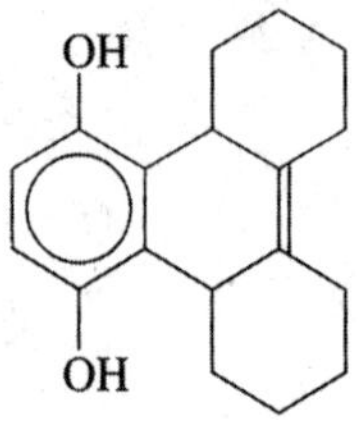

Problem 19.30 The ir OH stretching bands for the three isomeric nitrophenols in KBr pellets and in dilute CCl_4 solution are identical for the *ortho* but different for *meta* and *para* isomers. Explain.

In KBr (solid state), the OH for all three isomers is H-bonded. In CCl_4, the H-bonds of *meta* and *para* isomers, which are intermolecular, are broken. Their ir OH absorption bands shift to higher frequencies (3325–3520 cm^{-1}). There is no change in the absorption of the *ortho* isomer (3200 cm^{-1}), since intramolecular H-bonds are not broken upon dilution by solvent.

Problem 19.31 Show all major products for the following reactions:

(*a*) $CH_3-C_6H_4-OH + CHCl_3 + NaOH \xrightarrow{70°C}$ (*b*) $C_6H_5CH_2OOH + \text{acid} \longrightarrow$

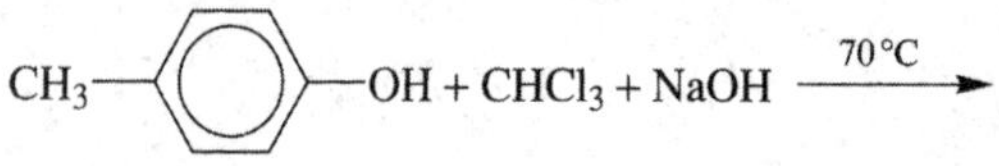

(*c*) $Br-C_6H_4-OCOCH_3 + C_2H_5-C_6H_4-OCOC_2H_5 \xrightarrow{AlCl_3}$

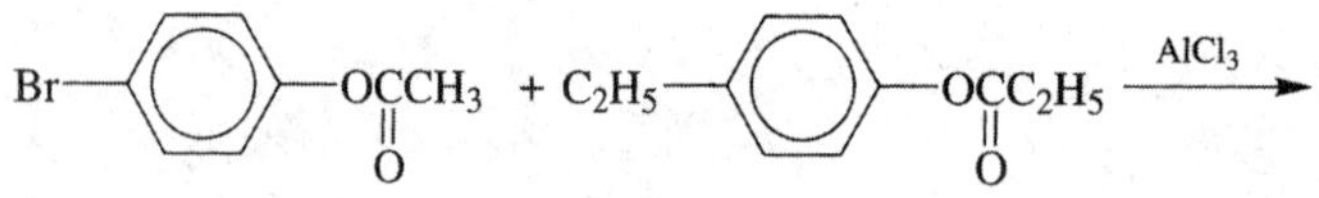

(*a*) Reimer-Tiemann reaction;

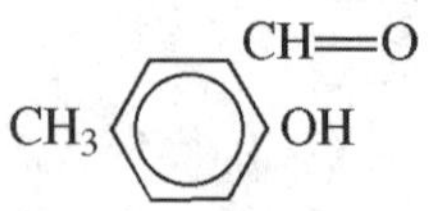

CH=O, CH_3, OH

(*b*) This is a 1,2-shift of phenyl to an electron-deficient O.

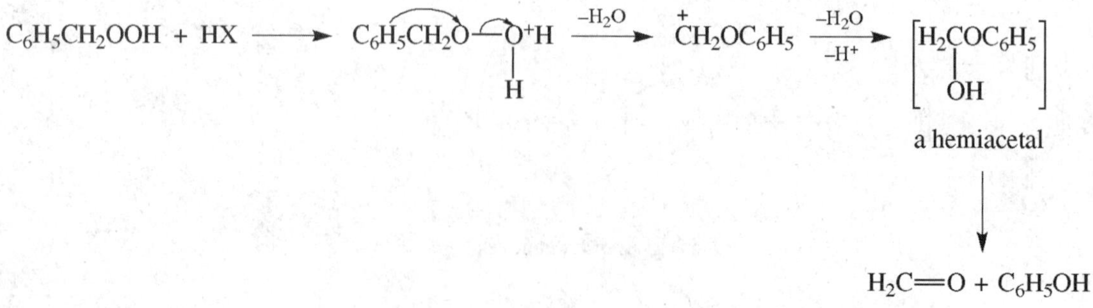

$$C_6H_5CH_2OOH + HX \longrightarrow C_6H_5CH_2O-\overset{+}{O}H(H) \xrightarrow{-H_2O} \overset{+}{C}H_2OC_6H_5 \xrightarrow[-H^+]{-H_2O} \left[\begin{matrix}H_2COC_6H_5\\ |\\ OH\end{matrix}\right]$$

a hemiacetal

$$\downarrow$$

$$H_2C{=}O + C_6H_5OH$$

(*c*) Intramolecular rearrangement gives two products, (A) and (B), and intermolecular rearrangement gives two more products, (C) and (D). These are Fries rearrangements.

(A) 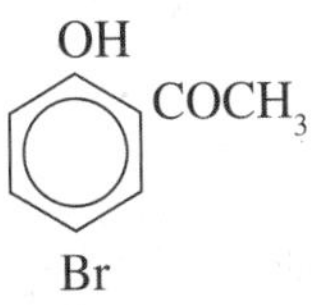 OH, $COCH_3$, Br (C) 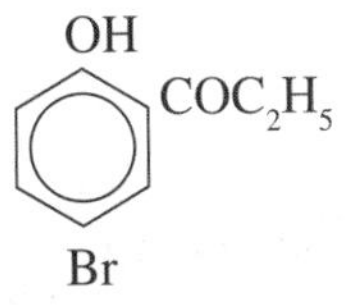 OH, COC_2H_5, Br (D) 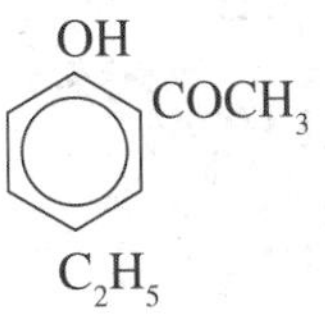 OH, $COCH_3$, C_2H_5 (B) 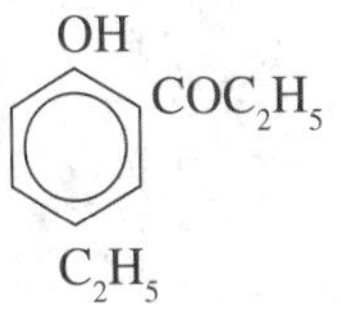 OH, COC_2H_5, C_2H_5

Problem 19.32 (*a*) Compound (A), C_7H_8O, is insoluble in aqueous $NaHCO_3$ but dissolves in NaOH. When treated with bromine water, (A) rapidly forms compound (B), $C_7H_5OBr_3$. Give structures for (A) and (B). (*b*) What would (A) have to be if it did not dissolve in NaOH?

(*a*) With four degrees of unsaturation, (A) has a benzene ring. From its solubility, (A) must be a phenol with a methyl substituent to account for the seventh C. Since a tribromo compound is formed, (A) must be *m*-cresol and (B) is 2,4,6-tribromo-3-methylphenol. (*b*) (A) could not be a phenol. It would have to be an ether, $C_6H_5OCH_3$ (anisole).

Problem 19.33 A compound ($C_{10}H_{14}O$) dissolves in NaOH but not in $NaHCO_3$ solution. It reacts with Br_2 in water to give $C_{10}H_{12}Br_2O$. The ir spectrum shows a broad band at 3520 cm^{-1} and a strong peak at 830 cm^{-1}. The proton nmr shows: a singlet at $\delta = 1.3$ (9 H), a singlet at $\delta = 4.9$ (1 H), a multiplet at $\delta = 7.0$ (4 H). Give the structure of the compound.

The level of acidity, the facile reaction with Br_2, the broad band at 3250 cm^{-1}, and the singlet at $\delta = 4.9$ suggest a phenolic compound. The multiplet at $\delta = 7.0$, the dibromination (not tribromination), and the strong band at 830 cm^{-1} indicate a *p*-substituted phenol. The singlet at $\delta = 1.3$ is typical of a *t*-butyl group. The compound is *p*-$HOC_6H_4C(CH_3)_3$.

Problem 19.34 **Dioxin** (2,3,7,8-tetrachlorodibenzo-*p*-dioxin):

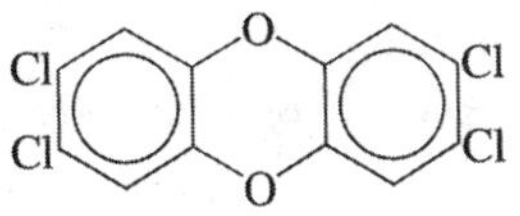

Cl, Cl, O, O, Cl, Cl

a very toxic compound, is a by-product of the manufacture of 2,4,5-trichlorophenol by treating 1,2,4,5-tetrachlorobenzene with NaOH. Suggest a mechanism for the formation of dioxin.

NaOH converts 2,4,5-trichlorophenol to its phenoxide anion, which undergoes aromatic nucleophilic displacement with unreacted 1,2,4,5-tetrachlorobenzene.

Cl Cl Cl Cl + $^-$O Cl Cl Cl ⟶ Cl Cl O Cl Cl Cl Cl

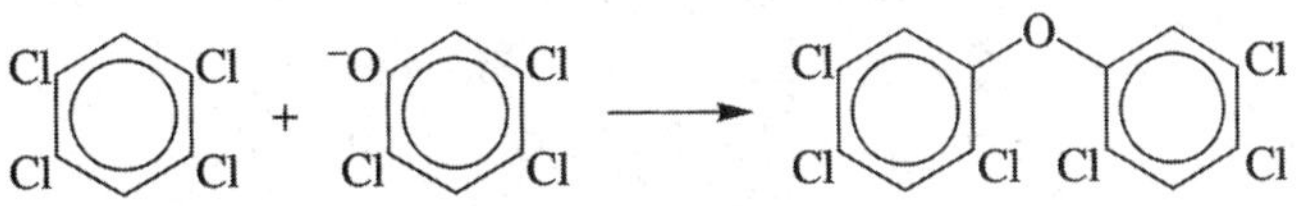

This product undergoes another aromatic nucleophilic displacement with OH^-, to give a phenoxide that reacts by an intramolecular displacement to give dioxin.

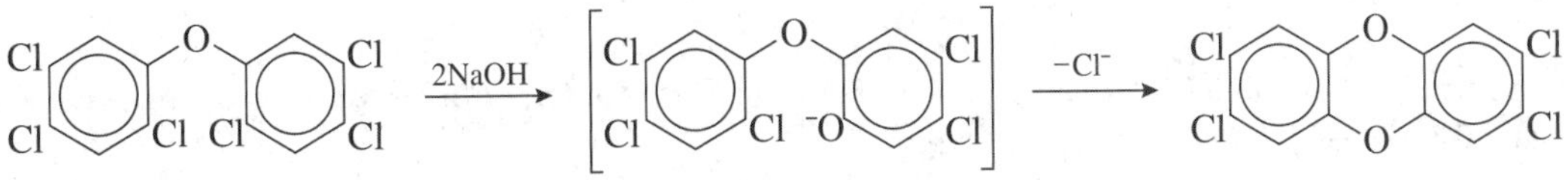

Cl Cl O Cl Cl Cl Cl $\xrightarrow{2NaOH}$ [Cl Cl O Cl $^-$O Cl Cl] $\xrightarrow{-Cl^-}$ Cl Cl O O Cl Cl

CHAPTER 20

Aromatic Heterocyclic Compounds

The chemistry of saturated heterocyclic compounds is characteristic of their functional group. For example, nitrogen compounds are amines, oxygen compounds are ethers, and sulfur compounds are sulfides. Differences in chemical reactivity are observed for three-membered rings, for instance, epoxides, whose enhanced reactivity is driven by the relief of their severe ring strain. This chapter discusses **heterocycles** that are aromatic and have unique chemical properties.

20.1 Five-Membered Aromatic Heterocycles with One Heteroatom

Nomenclature; Aromaticity

The **ring index system** combines (1) the prefix **oxa-** for O, **aza-** for N, or **thia**- for S; and (2) a stem for ring size and saturation or unsaturation. These are summarized in Table 20.1.

TABLE 20.1 Ring Index Heterocyclic Nomenclature

RING SIZE	STEM	
	SATURATED	UNSATURATED
3	irane	irene
4	etane	ete
5	olane	ole
6	ane	ine
7	epane	epine
8	ocane	ocin

The three most common five-membered aromatic rings are **furan**, with an O atom; **pyrrole**, with an N atom; and **thiophene**, with an S atom.

Problem 20.1 Name the following compounds, using (i) numbers and (ii) Greek letters.

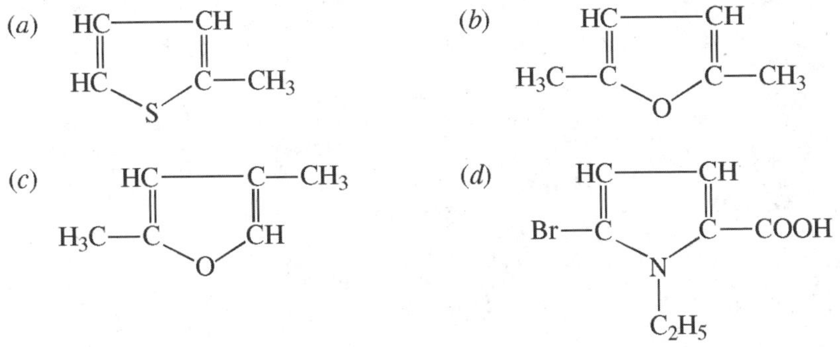

(*a*) HC—CH, HC, C—CH_3, S

(*b*) HC—CH, H_3C—C, C—CH_3, O

(*c*) HC—C—CH_3, H_3C—C, CH, O

(*d*) HC—CH, Br—C, C—COOH, N, C_2H_5

(*a*) 2-methylthiophene (2-methylthiole) or α-methylthiophene, (*b*) 2,5-dimethylfuran (2,5-dimethyloxole) or α,α'-dimethylfuran, (*c*) 2,4-dimethylfuran or α,β'-dimethylfuran (2,4-dimethyloxole), (*d*) 1-ethyl-5-bromo-2-pyrrolecarboxylic acid or *N*-ethyl-α-bromo-α'-pyrrolecarboxylic acid (*N*-ethyl-5-bromazole-2-carboxylic acid).

Problem 20.2 Write structures for (*a*) 2-benzoylthiophene, (*b*) 3-furansulfonic acid, (*c*) α,β'-dichloropyrrole.

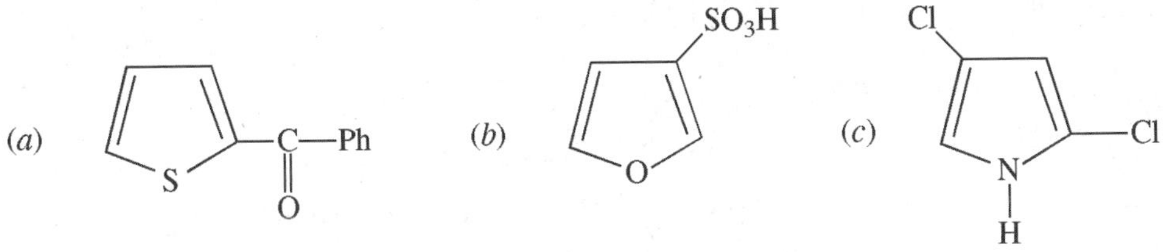

(*a*) S, C—Ph, O

(*b*) SO_3H, O

(*c*) Cl, Cl, N, H

Problem 20.3 Account for the aromaticity of furan, pyrrole, and thiophene, which are planar molecules with bond angles of 120°.

See Fig. 20.1. The four C's and the heteroatom Z use sp^2-hybridized atomic orbitals to form the σ bonds. When Z is O or S, one of the unshared pairs of e^-'s is in an sp^2 HO. Each C has a *p* orbital with one electron, and the heteroatom Z has a *p* orbital with two electrons. These five *p* orbitals are parallel to each other and overlap side by side to give a cyclic π system with six *p* electrons. These compounds are aromatic because six electrons fit Hückel's $4n + 2$ rule, which is extended to include heteroatoms.

It is noteworthy that typically these heteroatoms would use sp^3 HO's for bonding. The exceptional sp^2 HO's lead to a *p* AO for the cyclic aromatic π system.

Problem 20.4 Account for the following dipole moments: furan, 0.7D (away from O); tetrahydrofuran, 1.7 D (toward O).

In tetrahydrofuran, the greater electronegativity of O directs the moment of the C—O bond toward O. In furan, delocalization of an electron pair from O makes the ring C's negative and O positive; the moment is away from O. See Fig. 20.1.

Preparation

Problem 20.5 Pyrroles, furans, and thiophenes are made by heating 1,4-dicarbonyl compounds with $(NH_4)_2CO_3$, P_4O_{10}, and P_2S_5, respectively. Which are used to prepare (*a*) 3,4-dimethylfuran; (*b*) 2,5-dimethylthiophene; (*c*) 2,3-dimethylpyrrole?

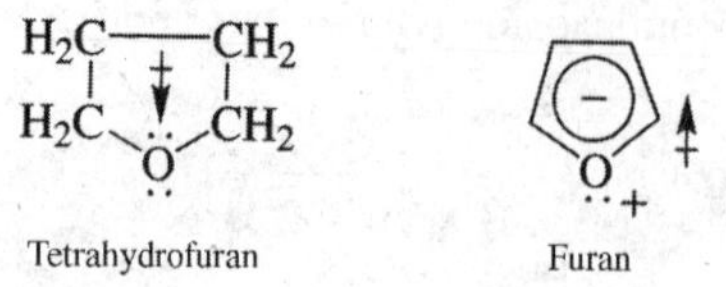

Tetrahydrofuran Furan

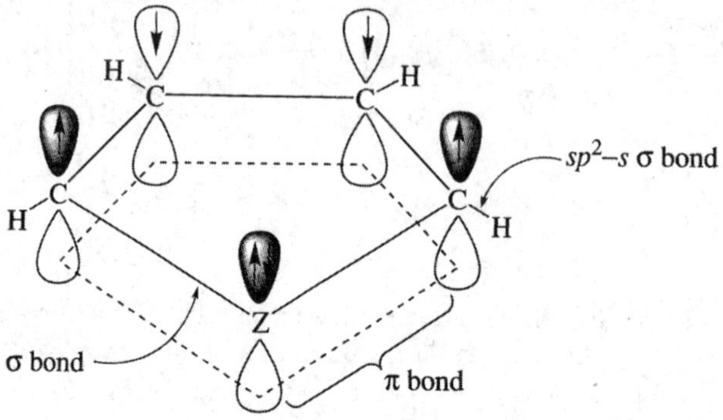

Figure 20.1

The carbonyl C's become the α C's in the heterocyclic compound.

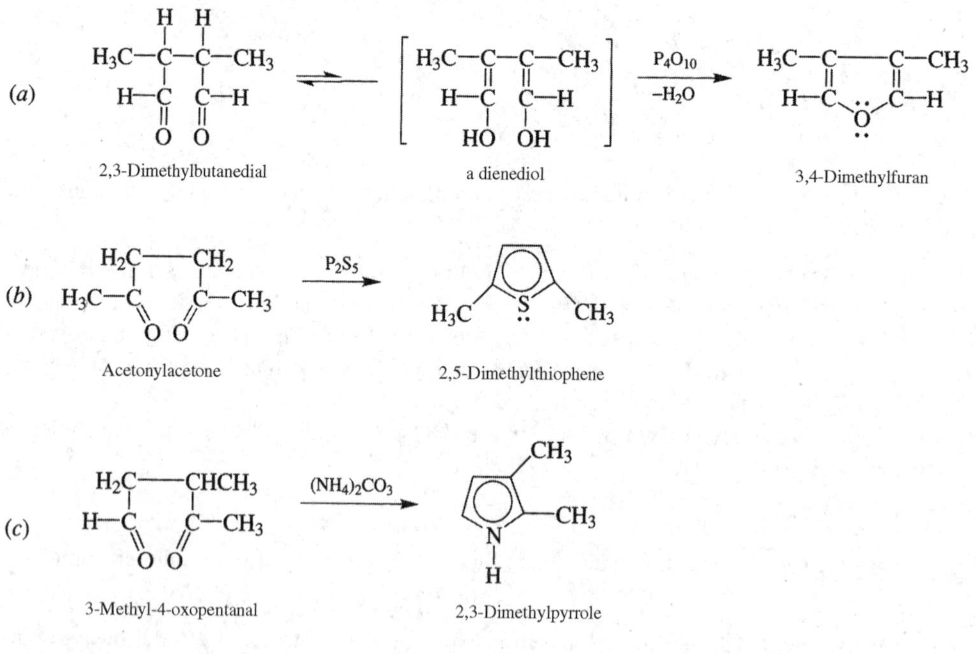

(*a*) 2,3-Dimethylbutanedial ⇌ [a dienediol] $\xrightarrow[-H_2O]{P_4O_{10}}$ 3,4-Dimethylfuran

(*b*) Acetonylacetone $\xrightarrow{P_2S_5}$ 2,5-Dimethylthiophene

(*c*) 3-Methyl-4-oxopentanal $\xrightarrow{(NH_4)_2CO_3}$ 2,3-Dimethylpyrrole

Problem 20.6 Prepare pyrrole from succinic anhydride.

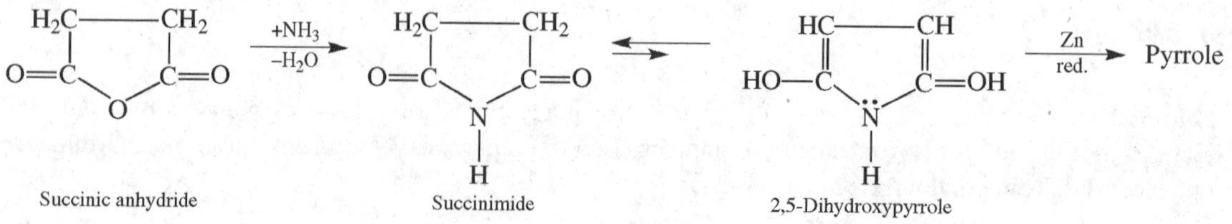

Succinic anhydride $\xrightarrow[-H_2O]{+NH_3}$ Succinimide ⇌ 2,5-Dihydroxypyrrole $\xrightarrow[\text{red.}]{Zn}$ Pyrrole

Problem 20.7 Identify compounds (A) through (D).

$$CH_3COOC_2H_5 + (A) \xrightarrow{NaOEt} C_6H_5COCH_2COOC_2H_5 \xrightarrow[I_2]{NaOEt} (B) \xrightarrow[2.\ H_3O^+]{1.\ dil.\ NaOH} (C) \xrightarrow{P_4H_{10}} (D)$$

(A) PhCOOEt

(B) $C_6H_5COCHCOOC_2H_5$ (via $C_6H_5COCH(I)COOC_2H_5$)
$\quad\quad C_6H_5COCHCOOC_2H_5$ (the two CH carbons joined by a bond)

(C) 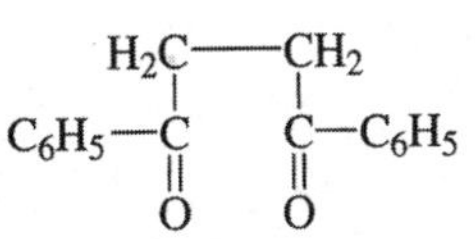 C_6H_5—C(=O)—CH_2—CH_2—C(=O)—C_6H_5

(D) 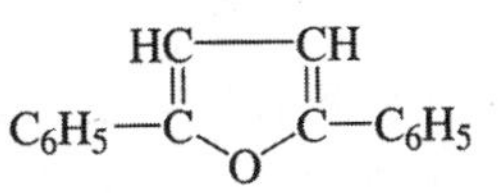 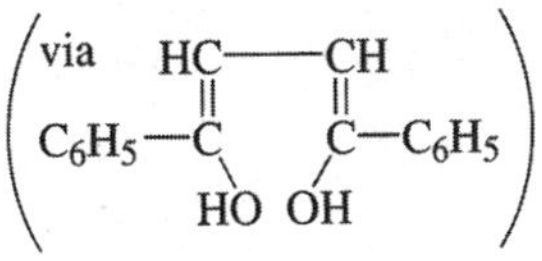 2,5-diphenylfuran: C_6H_5—C═CH—CH═C—C_6H_5 ring closed through O (via C_6H_5—C(OH)═CH—CH═C(OH)—C_6H_5)

Problem 20.8 Dilantin (5,5-diphenylhydantoin), an anticonvulsant drug used in the treatment of epileptic seizures, is a pyrrole with the molecular formula $C_{15}H_{12}N_2O_2$. What is the structural formula for Dilantin?

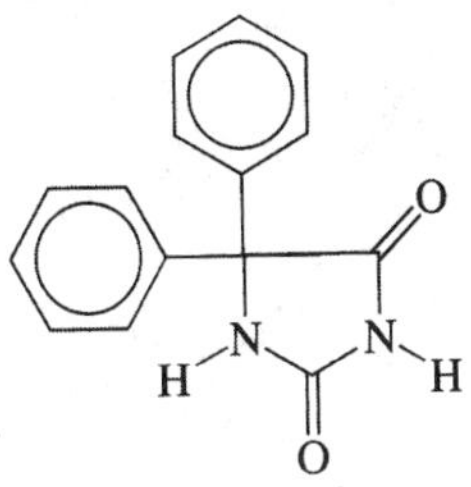

[Structure: 5,5-diphenylhydantoin — five-membered ring with C(Ph)₂, C=O, N–H, C=O, N–H]

Chemical Properties

Problem 20.9 (*a*) In terms of relative stability of the intermediate, explain why an electrophile (E^+) attacks the α rather than the β position of pyrrole, furan, and thiophene. (*b*) Why are these heterocyclics more reactive than C_6H_6 to E^+-attack?

(*a*) The transition state and the intermediate R^+ formed by α-attack is a hybrid of three resonance structures which possess less energy; the intermediate from β-attack is less stable and has more energy because it is a hybrid of only two resonance structures. I and II are also more stable allylic carbocations; V is not allylic.

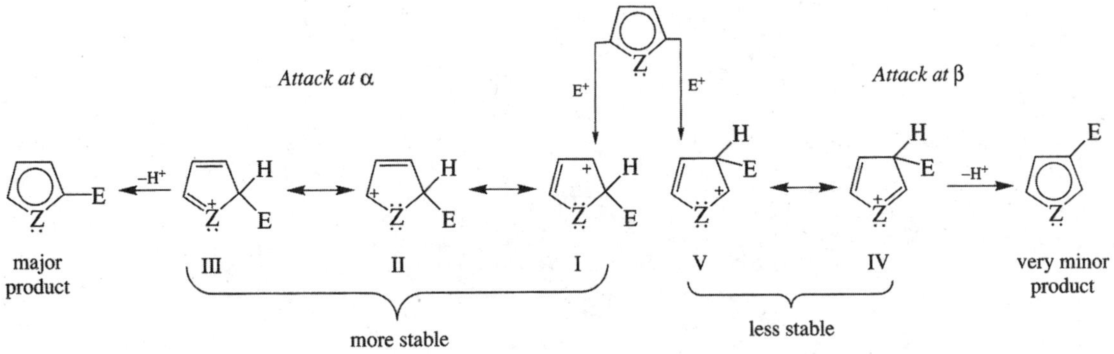

Attack at α: major product ← $-H^+$ ← III ↔ II ↔ I (more stable)

Attack at β: V ↔ IV (less stable) → $-H^+$ → very minor product

(*b*) This is ascribed to resonance structure III, in which Z has + charge and in which all ring atoms have an octet of electrons. These heterocyclics are as reactive as PhOH and $PhNH_2$.

Problem 20.10 Explain why pyrrole is not basic.

The unshared pair of electrons on N is delocalized and an "aromatic sextet." Adding an acid to N could prevent delocalization and destroy the aromaticity.

Problem 20.11 Give the type of reaction and the structures and names of the products obtained from: (*a*) furfural,

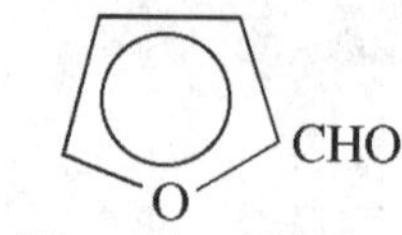

α-Furancarboxaldehyde

and concentrated aq. KOH; (*b*) furan with (i) $CH_3CO—ONO_2$ (acetyl nitrate), (ii) $(CH_3CO)_2O$ and BF_3 and then H_2O; (*c*) pyrrole with (i) SO_3 and pyridine, (ii) $CHCl_3$ and KOH, (iii) $PhN_2^+Cl^-$, (iv) Br_2 and C_2H_5OH; (*d*) thiophene and (i) H_2SO_4, (ii) $(CH_3CO)_2O$ and CH_3COONO_2, (iii) Br_2 in benzene.

(*a*) Cannizzaro reaction:

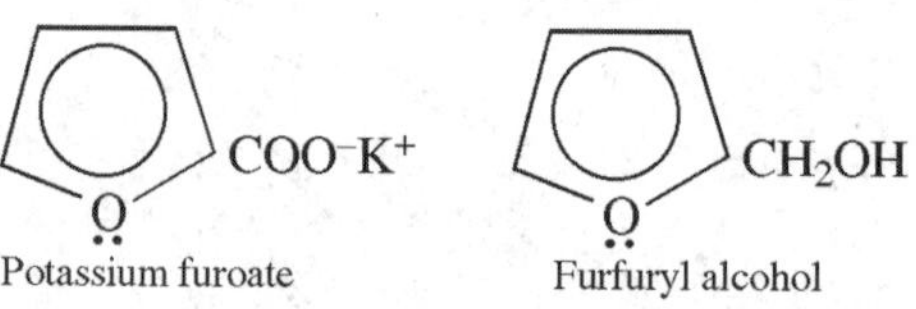

Potassium furoate Furfuryl alcohol

(*b*) (i) Nitration; 2-nitrofuran:

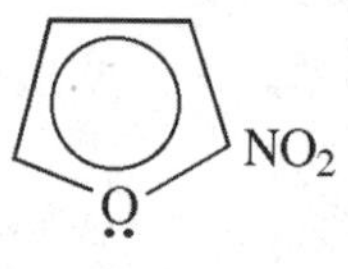

NO_2

(ii) Acetylation; 2-acetylfuran:

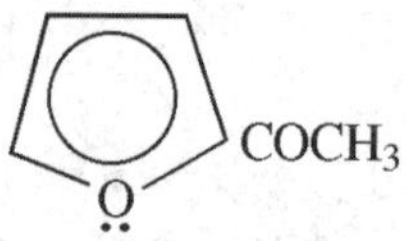

$COCH_3$

(*c*) (i) Sulfonation; 2-pyrrolesulfonic acid:

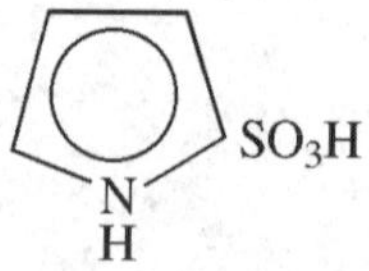

SO_3H (H_2SO_4 alone destroys ring).

(ii) Reimer-Tiemann formylation; 2-pyrrolecarboxaldehyde (2-formylpyrrole),

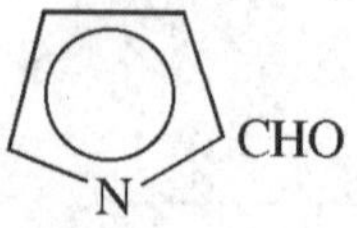

CHO

(iii) Coupling; 2-phenylazopyrrole:

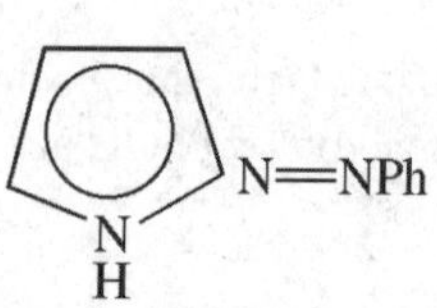

N═NPh

(iv) Bromination; 2,3,4,5-tetrabromopyrrole.

(*d*) (i) Sulfonation; thiophene-2-sulfonic acid:

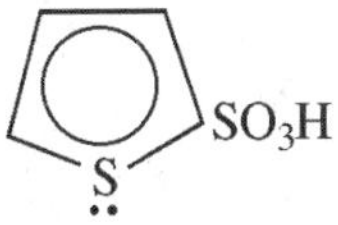

(ii) Nitration; 2-nitrothiophene:

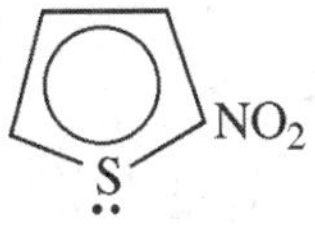

(iii) Bromination; 2,5-dibromothiophene. (Thiophene is less reactive than pyrrole and furan.)

Problem 20.12 Write structures for the mononitration products of the following compounds, and explain their formation: (*a*) 3-nitropyrrole, (*b*) 3-methoxythiophene, (*c*) 2-acetylthiophene, (*d*) 5-methyl-2-methoxythiophene, (*e*) 5-methylfuran-2-carboxylic acid.

(*a*) Nitration at C^5 to form 2,4-dinitropyrrole. After nitration, C^5 (*i*) becomes C^2, and C^3 becomes C^4. Nitration at C^2 (*ii*) would form an intermediate with a + on C^3, which has the electron-attracting —NO_2 group.

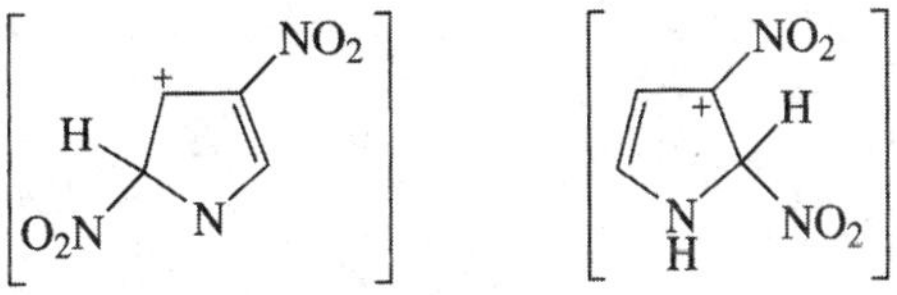

i *ii*

(*b*) 2-Nitro-3-methoxythiophene (α and *ortho* to OCH_3)

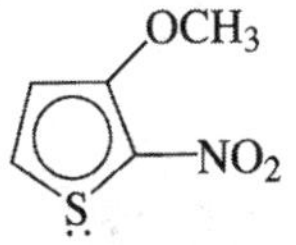

(*c*) 2-Acetyl-5-nitrothiophene (α-attack)

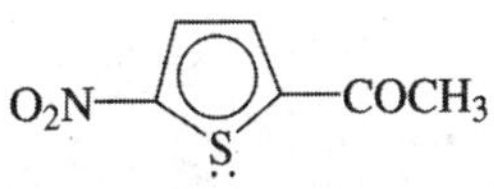

(*d*) 2-Methoxy-3-nitro-5-methylthiophene (*ortho* to OCH_3, a stronger activating group than CH_3)

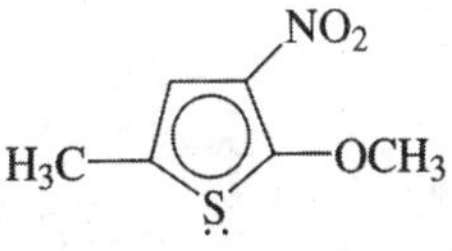

(*e*) 2-methyl-5-nitrofuran

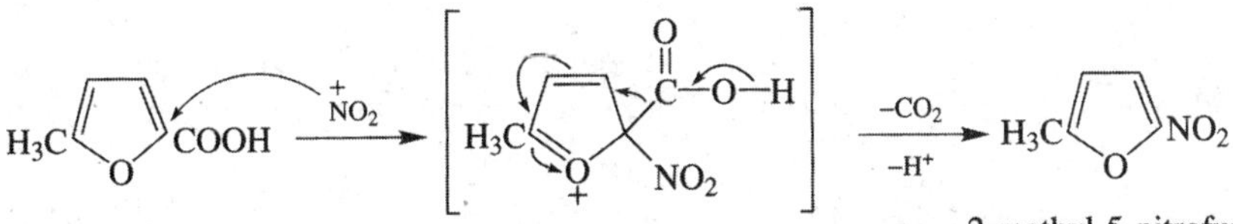

Attack of NO_2^+ at C^2, followed by elimination of CO_2 and H^+.

Problem 20.13 Give the Diels-Alder product for the reaction of furan and maleic anhydride.

Furan is the least aromatic of the five-membered ring heterocyclics and acts as a diene toward *strong* dienophiles.

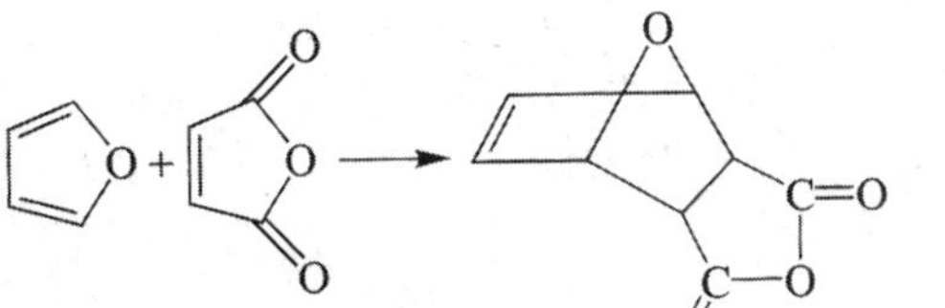

Furan Maleic anhydride

Problem 20.14 Give the products of reaction of pyrrole with (*a*) I_2 in aqueous KI; (*b*) CH_3CN + HCl, followed by hydrolysis; (*c*) CH_3MgI.

(*a*) 2,3,4,5-Tetraiodopyrrole (*b*) α-Acetylpyrrole [see Problem 19.21 (*c*)]

(*c*) $HC{=}CH$ / $HC{=}CH$ pyrrole ring with N—H + $H_3\overset{-}{C}{:}\overset{+}{Mg}I$ ⟶ pyrrolide anion + $\overset{+}{Mg}I + H_3C{:}H$

A₁ (stronger acid) B₂ (stronger base) B₁ (weaker base) A₂ (weaker acid)

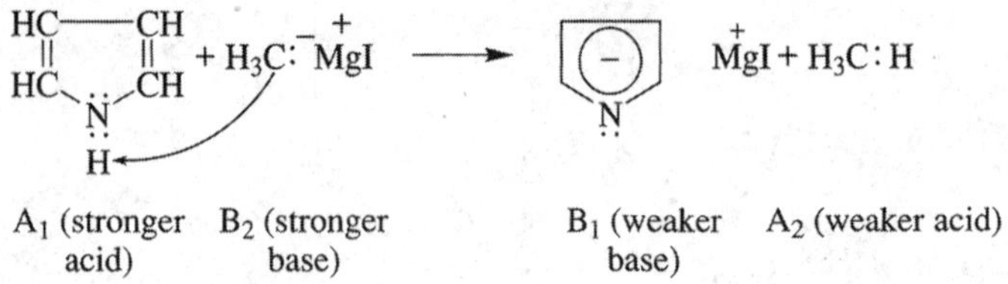

20.2 Six-Membered Heterocycles with One Heteroatom

The most important example of this category is **pyridine** (azabenzene), C_5H_5N:.

Problem 20.15 Write the structural formulas and give the names of the isomeric methylpyridines.

There are three isomers:

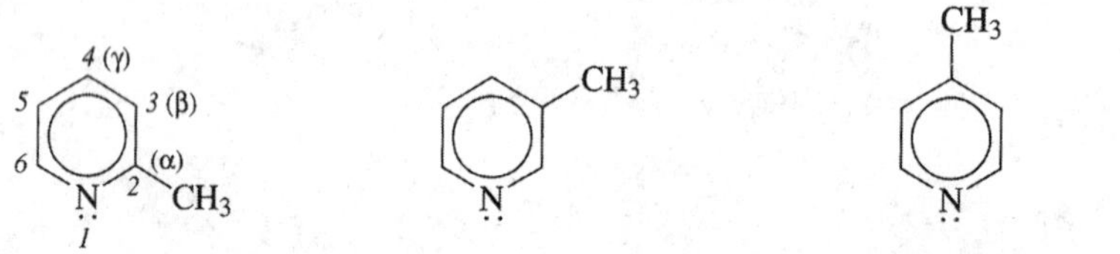

2- or α-Methylpyridine (α-Picoline) 3- or β-Methylpyridine (β-Picoline) 4- or γ-Methylpyridine (γ-Picoline)

Problem 20.16 (*a*) Account for the aromaticity of pyridine, a planar structure with 120° bond angles. (*b*) Is pyridine basic? Explain. (*c*) Explain why piperidine (azacyclohexane) is more basic than pyridine. (*d*) Write the equation for the reaction of pyridine with HCl.

(*a*) Pyridine (azabenzene) is the nitrogen analog of benzene, and both have the same *orbital* picture (Figs. 10.1 and 10.2). The three double bonds furnish six *p* electrons for the delocalized π system, in accordance with Hückel's rule.

(*b*) Yes. Unlike pyrrole, pyridine does not need the unshared pair of electrons on N for its aromatic sextet. The pair of electrons is available for bonding to acids.

(*c*) Piperidine: N ← sp^3-hybridized (less *s* character)

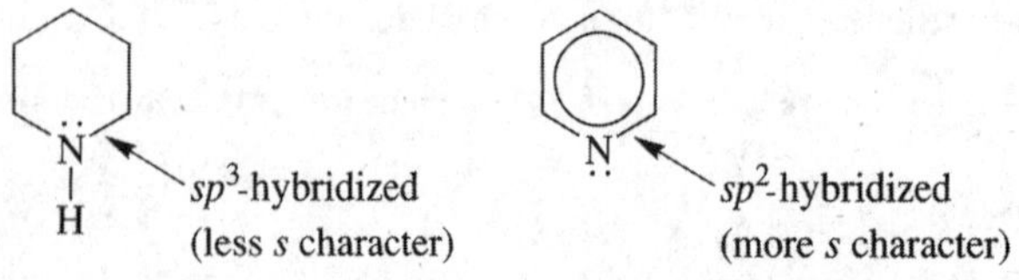

Pyridine: N ← sp^2-hybridized (more *s* character)

The fewer *s* characters in the orbital holding the unshared pair of electrons, the more basic the site.

(*d*) Pyridine N: —HCl→ $NH^+ Cl^-$

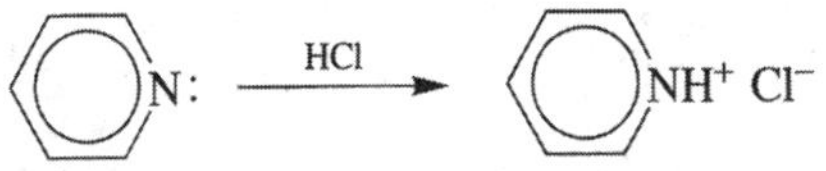

Pyridinum chloride
a salt

Problem 20.17 Explain why pyridine (*a*) undergoes electrophilic substitution at the β position, and (*b*) is less reactive than benzene.

(*a*) The R^+'s formed by attack of E^+ at the α or γ positions of pyridine have resonance structures (I, IV), with a positive charge on N having a sextet of electrons. These are high-energy structures.

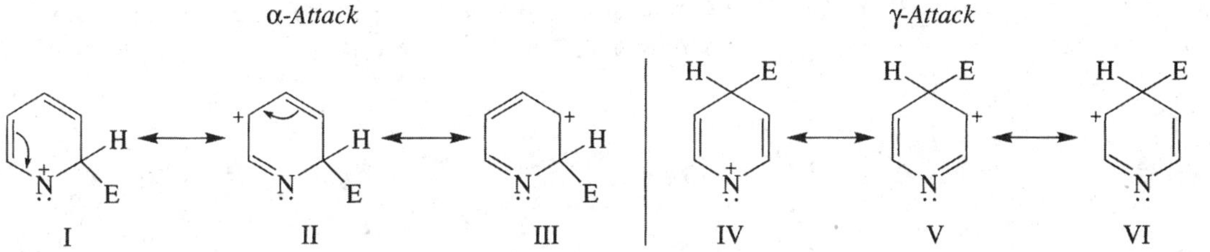

α-Attack γ-Attack

I II III IV V VI

With β-attack, the + charge in the intermediate is distributed only to C's. A + on C with six electrons is not as unstable as a + on N with six electrons, since N is more electronegative than C. β-Electrophilic substitution gives the more stable intermediate.

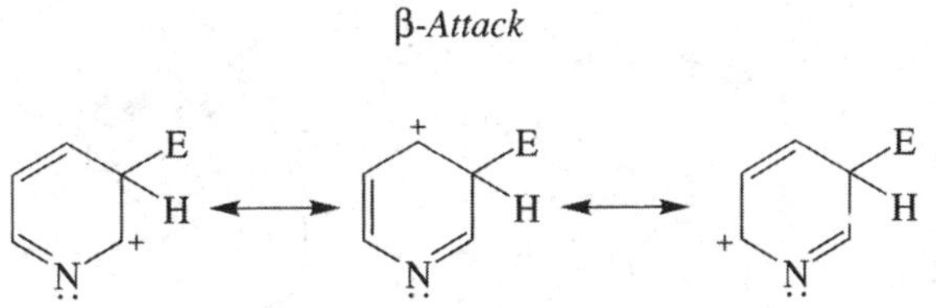

β-Attack

(*b*) N withdraws electrons by induction and destabilizes the R^+ intermediates formed from pyridine. Also, the N atom reacts with electrophiles to form a pyridinium cation, whose + charge decreases reactivity.

Problem 20.18 How do the ^{1}H nmr spectra of pyridine and benzene differ?

These are both aromatic compounds, and their ring-H signals are decidedly downfield. As all the H's of benzene are alike, one signal is observed. Pyridine gives three signals (not counting spin-spin coupling): $\delta = 8.5$ (two C^2 H's), $\delta = 7.06$ (two C^3 H's), and $\delta = 7.46$ (lone C^4 H). Notice that the C^2 H-signal is most downfield because the N is electron-withdrawing and less shielding.

Problem 20.19 Compare and explain the difference between pyridine and pyrrole with respect to reactivity toward electrophilic substitution.

Pyrrole is more reactive than pyridine because its intermediate is more stable. For both compounds, the intermediate has a + on N. However, the pyrrole intermediate is *relatively stable* because every atom has a *complete octet*, while the pyridine intermediate is *very unstable* because N has only *six* electrons.

Problem 20.20 Predict and account for the product obtained and conditions used in nitration of 2-aminopyridine.

The product is 2-amino-5-nitropyridine because substitution occurs preferentially at the sterically less hindered β position *para* to NH_2. The conditions are milder than those for nitration of pyridine, because NH_2 is activating.

Problem 20.21 Explain why (*a*) pyridine and $NaNH_2$ give α-aminopyridine, (*b*) 4-chloropyridine and NaOMe give 4-methoxypyridine, (*c*) 3-chloropyridine and NaOMe give no reaction.

Electron-attracting N facilitates attack by strong nucleophiles in α and γ positions. The intermediate is a carbanion stabilized by delocalization of – to the electronegative N. The intermediate carbanion readily reverts to a stable aromatic ring by ejecting an $H:^-$ in (*a*) or a $:\ddot{C}l:^-$ in (*b*).

(*a*) Pyridine $\xrightarrow[\text{heat}]{Na^+NH_2^-}$ [intermediate with NH₂, H, Na⁺] ⟶ 2-aminopyridine $-NH_2 + Na^+:H^-$

(*b*) 4-Chloropyridine $\xrightarrow{OMe^-}$ [MeO, Cl intermediate] ⟶ 4-methoxypyridine (OMe) $+ Cl^-$

(*c*) β-Nucleophilic attack does not give an intermediate with – on N.

3-Chloropyridine (Cl) + OMe^- ⟶✕ [Cl/OMe intermediate ⟷ Cl/OMe ⟷ Cl/OMe]

Problem 20.22 Account for the following orders of reactivity:

(*a*) Toward H_3O^+: 2,6-dimethylpyridine (2,6-lutidine) > pyridine

(*b*) Toward the Lewis acid BMe_3: pyridine > 2,6-lutidine

(*a*) Alkyl groups are electron-donating by induction and are base-strengthening. (*b*) BMe_3 is bulkier than an H_3O^+. The Me's at C^2 and C^6 flanking the N sterically inhibit the approach of BMe_3, causing 2,6-lutidine to be less reactive than pyridine. This is an example of **F-strain** (Front strain).

Problem 20.23 Pyridine *N*-oxide is converted to pyridine by PCl_5 or by zinc and acid. Use this reaction for the synthesis of 4-bromopyridine from pyridine.

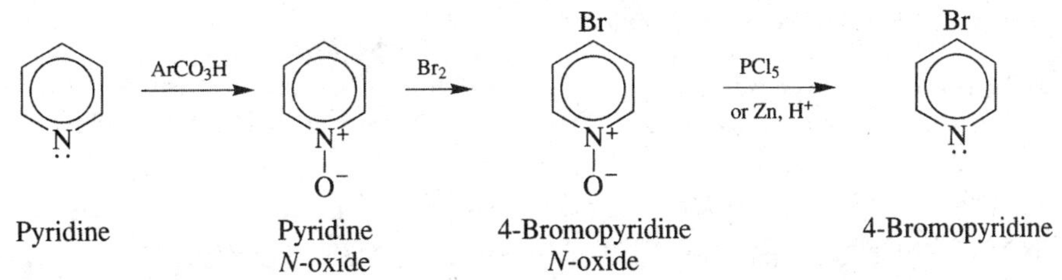

Pyridine $\xrightarrow{ArCO_3H}$ Pyridine *N*-oxide $\xrightarrow{Br_2}$ 4-Bromopyridine *N*-oxide $\xrightarrow[\text{or Zn, H}^+]{PCl_5}$ 4-Bromopyridine

Problem 20.24 Account for the fact that the CH_3's of α- and γ-picolines (methylpyridines) are more acidic than the CH_3 of toluene.

They react with strong bases to form resonance-stabilized anions with – on N.

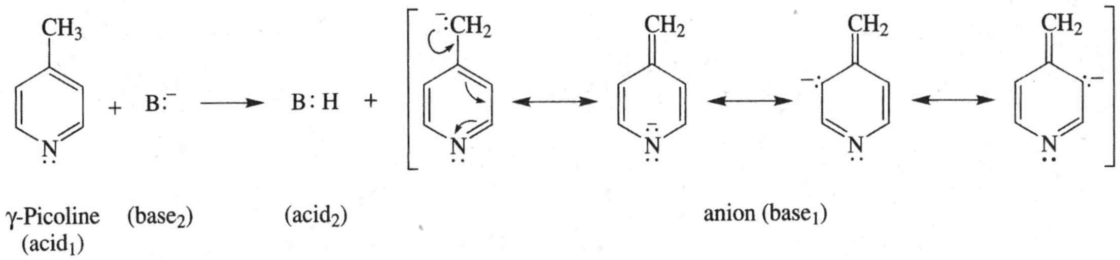

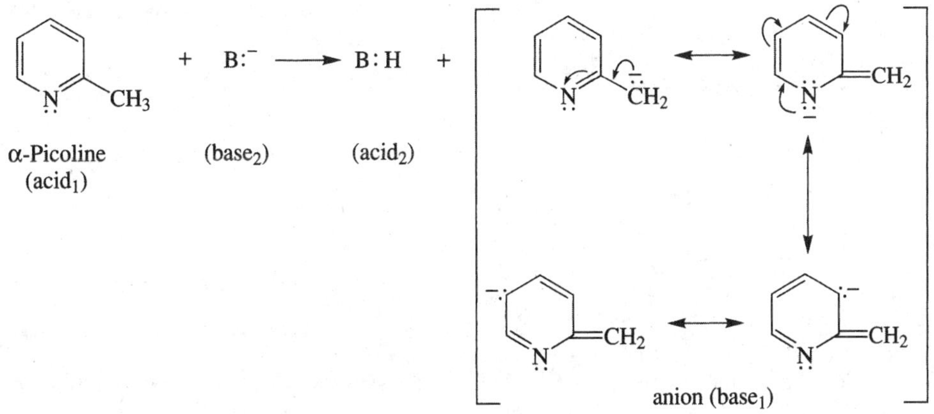

γ-Picoline (acid$_1$) + B:$^-$ (base$_2$) ⟶ B:H (acid$_2$) + [resonance structures] anion (base$_1$)

α-Picoline (acid$_1$) + B:$^-$ (base$_2$) ⟶ B:H (acid$_2$) + [resonance structures] anion (base$_1$)

Problem 20.25 From picolines, prepare (*a*) the vitamin niacin (3-pyridinecarboxylic acid), (*b*) the anti-tuberculosis drug isoniazide (4-pyridinecarboxylic acid hydrazide).

(*a*) 3-Picoline $\xrightarrow{KMnO_4}$ Niacin

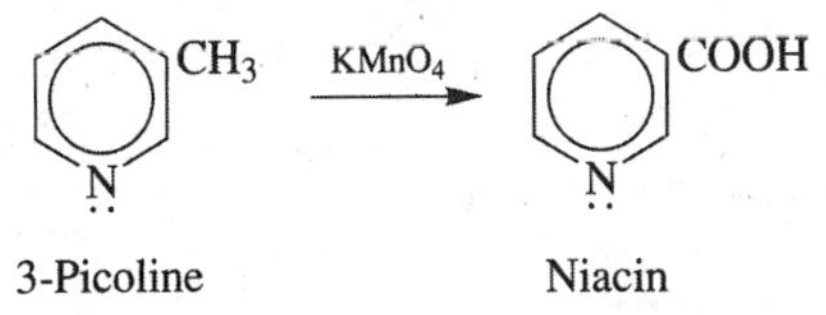

(*b*) 4-Picoline $\xrightarrow{KMnO_4}$ 4-Pyridine-carboxylic acid $\xrightarrow[\text{2. } H_2NNH_2]{\text{1. } SOCl_2}$ Isoniazide

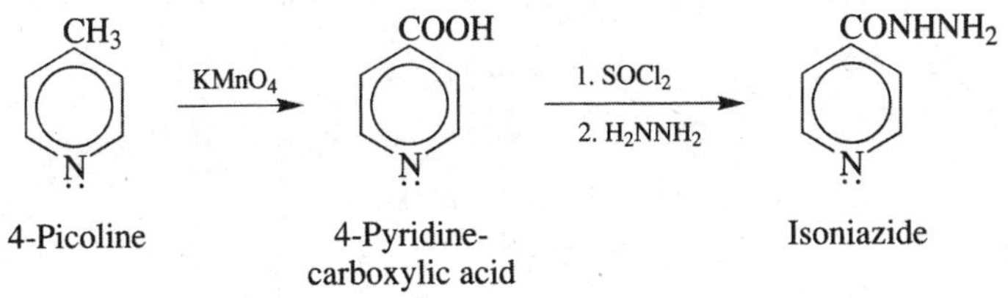

20.3 Compounds with Two Heteroatoms

We use the *oxa-aza-thia* system to name these compounds; the suffix **-ole** and **-ine** indicate five- and six-membered rings, respectively. When there is more than one kind of ring heteroatom, the atom of *higher* atomic number receives the *lower* number.

Problem 20.26 Name the following compounds:

(*a*) 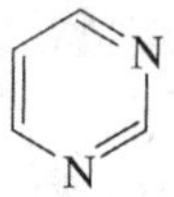 (*b*) 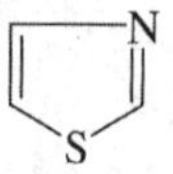 (*c*) 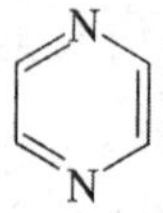 (*d*) 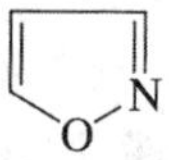 (*e*)

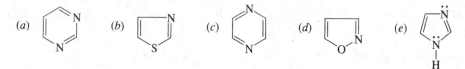

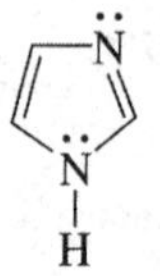

(*a*) 1,3-diazine (pyrimidine); (*b*) 1,3-thiazole; (*c*) 1,4-diazine (pyrazine); (*d*) 1,2-oxazole; (*e*) imidazole.

Three pyrimidines are among the constituents of nucleic acids:

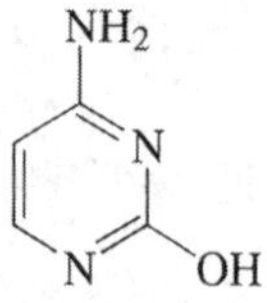

Cytosine 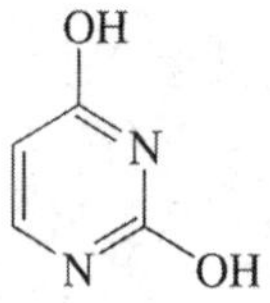 Uracil 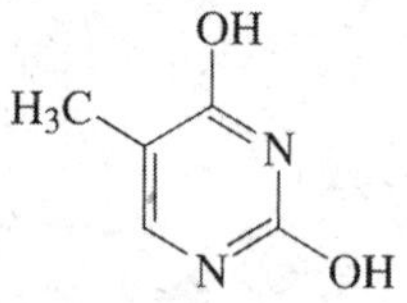 Thymine

Problem 20.27 Write the tautomeric structures of these pyrimidines.

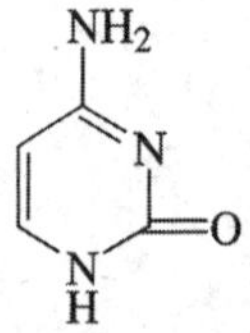

Cytosine 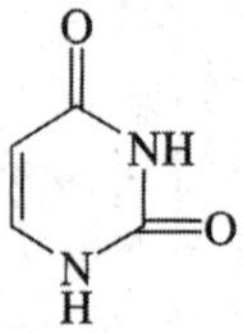 Uracil 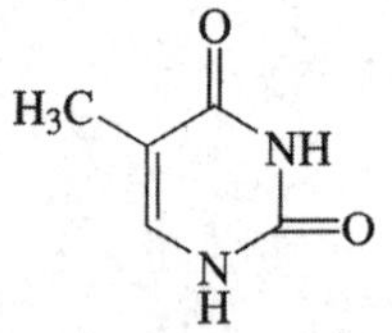 Thymine

Problem 20.28 (*a*) What makes imidazole (Problem 20.30*a*) aromatic? (*b*) Explain why imidazole, unlike pyrrole, is basic. Which N is the basic site?

(*a*) Imidazole is aromatic because it has a sextet of electrons from the two double bonds and the electron pair on the N bonded to H. (*b*) The proton acceptor is the N not bonded to H, because its lone pair is not part of the aromatic sextet.

20.4 Condensed Ring Systems

Many biologically important heterocyclic compounds have **fused (condensed)** ring systems. In particular, the purines adenine and guanine are found in DNA (with cytosine, 5-methylcytosine, and thymine) and also in RNA (with cytosine and uracil).

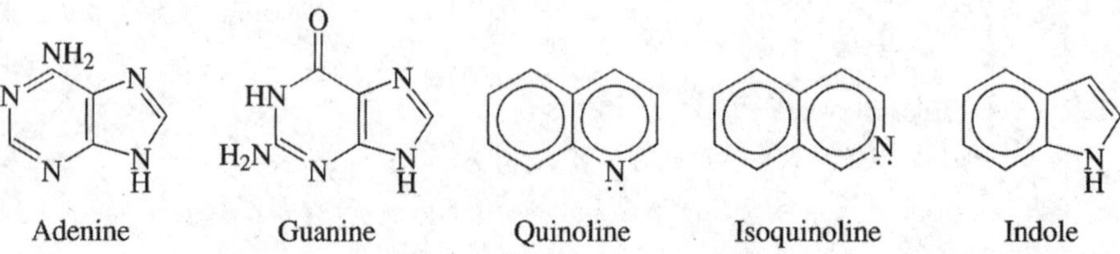

Adenine Guanine Quinoline Isoquinoline Indole

Quinoline (1-Azanaphthalene)

Problem 20.29 Which dicarboxylic acid is formed on oxidation of quinoline?

The pyridine ring is more stable than the benzene ring [Problem 20.17(*a*)].

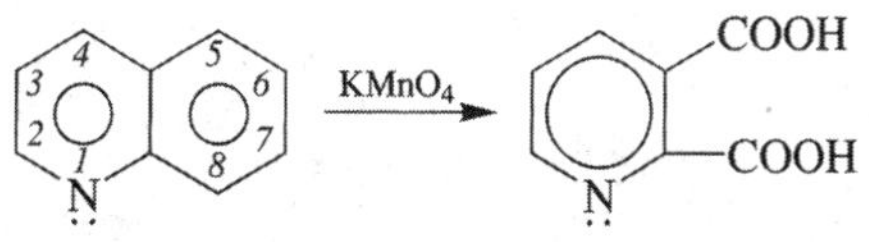

$$\text{Quinoline} \xrightarrow{KMnO_4} \text{Quinolinic acid}$$

Quinoline Quinolinic acid

Problem 20.30 Quinoline is prepared by the **Skraup reaction** of aniline, glycerol, and nitrobenzene. Suggest a mechanism involving Michael addition of aniline to an α,β-unsaturated aldehyde, ring closure, and then dehydration and oxidation.

The steps in the reaction are as follows:

(1) Dehydration of glycerol to acrolein (propenal).

$$\underset{\text{Glycerol}}{H_2COHCHOHCH_2OH} \xrightarrow{H_2SO_4} \underset{\text{Acrolein}}{H_2C{=}CHCHO} + 2H_2O$$

(2) Michael-type addition (Section 17.4).

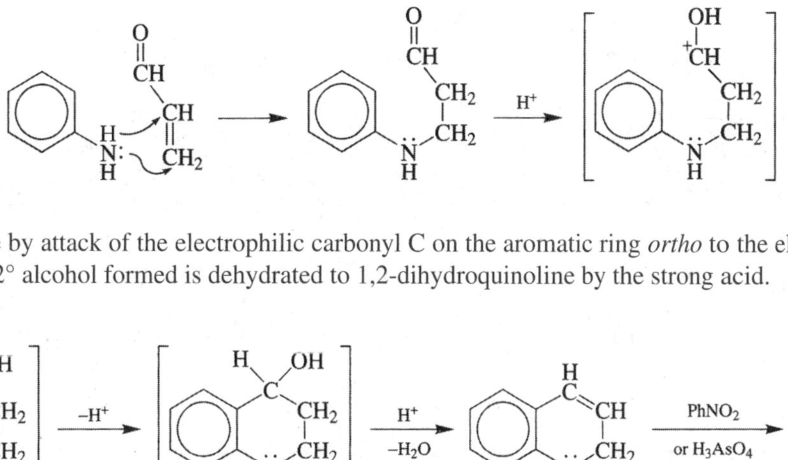

$$C_6H_5NH_2 + CH_2{=}CH{-}CH{=}O \longrightarrow C_6H_5NH{-}CH_2{-}CH_2{-}CH{=}O \xrightarrow{H^+} \left[C_6H_5NH{-}CH_2{-}CH_2{-}\overset{+}{C}H{-}OH\right]$$

(3) Ring closure by attack of the electrophilic carbonyl C on the aromatic ring *ortho* to the electron-releasing —NH. The 2° alcohol formed is dehydrated to 1,2-dihydroquinoline by the strong acid.

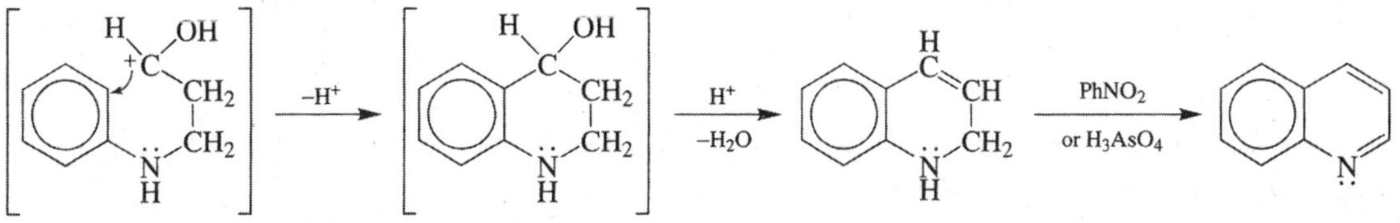

$$[\ ] \xrightarrow{-H^+} [\ ] \xrightarrow[-H_2O]{H^+} \text{1,2-Dihydroquinoline} \xrightarrow[\text{or } H_3AsO_4]{PhNO_2} \text{Quinoline}$$

1,2-Dihydroquinoline Quinoline

(4) $PhNO_2$ oxidizes the dihydroquinoline to the aromatic compound quinoline. $PhNO_2$ is reduced to $PhNH_2$, which then reacts with more acrolein. This often violent reaction is moderated by added $FeSO_4$.

Problem 20.31 Give structures for the products of reaction of quinoline with (*a*) HNO_3, H_2SO_4; (*b*) $NaNH_2$; (*c*) PhLi.

(*a*) 5-nitro- and 8-nitroquinoline; (*b*) 2-amino- and 4-aminoquinoline (like pyridine, quinoline undergoes nucleophilic substitution in the 2 and 4 positions); (*c*) 2-phenylquinoline.

Problem 20.32 Outline a mechanism for the **Bischler-Napieralski synthesis** of 1-methylisoquinoline from N-acetylphenylethylamine by reaction with strong acid and P_2O_5, and then oxidation of the dihydroisoquinoline intermediate.

The mechanism is similar to that of the Skraup synthesis (Problem 20.30) in that carbonyl O is protonated and the electrophilic C attacks the benzene ring in cyclization, to be followed by dehydration and oxidation.

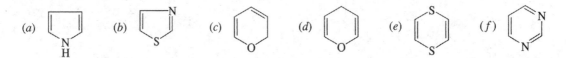

CH2 CH2 NH C O CH3 —H⁺→ CH2 CH2 NH +C HO CH3 → NH HO CH3 —$-H_2O$→ N CH3 —oxid.→ N CH3

SUPPLEMENTARY PROBLEMS

Problem 20.33 Supply systematic names for:

(*a*) C_6H_5 N O (*b*) Br N CH3 HOOC N N (*c*) H3C N CH3 S (*d*) N–N N S (*e*) N H

(*a*) 4-phenyl-1,2-oxazole, (*b*) 3-methyl-5-bromo-1,2,4-triazine-6-carboxylic acid, (*c*) 2,4-dimethyl-1,3-thiazole, (*e*) 1,2,3,4-thiatriazole, (*e*) 2,3-benzazole (indole).

Problem 20.34 Name the following compounds systematically:

(*a*) N H (*b*) N S (*c*) O (*d*) O (*e*) S S (*f*) N N

(*a*) azole (pyrrole), (*b*) 1,3-thiazole, (*c*) 2*H*-oxirine, (*d*) 4*H*-oxirine (pyran), (*e*) 1,4-dithiazine, (*f*) 1,3-diazine (pyrimidine).

2*H*- and 4*H*- are used in (*c*) and (*d*) to differentiate the position of the saturated sp^3 atom. Common names are given in parentheses.

Problem 20.35 Write structures for (*a*) oxirane, (*b*) 1,2-oxazole, (*c*) 1,4-diazine (pyrazine), (*d*) 1-thia-4-oxa-6-azocine, (*e*) 3*H*-1,2,4-triazole, (*f*) azepane.

(*a*) H_2C–CH_2 O (*c*) N N (*e*) N N N H

(*b*) N O (*d*) N O S (*f*) H N

Problem 20.36 How many thiophenyl-thiophenes (**bithienyls**) are possible?

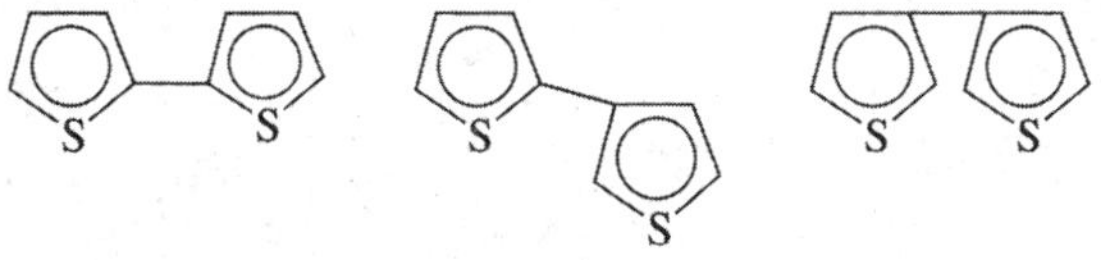

2,2′-Bithienyl 2,3′-Bithienyl 3,3′-Bithienyl

Problem 20.37 Identify the compounds represented by Roman numerals.

(*a*) Quinoline $\xrightarrow{C_6H_5\overset{O}{\overset{\|}{C}}O_2H}$ I $\xrightarrow{HNO_3}$ II $\xrightarrow{PCl_3}$ III

(*b*) Pyrrole (PyH) $\xrightarrow{p\text{-}HO_3SC_6H_4\overset{+}{N}_2X^-}$ I $\xrightarrow{Sn,HCl}$ II + III

(*c*) Furan (FuH) $\xrightarrow[(C_2H_5)_2O,\ BF_3]{(CH_3CO)_2O}$ I $\xrightarrow{NaOI}$ II $\xrightarrow{\text{fum. } H_2SO_4}$ III

	I	II	III
(*a*)	Quinoline *N*-oxide	4-Nitroquinoline *N* oxide	4-Nitroquinoline
(*b*)	2-PyN=$NC_6H_4SO_3H$-*p*	2-$PyNH_2$	*p*-$\overset{+}{N}H_3C_6H_4SO_3^-$
(*c*)	2-$FuCOCH_3$	2-$FuCOO^-Na^+$	5-HO_3S-2-FuCOOH

Problem 20.38 Prepare (*a*) 3-aminopyridine from *β*-picoline, (*b*) 4-aminopyridine from pyridine, (*c*) 8-hydroxyquinoline from quinoline, (*d*) 5-nitro-2-furoic acid from furfural, (*e*) 2-pyridylacetic acid from pyridine.

(*a*) 3-Methylpyridine $\xrightarrow[KMnO_4]{\text{oxid.}}$ 3-COOH-pyridine $\xrightarrow{SOCl_2}$ 3-COCl-pyridine $\xrightarrow{NH_3}$ 3-$CONH_2$-pyridine $\xrightarrow[OH^-]{NaOBr}$ 3-NH_2-pyridine

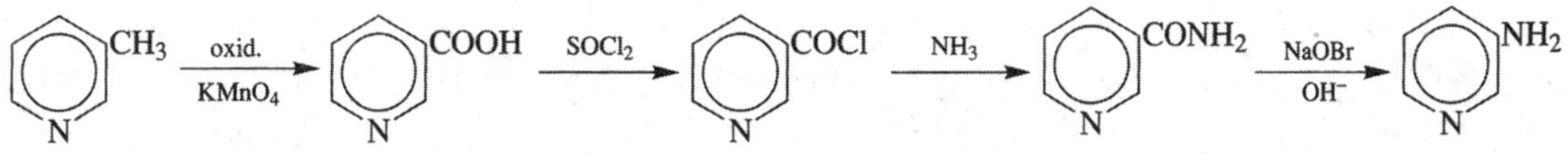

(*b*) Pyridine $\xrightarrow{PhCOOOH}$ pyridine N-oxide (N^+–O^-) $\xrightarrow[H_2SO_4]{HNO_3}$ 4-NO_2-pyridine N-oxide $\xrightarrow{PCl_3}$ 4-NO_2-pyridine $\xrightarrow[2.\ OH^-]{1.\ Sn,\ H^+}$ 4-NH_2-pyridine

(*c*) Quinoline $\xrightarrow[220\,°C]{H_2SO_4}$ Quinoline-8-sulfonic acid $\xrightarrow[2.\ NaOH,\ fuse;\ 3.\ H_3O^+]{1.\ OH^-}$ 8-hydroxyquinoline (OH, N)

(*d*) Furan-CHO $\xrightarrow{Ag(NH_3)_2^+}$ Furan-COOH $\xrightarrow{HNO_3}$ O_2N-furan-COOH

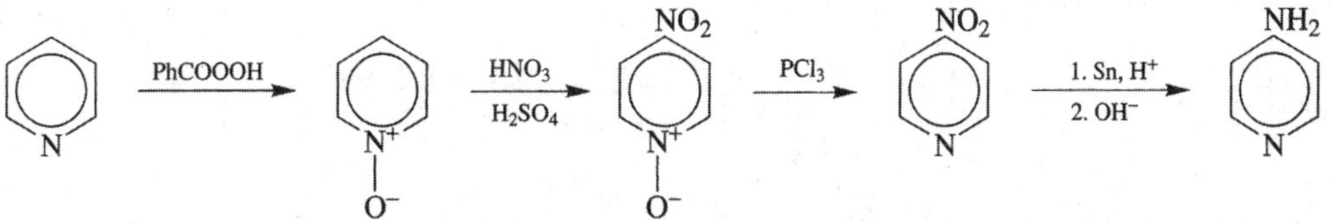

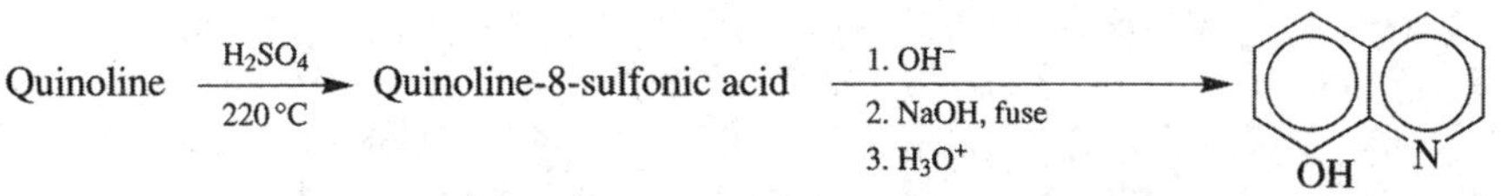

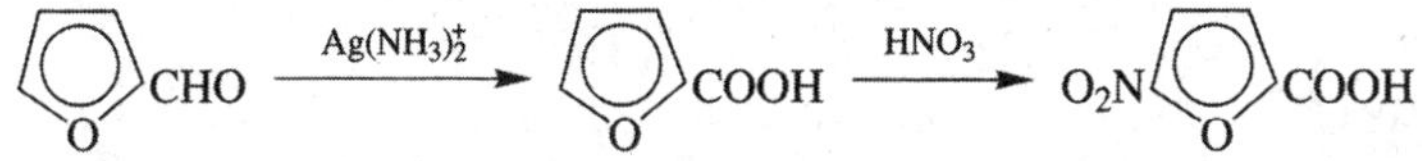

(COOH stabilizes the ring towards acid cleavage of the ether bond.)

(*e*) Pyridine $\xrightarrow{NH_2^-}$ 2-Aminopyridine $\xrightarrow[2.\ CuBr]{1.\ NaNO_2,\ H^+,\ 5^\circ C}$ 2-Bromopyridine

2-Bromopyridine + $[\bar{C}H(COOC_2H_5)_2]\ Na^+$ $\xrightarrow[\text{nucleophilic displacement}]{}$ 2-Py—$CH(COOC_2H_5)_2$ $\xrightarrow[3.\ \Delta,\ -CO_2]{1.\ NaOH\ \ 2.\ H_3O^+}$ 2-Py—CH_2COOH

Problem 20.39 (*a*) Explain why pyran [Problem 20.34(*d*)] is not aromatic. (*b*) What structural change would theoretically make it aromatic?

(*a*) There are six electrons available: four from the two π bonds and two from the O atom. However, C^4 is sp^3-hybridized and has no *p* orbital available for cyclic *p* orbital overlap. (*b*) Convert C^4 to a carbocation. C^4 would now be sp^2-hybridized and would have an empty *p* orbital for cyclic overlap.

Problem 20.40 How can pyridine and piperidine be distinguished by infrared spectroscopy?

Piperidine has an N—H bond absorbing at 3500 cm^{-1} and H—C(sp^3) stretch below 3000 cm^{-1}. Pyridine has no N—H; has H—C(sp^2) stretch above 3000 cm^{-1}; C═C and C═N stretches near 1600 and 1500 cm^{-1}, respectively; aromatic ring vibrations near 1200 and 1050 cm^{-1}; and C—H deformations at 750 cm^{-1}. The peak at 750 varies with substitution in the pyridine ring.

Problem 20.41 How can nmr spectroscopy distinguish among aniline, pyridine, and piperidine?

The NH_2 of aniline is electron-donating and shields the aromatic H's; their chemical shift is $\delta = 6.5–7.0$ (for benzene, the chemical shift is $\delta = 7.1$). The N of pyridine is electron-withdrawing and weakly shields the aromatic H's ($\delta = 7.5–8.0$). Piperidine is not aromatic and has no signals in these regions.

Problem 20.42 From pyridine (PyH), 2-picoline (2-PyMe), and any reagent without the pyridine ring, prepare (*a*) 2-acetylpyridine, (*b*) 2-vinylpyridine, (*c*) 2-cyclopropylpyridine, (*d*) 2-$PyCH_2CH_2CH_2COOH$, (*e*) 2-PyC(Me)═$CHCH_3$, (*f*) 2-pyridinecarboxaldehyde. Any synthesized compound can be used in ensuing steps.

(*a*) By a crossed-Claisen condensation

2-PyMe $\xrightarrow{KMnO_4}$ 2-PyCOOH $\xrightarrow[H^+]{EtOH}$ 2-PyCOOEt $\xrightarrow[NaOEt]{CH_3COOEt}$

2-$PyCOCH_2COOEt$ $\xrightarrow[heat]{OH^-}$ 2-$PyCOCH_2COO^-$ $\xrightarrow{H^+}$ [2-$PyCOCH_2COOH$] $\xrightarrow{-CO_2}$ 2-$PyCOCH_3$

(*b*) 2-$PyCOCH_3$ $\xrightarrow{NaBH_4}$ 2-$PyCHOHCH_3$ $\xrightarrow{P_4O_{10}}$ 2-PyCH═CH_2

(*c*) 2-PyCH═CH_2 + CH_2N_2 $\xrightarrow{uv}$ product

(*d*) By a Michael addition:

$$\text{2-PyCH═CH}_2 + (\text{EtOOC})_2\bar{\text{C}}\text{HNa}^+ \longrightarrow \text{2-Py}\ddot{\text{C}}\text{HCH}_2\text{CH(COOEt)}_2\text{Na}^+$$

This α-pyridylcarbanion is stabilized by charge delocalization to the ring N (Problem 20.30). Refluxing the salt in HCl causes decarboxylation and gives the pyridinium salt of the product, which is then neutralized with OH^-.

(*e*) Use the Wittig synthesis: 2-PyCOMe + Ph_3P═$CHCH_3$ → products (*trans* and *cis*).

(*f*) 2-PyCH═CH_2 $\xrightarrow[2.\ Zn,\ HOAc]{1.\ O_3}$ 2-PyCHO

Problem 20.43 (*a*) Account for the aromaticity of α- and γ-pyridones. (*b*) Explain why α-pyridone predominates over α-pyridinol, especially in the solid state. (The same is true for the γ-tautomers.) (*c*) How can ir spectroscopy show which tautomer predominates?

(*a*) See Fig. 20.2 for the cyclic aromatic extended π-bonding with six electrons. The carbonyl carbon furnishes an empty *p* AO, N furnishes a *p* AO with a pair of electrons, and each doubly bonded C furnishes a *p* AO with one electron.

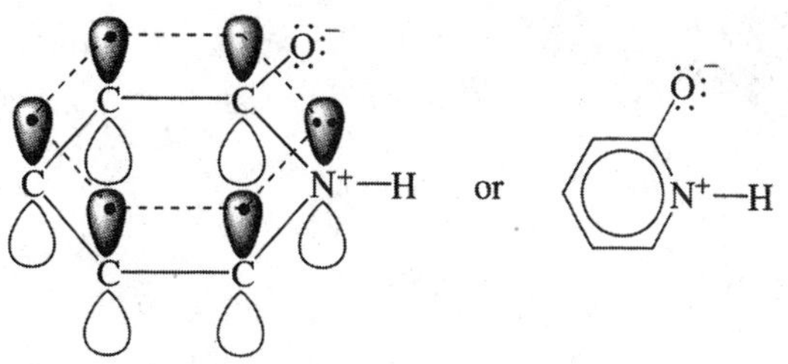

Figure 20.2

(*b*) The N—H forms a strong intermolecular H-bond with O of C═O. This bond, repeated throughout the crystalline solid, links molecules in endless helices.

(*c*) The ir spectra of the solid (and solution) show a strong C═O stretching band.

INDEX

PERIODIC TABLE

1 IA	2 IIA	3	4	5	6	7	8	9	10	11	12	13 IIIA	14 IVA	15 VA	16 VIA	17 VIIA	18 VIIIA
1 H 1.0079																	2 He 4.0026
3 Li 6.941	4 Be 9.0122											5 B 10.81	6 C 12.011	7 N 14.007	8 O 15.999	9 F 18.998	10 Ne 20.179
11 Na 22.989	12 Mg 24.305											13 Al 26.981	14 Si 28.086	15 P 30.974	16 S 32.06	17 Cl 35.453	18 Ar 39.948
19 K 39.098	20 Ca 40.08	21 Sc 44.956	22 Ti 47.88	23 V 50.941	24 Cr 51.996	25 Mn 54.938	26 Fe 55.847	27 Co 58.933	28 Ni 58.69	29 Cu 63.546	30 Zn 65.38	31 Ga 59.72	32 Ge 72.59	33 As 74.922	34 Se 78.96	35 Br 79.904	36 Kr 83.80
37 Rb 85.468	38 Sr 87.62	39 Y 88.906	40 Zr 91.22	41 Nb 92.905	42 Mo 95.94	43 Tc (98)	44 Ru 101.07	45 Rh 102.91	46 Pd 106.42	47 Ag 107.87	48 Cd 112.41	49 In 114.82	50 Sn 118.69	51 Sb 121.75	52 Te 127.60	53 I 126.90	54 Xe 131.29
55 Cs 132.91	56 Ba 137.33	57 *La 138.90	72 Hf 178.49	73 Ta 180.95	74 W 183.85	75 Re 186.21	76 Os 190.2	77 Ir 192.22	78 Pt 195.08	79 Au 196.97	80 Hg 200.59	81 Tl 204.38	82 Pb 207.2	83 Bi 208.98	84 Po (209)	85 At (210)	86 Rn (222)
87 Fr (223)	88 Ra 226.0	89 #Ac 227.03	104 Rf (261)	105 Db (262)	106 Sg (263)	107 Bh (262)	108 Hs (265)	109 Mt (266)	110 Uun (269)	111 Uuu (272)	112 Uub (277)						

*Lanthanides	58 Ce 140.12	59 Pr 140.91	60 Nd 144.24	61 Pm (145)	62 Sm 150.36	63 Eu 151.96	64 Gd 157.25	65 Tb 158.92	66 Dy 162.50	67 Ho 164.93	68 Er 167.26	69 Tm 168.93	70 Yb 173.04	71 Lu 174.97
#Actinides	90 Th 232.03	91 Pa 231.03	92 U 238.03	93 Np 237.05	94 Pu (244)	95 Am (243)	96 Cm (247)	97 Bk (247)	98 Cf (251)	99 Es (254)	100 Fm (257)	101 Md (257)	102 No (255)	103 Lr (256)